INTERPRETING ENGINEERING DRAWINGS

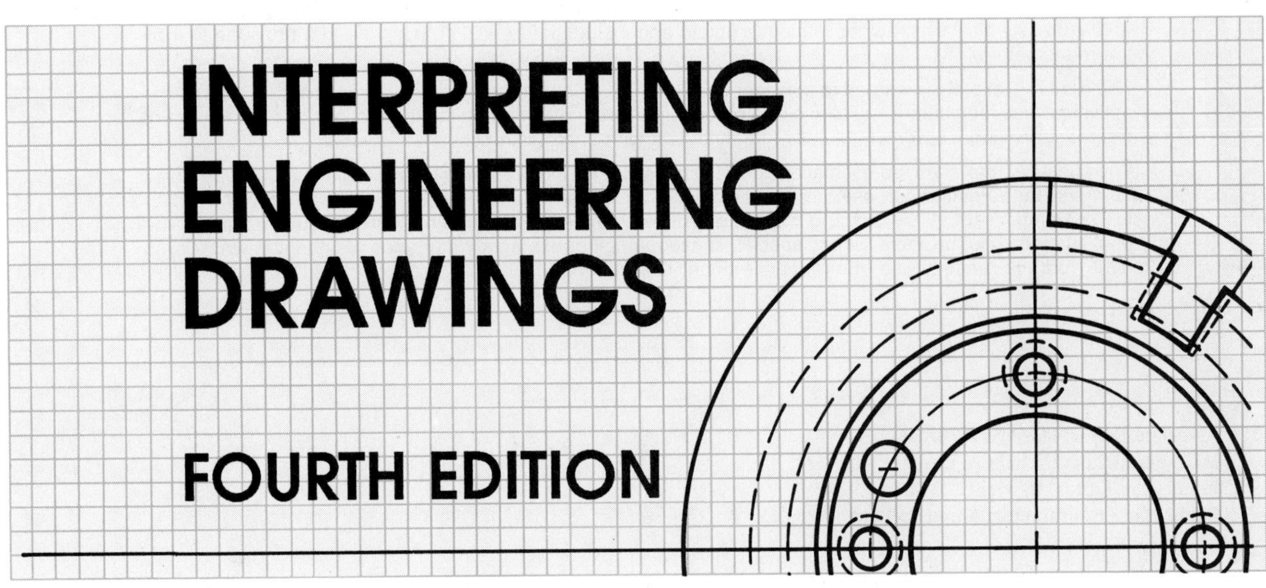

C. JENSEN • R. HINES

DELMAR PUBLISHERS INC.®

NOTICE TO THE READER

Publisher does not warrant or guarantee any of the products described herein or perform any independent analysis in connection with any of the product information contained herein. Publisher does not assume, and expressly disclaims, any obligation to obtain and include information other than that provided to it by the manufacturer.

The reader is expressly warned to consider and adopt all safety precautions that might be indicated by the activities described herein and to avoid all potential hazards. By following the instructions contained herein, the reader willingly assumes all risks in connection with such instructions.

The publisher makes no representations or warranties of any kind, including but not limited to, the warranties of fitness for particular purpose or merchantability, nor are any such representations implied with respect to the material set forth herein, and the publisher takes no responsibility with respect to such material. The publisher shall not be liable for any special, consequential or exemplary damages resulting, in whole or in part, from the readers' use of, or reliance upon, this material.

Delmar Staff
Associate Editor: Marjoria A. Bruce
Managing Editor: Barbara A. Christie
Production Editor: Lawrence T. Main
Publications Coordinator: Karen Seebald

For information, address Delmar Publishers Inc.
2 Computer Drive West, Box 15-015
Albany, New York 12212

10 9 8 7 6 5 4 3

Printed in the United States of America
Published simultaneously in Canada
by Nelson Canada,
A division of The Thomson Corporation

Library of Congress Cataloging in Publication Data

Jensen, Cecil Howard, 1925-
Interpreting engineering drawings / C. Jensen, R. Hines. – 4th ed.
p. cm.
Includes index.
ISBN 0-8273-3048-0 (pbk.). ISBN 0-8273-3049-9 (Instructor's guide)
1. Engineering drawings. I. Title.
T379.J45 1989 604.2'5–dc19 88-427

CONTENTS

PREFACE xiii

Unit 1 1
Basis for Interpreting Drawings 1
Third-Angle Projection 1
ISO Projection Symbol 4
Title Block 4
Drawing Standards 5
Visible Lines 5
Lettering on Drawings 5
Sketching 5
Assignments: A-1 Three Views Sketching 7
A-2 Sketching the Missing View 9
A-3 Sketching—Pictorial 10

Unit 2 11
Working Drawings 11
Dimensioning 11
Linear Units of Measurement 13
Choice of Dimensions 14
Basic Rules for Dimensioning 15
Information Shown on Assignment Drawings 16
Assignments: A-4M Corner Block 17
A-5 Step Bracket 18
A-6 Counter Clamp Bar 19
A-7 Dimensioning 20

Unit 3 21
The Drafting Office 21
Drawing Reproduction 23
Abbreviations Used on Drawings 24
Hidden Lines 24
Assignments: A-8 Matching—Pictorial to Orthographic 25
A-9 Matching—Pictorial to Orthographic 26
A-10 Sketching Three Views 27
A-11 Feed Hopper 28

Unit 4 30
Inclined Surfaces 30

Measurement of Angles 30
Assignments: A-12 Sketching Three Views 32
A-13 Sketching the Missing Views 33
A-14 Adapter 34
A-15M Guide Block 35
A-16 Baseplate 36

Unit 5 37
Circular Features 37
Center Lines 37
Dimensioning of Cylindrical Features 37
Dimensioning Cylindrical Holes 38
Repetitive Features and Dimensions 39
Drilling, Reaming, and Boring 40
Identifying Similarly Sized Features 40
Rounds and Fillets 40
Assignments: A-17 Sketching One- and Two-View Drawings 41
A-18 Coupling 42
A-19 Sketching the Missing Views 44

Unit 6 45
Drawing to Scale 45
SI (Metric Scales) 46
Machine Slots 46
Symmetrical Outlines 49
Reference Dimensions 49
Assignments: A-20 Reading Scales 50
A-21 Scale Measurement–Inch and Millimeter 51
A-22 Compound Rest Slide 52

Unit 7 54
Machining Symbols 54
Not-to-Scale Dimensions 55
Drawing Revisions 56
Break Lines 56
Assignment: A-23M Offset Bracket 58

Unit 8 60
Sectional Views 60
Types of Sections 61
Countersinks, Counterbores, and Spotfaces 62
Assignments: A-24 Sketching Full Sections 65
A-25 Slide Bracket 66
A-26 Sketching Half Sections 68

Unit 9 69

Chamfers 69
Undercuts 70
Tapers 70
Knurls 71
Intersection of Unfinished Surfaces 71
Assignments: A-27 Handle 73
A-28 Shaft Support 74

Unit 10 75
Selection of Views 75
One- and Two-View Drawings 75
Multiple Detail Drawings 75
Assignment: A-29 Centering Connector Details 76

Unit 11 78
Surface Texture 78
Surface Texture Symbol 79
Surface Texture Ratings 82
Control Requirements 83
Assignments: A-30 Hanger Details 86
A-31 Completion Test–Sketch Missing Views 88

Unit 12 89
Tolerances and Allowances 89
Definitions 89
Tolerancing Methods 90
Dimension Origin Symbol 91
Assignments: A-32 Inch Tolerances and Allowances 94
A-33M Millimeter Tolerances and Allowances 96

Unit 13 98
Inch Fits 98
Description of Fits 98
Standard Inch Fits 100
Assignment: A-34 Inch Fit Problems 103

Unit 14 104
Metric Fits 104
Assignments: A-35M Metric Fit Problems 109
A-36M Bracket 110

Unit 15 112
Threaded Fasteners 112
Threaded Assemblies 112
Inch Threads 115
Right- and Left-Handed Threads 115

Metric Threads 116
Assignments: A-37 Shaft Intermediate Support 119
A-38M Drive Support Details 120
A-39 Housing Details 122

Unit 16 124
Revolved and Removed Sections 124
Assignments: A-40 Shaft Supports 126
A-41 Terminal Block 128

Unit 17 129
Keys 129
Setscrews 130
Flats 131
Bosses and Pads 131
Rectangular Coordinate Dimensioning Without Dimension Lines 131
Rectangular Coordinate Dimensioning in Tabular Form 131
Assignments: A-42M Terminal Stud 133
A-43 Rack Details 134
A-44 Support Bracket 136

Unit 18 138
Primary Auxiliary Views 138
Assignment: A-45 Gear Box 140

Unit 19 141
Secondary Auxiliary Views 141
Assignments: A-46 Sketching Three Views 143
A-47 Completion of Views 144
A-48 Hexagon Bar Support 146

Unit 20 148
Development Drawings 148
Joints, Seams, and Edges 149
Sheet Metal Sizes 150
Straight Line Development 150
Stampings 150
Assignment: A-49 Letter Box 152

Unit 21 153
Arrangement of Views 153
Assignments: A-50 Mounting Plate 155
A-51 Index Pedestal 156
A-52M Contact Arm 158

Unit 22 159

Piping 159
Piping Drawings 161
Pipe Drawing Symbols 161
Assignments: A-53 Engine Starting Air System 166
A-54 Boiler Room 168

Unit 23 170
Bearings 170
Assignment: A-55 Corner Bracket 172

Unit 24 174
Steel Specifications 174
Drawings for Numerical Control 176
Assignments: A-56M Crossbar 177
A-57 Oil Chute 178

Unit 25 180
Castings 180
Assignments: A-58 Offset Bracket 183
A-59 Trip Box 184

Unit 26 186
Cast Irons 186
Casting Design 186
Cored Castings 188
Machining Lugs 189
Surface Coatings 189
Assignments: A-60 Auxiliary Pump Base 190
A-61M Slide Valve 192

Unit 27 193
Alignment of Parts and Holes 193
Partial Views 194
Naming of Views for Spark Adjuster 195
Drill Sizes 195
Assignment: A-62 Spark Adjuster 196

Unit 28 198
Broken-out and Partial Sections 198
Webs in Section 198
Ribs in Section 198
Spokes in Section 200
Assignments: A-63M Raise Block 202
A-64 Coil Frame 204

Unit 29 206

Pin Fasteners 206
Section Through Shafts, Pins, and Keys 210
Arrangement of Views of Drawing A-65M 210
Assignments: A-65M Spider 212
A-66 Hood 214

Unit 30 216
Chain Dimensioning 216
Base Line Dimensioning 216
Assignments: A-67 Interlock Base 218
A-68 Control Bracket 220
A-69M Contactor 222

Unit 31 223
Assembly Drawings 223
Bill of Materials (Items List) 223
Swivels and Universal Joints 225
Assignment: A-70 Universal Trolley 229

Unit 32 230
Structural Steel Shapes 230
Phantom Outlines 231
Conical Washers 232
Assignment: A-71 Four-wheel Trolley 235

Unit 33 236
Welding Drawings 236
Welding Symbols 236
Fillet Welds 240
Assignments: A-72 Fillet Welds 244
A-73 Shaft Support 245

Unit 34 246
Groove Welds 246
Supplementary Symbols 248
Assignments: A-74 Groove Welds 251
A-75 Base Skid 253

Unit 35 254
Other Basic Welds 254
Assignments: A-76 Plug, Slot and Spot Welds 261
A-77 Base Assembly 263
A-78 Seam and Flange Welds 264

Unit 36 265
Gears 265

Spur Gears 265
Assignments: A-79 Spur Gear Calculations 271
A-80 Spur Gear 272

Unit 37 274
Bevel Gears 274
Assignment: A-81 Miter Gear 276

Unit 38 278
Gear Trains 278
Assignments: A-82 Motor Drive Assembly 280
A-83 Gear Train Calculations 282

Unit 39 283
Cams 283
Assignments: A-84 Plate Cam 285
A-85 Cylindrical Feeder Cam 286

Unit 40 288
Antifriction Bearings 288
Retaining Rings 290
O-Ring Seals 290
Clutches 290
Belt Drives 290
Assignment: A-86 Power Drive 294

Unit 41 296
Ratchet Wheels 296
Assignment: A-87 Winch 298

Unit 42 300
Modern Engineering Tolerancing 300
Geometric Tolerancing 301
Form Tolerance 304
Assignment: A-88 Form Tolerancing—Straightness 308

Unit 43 309
Modifying Symbols 309
Straightness—RFS and MMC 311
Assignment: A-89 Straightness—RFS, MMC 313

Unit 44 314
Form Tolerances 314
Flatness 314
Circularity 315
Cylindricity 317

Assignment: A-90 Form Tolerancing–Flatness, Circularity, Cylindricity 319

Unit 45 320
Datums and the Three-Plane Method of Tolerancing 320
Datums for Geometric Tolerancing 320
Datum Identifying Symbol 321
Assignment: A-91M Datums and Datum Dimensioning 326

Unit 46 327
Orientation Tolerancing 327
Assignments: A-92 Orientation Tolerancing–Angularity, Perpendicularity, Parallelism 333
A-93 Orientation Tolerancing–Angularity, Perpendicularity, Parallelism 334

Unit 47 335
Tolerancing of Features by Position 335
Coordinate Tolerancing 336
Positional Tolerancing 337
Projected Tolerance Zone 344
Assignment: A-94 Positional Tolerancing 346

Unit 48 348
Datums for Tolerancing by Position 348
Datum Targets 349
Assignment: A-95 Datums for Positional Tolerancing 352

Unit 49 353
Profile of a Line 353
Profile of a Surface 354
Assignment: A-96 Profile of Lines and Surfaces 356

Unit 50 357
Correlative Tolerances 357
Assignments: A-97M Correlative Tolerances–Coplanarity, Symmetry, Concentricity, Runout 365
A-98 Housing 366
A-99M End Plate 368

APPENDIX 369

Table
1 Abbreviations and Symbols Used on Technical Drawings 369
2 Chart for Converting Inch Dimensions to Millimeters 370
3 Number and Letter–Size Drills 371
4 Metric Twist Drill Sizes 372

5 Unified and American (Inch) Threads 373
6 Metric Threads 374
7 Common Cap Screws 375
8 Hexagon-head Bolts and Cap Screws 376
9 Setscrews 377
10 Hexagon-head Nuts 378
11 Hex Flange Nuts 379
12 Common Washer Sizes 380
13 Square and Flat Stock Keys 382
14 Woodruff Keys 382
15 American Standard Wrought Steel Pipe 383
16 Wire and Sheet-metal Gages and Thicknesses 384
17 Running and Sliding Fits (Values in Thousandths of an Inch) 386
18 Locational Clearance Fits (Values in Thousandths of an Inch) 387
19 Locational Transition Fits (Values in Thousandths of an Inch) 388
20 Locational Interference Fits (Values in Thousandths of an Inch) 389
21 Force and Shrink Fits (Values in Thousandths of an Inch) 390
22 Preferred Hole Basic Metric Fits Description 391
23 Preferred Shaft Basis Metric Fits Description 392
24 Preferred Hole Basis Metric Fits (Values in Millimeters) 393
25 Preferred Shaft Basis Metric Fits (Values in Millimeters) 395
26 Metric Conversion Tables 397

INDEX 398

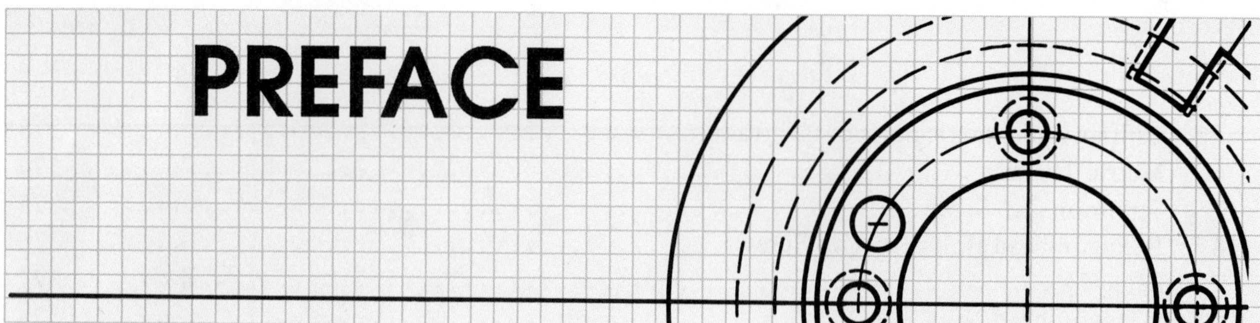

The fourth edition of INTERPRETING ENGINEERING DRAWINGS builds upon the success of previous editions in preparing students for careers in modern, technology-intensive industry. Now, more than ever, people entering industry, or those in industry who seek to upgrade their knowledge and skills, require educational materials that reflect the current state of technology.

INTERPRETING ENGINEERING DRAWINGS provides the necessary range of topics to ensure that readers who conscientiously study and practice will know how to interpret engineering drawings. The development of this skill is essential to successful employment. The text begins with the essential concepts of lines and views, proceeds through the rules of conventional dimensioning, covers specific machining practices, and concludes with extensive instruction in geometric dimensioning and tolerancing.

The features that have made the text such a valuable learning resource have been retained in the new edition. The topics build progressively from simple to complex. Numerous examples and illustrations are provided to show the applications of each concept described. Color is used extensively in the text to draw attention to important concepts and changes of state from one part or location to another. Tables are provided to summarize useful data for future reference. Similar topics are organized into discrete units of moderate length, followed by immediate application by way of assignments. Each assignment drawing contains questions that require the student to follow a logical thought process to arrive at the answer. Many questions require the application of basic mathematics to arrive at solutions. The Appendix contains 26 tables of machining data which serve as a ready reference for the solution of text problems as well as for on-the-job use.

The text consistently uses drafting practices based on the ANSI standard Y14.5M-1982. In recognition of the fact that certain industries require the use of the metric system of measurement, the principles of metric measurement are introduced early in the text and selected assignments throughout the text are dimensioned in metric units. In the Appendix, standard parts are shown in both U.S. Customary and metric sizes and standard metric fits are included.

CHANGES FOR THE FOURTH EDITION

Throughout the text, figures and assignment drawings were revised to correspond to the requirements of ANSI Y14.5M-1982. In addition to these general changes, a number of specific alterations were made.

- In early units, isometric and oblique pictorial drawings were added to assignments as an aid to visualization of the parts to assist students in drawing the missing views.
- In Unit 2, material was added on metric units of measurement and metric dimensioning. Beginning in this unit, selected assignments and assignment drawings were converted to metric units. Metric problems are identified by the letter "M" following the assignment number (such as "A-4M").
- Overall, the number of assignments was increased from 89 to 99, with a number of these added to early units to provide progressive sketching, dimensioning, and interpretative practice for both inch and millimeter dimensioned parts.
- Added to Unit 9 is a basic description of both circular tapers and flat tapers, including dimensioning of these tapers.

- In Unit 11, the use of surface texture symbols was updated, with the symbols modified to correspond to current practice; includes metric measurement in specifying finish.
- There is expanded coverage of tolerances and allowances in Unit 12, including limit dimensioning, dimension origin symbol, and the addition of millimeter tolerances.
- Unit 13 contains an expanded description of clearance, interference, and transition fits; this includes running and sliding fits, locational fits and drive and force fits; provides standard inch fits for all categories of fits; also introduces the basic hole system and the basic shaft system.
- NEW Unit 14 on metric fits covers the international tolerance grade, metric tolerance symbol, fit symbol, types and examples of millimeter fits (clearance, transition, and interference), the hole basis fits system, and the shaft basis fits system.
- Unit 15 contains updated standard thread conventions (ANSI and ISO) for internal and external threads, and an added section on metric threads.
- In Unit 20 (developments), there is added information on standard sheet metal sizes, for inch and millimeter dimensions, and a brief description of stamping (both forming and shearing).
- Unit 32, on structural shapes, provides a new figure showing examples of size designation for shapes, plates, bars, and tubes, in U.S. Customary and metric dimensions.
- Unit 33, welding drawings, was updated with clarification of the difference between the *weld* symbol and the *welding* symbol; all figures and assignments were revised to show current symbols and dimensioning practices; added content on the function of the tail of the welding symbol, multiple reference lines, and fillet welds.
- NEW Unit 34 on groove welds describes applications and dimensioning of various types of groove welds; also describes supplementary symbols–back and backing welds and the melt-through symbol.
- NEW Unit 35 covers plug welds, slot welds, spot welds, seam welds, and flange welds–applications and dimensioning. Overall, the coverage of welding symbol interpretation has been increased, with more assignments for student practice.
- Unit 36 on gears was expanded with more information on spur gears, including definitions of terms, spur gear calculations, and metric spur gears.
- NEW Unit 37 on bevel gears covers terminology and calculations.
- A separate unit on gear trains (Unit 38) was developed from existing content.
- The section on geometric tolerancing and true position (Units 42–50) was extensively updated and expanded to comply more exactly with the latest ANSI standards, including additional examples and figures to illustrate concepts and additional assignments for practice.
- The Appendix of tables was revised to include metric sizes as well as U.S. Customary sizes of parts. Metric fits were also added.

An Instructor's Guide accompanies the text and contains answers and solutions to all questions, problems and sketching exercises in the unit assignments.

ABOUT THE AUTHORS

Cecil H. Jensen, now retired, held the position of Technical Director of the McLaughlin Collegiate and Vocational Institute, Oshawa, Ontario, Canada, and has more than twenty-seven years of teaching experience in mechanical drafting. An active member of the Canadian Standards Association (CSA) Committee on Technical Drawings, Mr. Jensen has represented Canada at international (ISO) conferences on engineering drawing standards which took place in Oslo, Norway and Paris, France. He also represents Canada on the ANSI Y14.5M Committee on Dimensioning and Tolerancing. He is the successful author of numerous texts, including *Drafting Fundamentals, Engineering Drawing and Design,* and *Home Planning and Design.* Before he began teaching, Mr. Jensen spent several years in industrial design. He was also responsible for the supervision of evening courses and the teaching of selected courses for the General Motors Corporation apprentices in Oshawa, Canada.

Raymond D. Hines, now retired, contributes thirty-five years of experience in the field of electromechanical design to this text. As a Design Specialist with the General Electric Company, Guelph, Ontario, Canada, he was responsible for the internal design layout of the largest high-voltage power transformer built in North America. He is especially interested in the advancement of the drafting design profession through the latest methods of drawing presentations.

ACKNOWLEDGMENTS

The authors and staff at Delmar Publishers wish to express their appreciation to the following individuals for their thorough, professional reviews of the manuscript for the fourth edition.

Thomas Acuff, Industrial Technology Department, Black Hawk College, Moline, IL 61265
Erwin J. Seel, American Welding Society, Miami, FL 33135
Edward Wheaton, Aiken Technical College, Aiken, SC 29802
William Pearson, Black Hawk College, Moline, IL 61265
Ronald Butt, Waukesha County Technical Institute, Pewaukee, WI 53130
Brian Jacobs, Conestoga College, Kitchener, Ontario, Canada N2G 3W5

Appreciation is also expressed to the following instructors for their recommendations which guided the authors in developing the plan for the fourth edition.

John Phyllon, Milwaukee Area Technical College, Oak Creek, WI 53154
Ray Anderson, Schoolcraft College, Livonia, MI 48152
Donald Hartshorn, Columbus Technical Institute, Columbus, OH 43215
Gary Uthe, Anoka Vocational Technical Institute, Anoka, MN 55303
Robert J. Simon, Delta College, Bay City, MI 48706
James H. Rhea, Asheville-Buncombe Technical Institute, Asheville, NC 28801
Robert Thomas, Milwaukee Area Technical College, Milwaukee, WI 53203
Ken Franken, Linn Technical College, Linn, MO 65051

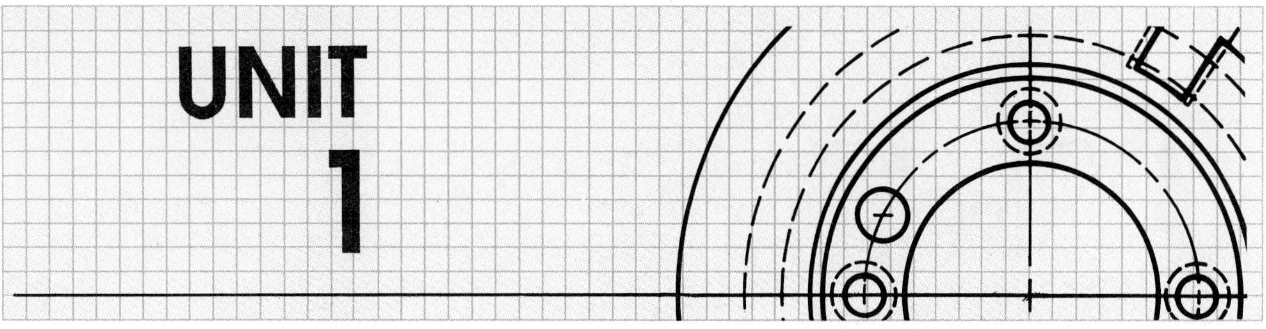

BASES FOR INTERPRETING DRAWINGS

Engineering or *technical drawings* furnish a description of the shape and size of an object. Other information necessary for the construction of the object is given in a way that renders it readily recognizable to anyone familiar with engineering drawings.

Pictorial drawings are similar to photographs, because they show objects as they would appear to the eye of the observer, figure 1-1. Such drawings, however, are not often used for technical designs since interior features and complicated detail are easier to understand and dimension on orthographic drawings. The drawings used in industry must clearly show the exact shape of objects. This usually cannot be accomplished in just one pictorial view, because many details of the object may be hidden or not clearly shown when the object is viewed from only one side.

For this reason, the drafter must show a number of views of the object as seen from different directions. These views, referred to as front view, top view, right-side view, and so forth, are systematically arranged on the drawing sheet and projected from one another, figure 1-2. This type of projection is called *orthographic projection.* The ability to understand and visualize an object from these views is essential in the interpretation of engineering drawings.

The principles of orthographic projection can be applied in four angles, or systems: first-, second-, third-, and fourth-angle projection. However, only two systems, first- and third-angle projection are used. *Third-angle projection* is used in the United States, Canada, and many other countries. *First-angle projection* is used mainly in Europe. Because world trade has brought about the exchange of both engineering drawings and products, drafters are now called upon to communicate in both types of orthographic projection, as well as pictorial representations.

THIRD-ANGLE PROJECTION

The third-angle system of projection is used almost exclusively on mechanical engineering drawings because it permits each feature of the object to be drawn in true proportion and without distortion along all dimensions.

Three views are usually sufficient to describe the shape of an object. The views most commonly used are the front, top, and right-side. In

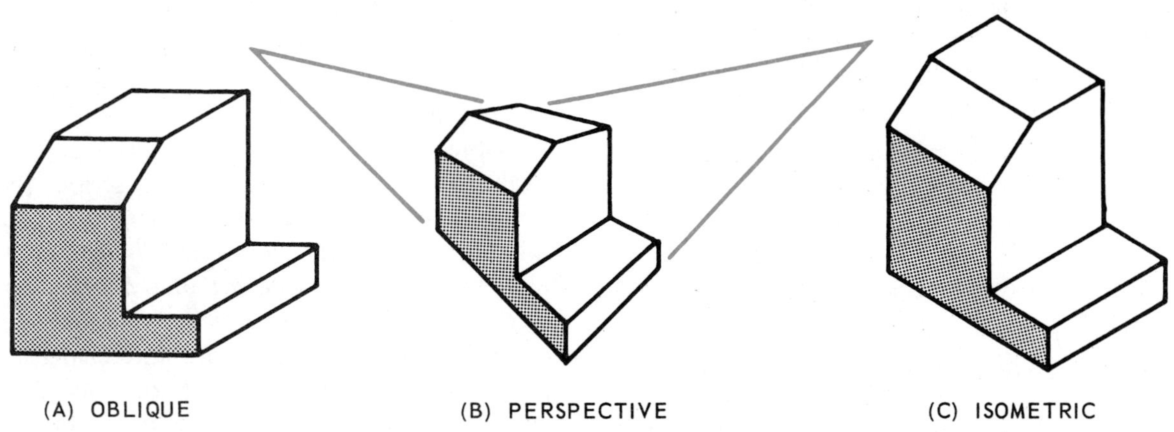

Fig. 1-1 Pictorial drawings

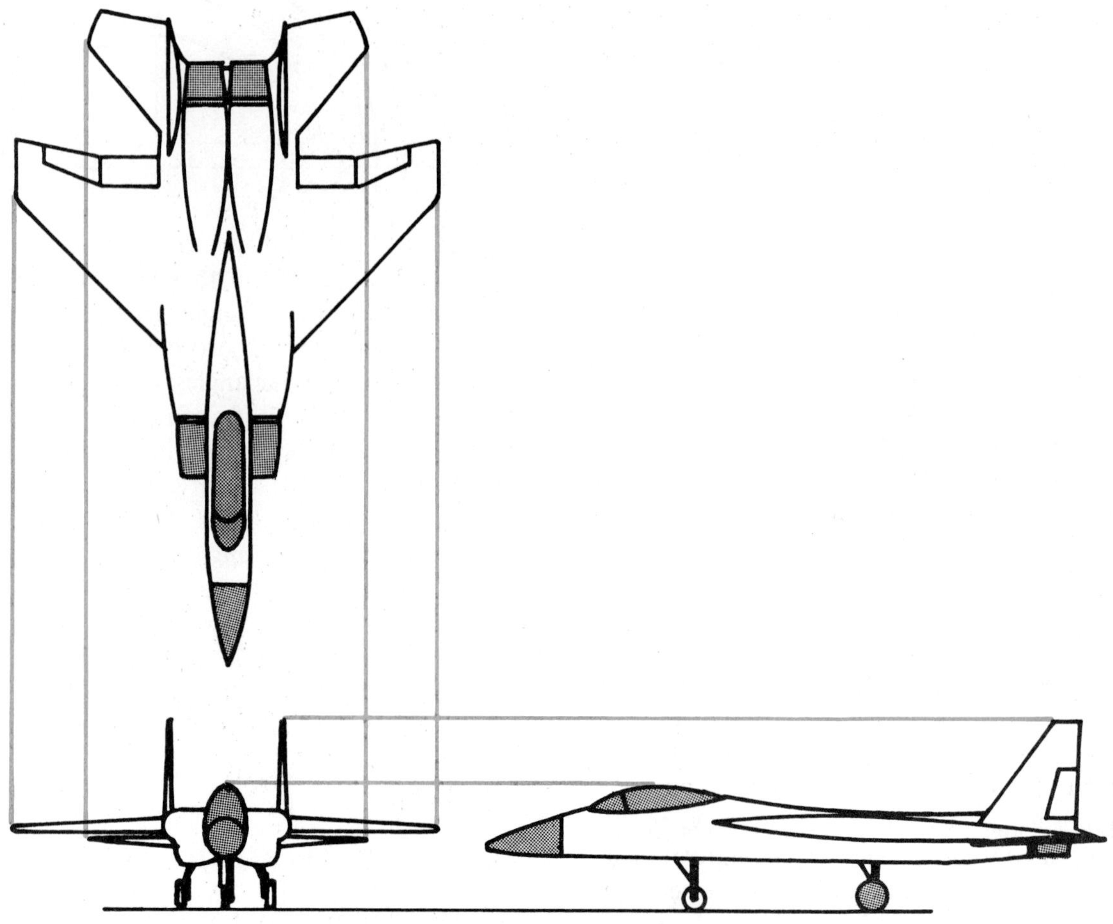

Fig. 1-2 Systematic arrangement of views

third-angle projection the object may be assumed to be enclosed in a glass box, figure 1-3(B). A view of the object drawn on each side of the box represents that which is seen when looking perpendicularly at each face of the box. If the box were unfolded as if hinged around the front face, the desired orthographic projection would result, figure 1-3(C) and (D). These views are identified by names as shown. With reference to the front view:

- the top view is placed above.
- the bottom view is placed underneath.
- the left view is placed on the left.
- the right view is placed on the right.
- the rear view is placed at the extreme left or right, whichever is convenient.

When looking at objects, we normally see them as three-dimensional; as having width, depth and height; or length, width and height. The choice of terms used depends on the shape and proportions of the object.

Spherical shapes, such as a basketball, would be described as having a certain *diameter* (one term).

Cylindrical shapes, such as a baseball bat, would have *diameter* and *length.* However, a hockey puck would have *diameter* and *thickness* (two terms).

Objects which are not spherical or cylindrical require *three* terms to describe their overall shape. The terms used for a car would probably be *length, width* and *height;* for a filing cabinet–*width, height* and *depth;* for a sheet of drawing paper–*length, width* and *thickness.* The terms used are interchangeable according to the *proportions* of the object being described, and the *position* it is in when being viewed. For example, a telephone pole

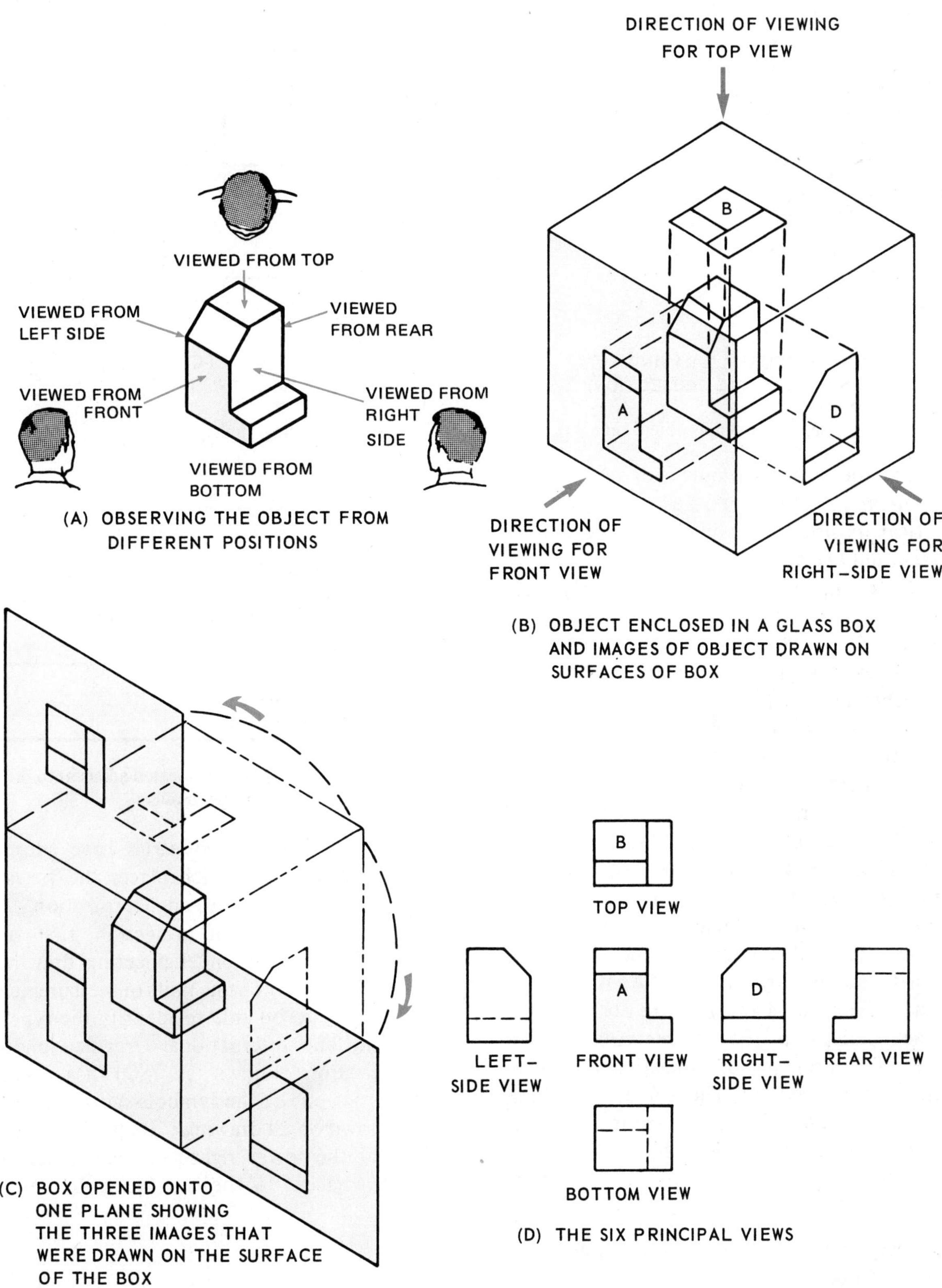
VIEWED FROM TOP
VIEWED FROM LEFT SIDE
VIEWED FROM REAR
VIEWED FROM FRONT
VIEWED FROM RIGHT SIDE
VIEWED FROM BOTTOM
(A) OBSERVING THE OBJECT FROM DIFFERENT POSITIONS
DIRECTION OF VIEWING FOR TOP VIEW
B
A
D
DIRECTION OF VIEWING FOR FRONT VIEW
DIRECTION OF VIEWING FOR RIGHT–SIDE VIEW
(B) OBJECT ENCLOSED IN A GLASS BOX AND IMAGES OF OBJECT DRAWN ON SURFACES OF BOX
(C) BOX OPENED ONTO ONE PLANE SHOWING THE THREE IMAGES THAT WERE DRAWN ON THE SURFACE OF THE BOX
B
TOP VIEW
A
D
LEFT–SIDE VIEW
FRONT VIEW
RIGHT–SIDE VIEW
REAR VIEW
BOTTOM VIEW
(D) THE SIX PRINCIPAL VIEWS

Fig. 1-3 Third-angle orthographic projection

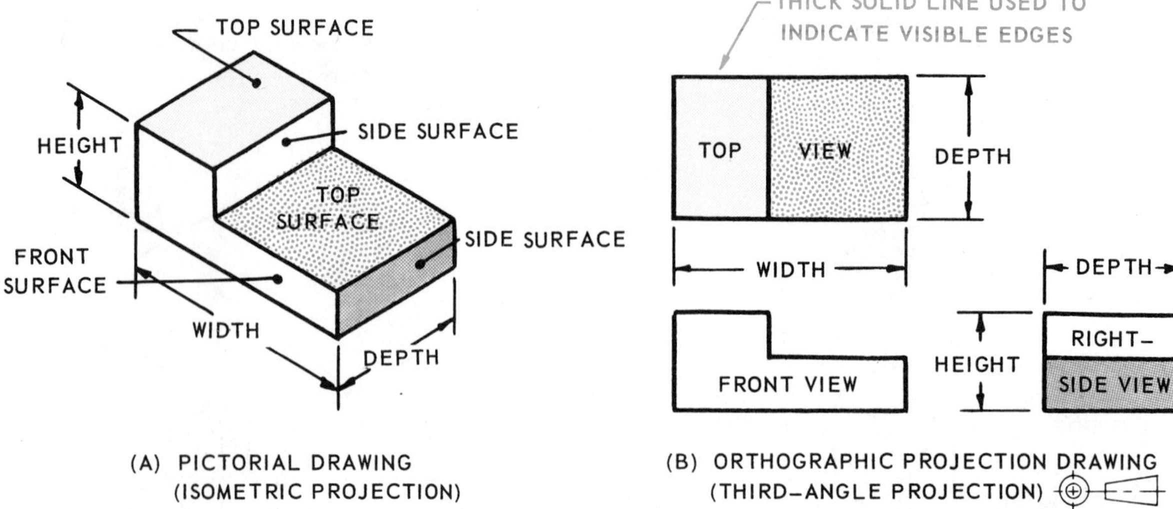

Fig. 1-4 A simple object shown in (A) pictorial form, and (B) orthographic projection

lying on the ground would be described as having *diameter* and *length,* but when placed in a vertical position, its dimensions would be *diameter* and *height.*

In general, distances from left to right are referred to as width or length, distances from front to back as depth or width, and vertical distances (except when very small in proportion to the others) as height.

On drawings, the multidimensional shape is represented by a view or views on the flat surface of the drawing paper. Seldom are more than three views necessary to completely describe the shape of an object. Therefore, the simple object shown in figure 1-4 can be used to illustrate the positions of these principal dimensions.

In figure 1-4, the object is shown in (A) pictorial form, and (B) orthographic projection. The orthographic drawing uses each view to represent the exact shape and size of the object and the relationship of the three views to one another. This principle of projection is used in all mechanical drawings. The isometric drawing shows the relationship of the front, top, and right-side surfaces in a single view.

ISO PROJECTION SYMBOL

Today, two systems of orthographic projection are used on engineering drawings. Third-angle orthographic projection is used by many countries, including the United States and Canada, and thus is used in this text. Most

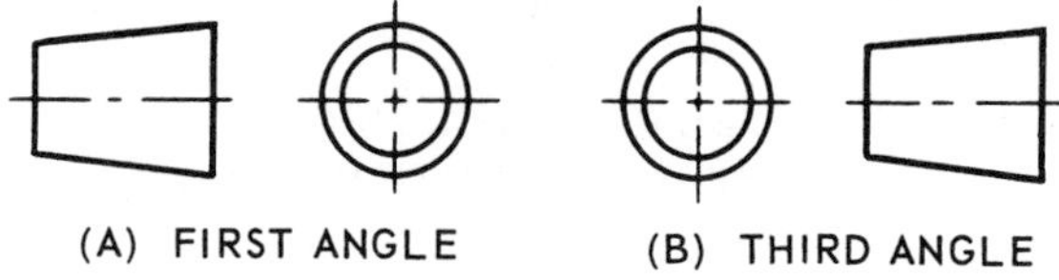

Fig. 1-5 ISO projection symbols

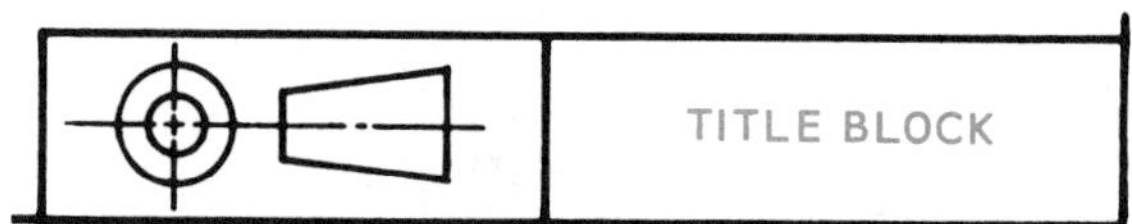

Fig. 1-6 The ISO symbol is located adjacent to the title block on the drawing

European and Asiatic countries have adopted first-angle projection which places the views in opposite positions to third-angle projection.

Since two types of projection, first- and third-angle, are used on engineering drawings, and since each has the same units of measurement, it is necessary to be able to identify the type of projection. The International Organization for Standardization (known as ISO) has recommended that one of the symbols shown in figure 1-5 be shown on all drawings. Its preferred location is in the lower right-hand corner of the drawing, adjacent to the title block, figure 1-6.

TITLE BLOCK

All drawings will have some form of title block, usually placed in the lower right-hand corner. A title block may contain such information as the

- name of the part.
- order number.
- date.
- drawing number.
- scale size used.
- name of the drafter.
- name of the drawing checker.
- material to be used.

ABCDEFGHIJKLMNOP
QRSTUVWXYZ
1234567890
VERTICAL

ABCDEFGHIJKLMNOP
QRSTUVWXYZ
1234567890
SLOPED

Fig. 1-7 Lettering for drawings

DRAWING STANDARDS

The drawings and information shown throughout this text are based on the *American National Standard Drafting Practices Y-14,* and any approved revisions. In some areas of drawing practice, such as in simplified drafting, national standards have not yet been established. The authors have, in such cases, adopted the practices used by leading industries in the United States.

VISIBLE LINES

A thick solid line is used to indicate the visible edges and corners of an object. Visible lines should stand out clearly in contrast to other lines, making the general shape of the object apparent to the eye.

LETTERING ON DRAWINGS

The most important requirements for lettering used on engineering drawings are legibility and reproducibility. These requirements are best met by the style of lettering known as standard uppercase Gothic, as shown in figure 1-7. Vertical lettering is preferred, but sloping style may be used, though never on the same drawing as vertical lettering. Suitable lettering size for notes and dimensions is .10 inch (in.). Larger characters are used for drawing titles and number. They are also used where it may be necessary to bring some part of the drawing to the attention of the reader.

SKETCHING

Sketching is a necessary part of the course of interpreting technical drawings since the skilled technician in the shop is frequently called upon to sketch and explain points to other people. Sketching also helps to develop a good sense of proportion and accuracy of observation.

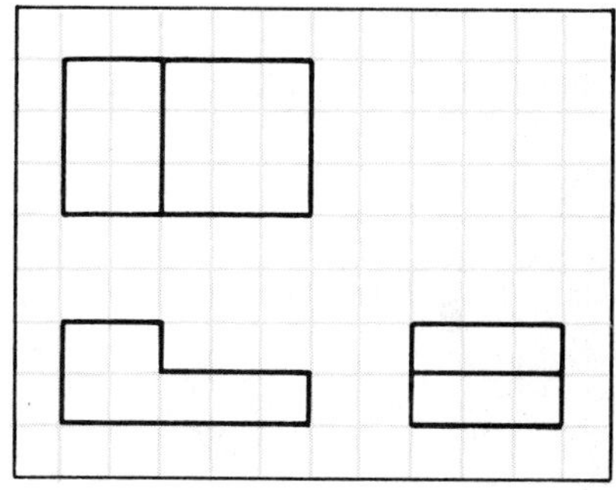

(A) COORDINATE SKETCHING PAPER USED FOR SKETCHING ORTHOGRAPHIC PROJECTION

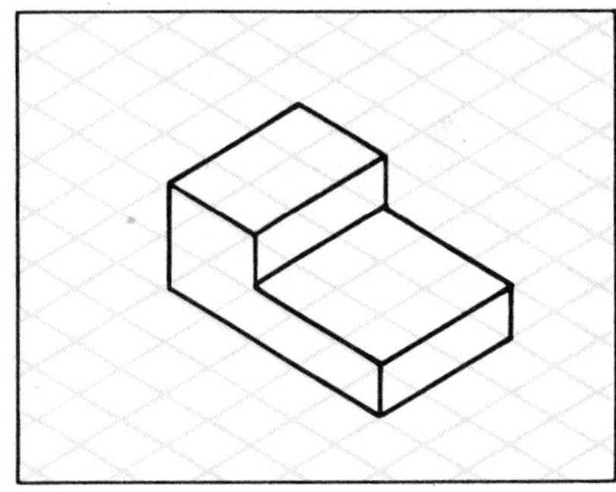

(B) ISOMETRIC SKETCHING PAPER USED FOR SKETCHING PICTORIAL DRAWINGS

Fig. 1-8 Sketching paper

The most common types of sketching paper are shown in figure 1-8. Each square on the paper may represent .10 in., .25 in., 1.00 in., one foot (ft.), or 10 mm of actual object length. Figure 1-8 illustrates the use of graph paper for sketching (A) orthographic projection, and (B) pictorial drawings. Figure 1-9 shows some simple objects drawn in orthographic projection.

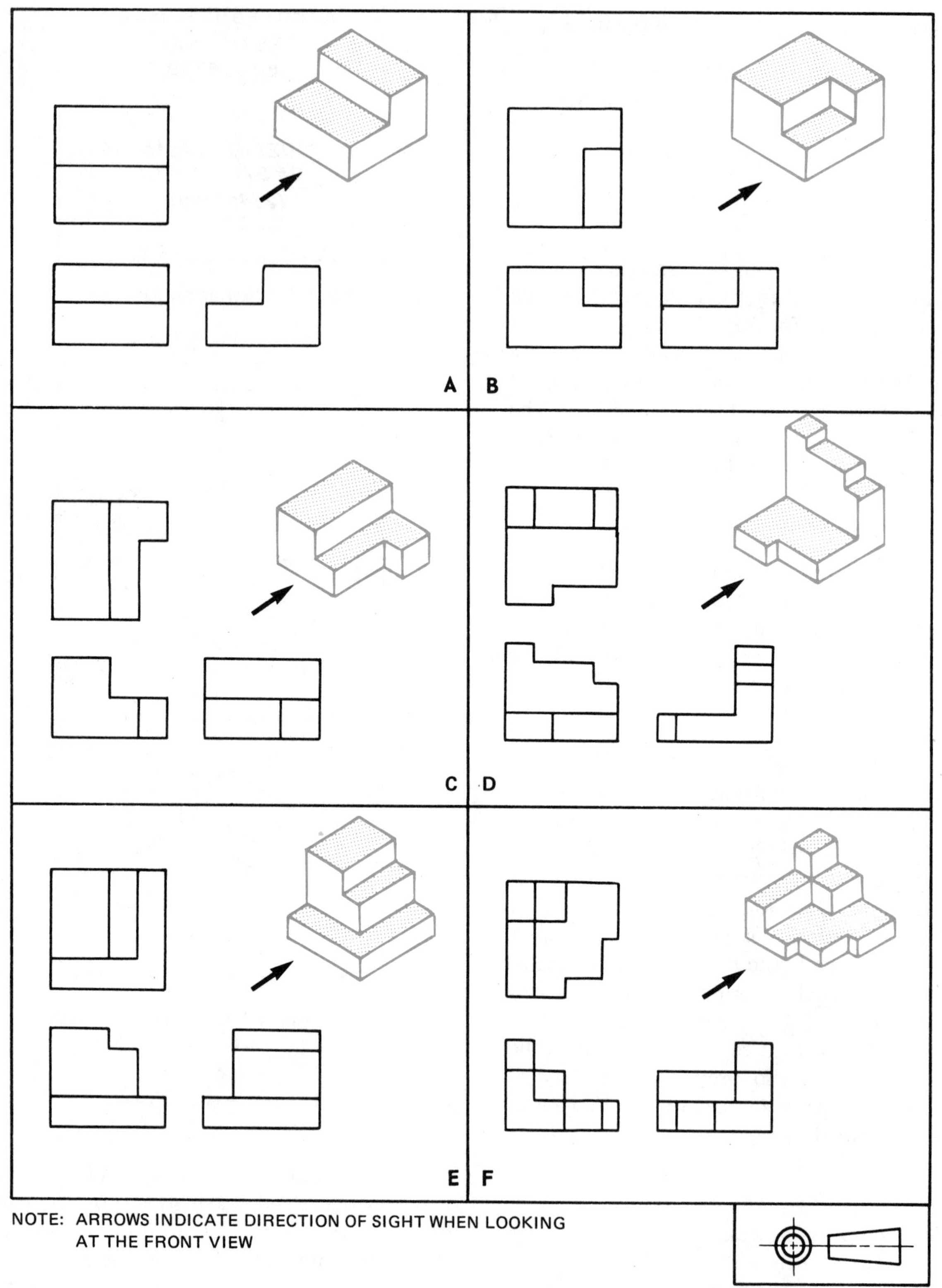
A
B
C
D
E
F
NOTE: ARROWS INDICATE DIRECTION OF SIGHT WHEN LOOKING AT THE FRONT VIEW

Fig. 1-9 Illustrations of simple objects drawn in orthographic projection

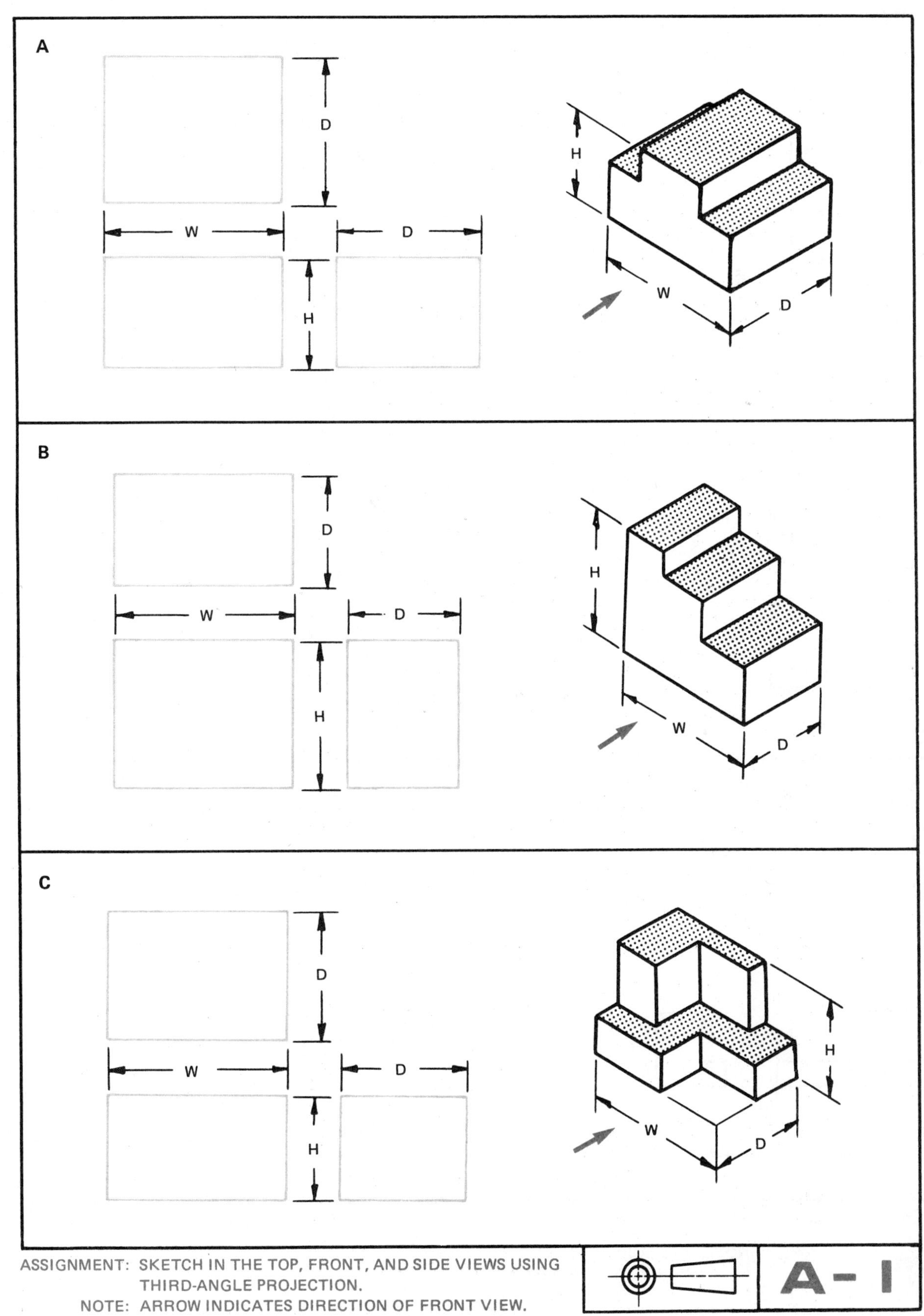

ASSIGNMENT: SKETCH IN THE TOP, FRONT, AND SIDE VIEWS USING THIRD-ANGLE PROJECTION.

NOTE: ARROW INDICATES DIRECTION OF FRONT VIEW.

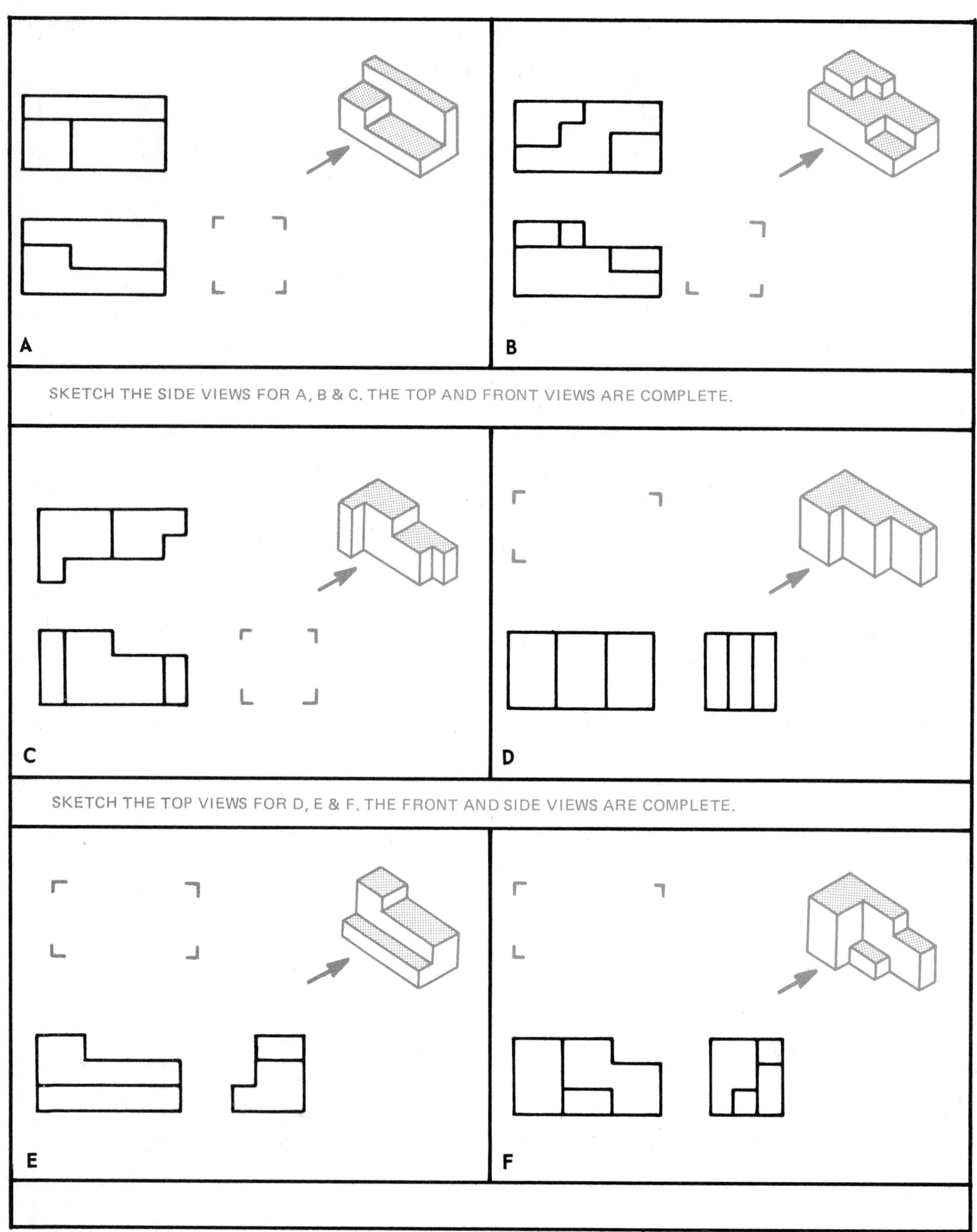

A
B
SKETCH THE SIDE VIEWS FOR A, B & C. THE TOP AND FRONT VIEWS ARE COMPLETE.
C
D
SKETCH THE TOP VIEWS FOR D, E & F. THE FRONT AND SIDE VIEWS ARE COMPLETE.
E
F

COMPLETION ASSIGNMENT:

NOTE: ALL SURFACES ARE VERTICAL OR HORIZONTAL.

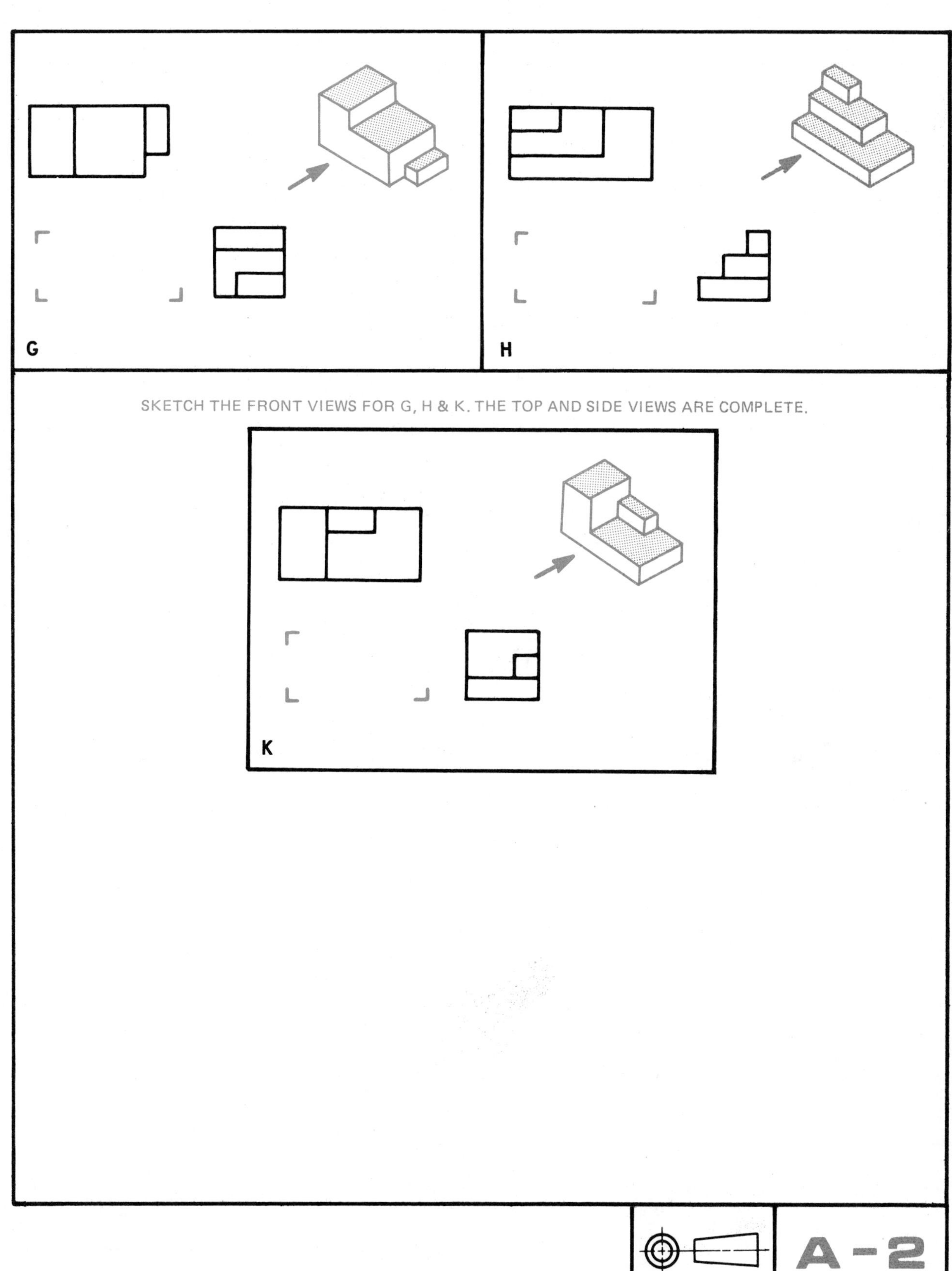
G
H
SKETCH THE FRONT VIEWS FOR G, H & K. THE TOP AND SIDE VIEWS ARE COMPLETE.
K
A-2

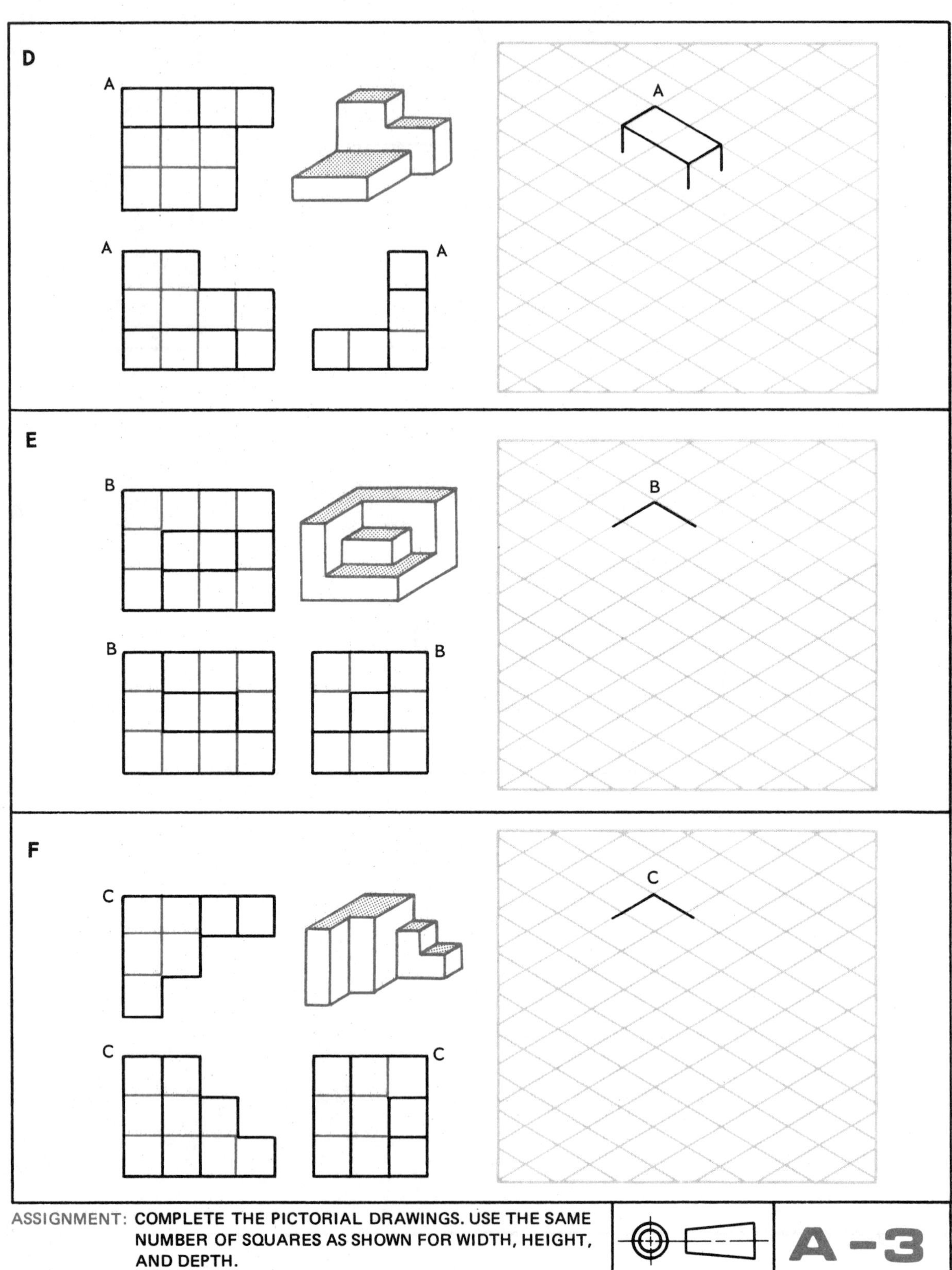

ASSIGNMENT: COMPLETE THE PICTORIAL DRAWINGS. USE THE SAME NUMBER OF SQUARES AS SHOWN FOR WIDTH, HEIGHT, AND DEPTH.

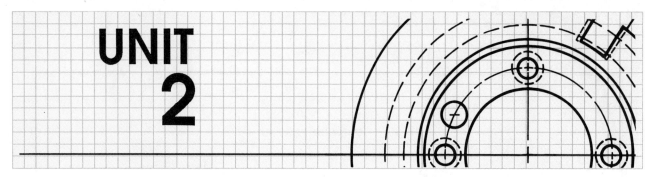

UNIT 2

WORKING DRAWINGS

A *working drawing,* sometimes referred to as a *detail* or *assembly drawing,* contains the complete information for the construction of an object or the assembly of a product. The information found on this drawing may be classified under three headings:

- **Shape.** The description of the shape of an object is indicated by the number and type of views and lines selected to completely show or describe the form of an object.
- **Dimensioning.** The description of the size and location of the shape of an object is indicated by lines and numerals used to specify the dimensions of an object.
- **Specifications.** Additional information such as general notes, material, heat treatment, machine finish, and applied finish are indicated by special entries written on the drawings. Such information is found on the drawing or in the title block.

DIMENSIONING

Dimensions are indicated on drawings by extension lines, dimension lines, leaders, arrowheads, figures, notes, and symbols. These lines and dimensions define such geometrical characteristics as distances, diameters, angles, and locations, figure 2-1. The lines used in dimensioning are thin in contrast to the outline of the object. The dimension must be clear and concise, permitting only one interpretation. In general, each surface, line, or point is located by only one set of dimensions. Exceptions to these rules are the two types of rectangular coordinate dimensioning discussed in Unit 17.

Placement of Dimensions

The two methods of dimensioning most commonly used are the *unidirectional system* and the *aligned system,* figure 2-2. The aligned system, which is read from the bottom and right side of the drawing, is becoming outdated. The method currently preferred is the unidirectional system which is read from the bottom only.

Dimension Lines

Dimension lines denote particular sections of the object. They should be drawn parallel to the section they define. Dimension lines terminate in arrowheads which touch an extension line and are broken in order to allow the insertion of the dimension. Unbroken lines are sometimes used as a simplified drafting practice, with the dimension placed above the line. Where space does not permit the insertion of the dimension line and the dimension between the extension lines, the dimension line may be placed outside the extension line. The dimension can also be placed outside the extension line if the space between the extension lines is limited. In restricted areas a dot may be used instead of two arrowheads. These methods are shown in figure 2-3.

Extension Lines

Extension lines denote the points or surfaces between which a dimension applies. They extend from object lines and are drawn perpendicular to the dimension lines (as shown in figures 2-1 and 2-2). A small gap (.03 to .06 in) is left between the extension line and the outline to which it refers.

Where extension lines cross arrowheads or dimension lines are close to arrowheads, a break in the extension line is permitted, figure 2-4.

Leaders

Leaders are used to direct dimensions or notes to the surface or points to which they

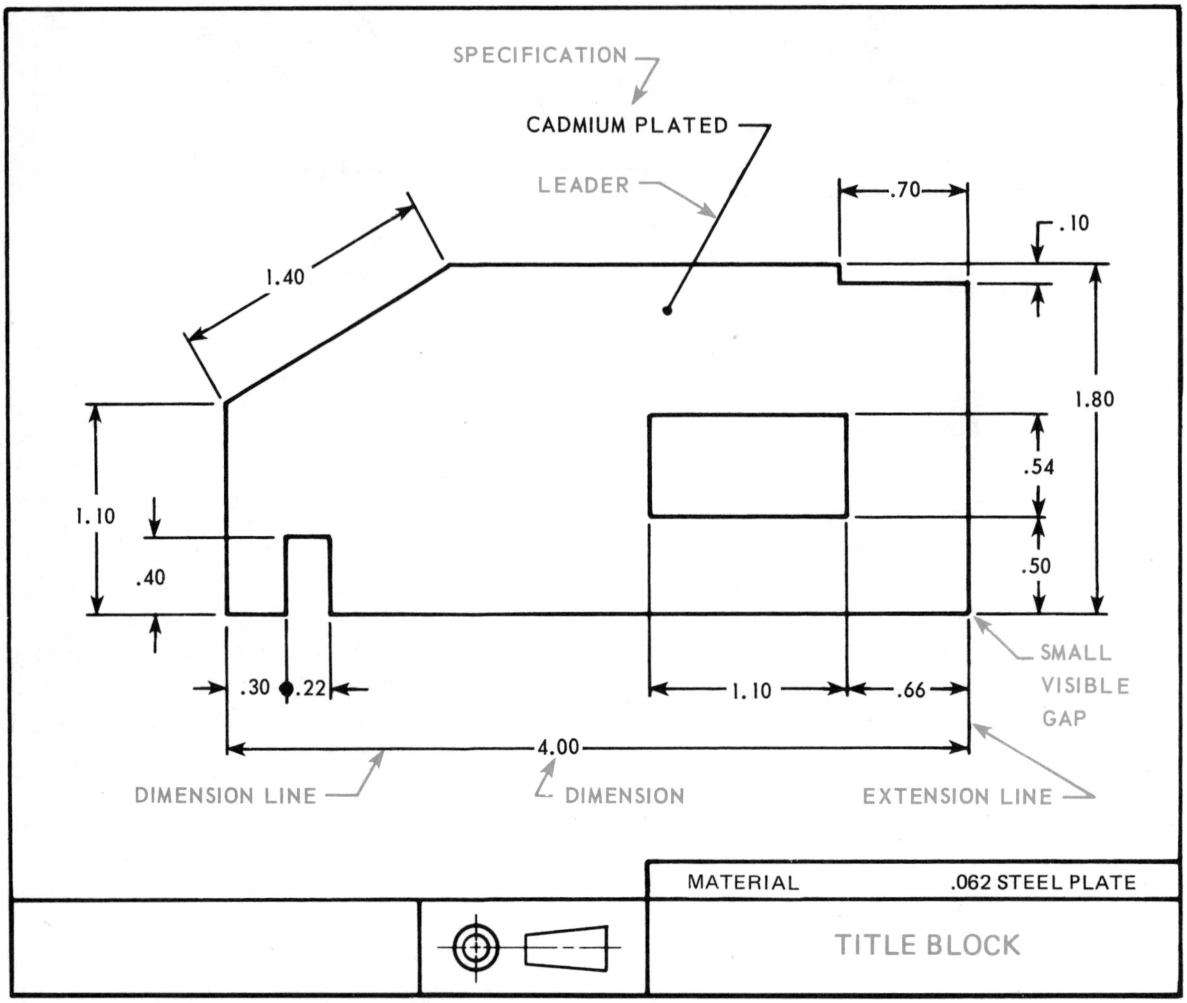

Fig. 2-1 Basic dimensioning elements

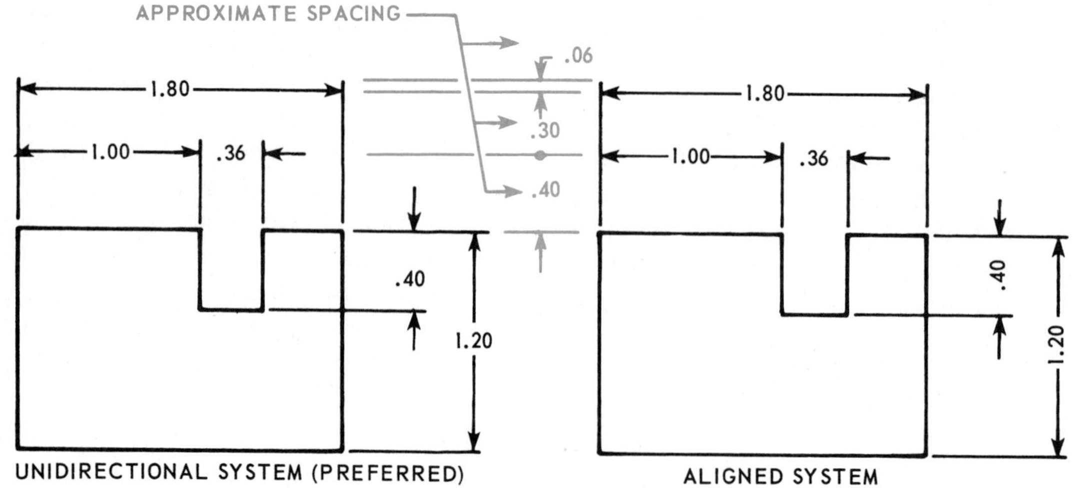

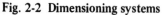

Fig. 2-2 Dimensioning systems

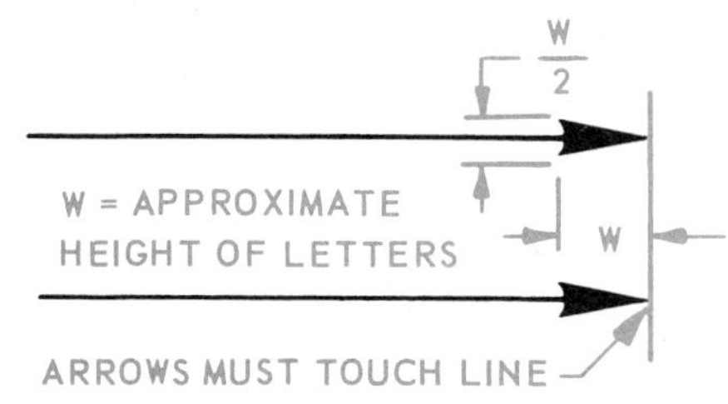

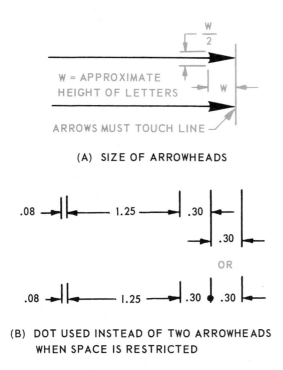

(A) SIZE OF ARROWHEADS

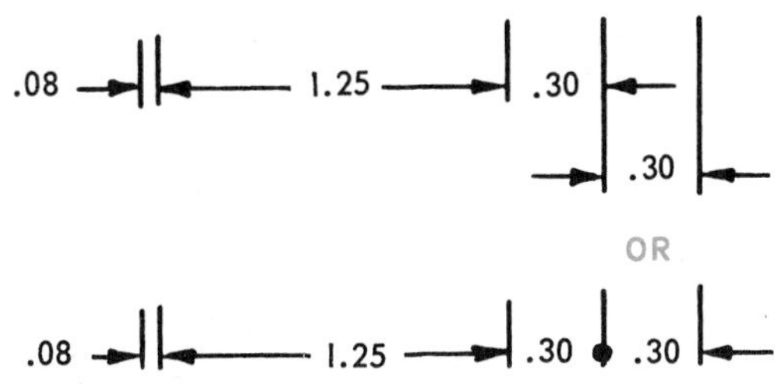

(B) DOT USED INSTEAD OF TWO ARROWHEADS WHEN SPACE IS RESTRICTED

Fig. 2-3 Arrowheads

apply, figure 2-5. A leader consists of a line with or without short horizontal bars adjacent to the note or dimension, and an inclined portion which terminates with an arrowhead touching the line or point to which it applies. A leader may terminate with a dot when it refers to a surface within the outline of a part.

NOTE: If by chance a dimension is omitted on a drawing, contact the drafting department. *Never* scale a drawing for a missing dimension.

LINEAR UNITS OF MEASUREMENT

Although the metric system of dimensioning is expected to become the official standard of measurement, most drawings in existence in North America today are dimensioned in inches or feet and inches. For this reason, drafters and persons involved in the reading of engineering drawings should be familiar with all the dimensioning systems they may encounter.

The dimensions used in this book are primarily decimal inch. However, metric dimensions are used very frequently. When metric units of measurement are used, the drawing prominently displays the word METRIC and a note stating the dimensions are in millimeters.

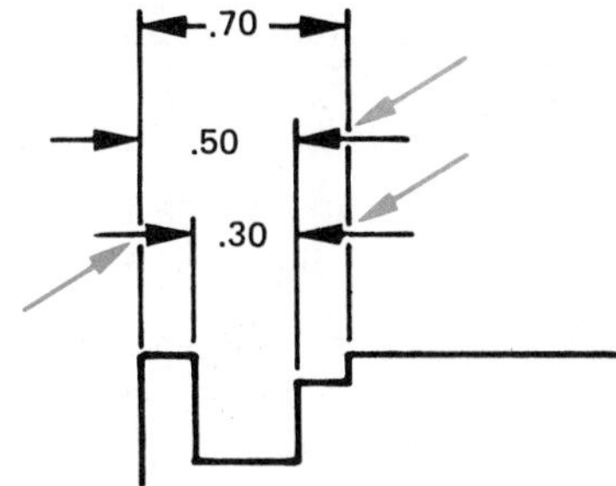

Fig. 2-4 Breaks in extension lines

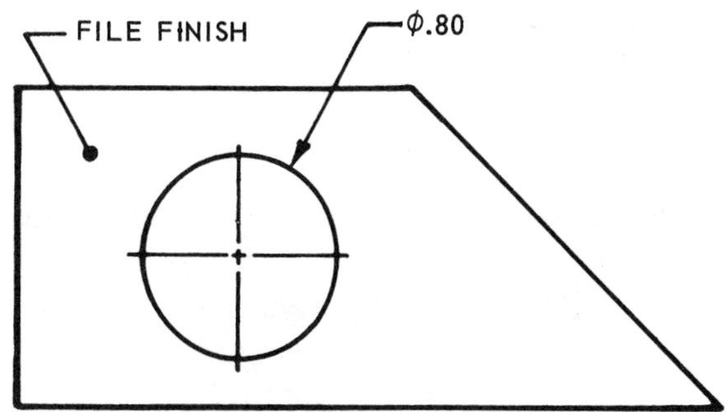

Fig. 2-5 Using leaders for dimensions and notes

Inch Units of Measurement

The Decimal-inch System (U.S. Customary). In the decimal-inch system, parts are designed in basic decimal increments, preferably .02 inch, and are expressed as two-place decimal numbers. Using the .02 module, the second decimal place (hundredths) is an even number or zero, figure 2-6. Sizes other than these, such as .25, are used when they are essential to meet design requirements. When greater accuracy is required, sizes are expressed as three- or four-place decimal numbers such as 1.875 or 4.5625.

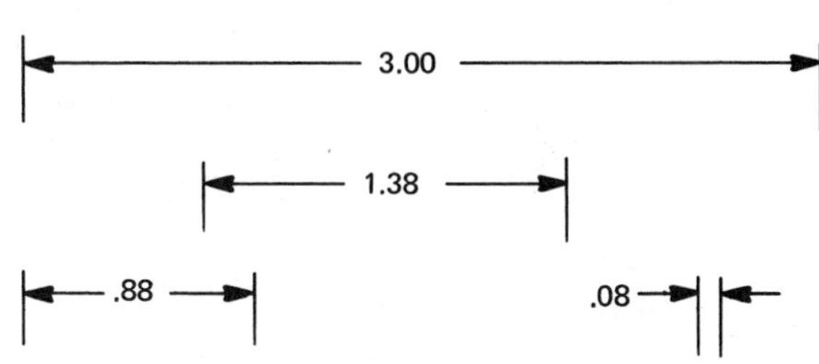

Fig. 2-6 Decimal-inch dimensioning

The Fractional-inch System. In this system, sizes are expressed in common fractions, the smallest division being 64ths, figure 2-7. Sizes other than common fractions are expressed as decimals.

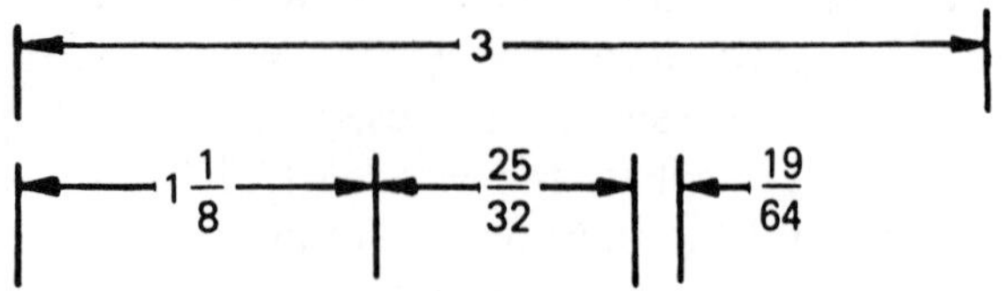

Fig. 2-7 Fractional-inch dimensioning

The Feet and Inches System. This system is used in architectural drawing and for large structural and installation drawings. Dimensions greater than 12 inches are given in feet and inches, with parts of an inch given as common fractions. Inch marks are not used, and a zero is added to indicate no full inches. A dash mark is placed between the foot and inch values, figure 2-8.

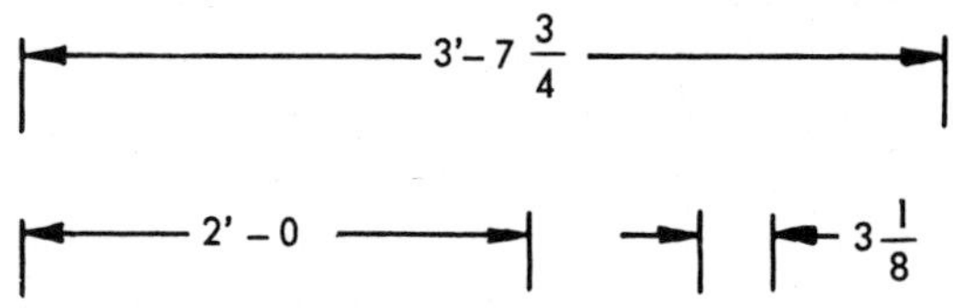

Fig. 2-8 Feet and inch dimensioning

SI (Metric) Units of Measurement

The standard metric units on engineering drawings are the millimeter for linear measure and the micrometer for surface roughness. For architectural drawings meter and millimeter units are used, figure 2-9.

In metric dimensioning, as in decimal-inch dimensioning, numerals to the right of the decimal point indicate the degree of precision.

Whole dimensions do not require a zero to the right of the decimal point.

2 not 2.0
10 not 10.0

A millimeter value of less than one is shown with a zero to the left of the decimal point.

0.2 not .2 or .20
0.26 not .26

Commas should not be used to separate groups of three numbers in metric values. A space should be used in place of the comma.

32 541 not 32,541
2.562 826 6 not 2.5628266

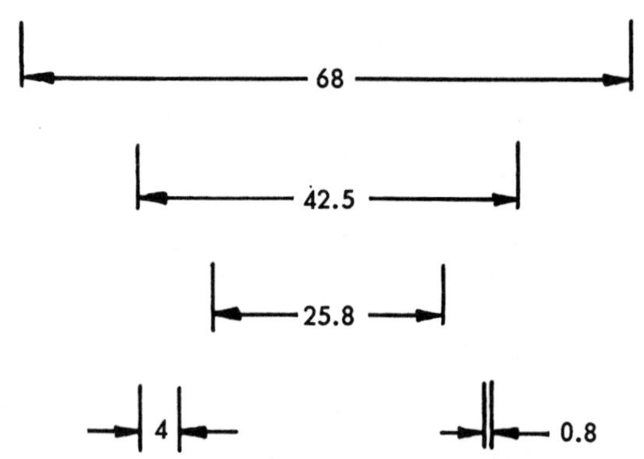

Fig. 2-9 Metric (millimeter) dimensioning

Identification. A metric drawing should include a general note, such as UNLESS OTHERWISE SPECIFIED, DIMENSIONS ARE IN MILLIMETERS. In addition, a metric drawing should be identified by the word METRIC prominentiy displayed near the title block.

CHOICE OF DIMENSIONS

The choice of the most suitable dimensions and dimensioning methods depends, to some extent, on whether the drawings are intended for unit production or mass production.

Unit production refers to applications in which each part is to be made separately, using general-purpose tools and machines. Details on custom-built machines, jigs, fixtures, and gauges required for the manufacture of production parts are made in this way. Frequently, only one of each part is required.

Mass production refers to parts produced in quantity, for which special tooling is usually provided. Most part drawings for manufactured products are considered to be for mass-produced parts.

Functional dimensioning should be expressed directly on the drawing, especially for mass-produced parts. This will result in the selection of datum features on the basis of function and assembly. For unit-produced parts, it is generally preferable to select datum features on the basis of manufacture and machining, figure 2-10.

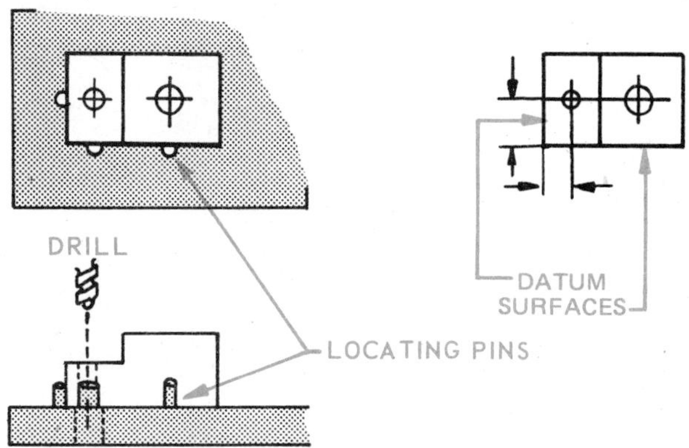

Fig. 2-10 Selection of datum surfaces for dimensioning

BASIC RULES FOR DIMENSIONING

- Place dimensions between the views when possible, figure 2-11(A).
- Place the dimension line for the shortest width, height, and depth, nearest the outline of the object, figure 2-11(B). Parallel dimension lines are placed in order of their size, making the longest dimension line the outermost line.
- Place dimensions near the view that best shows the characteristic contour or shape of the object, figure 2-11(C). In following this rule, dimensions will not always be between views.
- On large drawings, dimensions can be placed on the view to improve clarity.
- Use only one system of dimensions, either the unidirectional or the aligned, on any one drawing.

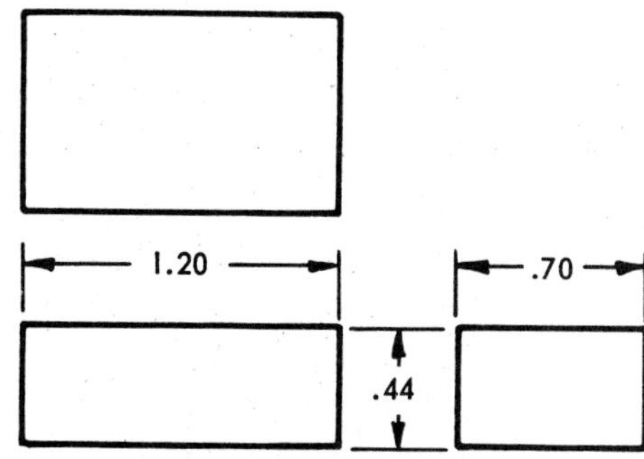

(A) PLACE DIMENSIONS BETWEEN VIEWS

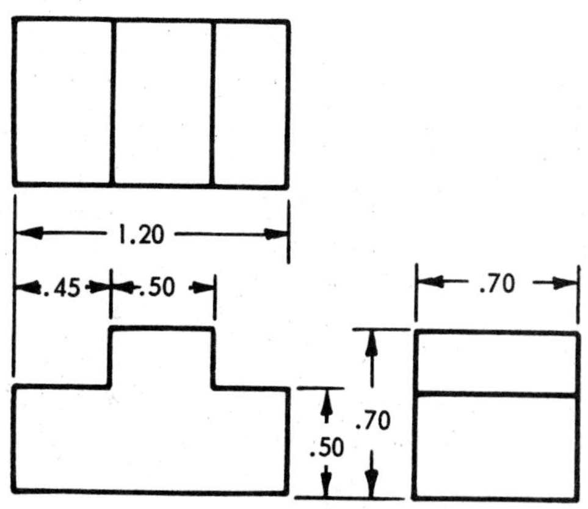

(B) PLACE SMALLEST DIMENSION NEAREST THE VIEW BEING DIMENSIONED

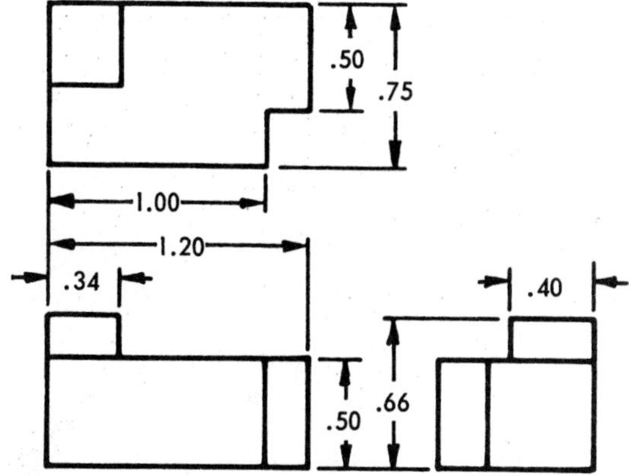

(C) DIMENSION THE VIEW THAT BEST SHOWS THE SHAPE

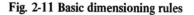

Fig. 2-11 Basic dimensioning rules

INFORMATION SHOWN ON ASSIGNMENT DRAWINGS

Circled numbers and letters shown in color are used to identify lines, distances, and surfaces on the drawing assignments so that questions may be asked about these features. To simplify the drawings, the actual working drawing is drawn in black. The information which is used in the developing of interpreting technical drawings is shown in color and would not appear on working drawings found in industry.

REFERENCES AND SOURCE MATERIALS

1. American National Standard Drafting Practices Y-14, and any approved revisions.

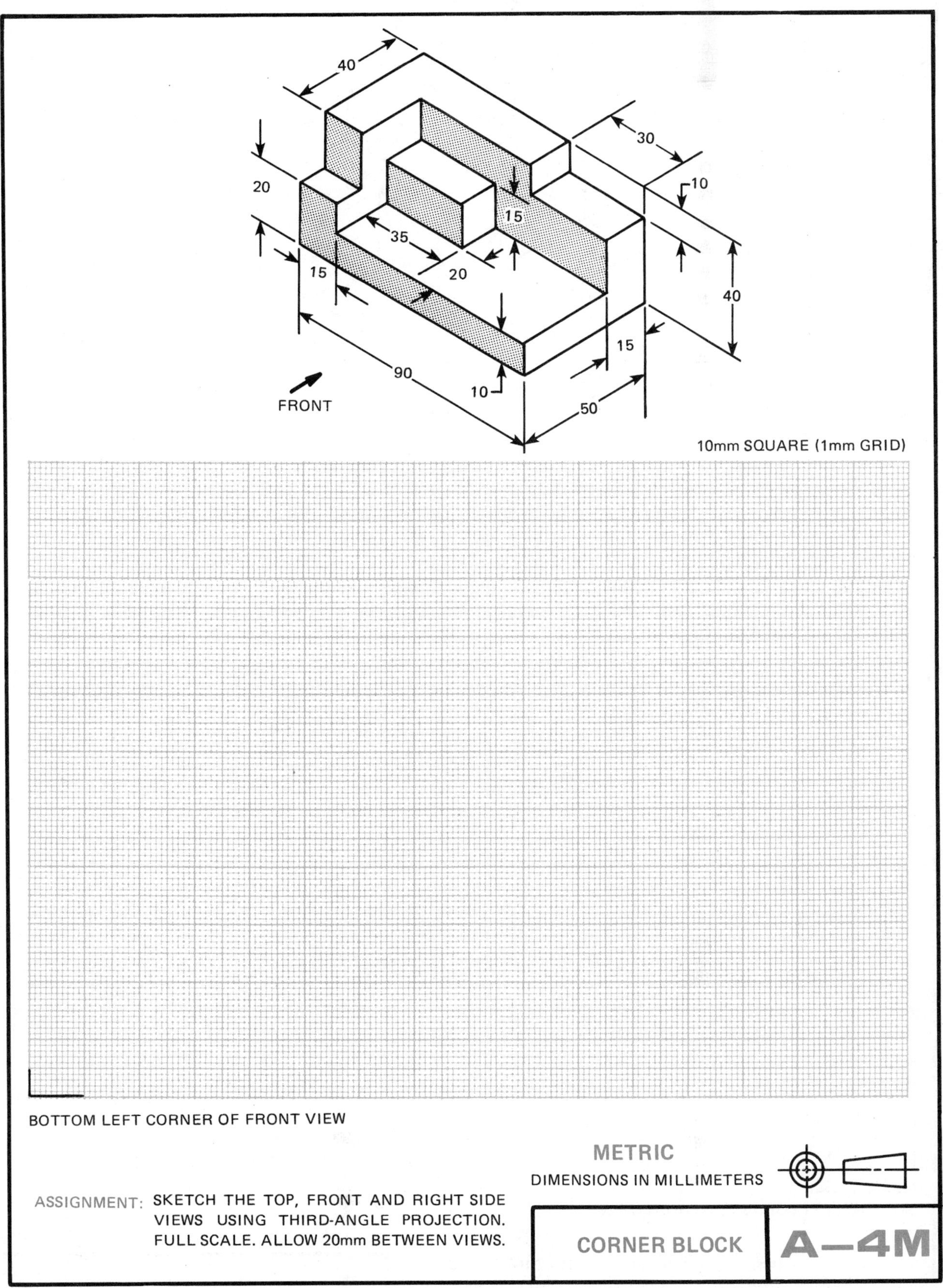
40
30
20
10
15
35
15
20
40
15
90
10
50
FRONT
10mm SQUARE (1mm GRID)
BOTTOM LEFT CORNER OF FRONT VIEW
METRIC
DIMENSIONS IN MILLIMETERS
ASSIGNMENT: SKETCH THE TOP, FRONT AND RIGHT SIDE VIEWS USING THIRD-ANGLE PROJECTION. FULL SCALE. ALLOW 20mm BETWEEN VIEWS.
CORNER BLOCK
A—4M

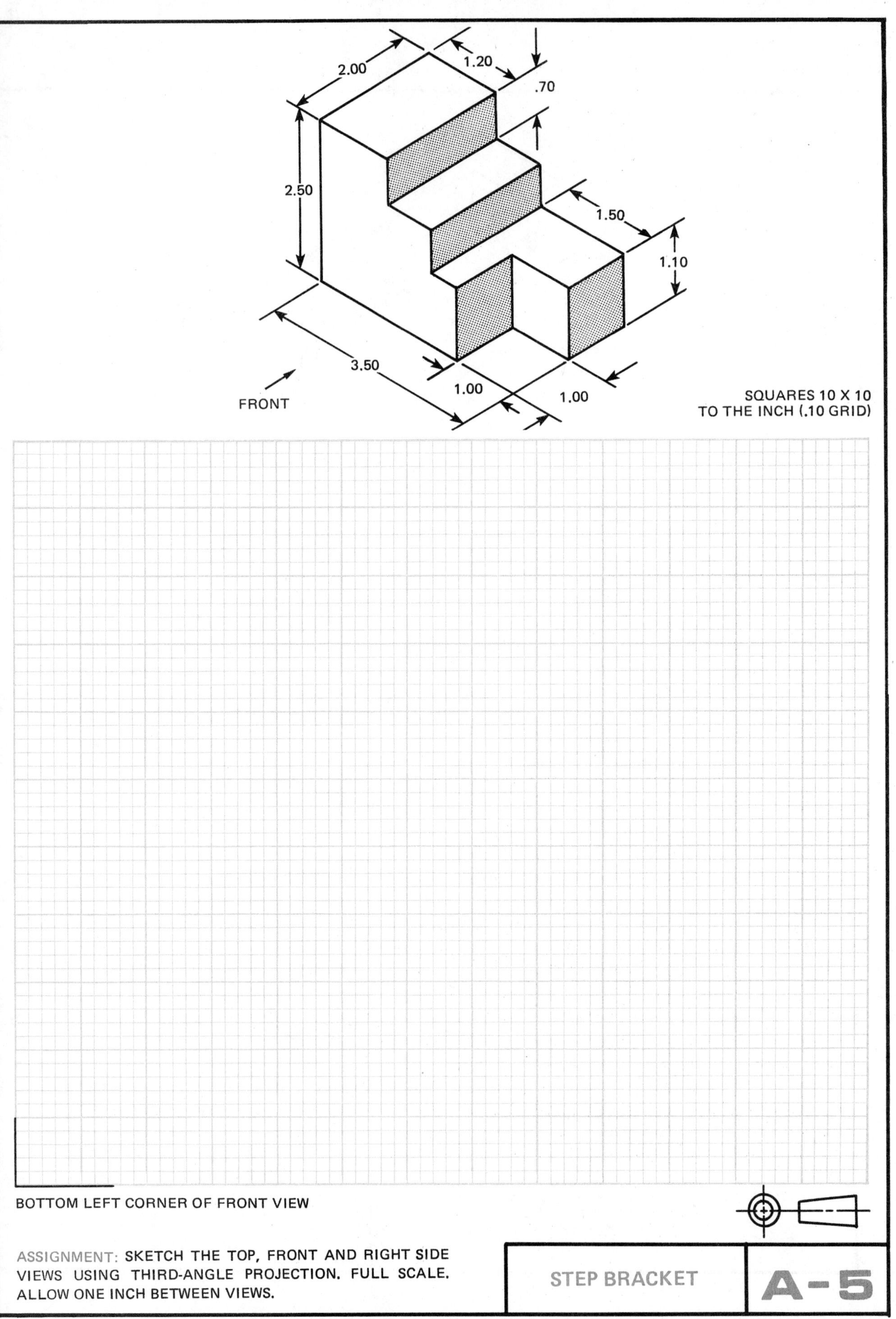

ASSIGNMENT: SKETCH THE TOP, FRONT AND RIGHT SIDE VIEWS USING THIRD-ANGLE PROJECTION. FULL SCALE. ALLOW ONE INCH BETWEEN VIEWS.

STEP BRACKET

A-5

QUESTIONS

1. What is the name of the object?
2. What is the drawing number?
3. How many pieces are to be made?
4. Of what material is the part made?
5. What is the overall width?
6. What is the overall depth?
7. What is the overall height?
8. Which line or surface in the side view represents surface (F) in the top view?
9. Which line or surface in the side view represents surface (E) in the top view?
10. Which line or surface in the side view represents surface (G) in the top view?
11. Which line or surface in the side view represents surface (L) of the front view?
12. What is the vertical height in the side view from the surface represented by line (P) to that represented by line (Q)?
13. What is the height of the step in the side view from the bottom of the part to the surface represented by surface (E)?
14. Which two dimensions (letters) in the top view represent distance V in the side view?
15. Which two dimensions (letters) in the top view represent distance W in the side view?
16. Which line or surface in the side view represents surface (M) in the front view?
17. What is the height of line (N)?
18. Which line or surface in the front view represents the surface (R) in the side view?
19. Which line or surface in the top view represents surface (L)?
20. Which line or surface in the front view represents surface (F)?
21. Which line or surface in the front view represents surface (E)?
22. Which line or surface in the top view represents surface (M)?
23. What type of line is (T)?
24. What type of line is (Y)?
25. What units of measurement are used on this drawing?
26. **Calculate dimensions B, C, D, and W.**

ANSWERS

1 ______
2 ______
3 ______
4 ______
5 ______
6 ______
7 ______
8 ______
9 ______
10 ______
11 ______
12 ______
13 ______
14 ______
15 ______
16 ______
17 ______
18 ______
19 ______
20 ______
21 ______
22 ______
23 ______
24 ______
25 ______
26 B ______
C ______
D ______
W ______

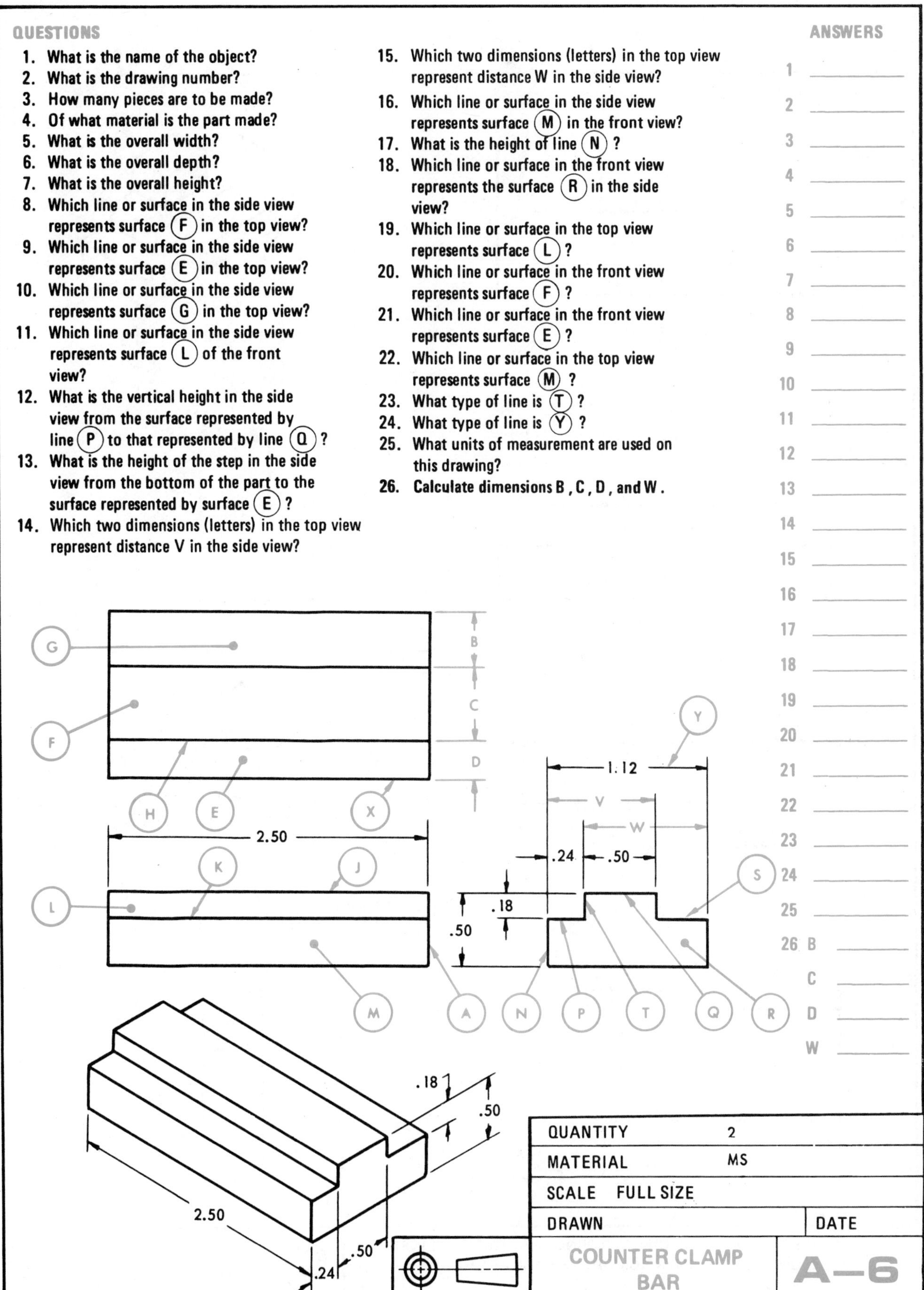

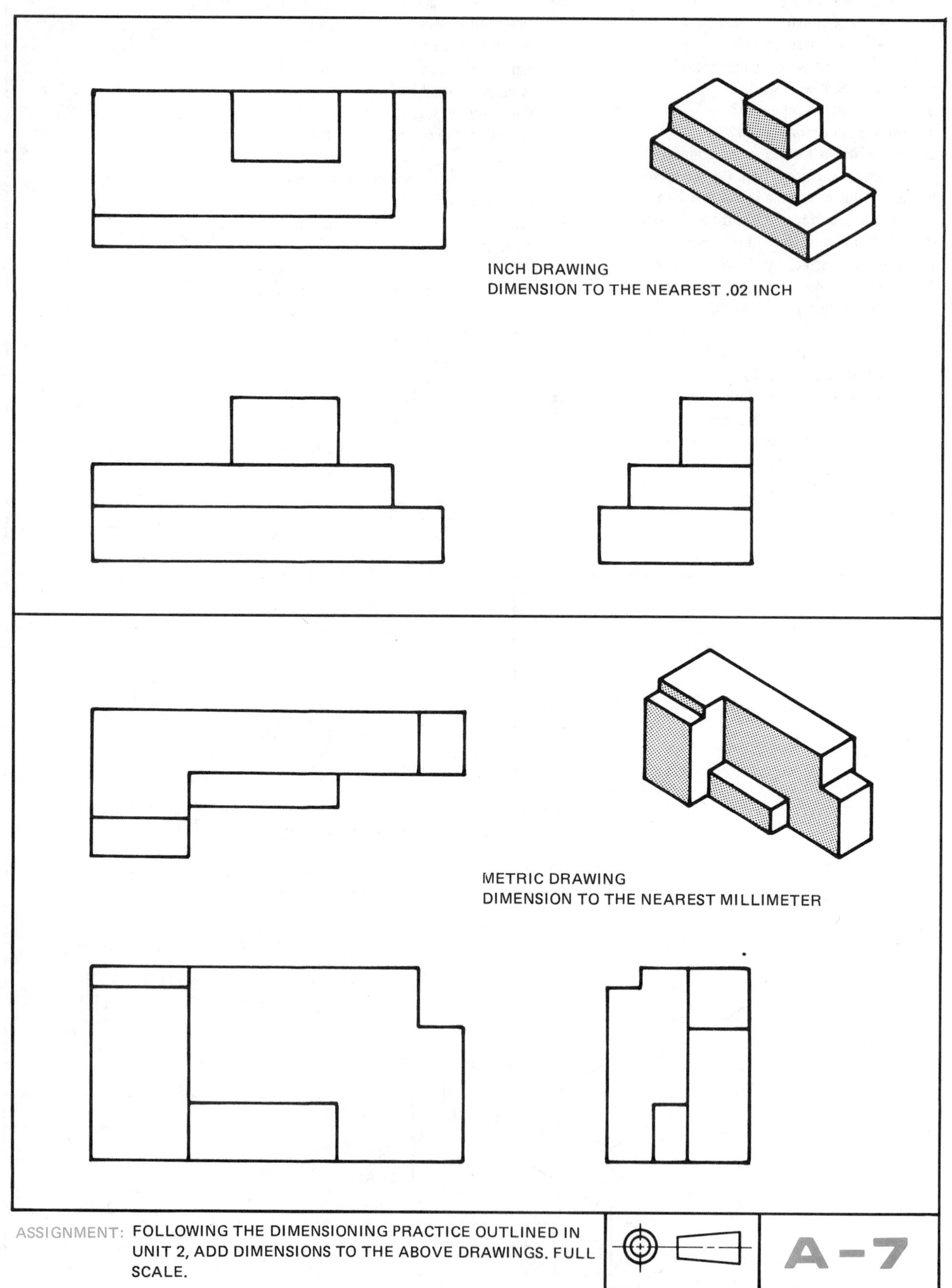
INCH DRAWING
DIMENSION TO THE NEAREST .02 INCH
METRIC DRAWING
DIMENSION TO THE NEAREST MILLIMETER
ASSIGNMENT: FOLLOWING THE DIMENSIONING PRACTICE OUTLINED IN UNIT 2, ADD DIMENSIONS TO THE ABOVE DRAWINGS. FULL SCALE.
A-7

UNIT 3

THE DRAFTING OFFICE

The drafting office is the starting point for all engineering work. Its product, the engineering drawing, is the main method of communication between all people concerned with the design and manufacture of parts.

Computer-Aided Drafting

The use of the computer in design and drafting is the most significant development to occur recently in these fields. The way in which drawings are prepared has been revolutionized, see figure 3-1. Producing engineering drawings on a computer is a process known as *computer-aided drafting,* commonly referred to as *CAD.* CAD has many advantages over manual drafting. Some major advantages are: (1) ability to rapidly access and change information on drawings; (2) cross reference related information in an easy manner; (3) increase productivity in the drafting room; and (4) make three-dimensional modelling economically possible.

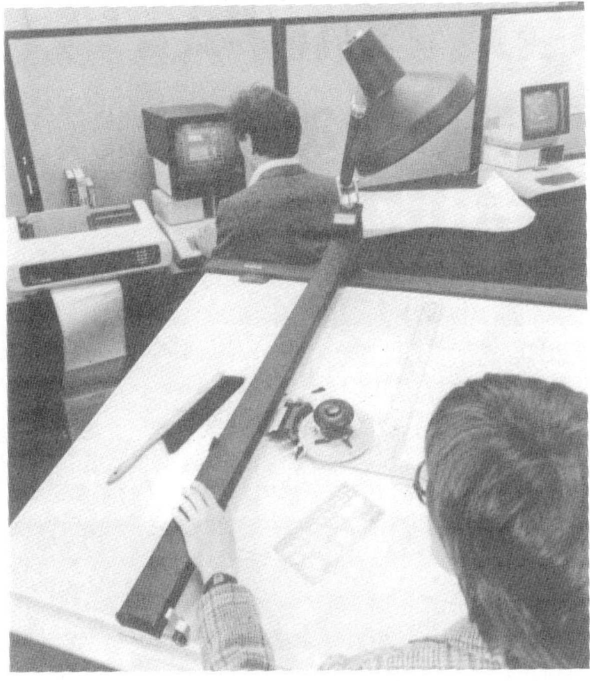

Fig. 3-1 The present and future state-of-the-art in drafting. (Cascade Graphics Development)

If information is to be sent directly to the fabricating machinery, the process is referred to as computer-aided drafting–computer-aided manufacturing *(CAD–CAM).* With the advent of CAD, the traditional drafting table has been, or is being, replaced with a station similar to that shown in figure 3-2.

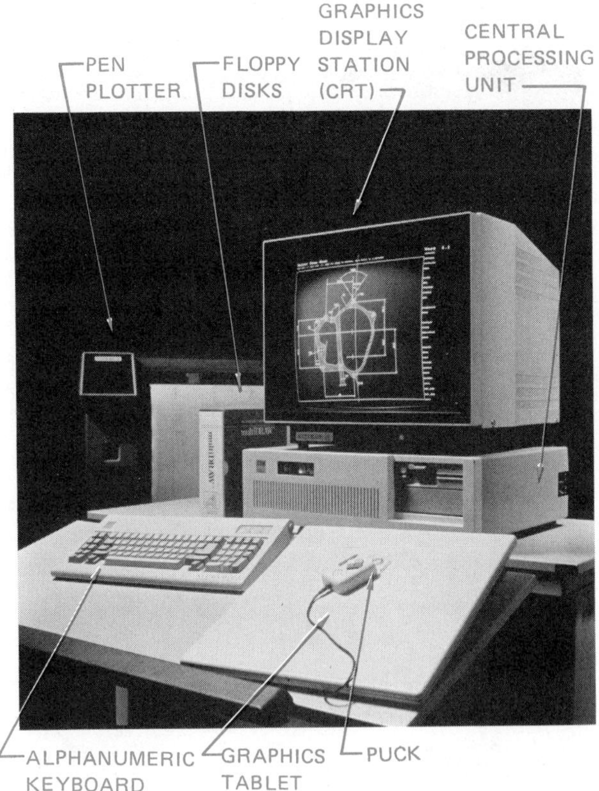

Fig. 3-2 A CAD work station (courtesy of Accugraph Inc.)

Whether preparing manual (traditional) or automated (computer-aided) drawings, the drafter must be familiar with the principles of drafting. Only the skill requirements and drafting equipment are different, figure 3-3. With manual drafting, skills in drawing lines and lettering are mandatory. With automated drafting, skills in drawing lines and lettering are not as critical. A cathode-ray terminal, a processing unit, a digitizer, and a plotter replace the manual drafting equipment.

(A) MANUAL DRAFTING

(B) COMPUTER-AIDED DRAFTING (CAD)

Fig. 3-3 Skill requirements for manual and computer-aided drafting

A CAD system by itself cannot create. A drafter must create the drawing, and thus a strong design and drafting background remains essential.

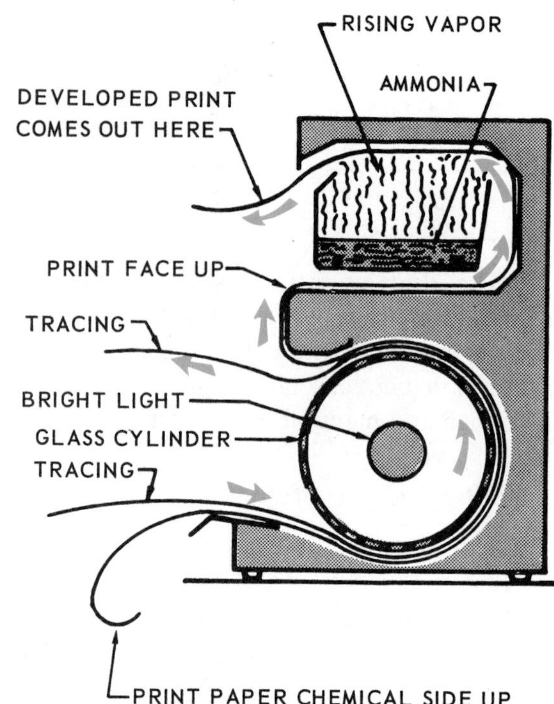

ROLLERS MOVE THE TRACING AND PRINT AROUND THE LIGHT, AND MOVE THE PRINT PAST THE RISING AMMONIA VAPOR.

(A) THE DIAZO PROCESS

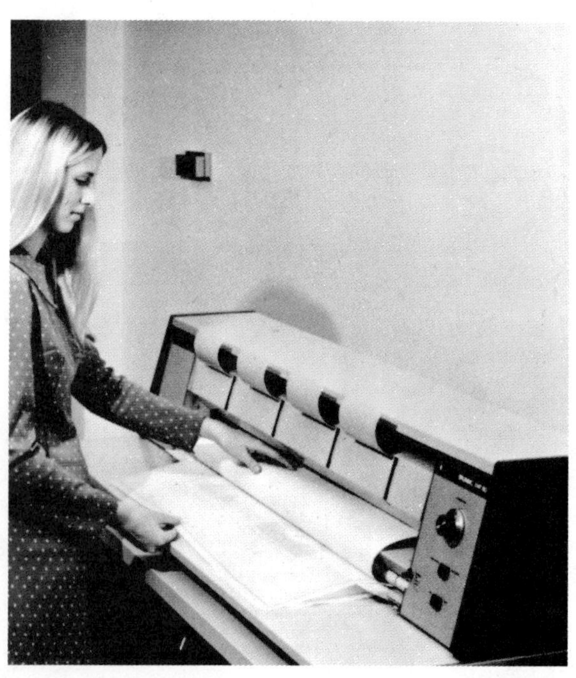

(B) A WHITEPRINT DIAZO MACHINE IN USE

Fig. 3-4 The diazo method of reproduction

DRAWING REPRODUCTION

The original drawings made by the drafter are kept in the engineering department at all times. Copies or prints of the originals are sent to the shop. For many years the most popular reproduction method was the blueprint. The demand for a faster and more versatile type of reproduction during World War II led to the introduction of the diazo process (white prints) of duplicating drawings, figure 3-4.

Today, the demand for more economical and versatile methods of reproducing drawings and the expense of maintaining filing systems has brought about the introduction of new methods of reproducing drawings, two of which are *microfilming* and *photoreproduction.*

The use of microfilm has been steadily increasing in recent years. Microfilm has a 35-mm image set into an aperture card, figure 3-5. Using this technique, the drawing may also be stored on a roll of 35-mm film.

Microfilm enables the drawing to be stored in a fraction of the floor space needed for filing the full-size paper originals. It can be referenced on the screen of a reader or, if a print is required, microfilm can be enlarged and reproduced on a reader-printer, figure 3-6.

Photoreproduction is one of the newest and most versatile methods for reproducing engineering drawings, figure 3-7. One such printer offers, in addition to full-scale printing, five reduction settings by which a drawing, or part of a drawing, can be reduced in size. Drawing reductions allow significant advantages: less paper consumption, lower handling and mailing costs, and smaller filing-space requirements.

Fig. 3-5 Aperture cards

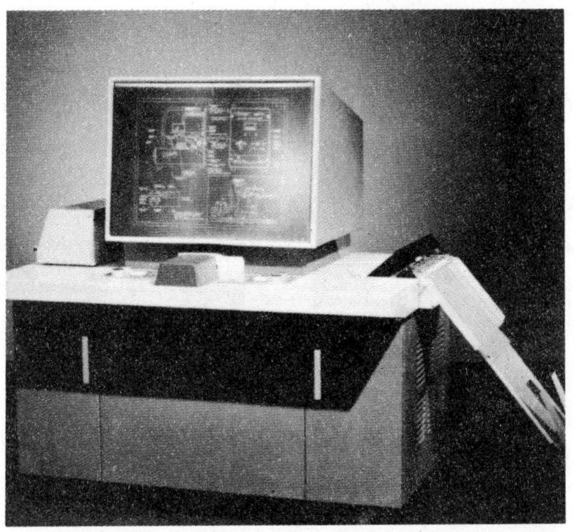

Fig. 3-6 Reader-printer for microfilm

Fig. 3-7 Photoreproduction process. This machine reproduces, reduces, folds, and sorts prints.

Another advantage is that since the drawing is photographed and prints or tracings can be made from the original work, the drafter can draw on practically any type of paper. However, in industrial practice, drawings are usually done on translucent sheets.

Because photoreproduction utilizes photography, it is suitable for scissors and paste-up drafting. This technique allows the drafter to incorporate previously printed or drawn material into his or her own work.

ABBREVIATIONS USED ON DRAWINGS

Abbreviations and symbols are used on drawings to conserve time and space, but only when their meaning is perfectly understood. A few of the more common abbreviations and symbols are shown in Table 1 of the appendix. A complete set of abbreviations for drawing use may be found in ANSI Y1.1, "Abbreviations For Use On Drawings And In Text."

HIDDEN LINES

Most objects drawn in engineering offices are complicated, and contain many surfaces and edges. Many features, such as lines and holes, cannot be seen when viewed from the outside of the piece. These hidden edges are shown on the drawings by a *hidden line,* a series of short dashes. The length of the dashes varies slightly in relationship to the size of the drawing. Hidden lines are usually required on the drawing to help show the true shape of the object, but may be omitted when not required to preserve the clarity of the drawing.

Hidden lines should always start and end with a dash except when such a dash would form a continuation of a visible detail line. Dashes should always join at corners. Illustrations of these hidden-line techniques are shown in figure 3-8.

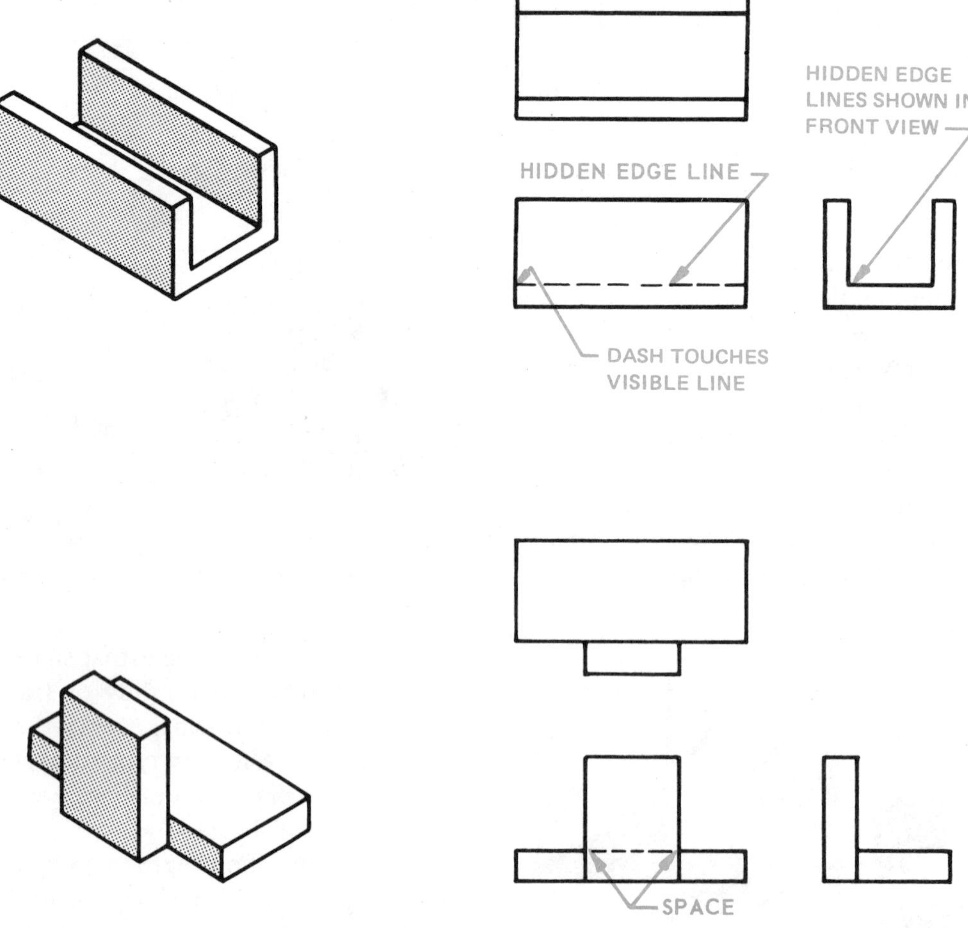

Fig. 3-8 Hidden lines

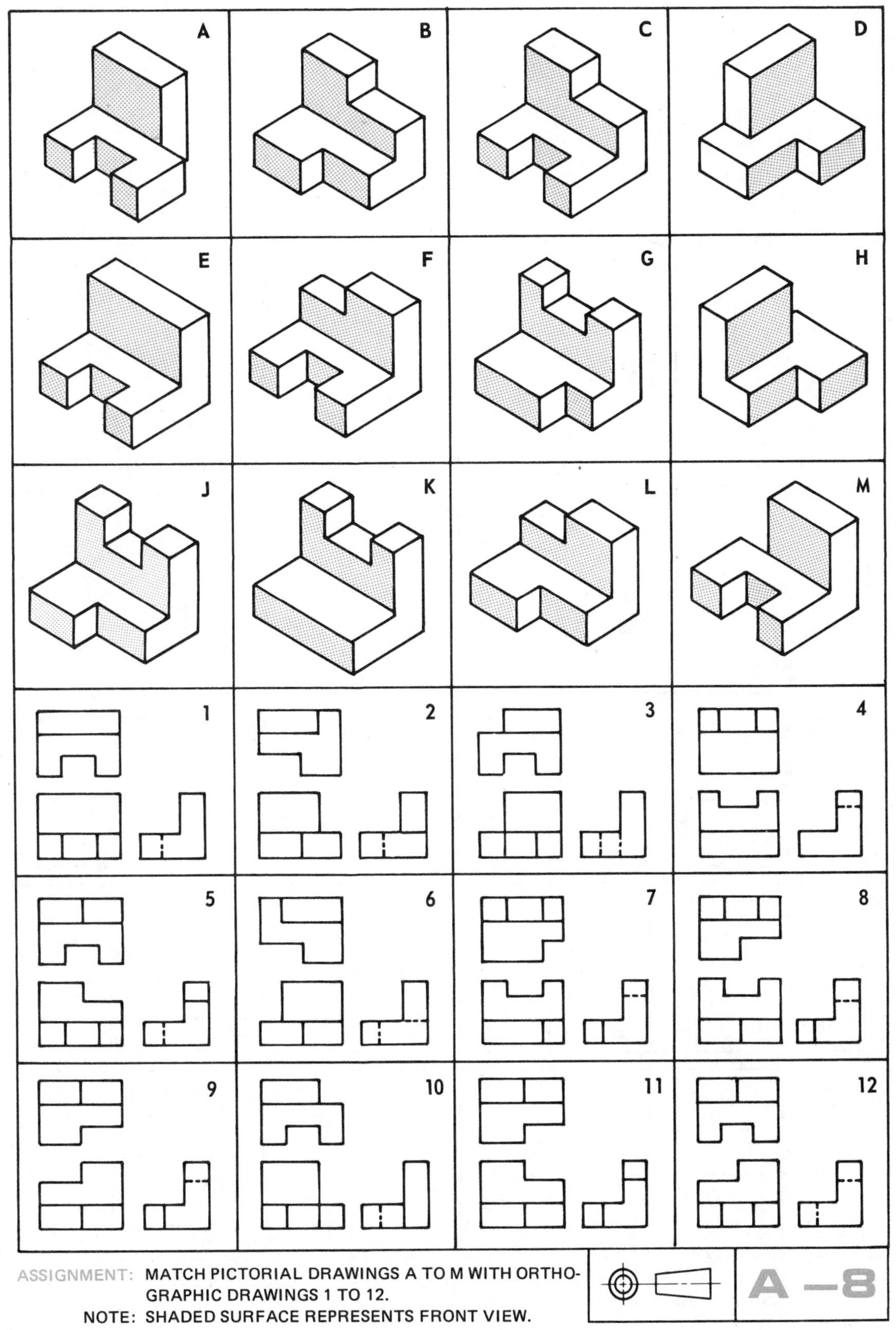

ASSIGNMENT: MATCH PICTORIAL DRAWINGS A TO M WITH ORTHOGRAPHIC DRAWINGS 1 TO 12.

NOTE: SHADED SURFACE REPRESENTS FRONT VIEW.

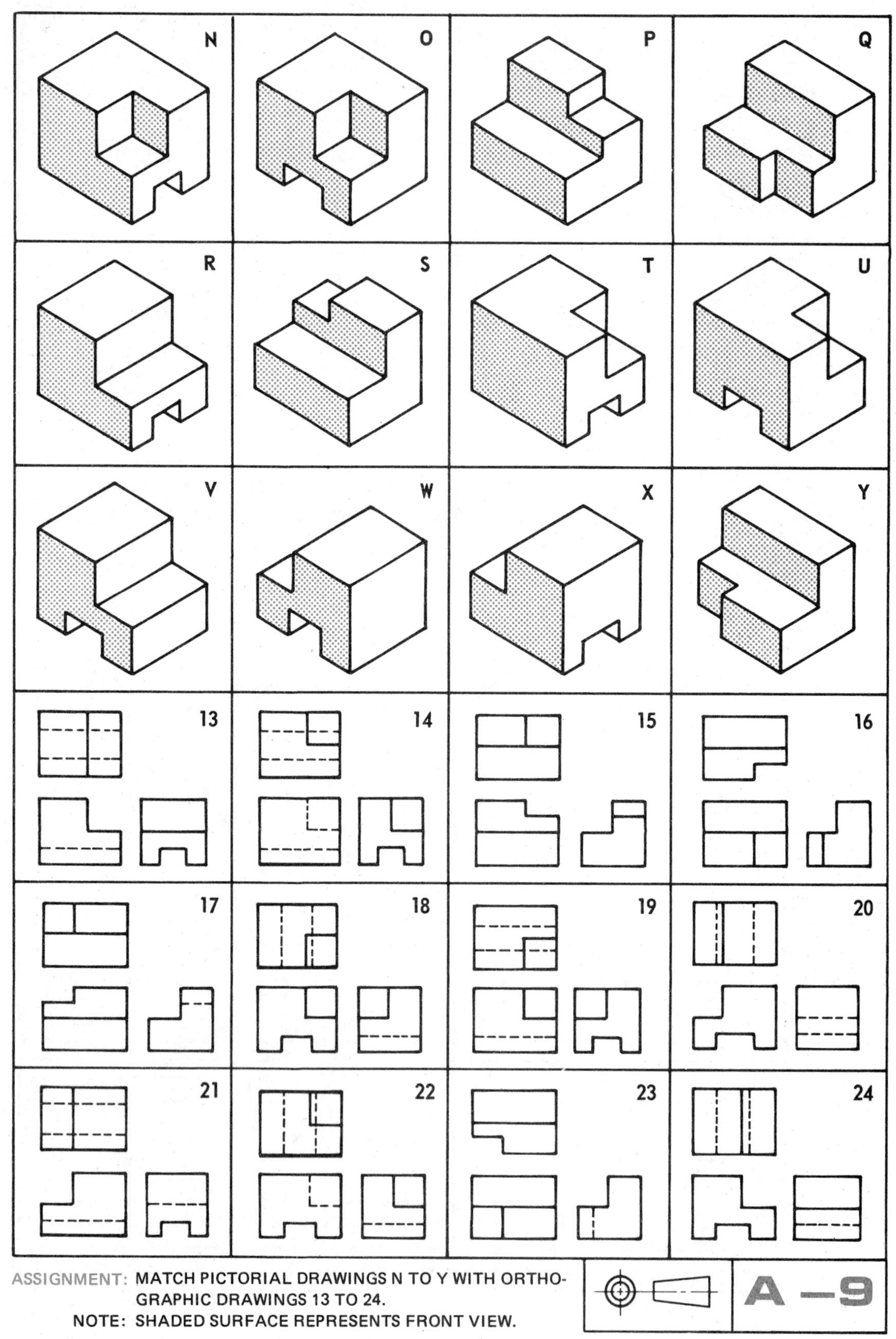

ASSIGNMENT: MATCH PICTORIAL DRAWINGS N TO Y WITH ORTHOGRAPHIC DRAWINGS 13 TO 24.

NOTE: SHADED SURFACE REPRESENTS FRONT VIEW.

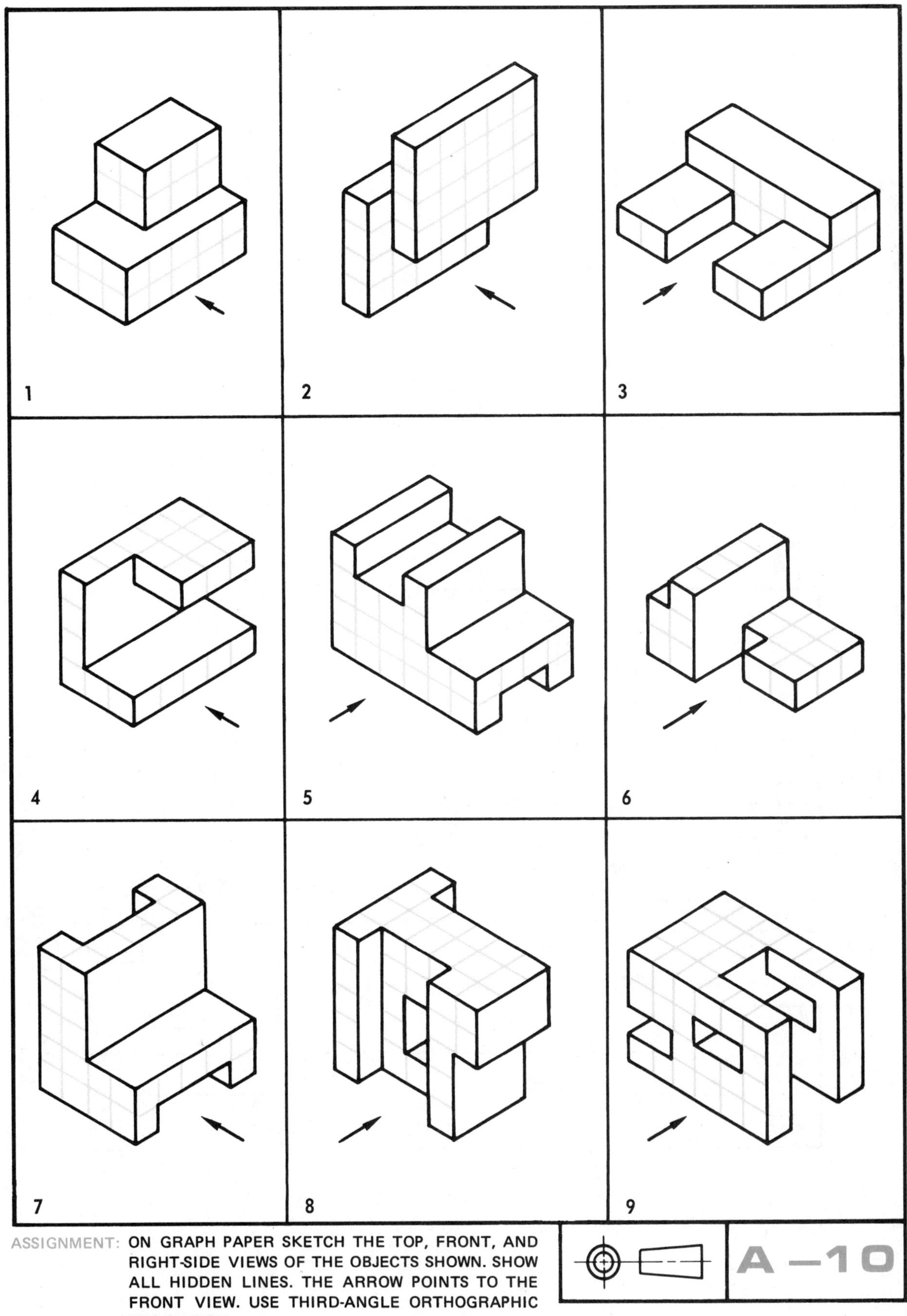

ASSIGNMENT: ON GRAPH PAPER SKETCH THE TOP, FRONT, AND RIGHT-SIDE VIEWS OF THE OBJECTS SHOWN. SHOW ALL HIDDEN LINES. THE ARROW POINTS TO THE FRONT VIEW. USE THIRD-ANGLE ORTHOGRAPHIC PROJECTION.

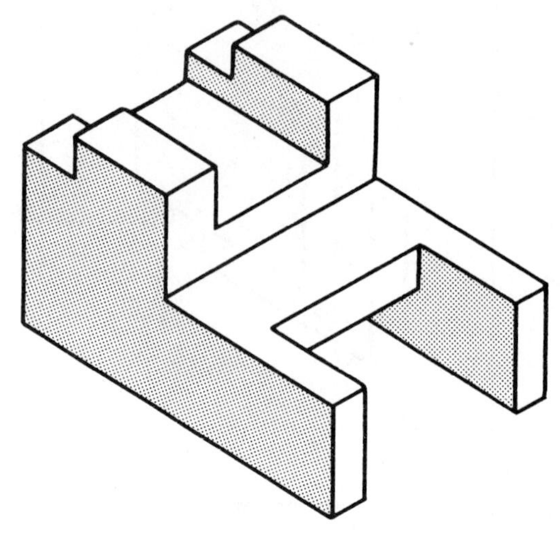
2
8
T
5
R
W
.25
4
.88
6
C
Y
G
S
9
3
Q
7
.25
3.00
1.40
.56
10
K
N
H
P
.40
U
D
B
E
.75
M
16
12
2.00
11
.44
A
.44
14
13
Z
.56
J
X
L
1.50
15
17
.44
V
F

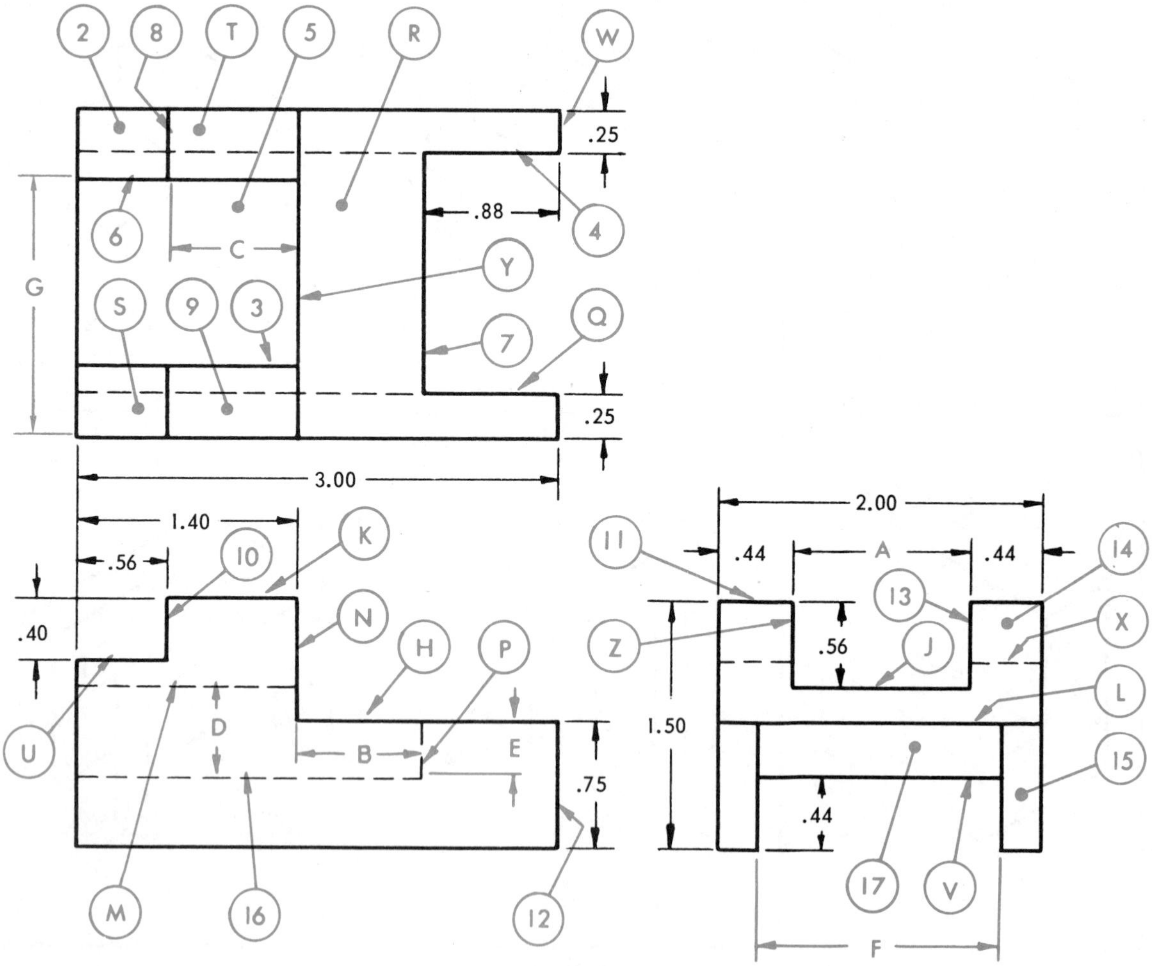

QUESTIONS

1. What is the object?
2. What is the drawing number?
3. How many castings are required?
4. What material is the part made of?
5. What is the overall width?
6. What is the overall height?
7. What is the overall depth?
8. Calculate distances A through G.
9. Which line in the top view represents surface (P)?
10. Which line in the side view represents surface (5)?
11. Which line in the side view represents surface (R)?
12. Which surfaces in the top view does line (K) of the front view represent?
13. Which surface in the top view does line (M) in the front view represent?
14. Which line in the side view represents the same surface represented by line (M) of the front view?
15. What kind, or type of line, is line (M)?
16. Which front view line does line (X) in the side view represent?
17. Which front view line does line (Y) in the top view represent?
18. Which line in the front view does surface (15) in the side view represent?
19. Which front view line represents surface (R) in the top view?
20. Which surface in the side view represents line (N) of the front view?
21. Which line in the side view represents surface (2)?
22. Which surface in the side view does line (P) represent?
23. Which surface in the top view does line (11) represent?
24. Which line in the side view does line (3) in the top view represent?
25. Which line in the side view does line (16) in the front view represent?
26. Which surface in the side view does line (W) represent?

ANSWERS

1 ______
2 ______
3 ______
4 ______
5 ______
6 ______
7 ______
8 A ______
B ______
C ______
D ______
E ______
F ______
G ______
9 ______
10 ______
11 ______
12 ______
13 ______
14 ______
15 ______
16 ______
17 ______
18 ______
19 ______
20 ______
21 ______
22 ______
23 ______
24 ______
25 ______
26 ______

QUANTITY 875	
MATERIAL MALLEABLE IRON	
SCALE NOT TO SCALE	
DRAWN BY	DATE
FEED HOPPER	A—11

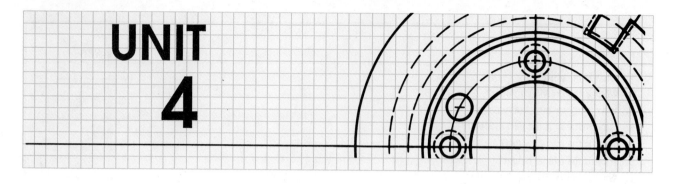

UNIT 4

INCLINED SURFACES

If the surfaces of an object lie in either a horizontal or a vertical position, the surfaces appear in their true shape in one of the three views, and these surfaces appear as a line in the other two views.

When a surface is sloped or inclined in only one direction, that surface is not seen in its true shape in the top, front, or side views. It is, however, seen in two views as a distorted surface. On the third view it appears as a line.

The true length of surfaces A and B in figure 4-1 is seen in the front view only. In the top and side views, only the width of surfaces A and B appears in its true size. The length of these surfaces is foreshortened.

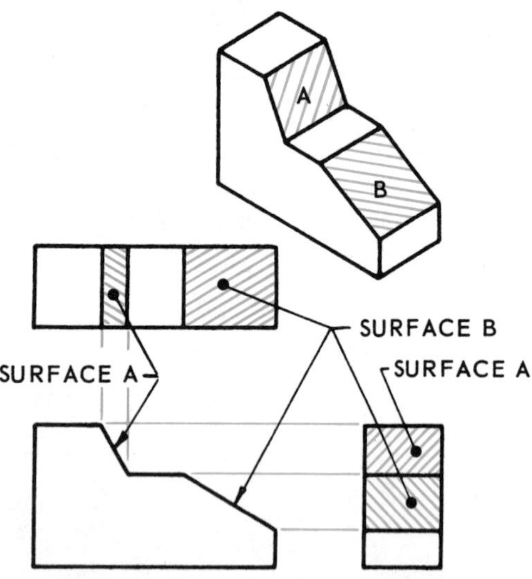

Fig. 4-1 Inclined surfaces

Where an inclined surface has important features that must be shown clearly and without distortion, an auxiliary or helper view must be used. These views are discussed in detail later in the book.

MEASUREMENT OF ANGLES

Some objects do not have all their features positioned in such a way that all surfaces can be in the horizontal and vertical plane at the same time. The design of the part may require that some of the lines in the drawing be shown in a direction other than horizontal or vertical. This will necessitate some lines to be drawn at an angle.

The amount of this divergence, or obliqueness, of lines may be indicated by either an offset dimension or an angle dimension, as shown in figure 4-2.

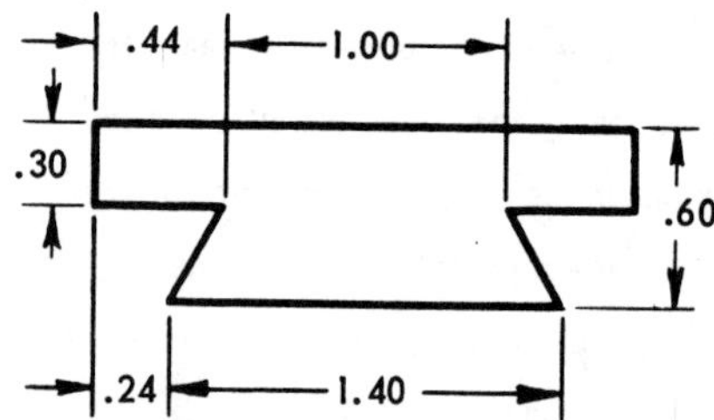

(A) LINEAR MEASUREMENTS

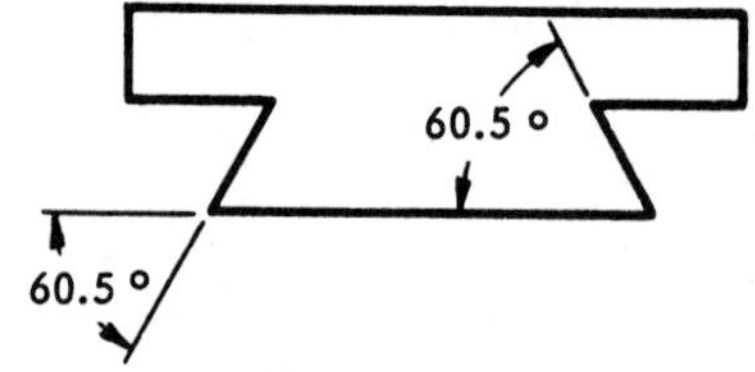

(B) ANGLE MEASUREMENTS USING DECIMAL DEGREES

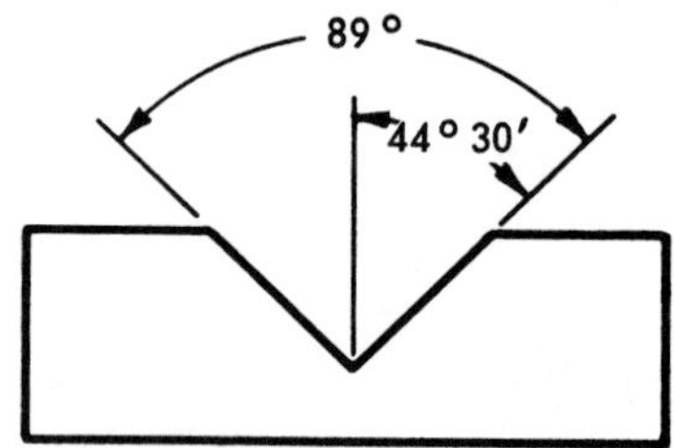

(C) ANGLE MEASUREMENTS USING DEGREES AND MINUTES

Fig. 4-2 Dimensioning angles

Angle dimensions may be expressed in degrees and decimal parts of a degree. They may also be expressed in degrees, minutes, and seconds. The former method is now preferred.

The symbols for degrees (°), minutes (′), and seconds (″) are included with the appropriate values. For example, 2°; 30°; 28°10′; 0°15′; 27°13′15″; 0°0′30″; 0.25°; 30°0′0″ ± 0°2′30″; and 2° ± 0.5° are all correct forms.

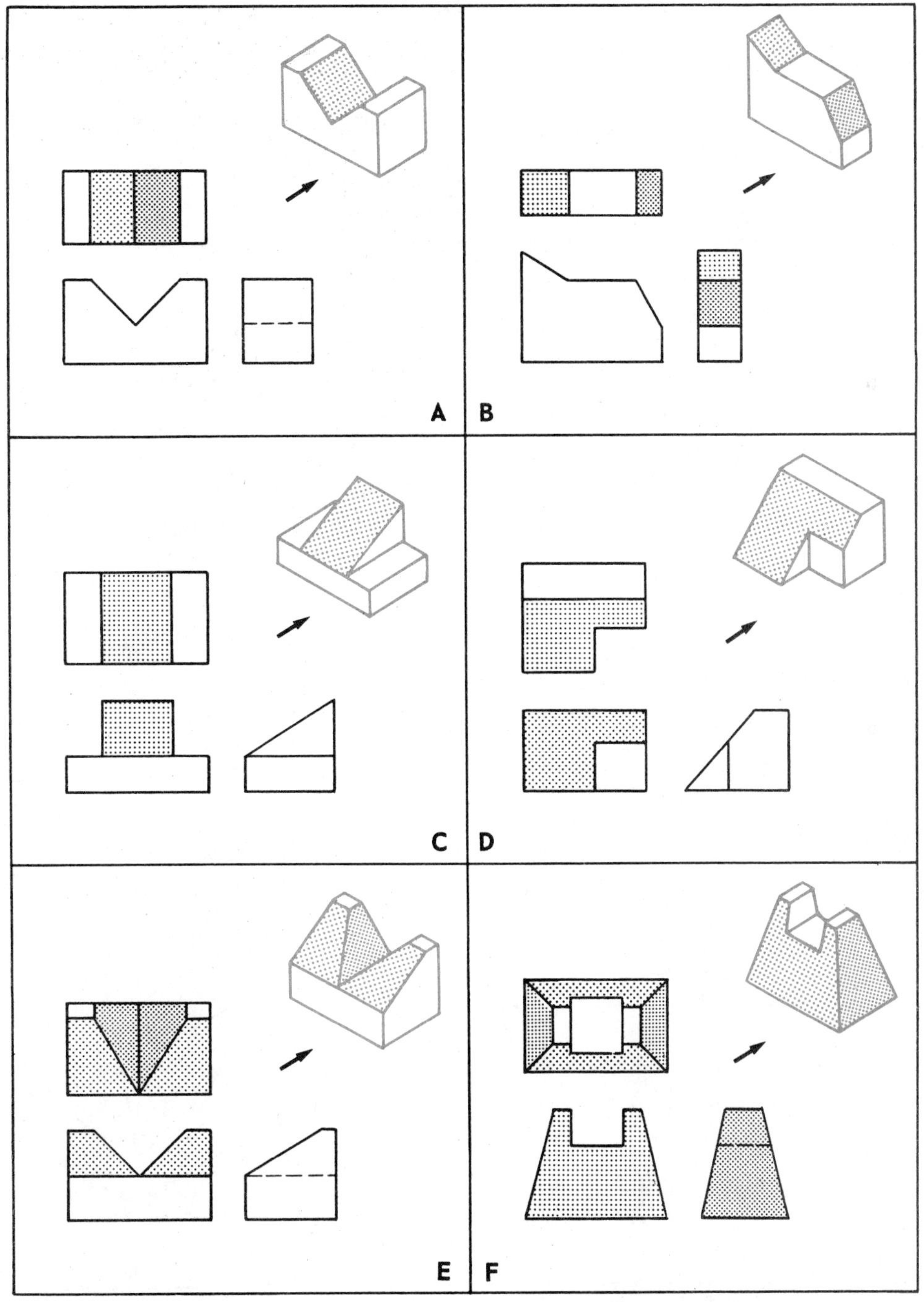

NOTE: ARROWS INDICATE DIRECTION OF SIGHT WHEN LOOKING AT THE FRONT VIEW.

Fig. 4-3 Illustrations of simple objects having inclined surfaces

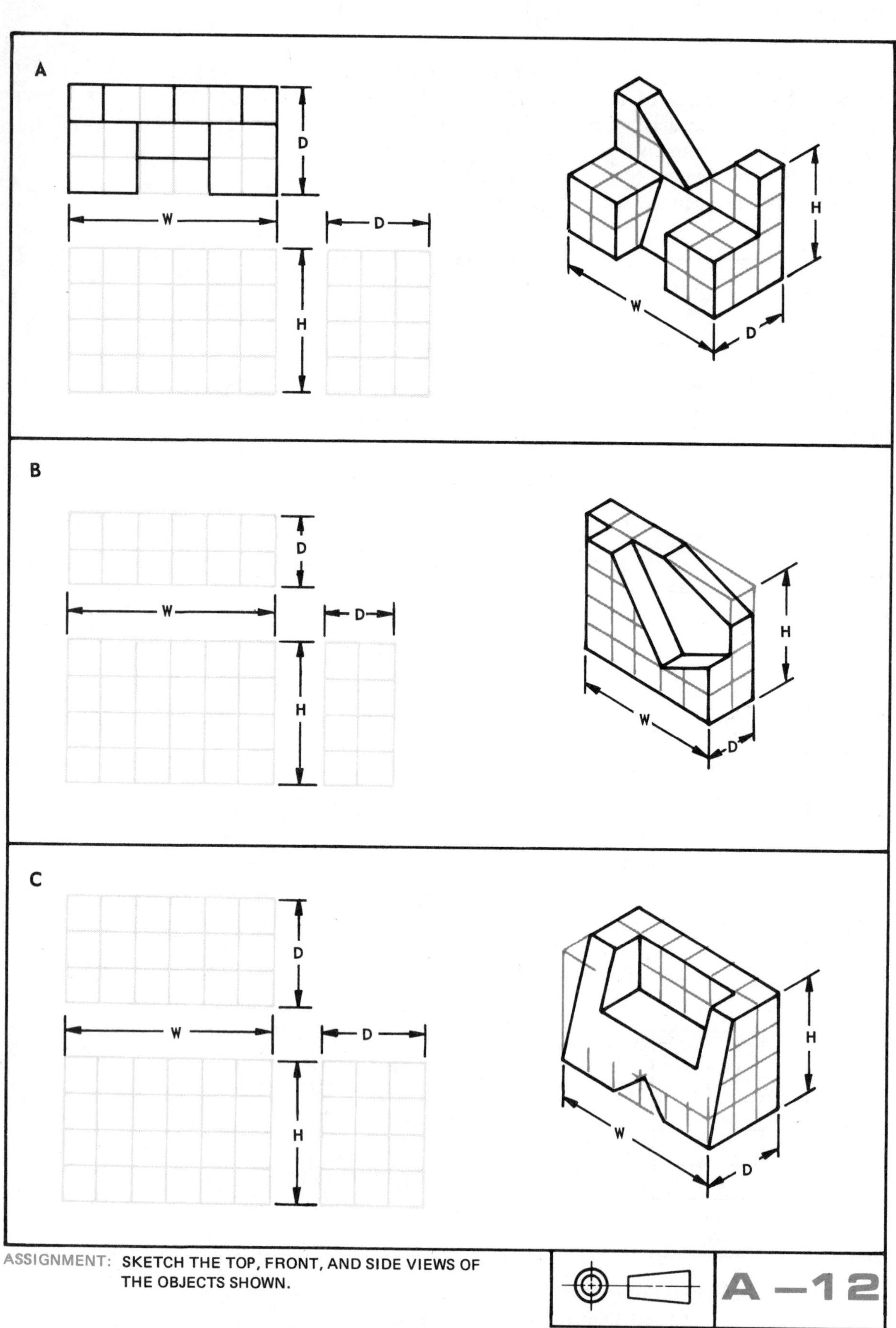

ASSIGNMENT: SKETCH THE TOP, FRONT, AND SIDE VIEWS OF THE OBJECTS SHOWN.

A — 12

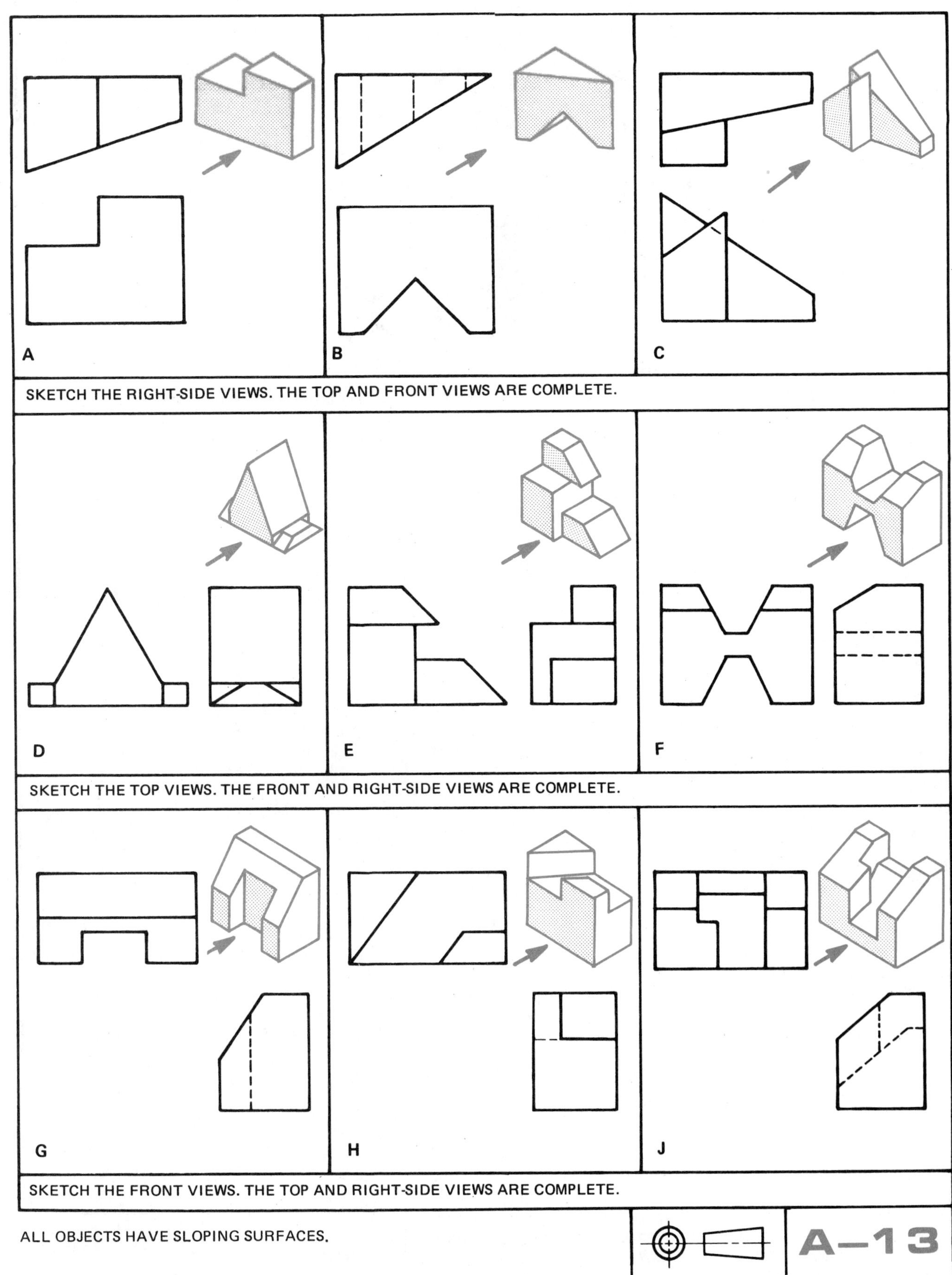
A
B
C
SKETCH THE RIGHT-SIDE VIEWS. THE TOP AND FRONT VIEWS ARE COMPLETE.
D
E
F
SKETCH THE TOP VIEWS. THE FRONT AND RIGHT-SIDE VIEWS ARE COMPLETE.
G
H
J
SKETCH THE FRONT VIEWS. THE TOP AND RIGHT-SIDE VIEWS ARE COMPLETE.

ALL OBJECTS HAVE SLOPING SURFACES.

A–13

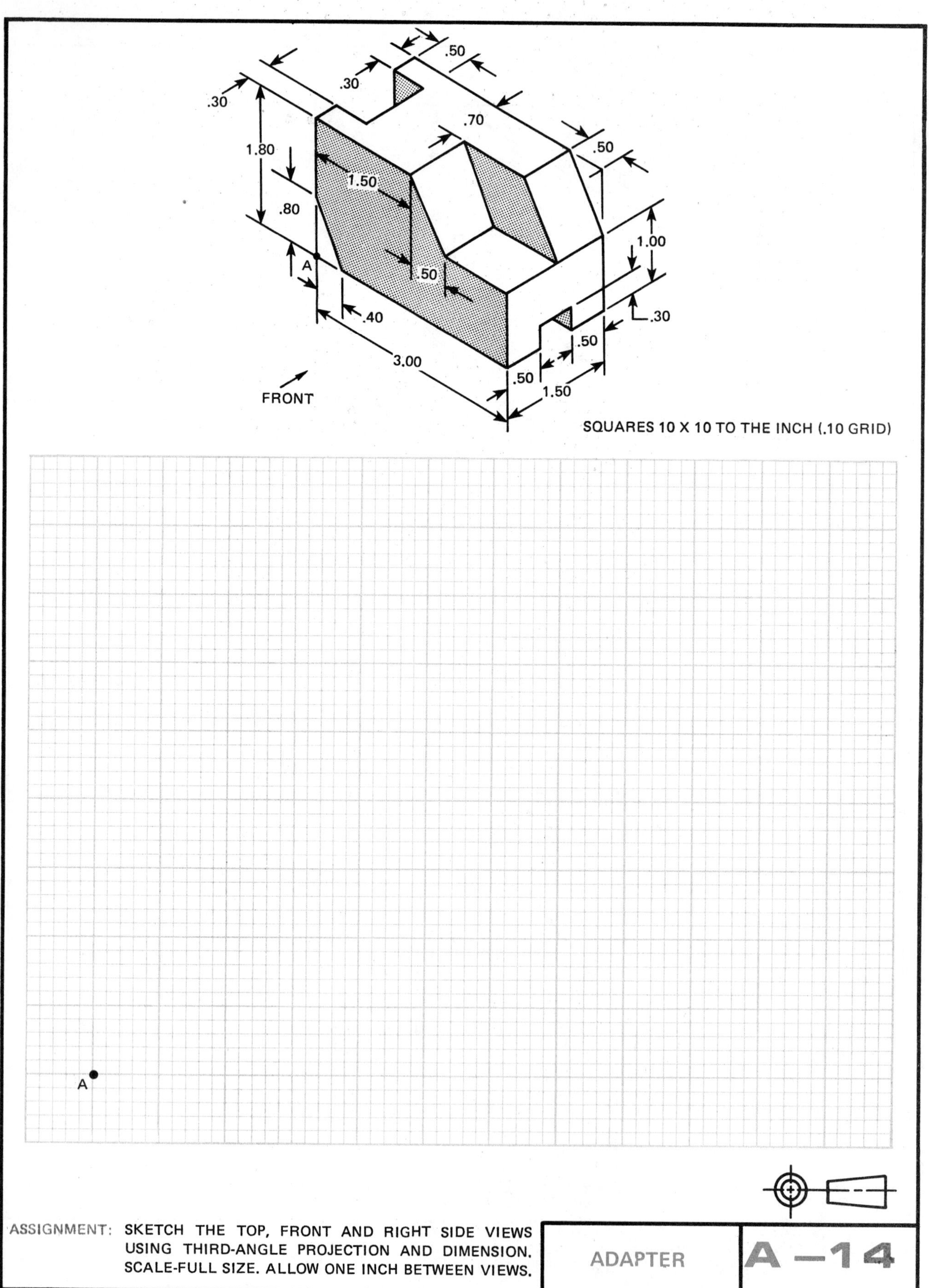
.50
.30
.30
.70
.50
1.80
1.50
.80
1.00
A
.50
.30
.40
.50
3.00
.50
1.50
FRONT
SQUARES 10 X 10 TO THE INCH (.10 GRID)
A
ASSIGNMENT: SKETCH THE TOP, FRONT AND RIGHT SIDE VIEWS USING THIRD-ANGLE PROJECTION AND DIMENSION. SCALE-FULL SIZE. ALLOW ONE INCH BETWEEN VIEWS.
ADAPTER
A — 14

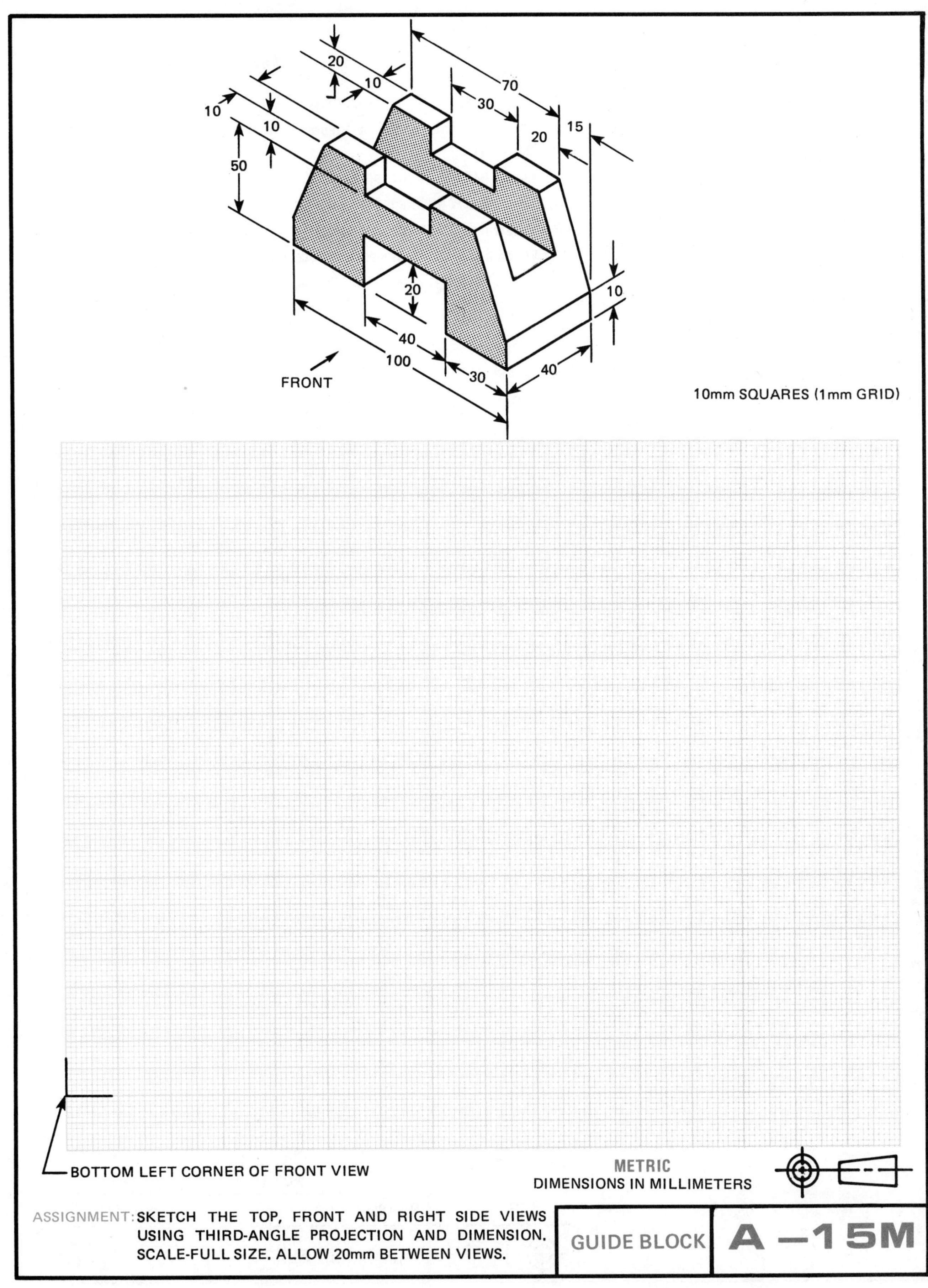

ASSIGNMENT: SKETCH THE TOP, FRONT AND RIGHT SIDE VIEWS USING THIRD-ANGLE PROJECTION AND DIMENSION. SCALE-FULL SIZE. ALLOW 20mm BETWEEN VIEWS.

GUIDE BLOCK	A —15M

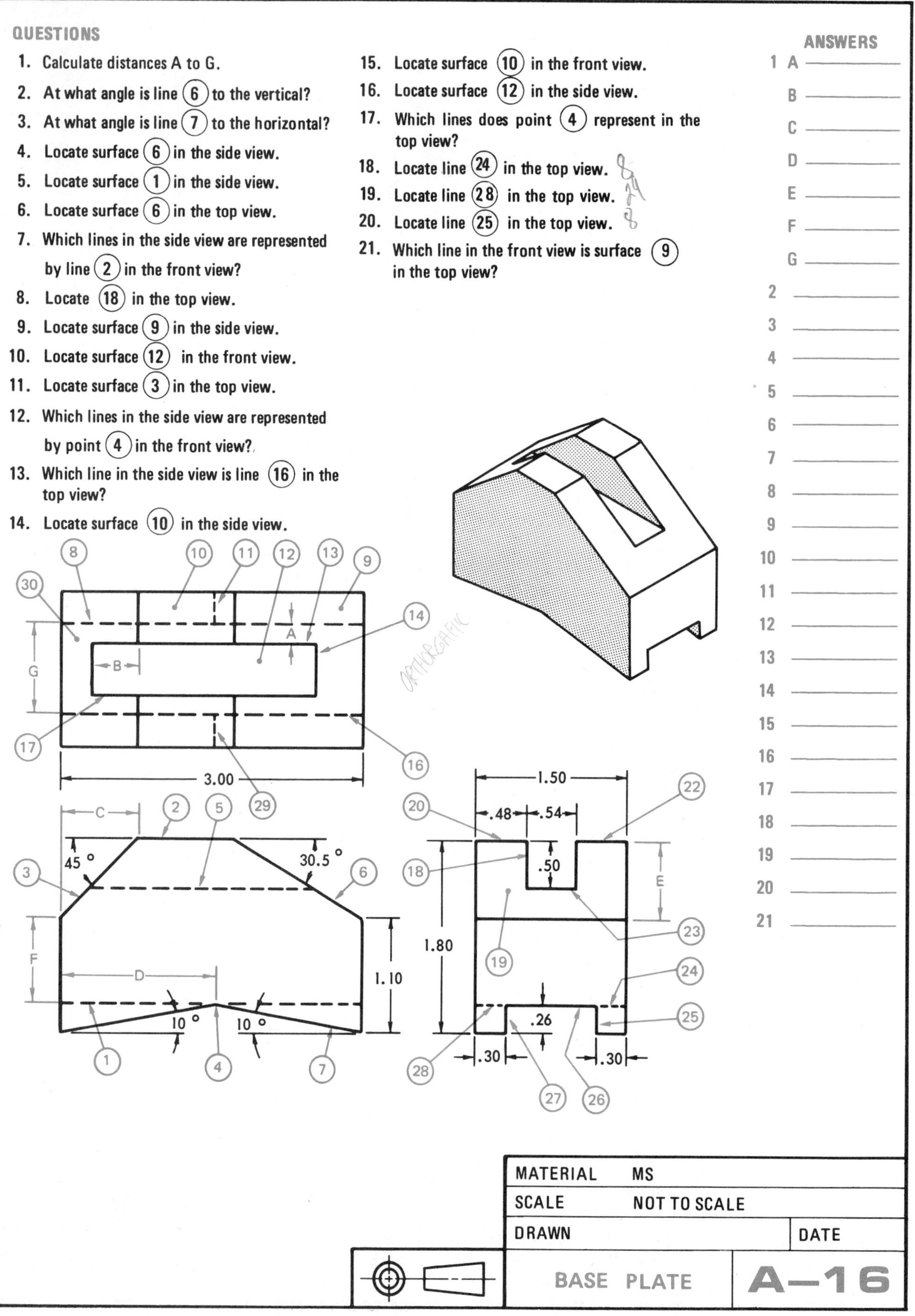

QUESTIONS

1. Calculate distances A to G.
2. At what angle is line (6) to the vertical?
3. At what angle is line (7) to the horizontal?
4. Locate surface (6) in the side view.
5. Locate surface (1) in the side view.
6. Locate surface (6) in the top view.
7. Which lines in the side view are represented by line (2) in the front view?
8. Locate (18) in the top view.
9. Locate surface (9) in the side view.
10. Locate surface (12) in the front view.
11. Locate surface (3) in the top view.
12. Which lines in the side view are represented by point (4) in the front view?
13. Which line in the side view is line (16) in the top view?
14. Locate surface (10) in the side view.
15. Locate surface (10) in the front view.
16. Locate surface (12) in the side view.
17. Which lines does point (4) represent in the top view?
18. Locate line (24) in the top view.
19. Locate line (28) in the top view.
20. Locate line (25) in the top view.
21. Which line in the front view is surface (9) in the top view?

ANSWERS

1 A ______
B ______
C ______
D ______
E ______
F ______
G ______
2 ______
3 ______
4 ______
5 ______
6 ______
7 ______
8 ______
9 ______
10 ______
11 ______
12 ______
13 ______
14 ______
15 ______
16 ______
17 ______
18 ______
19 ______
20 ______
21 ______

MATERIAL	MS	
SCALE	NOT TO SCALE	
DRAWN		DATE
BASE PLATE		A–16

UNIT 5

CIRCULAR FEATURES

Typical parts with circular features are illustrated in figure 5-1. Note that the circular feature appears circular in one view only and that no line is used to indicate where a curved surface joins a flat surface. Hidden circles, like hidden flat surfaces, are represented on drawings by a hidden line.

CENTER LINES

A *center line* is drawn as a thin broken line of long and short dashes, spaced alternately. Center lines may be used to indicate center points, axes of cylindrical parts, and axes of symmetry. Solid center lines are often used as a simplified drafting practice; however, the interrupted line is preferred. Center lines should project for a short distance beyond the outline of the part or feature to which they refer. They may be lengthened for use as extension lines for dimensioning purposes; in this case the extended portion is not broken.

In end views of circular features, the point of intersection of two center lines is shown by two intersecting short dashes. However, for very small circles a solid unbroken line is recommended. These techniques are shown in figure 5-2.

DIMENSIONING OF CYLINDRICAL FEATURES

Features shown as circles are normally dimensioned by one of the methods shown in

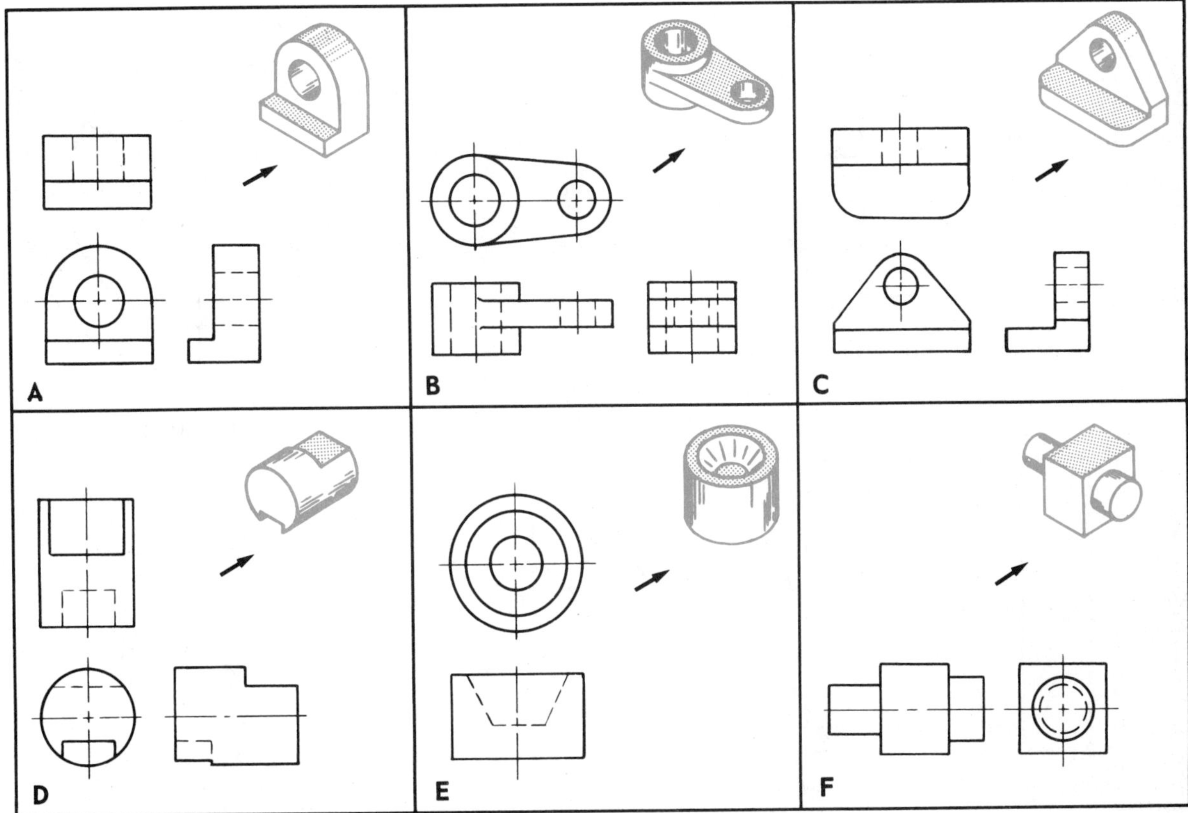

Fig. 5-1 Illustrations of simple objects having circular features

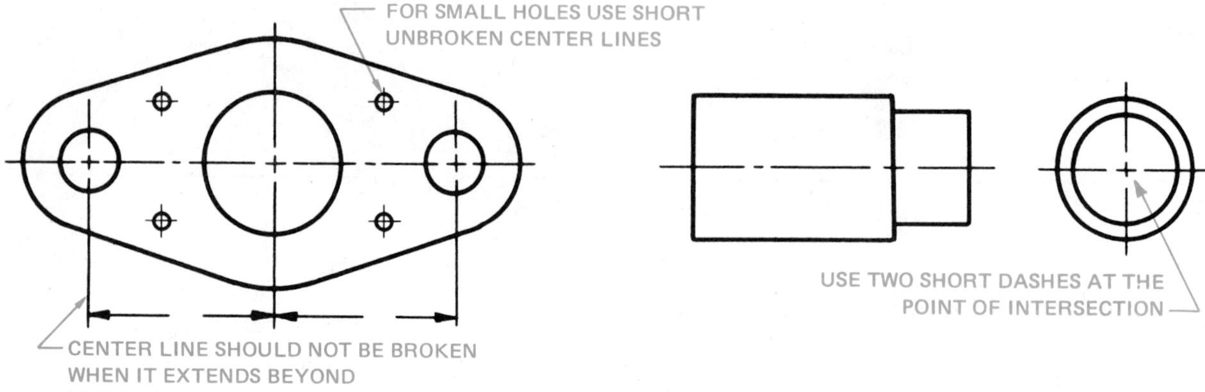

Fig. 5-2 Center line application

figure 5-3. Where the diameters of a number of concentric cylinders are to be given, it may be more convenient to show them on the side view. The diameter symbol ϕ should always precede the diametral dimension.

The radius of the arc is used in dimensioning a circular arc. The letter R is shown before the radius dimension to indicate that it is a radius. Approved methods for dimensioning arcs are shown in figure 5-4.

Previous ANSI drawing practice placed the diameter and radius symbols after the dimension.

DIMENSIONING CYLINDRICAL HOLES

Specification of the diameter with a leader, as shown in figure 5-5, is the preferred method for designating the size of small holes. For larger diameters, use one of the methods illustrated in figure 5-3. When the leader is used, the symbol ϕ

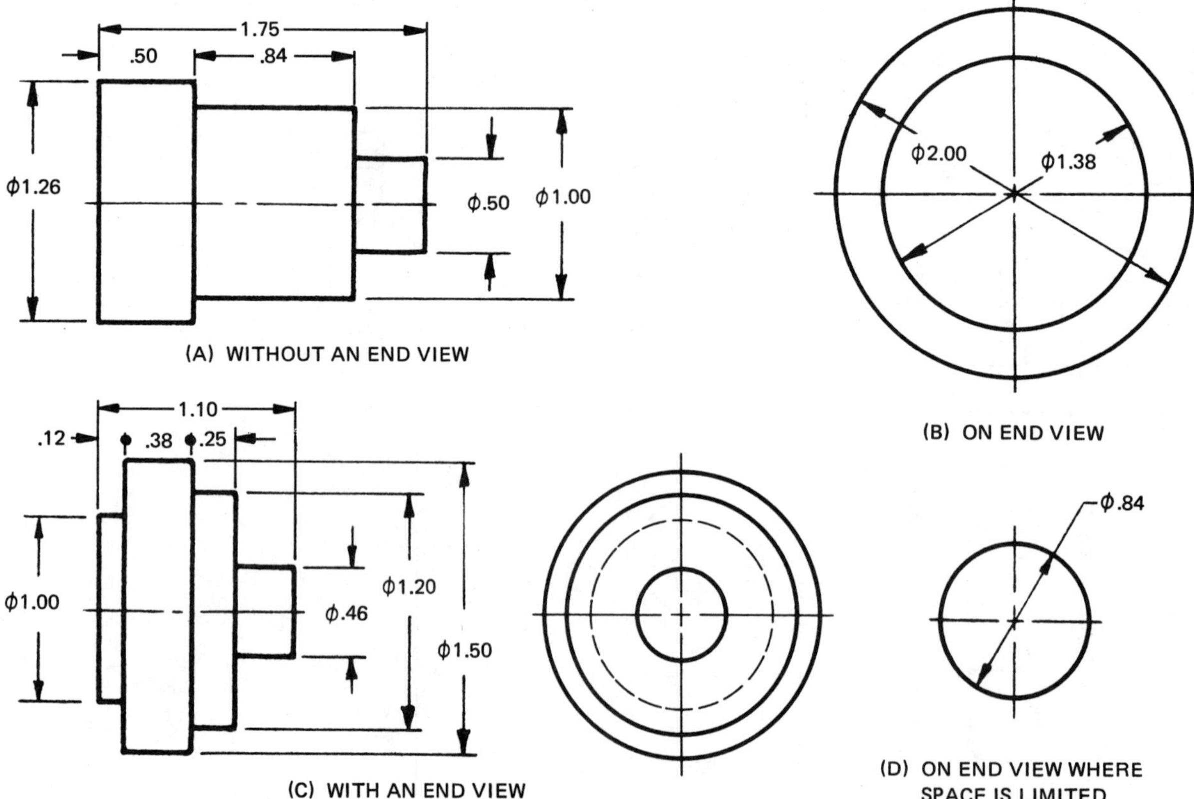

Fig. 5-3 Dimensioning diameters

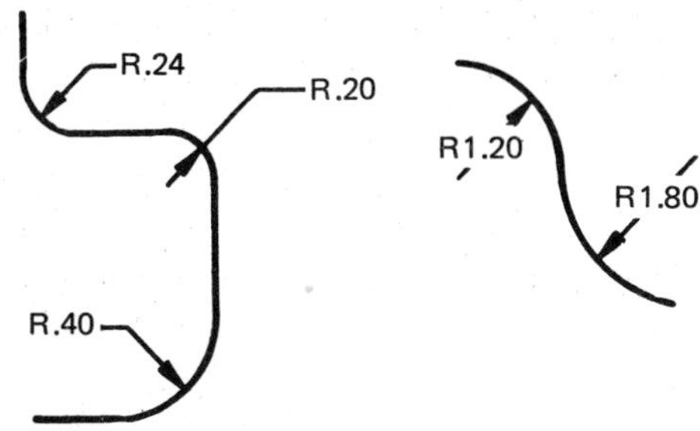

(A) DIMENSIONING RADII WHICH NEED NOT HAVE THEIR CENTER POINTS LOCATED

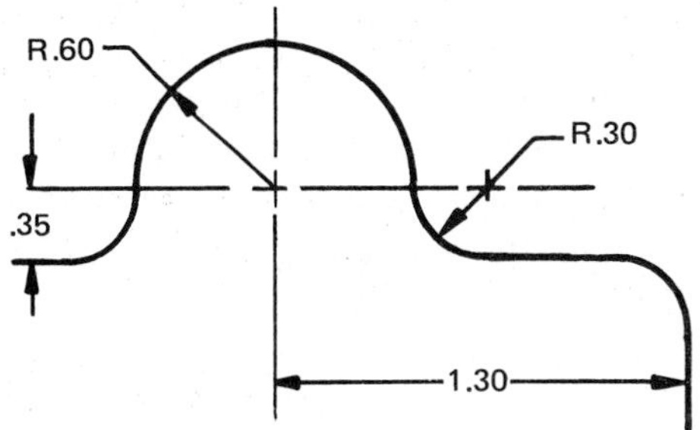

(B) RADII WITH LOCATED CENTERS

Fig. 5-4 Dimensioning radii

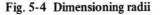

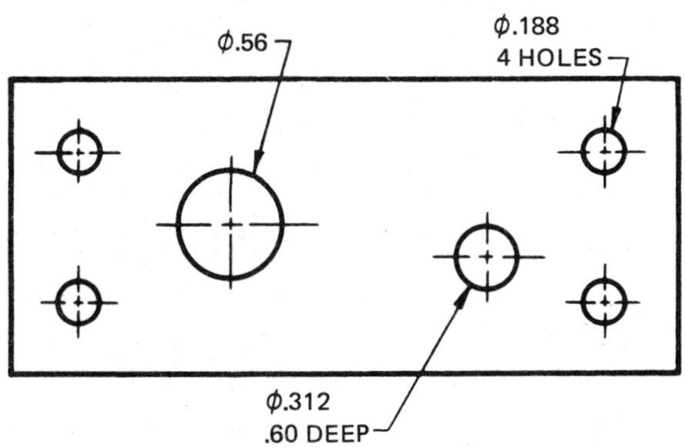

Fig. 5-5 Dimensioning cylindrical holes

precedes the size of the hole. The note end of the leader terminates in a short horizontal bar. When two or more holes of the same size are required, the number of holes is specified after the size. If a blind hole is required, the depth of the hole is included in the dimensioning note; otherwise, it is assumed that all holes shown are through holes. Terms such as *drill, ream, bore,* and so forth, should be avoided on drawings. The method of producing the hole should be left to the discretion of either the shop personnel or the planning department.

REPETITIVE FEATURES AND DIMENSIONS

Repetitive features and dimensions may be specified on a drawing by the use of an X in conjunction with the numeral to indicate the "number of times" or "places" they are required. A space is inserted between the X and the dimension, as shown in figure 5-6.

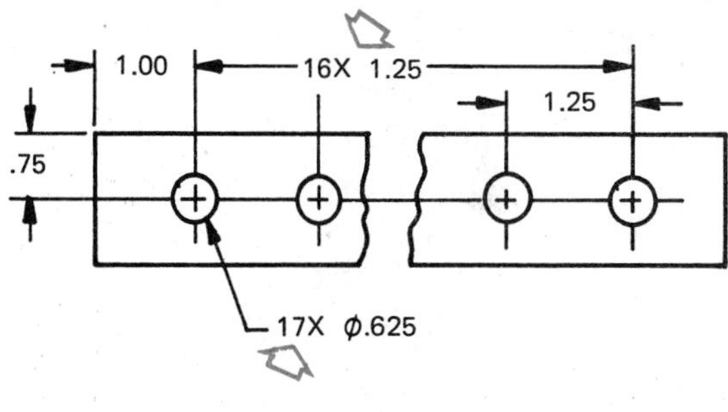

EXAMPLE 1

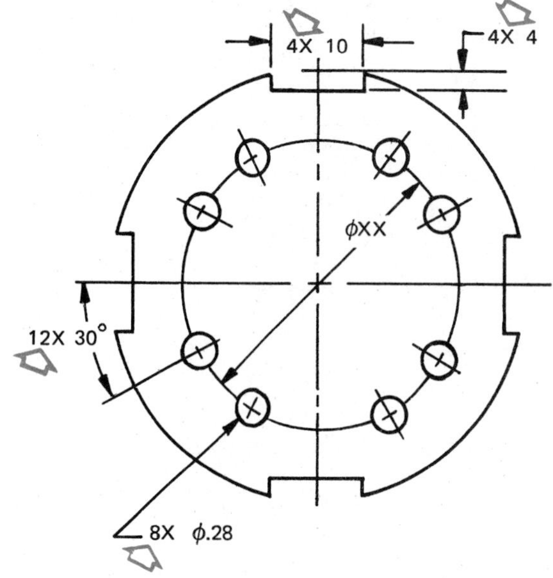

EXAMPLE 2

Fig. 5-6 Dimensioning repetitive detail

DRILLING, REAMING, AND BORING

Drilling is the process of using a drill to cut a hole through a solid, or to enlarge an existing hole. For some types of work, holes must be drilled smooth and straight, and to an exact size. In other work, accuracy of location and size of the hole are not as important.

When accurate holes of uniform diameter are required, they are first drilled slightly undersize and then reamed. *Reaming* is the process of sizing a hole to a given diameter with a reamer in order to produce a hole which is round, smooth, and straight.

Boring is one of the more dependable methods of producing holes which are round and concentric. The term *boring* refers to the enlarging of a hole by means of a boring tool. The use of reaming is limited to the sizes of available reamers. However, holes may be bored to any size desired.

The degree of accuracy to which a hole is to be machined is specified on the drawing.

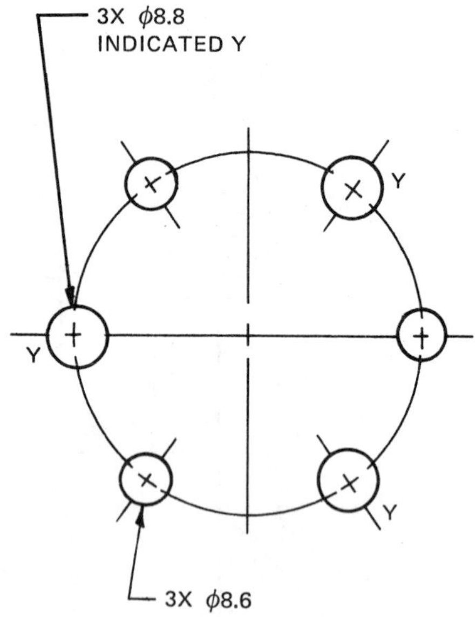

Fig. 5-7 Identifying similarly sized holes

IDENTIFYING SIMILARLY SIZED FEATURES

Where many similarly sized holes or features appear on a part, some form of identification may be desirable in order to insure the legibility of the drawing, figure 5-7.

ROUNDS AND FILLETS

A round, or radius, or chamfer is put on the outside of a piece to improve its appearance and to avoid forming a sharp edge that might chip off under a sharp blow or cause interference. A fillet is additional metal allowed in the inner intersection of two surfaces, figure 5-8. This increases the strength of the object. A general note, such as ROUNDS AND FILLETS R10 or ROUNDS AND FILLETS R10 UNLESS OTHERWISE SHOWN, is used on the drawing instead of individual dimensions.

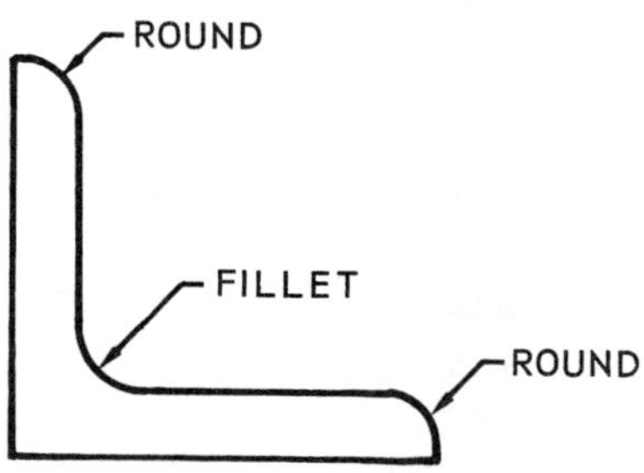

Fig. 5-8 Fillets and rounds

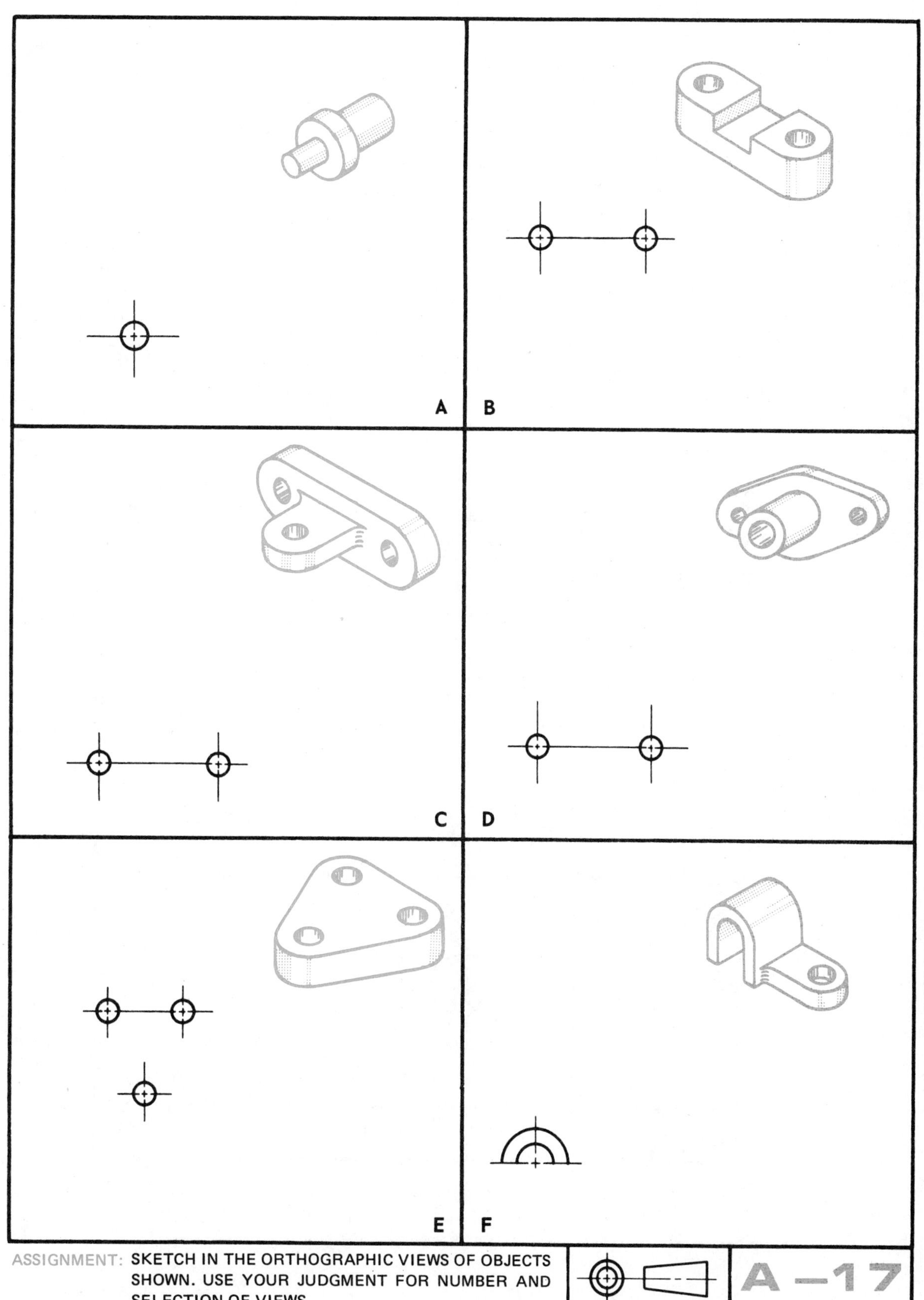

ASSIGNMENT: SKETCH IN THE ORTHOGRAPHIC VIEWS OF OBJECTS SHOWN. USE YOUR JUDGMENT FOR NUMBER AND SELECTION OF VIEWS.

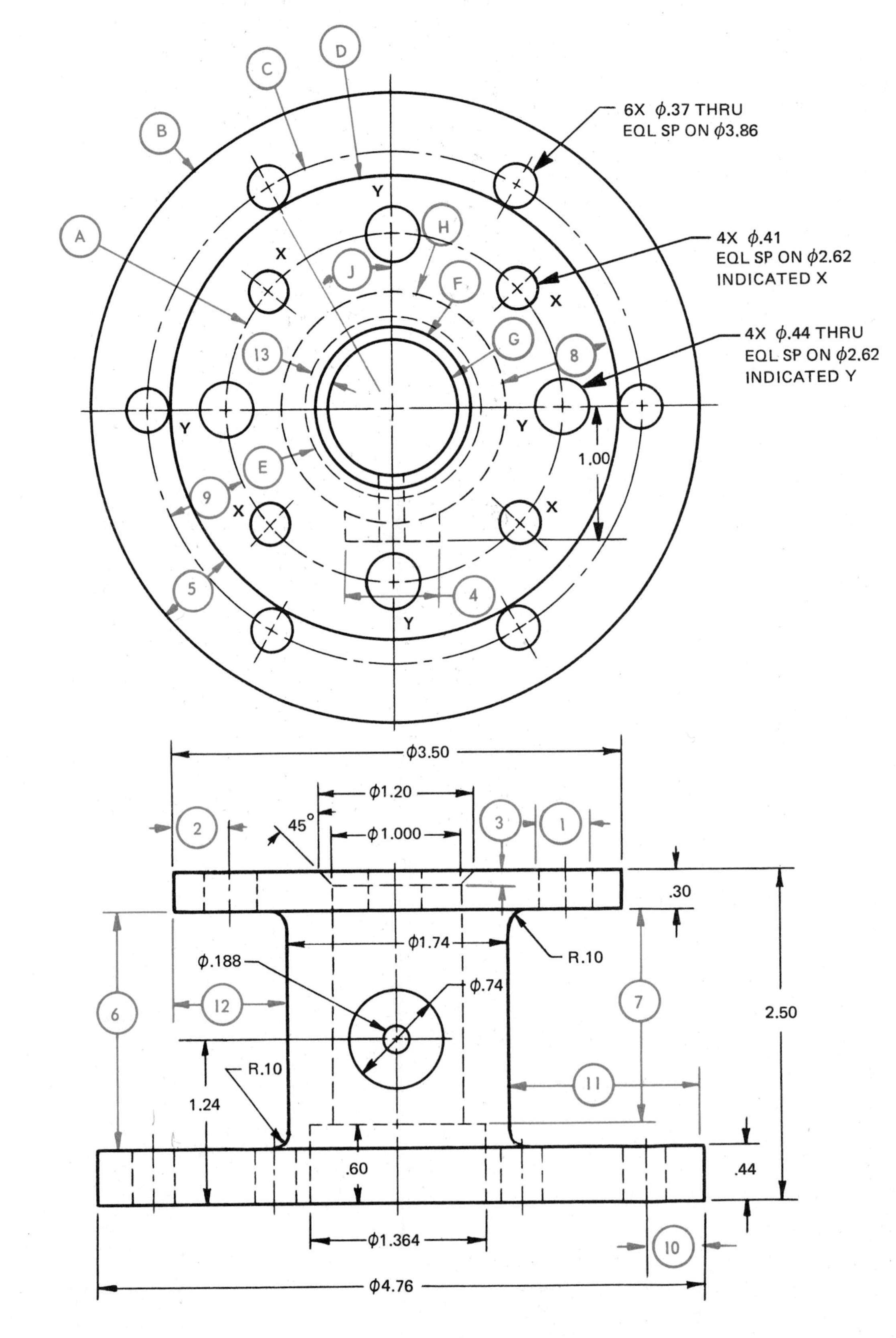
6X ϕ.37 THRU
EQL SP ON ϕ3.86
4X ϕ.41
EQL SP ON ϕ2.62
INDICATED X
4X ϕ.44 THRU
EQL SP ON ϕ2.62
INDICATED Y
1.00
ϕ3.50
ϕ1.20
45°
ϕ1.000
.30
ϕ1.74
R.10
ϕ.188
ϕ.74
2.50
R.10
1.24
.60
.44
ϕ1.364
ϕ4.76

QUESTIONS

1. What are the diameters of circles (A) to (H) ?
2. How many holes are in the bottom surface?
3. How many holes are in the top surface?
4. How deep is the Ø 1.000 hole from the top of the coupling?
5. What is angle (J) ?
6. How thick is the largest flange?
7. What size bolts would be used for the Y holes located on the top flange? Allow for .06 clearance. (Refer to the bolt sizes in the appendix.)
8. What size bolts would be used for the bottom flange? Allow for .06 clearance. (Refer to the bolt sizes in the appendix.)
9. Calculate distances (1) to (13) .

MATERIAL	GRAY IRON	
SCALE	NOT TO SCALE	
DRAWN		DATE
	COUPLING	A—18

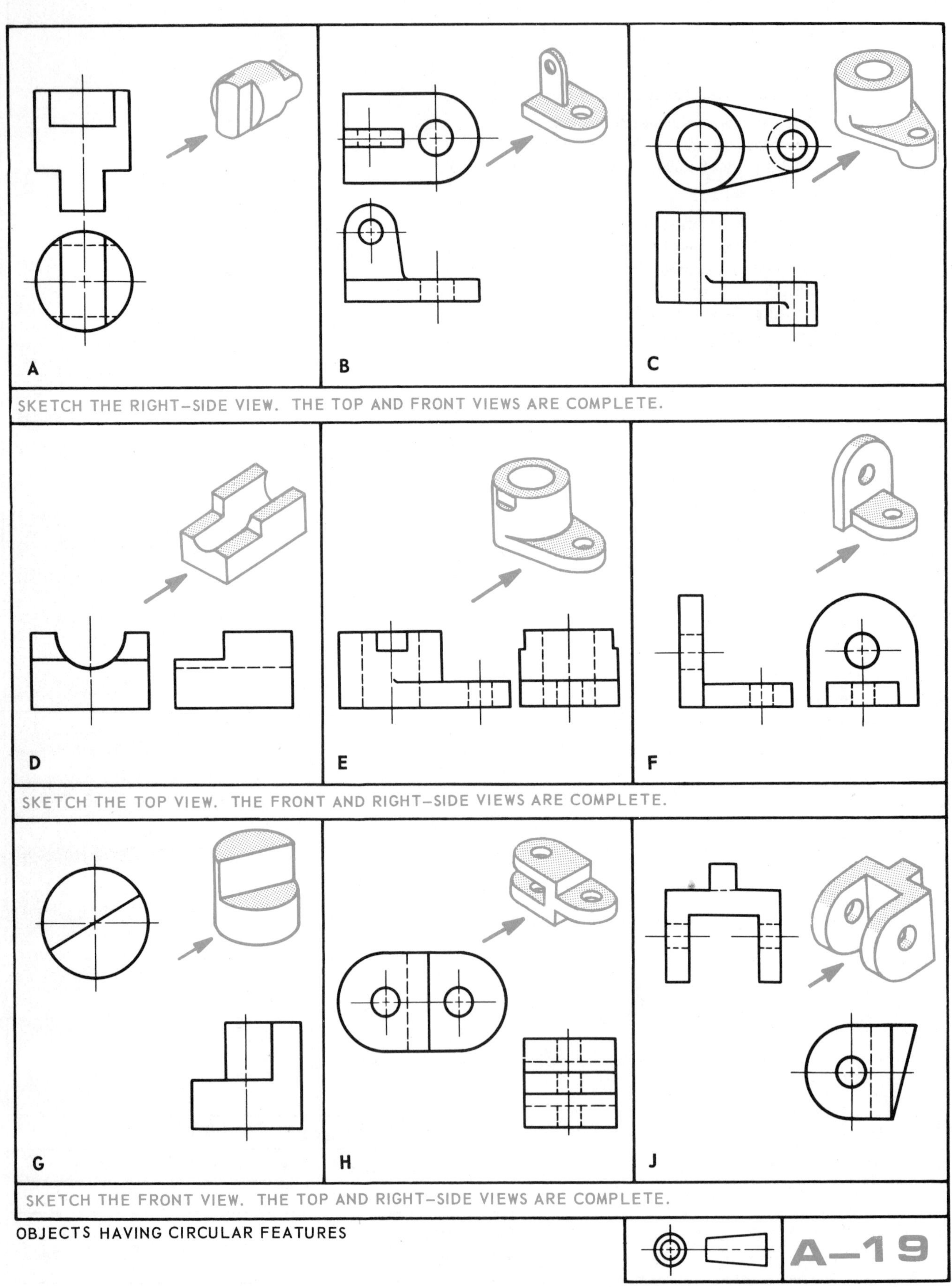
A
B
C
SKETCH THE RIGHT-SIDE VIEW. THE TOP AND FRONT VIEWS ARE COMPLETE.
D
E
F
SKETCH THE TOP VIEW. THE FRONT AND RIGHT-SIDE VIEWS ARE COMPLETE.
G
H
J
SKETCH THE FRONT VIEW. THE TOP AND RIGHT-SIDE VIEWS ARE COMPLETE.
A–19

OBJECTS HAVING CIRCULAR FEATURES

UNIT 6

DRAWING TO SCALE

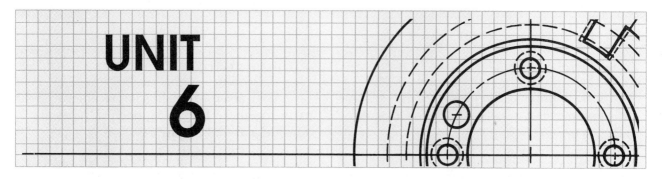

When objects are drawn at their actual size, the drawing is called *full-scale* or *scale 1:1.* Many objects, however, including buildings, ships, and airplanes, are too large to be drawn full-scale. Therefore, they must be drawn to a *reduced scale.* An example would be the drawing of a house to a scale of 1:48 (1/4″ = 1 foot) in the inch-foot scale.

Frequently, small objects, such as watch parts, are drawn larger than their actual size in order to clearly define their shapes. This is called drawing to an *enlarged scale.* The minute hand of a wrist watch, for example, could be drawn to scale 4:1.

Many mechanical parts are drawn to half scale, 1:2, and quarter scale, 1:4. Notice that the scale of the drawing is expressed in the form of a ratio. The left side of the ratio represents a unit of measurement of the size drawn. The right side represents the measurement of the actual object. Thus 1 unit of measurement on the drawing equals 4 units of measurement on the actual object. Many of the common scale ratios are shown in figure 6-1.

Drafting scales are constructed with a variety of combined scale ratios marked on their surfaces. This combination of scale ratios spares the drafter the inconvenience of calculating the size to be drawn when working to a scale other than full-size. The more common drafting scales are shown in figure 6-2.

TYPE OF SCALE		ENLARGED SCALE	SIZE AS	REDUCED SCALES
INCH SCALES	DECIMAL INCH	10:1 4:1 2:1	1:1	1:2 1:3 1:4
	FRACTIONAL INCH		1:1	1:2 1:4 3:4
	CIVIL ENGINEER'S SCALE		1:1	1:20 1:30 1:40 1:50 1:60
FOOT SCALES			1:1	$\frac{1}{8}$″ = 1 FOOT $\frac{1}{4}$″ = 1 FOOT 1″ = 1 FOOT 3″ = 1 FOOT
METRIC SCALE (MILLIMETERS		100:1 50:1 20:1 10:1 5:1 2:1	1:1	1:2 1:5 1:10 1:20 1:50 1:100

Fig. 6-1 Common scale ratios used on drawings

Inch and Foot Scales

The Inch Scales. There are three types of scales showing values that are equal to 1 inch. They are the *decimal-inch scale,* the *fractional-inch scale,* and the *civil engineer's scale.*

The civil engineer's scale has divisions of 10, 20, 30, 40, 50, 60, and 80 parts to the inch.

On fractional-inch scales, multipliers or divisors of 2, 4, 8, and 16 are used, offering such scales as full-size, half-size, quarter-size, and so forth.

The Foot Scales. Foot scales are used mostly in architectural work. The main difference between foot and inch scales is that in the foot scale each major division represents a foot, not an inch, and the end units are subdivided into inches or parts of an inch. The more common scales are the 1/8 inch = 1 foot; 1/4 inch = 1 foot; 1 inch = 1 foot; and 3 inches = 1 foot.

SI (METRIC SCALES)

Scale multipliers and divisors of 2, 5 and 10 are recommended, resulting in the scales shown in figures 6-1 and 6-2.

MACHINE SLOTS

Slots are used chiefly in machines to hold parts together. The two principal types are *T slots* and *dovetails*, figure 6-3.

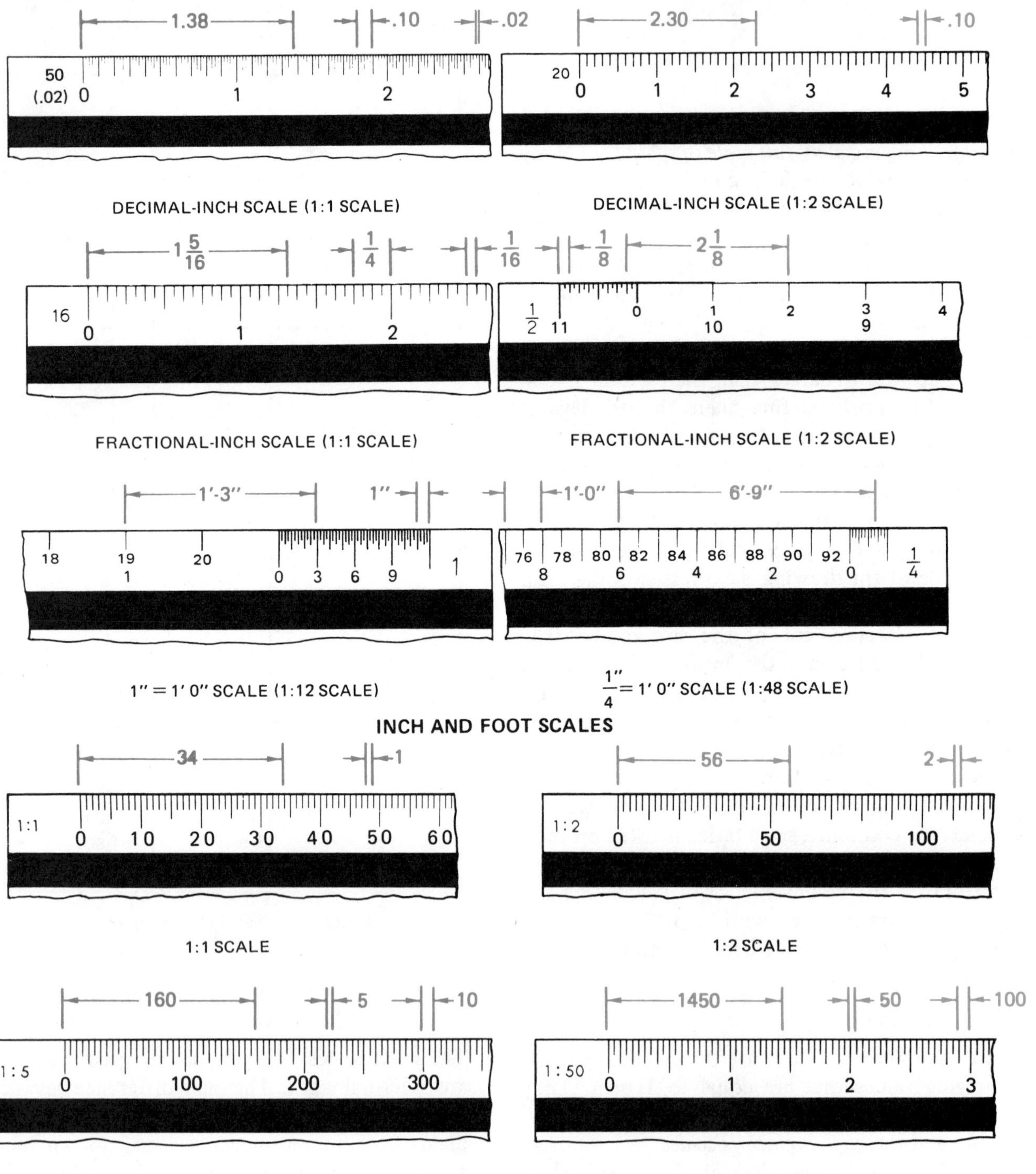

Fig. 6-2 Drafting scales

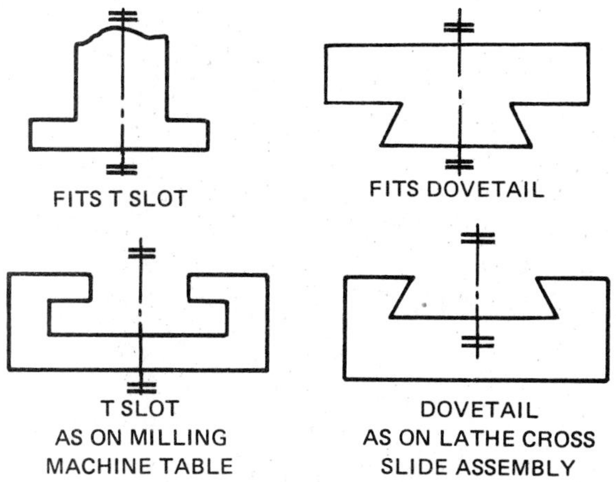

Fig. 6-3 Typical machine slots

A dovetail is a groove or slide whose sides are cut on an angle. This forms an interlocking joint between two pieces, enabling the slot to resist pulling apart in any direction other than along the lines of the dovetail slide itself.

The dovetail is commonly used in the design of slides, including lathe cross slides, milling machine table slides, and other sliding parts.

The two parts of a dovetail slide are shown in figure 6-3.

When dovetail parts are to be machined to a given width, they may be gauged by using accurately sized cylindrical rods or wires.

Dovetails are usually dimensioned as in figure 6-4. The dimensions limit the boundaries within which the machinist works.

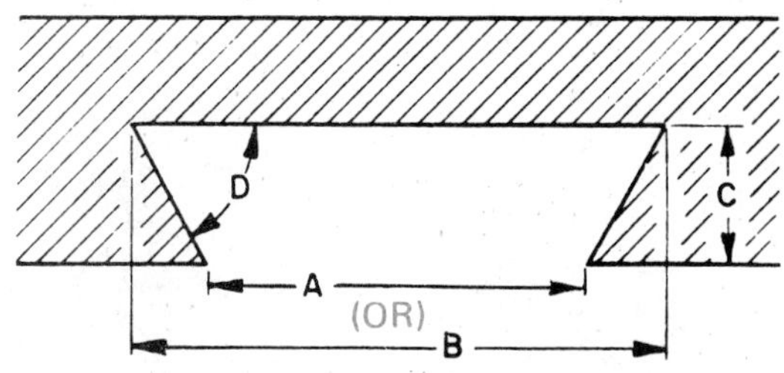

Fig. 6-4 Dimensions for dovetail

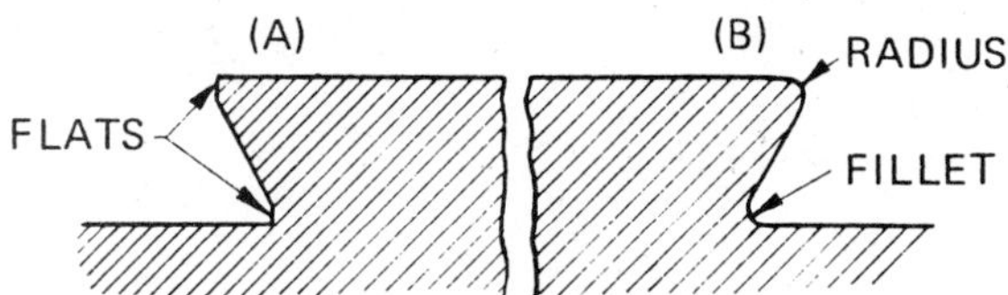

Fig. 6-5 Corners used on large dovetails

The edges of a dovetail are usually broken to remove the sharp corners. On large dovetails the external and internal corners are often machined as shown in figure 6-5 as at **A** or **B**.

Measuring Dovetails

Female dovetails may be measured by placing two accurate rods of known diameter against the sides and bottom of the dovetail as shown in figure 6-6. The distance **R** is then measured and checked against the computed value of **R** as found by the following formula:

$$R = M - [D\,(1 + \cot \frac{\text{angle } X}{2})\,]$$

Note: Diameter D should be slightly less than distance **L**.

Example:

Given:

Distance **M** = 3.000″
Diameter of rod **D** = .625″
Degrees in angle **X** = 55°

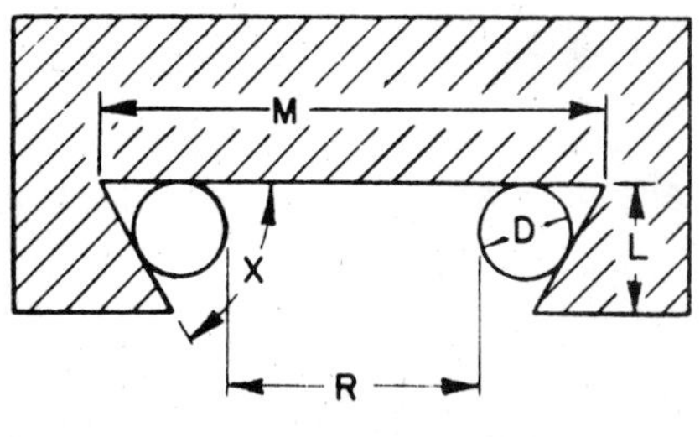

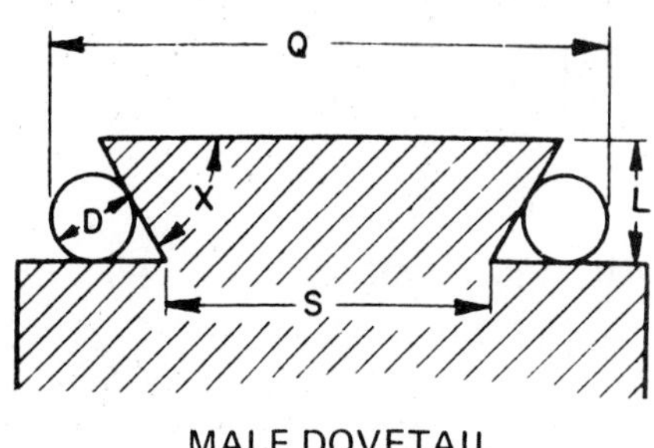

Fig. 6-6 Measuring dovetails

Then:

$$R = 3.000 - .625\ (1 + 1.921)$$

Combining: $R = 3.000 - 1.826$

$R = 1.174''$

Male dovetails may also be measured in a similar manner by placing two accurate rods of known diameter against the sides and bottom of the dovetail. The distance **Q** over the rods is measured and then checked against the computed value of **Q** as found by the following formula:

$$Q = D\ (1 + \cot \frac{\text{angle X}}{2}) + S$$

Example:

Given: Distance **S** = 2.000″
Diameter of rod **D** = .625″
Degrees in angle **X** = 55°

Then: $Q = .625\ (1 + 1.921) + 2$

Combining: $Q = (.625 \times 2.921) + 2$

$Q = 3.826''$

Other common slots produced by the milling machine are shown in figure 6-7.

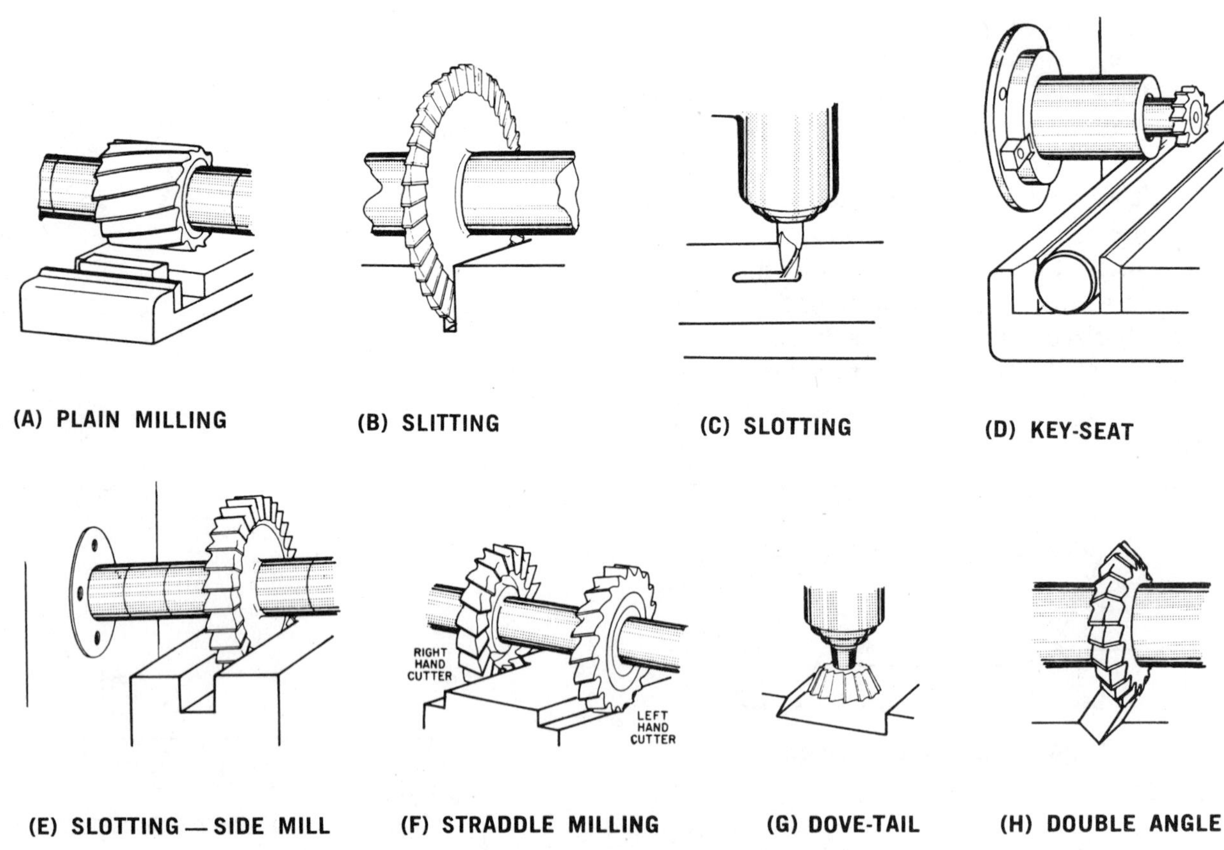

Fig. 6-7 Common milling machine operations

SYMMETRICAL OUTLINES

Symmetrical outlines or features may be indicated on a drawing by means of the symmetry symbol shown in figure 6-3. Two thick parallel lines are placed on the center lines above and below the feature. The use of this symbol means the part or feature is symmetrical about the center line or feature.

REFERENCE DIMENSIONS

When a reference dimension is shown on a drawing for information only and is not used for the manufacture of the part, it must be clearly labeled. The approved method for indicating reference dimensions on a drawing is the enclosure of the dimensions inside parentheses, figure 6-8. Previously, reference dimensions were indicated by placing the abbreviation REF after or below the dimension.

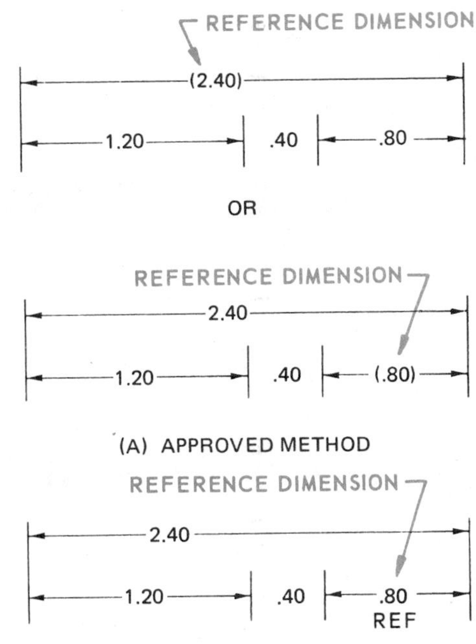

Fig. 6-8 Reference dimensions

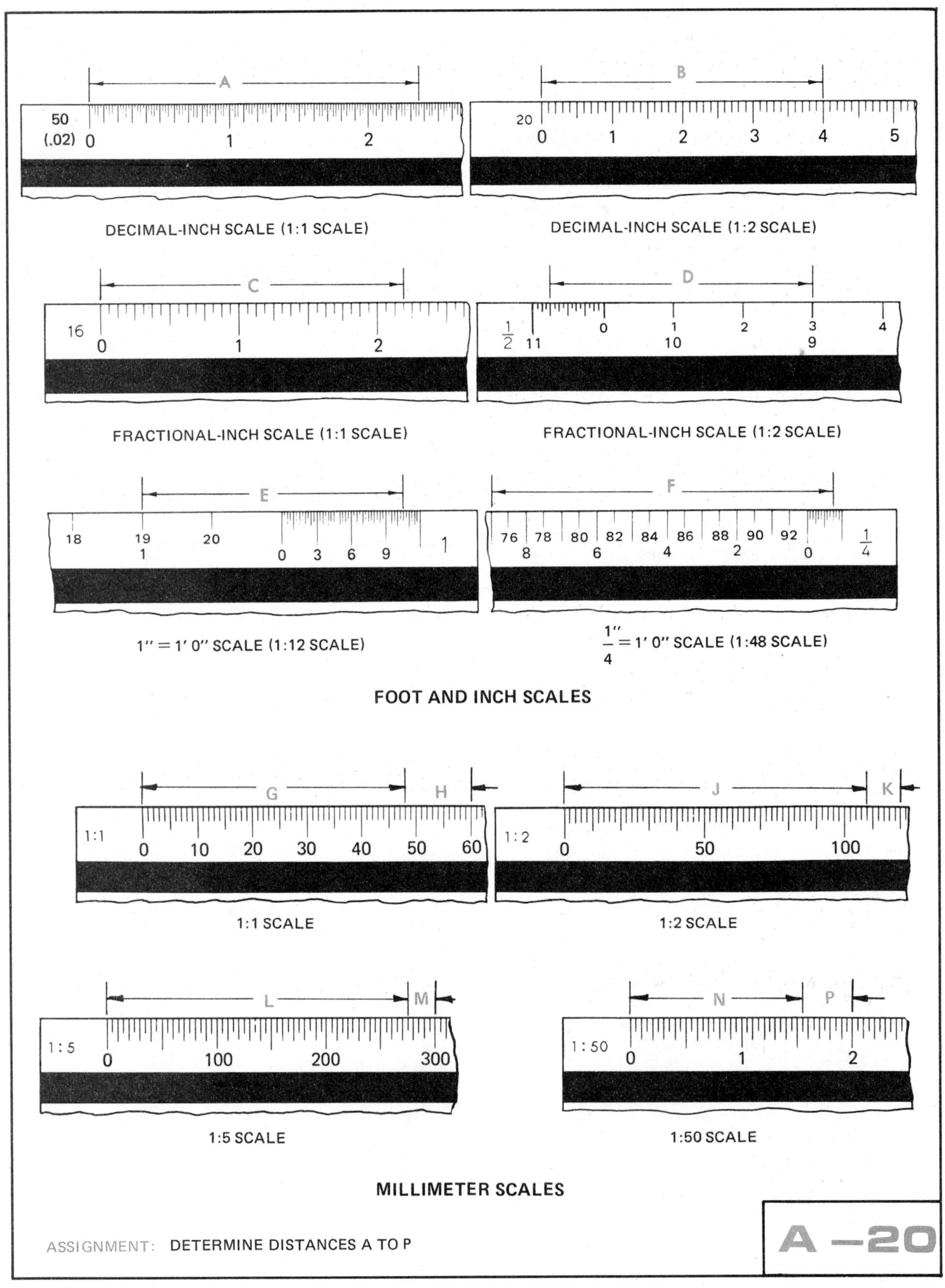

DECIMAL-INCH SCALE (1:1 SCALE)

DECIMAL-INCH SCALE (1:2 SCALE)

FRACTIONAL-INCH SCALE (1:1 SCALE)

FRACTIONAL-INCH SCALE (1:2 SCALE)

1″ = 1′ 0″ SCALE (1:12 SCALE)

$\frac{1''}{4}$ = 1′ 0″ SCALE (1:48 SCALE)

FOOT AND INCH SCALES

1:1 SCALE

1:2 SCALE

1:5 SCALE

1:50 SCALE

MILLIMETER SCALES

ASSIGNMENT: DETERMINE DISTANCES A TO P

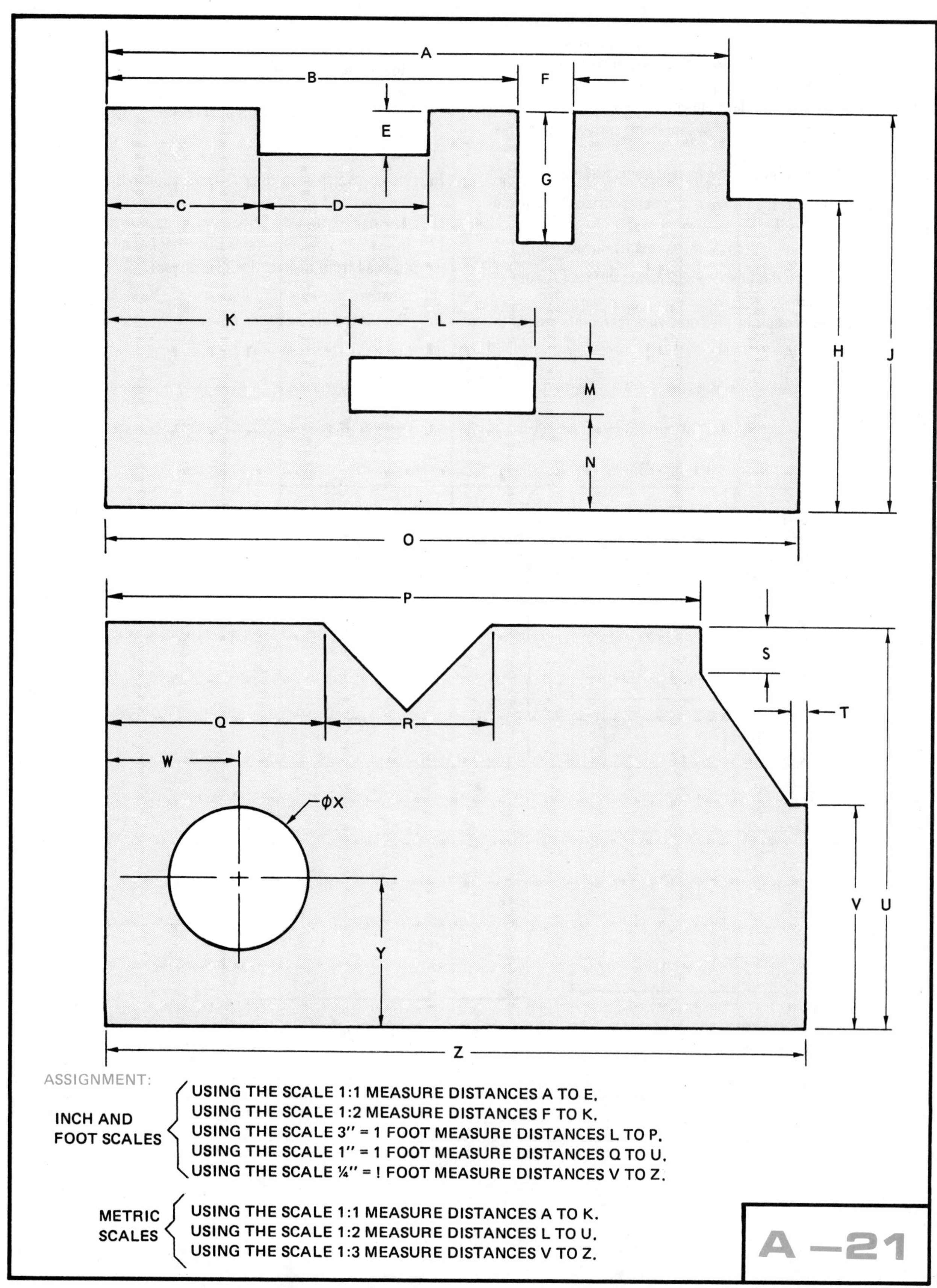

ASSIGNMENT:

INCH AND FOOT SCALES
- USING THE SCALE 1:1 MEASURE DISTANCES A TO E.
- USING THE SCALE 1:2 MEASURE DISTANCES F TO K.
- USING THE SCALE 3″ = 1 FOOT MEASURE DISTANCES L TO P.
- USING THE SCALE 1″ = 1 FOOT MEASURE DISTANCES Q TO U.
- USING THE SCALE ¼″ = ! FOOT MEASURE DISTANCES V TO Z.

METRIC SCALES
- USING THE SCALE 1:1 MEASURE DISTANCES A TO K.
- USING THE SCALE 1:2 MEASURE DISTANCES L TO U.
- USING THE SCALE 1:3 MEASURE DISTANCES V TO Z.

QUESTIONS

1. In which view is the shape of the dovetail shown?
2. In which view is the shape of the T slot shown?
3. How many rounds are shown in the top view?
4. In which view is a fillet shown?
5. Which line in the top view represents surface (R) of the side view?
6. Which line in the front view represents surface (R) ?
7. Which line in the top view represents surface (L) of the side view?
8. Which line in the front view represents surface (L) ?
9. Which line in the side view represents surface (A) on the top view?
10. Which dimension in the front view represents the width of surface (A) ?
11. What type of lines are (B), (J), and (K) ?
12. How far apart are the two hidden edge lines of the side view?
13. What dimension indicates how far line (J) is from the base of the slide?
14. How wide is the opening in the dovetail?
15. Which two lines in the top view indicate the opening of the dovetail?
16. At what angle to the horizontal is the dovetail cut?
17. In the side view, how far is the lower left edge of the dovetail from the left side of the piece?
18. What are the lengths of dimensions Y, V, and X ?
19. What is the height of the dovetail?

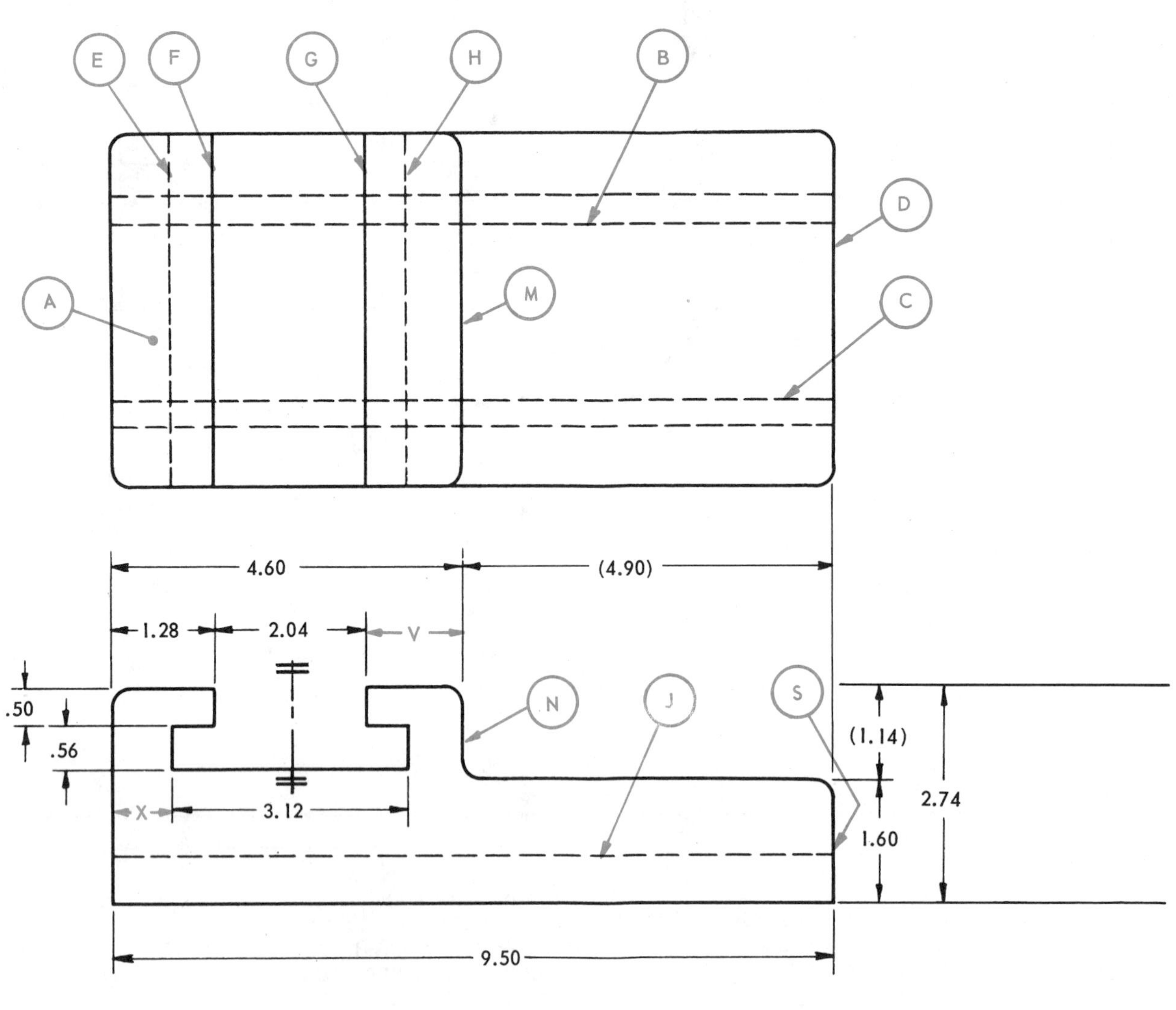

20. How much material remains between the surface represented by line (Q) and the top of the dovetail after the cut has been taken?
21. What is the vertical distance from the surface represented by line (Q) to that represented by line (T)?
22. Which dimension represents the distance between lines (F) and (G)?
23. What is the overall height of the T slot?
24. What is the distance between the bottom of the T slot and the top of the dovetail?
25. What is the width of the bottom of the T slot?
26. What is the height of the opening of the bottom of the T slot?
27. What is the horizontal distance from line (N) to line (S)?
28. What is the unit of measurement for the angles shown?
29. How many reference dimensions are shown on the drawing?
30. What is the size of the largest reference dimension?

ANSWERS

1		16	
2		17	
3		18 Y	
4		V	
5		X	
6		19	
7		20	
8		21	
9		22	
10		23	
11 (B)		24	
(J)		25	
(K)		26	
12		27	
13		28	
14		29	
15		30	

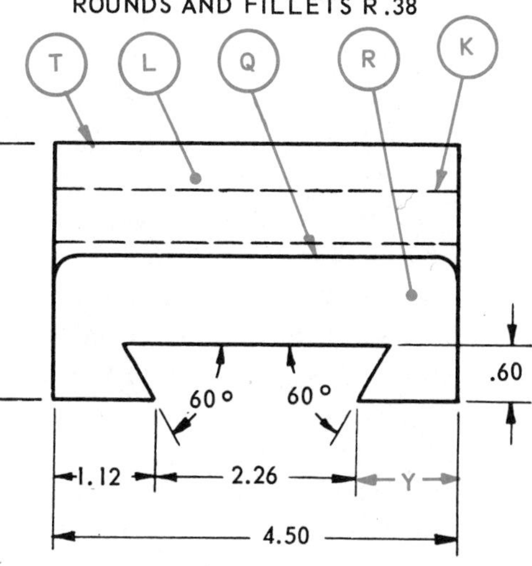

MATERIAL	GRAY IRON	
SCALE	NOT TO SCALE	
DRAWN		DATE
COMPOUND REST SLIDE		A–22

UNIT 7

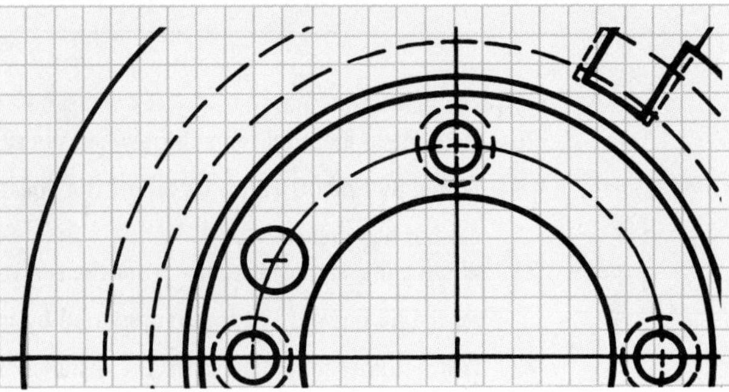

MACHINING SYMBOLS

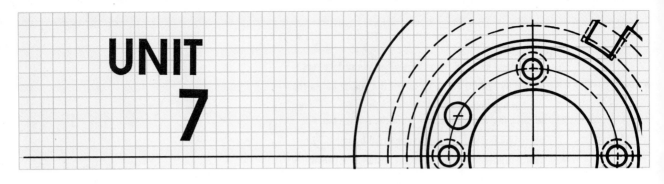

When preparing working drawings of parts to be cast or forged, the drafter must indicate part surfaces which require machining or finishing. This requirement is indicated by adding a machining allowance symbol to the surface or surfaces which must be finished by the removal of material, figure 7-1. Figure 7-2 shows the current machining symbol and those which were formerly used on drawings. This information is essential in order to alert the patternmaker and diemaker to provide extra metal on the casting or forging to allow for the finishing process. Depending on the material to be cast or forged, between .04 and .10 inch is usually allowed on small castings and forgings for each surface that requires finishing, figure 7-3.

Like dimensions, machining symbols are not duplicated on the drawing. They should be used on the same view as the dimensions that give the size or location of the surfaces. The symbol is placed on the line representing the surface or on a leader or an extension line locating the surface. The symbol and the inscription should be oriented so they may be read from the bottom or right-hand side of the drawing.

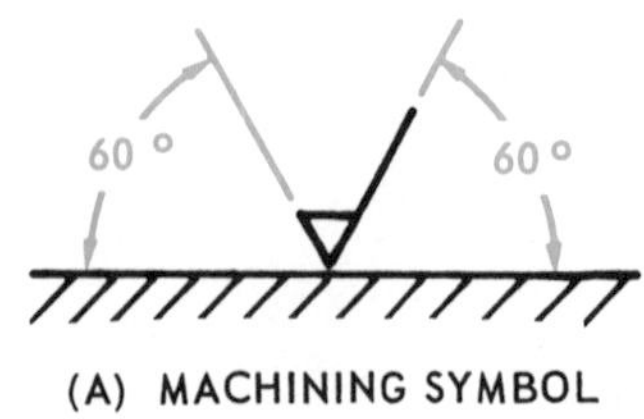

(A) MACHINING SYMBOL

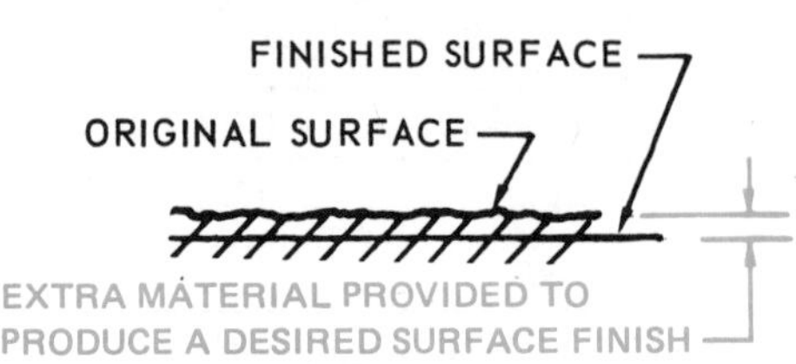

(B) MEANING

Fig. 7-1 Machining symbol

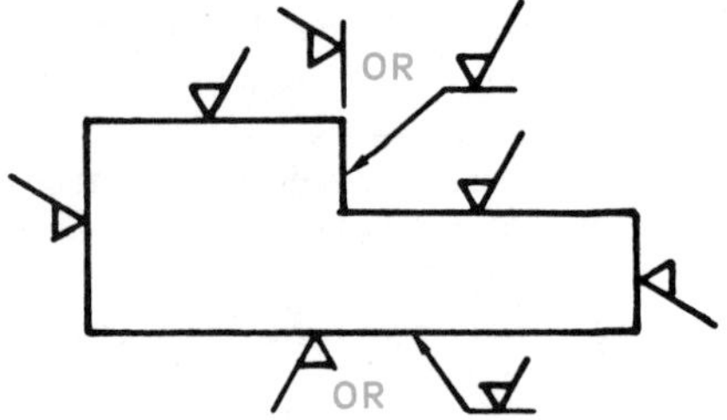

(A) RECOMMENDED SYMBOL

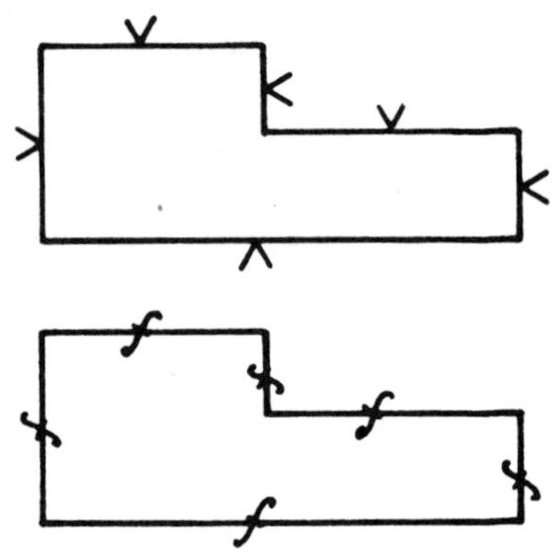

(B) FORMER MACHINING SYMBOLS

Fig. 7-2 Application of machining symbols

Where all the surfaces are to be machined, a general note, such as FINISH ALL OVER (FAO), may be used and the symbols on the drawings omitted.

The outdated machining symbols shown in figure 7-2(B) are found on many drawings in use today. When called upon to make changes or revisions on a drawing already in existence, a drafter must adhere to the drawing conventions shown on that drawing.

The machining symbol does not indicate surface-finish quality. The surface texture symbol which is defined and discussed in Unit 9 is used to signify the desired surface-finish quality.

Indicating Machining Allowance

When the value of the machining allowance must be specified, it is indicated to the left of

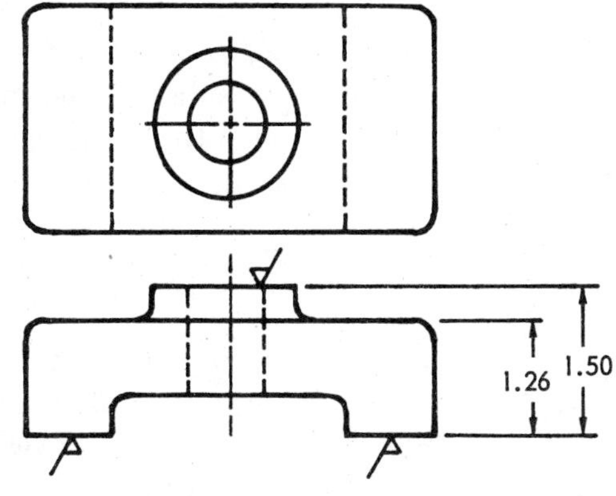

(A) FINISHED CAST PART

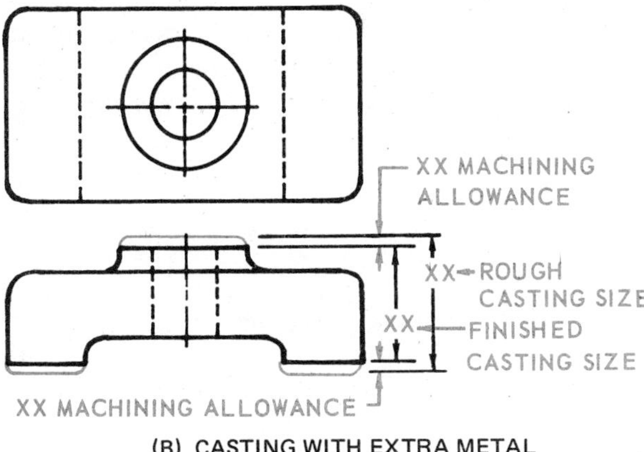

(B) CASTING WITH EXTRA METAL ALLOWED FOR MACHINING

Fig. 7-3 Allowances for machining

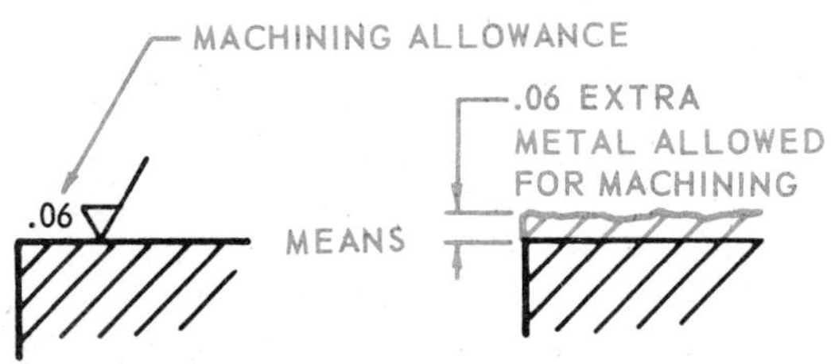

Fig. 7-4 Indicating the value of the machining allowance

the symbol, figures 7-4 and 7-5. This value is expressed in inches or millimeters depending on which units of measurement are used on the drawing.

Removal of Material Prohibited

A surface from which the removal of material is prohibited is indicated by the symbol shown in figure 7-6. This symbol indicates that

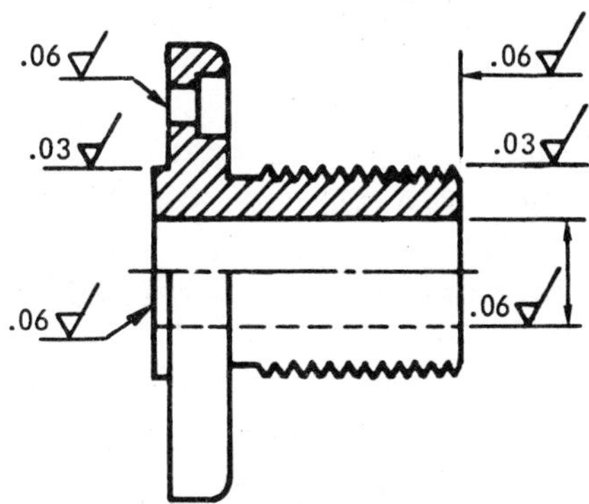

Fig. 7-5 The value of the machining allowance as shown on the drawing

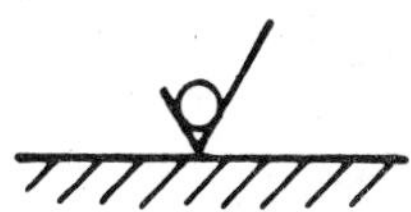

Fig. 7-6 Symbol for removal of material not permitted

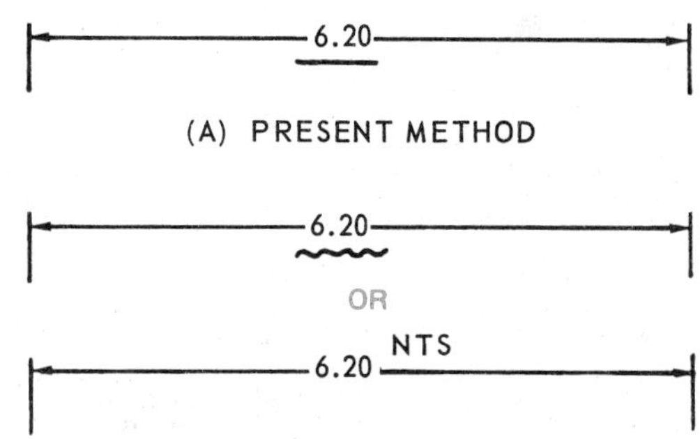

(B) FORMER METHODS FOUND ON EXISTING DRAWINGS

Fig. 7-7 Indicating dimensions that are not to scale

a surface must be left the way it is affected by a preceding manufacturing process, regardless of the removal of material or other changes.

NOT-TO-SCALE DIMENSIONS

When a dimension on a drawing is altered, making it not to scale, a straight freehand line is drawn below the dimension to indicate that the dimension is not drawn to scale, figure 7-7(A).

This is a change from earlier methods of indicating not-to-scale dimensions. Formerly, a wavy line below the dimension or the letters NTS beside the dimension were used to indicate not-to-scale dimensions, figure 7-7(B).

DRAWING REVISIONS

Drawing revisions are made to accommodate improved manufacturing methods, reduce costs, correct errors, and improve design. A clear record of these revisions must be registered on the drawing.

All drawings should carry a change or revision table located at the bottom or down the top right-hand side of the drawing. The revision number, enclosed in a circle or triangle, should be located near the revised dimension for easy identification, figure 7-8(A). The revision block should include a revision number or symbol, the date, the drafter's name or initials, and approval of the change. Should the drawing revision cause a dimension or dimensions to be different from the scale indicated, the dimensions which are not to scale should be indicated. Typical revision tables are shown in figure 7-8(B) and (C).

When many revisions are needed, a new drawing is often made. The words REDRAWN AND REVISED should appear in the revision column of the new drawing when this is done.

After a revision to a drawing is made, a new set of prints of that drawing is distributed to the appropriate departments and the prints of the original drawing are destroyed.

BREAK LINES

Break lines, as shown in figure 7-9, are used to shorten the view of long uniform sections. They are also used when only a partial view is required. Such lines are used on both detail and assembly drawings. The thin line with freehand zigzags is recommended for long breaks, and the jagged line for wood parts. The special breaks shown for cylindrical and tubular parts are useful when an end view is not shown; otherwise, the thick break line is adequate.

REFERENCES AND SOURCE MATERIAL

1. ANSI B46.1, "Surface Texture."

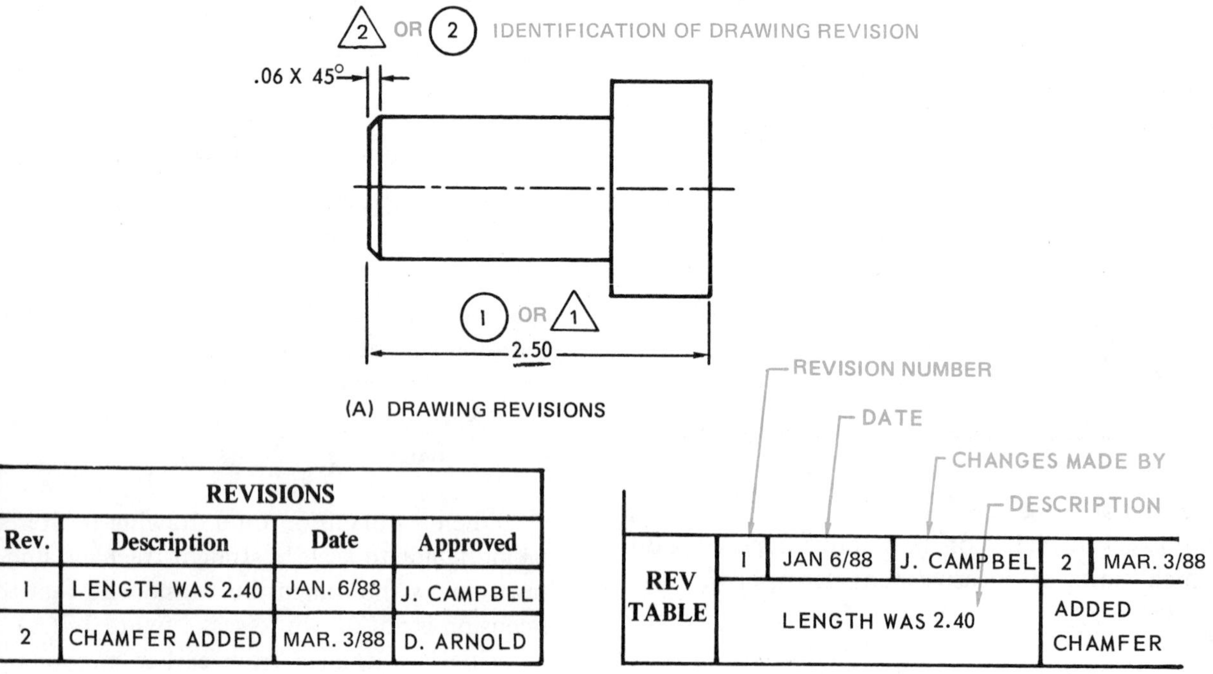

REVISIONS			
Rev.	Description	Date	Approved
1	LENGTH WAS 2.40	JAN. 6/88	J. CAMPBEL
2	CHAMFER ADDED	MAR. 3/88	D. ARNOLD

(B) TYPICAL VERTICAL REVISION BLOCK (SIZE MAY VARY)

(C) TYPICAL HORIZONTAL REVISION BLOCK (SIZE MAY VARY)

Fig. 7-8 Drawing revisions

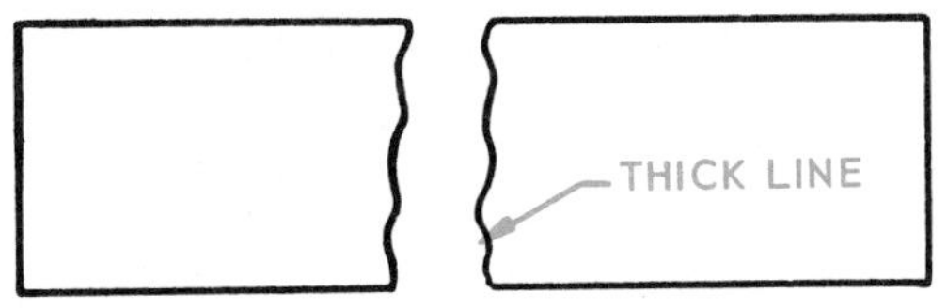

(A) SHORT BREAK – ALL SHAPES

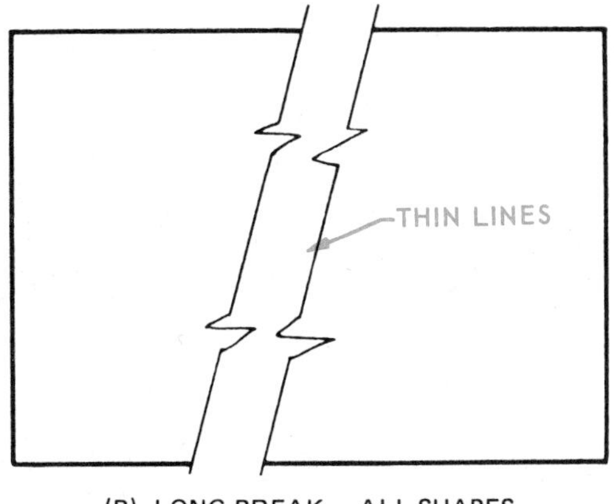

(B) LONG BREAK – ALL SHAPES

DRAWN FREEHAND OR WITH COMPASS

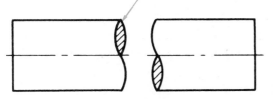

SOLID CYLINDER

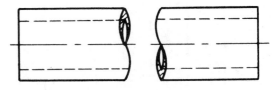

HOLLOW CYLINDER

(C) CYLINDERS – USEFUL WHEN END VIEW IS NOT SHOWN

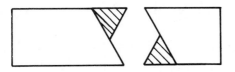

SOLID SQUARE AND RECTANGLES

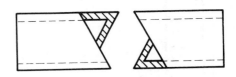

HOLLOW SQUARE AND RECTANGLES

(D) SQUARE AND RECTANGULAR

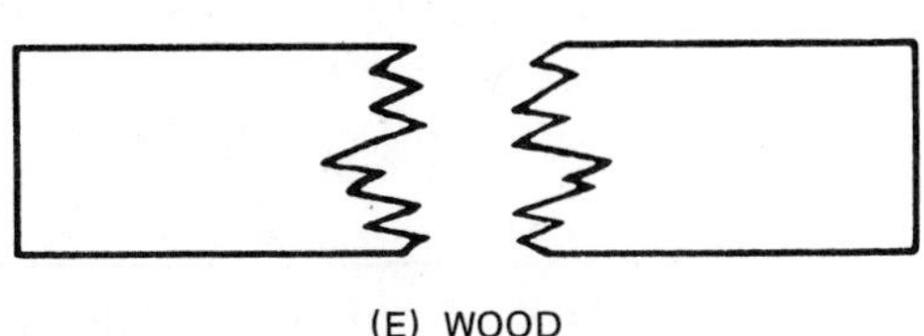

(E) WOOD

Fig. 7-9 Conventional break lines

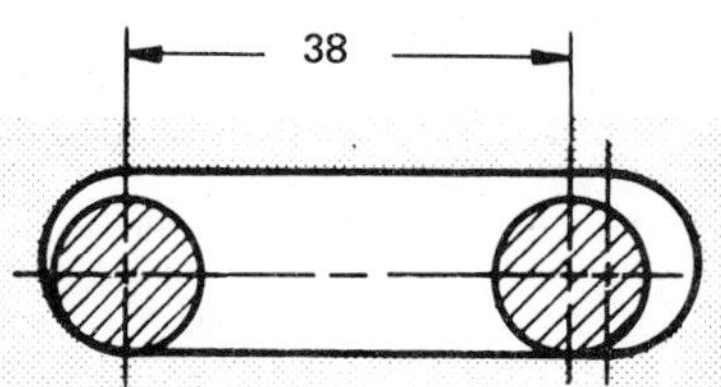

DETAIL OF 12mm BOLTS IN SLOT

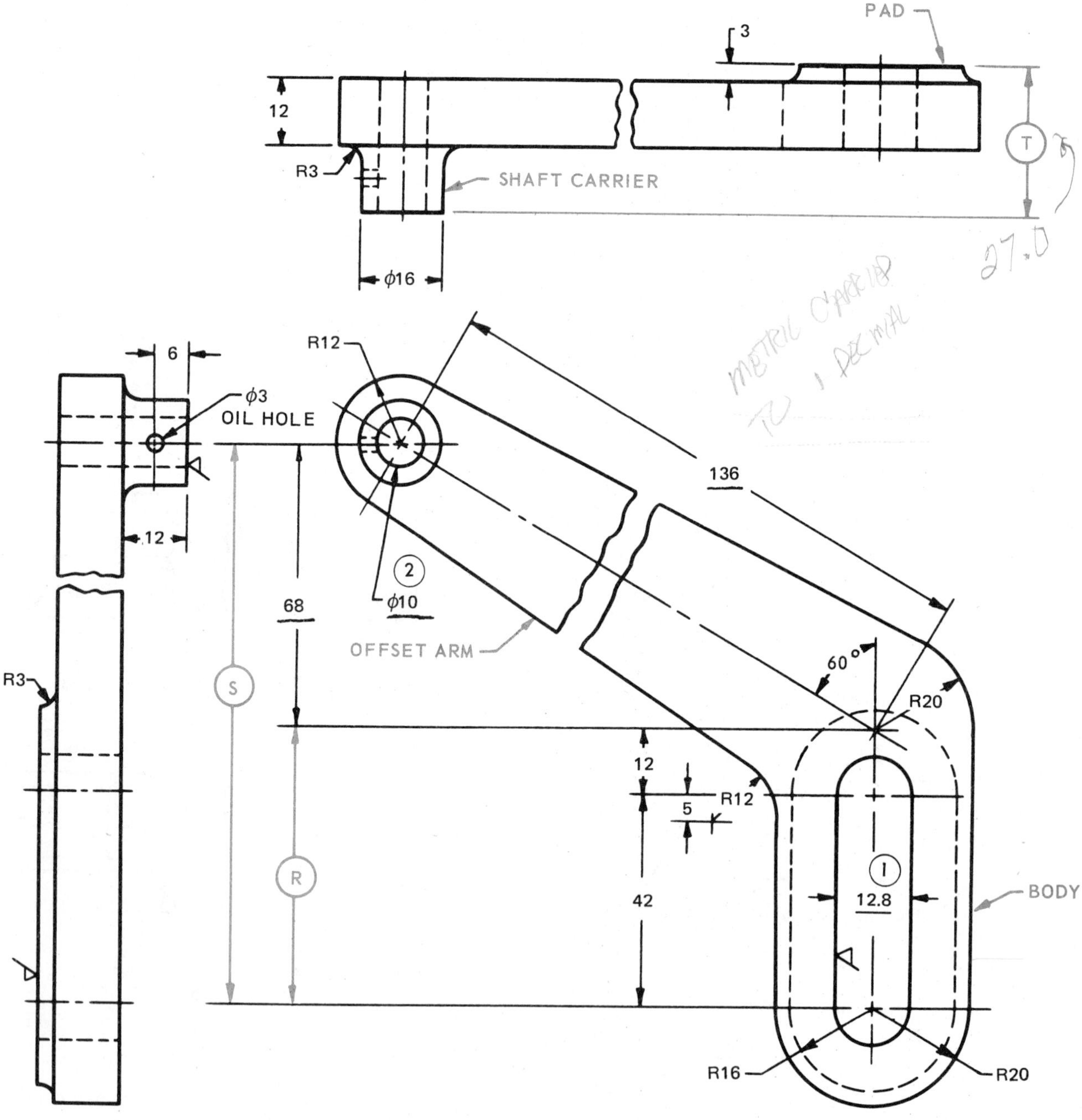

PAD
3
12
R3
SHAFT CARRIER
T
ϕ16

6
ϕ3
OIL HOLE
R12
136
12
2
ϕ10
68
OFFSET ARM
R3
S
60°
R20
12
5
R12
R
1
12.8
BODY
42
R16
R20

REVISIONS	1	JAN 12/88	K. DUNC	2	FEB. 3/88	M. SHOLAK			
	SLOT WIDTH WAS 16			HOLE WAS ϕ 8					

QUESTIONS

1. At what angle is the offset arm to the body of the piece?
2. What is the center-to-center measurement of the length of the offset arm?
3. Which radius forms the upper end of the offset arm?
4. Which radii form the lower end of the offset arm where it joins the body?
5. What is the width of the bolt slot in the body of the bracket?
6. What is the center-to-center length of this slot?
7. What was the slot width before revision?
8. Which radius forms the ends of the pad?
9. What is the overall length of this pad?
10. What is the overall width of the pad?
11. What is the radius of the fillet between the pad and the body?
12. What is the diameter of the shaft carrier body?
13. What is the diameter of the shaft carrier hole?
14. What is the distance from the face of the shaft carrier to the face of the pad?
15. What is the radius of the inside fillet between the arm and the body of the piece?
16. If M12 bolts are used in holding the bracket to the machine base, what is the clearance on each side of the slot?
17. If the center-to-center distance of the two M12 bolts which fit in the slot is 38mm, how much play is there lengthwise in the slot?
18. What size oil hole is in the shaft carrier?
19. How far is the center of the oil hole from the face of the shaft carrier?
20. How thick is the combined body and pad?
21. Calculate distances (R), (S), and (T).
22. The hole in the shaft carrier was revised. What is the difference in size between the new and old hole?
23. How many dimensions are not to scale?
24. If 2mm is allowed for each surface to be machined, what would be the overall thickness of the original casting?

ANSWERS

1 ____________
2 ____________
3 ____________
4 ____________
5 ____________
6 ____________
7 ____________
8 ____________
9 ____________
10 ____________
11 ____________
12 ____________
13 ____________
14 ____________
15 ____________
16 ____________
17 ____________
18 ____________
19 ____________
20 ____________
21 (R) ____________
(S) ____________
(T) ____________
22 ____________
23 ____________
24 ____________

DIMENSIONS IN MILLIMETERS

MATERIAL MI	
SCALE 1:1	
DRAWN	DATE
OFFSET BRACKET	A—23M

METRIC

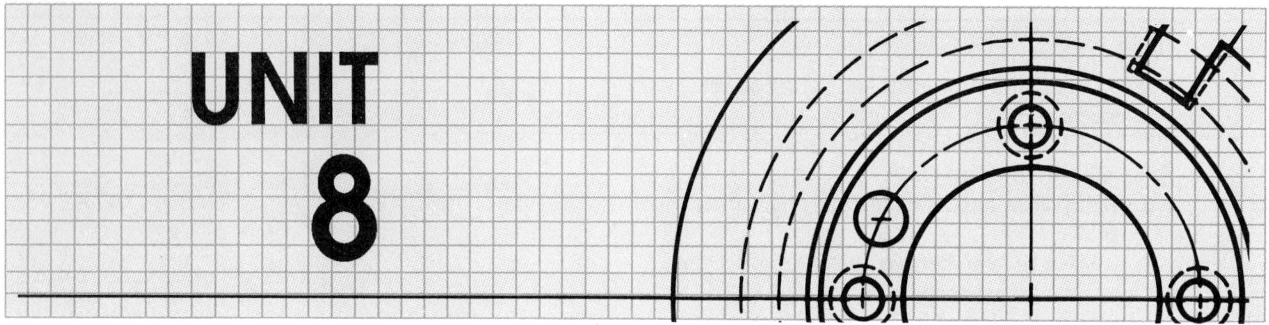

SECTIONAL VIEWS

Sectional views, commonly called *sections,* are used to show interior detail too complicated to be shown clearly and dimensioned by outside views and hidden lines. A sectional view is obtained by supposing the nearest part of the object has been cut or broken away on an imaginary cutting plane. The exposed or cut surfaces are identified by section lining or cross-hatching. Hidden lines and details behind the cutting-plane line are usually omitted unless they are required for clarity. It should be understood that only in the sectional view is any part of the object shown as having been removed.

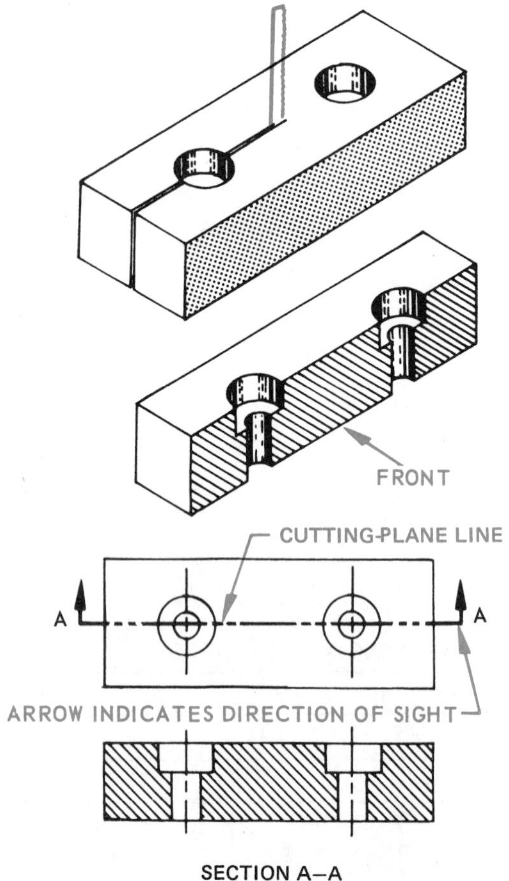

Fig. 8-1 A section drawing

A sectional view frequently replaces one of the regular views. For example, a regular front view is replaced by a front view in section, as shown in figure 8-1.

The Cutting-plane Line

A *cutting-plane line* indicates where the imaginary cutting takes place. The position of the cutting plane is indicated, when necessary, on a view of the object or assembly by a cutting-plane line, as shown in figure 8-2. The ends of the cutting-plane line are bent at 90 degrees and terminated by arrowheads to indicate the direction of sight for viewing the section. Cutting planes are not shown on sectional views. The cutting-plane line may be omitted when it corresponds to the center line of the part or when only one sectional view appears on a drawing.

If two or more sections appear on the same drawing, the cutting-plane lines are identified by two identical large, single-stroke, Gothic letters. One letter is placed at each end of the line. Sectional view subtitles are given when identification letters are used and appear directly below the view, incorporating the letters at each end of the cutting-plane line thus: SECTION A-A or, abbreviated, SECT A-A.

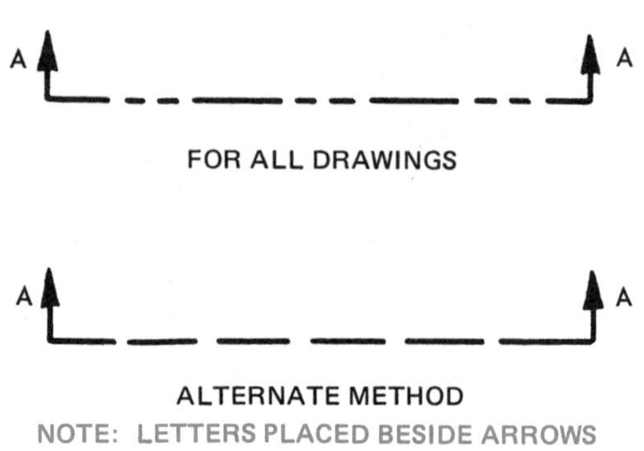

Fig. 8-2 Cutting-plane lines

Section Lining

Section lining indicates the surface that has been cut and makes it stand out clearly. Section lines usually consist of thin parallel lines, figure 8-3, drawn at an angle of approximately 45 degrees to the principal edges or axis of the part.

Since the exact material specifications for a part are usually given elsewhere, the general use section lining, figure 8-4, is recommended for general use.

When it is desirable to indicate differences in materials, other symbolic section lines are used, such as those shown in figure 8-4. If the part shape would cause section lines to be parallel or nearly parallel to one of the sides or features of the part, an angle other than 45 degrees is chosen.

The spacing of the hatching lines is uniform to give a good appearance to the drawing. The pitch, or distance, between lines varies from .06 to .18 inch, depending on the size of the area to be sectioned. Section lining is similar in direction and spacing in all sections of a single component.

Wood and concrete are the only two materials usually shown symbolically. When wood symbols are used, the direction of the grain is shown.

TYPES OF SECTIONS

Full Sections

When the cutting plane extends entirely through the object in a straight line and the front half of the object is theoretically removed,

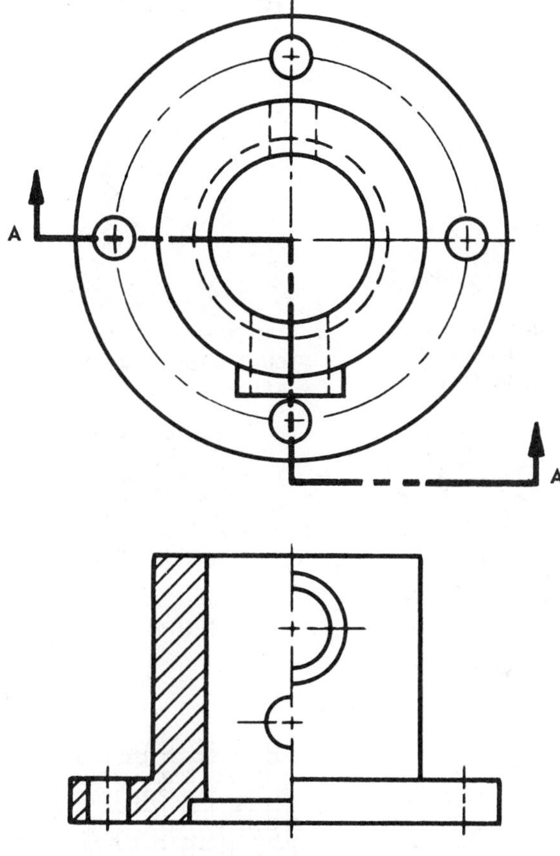

SECTION A–A

LETTERS, SUBTITLE, AND CUTTING-PLANE LINE USED WHEN MORE THAN ONE SECTION VIEW APPEARS ON A DRAWING OR WHEN THEY MAKE THE DRAWING CLEARER.

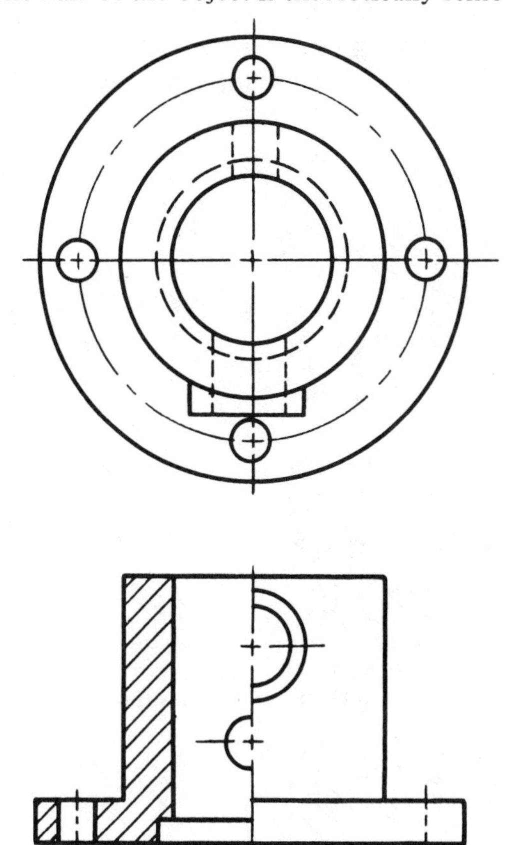

LETTERS, SUBTITLE, AND CUTTING-PLANE LINE MAY BE OMITTED WHEN THEY CORRESPOND WITH THE CENTER LINE OF THE PART AND WHEN THERE IS ONLY ONE SECTION VIEW ON THE DRAWING.

Fig. 8-3 Identification of cutting plane and section view

a *full section* is obtained, figure 8-5(B). This type of section is used for both detail and assembly drawings. When the cutting plane divides the object into two identical parts, it is not necessary to indicate its location. However, the cutting plane may be identified and indicated in the usual manner to increase clarity.

Half Sections

A symmetrical object or assembly may be drawn as a *half section,* figure 8-5(C), showing one half up to the center line in section and the other half in full view. A normal center line is used on the section view.

The half section drawing is not normally used where the dimensioning of internal diameters is required. This is because many hidden lines would have to be added to the portion showing the external features. This type of section is used mostly for assembly drawings where internal and external features are clearly shown and only overall and center-to-center dimensions are required.

COUNTERSINKS, COUNTERBORES, AND SPOTFACES

A *countersunk hole* is a conical depression cut in a piece to receive a countersunk type of flathead screw or rivet, as illustrated in figure 8-6. The size is usually shown by a note listing the diameter of the hole first, followed by the diameter of the countersink, the abbreviation CSK, and the angle. A *counterbored hole* is one which has been machined larger to a given depth to receive a fillister, hex-head, or similar type of bolt head. Counterbores are specified by a note giving the diameter of the hole first, followed by the counterbore diameter, the abbreviation CBORE, and depth of the counterbore. The counterbore and depth may also be indicated by

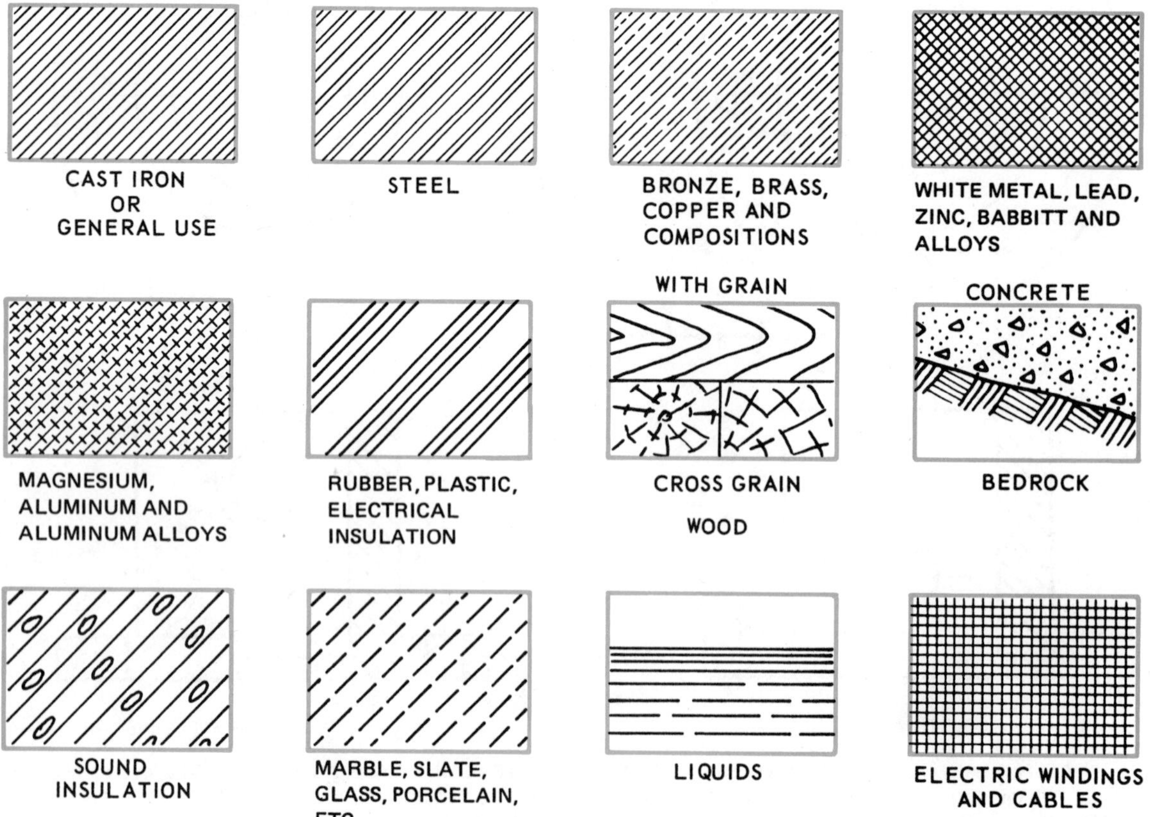

Fig. 8-4 Symbolic section lining

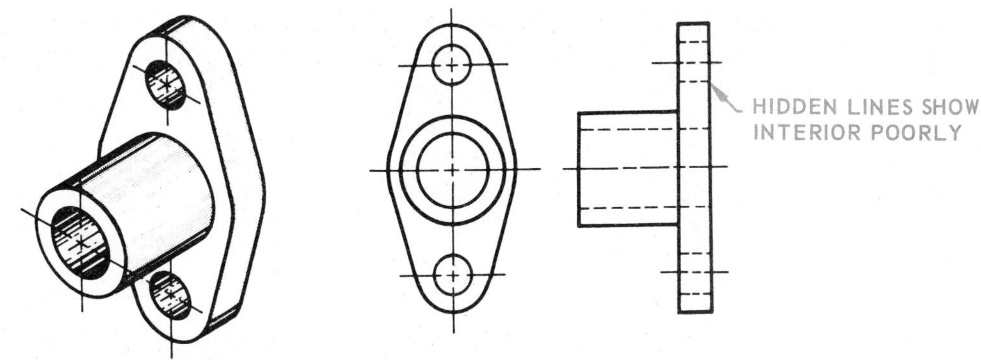

(A) SIDE VIEW NOT SECTIONED

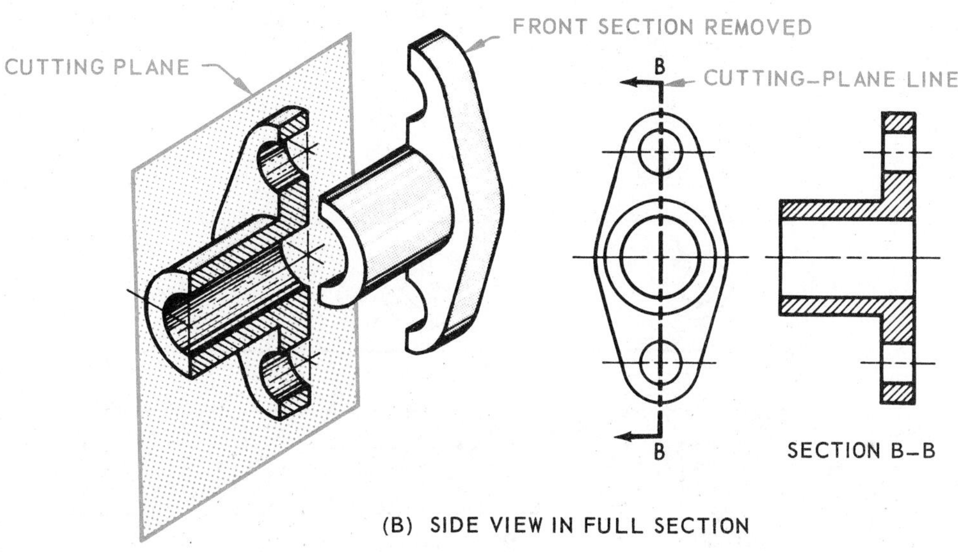

(B) SIDE VIEW IN FULL SECTION

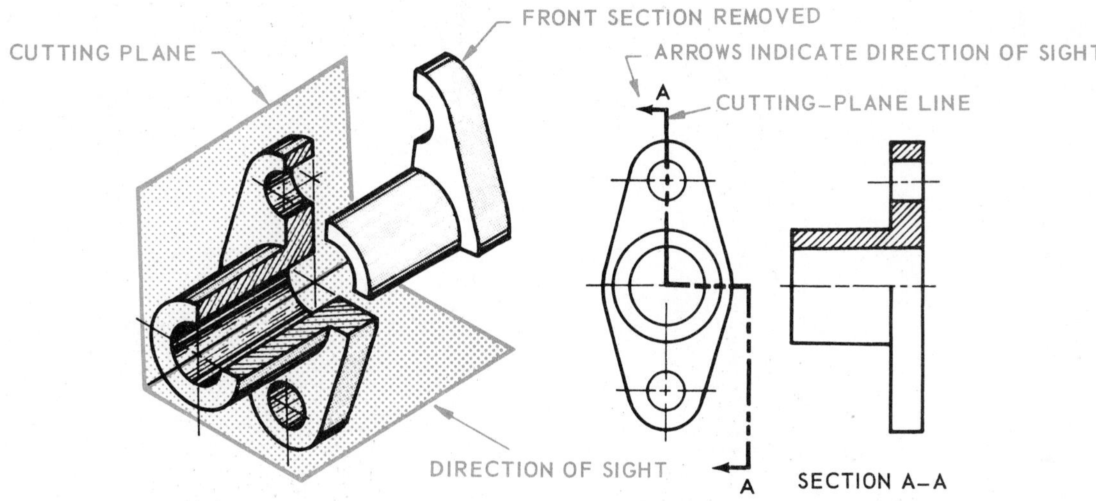

(C) SIDE VIEW IN HALF SECTION

Fig. 8-5 Full and half sections

direct dimensioning. A *spotface* is an area where the surface is machined just enough to provide a level seating surface for a bolt head, nut or washer. A spotface is specified by a note listing the diameter of the hole first, followed by the spotface diameter, and the abbreviation SFACE. The depth of the spotface is not usually given.

The symbolic means of indicating a counterbore or spotface, a countersink, and the depth of a feature are shown in figure 8-7. In each case, the symbol precedes the dimension.

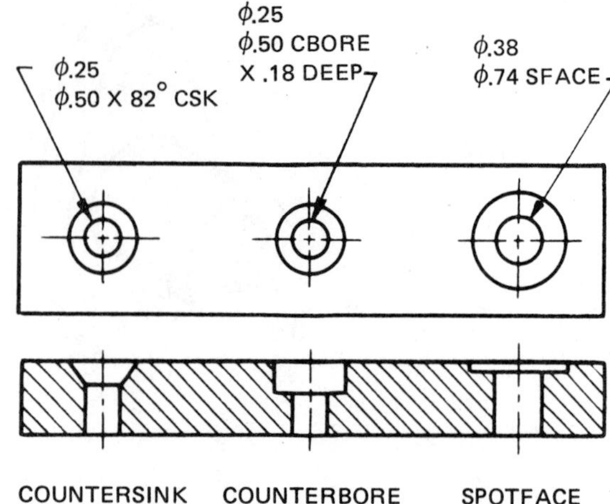

Fig. 8-6 Dimensioning countersinks, counterbores, and spotfaces

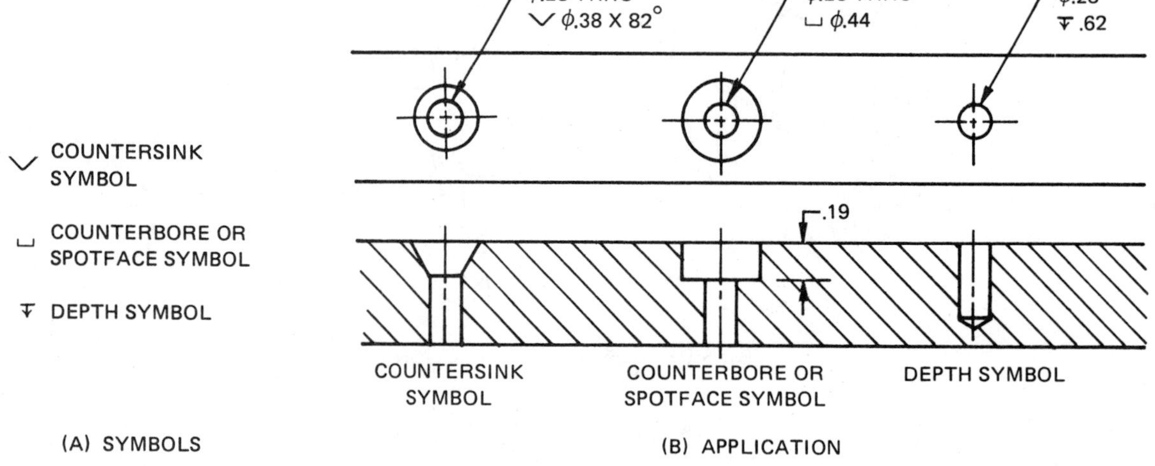

Fig. 8-7 Hole symbols

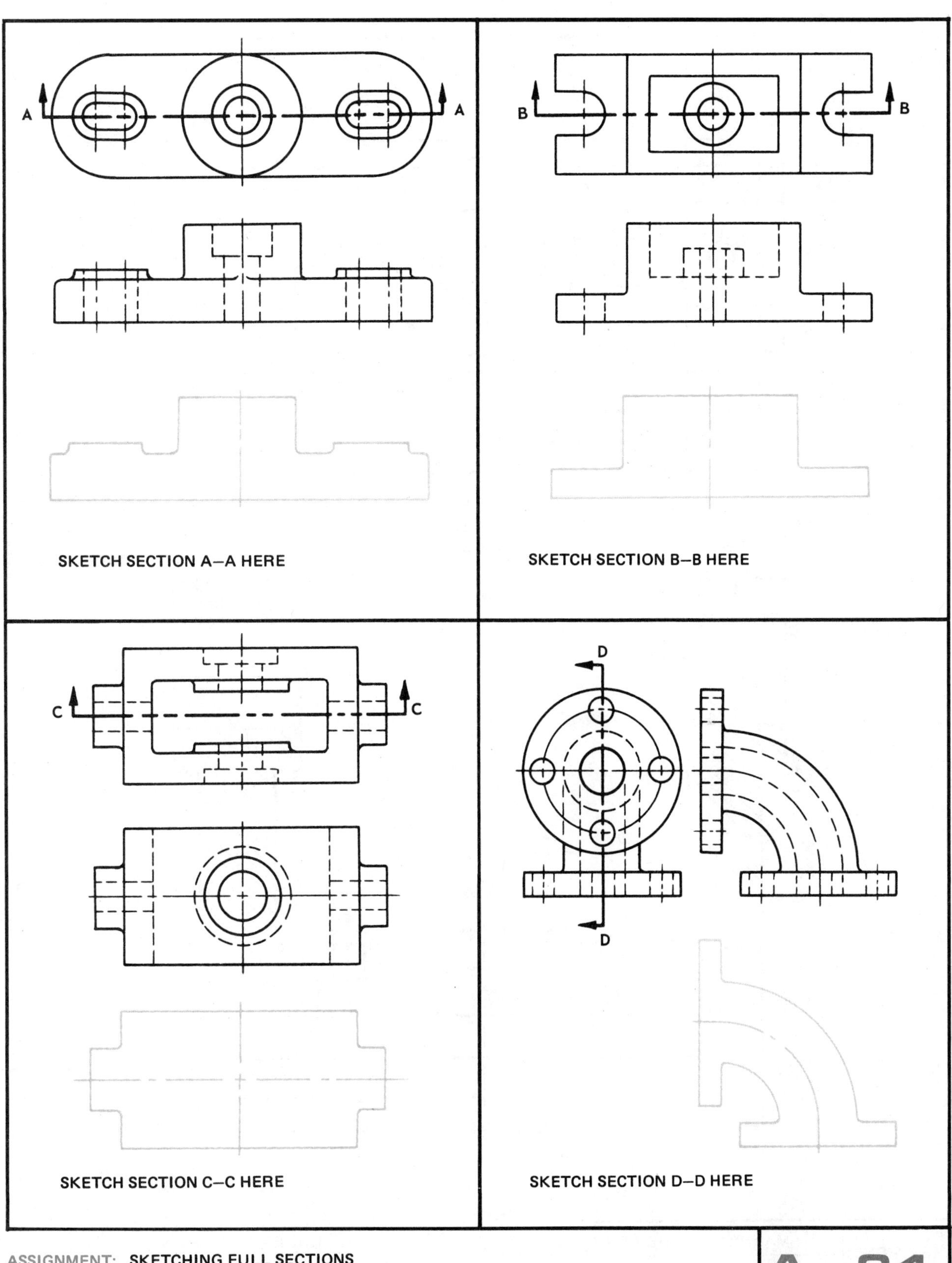

ASSIGNMENT: SKETCHING FULL SECTIONS

A –24

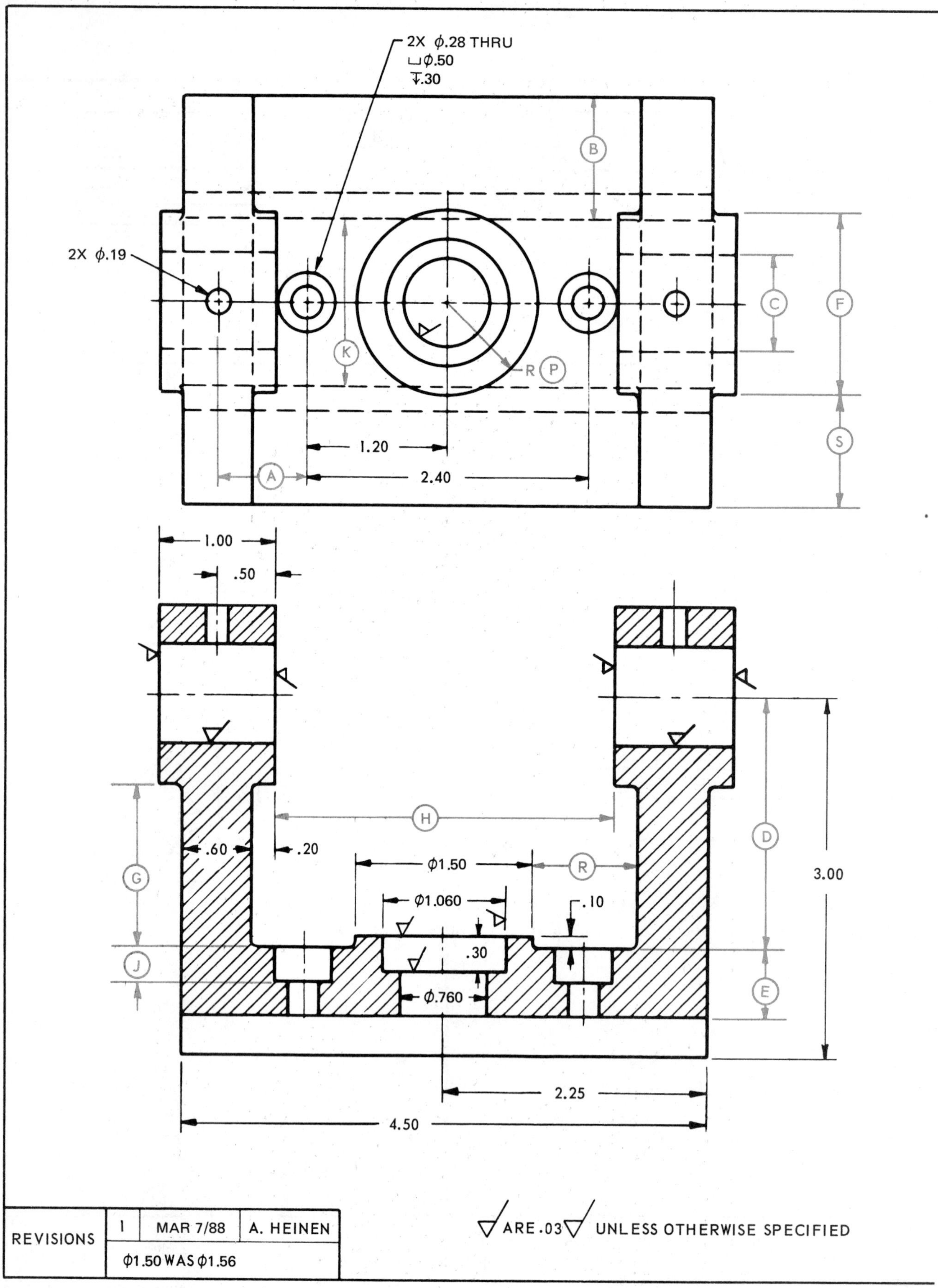

2X ϕ.28 THRU
⌴ϕ.50
↧.30
2X ϕ.19
B
C
F
K
R P
S
1.20
A
2.40
1.00
.50
H
.60
.20
G
ϕ1.50
R
D
3.00
ϕ1.060
.10
.30
J
ϕ.760
E
2.25
4.50
REVISIONS
1
MAR 7/88
A. HEINEN
ϕ1.50 WAS ϕ1.56
ARE .03 UNLESS OTHERWISE SPECIFIED

QUESTIONS

1. What is the overall width?
2. What is the overall height?
3. What is the center-to-center distance of the Ø.19 holes?
4. If .03 is allowed for each surface requiring finishing, calculate the width of the casting before machining.
5. At what angle to the vertical is the dovetail slot?
6. How many different surfaces require finishing?
7. Which type of lines in the top view represents the dovetail?
8. What was the original size of the Ø1.50?
9. How wide is the opening in the dovetail?
10. How high is the dovetail?
11. When was the Ø1.50 dimension altered?
12. How many reference dimensions are shown?
13. Calculate distances (A) to (S).

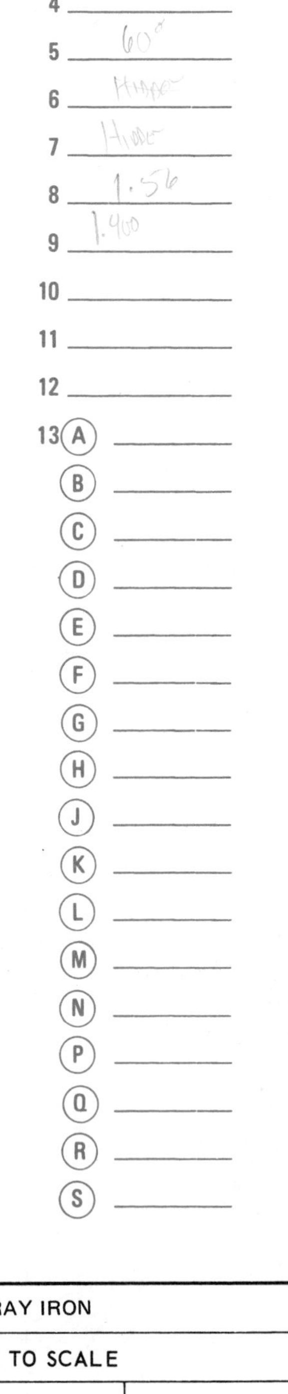

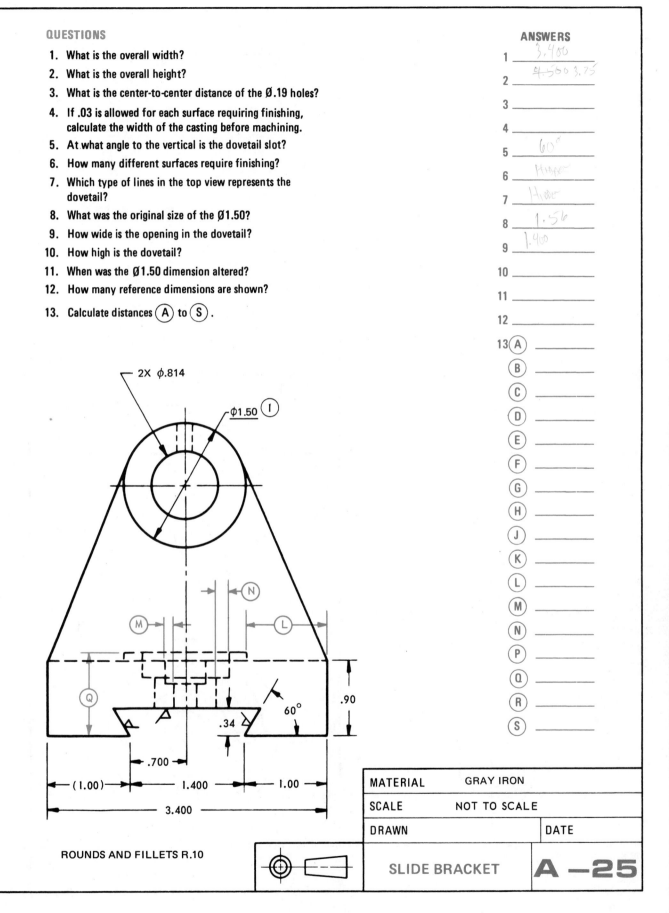

2X ϕ.814

ϕ1.50 (1)

(N) (M) (L) (Q)

60°

.90

.34

.700

(1.00) 1.400 1.00

3.400

ROUNDS AND FILLETS R.10

MATERIAL	GRAY IRON	
SCALE	NOT TO SCALE	
DRAWN		DATE
	SLIDE BRACKET	A–25

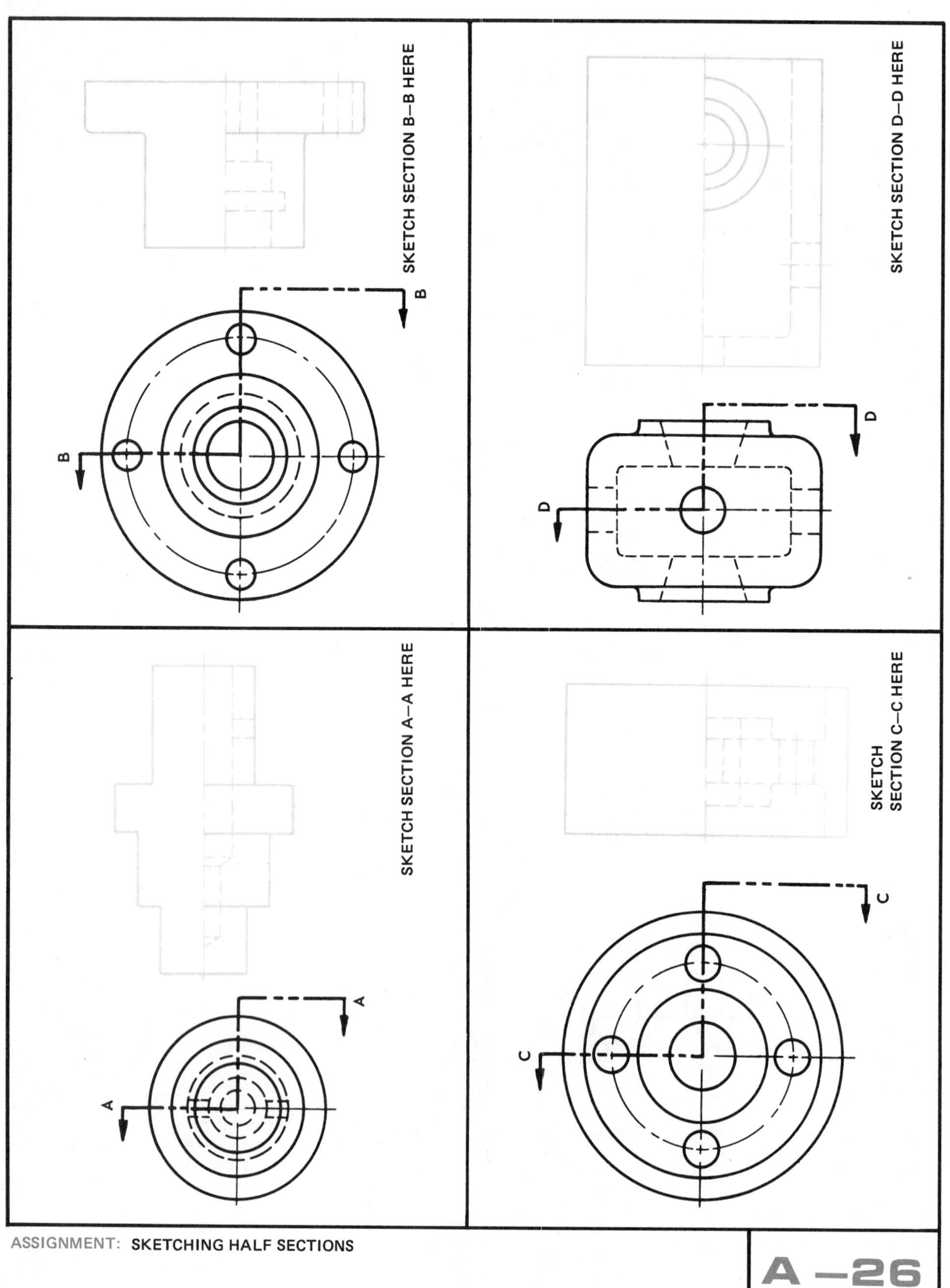

ASSIGNMENT: **SKETCHING HALF SECTIONS**

A –26

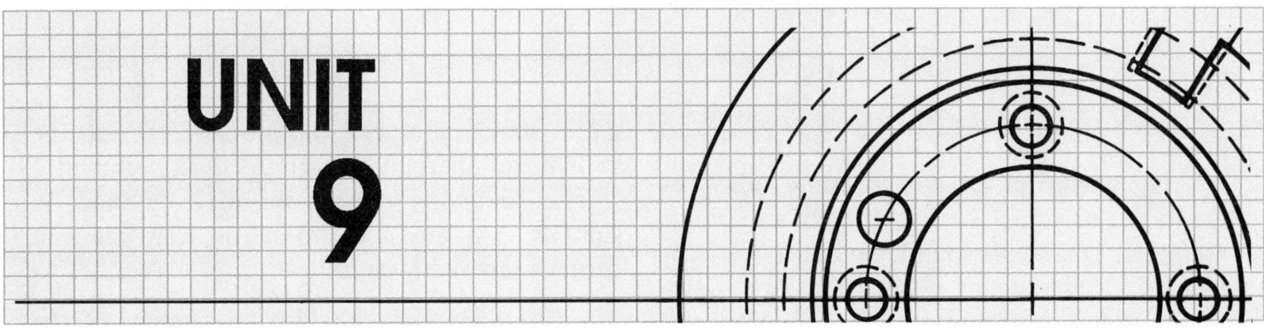

CHAMFERS

The process of *chamfering,* that is, cutting away the inside or outside corner of an object, is done to facilitate assembly, figure 9-1. The recommended method of dimensioning a chamfer is to give an angle and the linear length, or an angle and a diameter. For angles of 45 degrees only, a note form may be used. This method is permissible only with 45-degree angles because the size may apply to either the longitudinal or radial dimension. Chamfers are never measured along the angular surface.

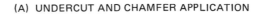

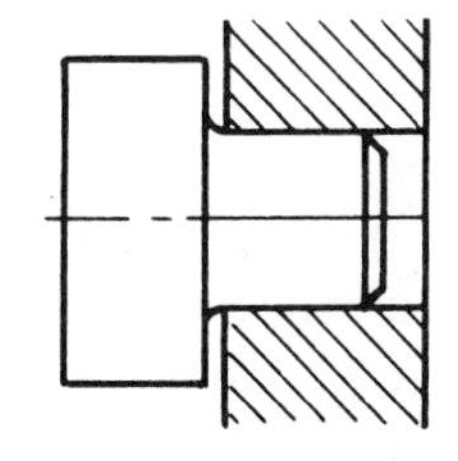

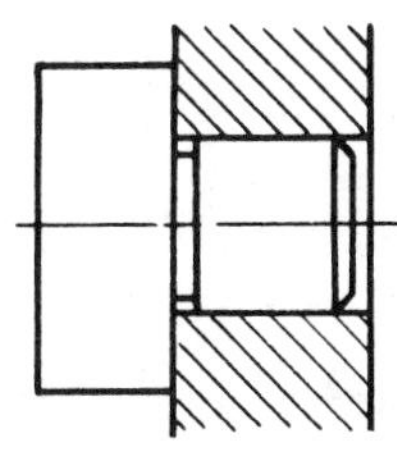

SAME PART WITH UNDERCUT
ADDED PERMITS PART
TO FIT FLUSH

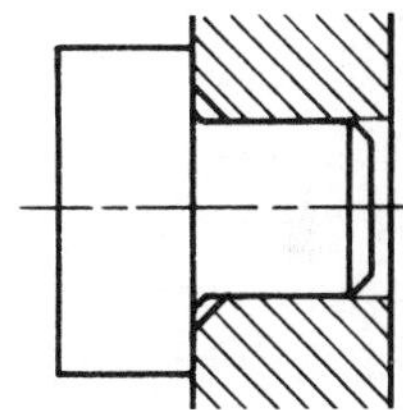

(A) UNDERCUT AND CHAMFER APPLICATION

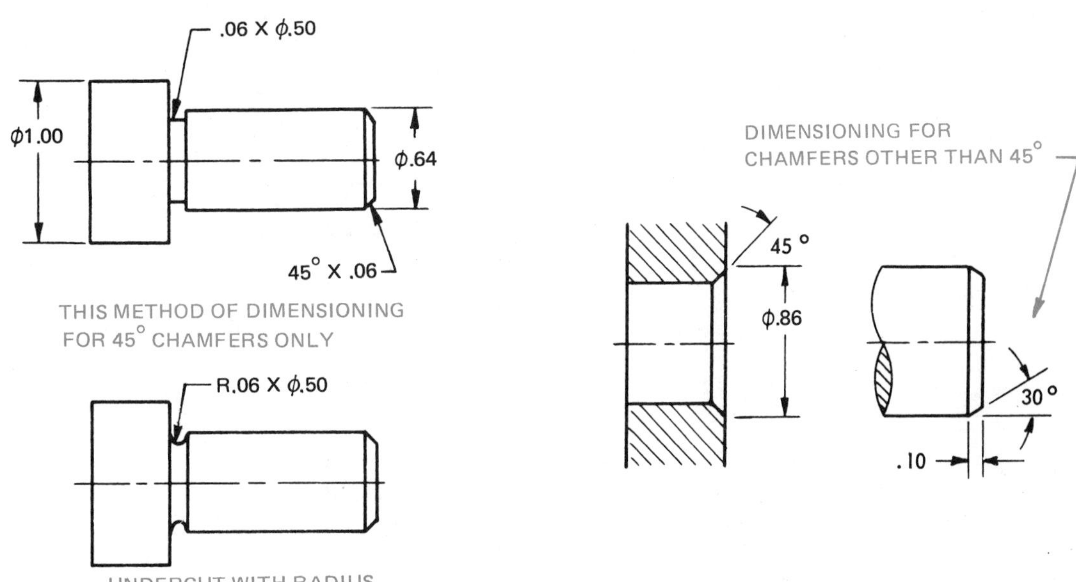

(B) DIMENSIONING CHAMFERS AND UNDERCUTS

Fig. 9-1 Chamfers and undercuts

UNDERCUTS

The operation of *undercutting,* also referred to as *necking,* is the cutting of a recess in a diameter. Undercuts permit two parts to join, figure 9-1. Undercutting is indicated on a drawing by a note listing the width first and then the diameter. If the radius is shown at the bottom of the undercut, it is assumed that the radius will be equal to half the width unless specified differently and the diameter will apply to the center of the undercut. Where the size of the neck is unimportant, the dimension may be omitted from the drawing.

TAPERS

Circular Tapers

Tapered shanks are used on many small tools such as drills, reamers, counterbores, and spotfaces to hold them accurately in the machine spindle. Taper means the difference in diameter or width in a given length. There are many standard tapers; the Morse taper and the Brown and Sharpe taper are the most common.

The following dimensions may be used, in suitable combinations, to define the size and form of tapered features:

- The diameter (or width) at one end of the tapered feature
- The length of the tapered feature
- The rate of taper
- The included angle
- The taper ratio

In dimensioning a taper by means of taper ratio, the circular taper symbol should precede the ratio figures, see figure 9-2.

Flat Tapers

Flat tapers (slopes) are used as locking devices such as taper keys and adjusting shims. The methods recommended for dimensioning flat tapers are shown in figure 9-3. The flat taper symbol should precede the ratio figures.

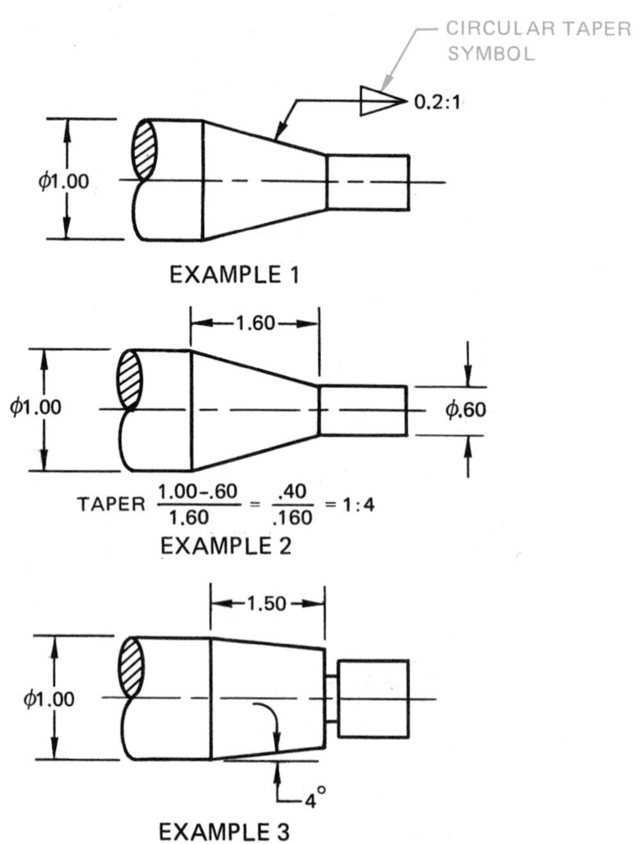

Fig. 9-2 Dimensioning circular tapers

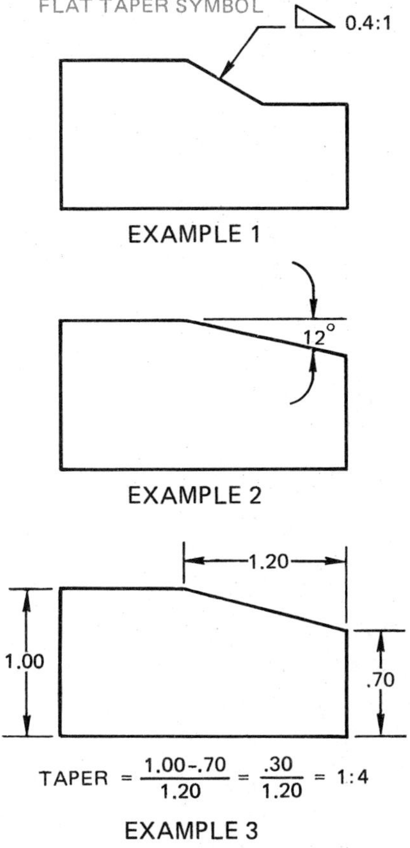

Fig. 9-3 Dimensioning flat tapers (slopes)

KNURLS

Knurling is the machining of a surface to create uniform depressions. Knurling permits a better grip. Knurling is shown on drawings as either a straight or diamond pattern. The pitch of the knurl may be specified. It is unnecessary to hatch in the whole area to be knurled if enough is shown to clearly indicate the pattern. Knurls are specified on the drawing by a note calling for the type and pitch. The length and diameter of the knurl are shown as dimensions, figure 9-4.

INTERSECTION OF UNFINISHED SURFACES

The intersection of unfinished surfaces that are rounded or filleted at the point of theoretical intersection is indicated by a line coinciding with the theoretical point of intersection. The need for this convention is shown by the examples in figure 9-5. For a large radius, figure 9-5(C), no line is drawn.

Members such as ribs and arms that blend into other features end in curves called *runouts.*

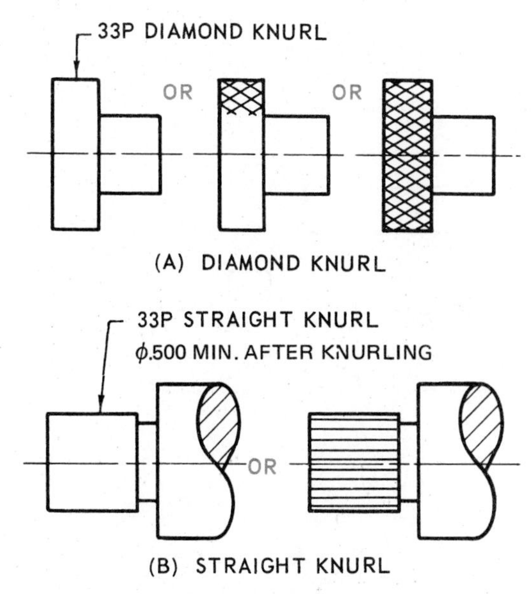

Fig. 9-4 Knurling

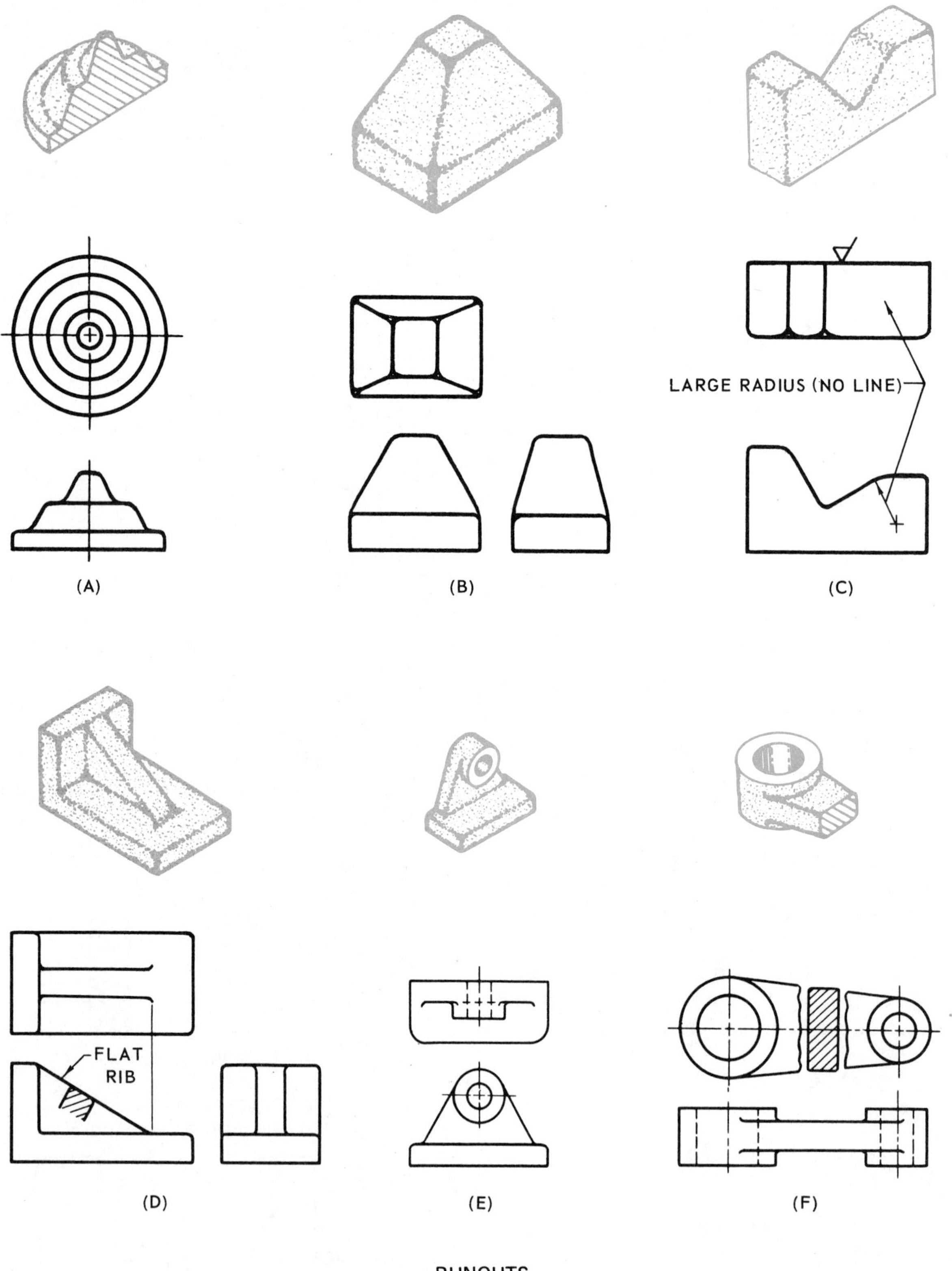

Fig. 9-5 Rounded and filleted intersections

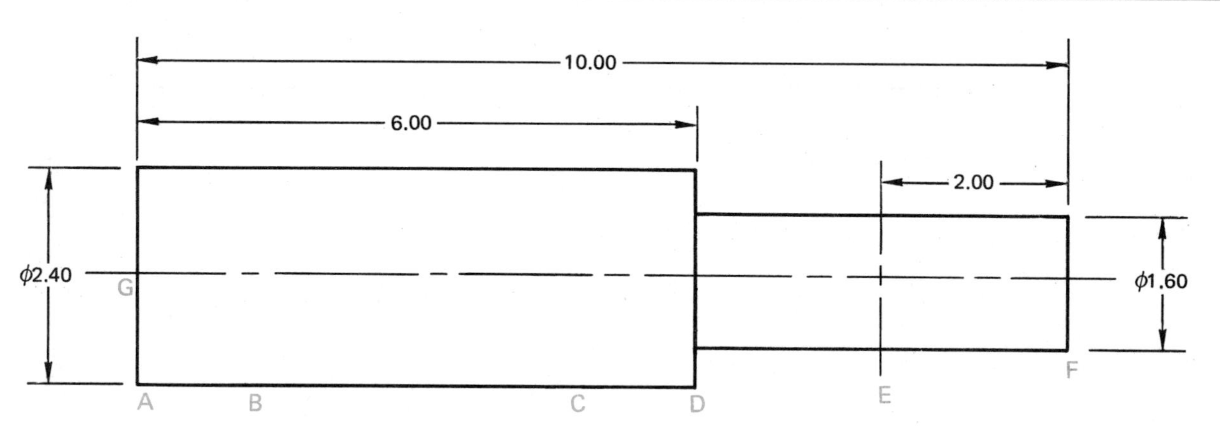

ASSIGNMENT: REDRAW THE HANDLE ON THE GRID SHOWN BELOW. SCALE IS 1:2. THE FOLLOWING FEATURES ARE TO BE ADDED AND DIMENSIONED USING SYMBOLS WHEREVER POSSIBLE AT THE LOCATIONS SHOWN.

A. 45° X .20 CHAMFER
B. 33P DIAMOND KNURL FOR 1.20 IN. STARTING .80 IN. FROM LEFT END
C. 0.1:1 CIRCULAR TAPER FOR 1.20-IN. LENGTH ON RIGHT END OF ϕ2.40 (SHOW THE CIRCULAR TAPER SYMBOL)
D. .20 X ϕ1.40 UNDERCUT ON ϕ1.60
E. ϕ.20 X .50 DEEP, 4 HOLES EQUALLY SPACED
F. 30° X .30 CHAMFER. THE .30 DIMENSION TAKEN HORIZONTALLY ALONG THE SHAFT
G. ϕ.60 HOLE, 1.50 DEEP

1.00 IN. SQUARES (.20 IN. GRID)

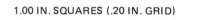

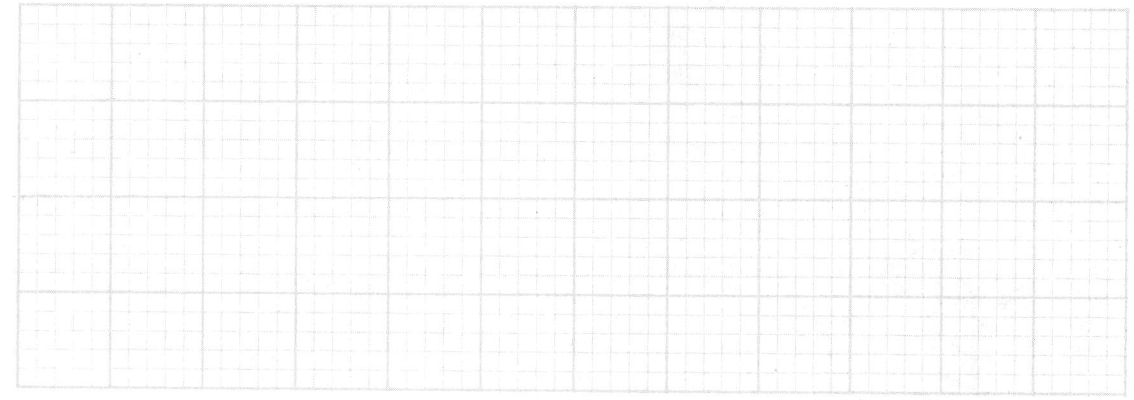

HANDLE	A–27

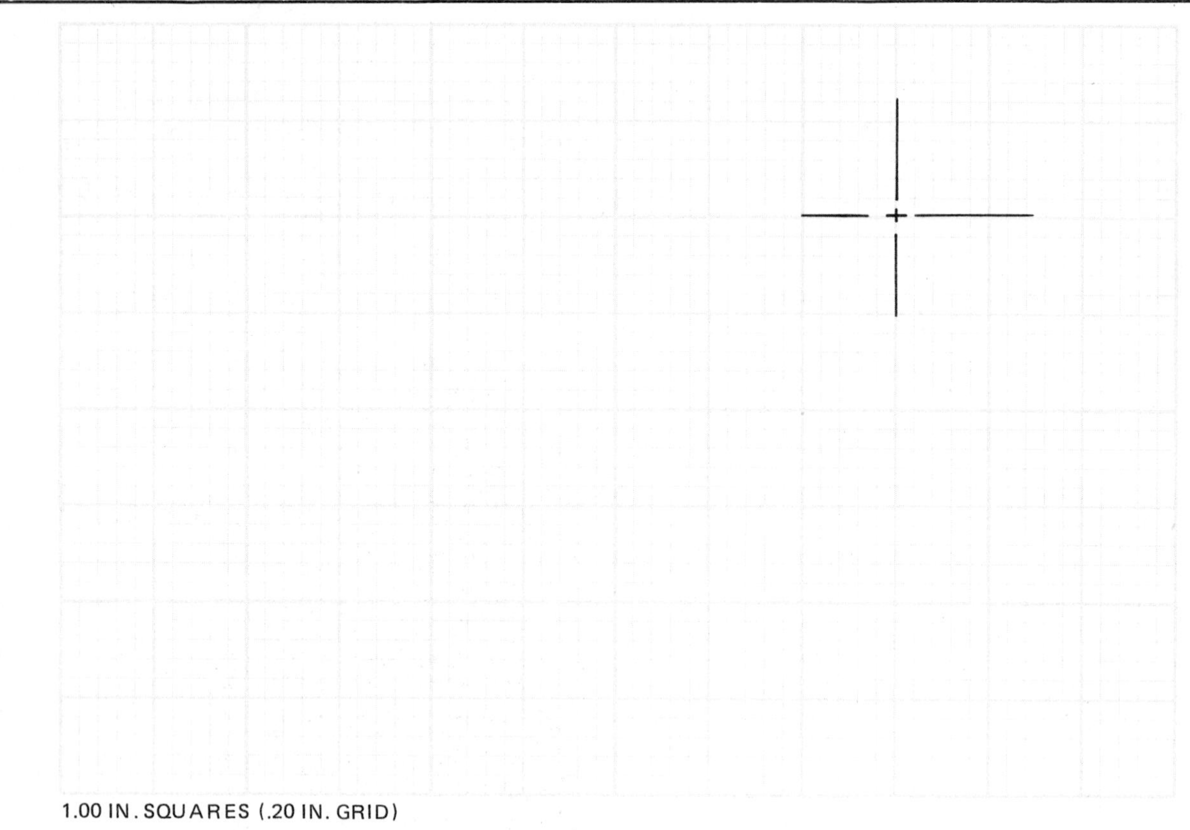

1.00 IN. SQUARES (.20 IN. GRID)

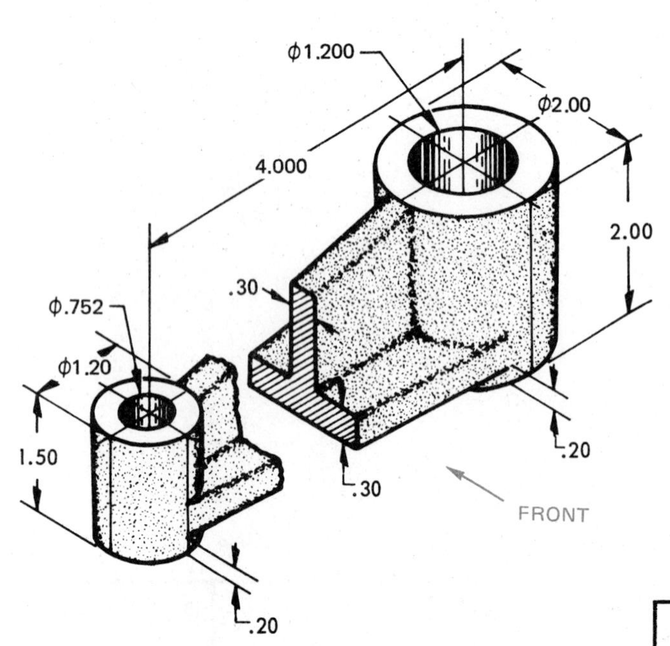

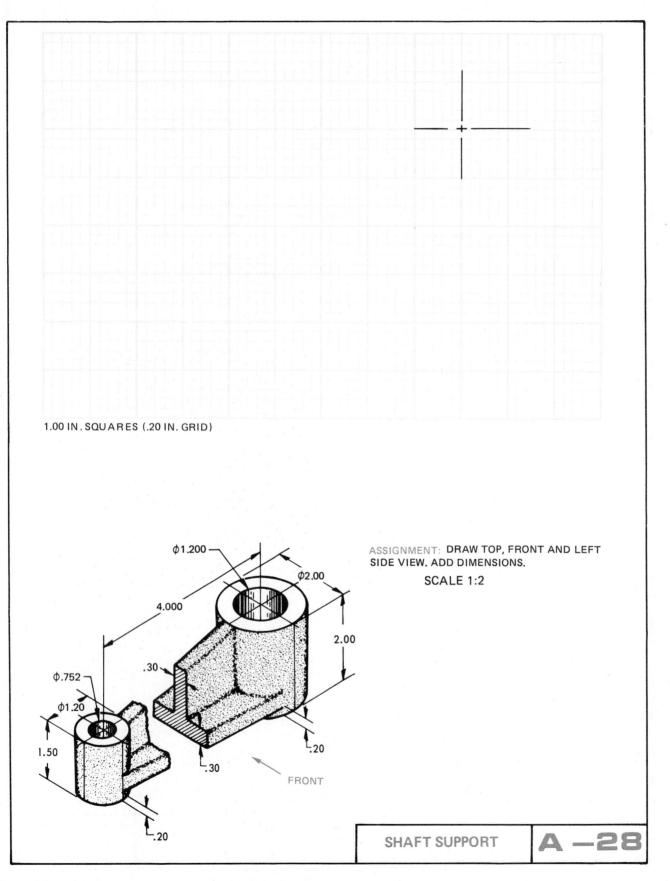

ASSIGNMENT: DRAW TOP, FRONT AND LEFT SIDE VIEW. ADD DIMENSIONS.

SCALE 1:2

SHAFT SUPPORT | A —28

UNIT 10

SELECTION OF VIEWS

ONE- AND TWO-VIEW DRAWINGS

Except for complex objects of irregular shapes, it is seldom necessary to draw more than three views, and for simple parts one- or two-view drawings will often suffice.

In one-view drawings the third dimension is expressed by a note or by symbols or abbreviations, such as ϕ, □, HEX ACR FLT, R, figure 10-1. The symmetry symbol shown at the ends of the center line indicates that the part is symmetrical. Frequently, the drafter will decide that only two views are necessary to explain the shape of an object fully, figure 10-2. One or two views usually show the shape of cylindrical objects adequately.

MULTIPLE DETAIL DRAWINGS

Details of parts may be shown on separate sheets, or they may be grouped together on one or more large sheets. Often the details of parts are grouped according to the department in which they are made. Metal parts to be fabricated in the machine shop may appear on one detail sheet while parts to be made in the wood shop may be grouped on another.

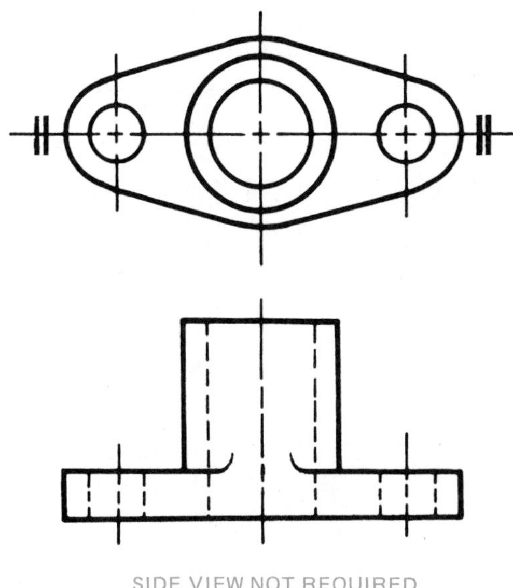

Fig. 10-2 Two-view drawing

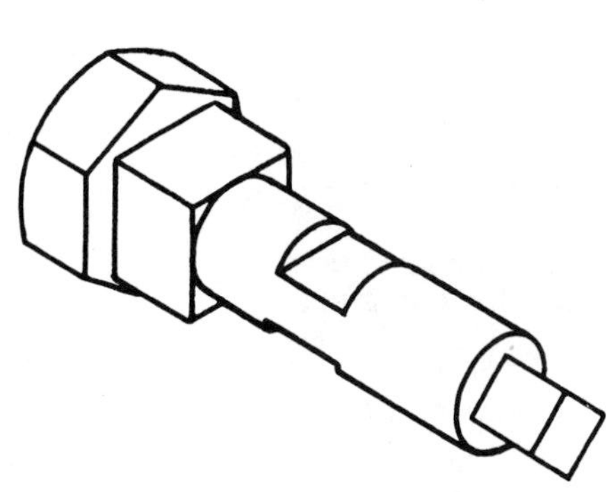

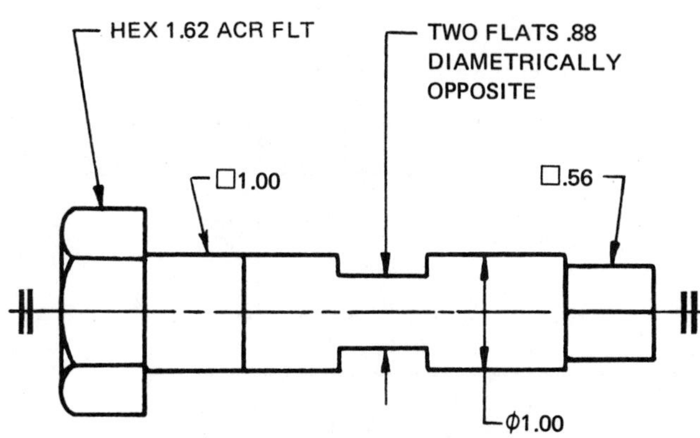

Fig. 10-1 One-view drawing of a turned part

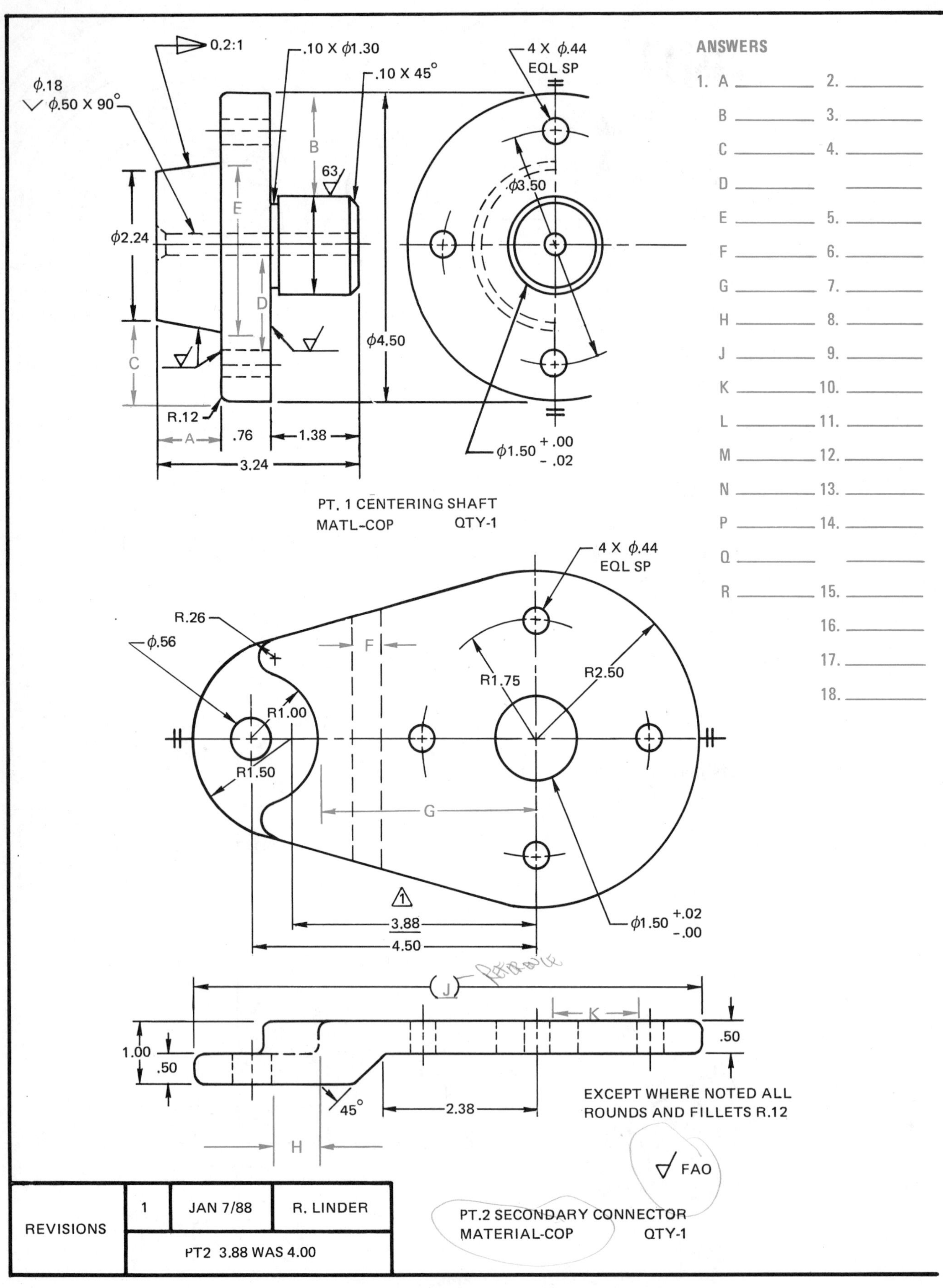
0.2:1
ϕ.18
ϕ.50 X 90°
.10 X ϕ1.30
.10 X 45°
4 X ϕ.44
EQL SP
B
63
ϕ3.50
E
ϕ2.24
D
C
ϕ4.50
R.12
A
.76
1.38
3.24
ϕ1.50 +.00 -.02
PT. 1 CENTERING SHAFT
MATL-COP QTY-1
ANSWERS
1. A
B
C
D
E
F
G
H
J
K
L
M
N
P
Q
R
2.
3.
4.
5.
6.
7.
8.
9.
10.
11.
12.
13.
14.
15.
16.
17.
18.
4 X ϕ.44
EQL SP
R.26
ϕ.56
F
R1.75
R2.50
R1.00
R1.50
G
1
3.88
4.50
ϕ1.50 +.02 -.00
J
K
1.00
.50
.50
45°
2.38
H
EXCEPT WHERE NOTED ALL
ROUNDS AND FILLETS R.12
FAO
REVISIONS
1
JAN 7/88
R. LINDER
PT2 3.88 WAS 4.00
PT.2 SECONDARY CONNECTOR
MATERIAL-COP QTY-1

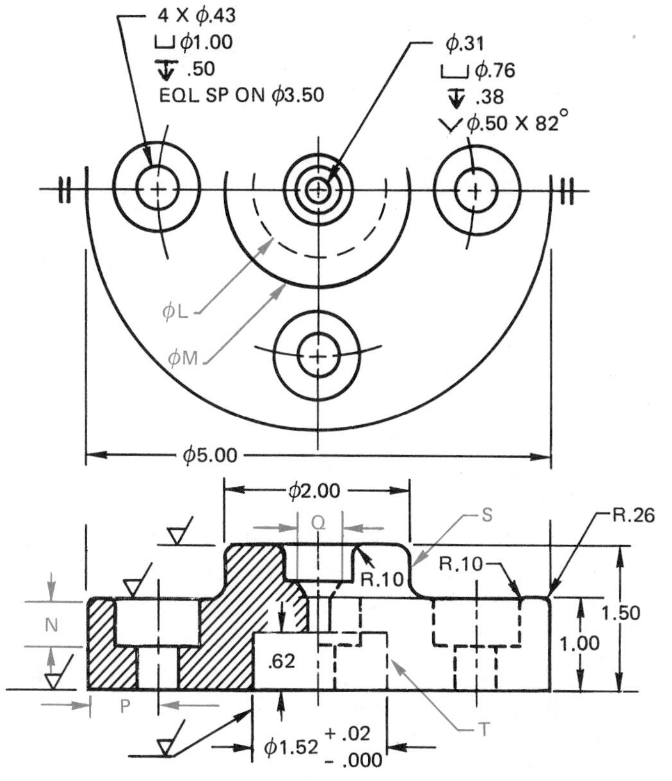

PT. 3 PRESSURE NUT
MATL COP QTY-1

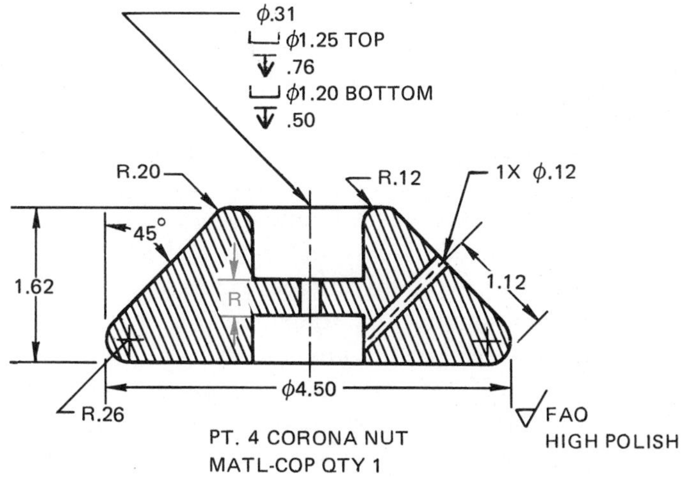

PT. 4 CORONA NUT
MATL-COP QTY 1

QUESTIONS

1. Calculate dimensions A to R. Use nominal sizes. There is no I or O.

Refer to Part 1

2. What is the length of the ϕ1.50 shaft? Do not include the undercut or chamfer.
3. What is the length of the ϕ.18 hole excluding the countersink?
4. Give two reasons why only half of the end view is drawn.
5. What is the diameter of the undercut?

Refer to Part 2

6. What does FAO mean?
7. What does the line under the 3.88 dimension indicate?
8. What does the abbreviation COP mean?
9. What do the parentheses around dimension J indicate?
10. What is the size of the fillets?

Refer to Part 3

11. How much machining allowance is provided for the bottom of the part?
12. What line in the front view represents line L in the top view?
13. What line in the top view represents line S in the front view?
14. What are the limits of the diameter of the counterbore in the bottom of the part?
15. What type of section view is shown?

Refer to Part 4

16. What is the diameter of the counterbore where the ϕ.12 hole terminates?
17. What is the distance between the center points of the .26 radii?
18. What type of section view is shown?

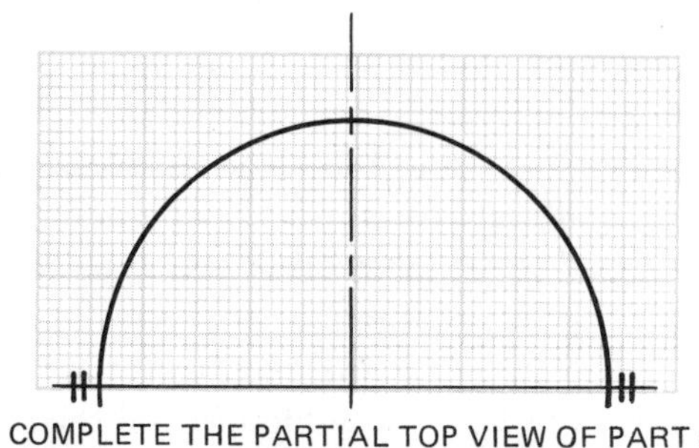

COMPLETE THE PARTIAL TOP VIEW OF PART 4

SKETCHING ASSIGNMENT

On the grid provided, complete the half top view of Part 4.

SCALE NTS	
DRAWN	DATE

CENTERING CONNECTOR DETAILS

A—29

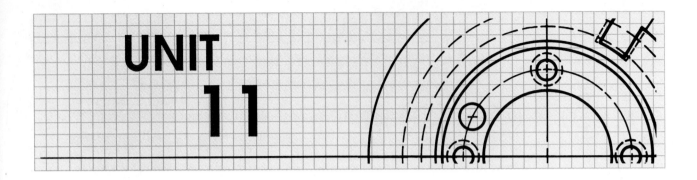

UNIT 11

SURFACE TEXTURE

The development of modern, high-speed machines has resulted in higher loadings and faster moving parts. To withstand these more severe operating conditions with minimum friction and wear, a particular surface texture is often essential. This requires the designer to accurately describe the needed texture (sometimes called *finish*) to the persons who are actually making the parts.

Rarely are entire machines designed and manufactured in one plant. They are usually designed in one location, manufactured in another, and perhaps assembled in a third.

All surface finish control begins in the drafting room. The designer is responsible for specifying the correct surface finish for maximum performance and service life at the lowest cost. In selecting the required surface finish for any particular part, the designer bases the decision on past experience with similar parts, field service data, and engineering tests. Many factors influence the designer's choice. These factors include the function of the parts, the type of loading, the speed and direction of movement, and the operating conditions. Also considered are such factors as the physical characteristics of both materials on contact, whether the part is subjected to stress reversals, the type and amount of lubricant, contaminants, and temperature.

The two principal reasons for surface finish control are friction reduction and the control of wear.

Whenever a lubricating film must be maintained between two moving parts, the surface irregularities must be small enough to prevent penetrating the oil film under even the most severe operating conditions. Such parts as bearings, journals, cylinder bores, piston pins, bushings, pad bearings, helical and worm gears, seal surfaces, and machine ways are objects where this condition must be fulfilled.

Surface finish is also important to the wear service of certain pieces subject to dry friction, such as machine tool bits, threading dies, stamping dies, rolls, clutch plates, and brake drums.

Smooth finishes are essential on certain high-precision pieces. In mechanisms such as injectors and high-pressure cylinders, smoothness and lack of waviness are essential to accuracy and pressure-retaining ability. Smooth finishes are also used on micrometer anvils, gauges, and gauge blocks, and other items.

Smoothness is often important for the visual appeal of the finished product. For this reason, surface finish is controlled on such articles as rolls, extrusion dies, and precision casting dies.

For gears and other parts, surface finish control may be necessary to insure quiet operation.

In cases where boundary lubrication exists or where surfaces are not compatible (for example, two hard surfaces running together), a certain amount of roughness or character of surface will assist in lubrication.

To meet the requirements for effective control of surface quality under diversified conditions, there is a system for accurately describing the surface.

Surfaces are usually very complex in character. Only the height, width, and direction of surface irregularities are covered in this section, since these are of practical importance in specific applications.

Surface Texture Definitions

The following terms relating to surface texture are illustrated in figure 11-1.

Microinch (μin). A microinch is one millionth of an inch (.000001 inch). For written specifications or reference to surface roughness requirements, microinches may be abbreviated as μin.

Micrometer (μm). A micrometer is one millionth of a meter (0.000 001 meters). For written specifications or reference to surface roughness requirements, micrometers may be abbreviated as μm.

Roughness. Roughness consists of the finer irregularities in the surface texture usually including those irregularities which result from the inherent action of the production process. These include traverse feed marks and other irregularities within the limits of the roughness-width cutoff.

Roughness Average (R_a). Roughness average is expressed in microinches, micrometers, or roughness grade numbers N1 to N12. The "N" series of roughness grade numbers is often used in lieu of the roughness average values to avoid misinterpretation when drawings are exchanged internationally.

Roughness Width. Roughness width is the distance parallel to the nominal surface between successive peaks or ridges which constitute the predominant pattern of the roughness. Roughness width is rated in inches or millimeters.

Roughness-width Cutoff. The greatest spacing of repetitive surface irregularities to be included in the measurement of average roughness height is the roughness-width cutoff. Roughness-width cutoff is rated in inches or millimeters and must always be greater than the roughness width in order to obtain the total roughness height rating.

Waviness. Waviness is the usually widely spaced component of surface texture and is generally spaced farther apart than the roughness-width cutoff. Waviness may result from machine or work deflections, vibration, chatter, heat treatment or warping strains. Roughness may be considered superimposed on a wavy surface. Although waviness is not currently in ISO standards, it is included as part of the surface texture symbol to follow present industrial practices in the United States.

Lay. The direction of the predominant surface pattern, which is ordinarily determined by the production method used, is the lay. Symbols for the lay are shown in figure 11-2.

Flaws. Flaws are surface irregularities occurring at one place or at relatively infrequent or widely varying intervals. Flaws include: cracks, blow holes, checks, ridges, scratches, and so forth. Unless otherwise specified, the effect of flaws is not included in the roughness height measurements.

SURFACE TEXTURE SYMBOL

The surface texture symbol, figure 11-3, denotes surface characteristics on the drawing. Roughness, waviness, and lay are controlled by

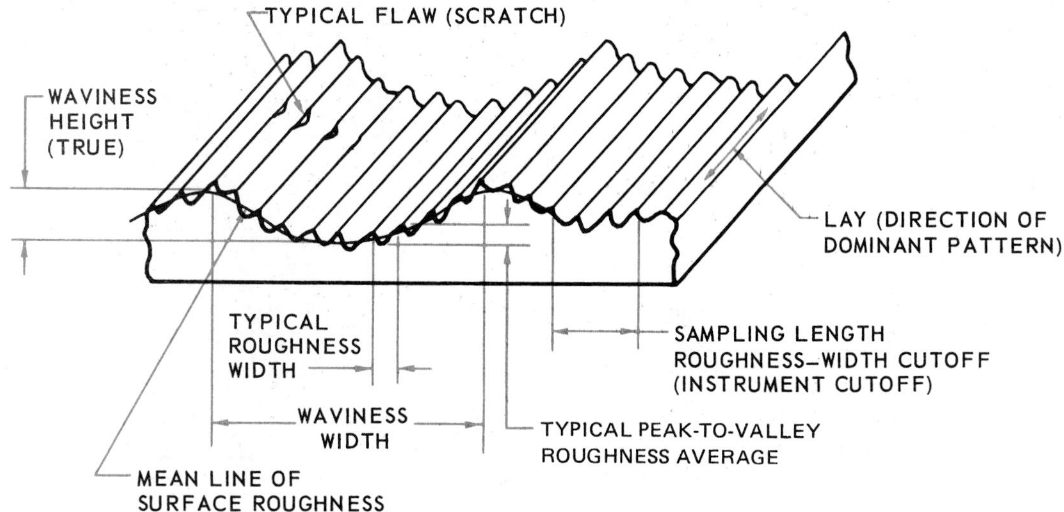

Fig. 11-1 Surface texture characteristics

SYMBOL	DESIGNATION	EXAMPLE
=	LAY PARALLEL TO THE LINE REPRESENTING THE SURFACE TO WHICH THE SYMBOL IS APPLIED	 DIRECTION OF TOOL MARKS
⊥	LAY PERPENDICULAR TO THE LINE REPRESENTING THE SURFACE TO WHICH THE SYMBOL IS APPLIED	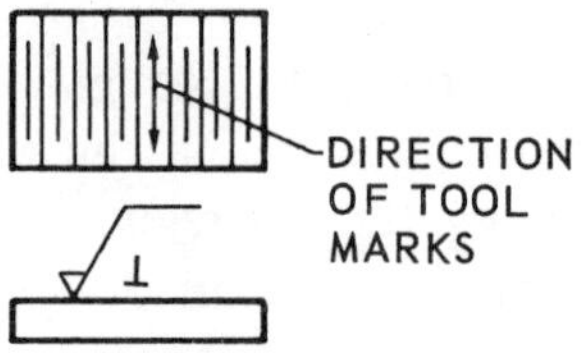 DIRECTION OF TOOL MARKS
X	LAY ANGULAR IN BOTH DIRECTIONS TO LINE REPRESENTING THE SURFACE TO WHICH THE SYMBOL IS APPLIED	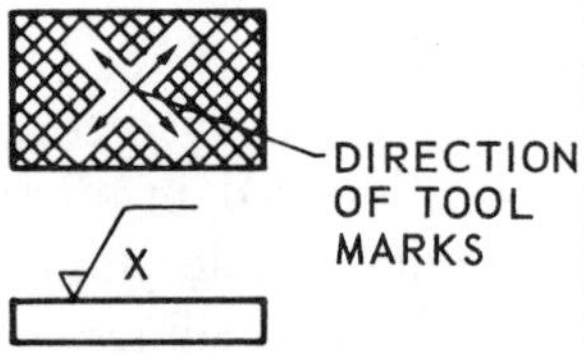 DIRECTION OF TOOL MARKS X
M	LAY MULTIDIRECTIONAL	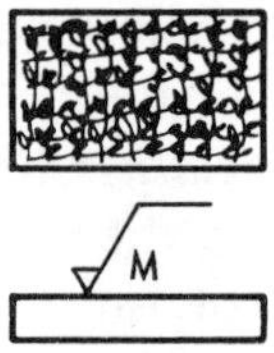 M
C	LAY APPROXIMATELY CIRCULAR RELATIVE TO THE CENTER OF THE SURFACE TO WHICH THE SYMBOL IS APPLIED	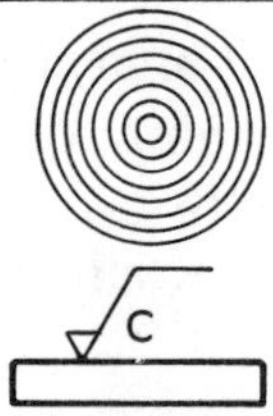 C
R	LAY APPROXIMATELY RADIAL RELATIVE TO THE CENTER OF THE SURFACE TO WHICH THE SYMBOL IS APPLIED	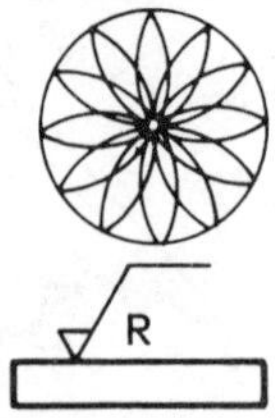 R
P	LAY NONDIRECTIONAL, PITTED, OR PROTUBERANT	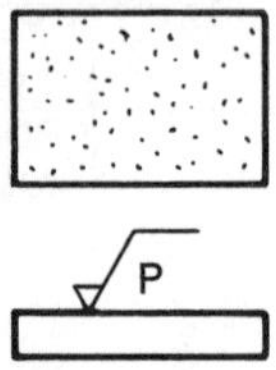 P

Fig 11-2 Lay symbols

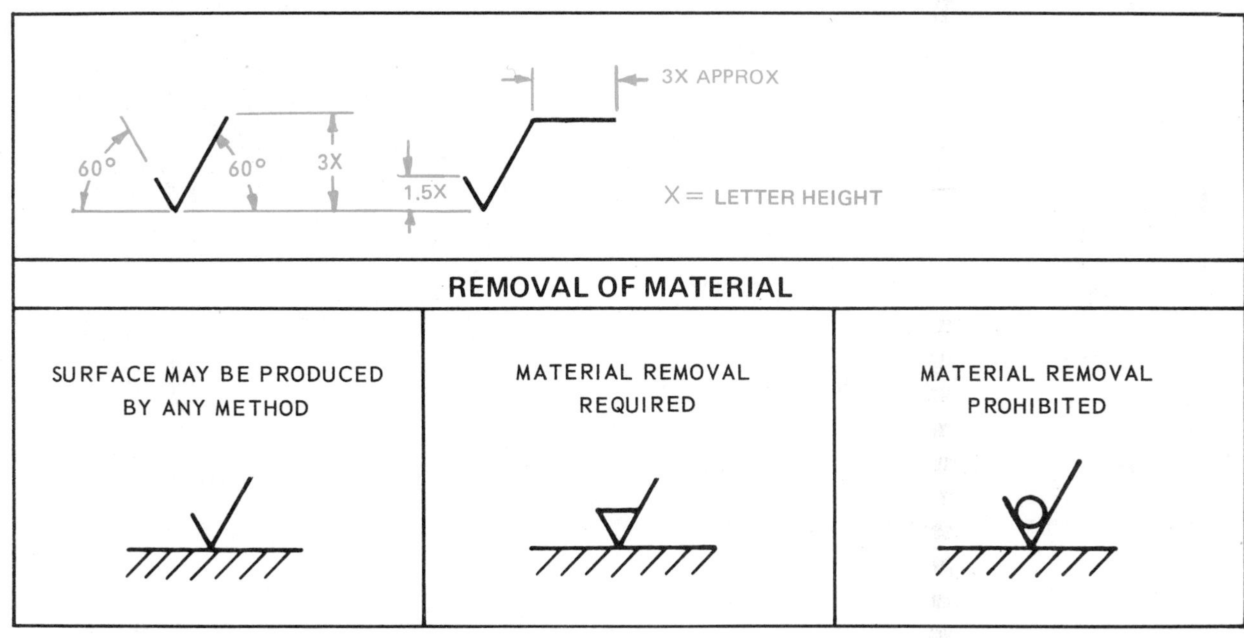

Fig 11-3 Basic surface texture symbol

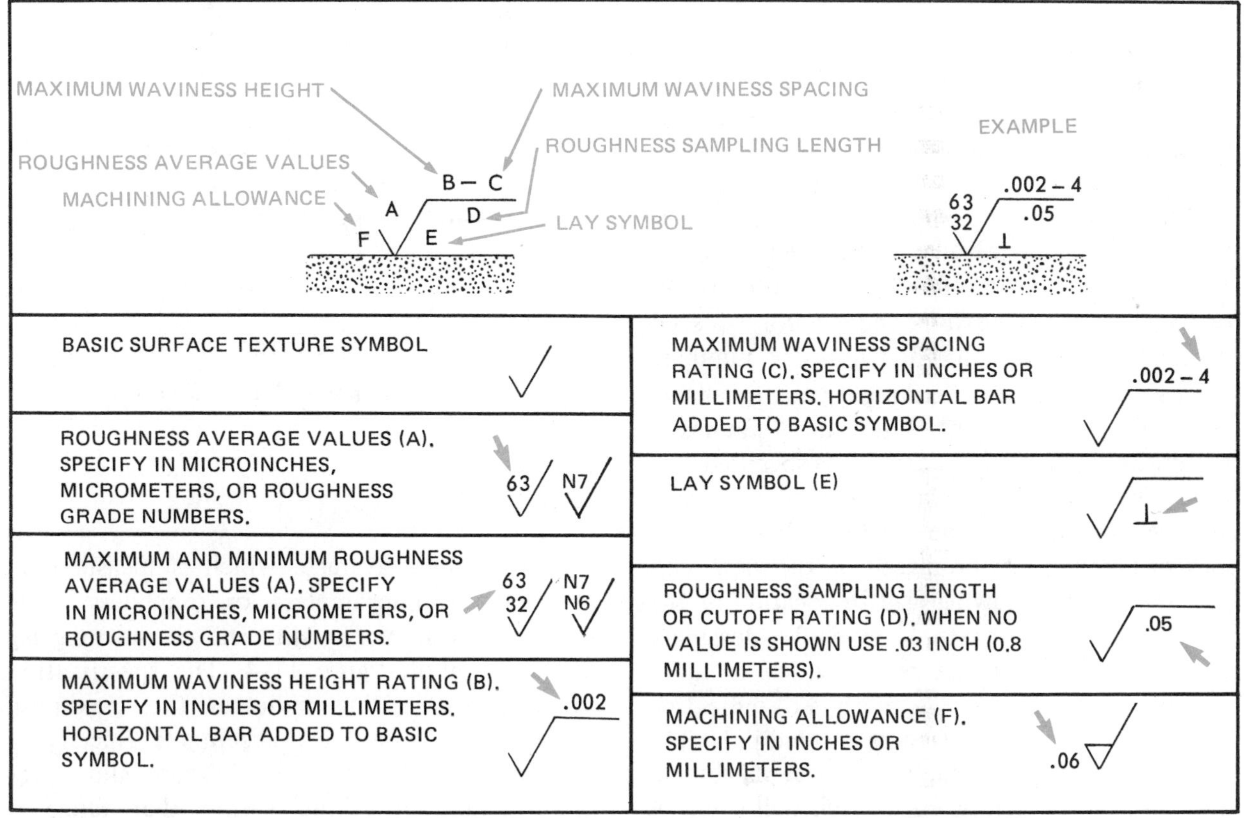

Fig. 11-4 Location of ratings and symbols on surface texture symbol

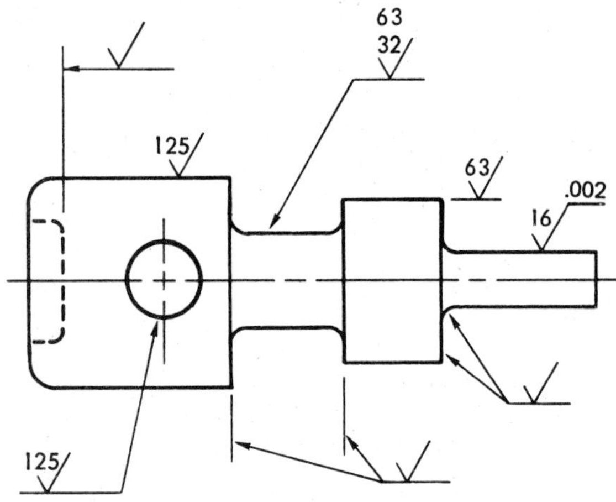

Fig. 11-5 Application of roughness values and waviness ratings

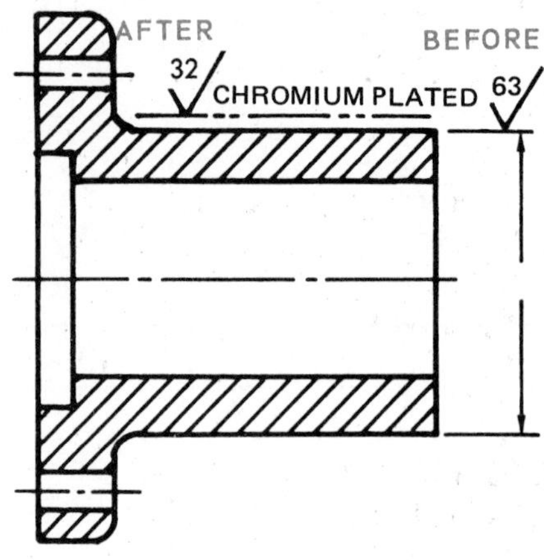

(A) INDICATING SURFACE TEXTURE BEFORE AND AFTER PLATING

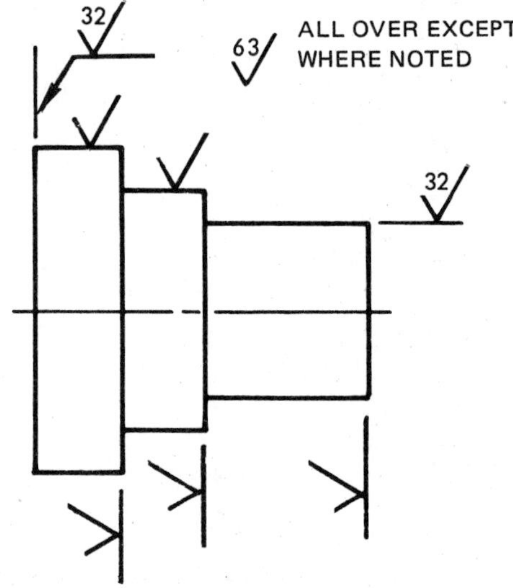

(B) A GENERAL NOTE BESIDE PART

Fig. 11-6 The use of notes with surface texture symbol

applying the desired values to the surface texture symbol, figure 11-4, or in a general note. The two methods may be used together. The point of the symbol should be on the line indicating the surface, on an extension line from the surface, or on a leader pointing either to the surface or extension line, figure 11-5. To be readable from the bottom, the symbol is placed in an upright position when notes or numbers are used. This means the long leg and extension line will be on the right. The symbol applies to the entire surface, unless otherwise specified.

Like dimensions, the symbol for the same surface should not be duplicated on other views. They should be placed on the view with the dimensions showing size or location of the surfaces. Surface texture symbols designate surface texture characteristics which includes machining of surfaces. The method of indicating machine finishes on surfaces is covered in Unit 7.

Where all the surfaces are to be machined, a general note such as FAO (finish all over) or √ ALL OVER may be used and the symbols on the part may be omitted.

SURFACE TEXTURE RATINGS

Roughness average, which is measured in microinches, micrometers or roughness grade numbers, is shown to the left of the long leg of the symbol, figure 11-4. The specification of only one rating indicates the maximum value; any lesser value is acceptable. Specifying two ratings indicates the minimum and maximum values. Anything within that range is acceptable. The maximum value is placed over the minimum.

Waviness height ratings are indicated in inches or millimeters and positioned as shown in figure 11-4. Any lesser value is acceptable.

Waviness spacing ratings are indicated in inches or millimeters positioned as shown in figure 11-4. Any lesser value is acceptable.

Lay symbols, indicating the directional pattern of the surface texture, are shown in figure 11-2. The symbol is on the right of the long leg of the symbol.

Roughness sampling length ratings are given in inches or millimeters and are located below the horizontal extension. Unless otherwise specified, roughness-width cutoff is .03 in (0.8 mm).

Notes

Usually, a note is used where a given roughness requirement applies to either the whole part or the major portion, or before or after plating. Examples are shown in figure 11-6.

CONTROL REQUIREMENTS

Surface texture control should be specified for surfaces where texture is a functional requirement. For example, most surfaces which have contact with a mating part have a certain texture requirement, especially for roughness. The drawing should reflect the texture necessary for optimum part function without depending upon the variables of machining practices.

Many surfaces do not need a specification of surface texture because the function is unaffected by the surface quality. Such surfaces should not receive surface quality designations because it could unnecessarily increase the product cost.

Figures 11-7, 11-8, and 11-9 show recommended roughness average ratings, the machining methods used to produce them, and the application of the ratings to the surface texture symbol.

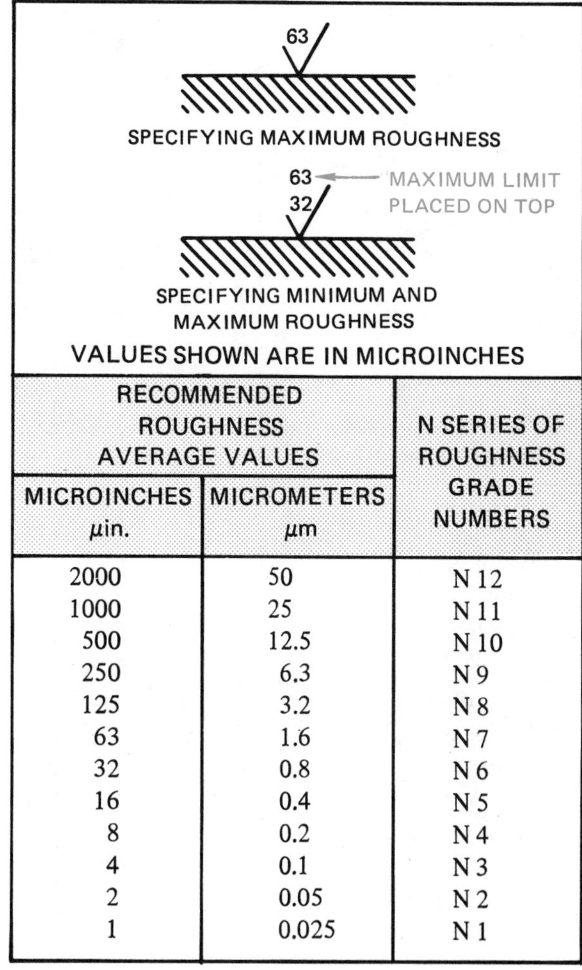

RECOMMENDED ROUGHNESS AVERAGE VALUES		N SERIES OF ROUGHNESS GRADE NUMBERS
MICROINCHES μin.	MICROMETERS μm	
2000	50	N 12
1000	25	N 11
500	12.5	N 10
250	6.3	N 9
125	3.2	N 8
63	1.6	N 7
32	0.8	N 6
16	0.4	N 5
8	0.2	N 4
4	0.1	N 3
2	0.05	N 2
1	0.025	N 1

Fig. 11-7 Recommended roughness average ratings

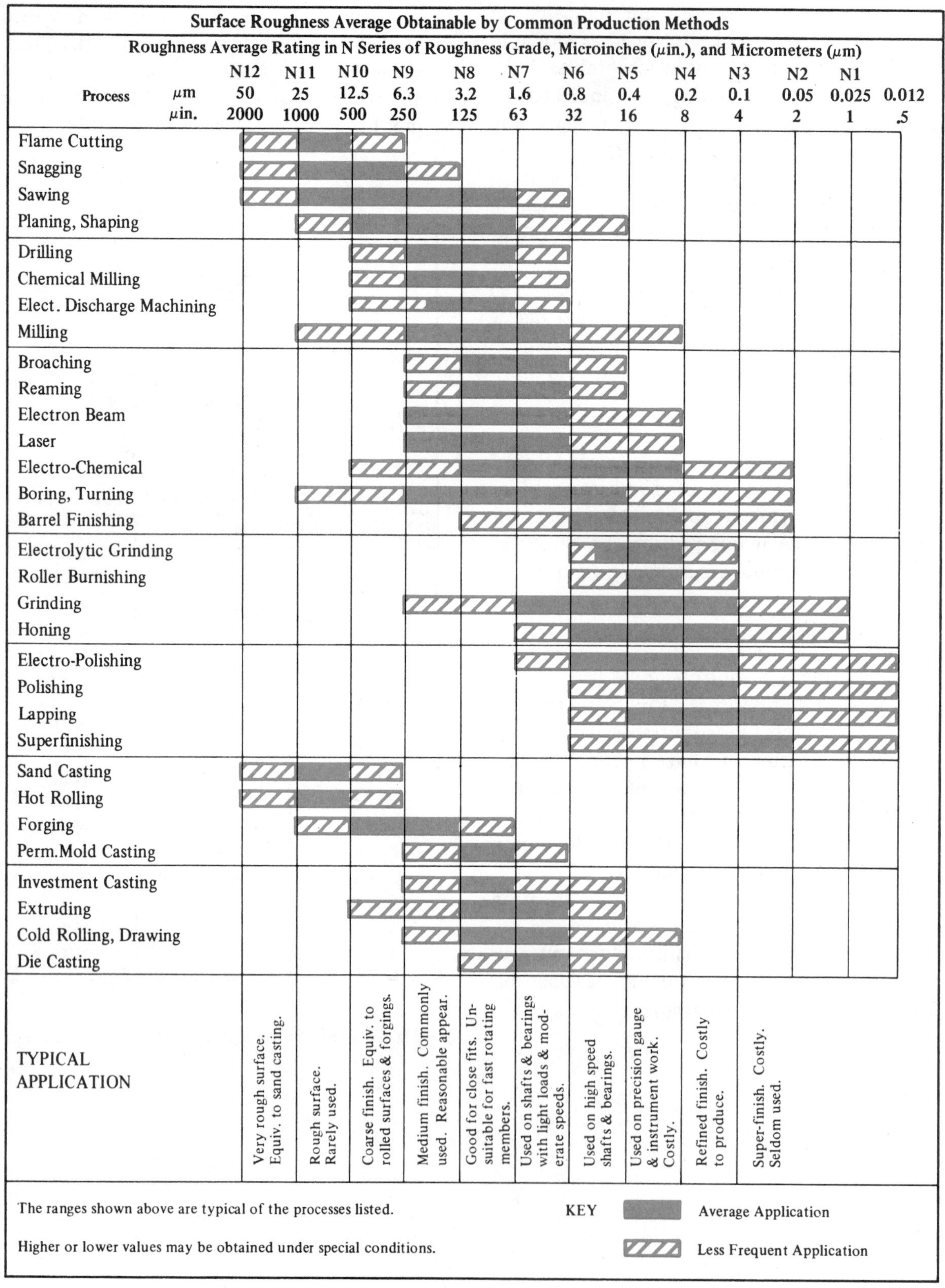

Surface Roughness Average Obtainable by Common Production Methods
Roughness Average Rating in N Series of Roughness Grade, Microinches (μin.), and Micrometers (μm)
N12 N11 N10 N9 N8 N7 N6 N5 N4 N3 N2 N1
Process μm 50 25 12.5 6.3 3.2 1.6 0.8 0.4 0.2 0.1 0.05 0.025 0.012
μin. 2000 1000 500 250 125 63 32 16 8 4 2 1 .5
Flame Cutting
Snagging
Sawing
Planing, Shaping
Drilling
Chemical Milling
Elect. Discharge Machining
Milling
Broaching
Reaming
Electron Beam
Laser
Electro-Chemical
Boring, Turning
Barrel Finishing
Electrolytic Grinding
Roller Burnishing
Grinding
Honing
Electro-Polishing
Polishing
Lapping
Superfinishing
Sand Casting
Hot Rolling
Forging
Perm.Mold Casting
Investment Casting
Extruding
Cold Rolling, Drawing
Die Casting
TYPICAL APPLICATION
Very rough surface. Equiv. to sand casting.
Rough surface. Rarely used.
Coarse finish. Equiv. to rolled surfaces & forgings.
Medium finish. Commonly used. Reasonable appear.
Good for close fits. Unsuitable for fast rotating members.
Used on shafts & bearings with light loads & moderate speeds.
Used on high speed shafts & bearings.
Used on precision gauge & instrument work. Costly.
Refined finish. Costly to produce.
Super-finish. Costly. Seldom used.
The ranges shown above are typical of the processes listed.
Higher or lower values may be obtained under special conditions.
KEY
Average Application
Less Frequent Application

Fig. 11-8 Surface roughness range for common production methods

MICROMETERS RATING	MICROINCHES RATING	APPLICATION
25	1000	Rough, low grade surface resulting from sand casting, torch or saw cutting, chipping, or rough forging. Machine operations are not required because appearance is not objectionable. This surface, rarely specified, is suitable for unmachined clearance areas on rough construction items.
12.5	500	Rough, low grade surface resulting from heavy cuts and coarse feeds in milling turning, shaping, boring, and rough filing, disc grinding and snagging. It is suitable for clearance areas on machinery, jigs, and fixtures. Sand casting or rough forging produces this surface.
6.3	250	Coarse production surfaces, for unimportant clearance and cleanup operations, resulting from coarse surface grind, rough file, disc grind, rapid feeds in turning, milling, shaping, drilling, boring, grinding, etc., where tool marks are not objectionable. The natural surfaces of forgings, permanent mold castings, extrusions, and rolled surfaces also produce this roughness. It can be produced economically and is used on parts where stress requirements, appearance, and conditions of operations and design permit.
3.2	125	The roughest surface recommended for parts subject to loads, vibration, and high stress. It is also permitted for bearing surfaces when motion is slow and loads light or infrequent. It is a medium commercial machine finish produced by relatively high speeds and fine feeds taking light cuts with sharp tools. It may be economically produced on lathes, milling machines, shapers, grinders, etc., or on permanent mold castings, die castings, extrusion, and rolled surfaces.
1.6	63	A good machine finish produced under controlled conditions using relatively high speeds and fine feeds to take light cuts with sharp cutters. It may be specified for close fits and used for all stressed parts, except fast rotating shafts, axles, and parts subject to severe vibration or extreme tension. It is satisfactory for bearing surfaces when motion is slow and loads light or infrequent. It may also be obtained on extrusions, rolled surfaces, die castings and permanent mold castings when rigidly controlled.
0.8	32	A high-grade machine finish requiring close control when produced by lathes, shapers, milling machines, etc., but relatively easy to produce by centerless, cylindrical, or surface grinders. Also, extruding, rolling or die casting may produce a comparable surface when rigidly controlled. This surface may be specified in parts where stress concentration is present. It is used for bearings when motion is not continuous and loads are light. When finer finishes are specified, production costs rise rapidly; therefore, such finishes must be analyzed carefully.
0.4	16	A high quality surface produced by fine cylindrical grinding, emery buffing, coarse honing, or lapping, it is specified where smoothness is of primary importance, such as rapidly rotating shaft bearings, heavily loaded bearing and extreme tension members.
0.2	8	A fine surface produced by honing, lapping, or buffing. It is specified where packings and rings must slide across the direction of the surface grain, maintaining or withstanding pressures, or for interior honed surfaces of hydraulic cylinders. It may also be required in precision gauges and instrument work, or sensitive value surfaces, or on rapidly rotating shafts and on bearings where lubrication is not dependable.
0.1	4	A costly refined surface produced by honing, lapping, and buffing. It is specified only when the design requirements make it mandatory. It is required in instrument work, gauge work, and where packing and rings must slide across the direction of surface grain such as on chrome plated piston rods, etc. where lubrication is not dependable.
0.05 0.025	2 1	Costly refined surfaces produced only by the finest of modern honing, lapping, buffing, and superfinishing equipment. These surfaces may have a satin or highly polished appearance depending on the finishing operation and material. These surfaces are specified only when design requirements make it mandatory. They are specified on fine or sensitive instrument parts or other laboratory items, and certain gauge surfaces, such as precision gauge blocks.

Fig. 11-9 Surface roughness description and application

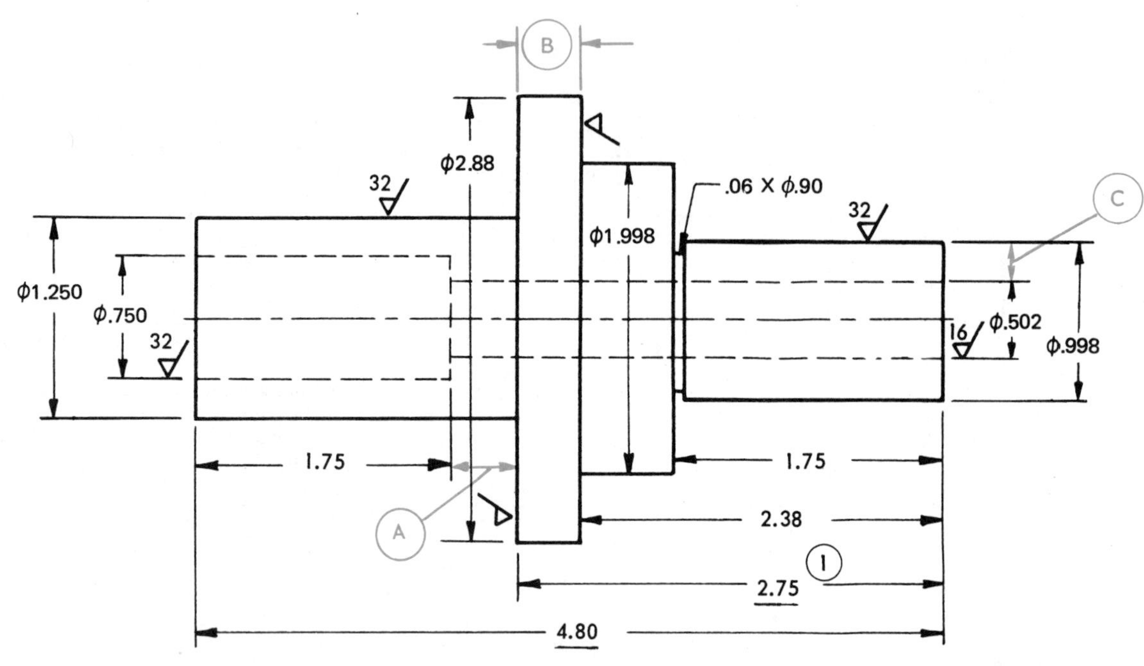

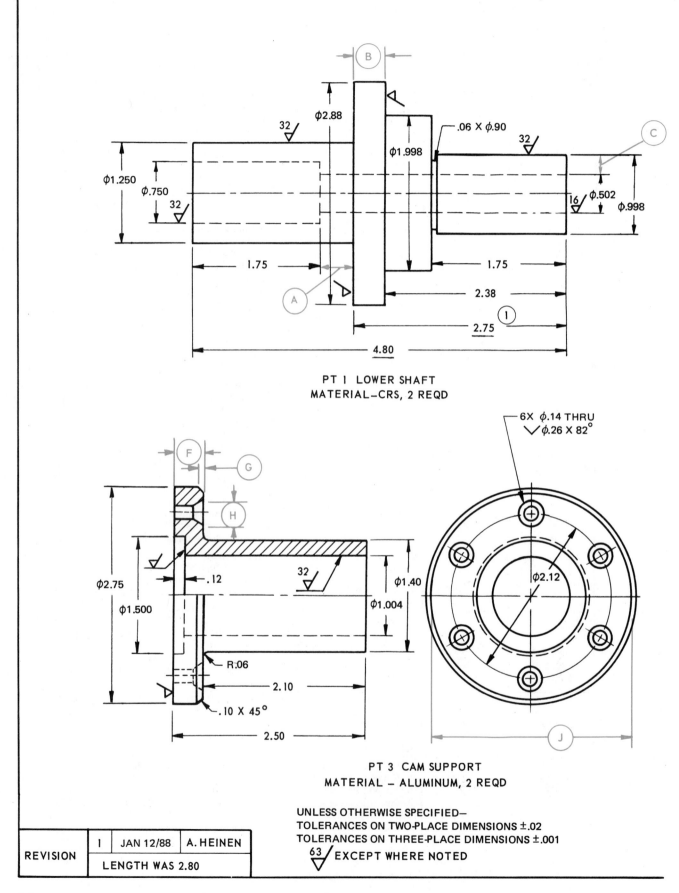

PT 1 LOWER SHAFT
MATERIAL–CRS, 2 REQD

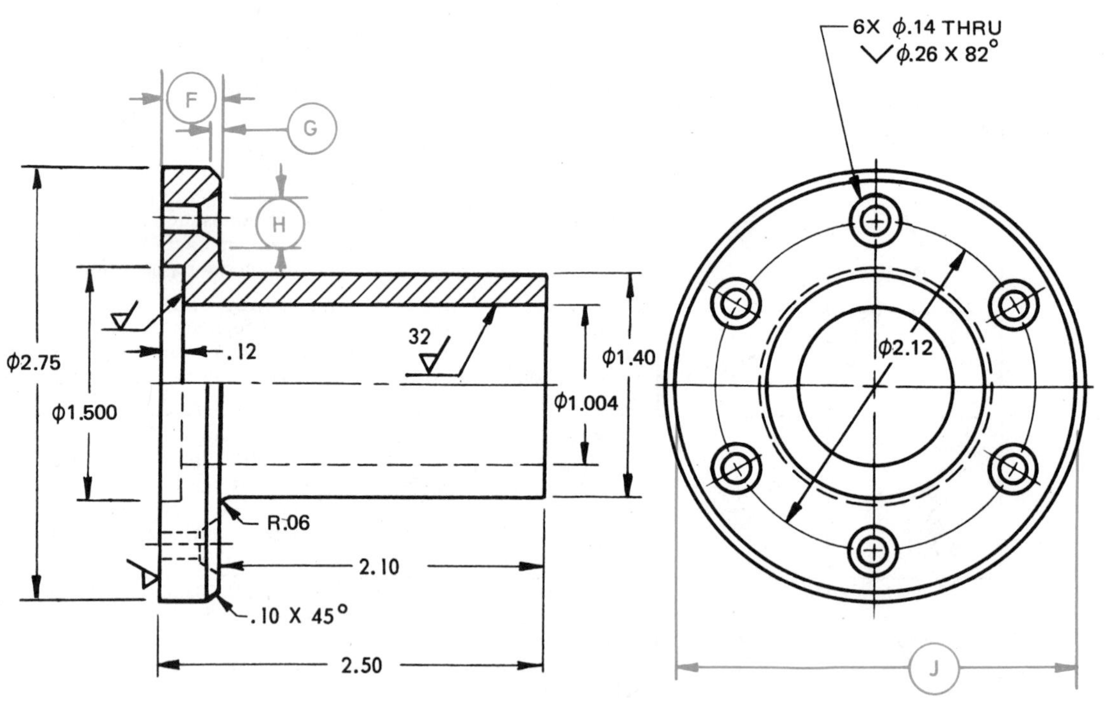

PT 3 CAM SUPPORT
MATERIAL – ALUMINUM, 2 REQD

UNLESS OTHERWISE SPECIFIED–
TOLERANCES ON TWO-PLACE DIMENSIONS ±.02
TOLERANCES ON THREE-PLACE DIMENSIONS ±.001
63 EXCEPT WHERE NOTED

REVISION	1	JAN 12/88	A. HEINEN
	LENGTH WAS 2.80		

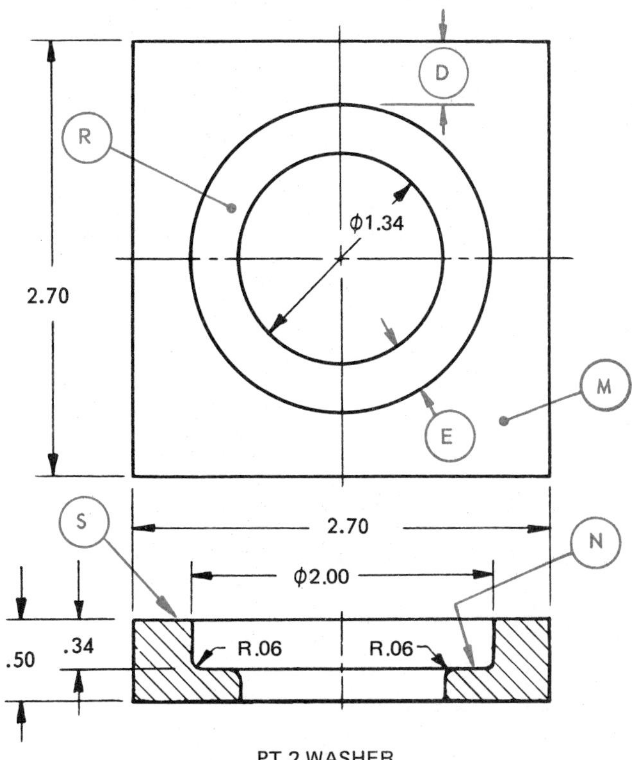

PT 2 WASHER
MATERIAL – MS. 4 REQD. FAO

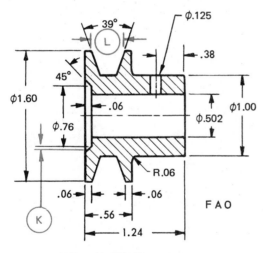

ϕ.125 39° L .38 45° ϕ1.60 .06 ϕ1.00 ϕ.502 ϕ.76 R.06 .06 .06 FAO .56 K 1.24

PT 4 V–BELT PULLEY
MATERIAL–CRS, 4 REQD

QUESTIONS

1. Calculate dimensions (A) to (L).

Referring to figure 9-8, what production methods would be suitable to produce the surface texture of

2. Ø.502 hole in Part 1?
3. Ø1.250 on Part 1?
4. Ø1.004 hole in Part 3?

Refer to Part 1

5. What is the length of the Ø.998 portion?
6. What was the original length of the 2.75 dimension?
7. How many hidden circles would be seen if the right end view were drawn?

Refer to Part 2

8. Which surface does (R) represent in the front view?
9. Which surface does (S) represent in the top view?
10. If the part that passes through the washer is Ø1.30, what is the clearance per side between the two parts?
11. **How many fillets are required?**

Refer to Part 3

12. How many degrees apart on the Ø2.12 are the Ø.14 holes?
13. Is the center line of the countersunk holes in the center of the flange?
14. What operation is performed to allow the head of the mounting screws to rest flush with the flange?
15. What type of section view is shown?
16. What is the amount and degree of chamfer?

Refer to Part 4

17. How deep is the Ø.125 hole?
18. How deep is the belt groove?
19. What does FAO mean?
20. What type of section view is shown?

ANSWERS

1 (A) ______
(B) ______
(C) ______
(D) ______
(E) ______
(F) ______
(G) ______
(H) ______
(J) ______
(K) ______
(L) ______
2 ______
3 ______
4 ______
5 ______
6 ______
7 ______
8 ______
9 ______
10 ______
11 ______
12 ______
13 ______
14 ______
15 ______
16 ______
17 ______
18 ______
19 ______
20 ______

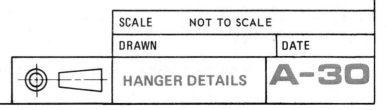

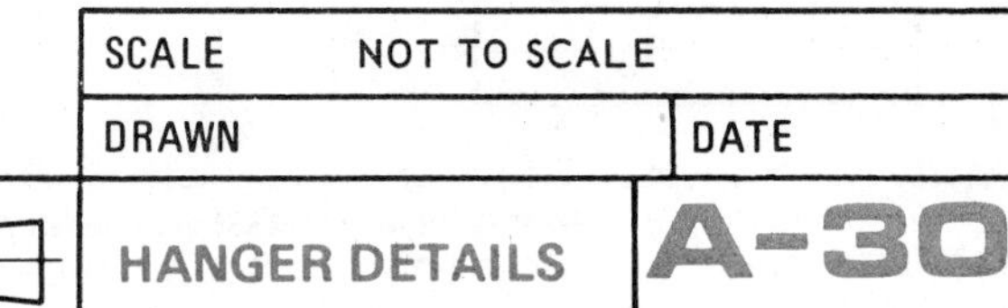

SCALE NOT TO SCALE
DRAWN
DATE
HANGER DETAILS
A-30

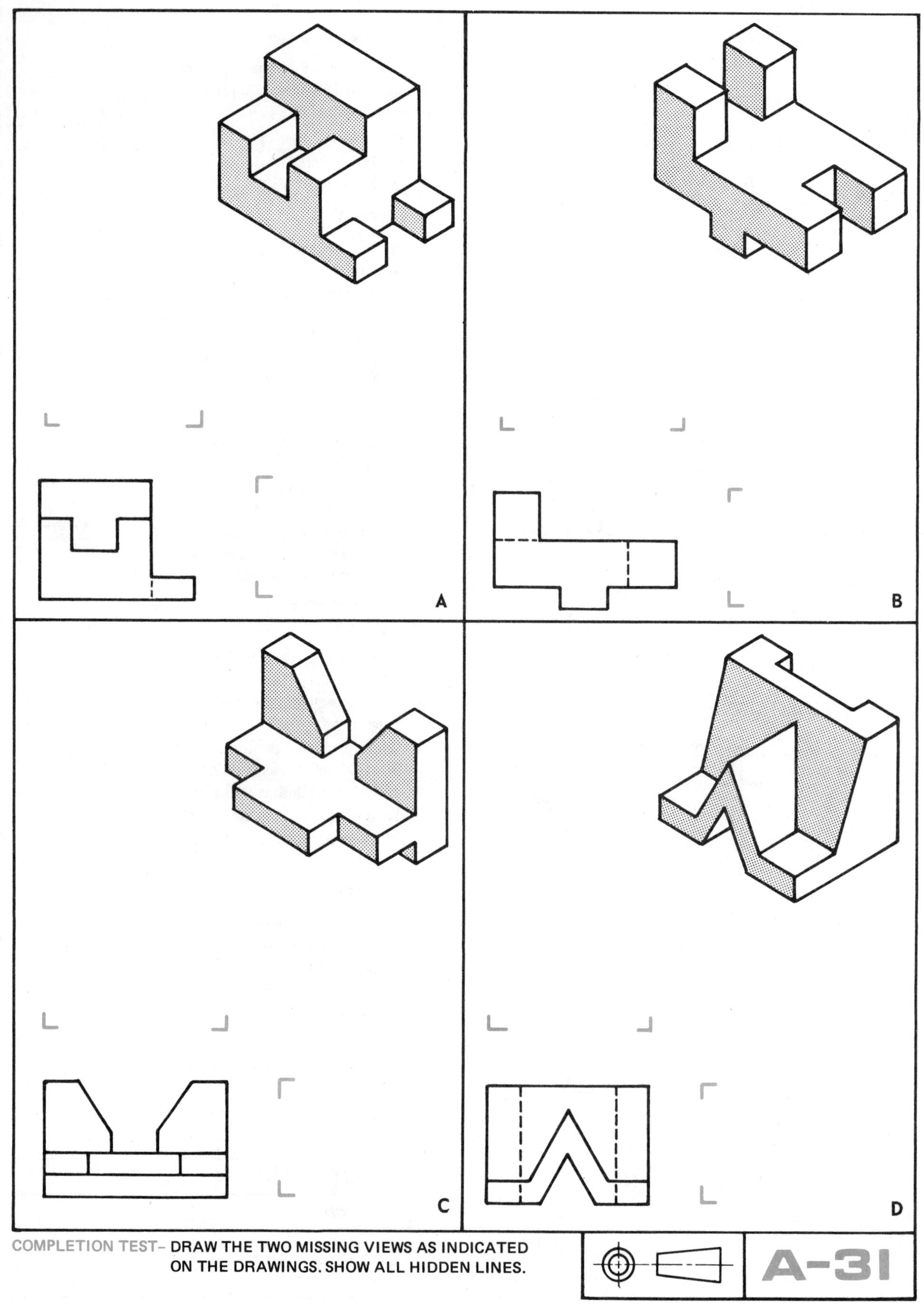

COMPLETION TEST– DRAW THE TWO MISSING VIEWS AS INDICATED ON THE DRAWINGS. SHOW ALL HIDDEN LINES.

A-31

UNIT 12

TOLERANCES AND ALLOWANCES

The history of engineering drawing as a means for the communication of engineering information spans a period of 6,000 years. It seems inconceivable that such an elementary practice as the tolerancing of dimensions, which is taken for granted today, was introduced for the first time about 80 years ago.

Apparently, engineers and workers realized only gradually that exact dimensions and shapes could not be attained in the shaping of physical objects. The skilled handicrafters of the past took pride in the ability to work to exact dimensions. This meant that objects were dimensioned more accurately than they could be measured. The use of modern measuring instruments would have shown the deviations from the sizes which were called exact.

It was soon realized that variations in the sizes of parts had always been present, that such variations could be restricted but not avoided, and that slight variation in the size which a part was originally intended to have could be tolerated without impairment of its correct functioning. It became evident that interchangeable parts need not be identical parts, but rather it would be sufficient if the significant sizes which controlled their fits lay between definite limits. Therefore, the problem of interchangeable manufacture developed from the making of parts to a supposedly exact size, to the holding of parts between two limiting sizes, lying so closely together than any intermediate size would be acceptable.

The concept of limits means essentially that a precisely defined basic condition (expressed by one numerical value or specification) is replaced by two limiting conditions. Any result lying between these two limits is acceptable. A workable scheme of interchangeable manufacture that is indispensable to mass production methods has been established.

DEFINITIONS

In order to calculate limits and tolerances, the following definitions should be clearly understood.

Basic Size. The basic size of a dimension is the theoretical size from which the limits for that dimension are derived, by the application of the allowance and tolerance.

Tolerances. The tolerance of a dimension is the total permissible variation in its size. The tolerance is the difference between the limits of size.

Limits of Size. The limits are the maximum and minimum sizes permitted for a specific dimension.

Allowance. An allowance is the intentional difference in size of mating parts. It is the minimum clearance (positive allowance) or maximum interference (negative allowance) between such parts. Fits between parts is covered in Units 13 and 14.

All dimensions on a drawing have tolerances. Some dimensions must be more exact than other dimensions and consequently have smaller tolerances.

When dimensions require a greater accuracy than the general note provides, individual tolerances or limits must be shown for that dimension.

Where limit dimensions are used and where either the maximum or minimum dimension has digits to the right of the decimal point, the other value should have the zeros added so that both the limits of size are expressed to the same number of decimal places.

When limit dimensions are used for diameter or radial features and the dimensions are placed one above the other, only one diameter or radius symbol is used and located at mid-height.

Where one limit alone is important and where any variations away from that limit in the other direction may be permitted, the MAX (maximum) or MIN (minimum) can be specified. Examples are depth of holes, corner radii, and chamfers. Figures 12-1 and 12-2 show applications of limit dimensioning.

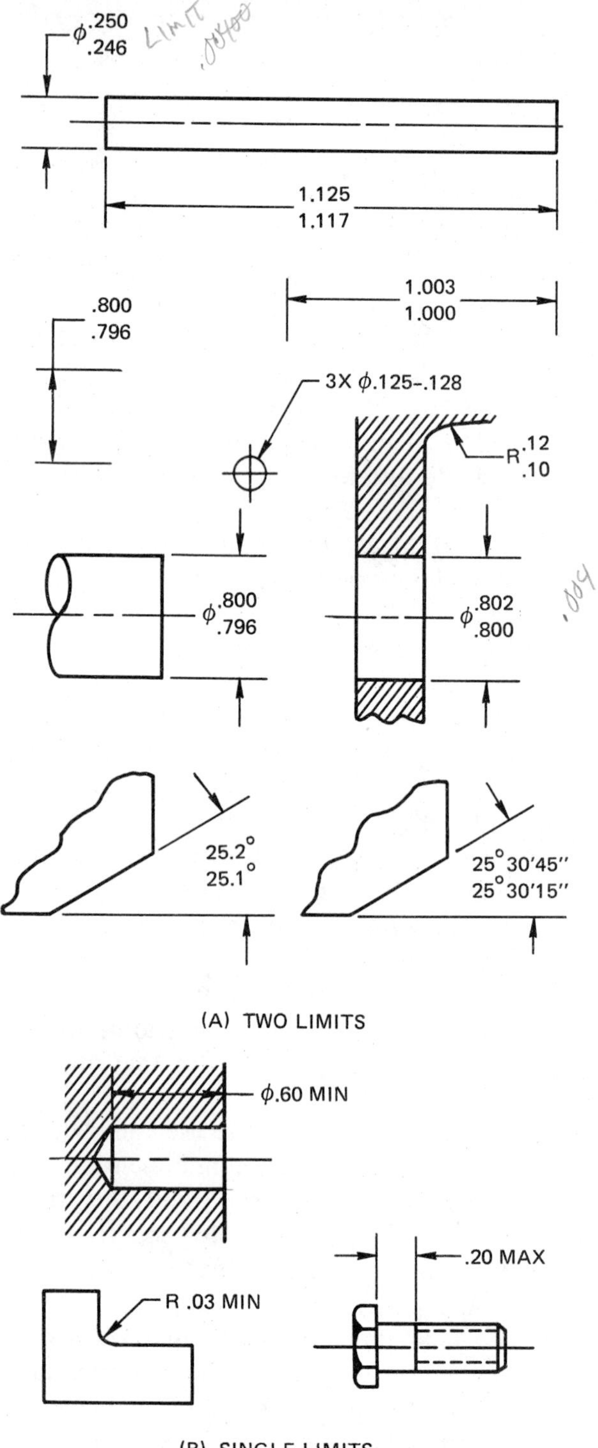

(A) TWO LIMITS

(B) SINGLE LIMITS

Fig. 12-1 Limit dimensioning

TOLERANCING METHODS

Dimensional tolerances are expressed in either of two ways: limit dimensioning or plus and minus tolerancing.

Limit Dimensioning

In the limiting dimensioning method, only the maximum and minimum dimensions are specified, figure 12-1. When placed above each other, the larger dimension is placed on top. When shown with a leader and placed in one line, the smaller size is shown first. A small dash separates the two dimensions. When limit dimensions are used for diameter or radial features, the ϕ or R symbol is centered midway between the two limits, figure 12-1(A).

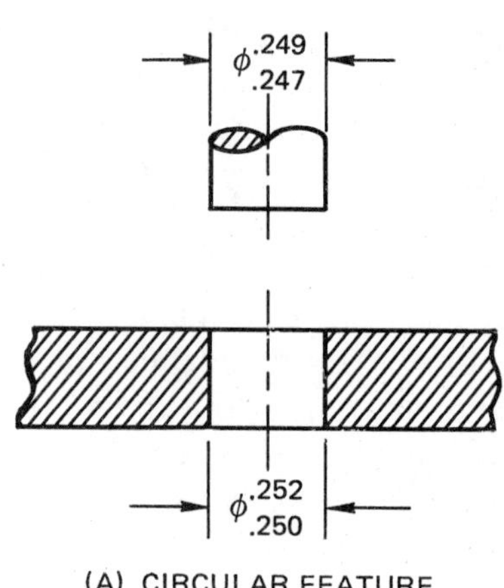

(A) CIRCULAR FEATURE

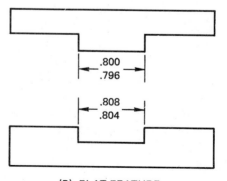

.800
.796
.808
.804

(B) FLAT FEATURE

Fig. 12-2 Limit dimensioning application

Plus and Minus Tolerancing

In this method the dimension of the specified size is given first and it is followed by a plus and minus tolerance expression. The tolerance can be bilateral or unilateral, figure 12-3.

A *bilateral tolerance* is a tolerance which is expressed as plus and minus values. These values need not be the same size.

A *unilateral tolerance* is one which applies in one direction from the specified size, so the permissible variation in the other direction is zero.

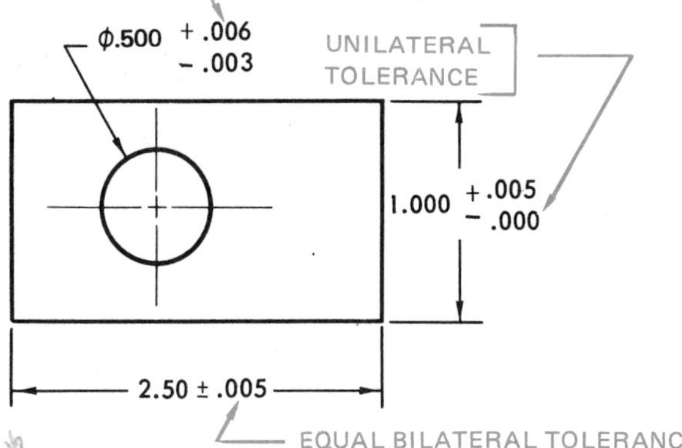

Fig. 12-3 Plus and minus tolerancing

Inch Tolerances

Where inch dimensions are used on the drawing, both limit dimensions or the plus and minus tolerance and its dimensions are expressed with the same number of decimal places.

Examples:

$.500 \quad .005$ not $.50 \pm .005$

$.500^{+.005}_{-.000}$ not $.500^{+.005}_{0}$

$25.0 \pm .2$ not $25 \pm .2$

General tolerance notes greatly simplify the drawing. The following examples illustrate the variety of applications in this system. The values given in the examples are typical only:

EXCEPT WHERE STATED OTHERWISE,
TOLERANCES ON DIMENSIONS ±.005

EXCEPT WHERE STATED OTHERWISE,
TOLERANCES ON FINISHED DIMENSIONS
TO BE AS FOLLOWS:

DIMENSION	TOLERANCE
UP TO 3.00	.01
OVER 3.00 TO 12.00	.02
OVER 12.00 TO 24.00	.04
OVER 24.00	.06

UNLESS OTHERWISE SPECIFIED
±.005 TOLERANCE ON MACHINED DIMENSIONS
±.04 TOLERANCE ON CAST DIMENSIONS
ANGULAR TOLERANCE ±30′

Millimeter Tolerances

Where millimeter dimensions are used on the drawings, the following applies:

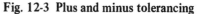

A. The dimension and its tolerance need not be expressed to the same number of decimal places.
Example:
15 ± 0.5 not 15.0 ± 0.5

B. Where unilateral tolerancing is used and either the plus or minus value is nil, a single zero is shown without a plus or minus sign.
Example:
$32^{\ 0}_{-0.02}$ or $32^{+0.02}_{\ 0}$

C. Where bilateral tolerancing is used, both the plus and minus values have the same number of decimal places, using zeros where necessary.
Example:
$32^{+0.25}_{-0.10}$ not $32^{+0.25}_{-0.1}$

DIMENSION ORIGIN SYMBOL

This symbol is used to indicate that a toleranced dimension between two features originates from one of these features, figures 12-4 and 12-5.

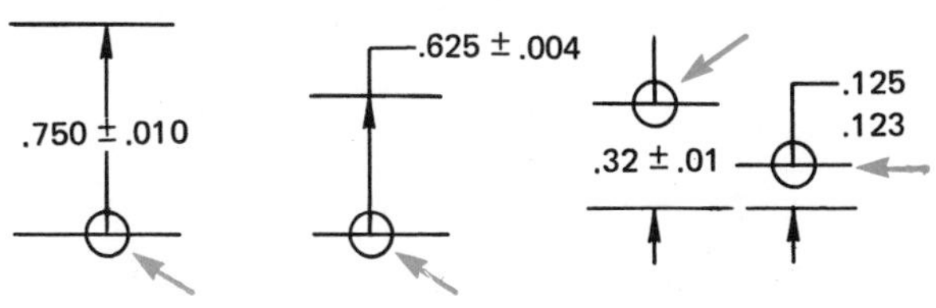

Fig. 12-4 Dimension origin symbol

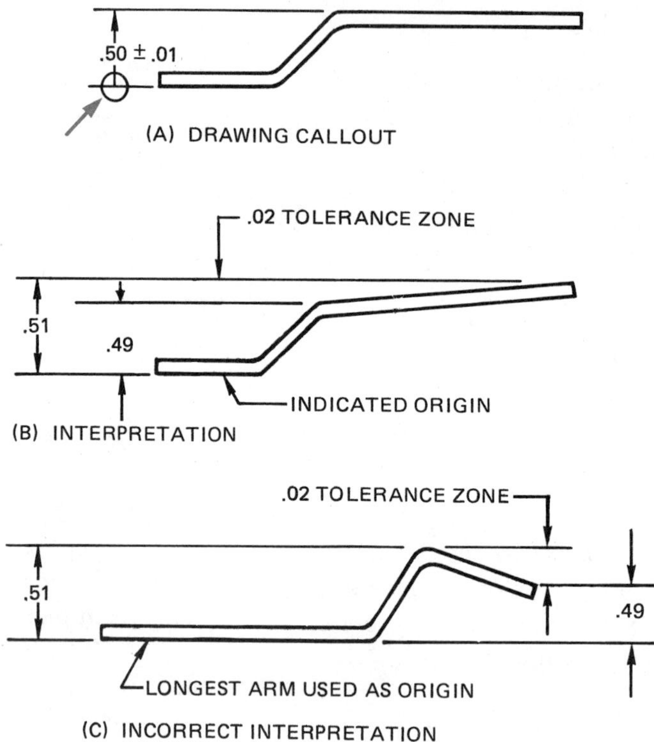

Fig. 12-5 Relating dimensional limits to an origin

PAGE 93 IS INTENTIONALLY BLANK. ASSIGNMENT DRAWING A-32 BEGINS ON PAGE 94.

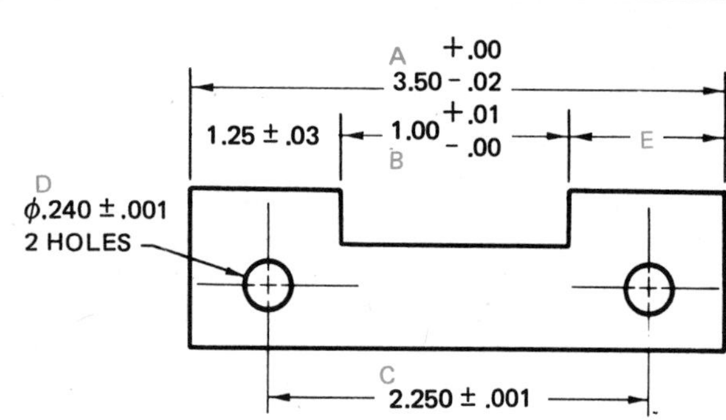

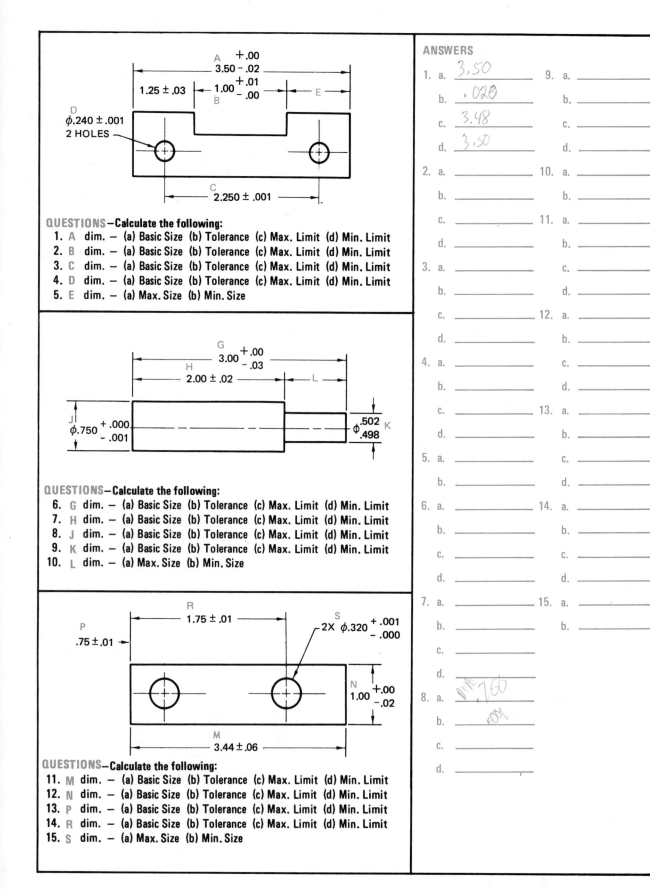

QUESTIONS—Calculate the following:

1. A dim. – (a) Basic Size (b) Tolerance (c) Max. Limit (d) Min. Limit
2. B dim. – (a) Basic Size (b) Tolerance (c) Max. Limit (d) Min. Limit
3. C dim. – (a) Basic Size (b) Tolerance (c) Max. Limit (d) Min. Limit
4. D dim. – (a) Basic Size (b) Tolerance (c) Max. Limit (d) Min. Limit
5. E dim. – (a) Max. Size (b) Min. Size

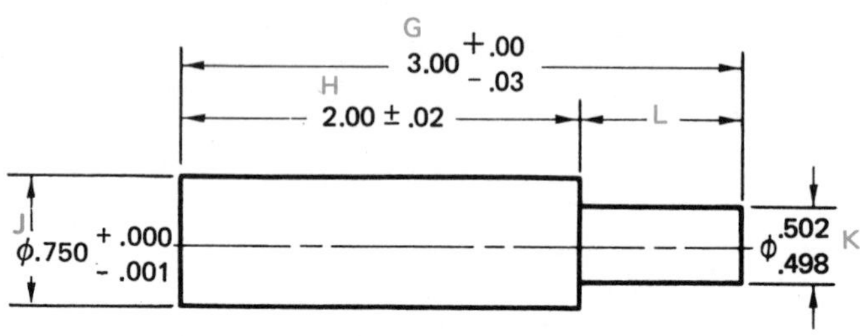

QUESTIONS—Calculate the following:

6. G dim. – (a) Basic Size (b) Tolerance (c) Max. Limit (d) Min. Limit
7. H dim. – (a) Basic Size (b) Tolerance (c) Max. Limit (d) Min. Limit
8. J dim. – (a) Basic Size (b) Tolerance (c) Max. Limit (d) Min. Limit
9. K dim. – (a) Basic Size (b) Tolerance (c) Max. Limit (d) Min. Limit
10. L dim. – (a) Max. Size (b) Min. Size

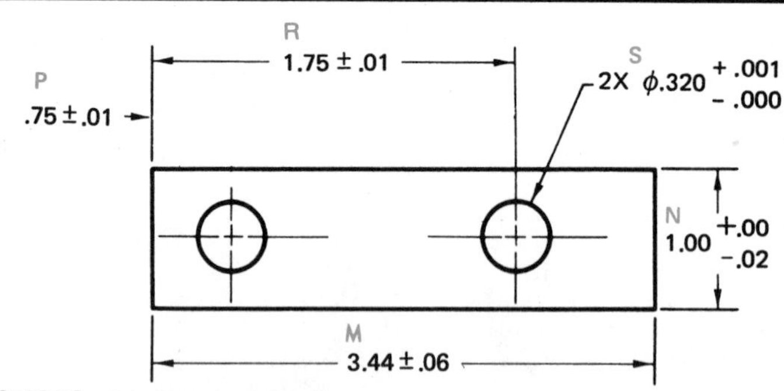

QUESTIONS—Calculate the following:

11. M dim. – (a) Basic Size (b) Tolerance (c) Max. Limit (d) Min. Limit
12. N dim. – (a) Basic Size (b) Tolerance (c) Max. Limit (d) Min. Limit
13. P dim. – (a) Basic Size (b) Tolerance (c) Max. Limit (d) Min. Limit
14. R dim. – (a) Basic Size (b) Tolerance (c) Max. Limit (d) Min. Limit
15. S dim. – (a) Max. Size (b) Min. Size

ANSWERS

1. a. 3.50 b. .020 c. 3.48 d. 3.50
2. a. ____ b. ____ c. ____ d. ____
3. a. ____ b. ____ c. ____ d. ____
4. a. ____ b. ____ c. ____ d. ____
5. a. ____ b. ____
6. a. ____ b. ____ c. ____ d. ____
7. a. ____ b. ____ c. ____ d. ____
8. a. DIA .750 b. ±.001 c. ____ d. ____
9. a. ____ b. ____ c. ____ d. ____
10. a. ____ b. ____
11. a. ____ b. ____ c. ____ d. ____
12. a. ____ b. ____ c. ____ d. ____
13. a. ____ b. ____ c. ____ d. ____
14. a. ____ b. ____ c. ____ d. ____
15. a. ____ b. ____

S ϕ .9992 / .9987

T ϕ 1.0008 / 1.0000

QUESTIONS

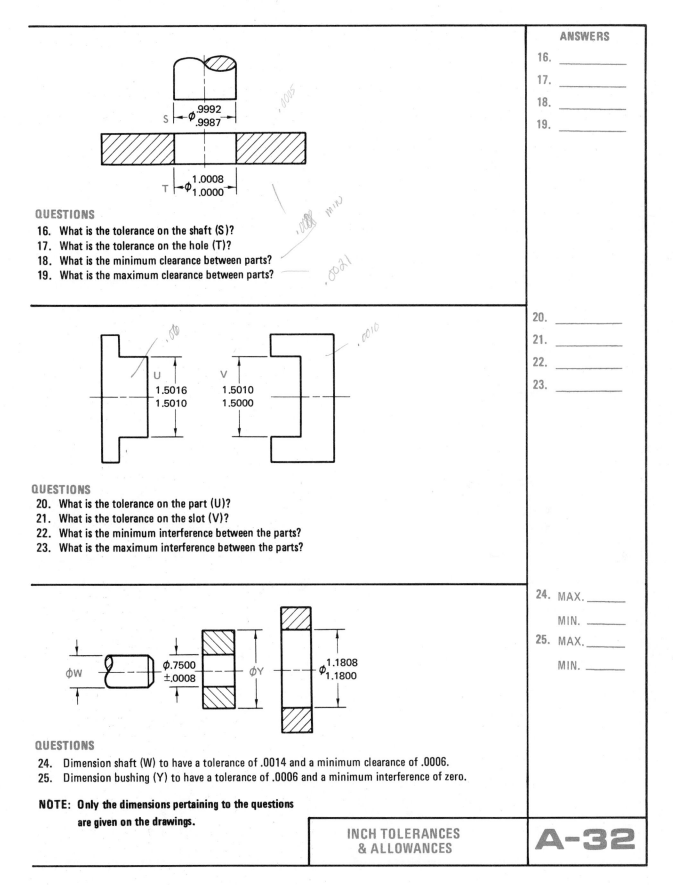

16. What is the tolerance on the shaft (S)?
17. What is the tolerance on the hole (T)?
18. What is the minimum clearance between parts?
19. What is the maximum clearance between parts?

ANSWERS

16. ________
17. ________
18. ________
19. ________

U 1.5016 / 1.5010

V 1.5010 / 1.5000

QUESTIONS

20. What is the tolerance on the part (U)?
21. What is the tolerance on the slot (V)?
22. What is the minimum interference between the parts?
23. What is the maximum interference between the parts?

20. ________
21. ________
22. ________
23. ________

ϕW

ϕ.7500 ±.0008

ϕY

ϕ 1.1808 / 1.1800

QUESTIONS

24. Dimension shaft (W) to have a tolerance of .0014 and a minimum clearance of .0006.
25. Dimension bushing (Y) to have a tolerance of .0006 and a minimum interference of zero.

24. MAX. ______

    MIN. ______

25. MAX. ______

    MIN. ______

NOTE: Only the dimensions pertaining to the questions are given on the drawings.

INCH TOLERANCES & ALLOWANCES

A-32

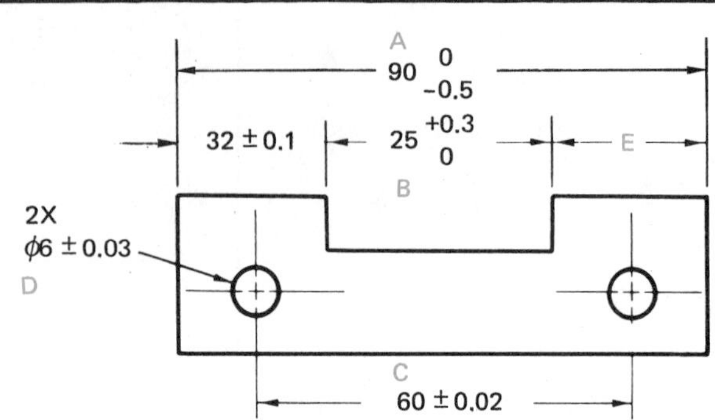

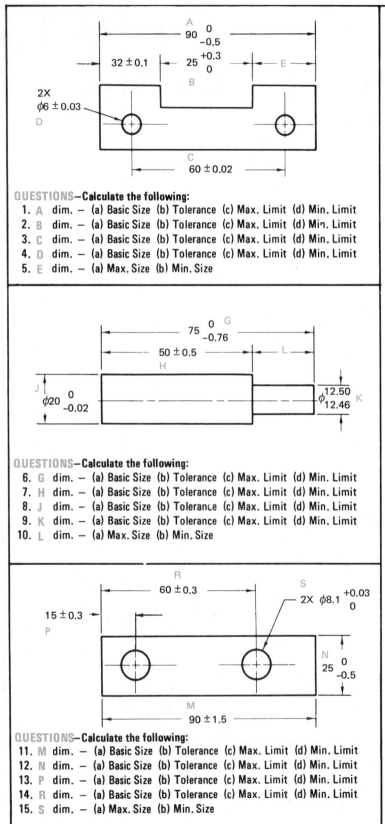

QUESTIONS—**Calculate the following:**

1. A dim. – (a) Basic Size (b) Tolerance (c) Max. Limit (d) Min. Limit
2. B dim. – (a) Basic Size (b) Tolerance (c) Max. Limit (d) Min. Limit
3. C dim. – (a) Basic Size (b) Tolerance (c) Max. Limit (d) Min. Limit
4. D dim. – (a) Basic Size (b) Tolerance (c) Max. Limit (d) Min. Limit
5. E dim. – (a) Max. Size (b) Min. Size

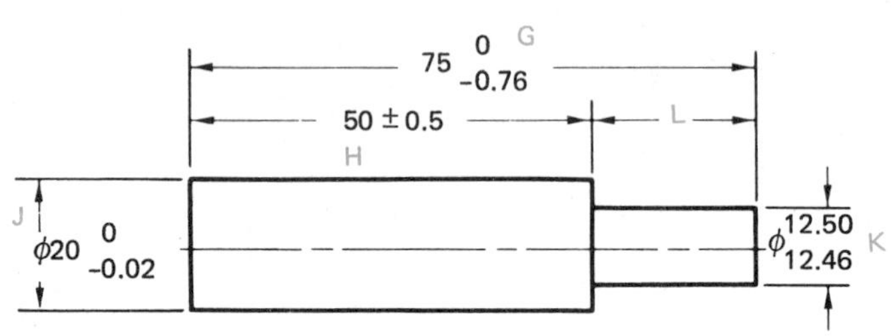

QUESTIONS—**Calculate the following:**

6. G dim. – (a) Basic Size (b) Tolerance (c) Max. Limit (d) Min. Limit
7. H dim. – (a) Basic Size (b) Tolerance (c) Max. Limit (d) Min. Limit
8. J dim. – (a) Basic Size (b) Tolerance (c) Max. Limit (d) Min. Limit
9. K dim. – (a) Basic Size (b) Tolerance (c) Max. Limit (d) Min. Limit
10. L dim. – (a) Max. Size (b) Min. Size

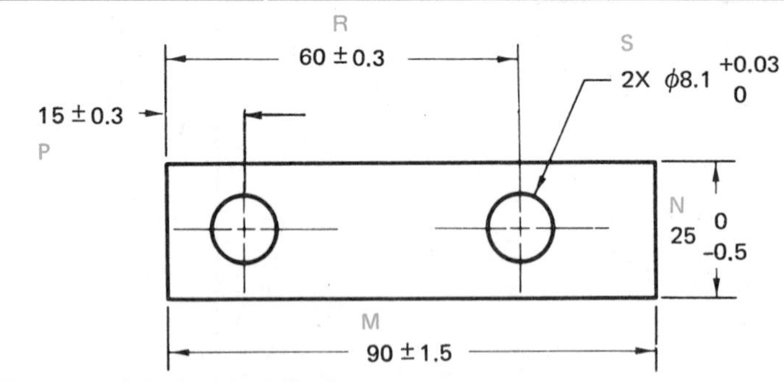

QUESTIONS—**Calculate the following:**

11. M dim. – (a) Basic Size (b) Tolerance (c) Max. Limit (d) Min. Limit
12. N dim. – (a) Basic Size (b) Tolerance (c) Max. Limit (d) Min. Limit
13. P dim. – (a) Basic Size (b) Tolerance (c) Max. Limit (d) Min. Limit
14. R dim. – (a) Basic Size (b) Tolerance (c) Max. Limit (d) Min. Limit
15. S dim. – (a) Max. Size (b) Min. Size

ANSWERS

1. a. ______ b. ______ c. ______ d. ______
2. a. ______ b. ______ c. ______ d. ______
3. a. ______ b. ______ c. ______ d. ______
4. a. ______ b. ______ c. ______ d. ______
5. a. ______ b. ______
6. a. ______ b. ______ c. ______ d. ______
7. a. ______ b. ______ c. ______ d. ______
8. a. ______ b. ______ c. ______ d. ______
9. a. ______ b. ______ c. ______ d. ______
10. a. ______ b. ______
11. a. ______ b. ______ c. ______ d. ______
12. a. ______ b. ______ c. ______ d. ______
13. a. ______ b. ______ c. ______ d. ______
14. a. ______ b. ______ c. ______ d. ______
15. a. ______ b. ______

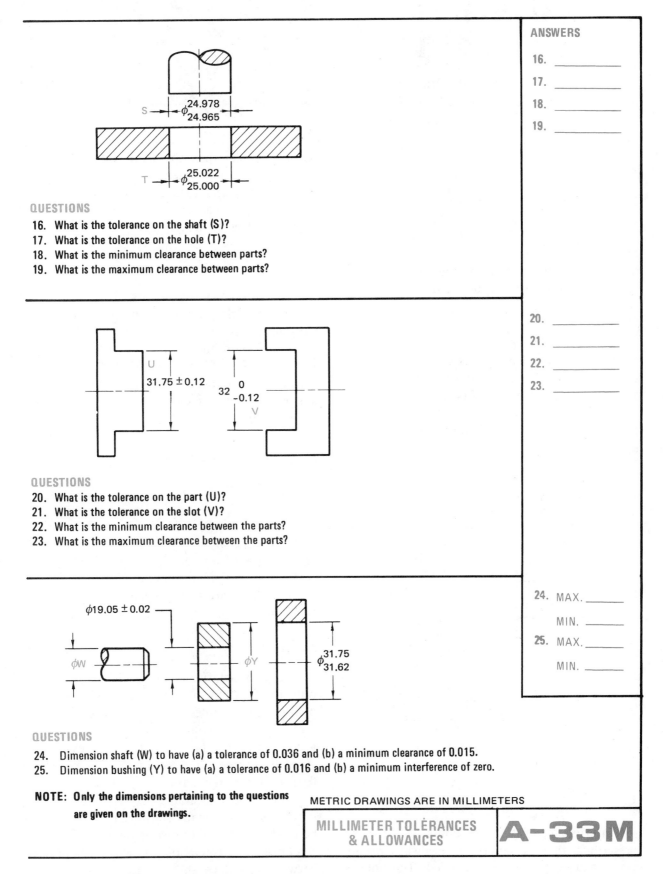

S ϕ 24.978 / 24.965

T ϕ 25.022 / 25.000

QUESTIONS

16. What is the tolerance on the shaft (S)?
17. What is the tolerance on the hole (T)?
18. What is the minimum clearance between parts?
19. What is the maximum clearance between parts?

ANSWERS

16. ________
17. ________
18. ________
19. ________

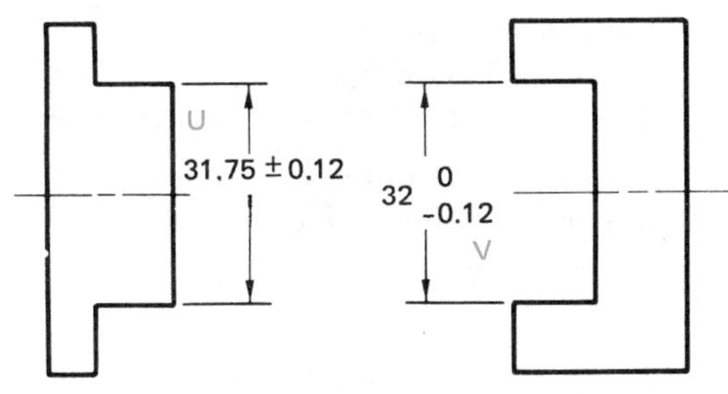

QUESTIONS

20. What is the tolerance on the part (U)?
21. What is the tolerance on the slot (V)?
22. What is the minimum clearance between the parts?
23. What is the maximum clearance between the parts?

20. ________
21. ________
22. ________
23. ________

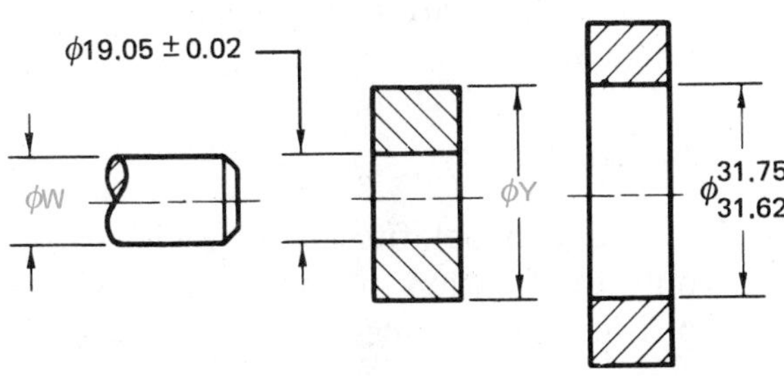

QUESTIONS

24. Dimension shaft (W) to have (a) a tolerance of 0.036 and (b) a minimum clearance of 0.015.
25. Dimension bushing (Y) to have (a) a tolerance of 0.016 and (b) a minimum interference of zero.

24. MAX. ______ MIN. ______
25. MAX. ______ MIN. ______

NOTE: Only the dimensions pertaining to the questions are given on the drawings.

METRIC DRAWINGS ARE IN MILLIMETERS

MILLIMETER TOLERANCES & ALLOWANCES	A-33M

UNIT 13

INCH FITS

Fit is the general term used to signify the range of tightness or looseness resulting from the application of a specific combination of allowances and tolerances in the design of mating parts. Fits are of three general types: clearance, interference, and transition. Figures 13-1 and 13-2 illustrate the three types of fits.

Clearance Fits

Clearance fits have limits of size prescribed so a clearance always results when mating parts are assembled. Clearance fits are intended for accurate assembly of parts and bearings. The parts can be assembled by hand because the hole is always larger than the shaft.

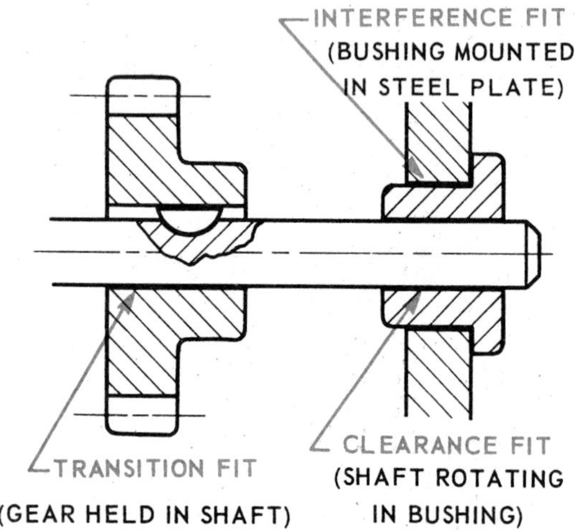

Fig. 13-1 Application of types of fits

Interference Fits

Interference fits have limits of size so prescribed that an interference always results when mating parts are assembled. The hole is always smaller than the shaft. Interference fits are for permanent assemblies of parts which require rigidity and alignment, such as dowel pins and bearings in castings. Parts are usually pressed together using an arbor press.

Transition Fits

Transition fits have limits of size indicating that either a clearance or an interference may result when mating parts are assembled. Transition fits are a compromise between clearance and interference fits. They are used for applications where accurate location is important, but either a small amount of clearance or interference is permissible.

DESCRIPTION OF FITS

Running and Sliding Fits

These fits, for which tolerances and clearances are given in the Appendix, represent a special type of clearance fit. These are intended to provide a similar running performance, with suitable lubrication allowance, throughout the range of sizes.

Locational Fits

Locational fits are intended to determine only the location of the mating parts; they may provide rigid or accurate location, as with interference fits, or some freedom of location, as with clearance fits. Accordingly, they are divided into three groups: clearance fits, transition fits, and interference fits.

Locational clearance fits are intended for parts that are normally stationary but which can be freely assembled or disassembled.

Locational transition fits are a compromise between clearance and interference fits, for ap-

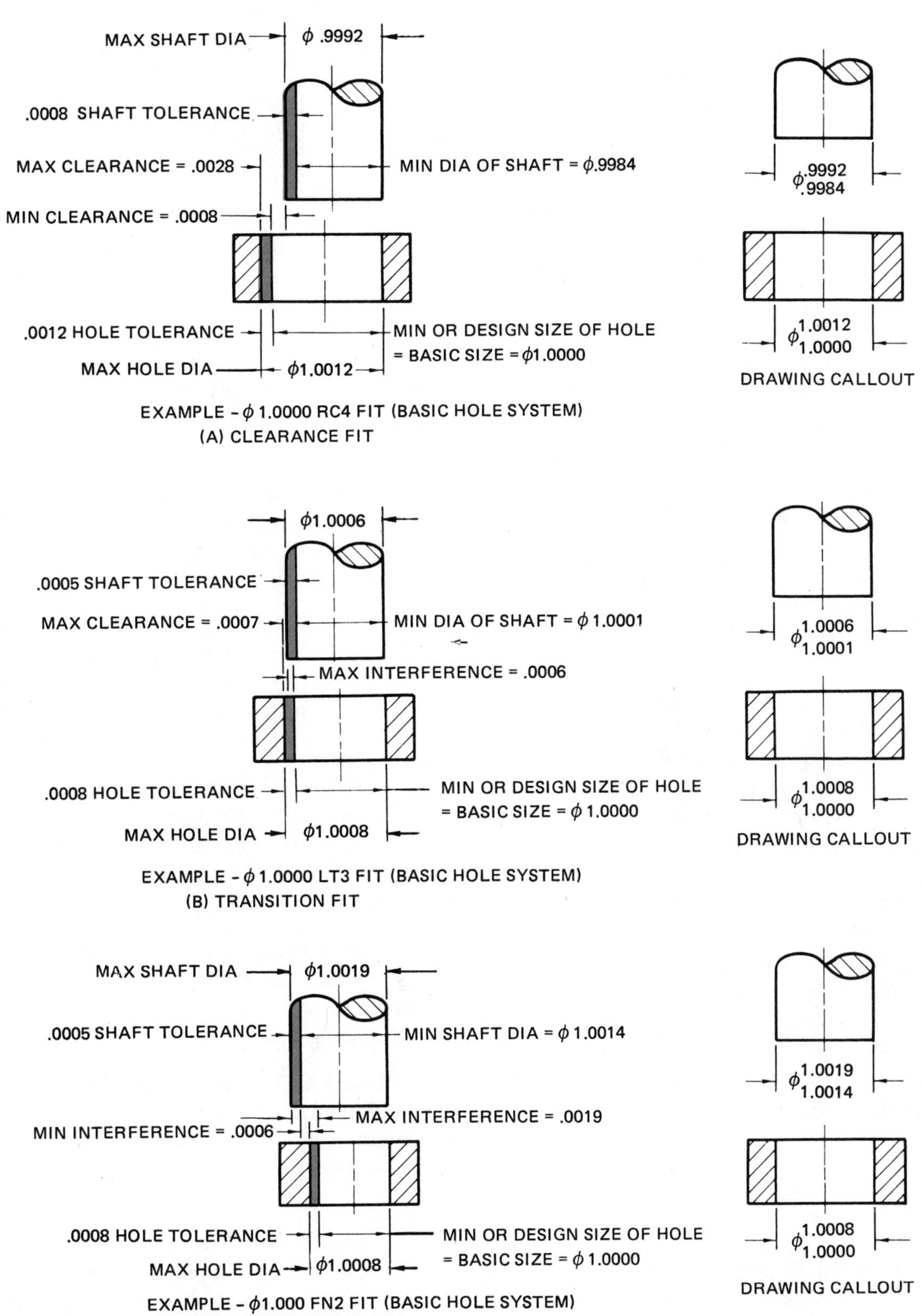

Fig. 13-2 Types and examples of inch fits

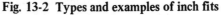

plication where accuracy of location is important but a small amount of either clearance or interference is permissible.

Locational interference fits are used where accuracy of location is of prime importance and for parts requiring rigidity and alignment.

Drive and Force Fits

Drive and force fits constitute a special type of interference fit, normally characterized by maintenance of constant bore pressures throughout the range of sizes. The interference therefore varies almost directly with diameter, and the difference between its minimum and maximum values is small to maintain the resulting pressures within reasonable limits.

STANDARD INCH FITS

Standard fits are designated for design purposes in specifications and on design sketches by means of the symbols shown in figure 13-3. These symbols, however, are not intended to be shown directly on shop drawings; instead the actual limits of size are determined, and these limits are specified on the drawings. The letter symbols used are as follows:

RC Running and sliding fit
LC Locational clearance fit
LT Locational transition fit
LN Locational interference fit
FN Force or shrink fit

These letter symbols are used in conjunction with numbers representing the class of fit; for example, FN4 represents a class 4, force fit.

Each of these symbols (two letters and a number) represents a complete fit, for which the minimum and maximum clearance or interference, and the limits of size for the mating parts, are given directly in Appendix Tables 17 through 21.

Running and Sliding Fits

RC1 Precision Sliding Fit. This fit is intended for the accurate location of parts that must assemble without perceptible play, for high precision work such as gages.

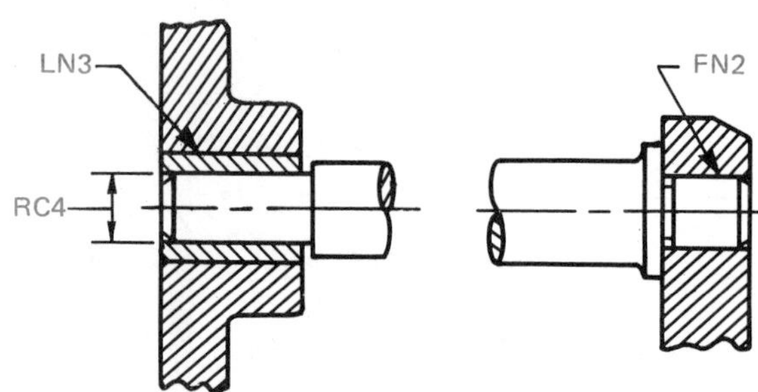

(A) SHAFT IN BUSHED HOLE (B) CRANK PIN IN CAST IRON

Fig. 13-3 Design sketches showing standard fits

RC2 Sliding Fit. This fit is intended for accurate location, but with greater maximum clearance than class RC1. Parts made to this fit move and turn easily but are not intended to run freely.

RC3 Precision Running Fit. This fit is about the closest fit which can be expected to run freely, and is intended for precision work for oil-lubricated bearings at slow speeds and light journal pressures.

RC4 Close Running Fit. This fit is intended chiefly as a running fit for grease or oil-lubricated bearings on accurate machinery with moderate surface speeds and journal pressures, where accurate location and minimum play are desired.

RC5 and RC6 Medium Running Fits. These fits are intended for higher running speeds and/or where temperature variations are likely to be encountered.

RC7 Free Running Fit. This fit is intended for use where accuracy is not essential, and/or where large temperature variations are likely to be encountered.

RC8 and RC9 Loose Running Fits. These fits are intended for use where materials made to commercial tolerances are involved such as cold-rolled shafting, tubing, etc.

Locational Clearance Fits

Locational clearance fits are intended for parts that are normally stationary, but can be freely assembled or disassembled. These are classified as follows:

LC1 to LC4. These fits have a minimum zero clearance, but in practice the probability is that the fit will always have a clearance.

LC5 and LC6. These fits have a small minimum clearance, intended for close location fits for nonrunning parts.

LC7 to LC11. These fits have progressively larger clearances and tolerances, and are useful for various loose clearances for assembly of bolts and similar parts.

Locational Transition Fits

Locational transition fits are a compromise between clearance and interference fits, for application where accuracy of location is important, but either a small amount of clearance or interference is permissible. These are classified as follows:

LT1 and LT2. These fits average a slight clearance, giving a light push fit.

LT3 and LT4. These fits average virtually no clearance, and are for use where some interference can be tolerated. These are sometimes referred to as an easy keying fit, and are used for shaft keys and ball race fits. Assembly is generally by pressure or hammer blows.

LT5 and LT6. These fits average a slight interference, although appreciable assembly force will be required.

Locational Interference Fits

Locational interference fits are used where accuracy of location is of prime importance, and for parts requiring rigidity and alignment with no special requirements for bore pressure. These are classified as follows:

LN1 and LN2. These are light press fits, with very small minimum interference, suitable for parts such as dowel pins, which are assembled with an arbor press in steel, cast iron, or brass. Parts can normally be dismantled and reassembled.

LN3. This is suitable as a heavy press fit in steel and brass, or a light press fit in more elastic materials and light alloys.

LN4 to LN6. While LN4 can be used for permanent assembly of steel parts, these fits are primarily intended as press fits for soft materials.

Force or Shrink Fits

Force or shrink fits constitute a special type of interference fit. The interference varies almost directly with diameter, and the difference between its minimum and maximum values is small to maintain the resulting pressures within reasonable limits. These fits are classified as follows:

FN1 Light Drive Fit. Requires light assembly pressure and produces more or less permanent assemblies. It is suitable for thin sections or long fits, or in cast-iron external members.

FN2 Medium Drive Fit. Suitable for heavier steel parts, or as a shrink fit on light sections.

FN3 Heavy Drive Fit. Suitable for heavier steel parts or as a shrink fit in medium sections.

FN4 and FN5 Force Fits. Suitable for parts which can be highly stressed.

Basic Hole System

In the basic hole system, which is recommended for general use, the basic size will be the design size for the hole, and the tolerance will be plus. The design size for the shaft will be the basic size minus the minimum clearance, or plus the maximum interference, and the tolerance will be minus, as given in the tables in the Appendix. For example, (see Table 17) for a 1-in. RC7 fit, values of +.0020, .0025, and −.0012 are given; hence, limits will be:

$$\text{Hole } \phi 1.0000 \begin{matrix} +.0020 \\ -.0000 \end{matrix}$$

$$\text{Shaft } \phi .9975 \begin{matrix} +.0000 \\ -.0012 \end{matrix}$$

Basic Shaft System

Fits are sometimes required on a basic shaft system, especially in cases where two or more fits are required on the same shaft. This is designated for design purposes by a letter S following the fit symbol; for example, RC7S.

Tolerances for holes and shaft are identical with those for a basic hole system, but the basic size becomes the design size for the shaft and the design size for the hole is found by adding the minimum clearance or subtracting the maximum interference from the basic size.

For example, for a 1-in. RC7S fit, values of +.0020, .0025, and −.0012 are given; therefore, limits will be:

Hole $\phi 1.0025 \begin{matrix} +.0020 \\ -.0000 \end{matrix}$

Shaft $\phi 1.000 \begin{matrix} +.0000 \\ -.0012 \end{matrix}$

REFERENCES AND SOURCE MATERIALS

1. ANSI Y14.5M "Dimensioning and Tolerancing."

ϕ A
ϕ B
ϕ.625 RC2

ϕ C
ϕ D
ϕ1.000 RC4

ϕ E
ϕ F
ϕ1.500 RC8

RUNNING AND SLIDING FITS

ϕ G
ϕ H
ϕ.625 LC5

ϕ J
ϕ K
ϕ1.125 LT3

ϕ L
ϕ M
ϕ1.375 LN2

LOCATIONAL FITS

ϕ N
ϕ P
ϕ.875 FN1

ϕ R
ϕ S
ϕ1.250 FN2

ϕ T
ϕ U
ϕ1.750 FN4

FORCE OR SHRINK FITS

ANSWERS

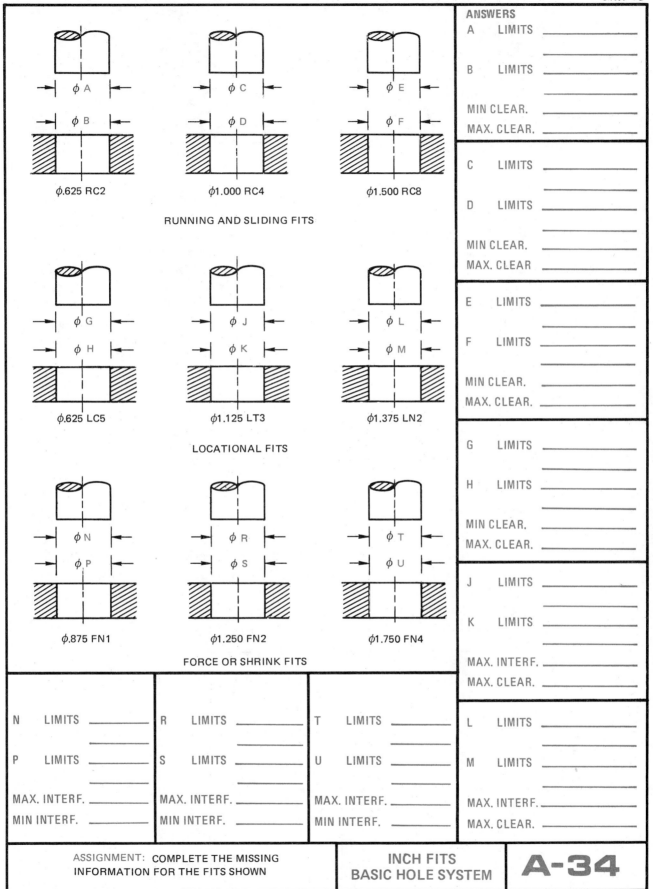

A LIMITS ______
______
B LIMITS ______
______
MIN CLEAR. ______
MAX. CLEAR. ______

C LIMITS ______
______
D LIMITS ______
______
MIN CLEAR. ______
MAX. CLEAR ______

E LIMITS ______
______
F LIMITS ______
______
MIN CLEAR. ______
MAX. CLEAR. ______

G LIMITS ______
______
H LIMITS ______
______
MIN CLEAR. ______
MAX. CLEAR. ______

J LIMITS ______
______
K LIMITS ______
______
MAX. INTERF. ______
MAX. CLEAR. ______

L LIMITS ______
______
M LIMITS ______
______
MAX. INTERF. ______
MAX. CLEAR. ______

N LIMITS ______
______
P LIMITS ______
______
MAX. INTERF. ______
MIN INTERF. ______

R LIMITS ______
______
S LIMITS ______
______
MAX. INTERF. ______
MIN INTERF. ______

T LIMITS ______
______
U LIMITS ______
______
MAX. INTERF. ______
MIN INTERF. ______

ASSIGNMENT: COMPLETE THE MISSING INFORMATION FOR THE FITS SHOWN

INCH FITS
BASIC HOLE SYSTEM

A-34

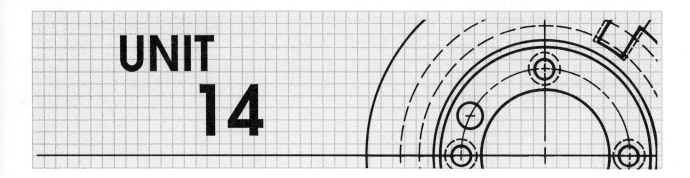

UNIT 14

METRIC FITS

The ISO (metric) system of limits and fits for mating parts is approved and adopted for general use in the United States. It establishes the designation symbols used to define specific dimensional limits on drawings.

The general terms "hole" and "shaft" can also be taken as referring to the space containing or contained by two parallel faces of any part, such as the width of slot, or the thickness of a key.

An "International Tolerance Grade" establishes the magnitude of the tolerance zone or the amount of part size variation allowed for internal and external dimensions alike. The smaller the grade number the smaller the tolerance zone. For general applications of IT grades see figure 14-1.

Grades 1 to 4 are very precise grades intended primarily for gage making and similar precision work, although grade 4 can also be used for very precise production work.

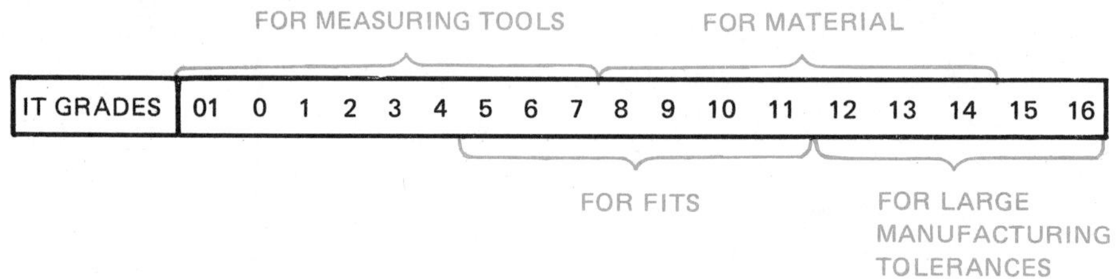

Fig. 14-1 Applications of international tolerance (IT) grades

Machining Processes	Tolerance Grades									
	4	5	6	7	8	9	10	11	12	13
Lapping & Honing	■	■								
Cylindrical Grinding		■	■	■						
Surface Grinding		■	■	■	■					
Diamond Turning		■	■	■						
Diamond Boring		■	■	■						
Broaching		■	■	■	■					
Reaming		■	■	■	■	■	■			
Turning				■	■	■	■	■	■	■
Boring					■	■	■	■	■	■
Milling							■	■	■	■
Planing & Shaping							■	■	■	■
Drilling							■	■	■	■

Fig. 14-2 Tolerancing grades for machining processes

Grades 5 to 16 represent a progressive series suitable for cutting operations, such as turning, boring, grinding, milling, and sawing. Grade 5 is the most precise grade, obtainable by the fine grinding and lapping, while 16 is the coarsest grade for rough sawing and machining.

Grades 12 to 16 are intended for manufacturing operations such as cold heading, pressing, rolling, and other forming operations.

As a guide to the selection of tolerances, figure 14-2 has been prepared to show grades which may be expected to be held by various manufacturing processes for work in metals. For work in other materials, such as plastics, it may be necessary to use coarser tolerance grades for the same process.

A fundamental deviation establishes the position of the tolerance zone with respect to the basic size. Fundamental deviations are expressed by "tolerance position letters." Capital letters are used for internal dimensions, and lower case letters for external dimensions.

Metric Tolerance Symbol

By combining the IT grade number and the tolerance position letter, the tolerance symbol is established which identifies the actual maximum and minimum limits of the part. The toleranced sizes are thus defined by the basic size of the part followed by the symbol composed of a letter and number, figure 14-3.

Hole basis fits have a fundamental deviation of "H" on the hole, and shaft basis fits have a fundamental deviation of "h" on the shaft. Normally, the hole basis system is preferred.

Fit Symbol

A fit is indicated by the basic size common to both components, followed by a symbol corresponding to each component, with the

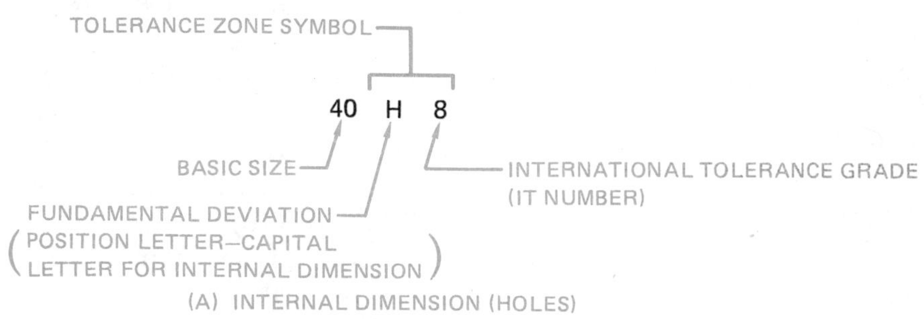

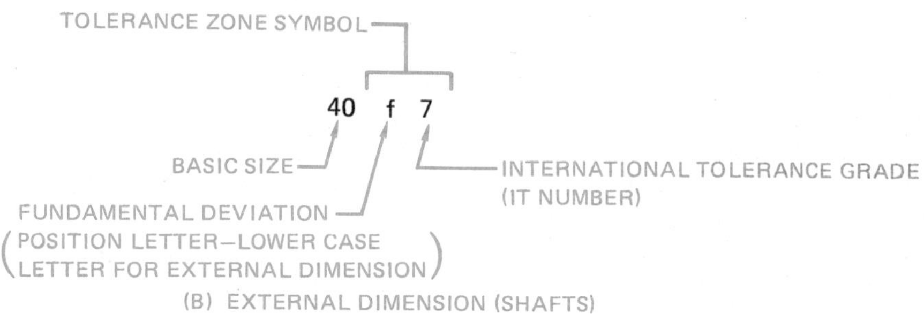

Fig. 14-3 Tolerance symbol (hole basis fit)

internal part symbol preceding the external part symbol, figure 14-4.

Figure 14-5 shows examples of three common fits.

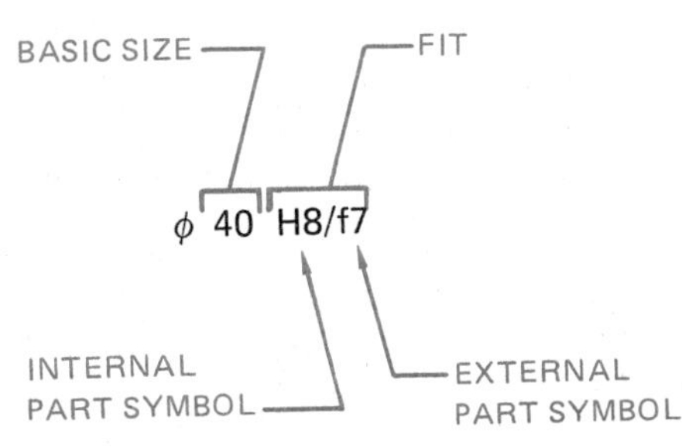

Fig. 14-4 Fit symbol

Hole Basis Fits System

In the hole basis fits system (see Tables 22 and 24 of the Appendix) the basic size will be the minimum size of the hole. For example, for a ϕ25 H8/f7 fit, which is a Preferred Hole Basis Clearance Fit, the limits for the hole and shaft will be as follows:

Hole limits = ϕ25.00–ϕ25.033
Shaft limits = ϕ24.959–ϕ24.980
Minimum clearance = 0.020
Maximum clearance = 0.074

If a ϕ25 H7/s6 Preferred Hole Basis Interference Fit is required, the limits for the hole and shaft will be as follows:

Hole limits = ϕ25.000–ϕ25.021
Shaft limits = ϕ25.035–ϕ25.048
Minimum interference = –0.014
Maximum interference = –0.048

Shaft Basis Fits System

Where more than two fits are required on the same shaft, the shaft basis fits system is recommended. Tolerances for holes and shafts are identical with those for a basic hole system, however, the basic size becomes the maximum shaft size. For example, for a ϕ16 C11/h11 fit, which is a Preferred Shaft Basis Clearance Fit, the limits for the hole and shaft will be as follows:

Hole limits = ϕ16.095–ϕ16.205
Shaft limits = ϕ15.890–ϕ16.000
Minimum clearance = 0.095
Maximum clearance = 0.315

Refer to Tables 23 and 25 of the Appendix.

Drawing Callout

The method shown in figure 14-6(a) is recommended when the system is first introduced. In this case, limit dimensions are specified and the basic size and tolerance symbol are identified as reference.

As experience is gained, the method shown in figure 14-6(b) may be used. When the system is established and standard tools, gages, and stock materials are available with size and symbol identification, the method shown in figure 14-6(c) may be used.

This would result in a clearance fit of 0.020–0.074 mm. A description of the preferred metric fits is shown in Tables 22 and 23 of the Appendix.

REFERENCES AND SOURCE MATERIALS

1. ANSI B4.2 "Preferred Metric Limits and Fits."

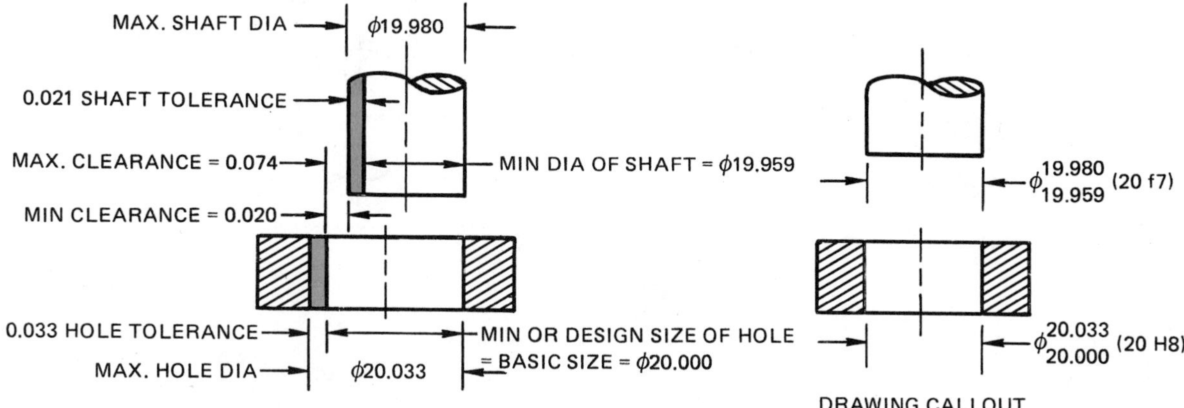

EXAMPLE—H8/f7 PREFERRED HOLE BASIS FIT FOR A ϕ20 HOLE (SEE APPENDIX, TABLE 24)

(A) CLEARANCE FIT

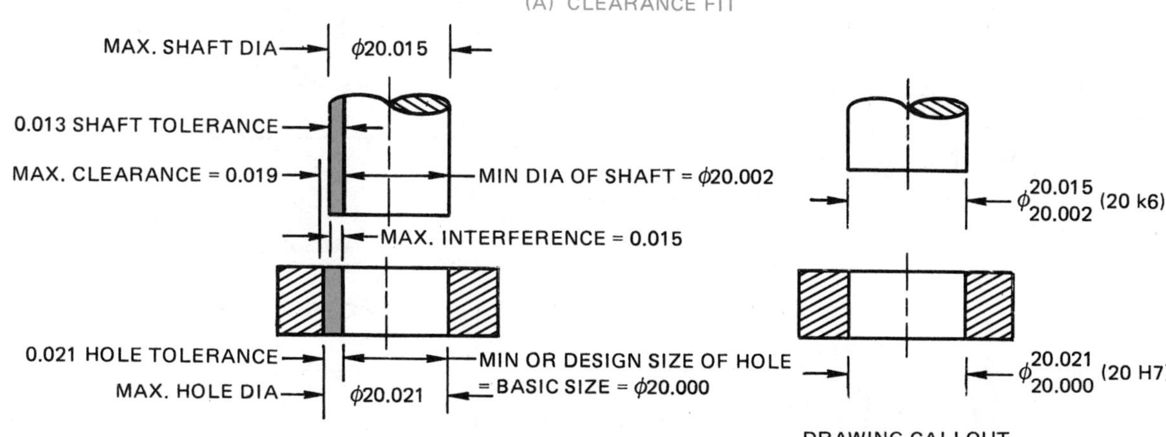

EXAMPLE—H7/k6 PREFERRED HOLE BASIS FIT FOR A ϕ20 HOLE (SEE APPENDIX, TABLE 24)

(B) TRANSITION FIT

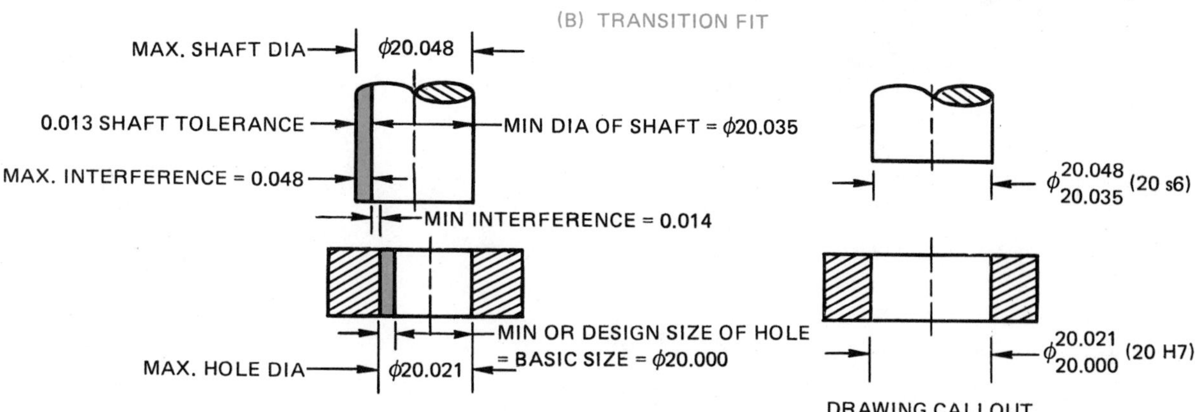

EXAMPLE—H7/s6 PREFERRED HOLE BASIS FIT FOR A ϕ20 HOLE (SEE APPENDIX, TABLE 24)

(C) INTERFERENCE FIT

Fig. 14-5 Types and examples of millimeter fits

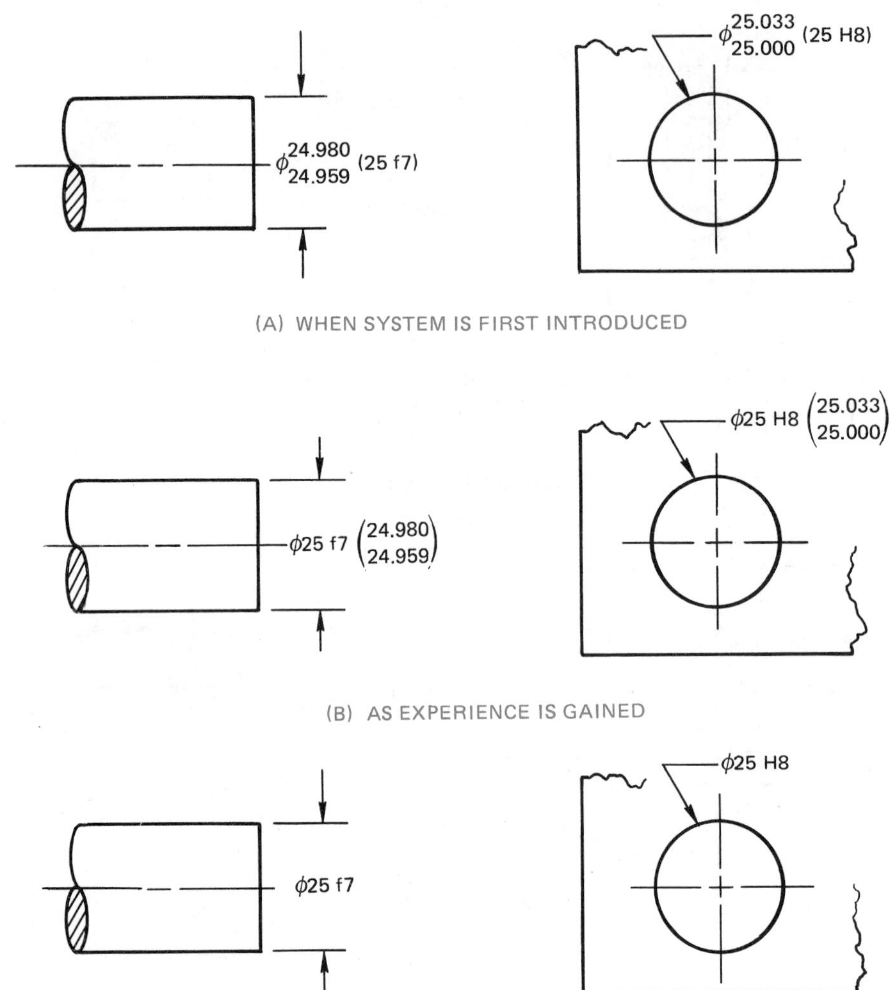

(C) WHEN SYSTEM IS ESTABLISHED

Fig. 14-6 Metric tolerance symbol shown on drawings

ϕA
ϕB
ϕ16 H7/g6

ϕC
ϕD
ϕ25 H9/d9

ϕE
ϕF
ϕ40 H11/c11

RUNNING AND SLIDING FITS

ϕG
ϕH
ϕ20 H7/h6

ϕJ
ϕK
ϕ30 H7/k6

ϕL
ϕM
ϕ45 H7/p6

LOCATIONAL FITS

ϕN
ϕP
ϕ16 H7/s6

ϕR
ϕS
ϕ25 H7/u6

ϕT
ϕU
ϕ40 H7/n6

FORCE OR SHRINK FITS

ANSWERS

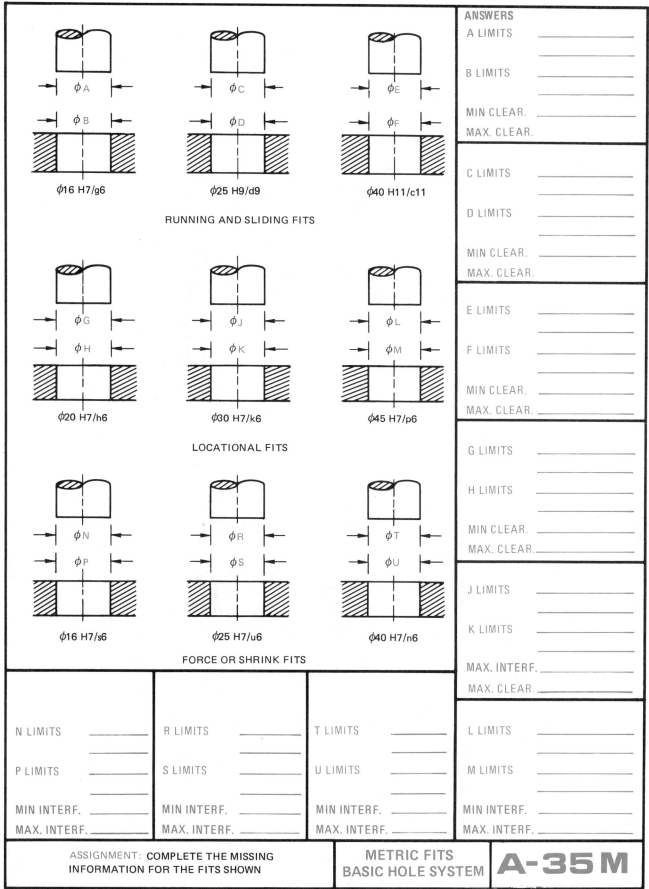

A LIMITS ________
____
B LIMITS ________
____
MIN CLEAR. ________
MAX. CLEAR.

C LIMITS ________
____
D LIMITS ________
____
MIN CLEAR. ________
MAX. CLEAR.

E LIMITS ________
____
F LIMITS ________
____
MIN CLEAR. ________
MAX. CLEAR. ________

G LIMITS ________
____
H LIMITS ________
____
MIN CLEAR. ________
MAX. CLEAR. ________

J LIMITS ________
____
K LIMITS ________
____
MAX. INTERF. ________
MAX. CLEAR. ________

N LIMITS ________
____
P LIMITS ________
____
MIN INTERF. ________
MAX. INTERF. ________

R LIMITS ________
____
S LIMITS ________
____
MIN INTERF. ________
MAX. INTERF. ________

T LIMITS ________
____
U LIMITS ________
____
MIN INTERF. ________
MAX. INTERF. ________

L LIMITS ________
____
M LIMITS ________
____
MIN INTERF. ________
MAX. INTERF. ________

ASSIGNMENT: COMPLETE THE MISSING INFORMATION FOR THE FITS SHOWN

METRIC FITS
BASIC HOLE SYSTEM

A-35M

NOTE: EXCEPT WHERE NOTED
TOLERANCES ON DIMENSIONS ±0.5
6.3 EXCEPT WHERE NOTED
ROUNDS AND FILLETS R.5

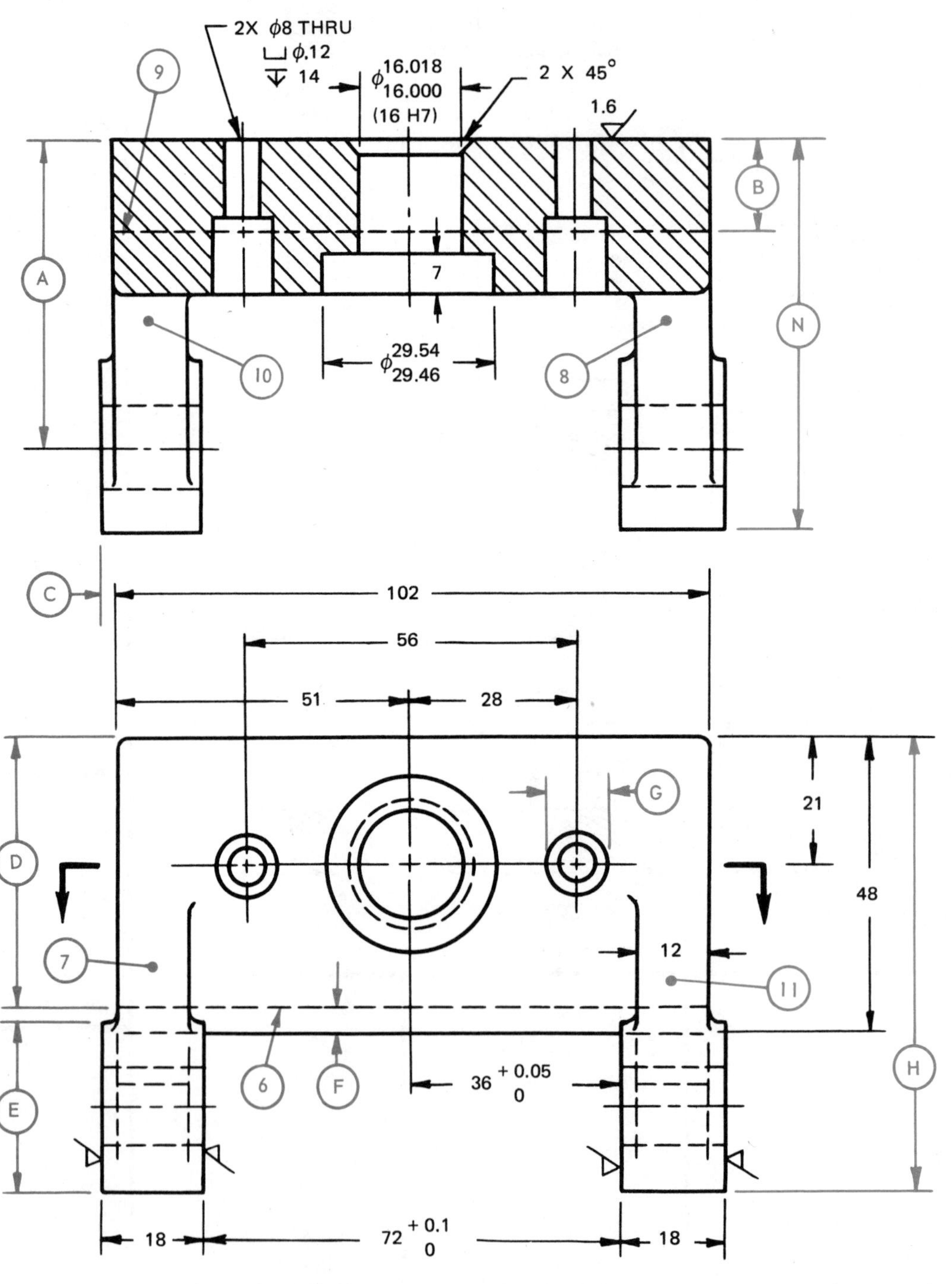

QUESTIONS

1. How many surfaces are to be finished?
2. Except where noted otherwise, what is the tolerance on all dimensions?
3. What is the tolerance on the ϕ12.000–12.018 holes?
4. What are the limit dimensions for the 40.64 dimension shown on the side view?
5. What are the limit dimensions for the 26 dimension shown on the side view?
6. What is the maximum distance permissible between the centers of the ϕ8 hole?
7. Express the ϕ12.000–12.018 holes as a plus-and-minus tolerance dimension.
8. How many surfaces require a 6.3 finish?
9. What are the limit dimensions for the 7 dimension shown on the top view?
10. Locate surfaces (4) on the top view.
11. How many bosses are there?
12. Locate line (3) in the top view.
13. Locate line (6) in the side view.
14. Which surfaces in the front view indicate line (4) in the side view?
15. Calculate nominal distances (A) to (N).

ANSWERS

1. ____________
2. ____________
3. ____________
4. ____________
5. ____________
6. ____________
7. ____________
8. ____________
9. ____________
10. ____________
11. ____________
12. ____________
13. ____________
14. ____________

15. (A) ____________
(B) ____________
(C) ____________
(D) ____________
(E) ____________
(F) ____________
(G) ____________
(H) ____________
(J) ____________
(K) ____________
(L) ____________
(M) ____________
(N) ____________

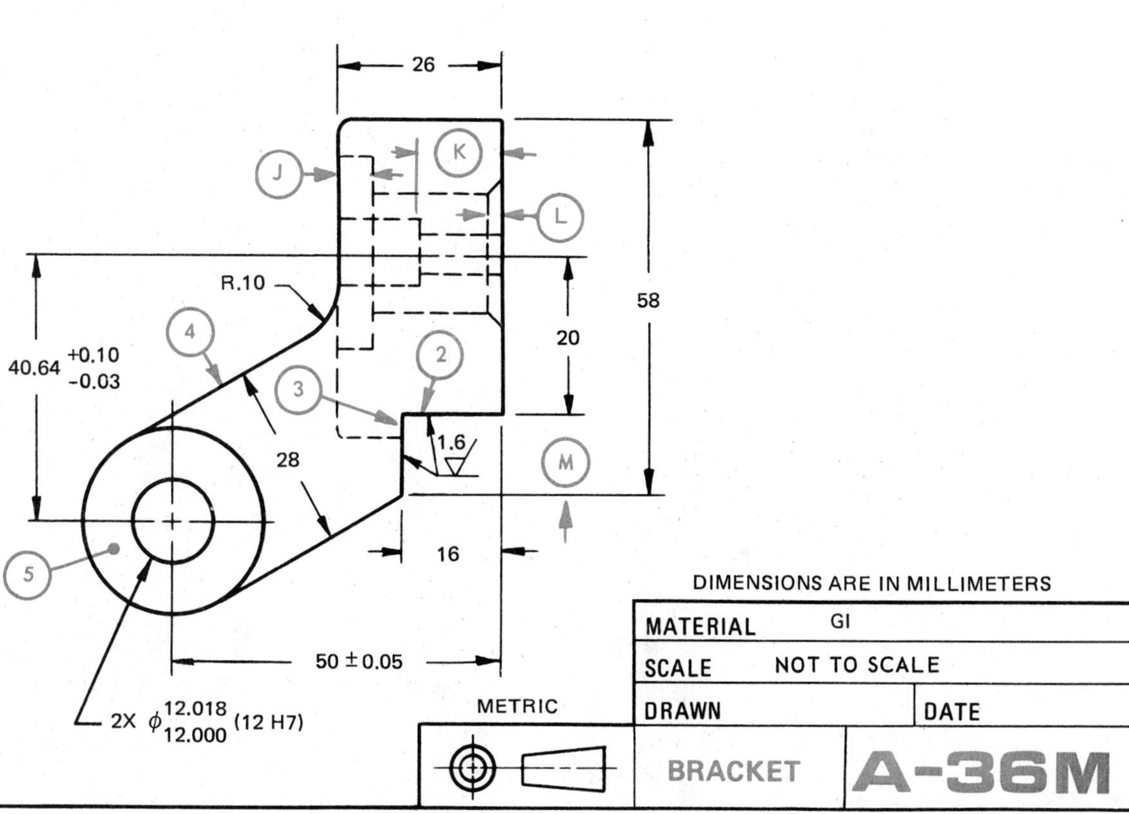

UNIT 15

THREADED FASTENERS

Fastening devices are vital to most phases of industry. They are used in assembling manufactured products, the machines and devices used in the manufacturing processes, and in the construction of all types of buildings.

There are two basic types of fasteners: semipermanent and removable. Rivets are semipermanent fasteners; bolts, screws, studs, nuts, pins, and keys are removable fasteners. With the progress of industry, fastening devices have become standardized. A thorough knowledge of the design and graphic representation of the common fasteners is essential for interpreting engineering drawings, figure 15-1.

Thread Representation

True representation of a screw thread is seldom provided on working drawings because of the time involved and the drawing cost. Three types of conventions are generally used for screw thread representation: detailed, schematic, and simplified representation. Simplified representation is used whenever it will clearly indicate the requirements. Schematic and detailed representations require more drafting time, but they are sometimes used to avoid confusion with other parallel lines or to more clearly portray particular aspects of threads. One method is generally used within any one drawing. When required, however, all three methods may be used. ANSI and ISO thread representation vary slightly, figures 15-2 and 15-3.

THREADED ASSEMBLIES

Any of the thread conventions shown here may be used for assemblies of threaded parts, and two or more methods may be used on the same drawing, figures 15-2 and 15-3. In sectional views, the externally threaded part is always shown covering the internally threaded part, figure 15-4.

Thread Standards

With the progress and growth of industry, there is a growing need for uniform, interchangeable, threaded fasteners. Aside from the threaded forms previously mentioned, the pitch of the thread and the major diameters are factors affecting standards.

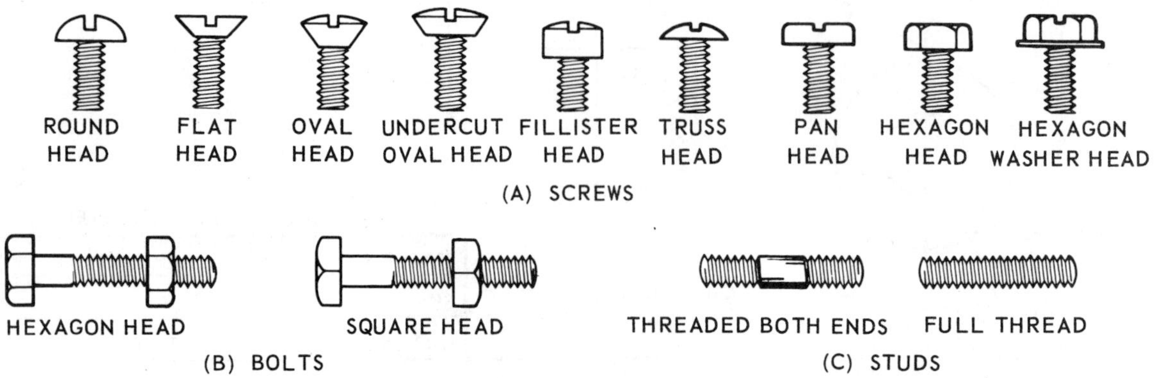

Fig. 15-1 Threaded fasteners

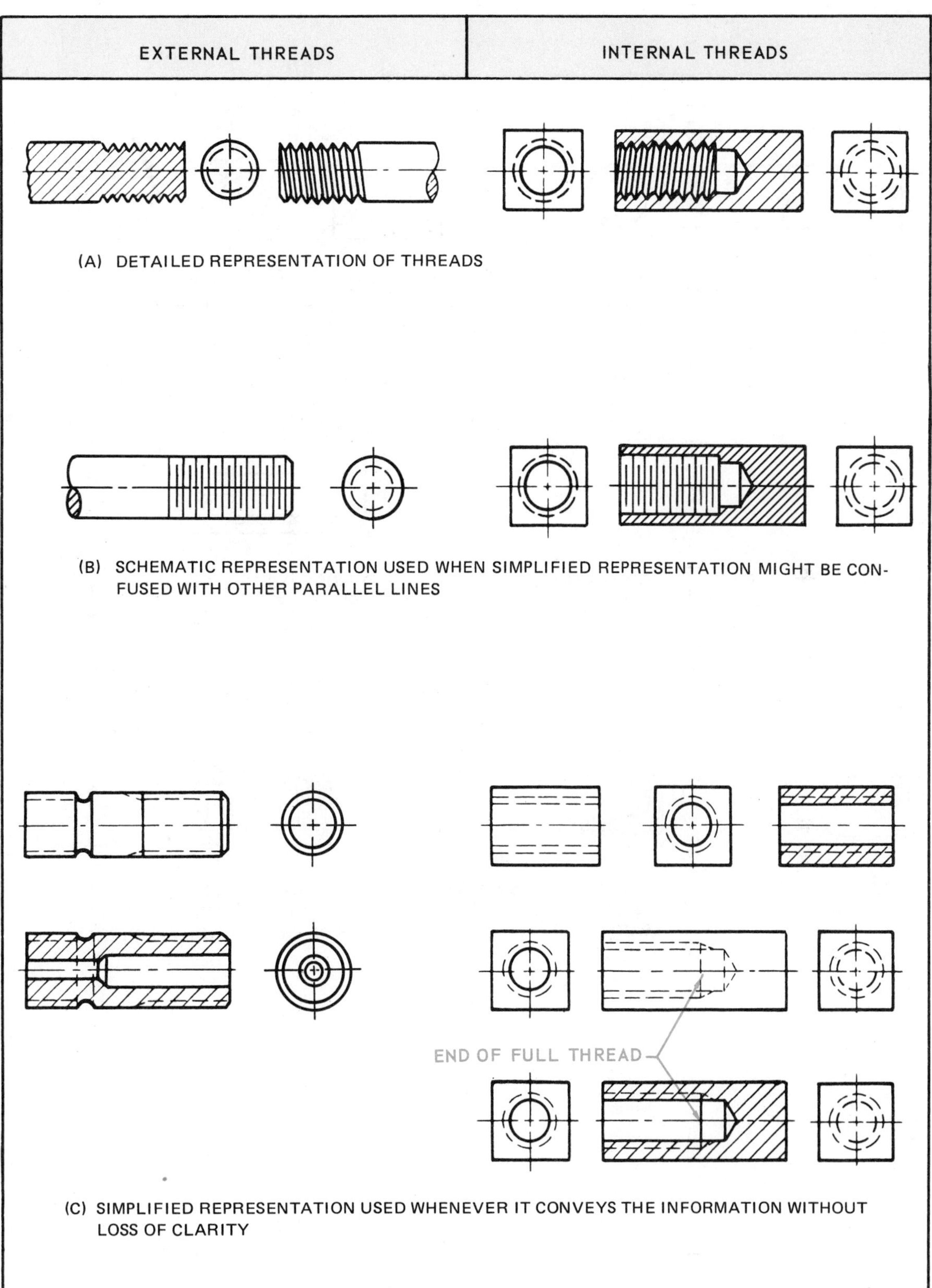

Fig. 15-2 Standard thread conventions (ANSI)

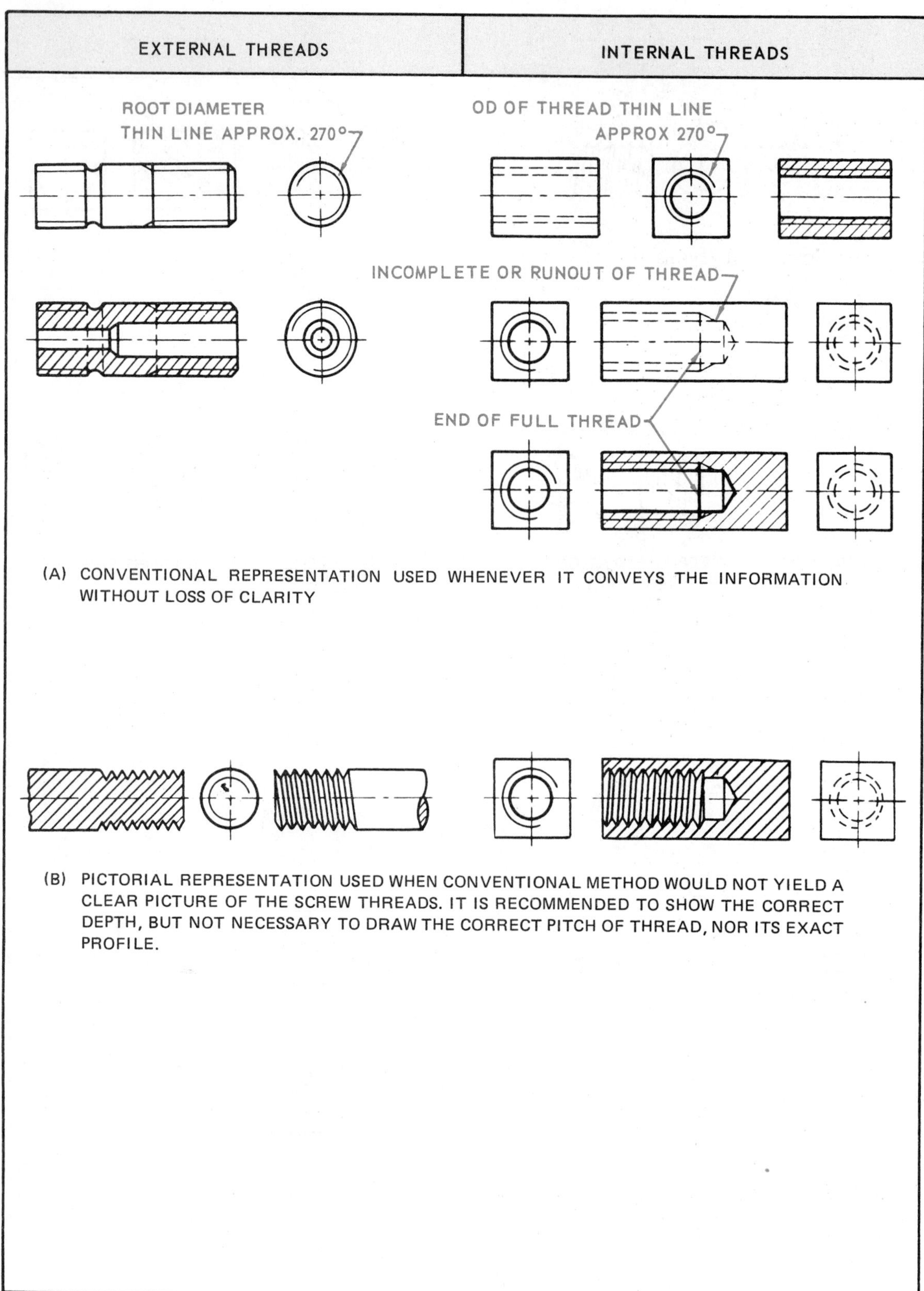
EXTERNAL THREADS
INTERNAL THREADS
ROOT DIAMETER
THIN LINE APPROX. 270°
OD OF THREAD THIN LINE
APPROX 270°
INCOMPLETE OR RUNOUT OF THREAD
END OF FULL THREAD
(A) CONVENTIONAL REPRESENTATION USED WHENEVER IT CONVEYS THE INFORMATION WITHOUT LOSS OF CLARITY
(B) PICTORIAL REPRESENTATION USED WHEN CONVENTIONAL METHOD WOULD NOT YIELD A CLEAR PICTURE OF THE SCREW THREADS. IT IS RECOMMENDED TO SHOW THE CORRECT DEPTH, BUT NOT NECESSARY TO DRAW THE CORRECT PITCH OF THREAD, NOR ITS EXACT PROFILE.

Fig. 15-3 Standard thread conventions (ISO)

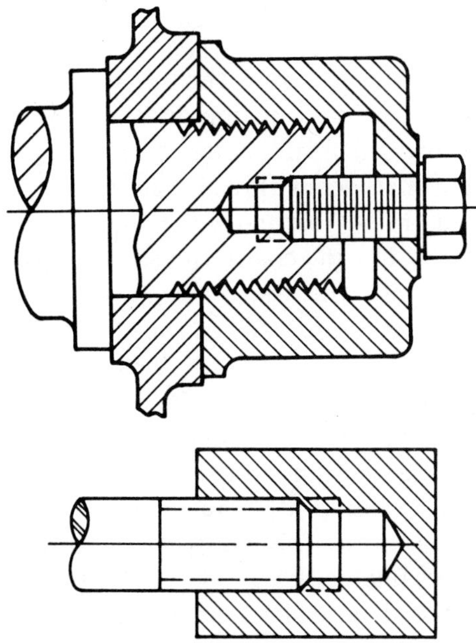

Fig. 15-4 Showing threads on assembly drawings

INCH THREADS

Until 1976, nearly all threaded assemblies in North America were designed using inch-sized threads. In this system the pitch is equal to the distance between corresponding points on adjacent threads and is expressed as:

$$\text{Pitch} = \frac{1}{\text{Number of threads per inch}}$$

The number of threads per inch is set for different diameters in a thread "series." For the Unified National system there are the coarse thread series and the fine thread series.

There is also an extra-fine thread series, UNEF, for use where a small pitch is wanted, such as on thin-walled tubing. For special work and diameters larger than those specified in the coarse and fine series, the Unified Thread system has three series that allow the same number of threads per inch regardless of the diameter. These are the 8-thread series, the 12-thread series, and the 16-thread series.

Three classes of external thread (Classes 1A, 2A, and 3A) and three classes of internal thread (Classes 1B, 2B, and 3B) are provided.

The general characteristics and uses of the classes are:

Classes 1A and 1B. These classes produce the loosest fit, that is, the most play in assembly. They are useful for work where ease of assembly and disassembly is essential, such as for some automotive work and for stove bolts and other rough bolts and nuts.

Classes 2A and 2B. These classes are designed for the ordinary good grade of commercial products including machine screws and fasteners and most interchangeable parts.

Classes 3A and 3B. These classes are intended for exceptionally high-grade commercial products needing a particularly close or snug fit. Classes 3A and 3B are used only when the high cost of precision tools and machines is warranted.

Thread Designation

Thread designation for both external and internal inch threads is expressed in this order: diameter (nominal or major diameter), number of threads per inch, thread form and series, and class of fit, figure 15-5. Although not recommended in ANSI standards, some industries, when specifying threads on a drawing, show only the nominal diameter and the number of threads.

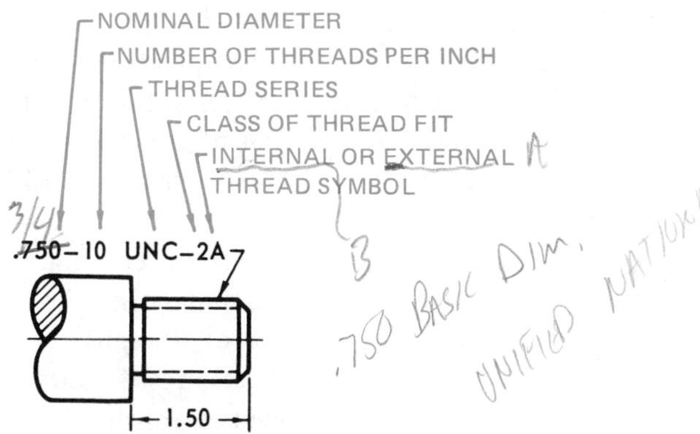

Fig. 15-5 Specifications for inch threads

RIGHT- AND LEFT-HANDED THREADS

Unless designated differently, threads are assumed to be right hand. A bolt being threaded into a tapped hole would be turned in a right-hand (clockwise) direction. For some special uses, such as turnbuckles, left-hand threads are required. When such a thread is necessary, the letters "LH" are added after the thread designation.

Typical hole and thread callouts are shown in figure 15-8.

METRIC THREADS

Metric threads are grouped into diameter-pitch combinations differentiated by the pitch applied to specified diameters. The pitch for metric threads is the distance between corresponding points on adjacent teeth. In addition to a coarse and fine pitch series, a series of constant pitches is available.

For each of the two main thread elements – pitch diameter and crest diameter – there are numerous tolerance grades. The number of the tolerance grade reflects the tolerance size. For example: Grade 4 tolerances are smaller than Grade 6 tolerances; Grade 8 tolerances are larger than Grade 6 tolerances.

In each case, Grade 6 tolerances should be used for medium quality length of engagement applications. The tolerance grades below Grade 6 are intended for applications involving fine quality and/or short lengths of engagement. Tolerance grades above Grade 6 are intended for coarse quality and/or long periods of engagement.

In addition to the tolerance grade, positional tolerance is required. The positional tolerance defines the maximum-material limits of the pitch and crest diameters of the external and internal threads and indicates their relationship to the basic profile.

In conformance with current coating (or plating) thickness requirements and the demand for ease of assembly, a series of tolerance positions reflecting the application of varying amounts of allowance has been established:

For External Threads:

Tolerance position "e" (large allowance)
Tolerance position "g" (small allowance)
Tolerance position "h" (no allowance)

For Internal Threads:

Tolerance position "G" (small allowance)
Tolerance position "H" (no allowance)

Thread Designation

ISO metric screw threads are defined by nominal size (basic major diameter) and pitch, both expressed in millimeters. An "M" specifying an ISO metric screw thread precedes the nominal size and an "X" separates the nominal size from the pitch. For the coarse thread series, the pitch is shown only when the dimension for the length of thread is required. When specifying the length of thread, an "X" is used to separate the length of thread from the rest of the designations. For external threads, the length of thread may be given as a dimension on the drawing.

For example, a 10 mm diameter, 1.25 pitch, fine thread series is expressed as M10 X 1.25. A 10 mm diameter, 1.5 pitch, coarse thread series is expressed as M10; the pitch need not be

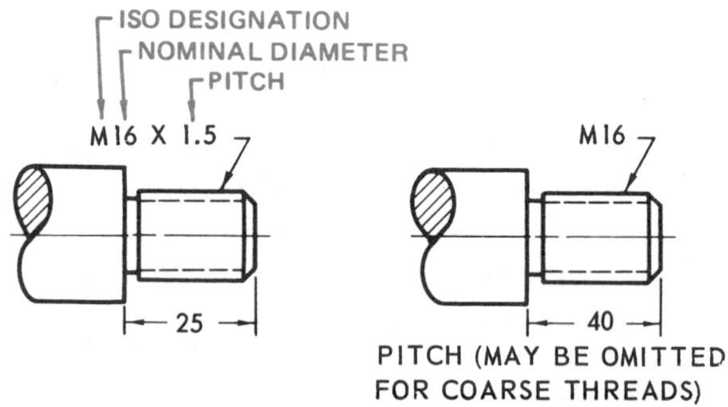

(A) BASIC THREAD CALLOUT

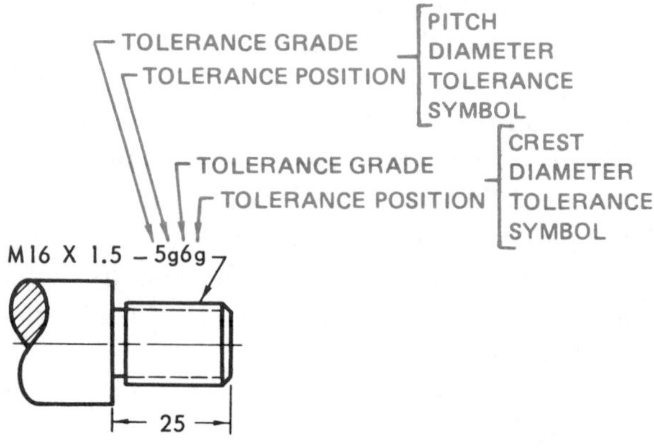

(B) ADDITIONAL THREAD CALLOUT

Fig. 15-6 Specifications for metric threads

shown unless the length of thread is required. If the latter thread was 25 mm long and this information was required on the drawing, the thread callout would be M10 × 1.5 × 25.

In addition to the basic designation, complete designation for an ISO metric screw thread includes a tolerance class identification. A dash separates the tolerance class designation from the basic designation and includes the symbol for the pitch diameter tolerance followed immediately by the symbol for crest diameter tolerance. Each of these symbols consists of a numeral indicating the grade tolerance followed by a letter indicating the tolerance position (a capital letter for internal threads and lowercase letter for external threads). Where the pitch and crest diameter symbols are identical, the symbol is necessary only once. Figure 15-6 illustrates the labeling of metric threads.

Figure 15-7 shows a comparison of metric threads and inch threads.

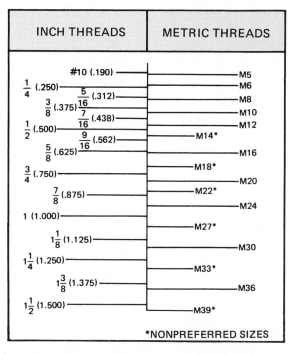

INCH THREADS	METRIC THREADS
#10 (.190)	M5
1/4 (.250)	M6
5/16 (.312)	M8
3/8 (.375)	M10
7/16 (.438)	M12
1/2 (.500)	M14*
9/16 (.562)	M16
5/8 (.625)	M18*
3/4 (.750)	M20
7/8 (.875)	M22*
1 (1.000)	M24
1 1/8 (1.125)	M27*
1 1/4 (1.250)	M30
1 3/8 (1.375)	M33*
1 1/2 (1.500)	M36
	M39*

*NONPREFERRED SIZES

Fig. 15-7 Size comparison of inch and metric threads

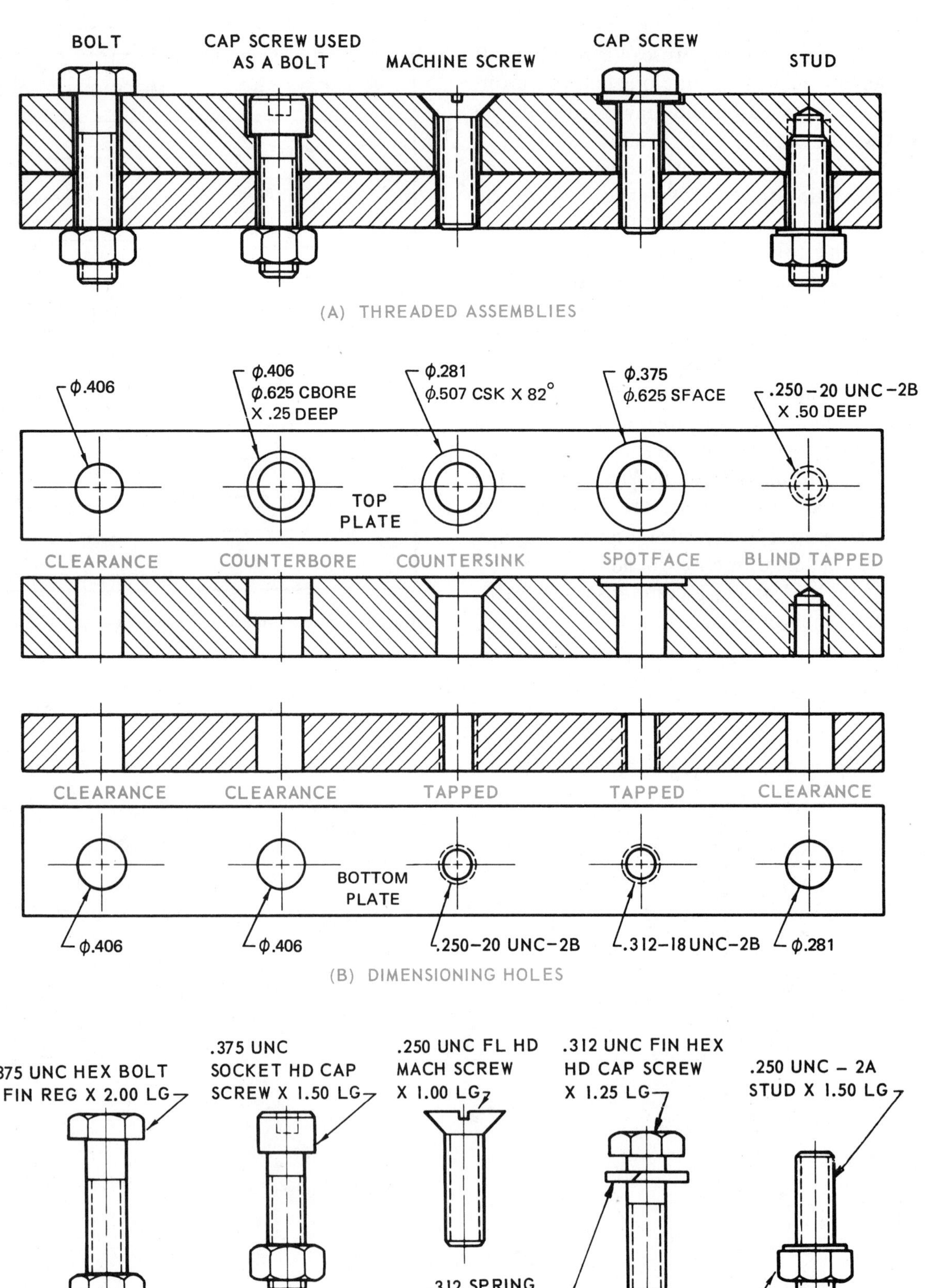

BOLT
CAP SCREW USED AS A BOLT
MACHINE SCREW
CAP SCREW
STUD
(A) THREADED ASSEMBLIES
ϕ.406
ϕ.406
ϕ.625 CBORE
X .25 DEEP
ϕ.281
ϕ.507 CSK X 82°
ϕ.375
ϕ.625 SFACE
.250 – 20 UNC –2B
X .50 DEEP
TOP
PLATE
CLEARANCE
COUNTERBORE
COUNTERSINK
SPOTFACE
BLIND TAPPED
CLEARANCE
CLEARANCE
TAPPED
TAPPED
CLEARANCE
BOTTOM
PLATE
ϕ.406
ϕ.406
.250–20 UNC–2B
.312–18UNC–2B
ϕ.281
(B) DIMENSIONING HOLES
.375 UNC HEX BOLT
FIN REG X 2.00 LG
.375 UNC
SOCKET HD CAP
SCREW X 1.50 LG
.250 UNC FL HD
MACH SCREW
X 1.00 LG
.312 UNC FIN HEX
HD CAP SCREW
X 1.25 LG
.250 UNC – 2A
STUD X 1.50 LG
.312 SPRING
LOCKWASHER
.250 UNC HEX NUT,
WASHER FACE
.375 UNC HEX NUT, REG
(C) DESCRIPTION OF FASTENERS

Fig. 15-8 Common threaded fasteners

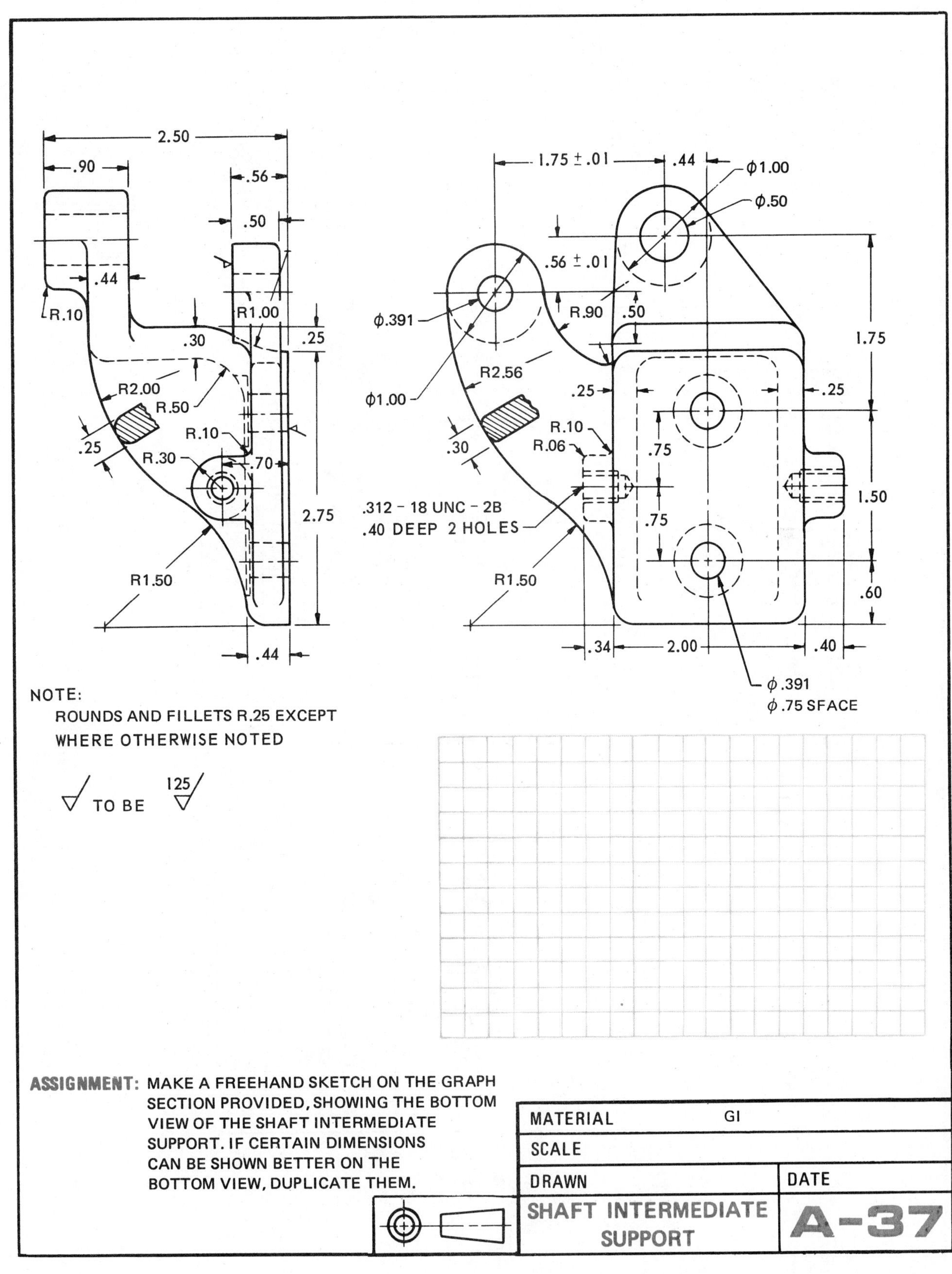
2.50
.90
.56
.50
.44
R.10
R1.00
.30
.25
R2.00
R.50
R.10
.25
R.30
.70
2.75
R1.50
.44
1.75 ± .01
.44
ϕ1.00
ϕ.50
.56 ± .01
ϕ.391
R.90
.50
1.75
R2.56
ϕ1.00
.25
.25
R.10
R.06
.30
.75
.75
1.50
.312 - 18 UNC - 2B
.40 DEEP 2 HOLES
R1.50
.60
.34
2.00
.40
ϕ.391
ϕ.75 SFACE
NOTE:
ROUNDS AND FILLETS R.25 EXCEPT
WHERE OTHERWISE NOTED
TO BE
125
ASSIGNMENT: MAKE A FREEHAND SKETCH ON THE GRAPH SECTION PROVIDED, SHOWING THE BOTTOM VIEW OF THE SHAFT INTERMEDIATE SUPPORT. IF CERTAIN DIMENSIONS CAN BE SHOWN BETTER ON THE BOTTOM VIEW, DUPLICATE THEM.
MATERIAL GI
SCALE
DRAWN
DATE
SHAFT INTERMEDIATE SUPPORT
A-37

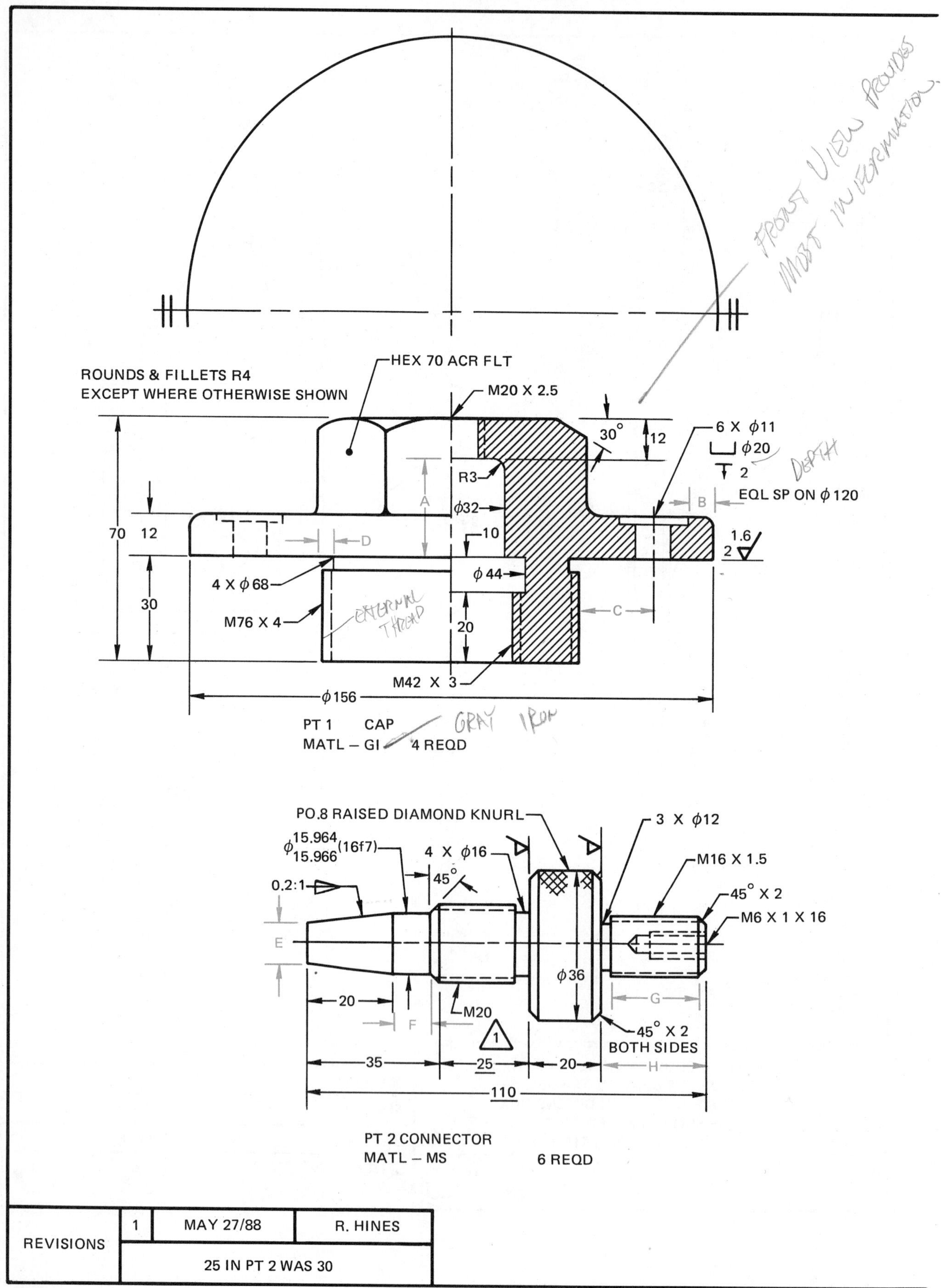

REVISIONS	1	MAY 27/88	R. HINES
	25 IN PT 2 WAS 30		

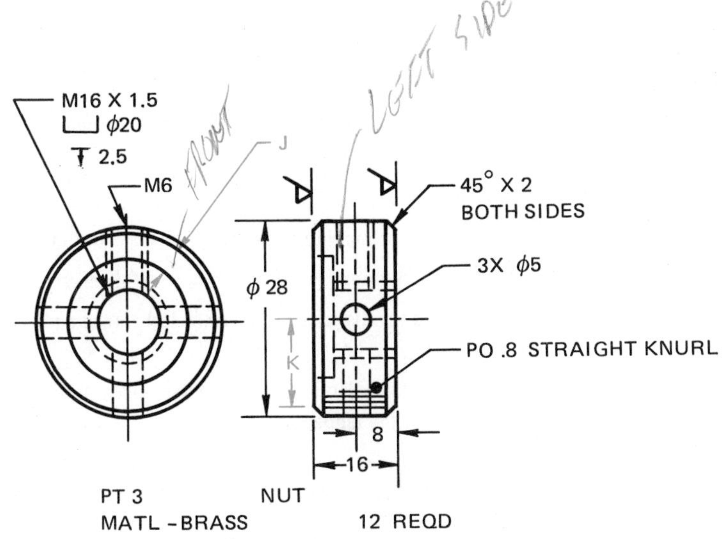

ANSWERS

1. A. ________
   B. ________
   C. ________
   D. ________
   E. ________
   F. ________
   G. ________
   H. ________
   J. ________
   K. ________
2. ___7___
3. ___1___
4. ___2___
5. ________
6. ________
7. ________
8. ________
9. ________
10. ________
11. ________
12. ________
13. ________
14. ________
15. ________
16. ________
17. ________
18. ________
19. ________

QUESTIONS

1. Calculate dimensions A to K.

Refer to Part 1

2. How many holes are in the part?
3. How many external threads are on the part?
4. How many internal threads are on the part?
5. What is the pitch for the M20 thread?
6. How many complete threads are on the M20 hole?
7. How many complete threads are on the M42 section?
8. What is the diameter of the spotface?
9. What is the angle between the ϕ11 holes?
10. How much extra metal was provided on the surface that required machining?

Refer to Part 2

11. What is the pitch on the internal thread?
12. What is the pitch on the M20 thread?
13. What provides for better gripping when rotating the part?
14. How many undercuts are shown?
15. How many chamfers are required?

Refer to Part 3

16. How many holes are in the part?
17. What is the depth of the counterbore?
18. How many full threads has the M16 hole?
19. What surface finish is required for the sides of the knurled portion?

SKETCHING ASSIGNMENT

Sketch the top half of Part 1 in the space provided.

NOTE: UNLESS OTHERWISE SHOWN:
– TOLERANCES ON DIMENSIONS ±0.5
– TOLERANCES ON ANGLES ±30′
SURFACES ✓ TO BE 3.2✓

METRIC

DIMENSIONS ARE IN MILLIMETERS

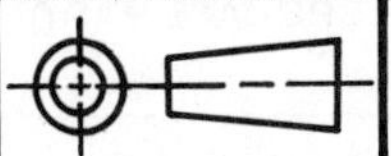

DRIVE SUPPORT DETAILS

A-38M

QUESTIONS

1. Calculate distances (A) to (J).
2. How many threaded holes or shafts are shown?
3. How many chamfers are shown?
4. How many necks are shown?
5. How many surfaces require finishing?

Refer to Part 1

6. What are the limit sizes for the Ø.64 dimension?
7. What operation provides better gripping when turning the clamping nut?
8. What is the tap drill size?
9. Is the thread right hand or left hand?
10. What is the depth of the threads?

Refer to Part 2

11. What is the length of thread?
12. Is the thread pitch fine or coarse?
13. What heat treatment does the part undergo?
14. How many threads are there in a one inch length?

Refer to Part 3

15. What is the maximum permissible width of the part?
16. What size thread is cut on the outside of the piece?
17. What is the distance between the last thread and the flange?
18. What is the tolerance on the center-to-center distance between the tapped holes?
19. What is the smallest diameter to which the hole through the stuffing box can be made?
20. What are the limits for the Ø2.000 dimension?

NOTE: EXCEPT WHERE NOTED –

TOLERANCE ON TWO-PLACE DIMENSIONS ±.02

TOLERANCE ON THREE-PLACE DIMENSIONS ±.005

TOLERANCE ON ANGLES ±0.5°

32/ EXCEPT WHERE NOTED

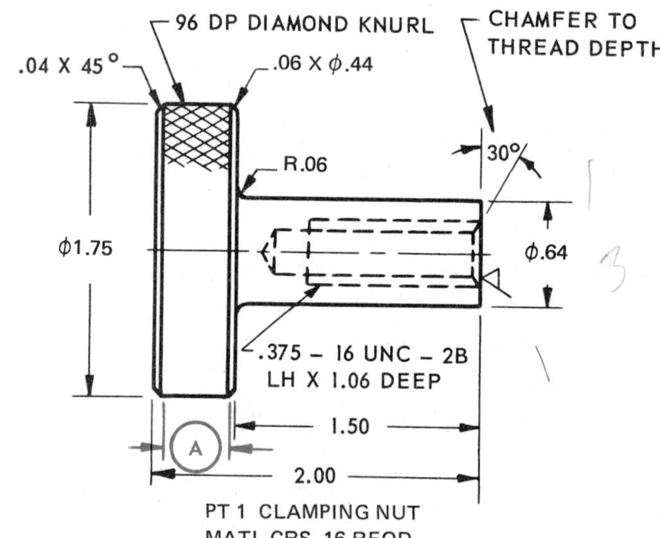

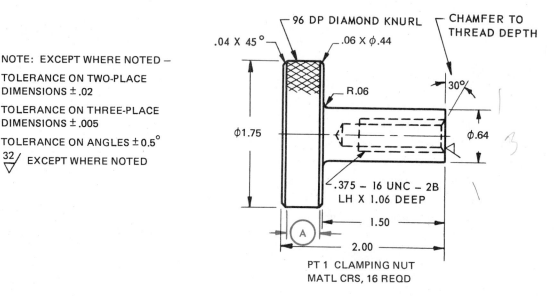

PT 1 CLAMPING NUT
MATL CRS, 16 REQD

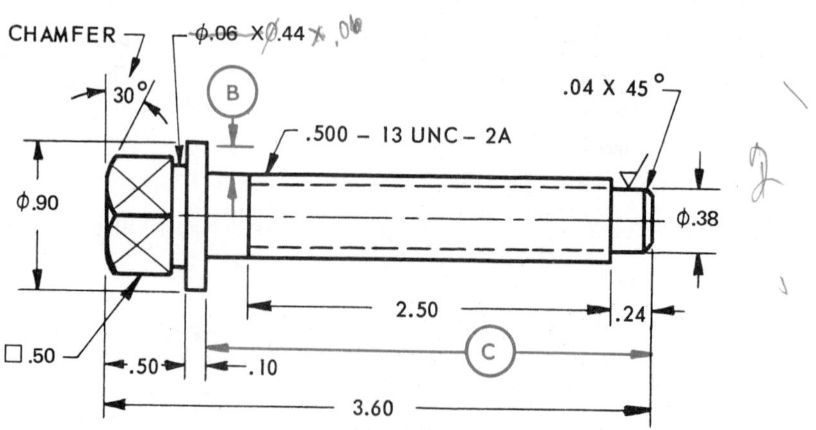

PT 2 ADJUSTING SCREW
MATERIAL CRS CASE HARDEN, 8 REQD

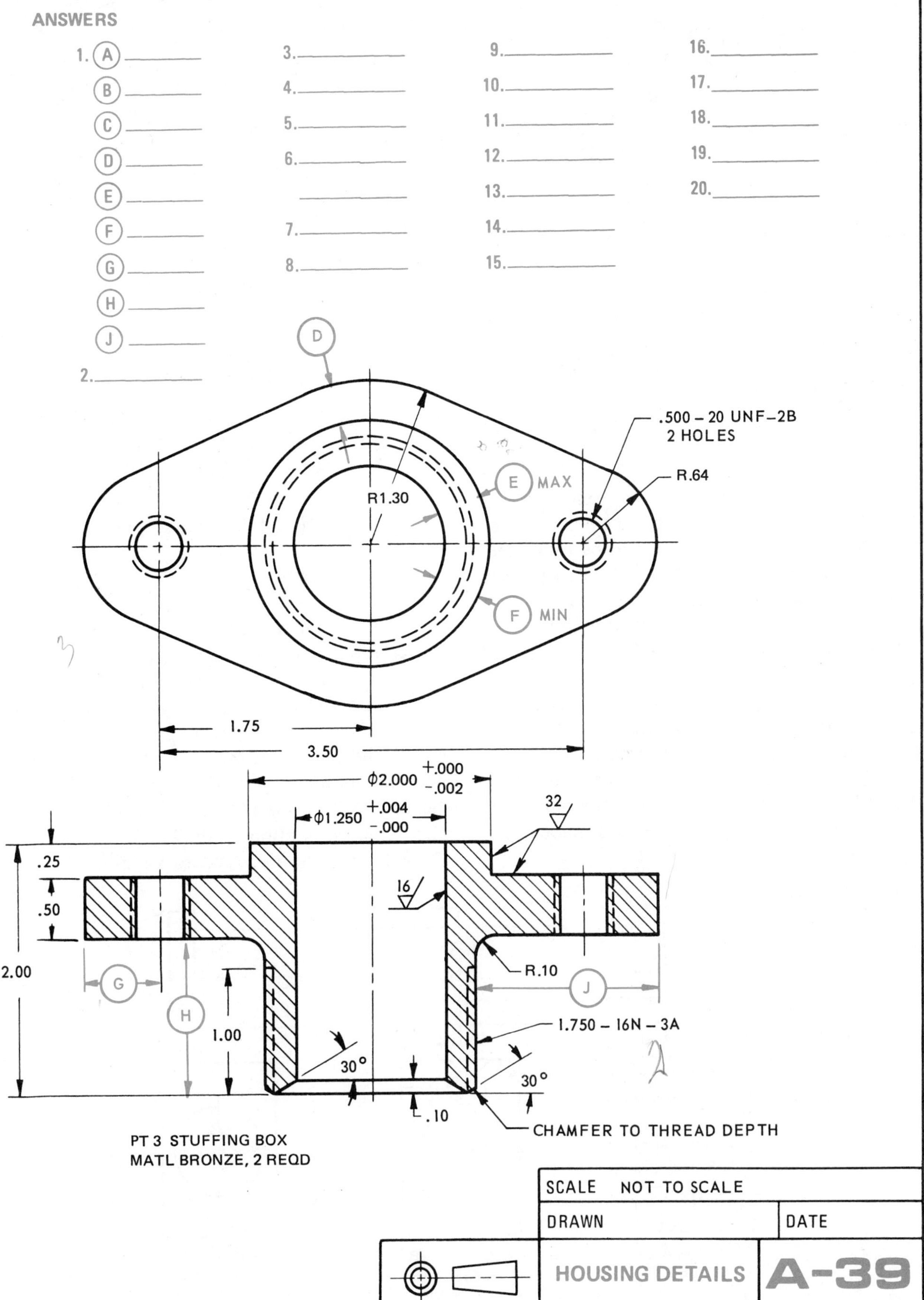

ANSWERS
1. (A)
(B)
(C)
(D)
(E)
(F)
(G)
(H)
(J)
2.
3.
4.
5.
6.
7.
8.
9.
10.
11.
12.
13.
14.
15.
16.
17.
18.
19.
20.
D
.500 – 20 UNF–2B
2 HOLES
R.64
R1.30
E MAX
F MIN
1.75
3.50
ϕ2.000 +.000 –.002
ϕ1.250 +.004 –.000
32
.25
.50
16
2.00
G
H
R.10
J
1.00
1.750 – 16N – 3A
30°
30°
.10
CHAMFER TO THREAD DEPTH
PT 3 STUFFING BOX
MATL BRONZE, 2 REQD
SCALE NOT TO SCALE
DRAWN
DATE
HOUSING DETAILS
A-39

UNIT 16

REVOLVED AND REMOVED SECTIONS

Revolved and removed sections are used to show the cross-sectional shape of ribs, spokes, or arms, when the shape is not obvious in the regular views. End views are often not needed when a revolved section is used.

Revolved Sections

For a *revolved section* a center line is drawn through the shape on the plane to be described, the part is imagined to be rotated 90 degrees, and the view that would be seen when rotated is superimposed on the view. If the revolved section does not interfere with the view on which it is revolved, then the view is not broken unless it would facilitate clearer dimensioning. When the revolved section interferes or passes through lines on the view on which it is revolved, the view is usually broken. Often the break is used to shorten the length of the object. When superimposed on the view, the outline of the revolved section is a thin continuous line, figure 16-1.

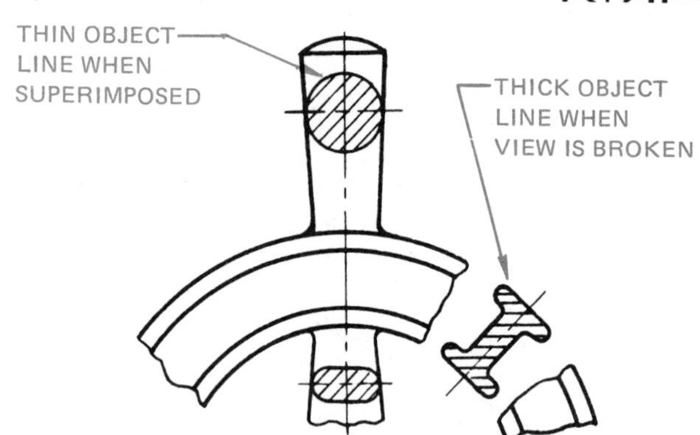

Fig. 16-1 Revolved section

Removed Sections

The *removed section* differs from the revolved section in that the section is removed to an open area on the drawing instead of being drawn directly on the view. Whenever practical, sectional views should be projected perpendicular to the cutting plane and be placed in the normal position for third-angle projection, figures 16-2, 16-3, 16-4 and 16-5. Frequently, the removed section is drawn to an enlarged scale for clarification and easier dimensioning.

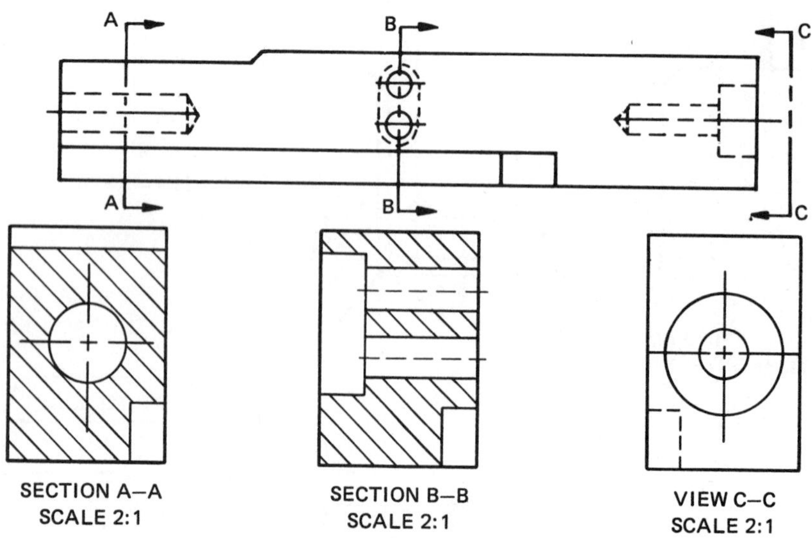

Fig. 16-2 Removed sections and removed view

Removed sections of symmetrical parts are placed on the extension of the center line where possible.

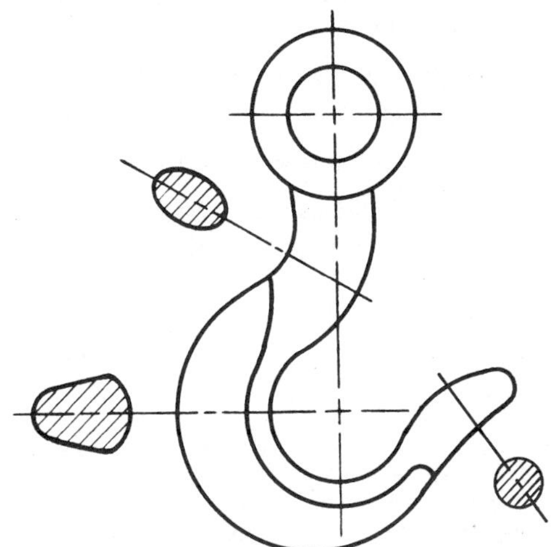

Fig. 16-3 Removed section of crane hook

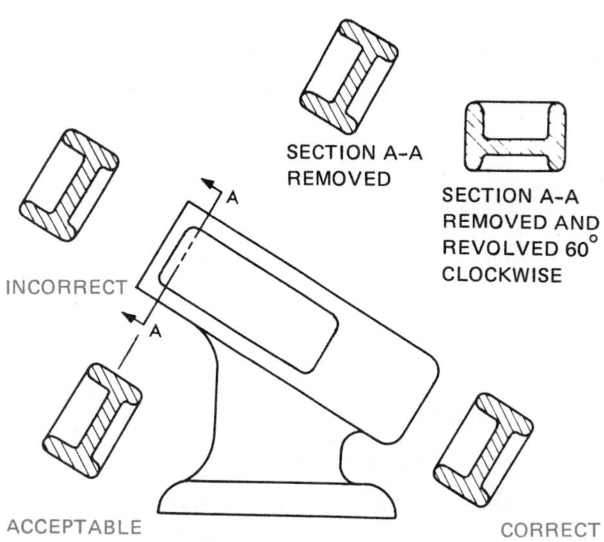

Fig. 16-4 Placement of sectional views

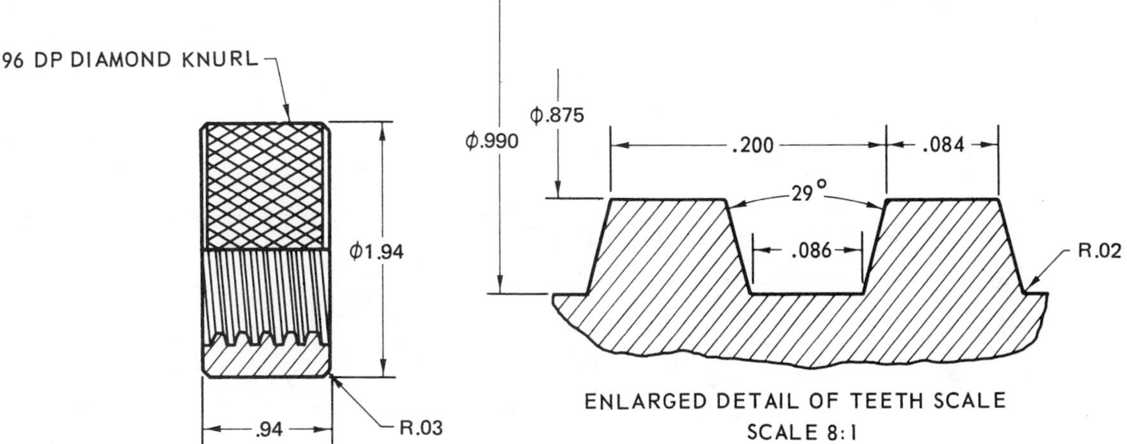

Fig. 16-5 Removed section of nut

QUESTIONS

1. Calculate distances (A) to (K) .
2. How many surfaces require finishing?
3. What tolerance is permitted on dimensions which do not specify a certain tolerance?
4. What type of section view is used on part 1?
5. What type of section view is used on part 2?
6. How many holes are there?

Refer to Part 1

7. What is the maximum center-to-center distance between the holes?
8. Which surface finish is required?
9. Using the smallest permissible hole size as the basic size, replace the limit dimensions for the larger hole with plus-and-minus tolerances.
10. What is the maximum permissible wall thickness at the larger hole?
11. What would be the minimum permissible wall thickness at the smaller hole?

Refer to Part 2

12. Name two machines that could produce the type of finish required for the slot.
13. What is the nominal size machine screw used in the counterbored holes?
14. What type of machine screw would be used?
15. What is the distance between the center of the counterbored holes and the center of the slot?

EXCEPT WHERE STATED OTHERWISE:
TOLERANCES ON
TWO-DECIMAL DIMENSIONS ±.02
THREE-DECIMAL DIMENSIONS ±.005
ROUNDS AND FILLETS R.10
MACHINE FINISH 32

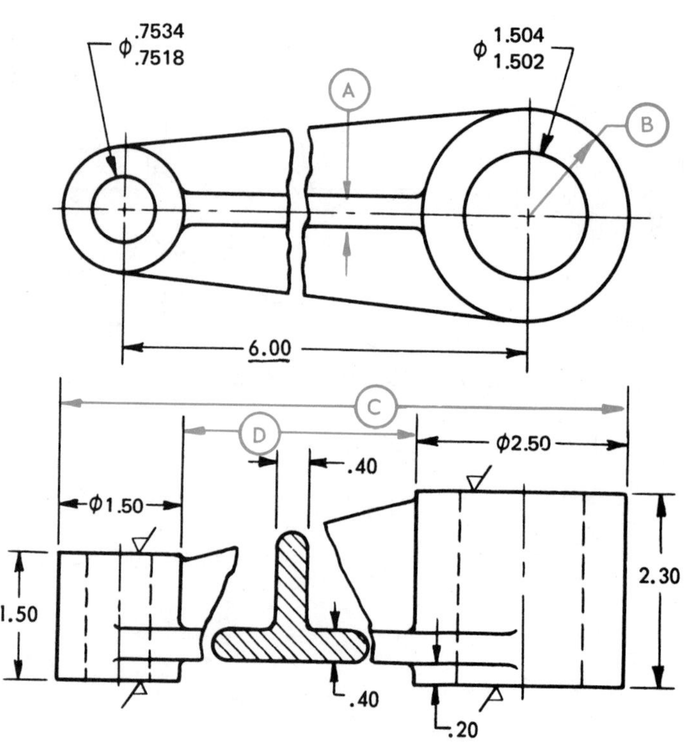

PT 1 SHAFT SUPPORT
MATL – MI

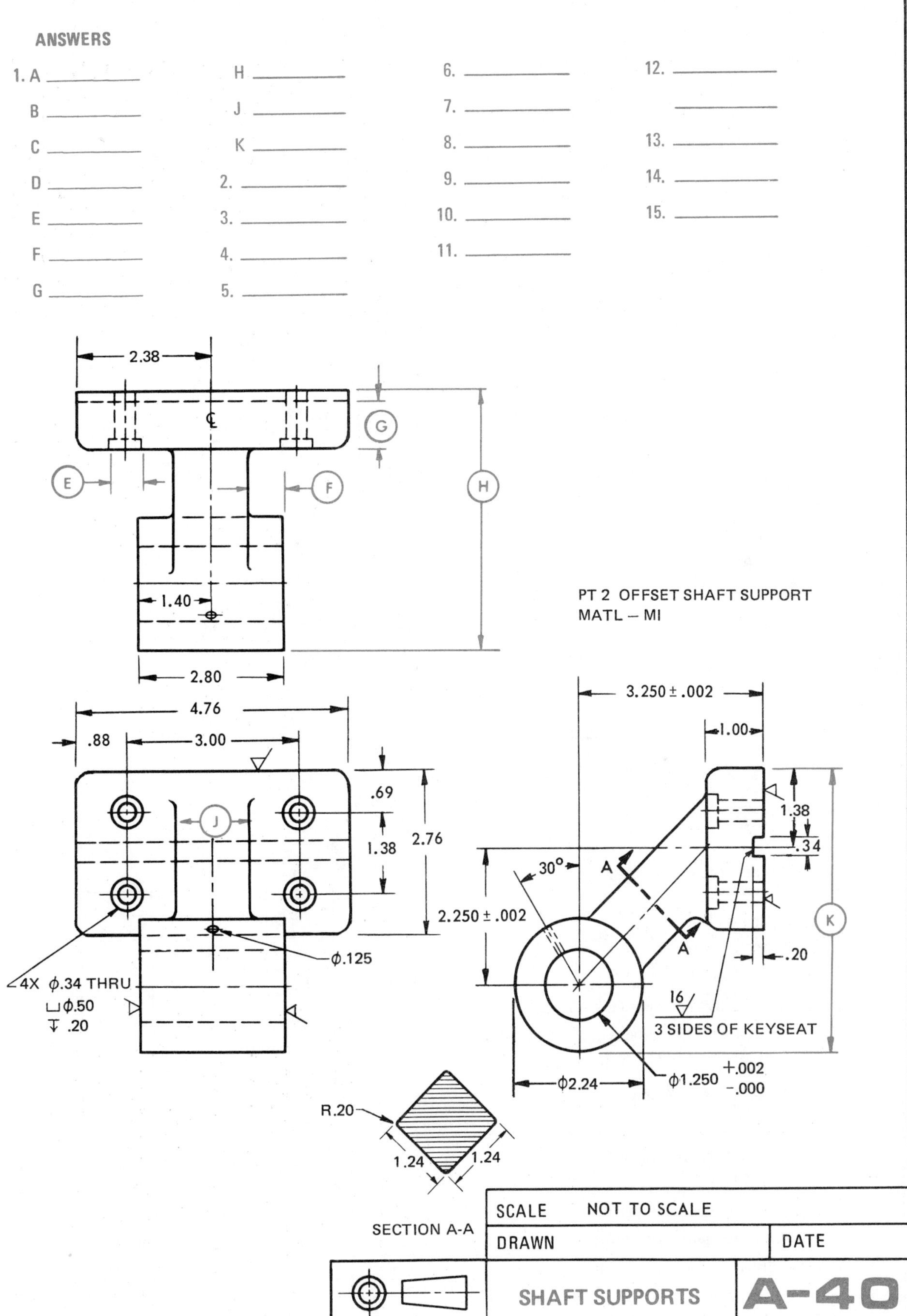
ANSWERS
1. A
B
C
D
E
F
G
H
J
K
2.
3.
4.
5.
6.
7.
8.
9.
10.
11.
12.
13.
14.
15.
2.38
℄
G
E
F
H
1.40
2.80
4.76
.88
3.00
.69
1.38
2.76
J
4X ϕ.34 THRU
⌴ϕ.50
↧ .20
ϕ.125
PT 2 OFFSET SHAFT SUPPORT
MATL – MI
3.250 ± .002
1.00
1.38
.34
30°
A
A
2.250 ± .002
K
.20
16
3 SIDES OF KEYSEAT
ϕ2.24
ϕ1.250 +.002 -.000
R.20
1.24
1.24
SECTION A-A
SCALE NOT TO SCALE
DRAWN
DATE
SHAFT SUPPORTS
A-40

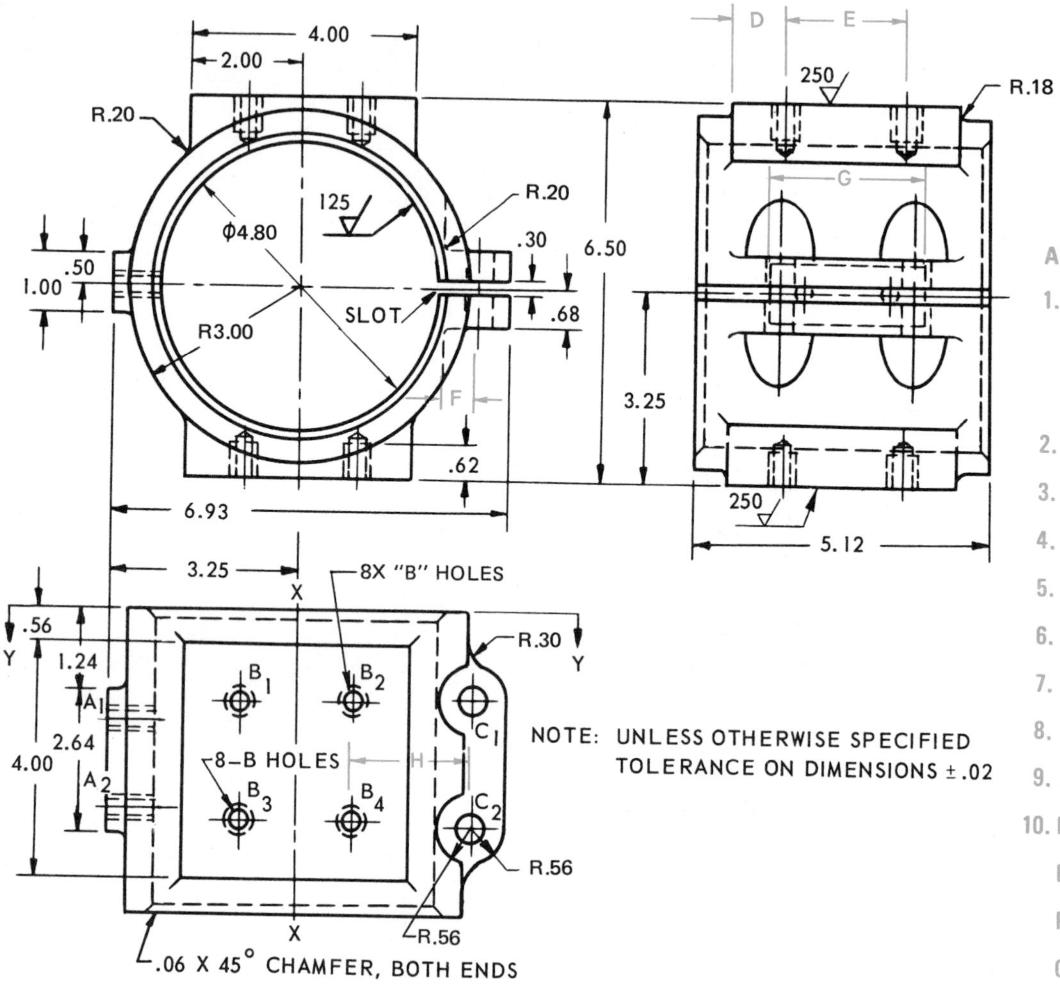

8X "B" HOLES

.06 X 45° CHAMFER, BOTH ENDS

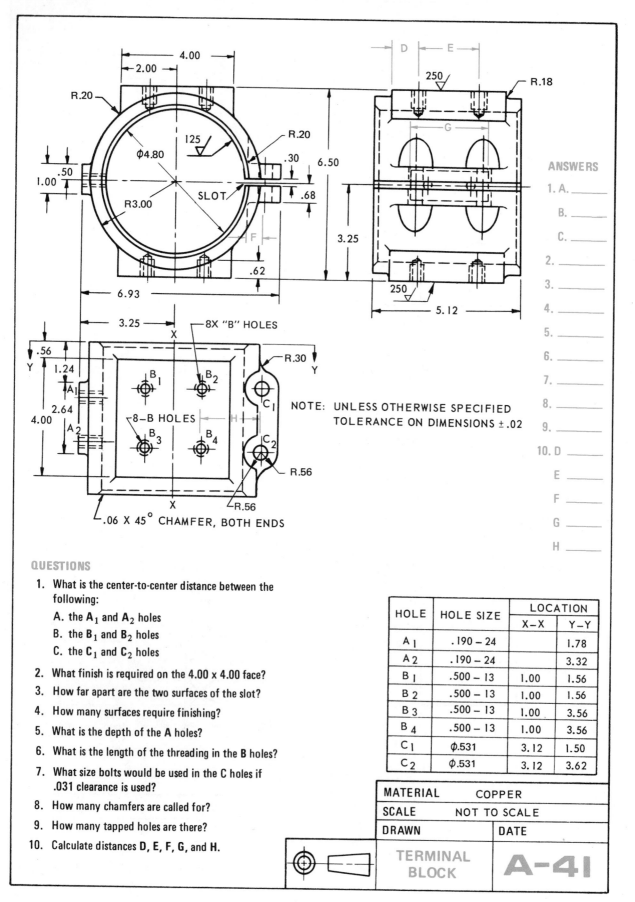

NOTE: UNLESS OTHERWISE SPECIFIED TOLERANCE ON DIMENSIONS ±.02

ANSWERS

1. A. ______
   B. ______
   C. ______
2. ______
3. ______
4. ______
5. ______
6. ______
7. ______
8. ______
9. ______
10. D ______
    E ______
    F ______
    G ______
    H ______

QUESTIONS

1. What is the center-to-center distance between the following:
   A. the A_1 and A_2 holes
   B. the B_1 and B_2 holes
   C. the C_1 and C_2 holes
2. What finish is required on the 4.00 x 4.00 face?
3. How far apart are the two surfaces of the slot?
4. How many surfaces require finishing?
5. What is the depth of the A holes?
6. What is the length of the threading in the B holes?
7. What size bolts would be used in the C holes if .031 clearance is used?
8. How many chamfers are called for?
9. How many tapped holes are there?
10. Calculate distances D, E, F, G, and H.

HOLE	HOLE SIZE	LOCATION	
		X–X	Y–Y
A_1	.190 – 24		1.78
A_2	.190 – 24		3.32
B_1	.500 – 13	1.00	1.56
B_2	.500 – 13	1.00	1.56
B_3	.500 – 13	1.00	3.56
B_4	.500 – 13	1.00	3.56
C_1	ϕ.531	3.12	1.50
C_2	ϕ.531	3.12	3.62

MATERIAL	COPPER
SCALE	NOT TO SCALE
DRAWN	DATE

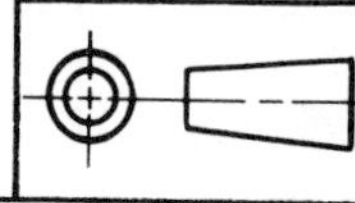

TERMINAL BLOCK

A-41

UNIT 17

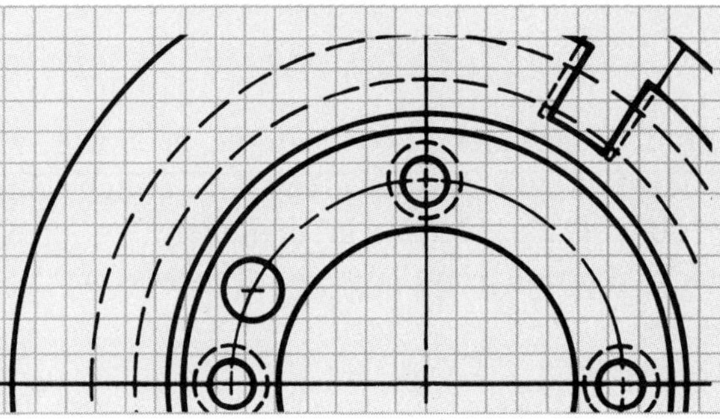

KEYS

A *key* is a piece of metal lying partly in a groove in the shaft, and extending into another groove in the hub, figure 17-1. These grooves are called *keyseats.* Keys are used to secure gears, pulleys, cranks, handles, and similar machine parts to shafts, so that the motion of the part is transmitted to the shaft or the motion of the shaft to the part, without slippage. The key is also a safety feature. Because of its size, when overloading occurs, the key slows or breaks before the part or shaft breaks.

Common key types are square, flat and Woodruff. Tables in the appendix give standard square and flat key sizes recommended for various shaft diameters and the necessary dimensions for Woodruff keys.

The *Woodruff key* is semicircular and fits into a semicircular keyseat in the shaft and a rectangular keyseat in the hub. Woodruff keys currently are available only in inch sizes, and are identified by a three- or four-digit key number. The last two numbers give the nominal diameter in eighths of an inch. The digit or digits preceding the last two digits gives the key width in thirty-

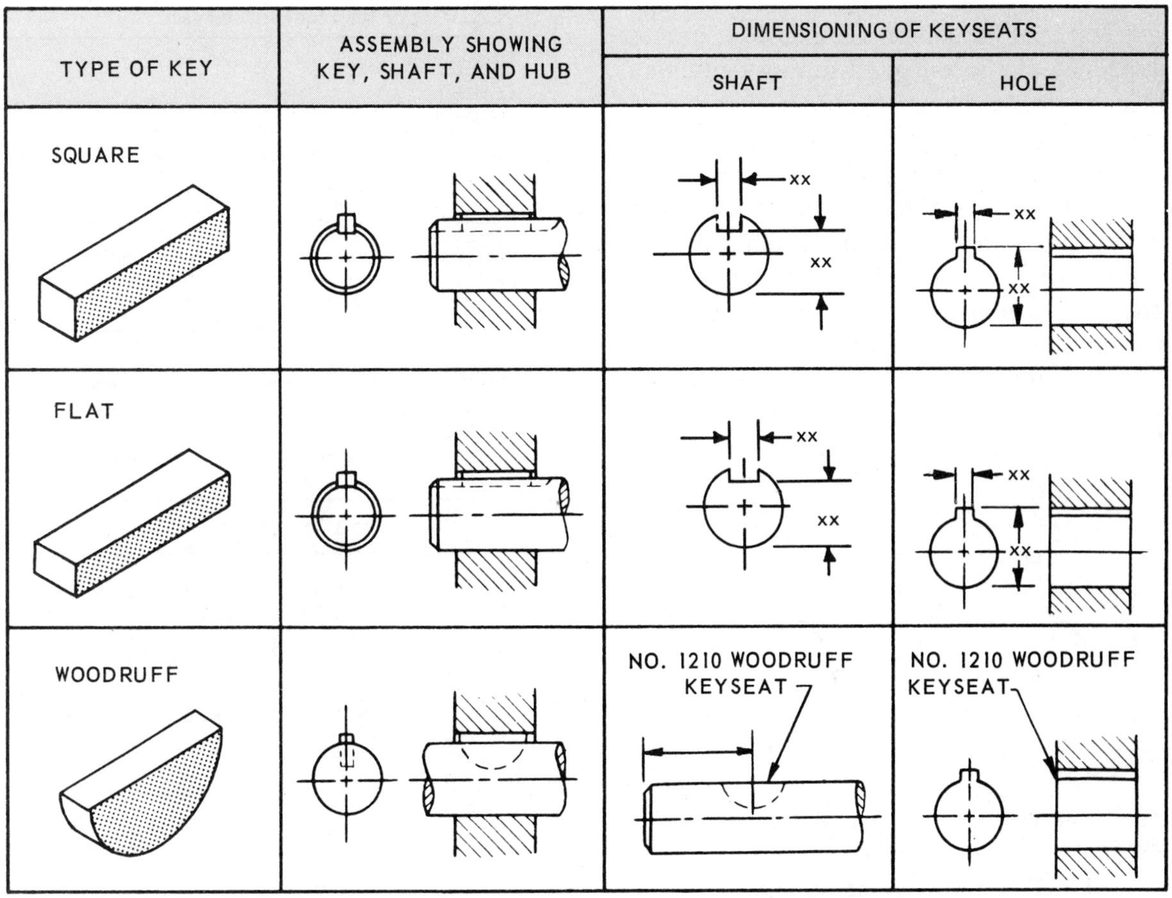

TYPE OF KEY	ASSEMBLY SHOWING KEY, SHAFT, AND HUB	DIMENSIONING OF KEYSEATS	
		SHAFT	HOLE
SQUARE		XX XX	XX XX
FLAT		XX XX	XX XX
WOODRUFF		NO. 1210 WOODRUFF KEYSEAT	NO. 1210 WOODRUFF KEYSEAT

Fig. 17-1 Common keys

second of an inch. For example, a No. 1210 Woodruff key indicates a key 12/32 inch wide by 10/8 (1.25) inches in diameter. Woodruff keys are currently available in inch sizes only.

Dimensioning of Keyseats

Keyseat dimensions are usually given in limit dimensions to ensure proper fits and are located from the opposite side of the hole or shaft, figure 17-2. Alternatively, for unit production, where the machinist is expected to fit the key into the keyseat, a leader pointing to the keyseat, specifying first the width and then the depth of the keyseat, may be used. See Part 2, Rack Details, drawing A-42.

SETSCREWS

Setscrews are used as semipermanent fasteners to hold a collar, sheave, or gear on a shaft against rotational or translational forces. In contrast to most fastening devices, the setscrew is essentially a compression device. Forces developed by the screw point during tightening produce a strong clamping action that resists relative motion between assembled parts. The basic problem in setscrew selection is finding the best combination of setscrew form, size, and point style providing the required holding power.

Setscrews are categorized by their forms and the desired point style, figure 17-3. Each of the standardized setscrew forms is available in a variety of point styles. Selection of a specific form or point is influenced by functional as well as other considerations.

The selection of the type of driver and thus, the setscrew form, is usually determined by factors other than tightening. Despite higher tightening ability, the protrusion of the square head is a major disadvantage. Compactness, weight saving, safety, and appearance may dictate the use of flush-seating socket or slotted headless forms.

The conventional approach to setscrew selection is usually based on a rule-of-thumb

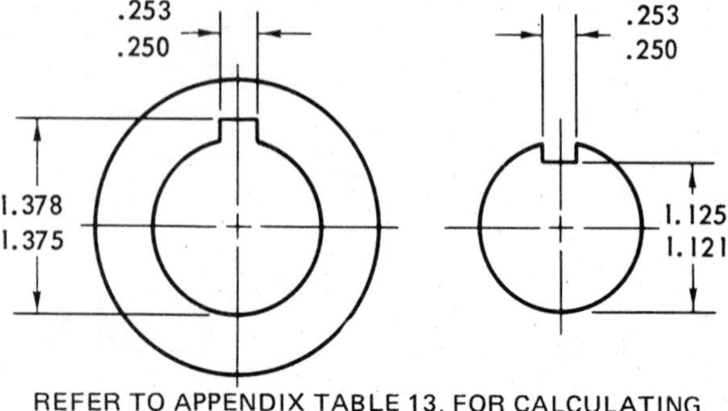

Fig. 17-2 Dimensioning keyseats

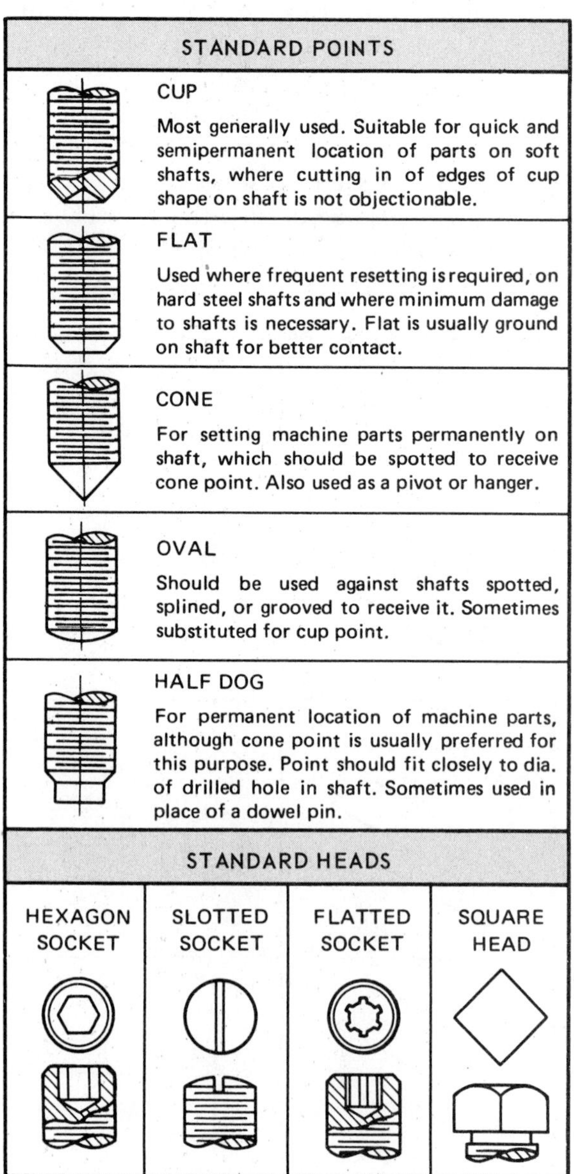

Fig. 17-3 Setscrews

procedure: the setscrew diameter should be roughly equal to one-half the shaft diameter. This rule of thumb often gives satisfactory results, but it has a limited range of usefulness. When a setscrew and key are used together, the screw diameter should be equal to the width of the key.

FLATS

A *flat* is a slight depression usually cut on a shaft to serve as a surface on which the end of a setscrew can rest when holding an object in place, figure 17-4.

BOSSES AND PADS

A *boss* is a relatively small cylindrical projection above the surface of an object. A *pad* is a slight, noncircular projection above the surface of an object, figure 17-5. Bosses and pads are used to provide additional clearance or strength in the area where they are used and allow machining over a smaller area.

RECTANGULAR COORDINATE DIMENSIONING WITHOUT DIMENSION LINES

To avoid having many dimensions extending away from the object, *rectangular coordinate dimensioning* without dimension lines may be used, as shown in figure 17-6. In this system, the "zero" lines are used as reference lines and each of the dimensions shown without arrowheads indicates the distance from the zero line. There is never more than one zero line in each direction.

This type of dimensioning is particularly useful when such features are produced on a general-purpose machine, such as a jig borer, a tape-controlled drill, or a turret-type press. Drawings for numerical control are covered in Unit 24.

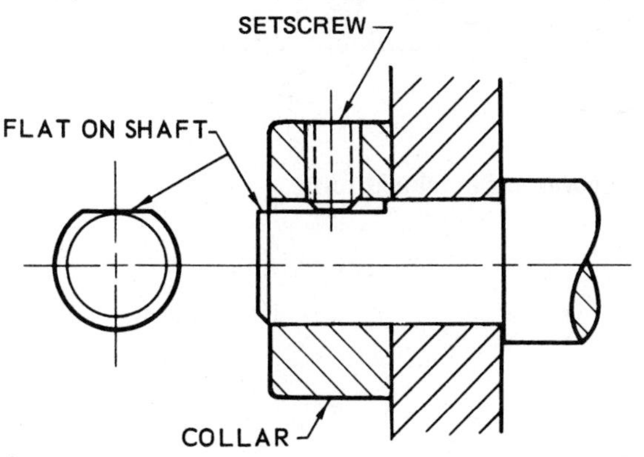

Fig. 17-4 Flat and setscrew application

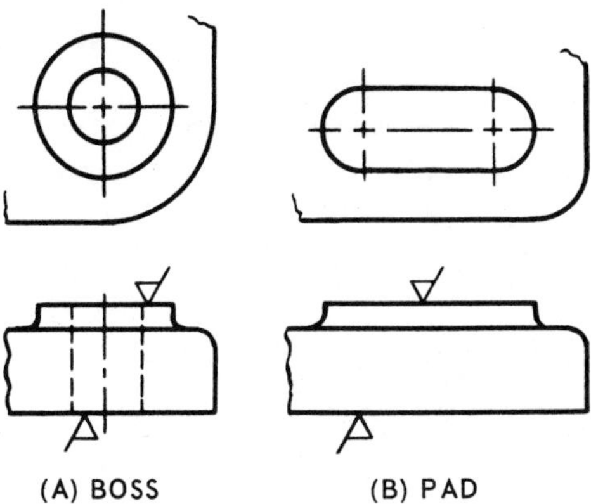

Fig. 17-5 Bosses and pads

RECTANGULAR COORDINATE DIMENSIONING IN TABULAR FORM

When there are many holes or repetitive features, such as in a chassis or a printed circuit board, and where the multitude of center lines would make a drawing difficult to read, rectangular coordinate dimensioning in tabular form is recommended. In this system each hole or feature is assigned a letter or a letter and numeral subscript. The feature dimensions and the feature location along the X, Y and Z axes are given in a table as shown in figure 17-7.

REFERENCES AND SOURCE MATERIALS

1. ANSI Y14.5M, "Dimensioning and Tolerancing."

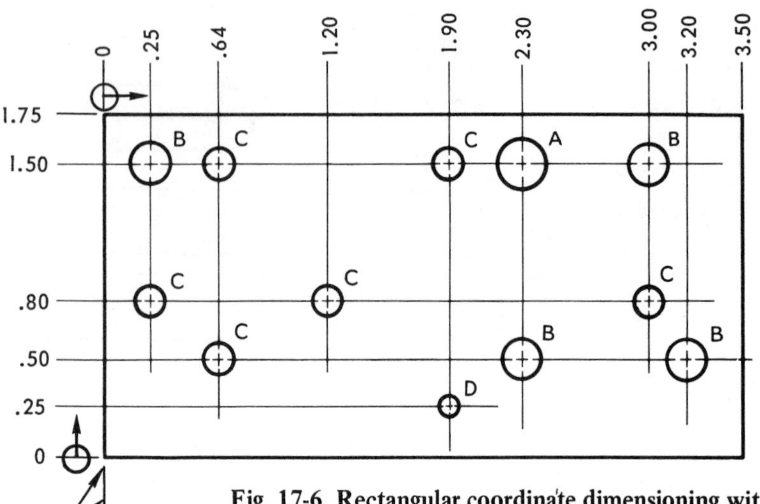

HOLE SYMBOL	HOLE ϕ
A	.250
B	.188
C	.156
D	.125

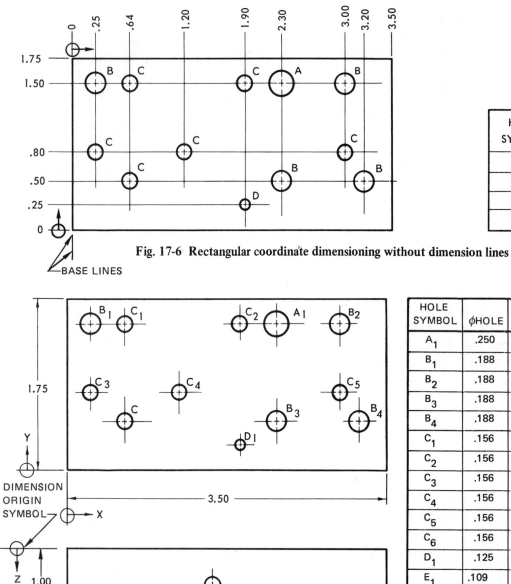

Fig. 17-6 Rectangular coordinate dimensioning without dimension lines

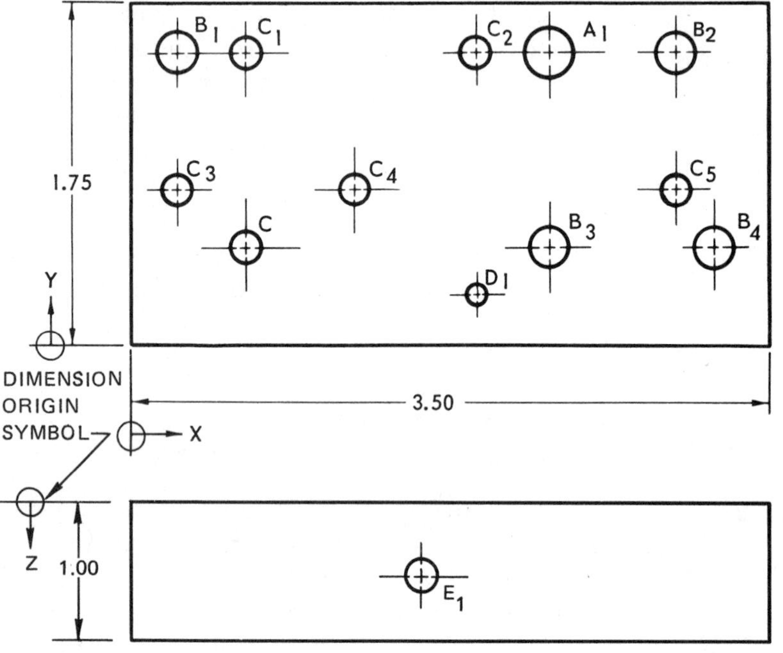

HOLE SYMBOL	ϕHOLE	LOCATION X	Y	Z
A_1	.250	2.30	1.50	.62
B_1	.188	.25	1.50	THRU
B_2	.188	3.00	1.50	THRU
B_3	.188	2.30	.50	THRU
B_4	.188	3.20	.50	THRU
C_1	.156	.64	1.50	THRU
C_2	.156	1.90	1.50	THRU
C_3	.156	.25	.80	THRU
C_4	.156	1.20	.80	THRU
C_5	.156	3.00	.80	THRU
C_6	.156	.64	.50	THRU
D_1	.125	1.90	.25	.50
E_1	.109	1.75	1.00	.75

Fig. 17-7 Rectangular coordinate dimensioning in tabular form

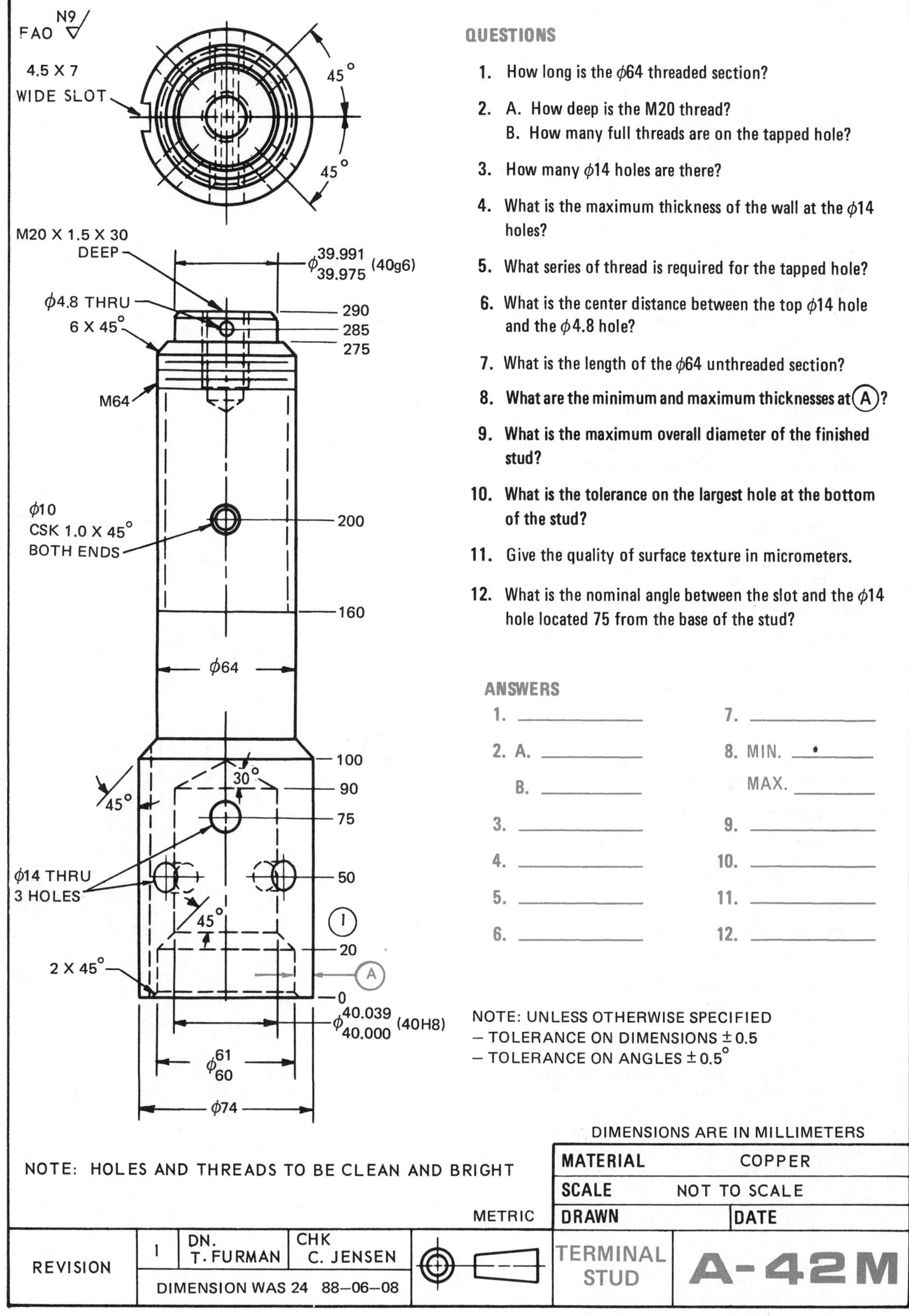

QUESTIONS

1. How long is the $\phi 64$ threaded section?
2. A. How deep is the M20 thread?
   B. How many full threads are on the tapped hole?
3. How many $\phi 14$ holes are there?
4. What is the maximum thickness of the wall at the $\phi 14$ holes?
5. What series of thread is required for the tapped hole?
6. What is the center distance between the top $\phi 14$ hole and the $\phi 4.8$ hole?
7. What is the length of the $\phi 64$ unthreaded section?
8. What are the minimum and maximum thicknesses at (A)?
9. What is the maximum overall diameter of the finished stud?
10. What is the tolerance on the largest hole at the bottom of the stud?
11. Give the quality of surface texture in micrometers.
12. What is the nominal angle between the slot and the $\phi 14$ hole located 75 from the base of the stud?

ANSWERS

1. ______
2. A. ______
   B. ______
3. ______
4. ______
5. ______
6. ______
7. ______
8. MIN. ______
   MAX. ______
9. ______
10. ______
11. ______
12. ______

QUESTIONS

Refer to Part 1

1. What was the original length of the shaft?
2. What symbol is used to indicate that the 1.90 dimension is not to scale?
3. At how many places are threads being cut?
4. Specify for any left-hand threads the thread diameter and the number of threads per inch.
5. What is the pitch for the 1.000 thread?
6. What is the length of that portion of the shaft which includes the (A) .875 thread (include chamfer), (B) 1.250 thread (do not include undercut), and (C) 1.000 thread (include chamfer)?
7. What distance is there between the last thread and the shoulder of the Ø.875 portion of the shaft?
8. How many dimensions have been changed?
9. What type of section view is used?
10. What is the largest size to which the Ø1.250 shaft can be turned?
11. What is the minimum permissible size for dimension A?

Refer to Part 2

12. How many holes are there?
13. What are the overall width and depth dimensions of the base?
14. How many surfaces are to be finished?
15. What is the diameter of the bosses on the base?
16. How wide is the pad on the upright column?
17. What is the depth of the keyseat?
18. How far does the horizontal hole overlap the vertical hole? (Use maximum sizes of holes and minimum center-to-center distances.)
19. How much material was added when the change to the bosses was made?
20. Which scale was used on this drawing?
21. What is the maximum permissible center-to-center distance of the two large holes?
22. If limit dimensions were to replace the keyseat dimensions shown, what would they be? Refer to Table 13 of the Appendix.

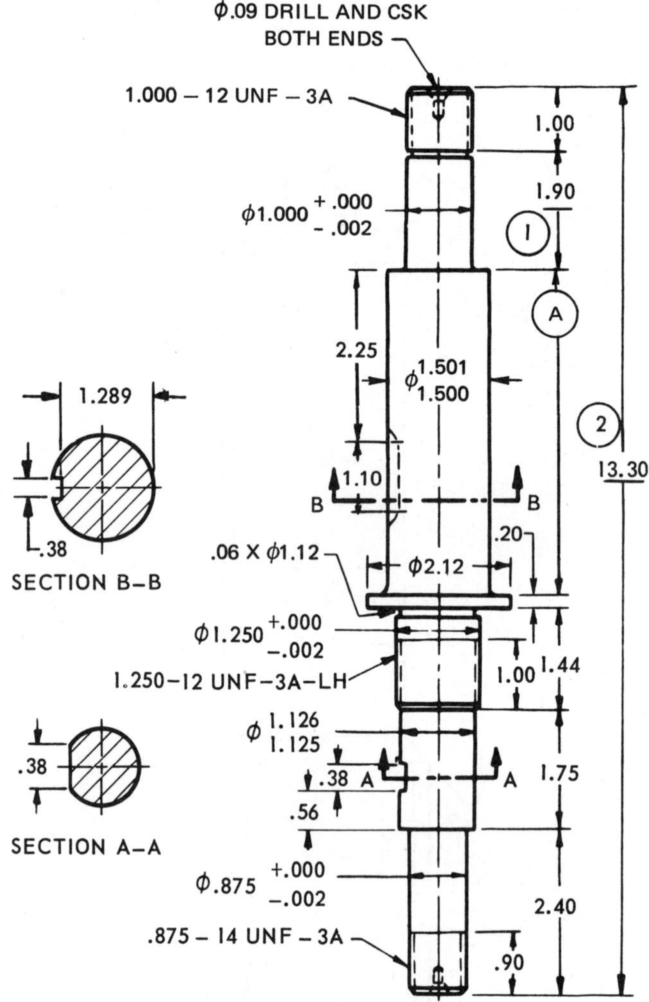

CHAMFER STARTING END OF ALL THREADS 45° TO THREAD DEPTH

NOTE: ALL FILLETS R.10

PT 1 SPINDLE SHAFT, SCALE 1:2, MATL – CRS
2 REQD

REVISIONS	1	88–01–12	R.H.	2	88–08–12	C.J.	3	88–12–15	G.H.
	1.90 WAS 2.00			13.30 WAS 13.40			.80 WAS .75		

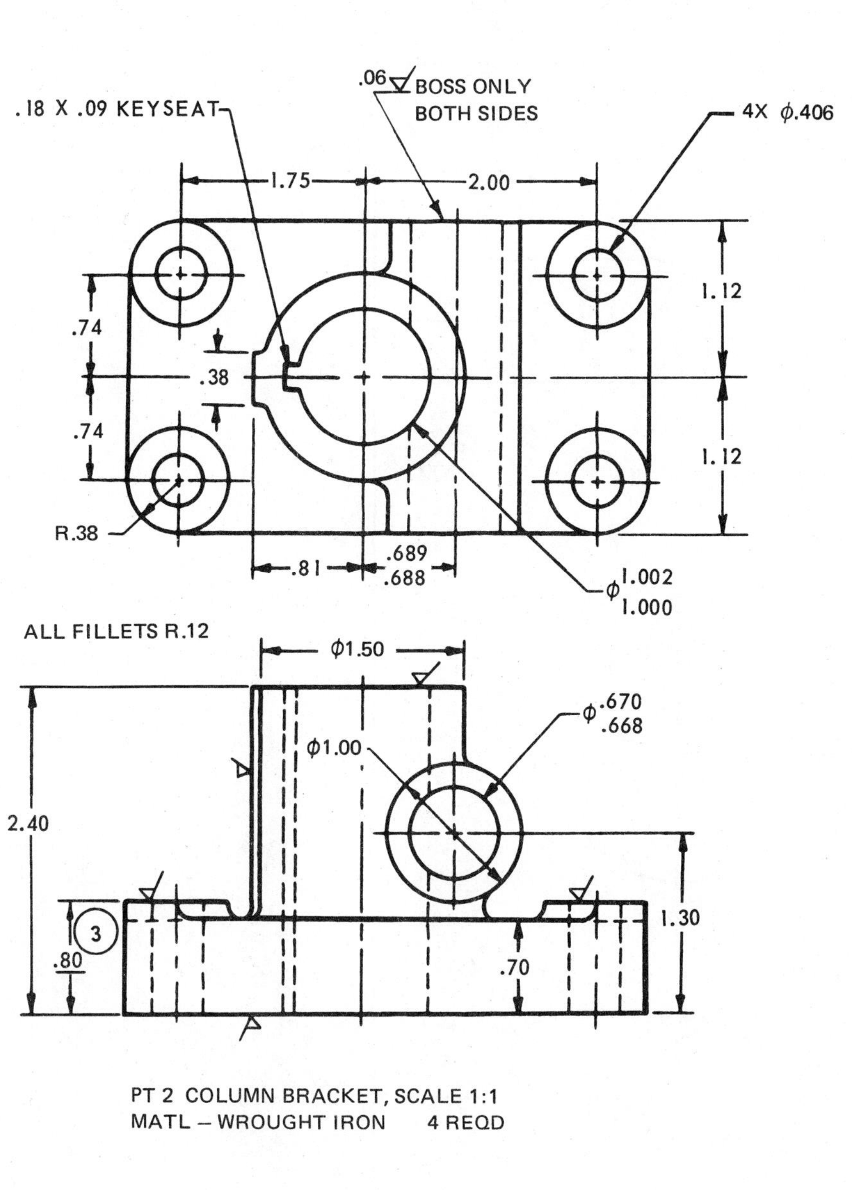

PT 2 COLUMN BRACKET, SCALE 1:1
MATL – WROUGHT IRON 4 REQD

ANSWERS

1. ______
2. ______
3. ______
4. ______
5. ______
6. A. ______
B. ______
C. ______
7. ______
8. ______
9. ______
10. ______
11. ______
12. ______
13. W ______
D ______
14. ______
15. ______
16. ______
17. ______
18. ______
19. ______
20. ______
21. ______
22. W ______
M ______

UNLESS OTHERWISE SPECIFIED
TOLERANCE ON DIMENSIONS ±.02

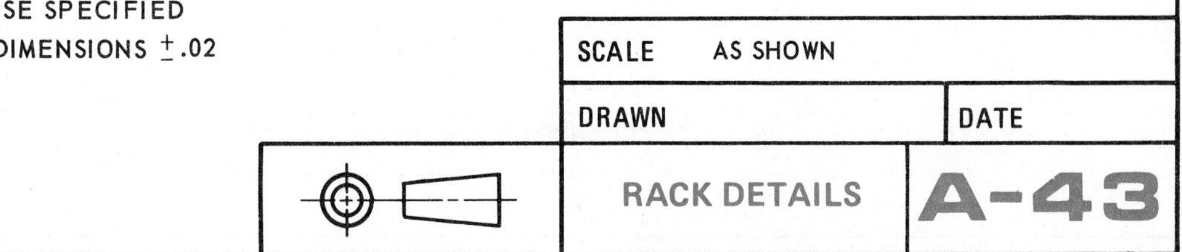

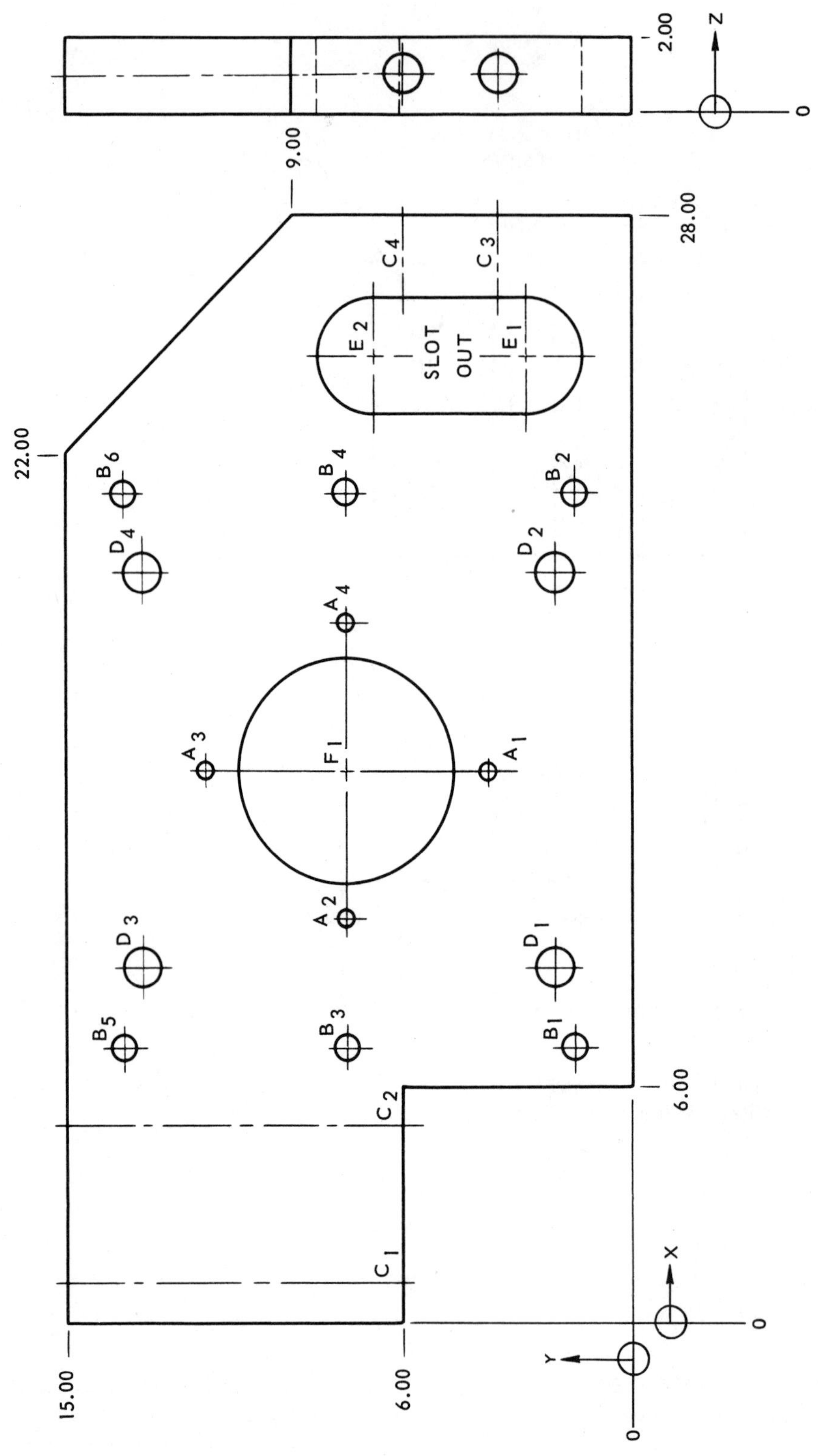
2.00
Z
0
9.00
28.00
C_4
C_3
E_2
E_1
SLOT
OUT
22.00
B_6
B_4
B_2
D_4
D_2
A_4
A_3
F_1
A_1
A_2
D_3
D_1
B_5
B_3
B_1
6.00
C_2
C_1
X
0
Y
15.00
6.00
0
NOTE: TOLERANCE ON LINEAR DIMENSIONS ±.02

QUESTIONS

ANSWERS

1. What is the width of the part?
2. What is the height of the part?
3. What are the width and height of the chamfer on the corner?
4. What is the width of slot E?
5. What is the height of slot E?
6. How much wood is left between slot E and the right-hand edge of the part?
7. What are the center distances between holes
   (A) B_5 and B_6
   (B) D_2 and D_4
   (C) B_1 and B_5
   (D) D_3 and D_4
   (E) A_1 and A_3
   (F) C_1 and C_2
   (G) B_3 and A_4?
8. How much wood is left between holes
   (A) A_1 and E_1
   (B) A_2 and B_3
   (C) C_1 and C_2
   (D) B_3 and B_4?

1. ______
2. ______
3. W ______
   H ______
4. ______
5. ______
6. ______
7. A. ______
   B. ______
   C. ______
   D. ______
   E. ______
   F. ______
   G. ______
8. A. ______
   B. ______
   C. ______
   D. ______

HOLE SYMBOL	HOLE DIA.	LOCATION		
		X	Y	Z
A_1	.375	14.00	3.75	
A_2	.375	10.25	7.50	
A_3	.375	14.00	11.25	
A_4	.375	17.75	7.50	
B_1	.625	7.00	1.50	
B_2	.625	21.00	1.50	
B_3	.625	7.00	7.50	
B_4	.625	21.00	7.50	
B_5	.625	7.00	13.50	
B_6	.625	21.00	13.50	
C_1	.812	1.00		1.00
C_2	.812	5.00		1.00
C_3	.812		3.50	1.00
C_4	.812		6.00	1.00
D_1	1.000	9.00	2.00	
D_2	1.000	19.00	2.00	
D_3	1.000	9.00	13.00	
D_4	1.000	19.00	13.00	
E_1	3.000	24.50	2.75	
E_2	3.000	24.50	6.75	
F_1	5.688	14.00	7.50	

NOTE: ALL HOLES THRU UNLESS OTHERWISE SPECIFIED

MATERIAL	DRY MAPLE
SCALE	NOT TO SCALE
DRAWN	DATE
SUPPORT BRACKET	A-44

UNIT 18

PRIMARY AUXILIARY VIEWS

Many objects have surfaces that are perpendicular to only one plane of projection. These surfaces are referred to as *inclined sloped surfaces.* In the remaining two orthographic views such surfaces appear to be foreshortened and their true shape is not shown, figure 18-1. When an inclined surface has important characteristics that should be shown clearly and without distortion, an auxiliary view is used to completely explain the shape of the object.

For example, figure 18-2 clearly shows why an auxiliary view is required. The circular features on the sloped surface on the front view

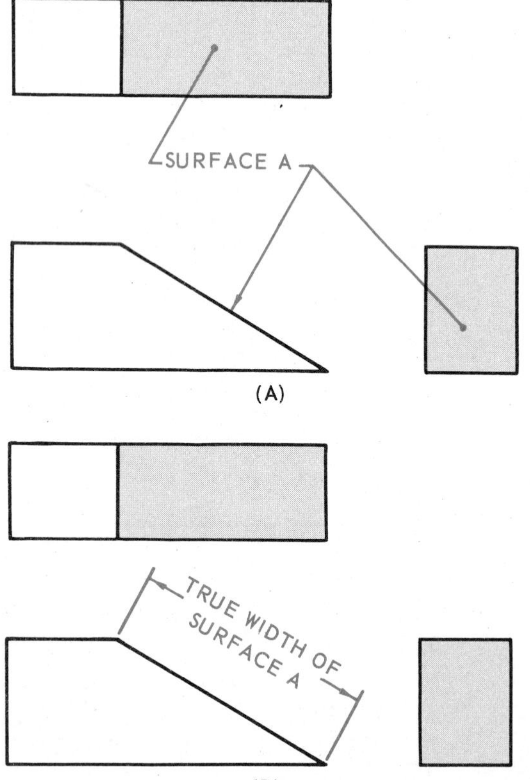

Fig. 18-1 (A) Surface A is perpendicular only to the front plane of projection. (B) Therefore, only the front view shows the true width of surface A, but no view shows the true shape.

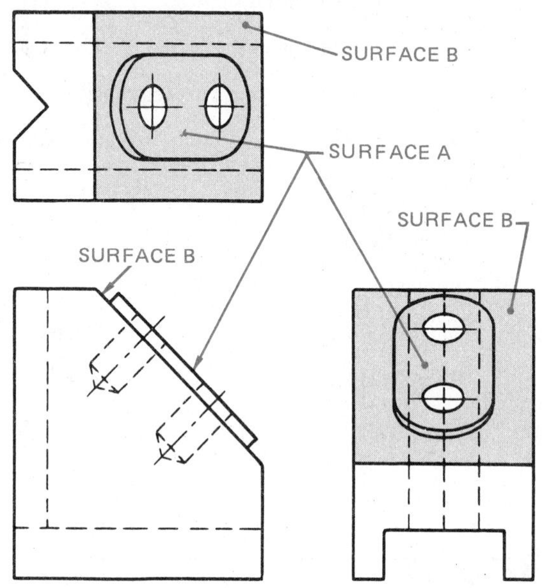

(A) REGULAR VIEWS DO NOT SHOW TRUE FEATURES OF SURFACES A AND B.

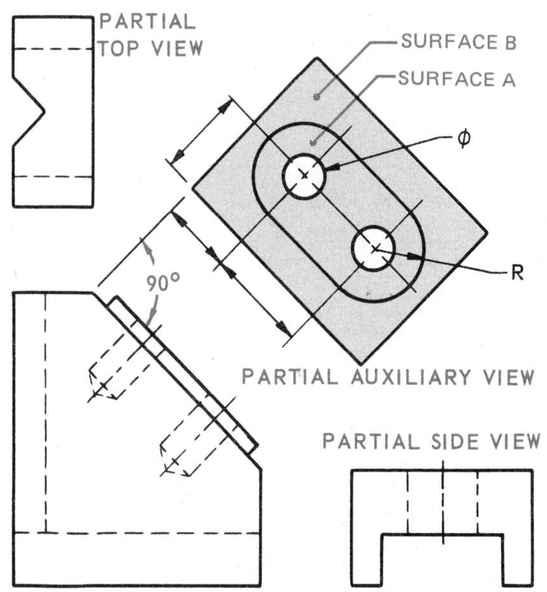

(B) AUXILIARY VIEW ADDED TO SHOW TRUE FEATURES OF SURFACES A AND B.

Fig. 18-2 The need for auxiliary views

cannot be seen in their true shape on either the top or side views. The auxiliary view is the only view which shows the actual shape of these features, figure 18-3. Note that only the sloped surface details are shown. Background detail is often omitted on auxiliary views and regular views to simplify the drawing and avoid confusion. A break line is used to signify the break in an incomplete view. The break line is not required if only the exact surface is drawn for either an auxiliary view or a partial regular view. Dimensions for the detail on the inclined face are placed on the auxiliary view, showing the detail in its true shape.

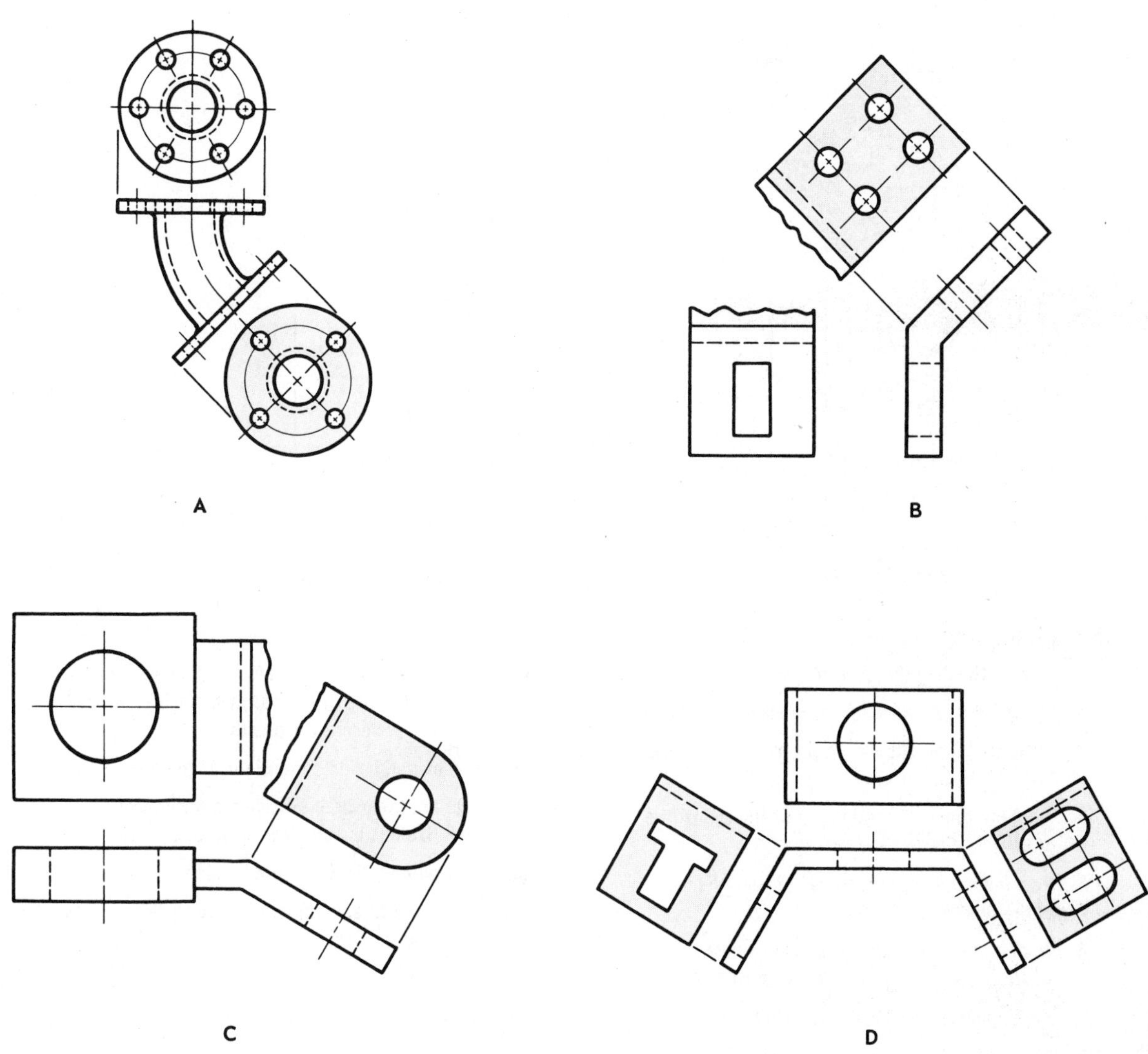

NOTE: CONVENTIONAL BREAK OF PROJECTED SURFACE ONLY NEED BE SHOWN ON PARTIAL VIEWS

Fig. 18-3 Examples of auxiliary-view drawings

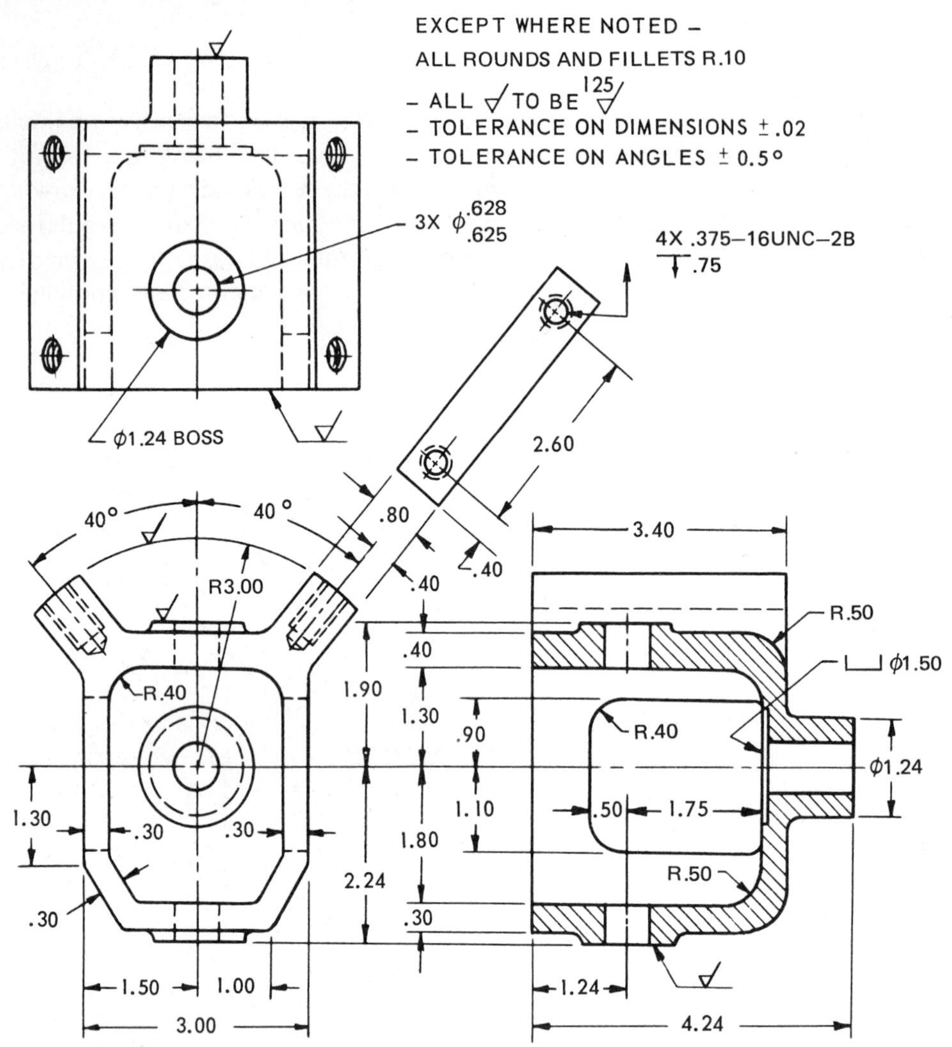

EXCEPT WHERE NOTED –
ALL ROUNDS AND FILLETS R.10
– ALL $\sqrt{}$ TO BE $\overset{125}{\sqrt{}}$
– TOLERANCE ON DIMENSIONS ±.02
– TOLERANCE ON ANGLES ± 0.5°

3X ϕ$\begin{matrix}.628\\.625\end{matrix}$

4X .375–16UNC–2B
↧.75

ϕ1.24 BOSS

ANSWERS

1. ______
2. ______
3. W ______
   H ______
4. W ______
   D ______
5. ______
6. ______
7. ______
8. ______
9. ______
10. A. ______
    B. ______
    C. ______
11. H ______
    W ______
    D ______
12. H ______
    W ______
    D ______
13. ______
14. ______

QUESTIONS

1. What is the diameter of the bosses?
2. What is the tolerance on the holes in the bosses?
3. What are the width and height of the cutouts in the sides of the box?
4. What are the width and depth of the legs of the box?
5. How many degrees are there between the legs of the box?
6. What is the maximum surface roughness in micro-inches permitted on the machined surfaces?
7. How many surfaces are machined?
8. What would be the inside diameter of the mating part that this box fits into?
9. If the mating part is .40 thick, and a flat washer and lockwasher are used, what diameter and length of socket head cap screws would be used to fasten the parts together? (See Table 7 in the Appendix.) Note: Lengths available in .25 in. increments.
10. What is the thickness of (A) the side walls of the box; (B) the top of the box; (C) the bottom of the box? Disregard the bosses.
11. Give overall inside dimensions of the box.
12. What are the overall outside dimensions of the box? (Do not include legs or bosses.)
13. Of what material is the box **made?**
14. How many screws are required to fasten the box to the mating part?

MATERIAL	GRAY IRON	
SCALE	NOT TO SCALE	
DRAWN		DATE

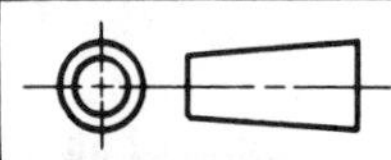

GEAR BOX

A-45

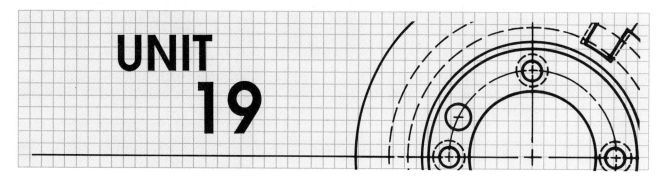

UNIT 19

SECONDARY AUXILIARY VIEWS

As mentioned earlier, auxiliary views show the true lengths of lines and the true shapes of surfaces which cannot be described in the ordinary views.

A primary auxiliary view is drawn by projecting lines from a regular view where the inclined surface appears as an edge.

The auxiliary view in figure 19-1(A), which shows the true projection and true shape of surface X, is called a *primary auxiliary view* because it is projected directly from the regular front view.

Some surfaces are inclined so that they are not perpendicular to any of the three viewing planes. In this case, they appear as a surface in all three views, but never in their true shape. These are referred to as *oblique surfaces,* figure 19-2. Because the oblique surface is not perpendicular to the viewing planes, it cannot be parallel to them. Consequently, the surface appears foreshortened in all three views. If a true view is required for this surface, two auxiliary views, a primary and a secondary view, must be drawn.

The chamfered corners **Z** have threaded holes on the oblique surfaces. To show the true shape and location of the threaded holes **M** a second auxiliary view must be shown, as at **(C)**. This auxiliary view is projected from the first or primary auxiliary view, and is known as a

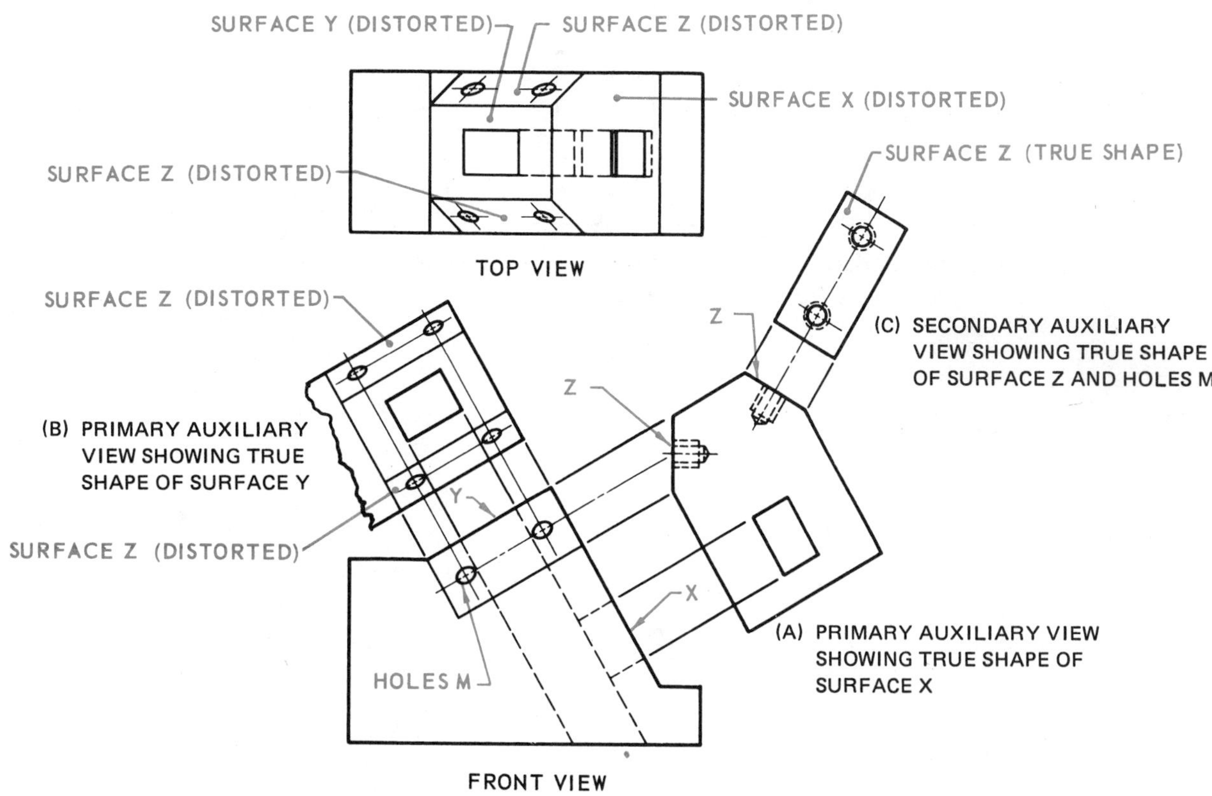

Fig. 19-1 Primary and secondary auxiliary views

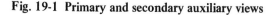

secondary auxiliary view. The view at (**B**) is a primary auxiliary view because it is projected from one of the regular views. Notice that the side view is not drawn since the auxiliary views provide the information usually shown on this view.

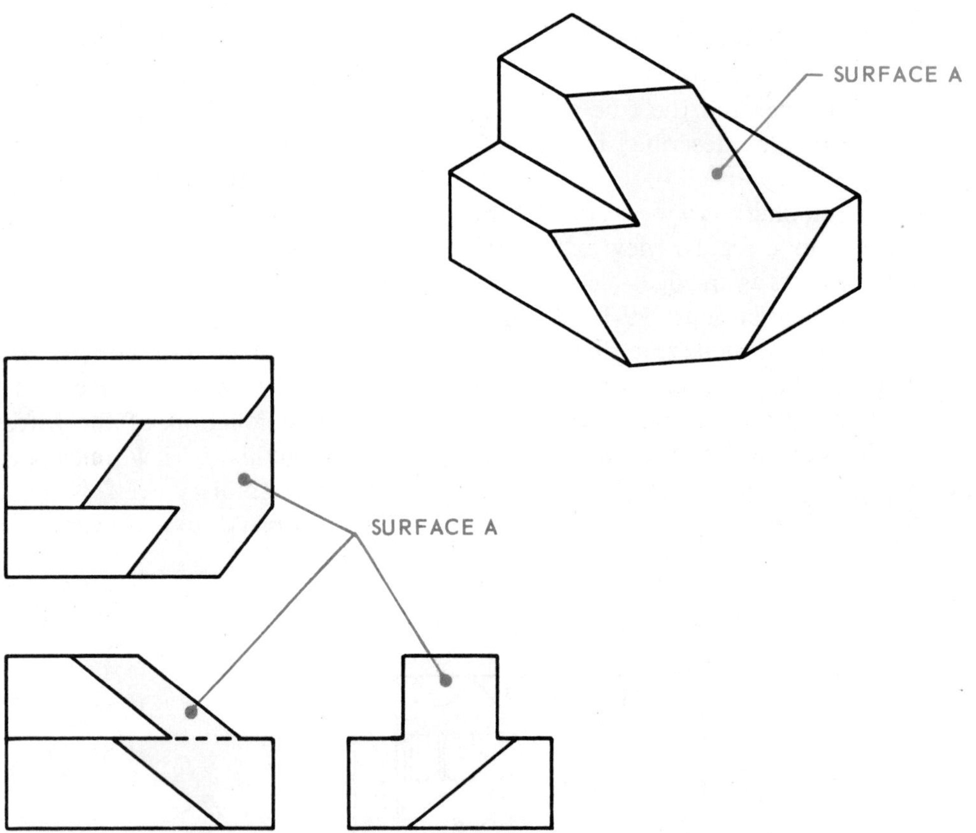

Fig. 19-2 Surface A is not perpendicular to any of the three viewing planes and is therefore considered oblique

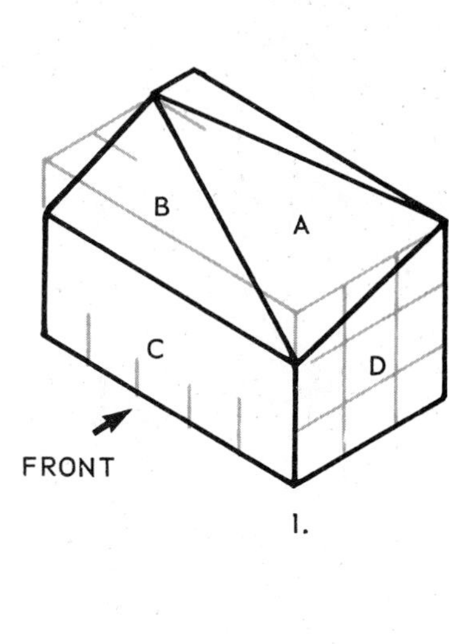

1.

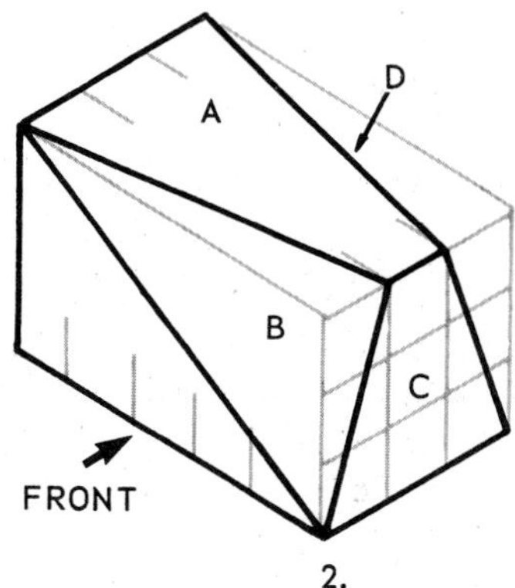

2.

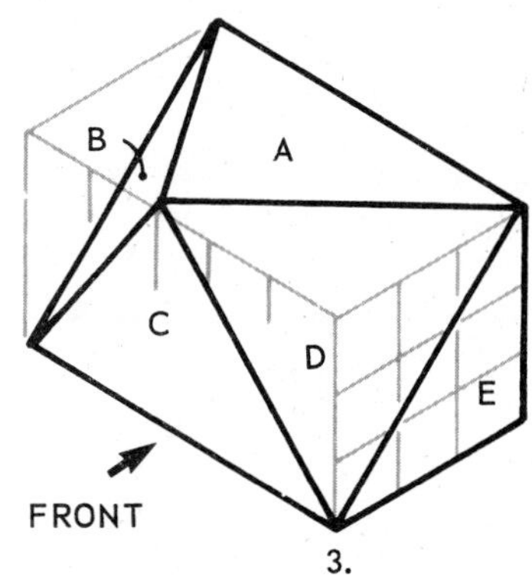

3.

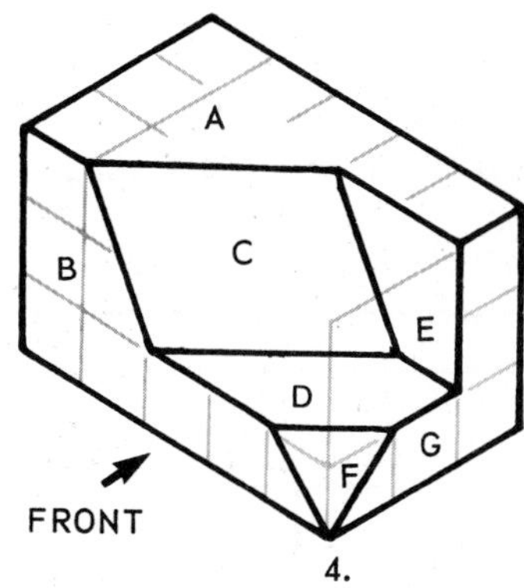

4.

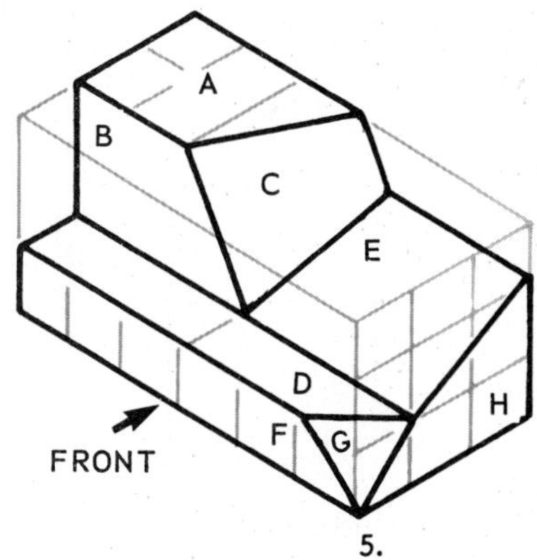

5.

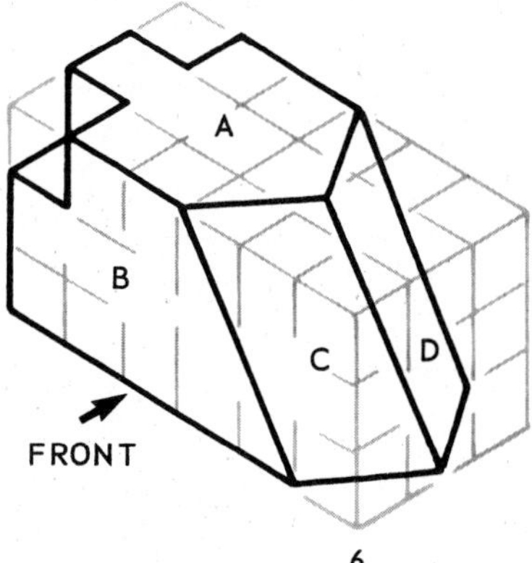

6.

ASSIGNMENT: USING THE SAME NUMBER OF SQUARES, DRAW ON GRAPH PAPER THE FRONT, TOP, AND SIDE VIEWS. IDENTIFY THE OBLIQUE SURFACES ON THE THREE VIEWS WITH THE APPROPRIATE LETTERS.	A-46

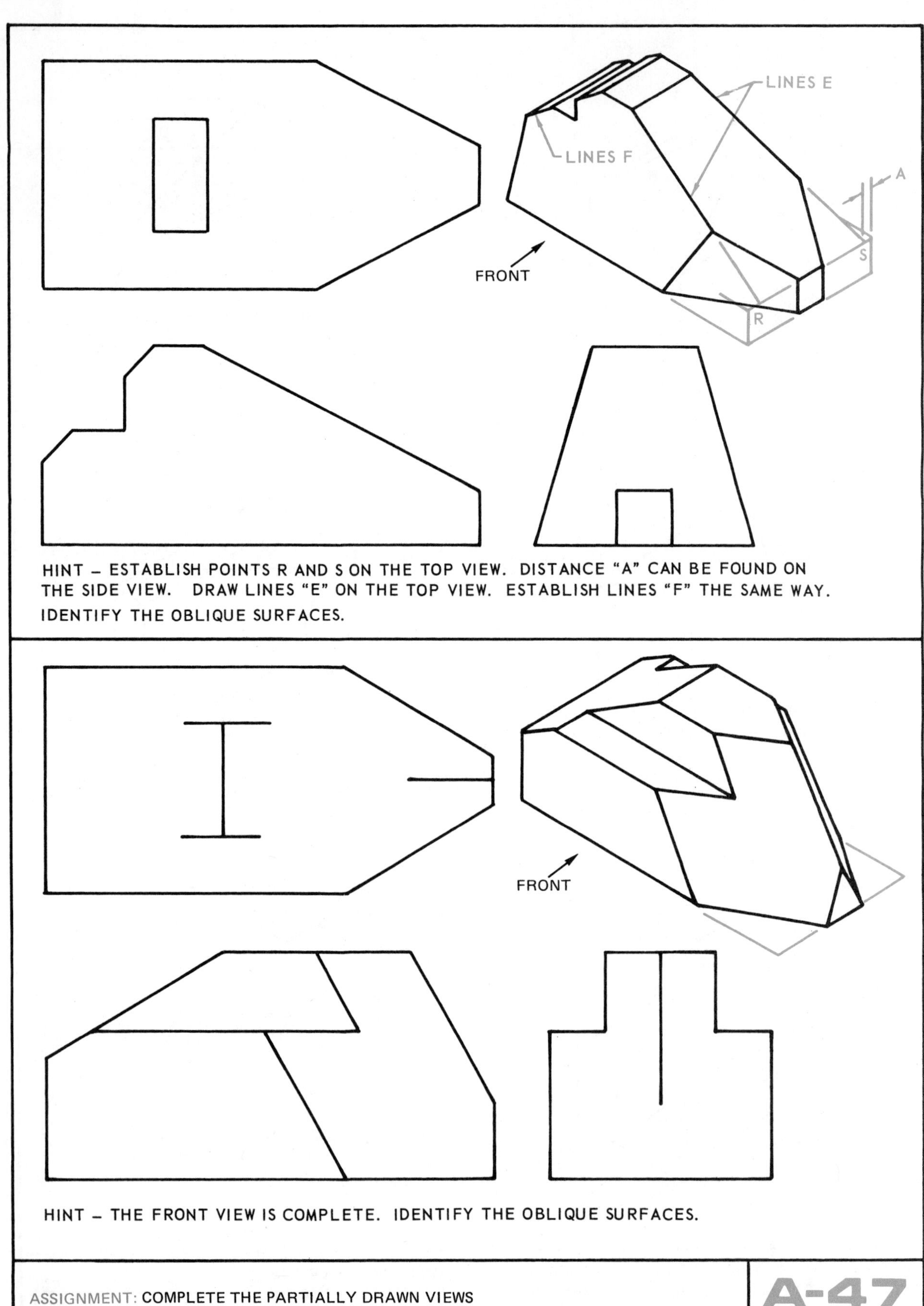
LINES E
LINES F
A
S
R
FRONT
HINT – ESTABLISH POINTS R AND S ON THE TOP VIEW. DISTANCE "A" CAN BE FOUND ON THE SIDE VIEW. DRAW LINES "E" ON THE TOP VIEW. ESTABLISH LINES "F" THE SAME WAY. IDENTIFY THE OBLIQUE SURFACES.
FRONT
HINT – THE FRONT VIEW IS COMPLETE. IDENTIFY THE OBLIQUE SURFACES.
ASSIGNMENT: COMPLETE THE PARTIALLY DRAWN VIEWS
A-47

PAGE 145 IS INTENTIONALLY BLANK. ASSIGNMENT DRAWING A-48 IS ON THE NEXT PAGE.

QUESTIONS

1. Which other view(s) show the true height of the .62 dimension shown in the front view?
2. Which other view(s) show the true width of the .60 dimension shown in the primary auxiliary view?
3. What view(s) show the true (A) height, (B) depth, (C) width of the top portion of the support?
4. How many surfaces require finishing?
5. List the surface(s) or feature(s) which are shown in their true shape or size in the primary auxiliary view but are distorted in all other views.
6. List the surface(s) or feature(s) which are shown in their true shape or size in the secondary auxiliary view but are distorted in all other views.
7. What would be the thickness of the base (26) before machining?
8. If the hexagon bar support was fastened to another member, –

   (Refer to the Appendix when necessary.)

   (A) How many cap screws would be used?

   (B) What would be the cap screw size?

   (C) If the supporting member had tapped holes which were .80 deep and flat washers were used under the cap screw heads; what would be the cap screw length?

   (D) If the cap screws were of the fine thread series, how would you call out these cap screws?

   (E) What would be the I.D. and O.D. of the flat washers used under these cap screws?

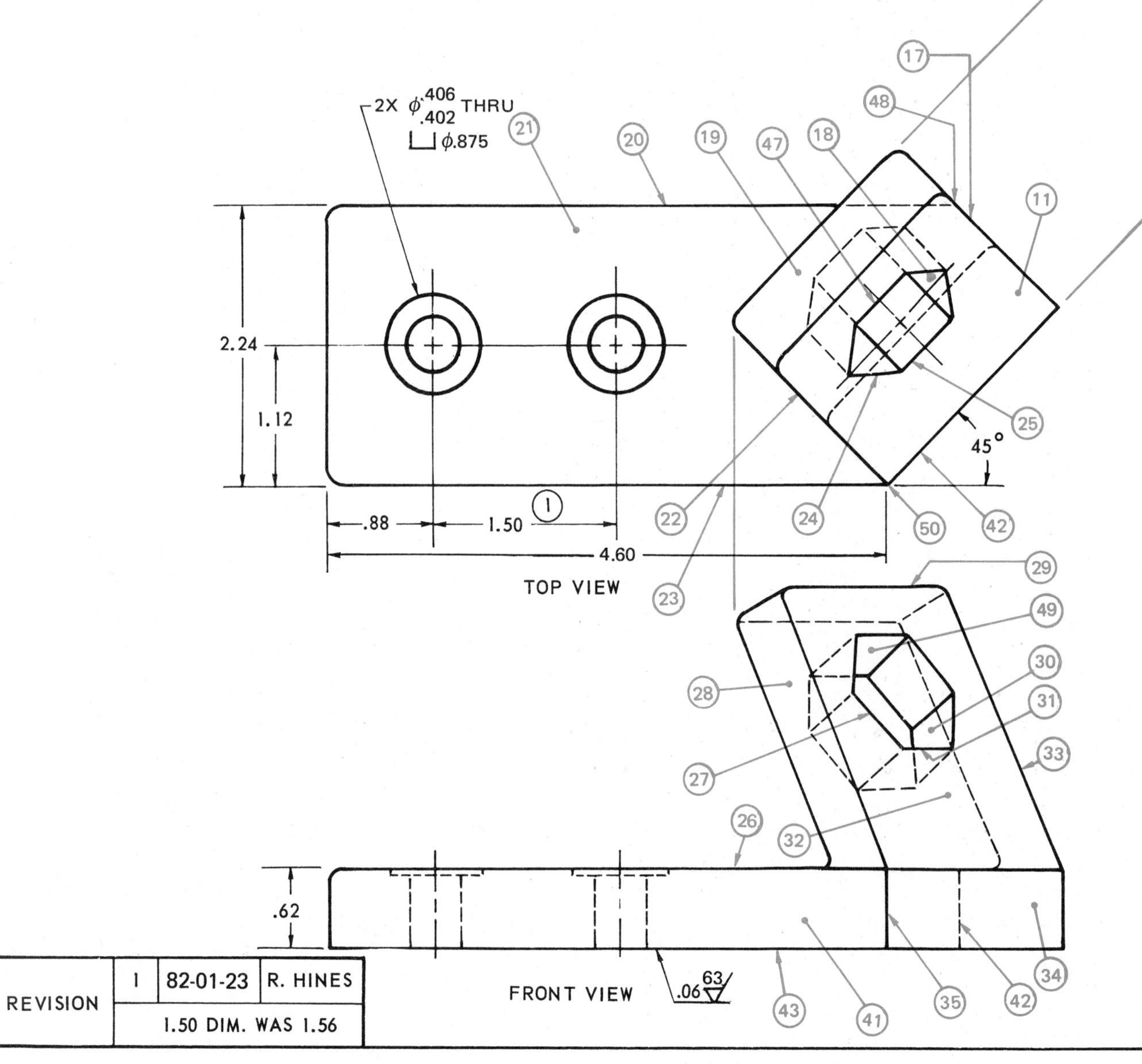

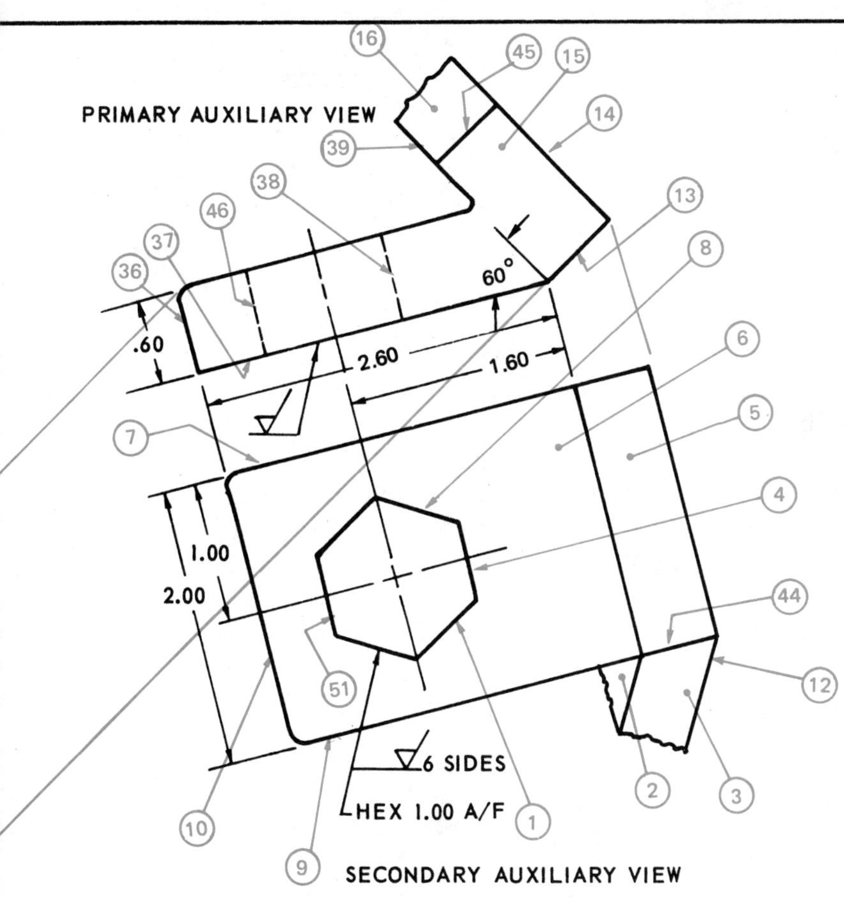

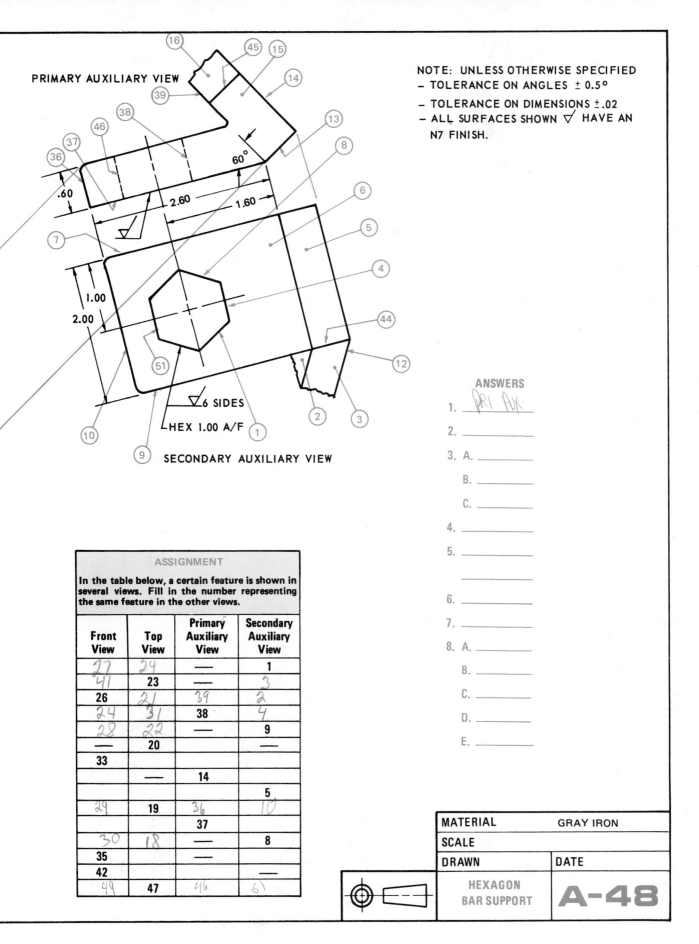

NOTE: UNLESS OTHERWISE SPECIFIED
- TOLERANCE ON ANGLES ± 0.5°
- TOLERANCE ON DIMENSIONS ±.02
- ALL SURFACES SHOWN ∀ HAVE AN N7 FINISH.

ANSWERS

1. PRI AUX
2. ______
3. A. ______
   B. ______
   C. ______
4. ______
5. ______
   ______
6. ______
7. ______
8. A. ______
   B. ______
   C. ______
   D. ______
   E. ______

ASSIGNMENT

In the table below, a certain feature is shown in several views. Fill in the number representing the same feature in the other views.

Front View	Top View	Primary Auxiliary View	Secondary Auxiliary View
27	24	—	1
41	23	—	3
26	21	39	2
24	31	38	4
28	22	—	9
—	20		—
33			
	—	14	
			5
29	19	36	10
		37	
30	18	—	8
35		—	
42			—
49	47	46	51

MATERIAL	GRAY IRON
SCALE	
DRAWN	DATE
HEXAGON BAR SUPPORT	A-48

UNIT 20

DEVELOPMENT DRAWINGS

Many objects, such as metal and cardboard boxes, duct work for heating, funnels and eaves-troughing, are made from flat sheet material that is cut so that, when folded, formed or rolled, it will take the shape of the object. Since a definite shape and size is desired, a regular orthographic drawing of the object is first made; then a development drawing is made to show the complete surface or surfaces laid out in a flat plane, figure 20-1.

A development drawing is sometimes referred to as a pattern drawing, because the layout, when made of heavy cardboard, metal, or wood, is used as a pattern for tracing out the developed shape on flat material. Such patterns are used extensively in sheet metal shops.

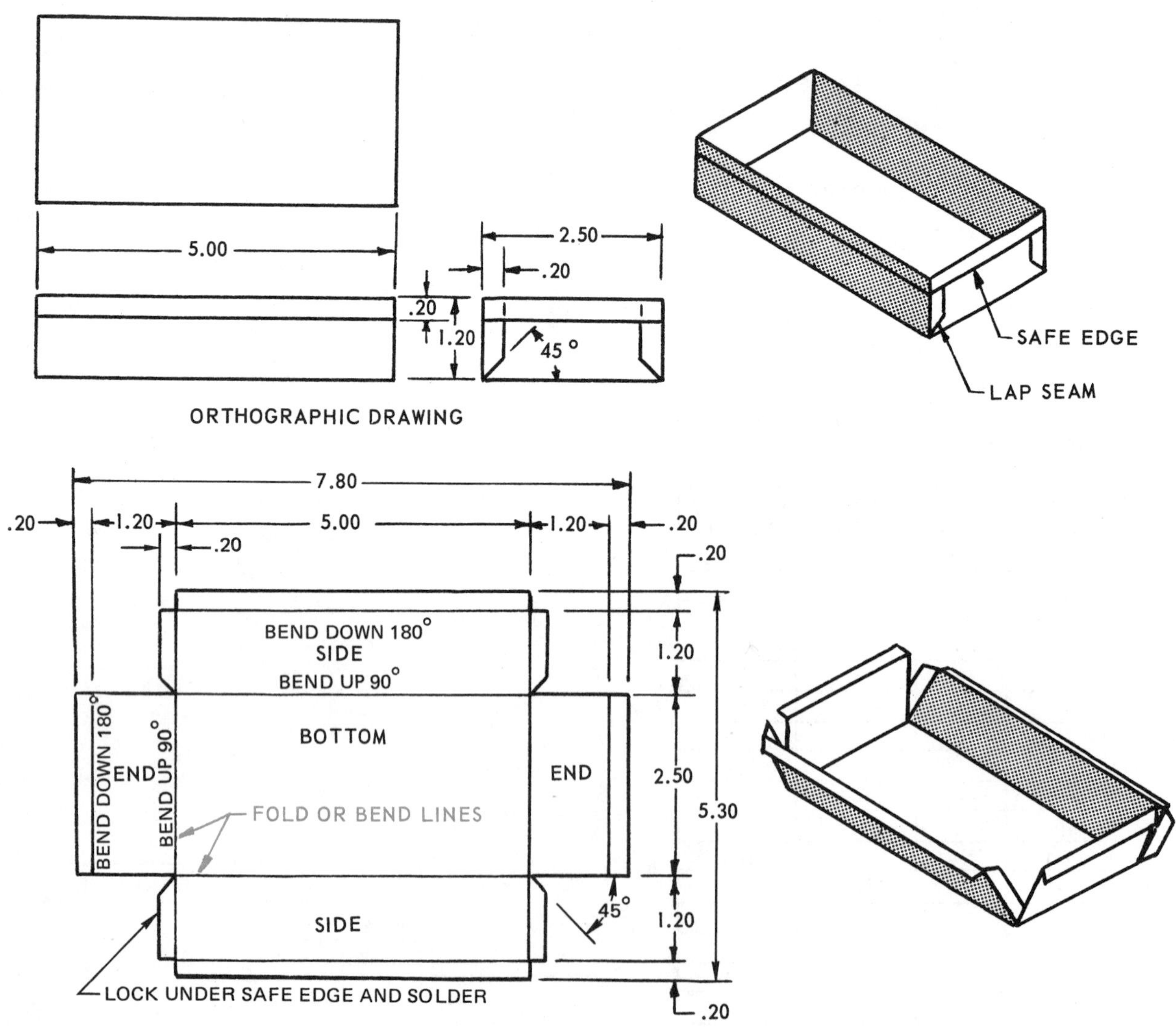

Fig. 20-1 Development drawing with a complete set of folding instructions

JOINTS, SEAMS, AND EDGES

Additional material is required for assembly and design purposes. When two or more pieces of material or surfaces are joined together extra material must be provided for the joint or seam. The type of joint or seam for joining metal, figure 20-2, is dependent on design criteria such as strength, waterproofing and appearance. Rivets or solder can be added to these joints if required. Exposed edges of metal parts may also be reinforced by the addition of extra material for hemming or for containing a wire. A round metal wastebasket, figure 20-3, is one example where extra material is provided for the joint, seam and edge.

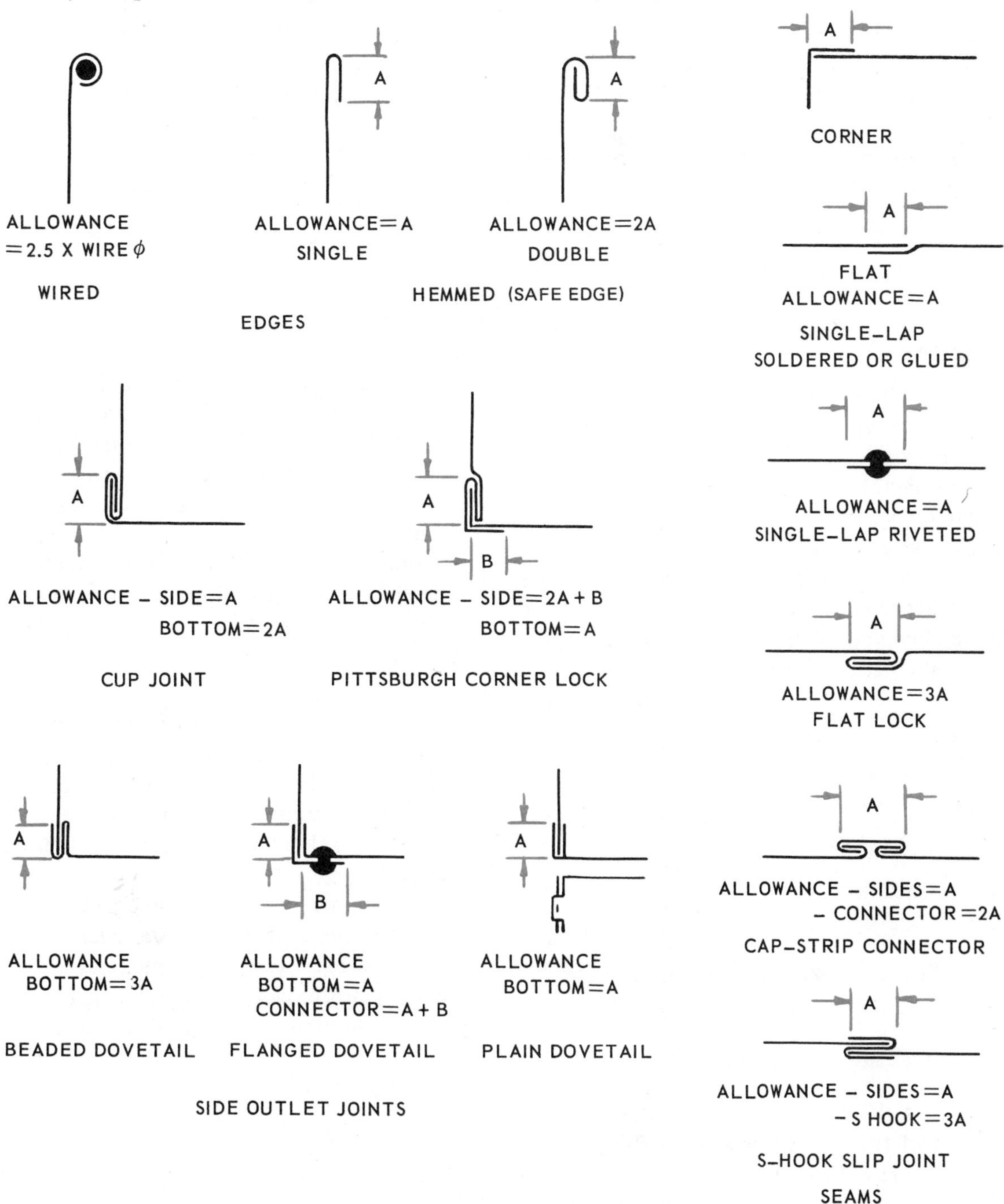

Fig. 20-2 Joints, seams, and edges

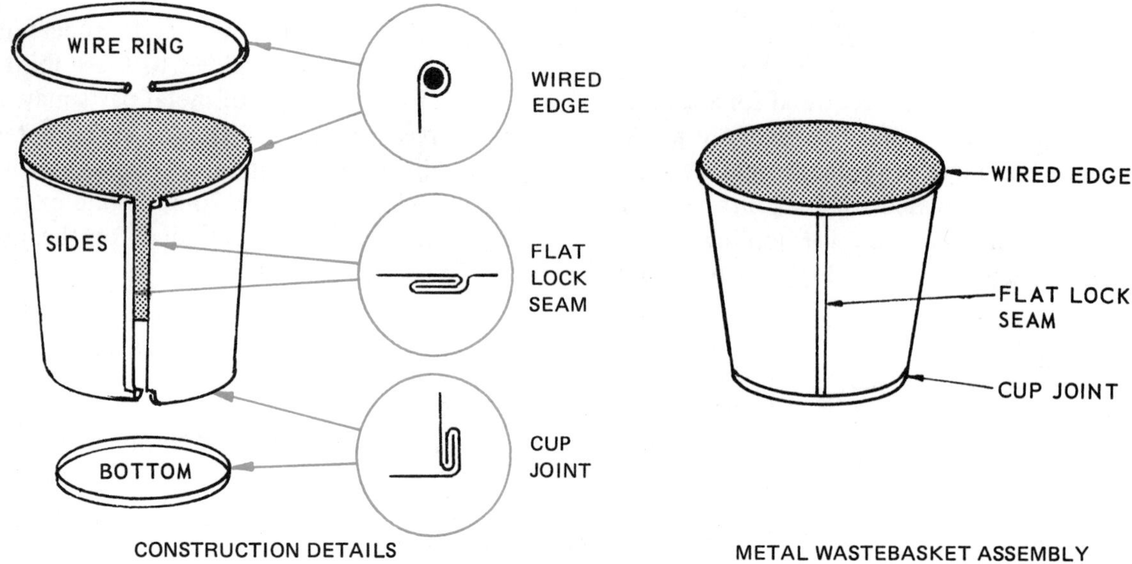

Fig. 20-3 Wastebasket construction

SHEET METAL SIZES

Metal thicknesses up to .25 in. (6 mm) are usually designated by a series of gage numbers, the more common gages are shown in Table 16 of the Appendix. Metal .25 in. and over is given in inch or millimeter sizes. In calling for the material size of sheet metal developments, customary practice is to give the gage number, type of gage, and its inch or millimeter equivalent in brackets followed by the developed width and length, figure 20-4.

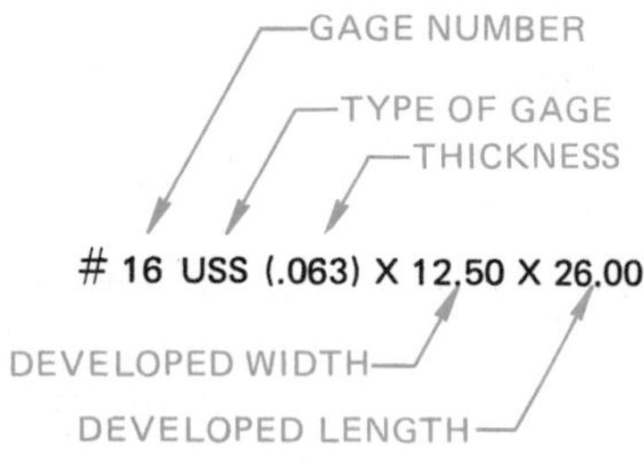

Fig. 20-4 Callout of sheet metal material

STRAIGHT LINE DEVELOPMENT

This is the term given to the development of an object that has surfaces on a flat plane of projection. The true size of each side of the object is known, and these sides can be laid out in successive order. Figure 20-1 shows the development of a simple rectangular box having a bottom and four sides. Note that in the development of the box an allowance is made for lap seams at the corners for a folded edge. All lines for each surface are parallel or perpendicular to the other surfaces. The bottom corners of the lap joints are chamfered to facilitate assembly. The fold lines on the development are shown as thin unbroken lines.

STAMPINGS

Stamping is the art of pressworking sheet metal to change its shape by the use of punches and dies. It may involve punching out a hole or the product itself from a sheet of metal. It may also involve bending or forming, figures 20-5 and 20-6.

Stamping may be divided into two general classifications: *forming* and *shearing.*

Forming. Forming includes stampings made by forming sheet metal to the shape desired without cutting or shearing the metal. For thicker sheet metal plates, bending allowances must be taken into consideration.

Shearing. Shearing includes stampings made by shearing the sheet metal either to change the outline or to cut holes in the interior of the part. Punching forms a hole or opening in the part.

Fig. 20-5 Punch press used to punch and form metal. (Courtesy of Whitney Metal Tool Co.)

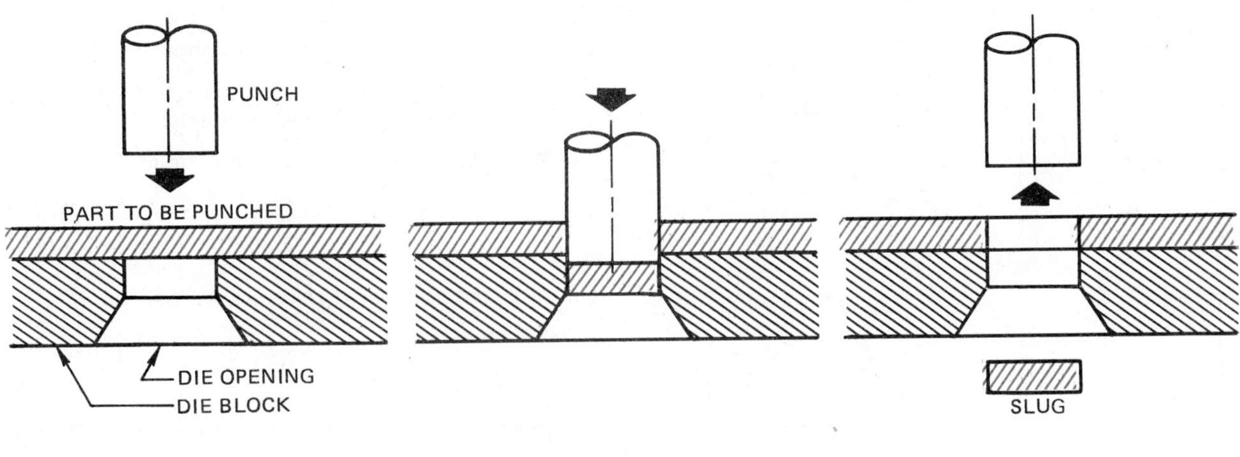

Fig. 20-6 Punching a hole in sheet metal

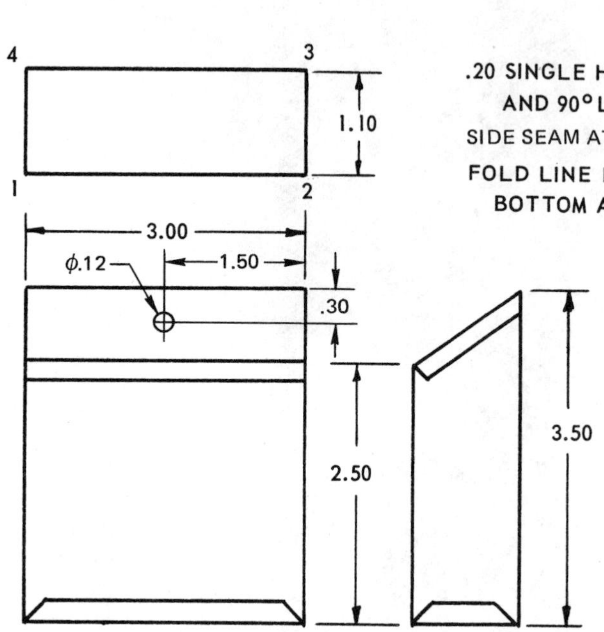

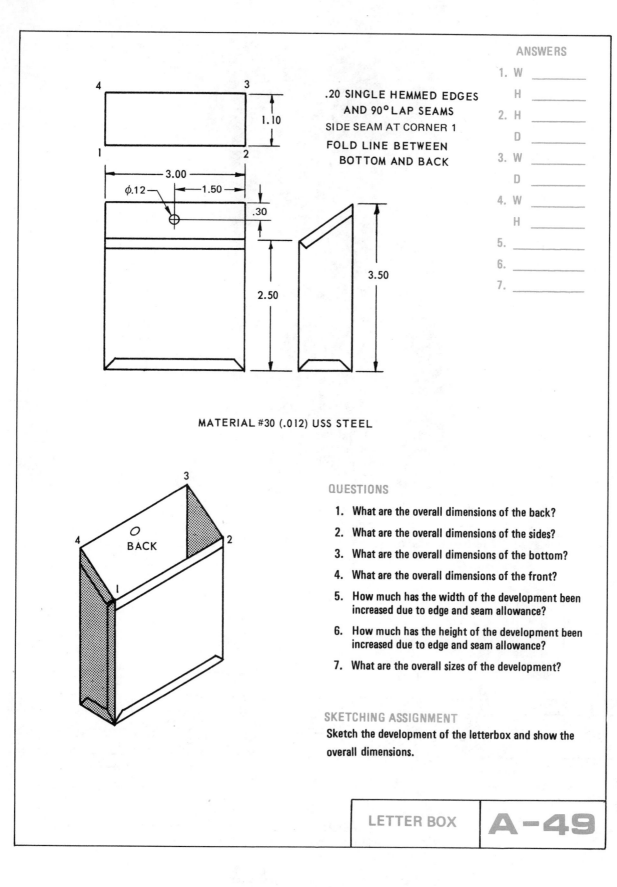

.20 SINGLE HEMMED EDGES
AND 90° LAP SEAMS
SIDE SEAM AT CORNER 1
FOLD LINE BETWEEN
BOTTOM AND BACK

ANSWERS

1. W ________
   H ________
2. H ________
   D ________
3. W ________
   D ________
4. W ________
   H ________
5. ________
6. ________
7. ________

MATERIAL #30 (.012) USS STEEL

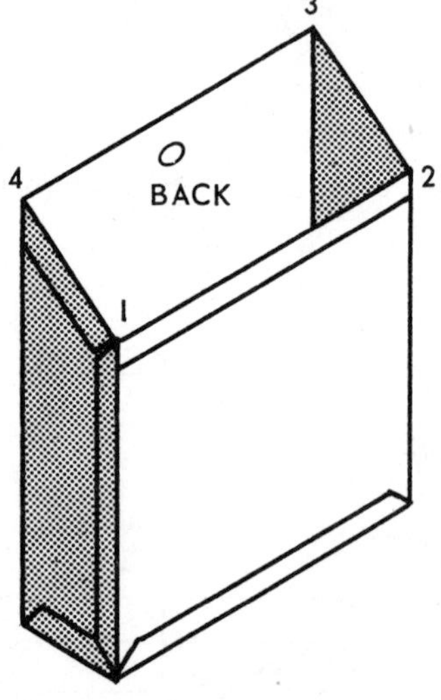

QUESTIONS

1. What are the overall dimensions of the back?
2. What are the overall dimensions of the sides?
3. What are the overall dimensions of the bottom?
4. What are the overall dimensions of the front?
5. How much has the width of the development been increased due to edge and seam allowance?
6. How much has the height of the development been increased due to edge and seam allowance?
7. What are the overall sizes of the development?

SKETCHING ASSIGNMENT

Sketch the development of the letterbox and show the overall dimensions.

LETTER BOX	A-49

UNIT 21

ARRANGEMENT OF VIEWS

The shape of an object and its complexity influence the possible choices and arrangement of views for that particular object. Since one of the main purposes of making drawings is to furnish the worker with enough information to be able to make the object, only the views which will aid in the interpretation of the drawing should be drawn.

The drafter chooses the view of the object which gives the viewer the clearest idea of the purpose and general contour of the object, and then calls this the front view. This choice of the front view may have no relationship to the actual front of the piece when it is used. The front view does not have to be the actual front of the object itself.

The selection of views to best describe the mounting plate shown in drawing A-50 is shown in figure 21-1. These views could be called front, right side, and bottom views as illustrated in Arrangement A. These views could also be designated top, front, and right side views, as shown in Arrangement B. Note that in Arrangement B the right side view is projected from the top view, and not the front view as shown in Arrangement A. The designation of names is not of major importance. What is important is that these views, in the opinion of the drafter or designer, give the necessary information in the most understandable way.

A variety of arrangements and naming of views to describe the index pedestal in drawing A-51 is shown in figure 21-2. Arrangements A and B are identical except for the naming of the views. While Arrangements C and D are acceptable, they are not as easily read.

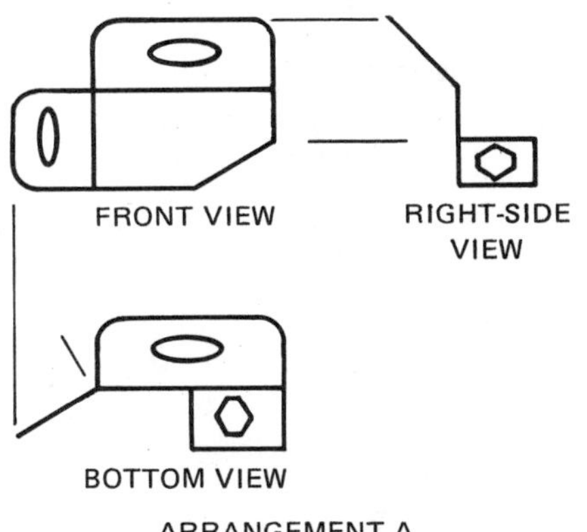

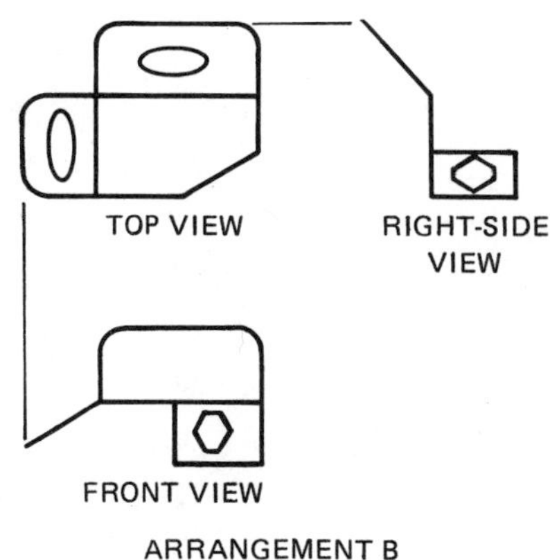

Fig. 21-1 Naming of views for mounting plate, drawing A-50

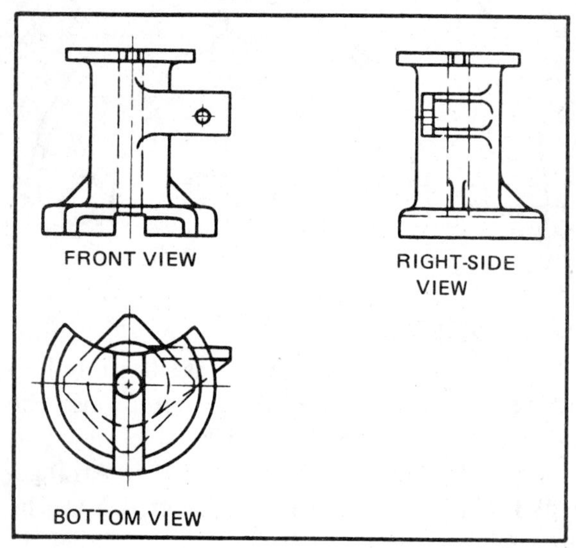

ARRANGEMENT A

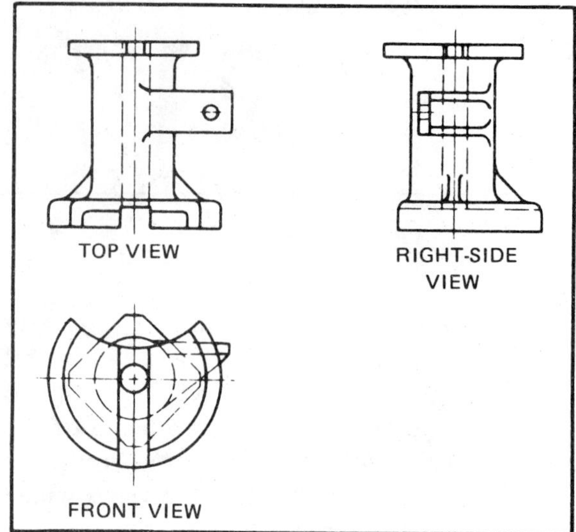

ARRANGEMENT B

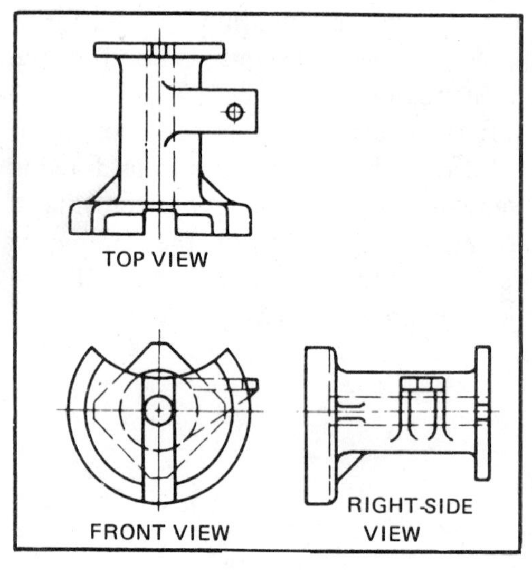

ARRANGEMENT C

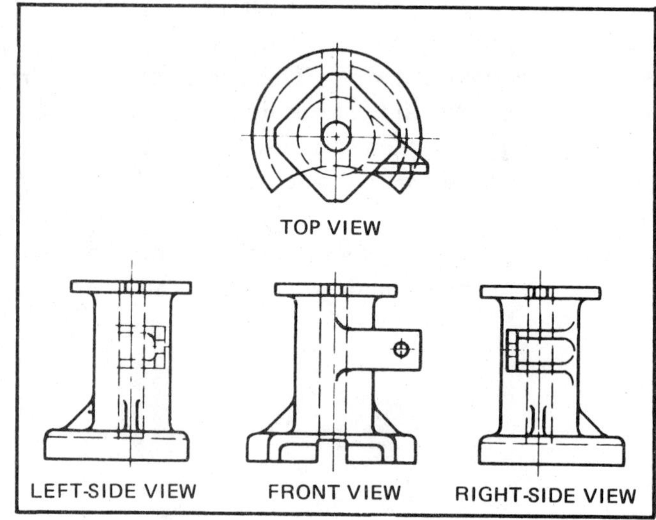

ARRANGEMENT D

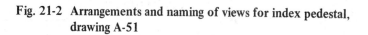

Fig. 21-2 Arrangements and naming of views for index pedestal, drawing A-51

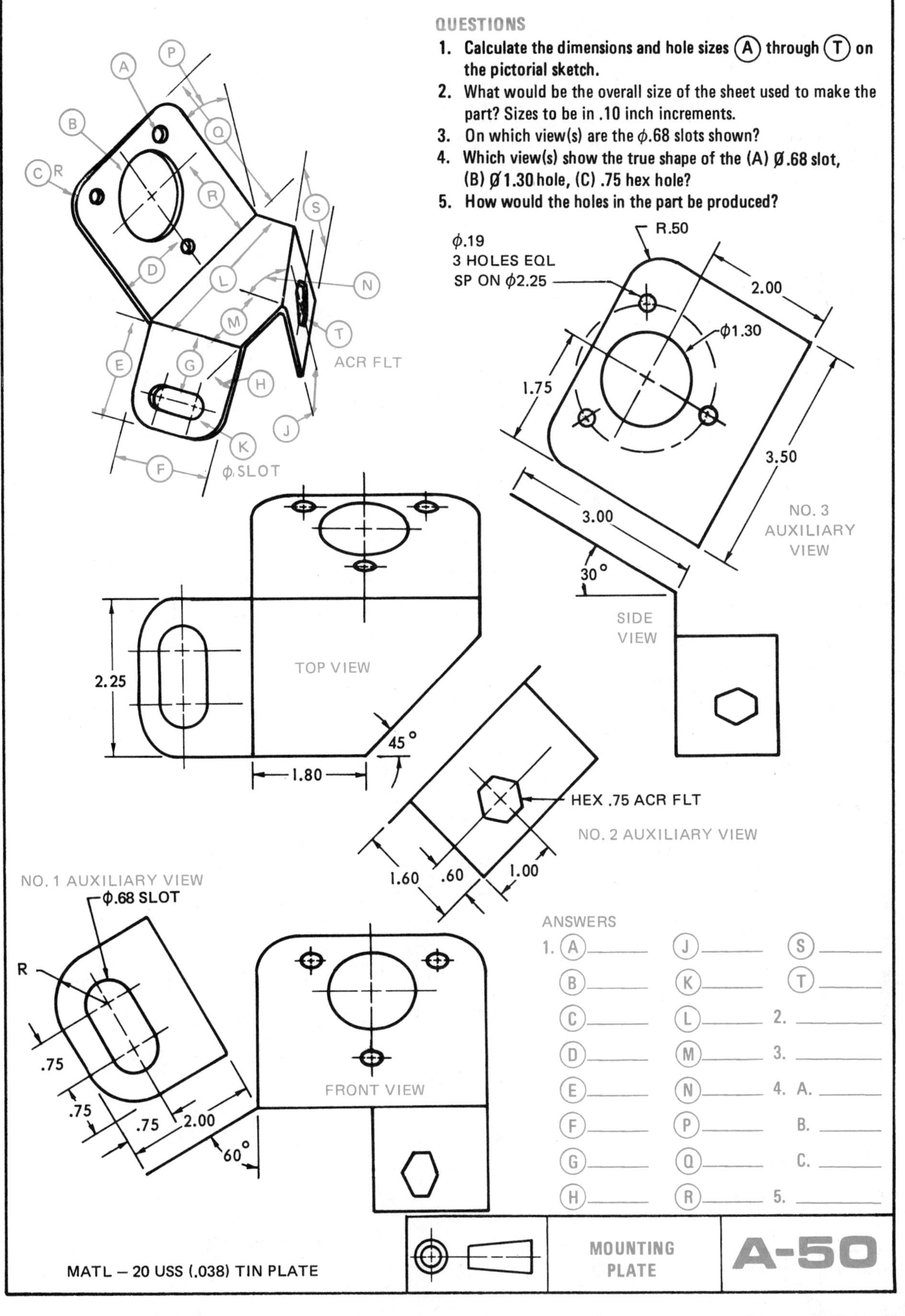

QUESTIONS

1. Calculate the dimensions and hole sizes (A) through (T) on the pictorial sketch.
2. What would be the overall size of the sheet used to make the part? Sizes to be in .10 inch increments.
3. On which view(s) are the ϕ.68 slots shown?
4. Which view(s) show the true shape of the (A) Ø.68 slot, (B) Ø1.30 hole, (C) .75 hex hole?
5. How would the holes in the part be produced?

ANSWERS

1. (A) ____ (B) ____ (C) ____ (D) ____ (E) ____ (F) ____ (G) ____ (H) ____ (J) ____ (K) ____ (L) ____ (M) ____ (N) ____ (P) ____ (Q) ____ (R) ____ (S) ____ (T) ____
2. ____
3. ____
4. A. ____ B. ____ C. ____
5. ____

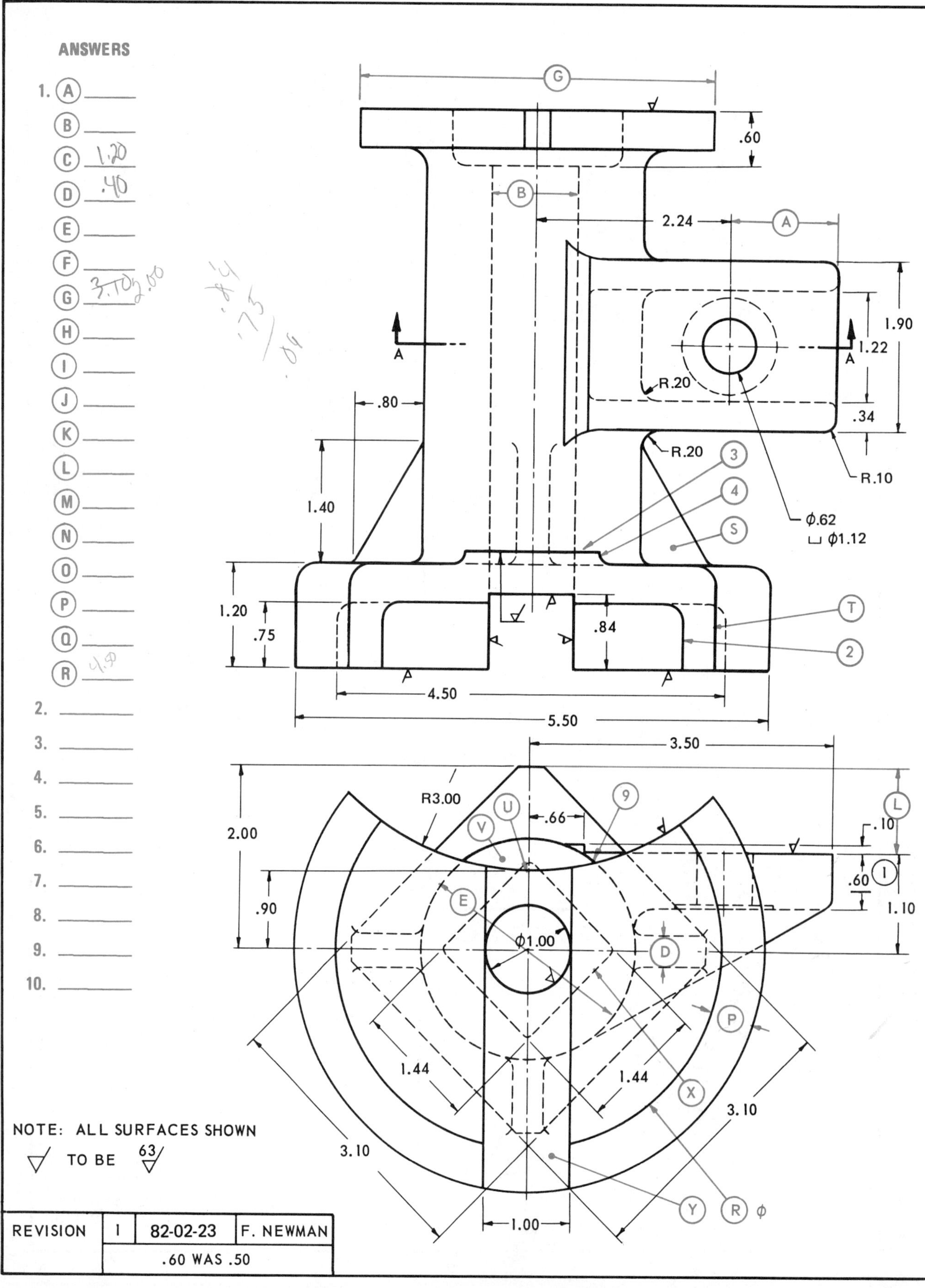
ANSWERS
1. A
B
C 1.20
D .40
E
F
G 3.10 2.00
H
I
J
K
L
M
N
O
P
Q
R 4.50
2.
3.
4.
5.
6.
7.
8.
9.
10.
NOTE: ALL SURFACES SHOWN
TO BE 63
REVISION
1
82-02-23
F. NEWMAN
.60 WAS .50
.60
2.24
1.90
1.22
R.20
.80
.34
R.20
R.10
1.40
ϕ.62
⌴ ϕ1.12
1.20
.75
.84
4.50
5.50
3.50
R3.00
.66
.10
2.00
.60
.90
1.10
ϕ1.00
1.44
1.44
3.10
3.10
1.00

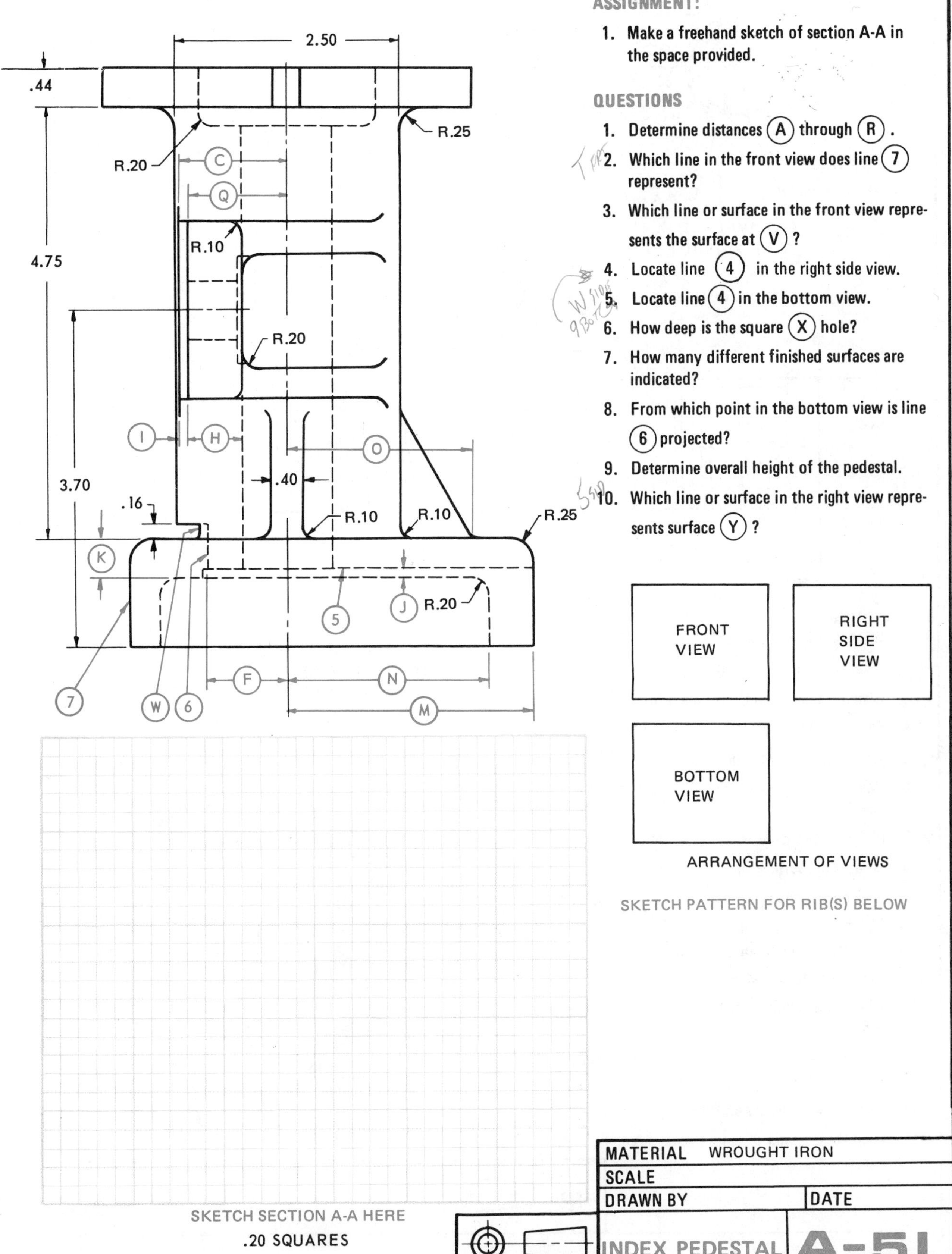

ASSIGNMENT:

1. Make a freehand sketch of section A-A in the space provided.

QUESTIONS

1. Determine distances (A) through (R).
2. Which line in the front view does line (7) represent?
3. Which line or surface in the front view represents the surface at (V)?
4. Locate line (4) in the right side view.
5. Locate line (4) in the bottom view.
6. How deep is the square (X) hole?
7. How many different finished surfaces are indicated?
8. From which point in the bottom view is line (6) projected?
9. Determine overall height of the pedestal.
10. Which line or surface in the right view represents surface (Y)?

MATERIAL WROUGHT IRON	
SCALE	
DRAWN BY	DATE
INDEX PEDESTAL	A-51

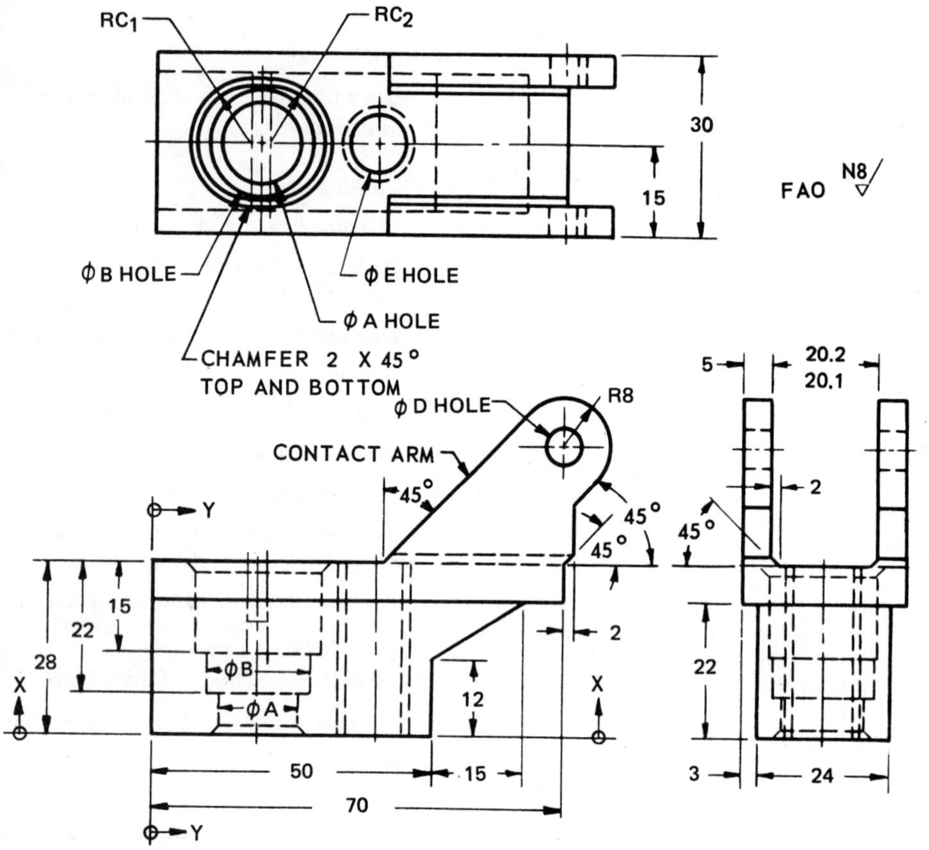

ANSWERS

1. ______
2. ______
3. ______
4. ______
5. ______
6. ______
7. ______
8. ______
9. ______
10. ______
11. ______
12. ______
13. ______
14. ______
15. ______

QUESTIONS

1. What is the overall width?
2. What is the overall height?
3. Give the chamfer angle for the **C** hole.
4. What is the distance from **Y-Y** to **E** hole?
5. What is the distance from **X-X** to **D** hole?
6. How many complete threads does the tapped hole have?
7. Which thread series is the tapped hole?
8. What is the surface finish in micrometers?
9. How deep is the **C** counterbore from the top of the surface?
10. How long is the **C** counterbored hole?
11. Give the distance between the ℄ of C_1 and C_2 radii.
12. What is the nominal thickness of the contact arm at **D** hole?
13. What is the tolerance on the distance between the contact arms?
14. What is the tolerance on **D** hole?
15. What is the center distance between **A** and **E** holes?

NOTE: UNLESS OTHERWISE SPECIFIED
- TOLERANCE ON DIMENSIONS ±0.5
- TOLERANCE ON ANGLES ±0.5°

HOLE SYMBOL	HOLE SIZE	LOCATION	
		X–X	Y–Y
A	ϕ13.5		18
B	ϕ17		18
C_1	R9		16
C_2	R9		20
D	ϕ6.5-6.6	50	70
E	M12 X 1.25		38

DIMENSIONS ARE IN MILLIMETERS

MATERIAL	MANGANESE BRONZE
SCALE	NOT TO SCALE
DRAWN	DATE

METRIC

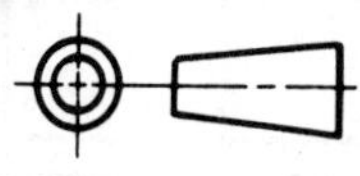

CONTACT ARM

A-52M

UNIT 22

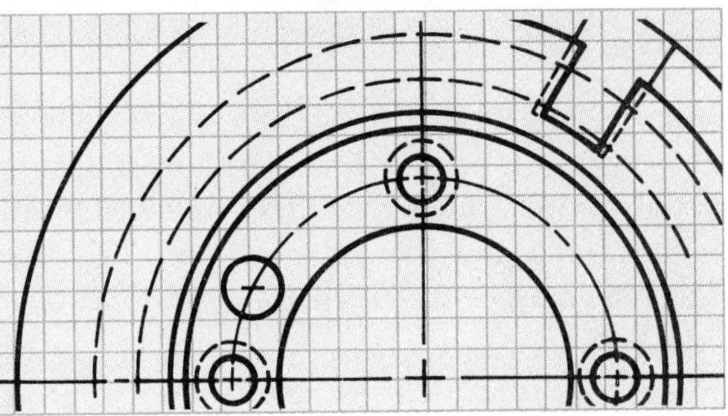

PIPING

Until one hundred years ago, water was the only important fluid conveyed from place to place through pipe. Today nearly every conceivable fluid is handled in pipe during its production, processing, transportation, or utilization. During the age of atomic energy and rocket power liquid metals, sodium, and nitrogen have been added to the list of more common fluids such as oil, water, and acids being transported through pipe. Many gases are also being stored and delivered through piping systems.

Pipe is also used for hydraulic and pneumatic mechanisms and used extensively for the controls of machinery and other equipment. Piping is also used as a structural element for columns and handrails.

The nominal size of pipe and the inside diameter, outside diameter, and wall thickness are given in inches.

Kinds of Pipe

Steel and Wrought Iron Pipe. This pipe carries water, steam, oil, and gas and is commonly used under high temperatures and pressures. Standard steel or cast iron pipe is specified by the nominal diameter, which is always less than the actual inner diameter (ID) of the pipe. Until recently, this pipe was available in only three weights – standard, extra strong, and double extra strong, figure 22-1. In order to use common fittings with these different pipe weights, the outside diameter (OD) of each of the different pipes remained the same. The extra metal was added to the ID to increase the wall thickness of the extra strong and double extra strong pipe.

The demand for a greater variety of pipe for use under increased pressure and temperature led to the introduction of ten different pipe weights, each designated by a schedule number. Standard pipe is now called schedule 40 pipe. Extra strong pipe is schedule 80.

Cast Iron Pipe. This is often installed underground to carry water, gas and sewage.

Seamless Brass and Copper Pipe. These pipes are used extensively in plumbing because of their ability to withstand corrosion.

Copper Tubing. This is used in plumbing and heating and where vibration and misalignment are factors, such as in automotive, hydraulic, and pneumatic design.

Plastic Pipe. This pipe or tubing, because of its resistance to corrosion and chemicals, is often used in the chemical industry. It is easily installed. However, it is not recommended where heat or pressure is a factor.

Pipe Joints and Fittings

Parts joined to pipe are called *fittings.* They may be used to change size or direction and to join or provide branch connections. There are three general classes of fittings: screwed, welded, and flanged. Other methods such as soldering, brazing, and gluing are used for cast iron pipe, and copper and plastic tubing.

Pipe fittings are specified by the nominal pipe size, the name of the fitting, and the material. Some fittings, for example, tees, crosses, and elbows, connect different sizes of pipe. These are called *reducing fittings.* Their nominal

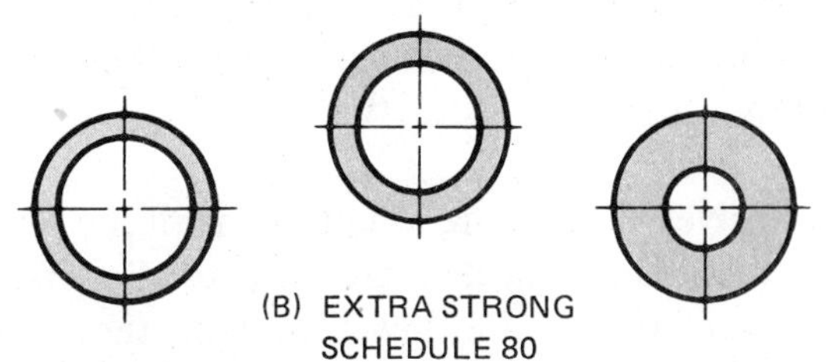

Fig. 22-1 Comparison of wall thicknesses

pipe sizes must be specified. The largest opening of the through run is given first, followed by the opposite end and the outlet. Figure 22-2 illustrates the method of designating sizes of reducing fittings.

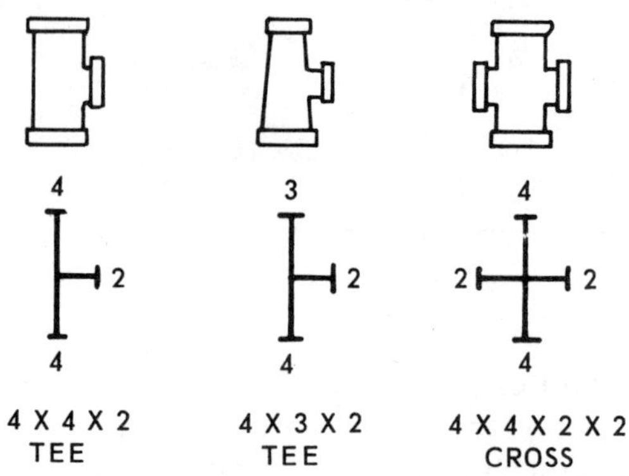

Fig. 22-2 Order of specifying the openings of reduced fittings

Screwed Fittings. Screwed fittings are generally used on small pipe design of 2.50 inch nominal pipe size or less.

There are two types of American Standard Pipe Thread: tapered and straight. The tapered thread is more common. Straight threads are used for special applications which are listed in the ANSI Handbook.

Tapered threads are designated on drawings as NPT (National Pipe Thread) or whichever standard is used and may be drawn either with or without the taper, figure 22-3. When drawn in tapered form, the taper is exaggerated. Straight pipe threads are designated on drawings as NPTS and standard thread symbols are used. Pipe threads are assumed to be tapered unless specified otherwise.

Welded Fittings. Welded fittings are used where connections will be permanent and on high pressure and temperature lines. The ends of the pipe and pipe fittings are usually bevelled to accommodate the weld.

Flanged Fittings. Flanged joint fittings provide a quick way to disassemble pipe. Flanges are attached to the pipe ends by welding, screwing, or lapping.

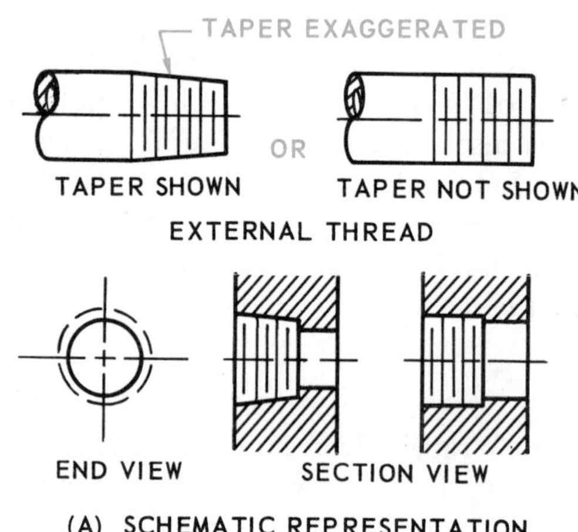

Valves

Valves are used in piping systems to stop or regulate the flow of fluids and gases. The following information describes a few of the more common types.

Gate Valves. These are used to control the flow of liquids. The wedge, or gate, lifts to allow full, unobstructed flow and lowers to stop the flow. They are generally used where operation is infrequent and are not intended for throttling or close control.

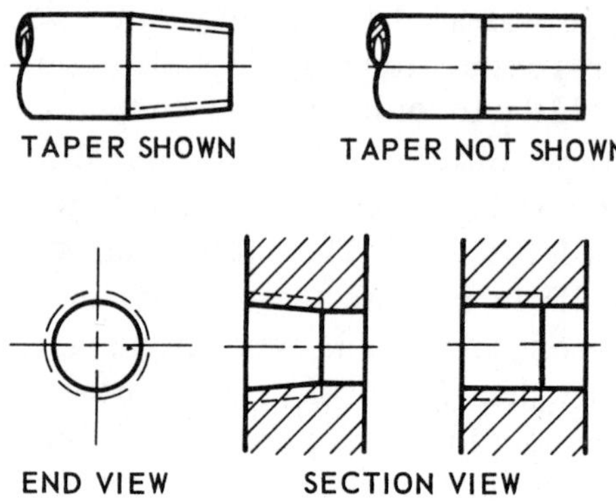

Fig. 22-3 Pipe thread conventions

Globe Valves. These are used to control the flow of liquids or gases. The design of the globe valve produces two changes in the direction of flow, slightly reducing the pressure in the system. The globe valve is recommended for the control of air, steam, gas, and other compressibles where instantaneous on-and-off operation is essential.

Check Valves. Check valves permit flow in one direction, but check all reverse flow. They are operated by pressure and velocity of line flow alone and have no external means of operation.

PIPING DRAWINGS

Piping drawings show the size and location of pipes, fittings, and valves. Because of the detail required to accurately describe these items, there is a set of symbols to indicate them on drawings.

There are two types of piping drawings in use, single-line and double-line drawings, figure 22-4. Double-line drawings take longer to draw and are therefore not recommended for production drawings. They are, however, suitable for catalogs and other applications where the appearance is more important than the extra drafting time.

Single-Line Drawings

Single-line drawings, also known as simplified representation, of pipe lines provide substantial savings without loss of clarity or reduction of comprehensiveness of information. Therefore, the simplified method is used whenever possible.

Single-line piping drawings use a single line to show the arrangement of the pipe and fittings. The center line of the pipe, regardless of pipe size, is drawn as a thick line to which symbols are added. The size of the symbol is left to the discretion of the drafter. When pipe lines carry different liquids, such as cold or hot water, a coded line symbol is often used.

Drawing Projection

Two methods of projection are used, orthographic and isometric, figure 22-5. Orthographic projection is recommended for the representation of single pipes which are either straight or bent in one plane only. However, this method is also used for more complicated pipings.

Isometric projection is recommended for all pipes bent in more than one plane and for assembly and layout work because the finished drawing is easier to understand.

Crossings. The crossing of pipes without connections is usually drawn without interrupting the line representing the hidden line, figure 22-6(A). But, when it is desirable to indicate that one pipe must pass behind the other, the line representing the pipe farthest away from the viewer will be shown with a break or interruption where the other pipe passes in front of it.

Connections. Permanent connections or junctions, whether made by welding or other processes, are indicated on the drawing by a heavy dot, figure 22-6(F). A general note or specification may indicate the process used.

Detachable connections or junctions are indicated by a single thick line. Specifications, a general note, or the Bill of Material will indicate the type of fitting, for example, flanges, union, or coupling. The specifications will also indicate whether the fittings are flanged, threaded, or welded.

PIPE DRAWING SYMBOLS

If specific symbols are not standardized, fittings such as tees, elbows, crosses, etc., are not specially drawn, but are represented, like pipe, by a continuous line. The circular symbol for a tee or elbow may be used when necessary to indicate whether the piping is viewed from the front or back, as shown in figure 22-6(H). Elbows on isometric drawings may be shown without the radius. However, if this method is used, the direction change of the piping must be shown clearly.

Adjoining Apparatus. If needed, adjoining apparatus, such as tanks, machinery, etc., not belong-

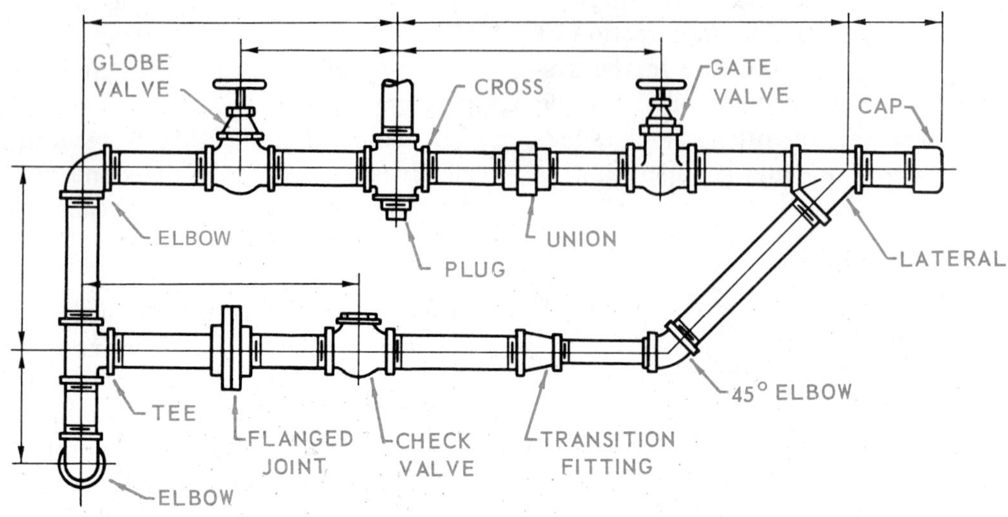

(A) DOUBLE-LINE DRAWING

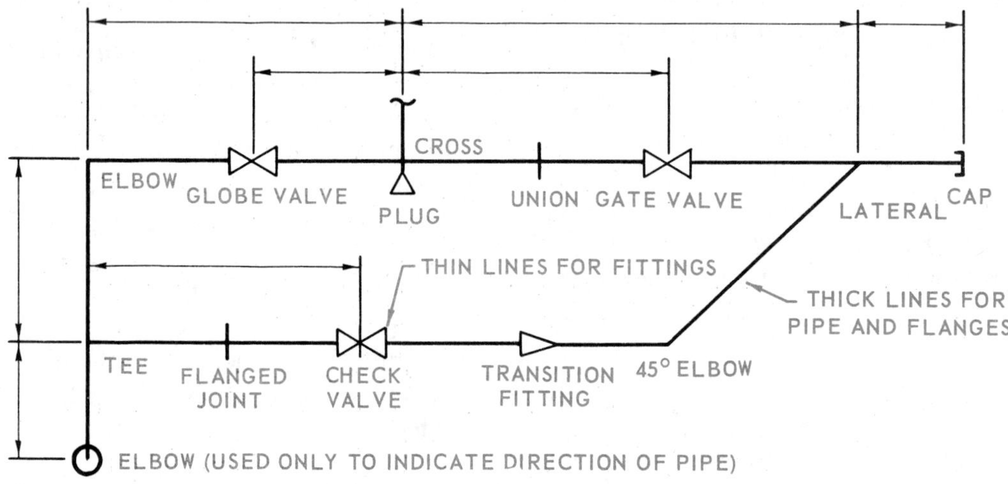

(B) SINGLE-LINE DRAWING

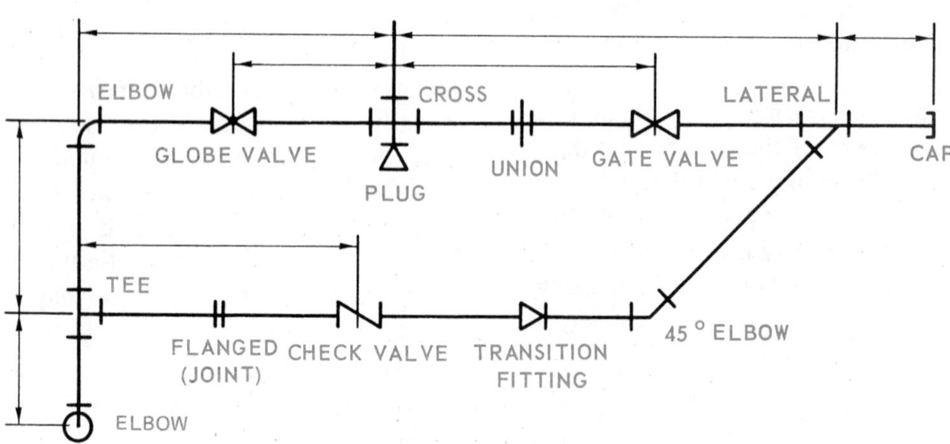

(C) FORMER SINGLE-LINE DRAWING SYMBOL

Fig. 22-4 Piping drawing symbols

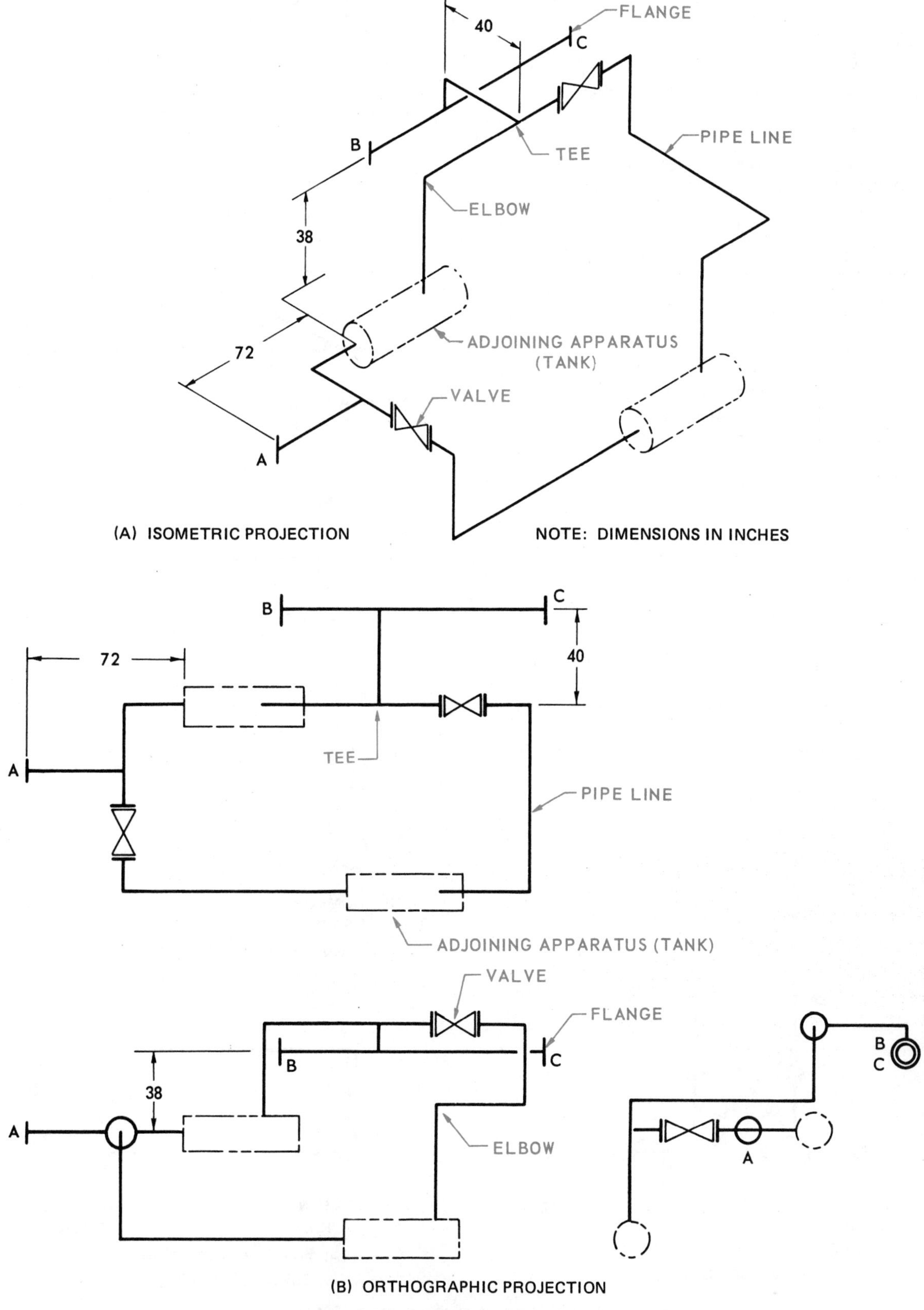

Fig. 22-5 Single-line piping drawing

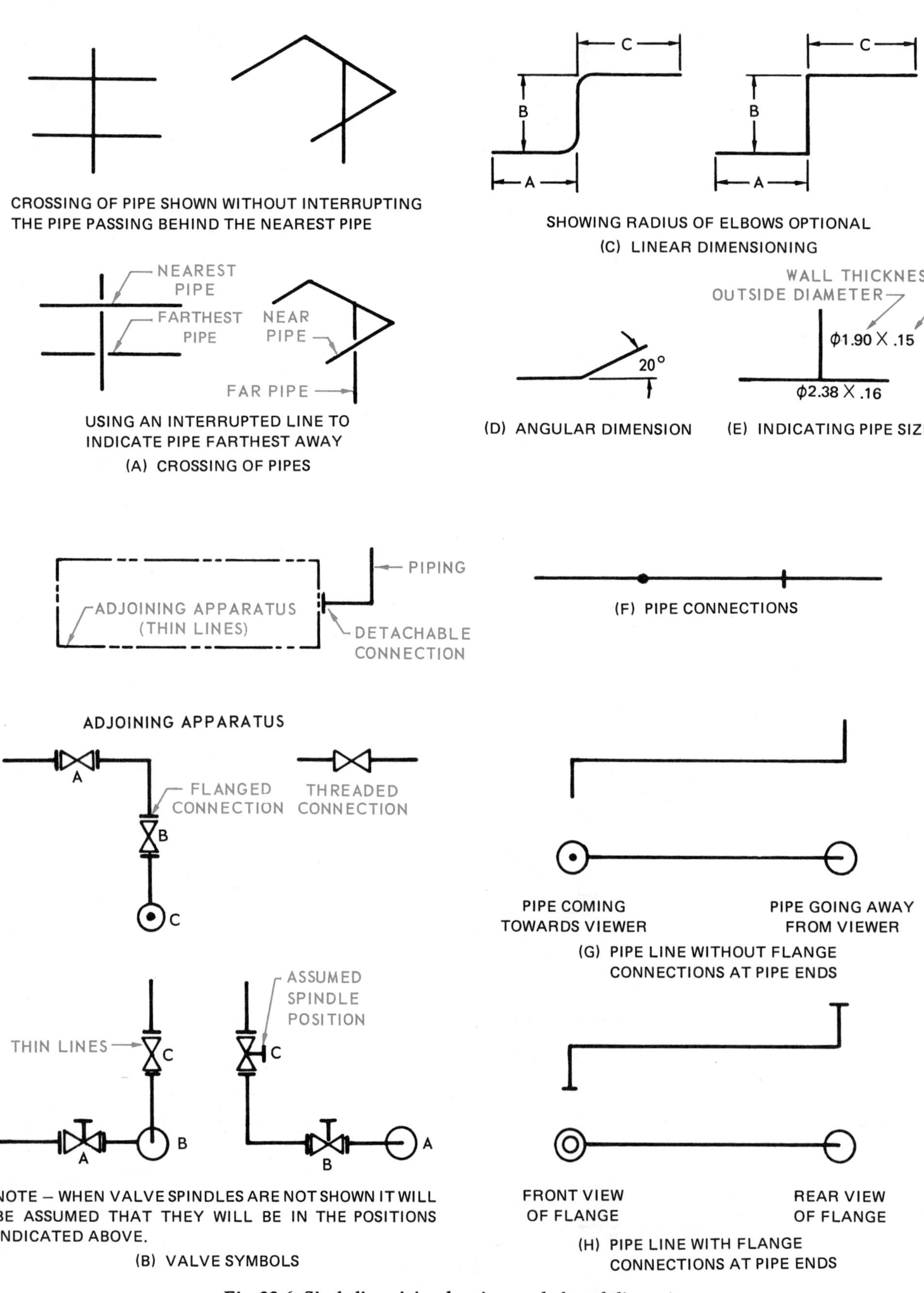
C
B
A
CROSSING OF PIPE SHOWN WITHOUT INTERRUPTING THE PIPE PASSING BEHIND THE NEAREST PIPE
SHOWING RADIUS OF ELBOWS OPTIONAL
(C) LINEAR DIMENSIONING
NEAREST PIPE
FARTHEST PIPE
NEAR PIPE
FAR PIPE
USING AN INTERRUPTED LINE TO INDICATE PIPE FARTHEST AWAY
(A) CROSSING OF PIPES
WALL THICKNESS
OUTSIDE DIAMETER
20°
ϕ1.90 X .15
ϕ2.38 X .16
(D) ANGULAR DIMENSION
(E) INDICATING PIPE SIZE
PIPING
ADJOINING APPARATUS (THIN LINES)
DETACHABLE CONNECTION
(F) PIPE CONNECTIONS
ADJOINING APPARATUS
FLANGED CONNECTION
THREADED CONNECTION
PIPE COMING TOWARDS VIEWER
PIPE GOING AWAY FROM VIEWER
(G) PIPE LINE WITHOUT FLANGE CONNECTIONS AT PIPE ENDS
ASSUMED SPINDLE POSITION
THIN LINES
NOTE – WHEN VALVE SPINDLES ARE NOT SHOWN IT WILL BE ASSUMED THAT THEY WILL BE IN THE POSITIONS INDICATED ABOVE.
(B) VALVE SYMBOLS
FRONT VIEW OF FLANGE
REAR VIEW OF FLANGE
(H) PIPE LINE WITH FLANGE CONNECTIONS AT PIPE ENDS

Fig. 22-6 Single-line piping drawing symbols and dimensions

ing to the piping itself, are shown by outlining them with a thin phantom line.

Dimensioning

- Dimensions for pipe and pipe fittings are always given from center-to-center of pipe and to the outer face of the pipe end or flange, figure 22-6(C).
- Individual pipe lengths are usually cut to suit by the pipe fitter. However, the total length of pipe required is usually called for in the Bill of Material.
- Pipe and fitting sizes and general notes are placed on the drawing beside the part concerned or, where space is restricted, with a leader.
- A Bill of Material is usually provided with the drawing.
- Pipes with bends are dimensioned from vertex to vertex.
- Radii and angles of bends are indicated as shown in figure 22-6(C) and (D). Whenever possible, the smaller of the supplementary angles is specified.
- The outer diameter and wall thickness of the pipe are indicated on the line representing the pipe, or in the Bill of Material, general note, or specifications, figure 22-6(E).

Orthographic Piping Symbols

Pipe Symbols. If flanges are not attached to the ends of the pipe lines when drawn in orthographic projection, pipe line symbols indicating the direction of the pipe are required. If the pipe line direction is toward the front (or viewer), it is shown by two concentric circles, the smaller one of which is a large solid dot, figure 22-6(G). If the pipe line direction is toward the back (or away from the viewer), it will be shown by one solid circle. No extra lines are required on the other views.

Flange Symbols. Irrespective of their type and sizes, flanges are to be represented by:

- two concentric circles for the front view,
- one circle for the rear view,
- a short stroke for the side view,

while using lines of equal thickness as chosen for the representation of pipes, figure 22-6(H).

Valve Symbols. Symbols representing valves are drawn with continuous thin lines (not thick lines as for piping and flanges). The valve spindles should only be shown if it is necessary to define their positions. It will be assumed that unless otherwise indicated, the valve spindle is in the position shown in figure 22-6(B).

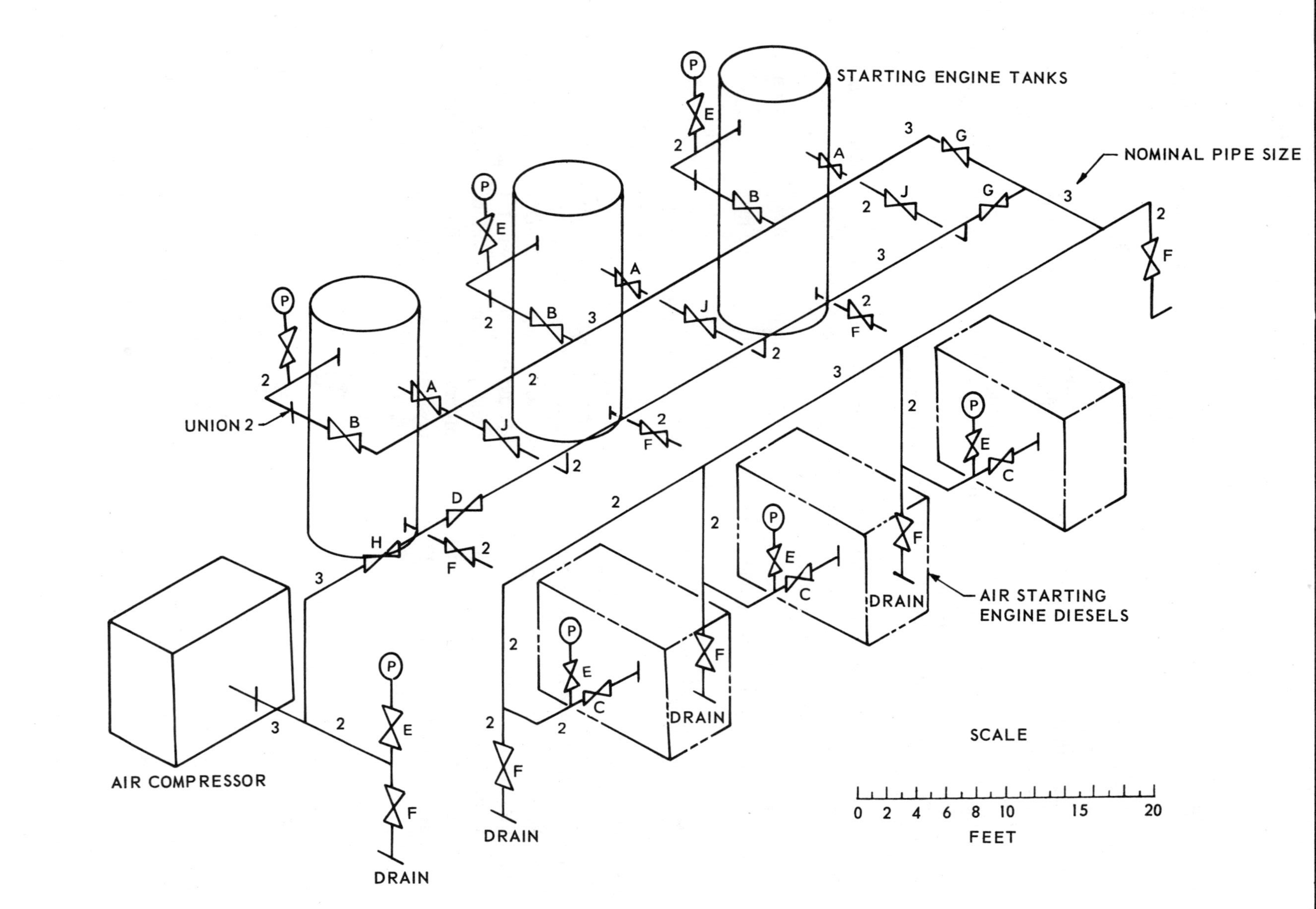

STARTING ENGINE TANKS
NOMINAL PIPE SIZE
UNION 2
AIR STARTING
ENGINE DIESELS
AIR COMPRESSOR
DRAIN
DRAIN
DRAIN
DRAIN
SCALE
0 2 4 6 8 10 15 20
FEET

Safety valves are provided for the compressor and the air storage tanks. Check valves are installed on the air storage tank feed lines and the compressor discharge lines to prevent accidental discharge of the tanks.

Piping is arranged so that the compressor will either fill the storage tanks and/or pump directly to the engines. Any of the three storage tanks may be used for starting, and pressure gauges indicate their readiness. The engines are fitted with quick-opening valves to admit air quickly at full pressure and shut it off at the instant rotation is obtained. A bronze globe valve is installed to permit complete shut-down of the engine for repairs, and regulation of the flow of air. Drains are provided at low points to remove condensate from the air storage tanks, lines, and engine feeds.

Globe valves are recommended throughout this hookup except on the main shutoff lines where gate valves are used because of infrequent operation.

ASSIGNMENT:

Complete the Bill of Material calling for all valves and fittings.

QUESTIONS (Use scale provided where necessary):

1. What size pipe is used for the main feed line?
2. Disregarding the length of the valves and fittings, what is the total approximate length of
   (a) the 3-inch pipe? (b) the 2-inch pipe?
3. What is the approximate center line to center line spacing of the diesel engines?
4. Calculate the approximate number of cubic feet of each of the air tanks.
5. Why are diesel engines used instead of public power supply?
6. What is the purpose of the air compressor?
7. Why is a gate valve used instead of a globe valve in the main line shutoff?
8. State the uses of the following parts.
   (a) Valve C (c) Gauge P (e) Valve D
   (b) Valve G (d) Valve F (f) Valve H

				14
				13
				12
				11
				10
				9
				8
				7
				6
				5
				4
				3
				2
				1
QTY	ITEM	MAT'L	DESCRIPTION	PT NO

ANSWERS

1. ________ 8. a. ________
2. a. ________ ________
   b. ________ b. ________
3. ________ ________
4. ________ c. ________
5. ________ ________
   ________ d. ________
6. ________ ________
   ________ e. ________
7. ________ ________
   ________ f. ________
   ____

CODE	VALVE	SERVICE
A	BRONZE GLOBE	AIR STORAGE TANK FEED LINES
B	BRONZE GLOBE	AIR STORAGE TANK DISCHARGE LINES
C	BRONZE GLOBE	DIESEL ENGINE SHUTOFF CONTROL
D	BRONZE GLOBE	AIR COMPRESSOR DISCHARGE
E	BRONZE GLOBE	PRESSURE GAUGE SHUTOFF
F	BRONZE GLOBE	DRAIN VALVES
G	SPINDLE GATES	MAIN LINE SHUTOFF
H	BRONZE CHECK	AIR COMPRESSOR CHECK
J	BRONZE CHECK	AIR STORAGE TANK FEED LINES
P	PRESSURE GAUGE	DISCHARGE OR FEED LINES

	SCALE	
	DRAWN	DATE
ISOMETRIC PROJECTION	**ENGINE STARTING AIR SYSTEM**	**A-53**

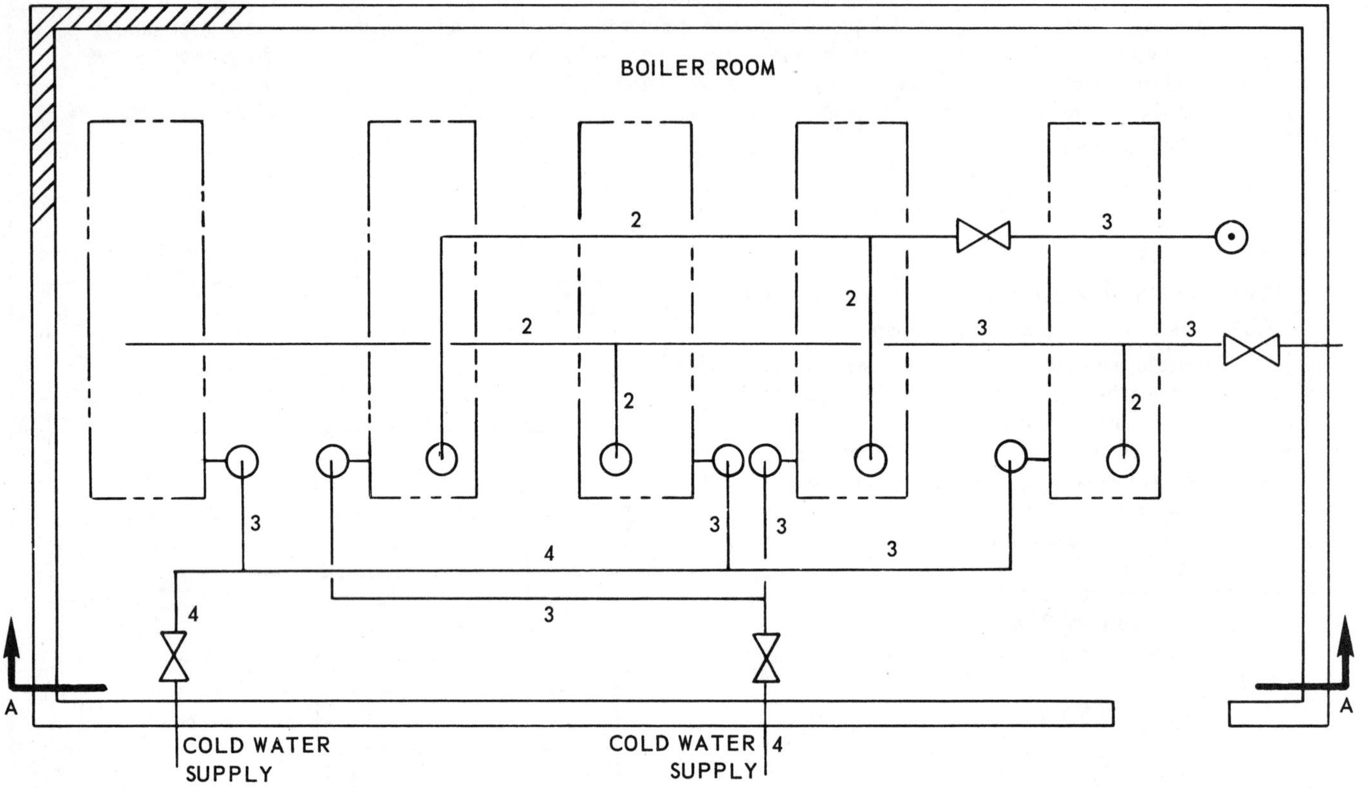

BOILER ROOM
2
3
2
2
3
3
2
2
2
3
3
3
4
3
4
3
A
A
COLD WATER SUPPLY
COLD WATER SUPPLY
4

PLAN VIEW OF BOILER ROOM

2ND FLOOR
HOT WATER MAINS
1 BOILER
2 BOILER
3 BOILER
4 BOILER
5 BOILER
IST FLOOR

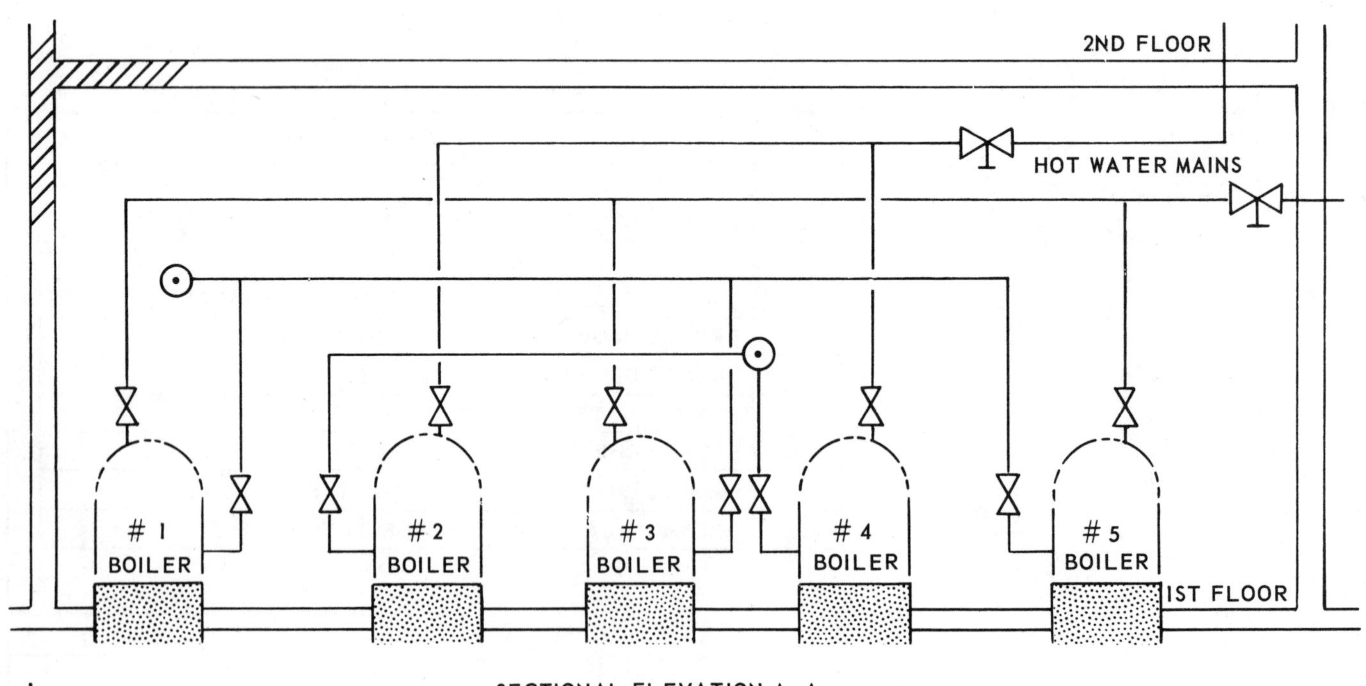

SECTIONAL ELEVATION A–A

ASSIGNMENT

1. Complete the isometric view of the Boiler Room piping shown in orthographic projection on the opposite page. It is not necessary to show the walls and floors.
2. On a separate sheet, make up a Bill of Material calling for all valves, fittings, and pipe. (Give the total approximate length of each size of pipe.)
3. From the Bill of Material, add part numbers to the isometric drawing for the valves and fittings only.

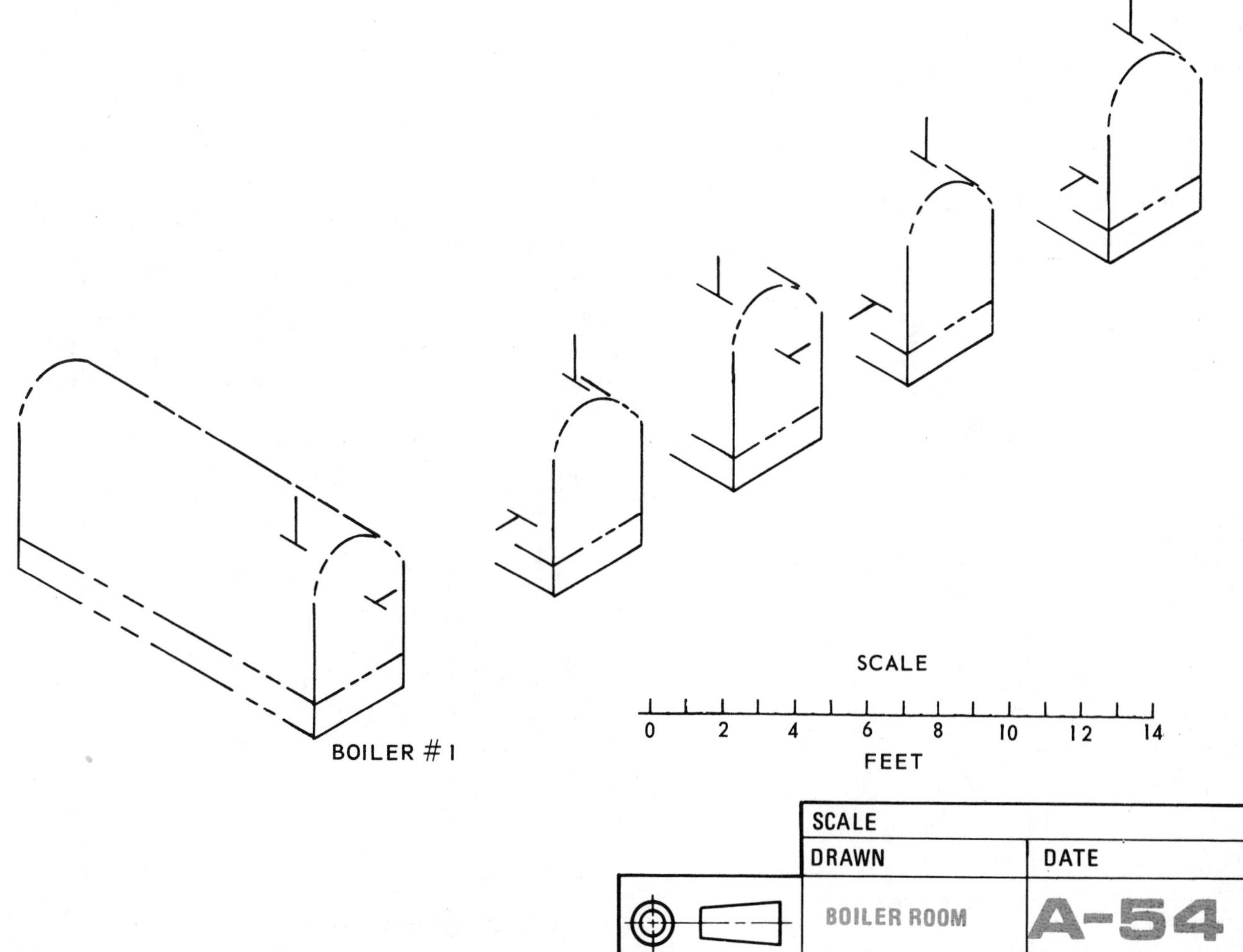

UNIT 23

BEARINGS

All rotating machinery parts are supported by *bearings*. Each bearing type and style has its particular advantages and disadvantages. Bearings are classified into two groups: plain bearings and antifriction bearings.

Plain Bearings

Plain bearings have many uses. They are available in a variety of shapes and sizes, figure 23-1. Because of their simplicity, plain bearings are versatile. There are several plain bearing categories. The most common are journal (sleeve) bearings and thrust bearings. These are available in a variety of standard sizes and shapes.

Journal or Sleeve Bearings. Journal bearings are the simplest and most economical means of supporting moving parts. *Journal bearings* are usually made of one or two pieces of metal enclosing a shaft. They have no moving parts. The journal is the supporting portion of the shaft.

Speed, mating materials, clearances, temperature, lubrication and type of loading affect the performance of bearings. The maintenance of an oil film between the bearing surfaces is important. The oil film reduces friction, dissipates heat, and retards wear by minimizing metal-to-metal contact, figure 23-2. Starting and stopping are the most critical periods of operation, because the load may cause the bearing surfaces to touch each other.

The shaft should have a smooth finish and be harder than the bearing material. The bearing will perform best with a hard, smooth shaft. For practical reasons, the length of the bearing should be between one and two times the shaft diameter. The outside diameter should be approximately 25 percent larger than the shaft diameter.

Cast bronze and porous bronze are usually used for journal bearings.

Bearings are sometimes split. This design feature facilitates assembly and permits adjustment and replacement of worn parts. Split bearings allow the shaft to be set in one half of

(A) JOURNAL TYPE

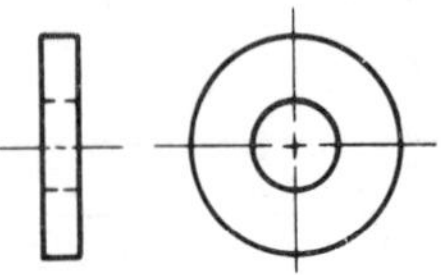

(B) THRUST TYPE

Fig. 23-1 Plain bearings

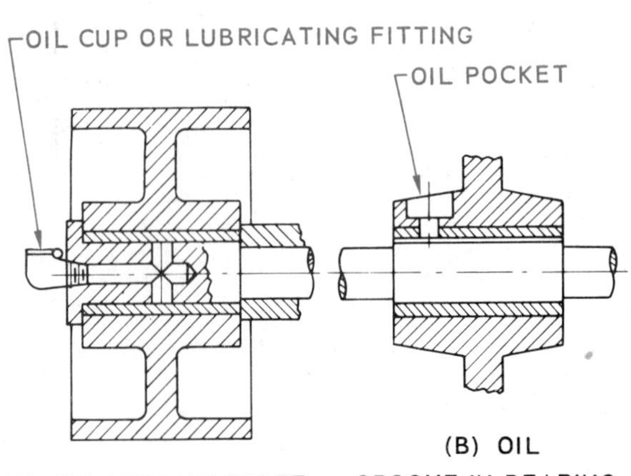

Fig. 23-2 Common methods of lubricating journal bearings

the bearing while the other half, or cover, is later secured in position, figure 23-3.

If the bearings shown in figure 23-3 are to be made from two parts, they must be fastened together before the hole is bored or reamed. This will facilitate the machining operation and make a perfectly round bearing. For an incorrectly assembled bearing, see figure 23-4.

One method that gives longer life to the bearing is the insertion of very thin strips of metal between the base and cover halves before boring. These thin strips of varying thickness are called *shims,* figure 23-5.

When a bearing is shimmed, the same number of pieces of corresponding thickness are used on both sides of the bearing.

As the hole wears, one or more pairs of these shims may be removed for wear compensation.

Thrust Bearings. Plain thrust bearings or thrust washers are available in various materials, including: sintered metal, plastic, woven TFE fabric on steel backing, sintered Teflon-bronze-lead on metal backing, aluminum alloy on steel, aluminum alloy, and carbon-graphite.

Antifriction Bearings

Ball, roller and needle bearings are classified as antifriction bearings as friction has been reduced to a minimum. These are covered in detail in Unit 40.

REFERENCES AND SOURCE MATERIAL

1. A.O. De Hart, "Basic Bearing Types" *Machine Design,* 40, No. 14
2. W.A. Glaeser, "Plain Bearings" *Machine Design,* 40, No. 14

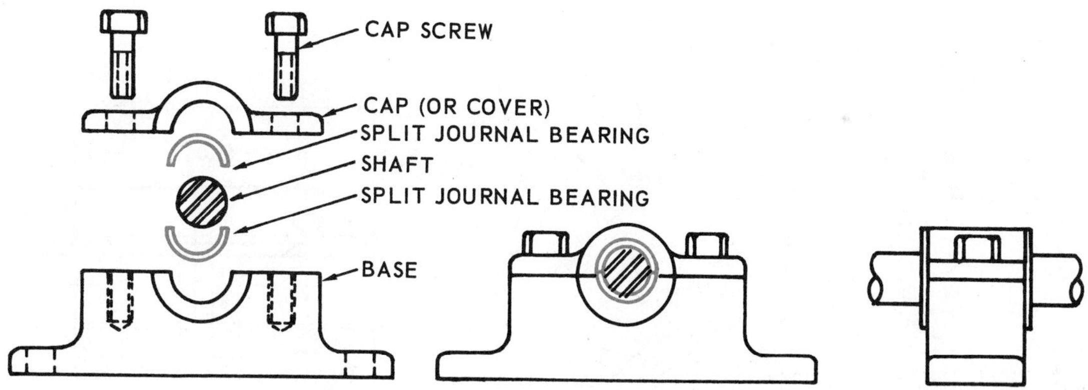

Fig. 23-3 Pillow block with split journal bearing

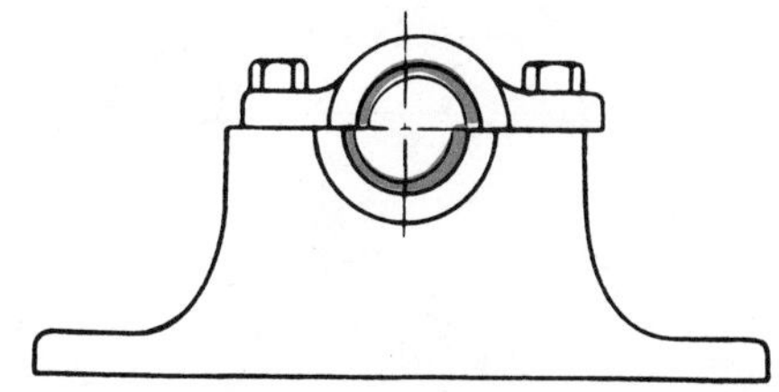

Fig. 23-4 Bearing halves incorrectly matched

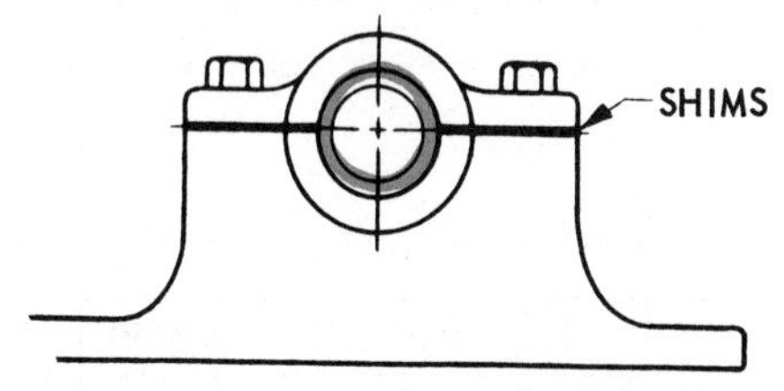

Fig. 23-5 "Shimmed" bearing

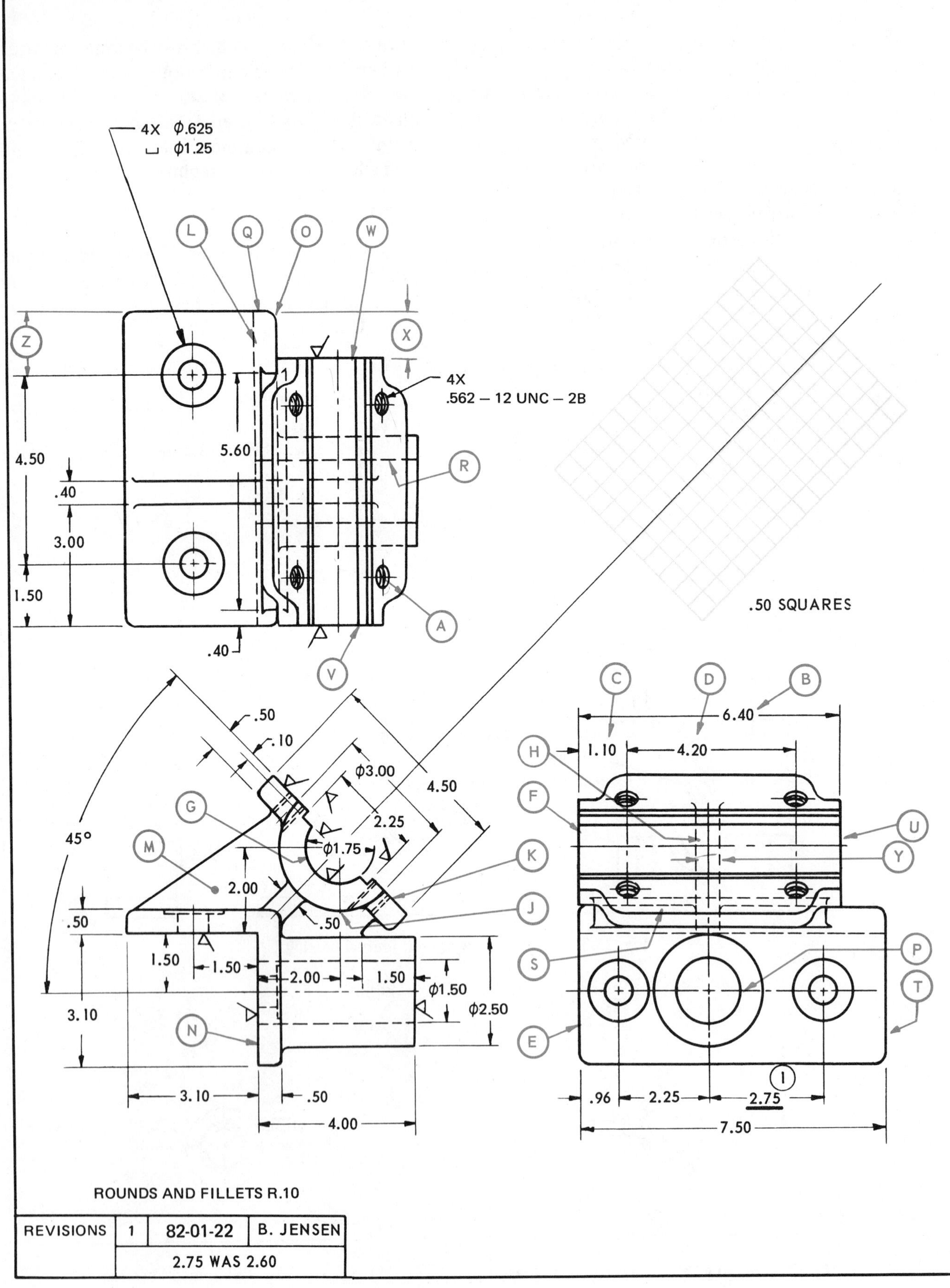

REVISIONS	1	82-01-22	B. JENSEN
	2.75 WAS 2.60		

QUESTIONS

1. How many definite finished surfaces are on the casting? (Note—two or more surfaces could lie on one plane.)
2. What is the size of (P) hole?
3. What size spotface is used on the mounting holes?
4. What are the number and size of the mounting holes?
5. Note that surface (E) is not finished, but surface (F) is to be finished. Allowing .06 in. for finishing, what would be the depth of the rough casting?
6. Would (G) hole be bored before or after bearing cap is assembled?
7. Which line in the side view shows surface (M) ?
8. In which view, and by what line, is the surface represented by (N) shown?
9. Which line or surface in the side view shows the projection of point (J) ?
10. Which point or surface in the side view does line (R) represent?
11. Locate surface (T) in the top view.
12. Locate surface (U) in the top view.
13. Locate (V) in the side view.
14. What are dimensions (X), (Y), and (Z) ?
15. What size is the round (O) ?
16. Determine dimension (A) and place it correctly on the sketch of the auxiliary view.
17. Place dimensions (B), (C), and (D) correctly on the sketch of the auxiliary view.

ANSWERS

1 __________
2 __________
3 __________
4 __________
5 __________
6 __________
7 __________
8 __________
9 __________
10 __________
11 __________
12 __________
13 __________
14 (X) __________
(Y) __________
(Z) __________
15 __________
16 __________

SKETCHING ASSIGNMENT

MAKE A FREEHAND SKETCH IN THE GRAPH AREA SHOWN AT LEFT, SHOWING THE AUXILIARY VIEW OF THE SPLIT BEARING SURFACES (K) AND (G). SEE QUESTIONS 16 AND 17.

NOTE: UNLESS OTHERWISE SPECIFIED:

- TOLERANCE ON DIMENSIONS ±.02
- TOLERANCE ON ANGLES ±0.5°
- ∀ TO BE 63∀

MATERIAL	GRAY IRON	
SCALE	NOT TO SCALE	
DRAWN		DATE
CORNER BRACKET		A-55

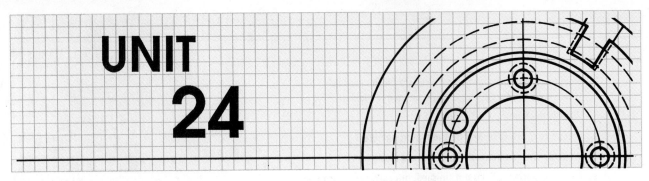

UNIT 24

STEEL SPECIFICATIONS

Carbon steels are the workhorses of product design. They account for over 90 percent of total steel production. More carbon steels are used in product manufacturing than all other metals combined. Far more research is going into carbon steel metallurgy and manufacturing technology than into all other steel mill products. Various technical societies and trade associations have issued the specifications covering the composition of metals. They serve as a selection guide and are a way for the buyer to conveniently specify certain known and recognized requirements. The main technical societies and trade associations concerned with metal identification in the United States are the American Iron and Steel Institute (AISI), and the Society of Automotive Engineers (SAE).

SAE and AISI Systems of Steel Identification

The specifications for steel bar are based on a number code indicating the composition of each type of steel covered. They include both plain carbon and alloy steels. The code is a 4-number system. The first two figures indicate the alloy series and the last two figures the carbon content in hundredths of a percent, figure 24-1. Therefore, the figure XX15 indicates 0.15 of 1 percent carbon.

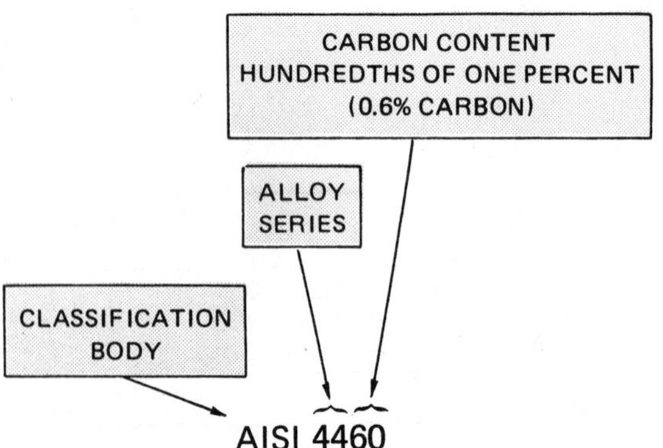

Fig. 24-1 Steel designation system

For example, AISI 4830 is a molybdenum-nickel steel containing 0.2–0.3 percent molybdenum, 3.25–3.75 percent nickel and 0.3 percent carbon. In addition to the 4-number designation the suffix "H" is used for steels produced to specify hardenability limits, and the prefix "E" indicates a steel made by the basic electric-furnace method.

Originally, the second figure indicated the percentage of the major alloying element present. This was true of many of the alloy steels. However, this had to be varied in order to classify all the steels that became available.

Alloying materials are added to steel to improve properties such as strength, hardness, machinability, corrosion resistance, electrical conductivity, and ease of forming. Figure 24-2 lists many types of steel and their uses.

Effect of Alloys on Steel

Carbon. Increasing the carbon content increases the tensile strength and hardness.

Sulphur. When sulphur content is over 0.06 percent the metal tends toward *red shortness* (brittleness in steel when it is red hot). Free cutting steel, for threading and screw machine work, is obtained by increasing sulphur content from about 0.075 to 0.1 percent.

Phosphorus. Phosphorus produces brittleness and general cold shortness. It also strengthens low carbon steel, increases resistance to corrosion, and improves machinability.

Manganese. Manganese is added during the making of steel to prevent red shortness and increase hardenability.

Chromium. Chromium increases hardenability, corrosion resistance, oxidation, and abrasion.

Nickel. Nickel strengthens and toughens ferrite and pearlite steels.

Silicon. Silicon is used as a general purpose deoxidizer. It strengthens low-alloy steels and increases hardenability.

Copper. Copper is used to increase atmospheric corrosion resistance.

Molybdenum. Molybdenum increases hardenability and coarsening temperature.

Type of Steel	AISI Symbol	Principal Properties	Common Uses
Carbon Steels			
– Nonresulfurized (Basic Carbon Steel)	10XX		
– Plain Carbon	10XX		
– Low Carbon Steel (0.06% to 0.2% Carbon)	1006 to 1020	Toughness and Less Strength	Chains, Rivets, Shafts, Pressed Steel Products
– Medium Carbon Steel (0.2% to 0.5% Carbon)	1020 to 1050	Toughness and Strength	Gears, Axles, Machine Parts, Forgings, Bolts and Nuts
– High Carbon Steel (Over 0.5% Carbon)	1050 and over	Less Toughness and Greater Hardness	Saws, Drills, Knives, Razors, Finishing Tools, Music Wire
– Resulfurized (Free Cutting)	11XX	Improves Machinability	Threads, Splines, Machined Parts
– Phosphorized	12XX	Increases Strength and Hardness but Reduces Ductility	
– Manganese Steels	13XX	Improves Surface Finish	
Molybdenum Steels			
0.15% – 0.30% Mo	40XX	High Strength	Axles, Forgings, Gears, Cams, Mechanism Parts
0.08% – 0.35% Mo 0.4% – 1.1% Cr	41XX		
1.65% – 2.00% Ni 0.2% – 0.3% Mo 0.4% – 0.9% Cr	43XX		
0.45% – 0.60% Mo	44XX		
0.7% – 2.0% Ni 0.15% – 0.30% Mo	46XX		
0.9% – 1.2% Ni 0.15% – 0.40% Mo 0.35% – 0.55% Cr	47XX		
3.25% – 3.75% Ni 0.2% – 0.3% Mo	48XX		
Chromium Steels			
0.3% – 0.5% Cr	50XX	Hardness, Great Strength and Toughness	Gears, Shafts, Bearings Springs, Connecting Rods
0.70% – 1.15% Cr	51XX		
1.00% C 0.90% – 1.15% Cr	E51100		
1.00% C 0.90% – 1.15% Cr	E52100		
Chromium Vanadium Steels			
0.5% – 1.1% Cr 0.10% – 0.15% V	61XX	Hardness and Strength	Punches and Dies, Piston Rods, Gears, Axles
Nickel - Chromium - Molybdenum Steels			
0.4% – 0.7% Ni 0.4% – 0.6% Cr 0.15% – 0.25% Mo	86XX	Rust Resistance, Hardness and Strength	Food Containers, Surgical Equipment
0.4% – 0.7% Ni 0.4% – 0.6% Cr 0.2% – 0.3% Mo	87XX		
0.4% – 0.7% Ni 0.4% – 0.6% Cr 0.3% – 0.4% Mo	88XX		
Silicon Steels			
1.8% – 2.2% Si	92XX	Springiness and Elasticity	Springs

Fig. 24-2 Designations, uses, and properties of steel

Boron. Boron increases hardenability of lower carbon steels and has better machinability than standard alloy steels.

Vanadium. Vanadium elevates coarsening temperatures, increases hardenability, and is a strong deoxidizer.

DRAWINGS FOR NUMERICAL CONTROL

Numerical control is a means of directing some or all of the functions of a machine automatically from instructions. The instructions are generally stored on tape and are fed to the controller through a tape reader. The controller interprets the coded instructions and directs the machine through the required operations.

It has been established that because of the consistent high accuracy of numerically controlled machines, and because human errors have been almost entirely eliminated, scrap has been considerably reduced.

Another area where numerically controlled machines are better is in the quality or accuracy of the work. In many cases a numerically controlled machine can produce parts more accurately at no additional cost, resulting in reduced assembly time and better interchangeability of parts. This latter fact is especially important when spare parts are required.

Computer-aided design and computer-aided manufacturing techniques are now widely used in conjunction with numerical control processes in industry.

Dimensioning for Numerical Control

Common guidelines have been established that enable dimensioning and tolerancing practices to be used effectively in delineating parts for both numerical control and conventional fabrication.

Coordinate System

The numerical control concept is based on the system of rectangular or Cartesian coordinates in which any position can be described in terms of distance from an origin point along either two or three mutually perpendicular axes. Two dimensional coordinates (x, y) define points in a plane. See figure 24-3, and drawings A-56M and A-57.

The x-axis is horizontal and is considered the first and basic reference axis. Distances to the right of the zero x-axis are considered positive x values and to the left of the zero x-axis as negative x values.

The y-axis is vertical and perpendicular to the x-axis in the plane of a drawing showing xy relationships. Distances above the zero y-axis are considered positive y values and below the zero y-axis as negative y values. The position where the x and y axes cross is called the origin, or zero point. For additional information on coordinate dimensioning, refer to Unit 17.

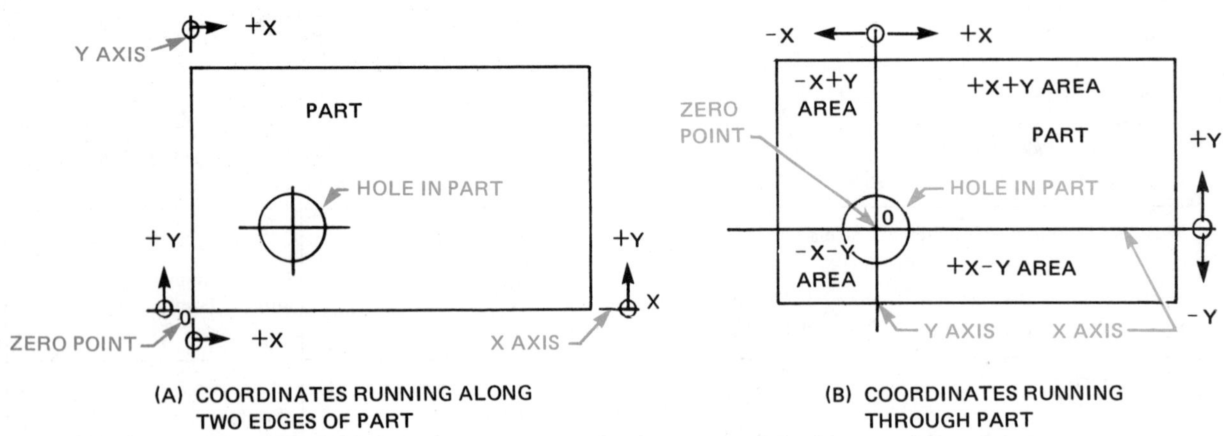

Fig. 24-3 Coordinate systems

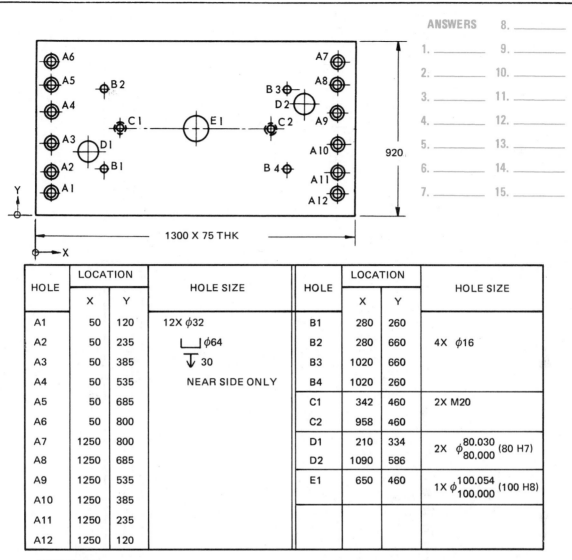

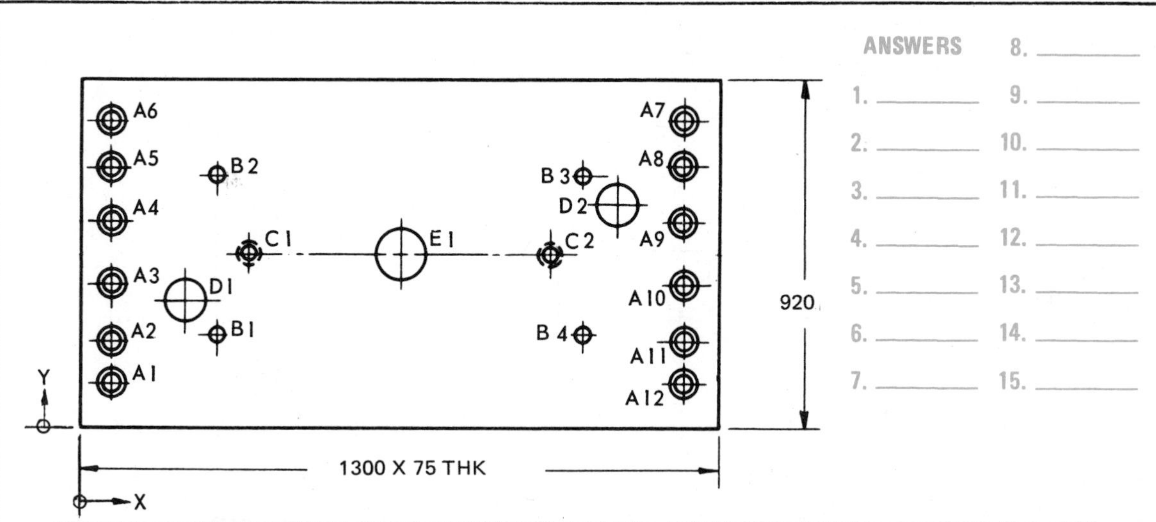

ANSWERS

1. ______
2. ______
3. ______
4. ______
5. ______
6. ______
7. ______
8. ______
9. ______
10. ______
11. ______
12. ______
13. ______
14. ______
15. ______

HOLE	LOCATION		HOLE SIZE	HOLE	LOCATION		HOLE SIZE
	X	Y			X	Y	
A1	50	120	12X ϕ32	B1	280	260	
A2	50	235	⌴ ϕ64	B2	280	660	4X ϕ16
A3	50	385	↧ 30	B3	1020	660	
A4	50	535	NEAR SIDE ONLY	B4	1020	260	
A5	50	685		C1	342	460	2X M20
A6	50	800		C2	958	460	
A7	1250	800		D1	210	334	2X $\phi^{80.030}_{80.000}$ (80 H7)
A8	1250	685		D2	1090	586	
A9	1250	535		E1	650	460	1X $\phi^{100.054}_{100.000}$ (100 H8)
A10	1250	385					
A11	1250	235					
A12	1250	120					

NOTE: UNLESS OTHERWISE SPECIFIED, TOLERANCE ON DIMENSIONS ±0.5

QUESTIONS

1. What is the thickness of the material?
2. What type of steel is specified?
3. What indicates that the drawing is not to scale?
4. How many counterbored holes are there?
5. What is the diameter of the CBORE?
6. How many **D** holes are there?
7. Which letter specifies the ϕ100 hole?
8. How many different sized holes are specified?
9. What is the total number of tapped holes?
10. What is the thread pitch of the **C** holes?
11. What is the tolerance given for the **E** hole?
12. What is the high limit of the **E** hole?
13. What is the maximum size for the **D** holes?
14. Is the **E** hole on the center of the part?
15. What is the center distance between the **C** holes?

NOTE: ALL DIMENSIONS ARE IN MILLIMETERS

MATERIAL	SAE 1020
SCALE	NOT TO SCALE
DRAWN	DATE
CROSSBAR	A-56M

METRIC

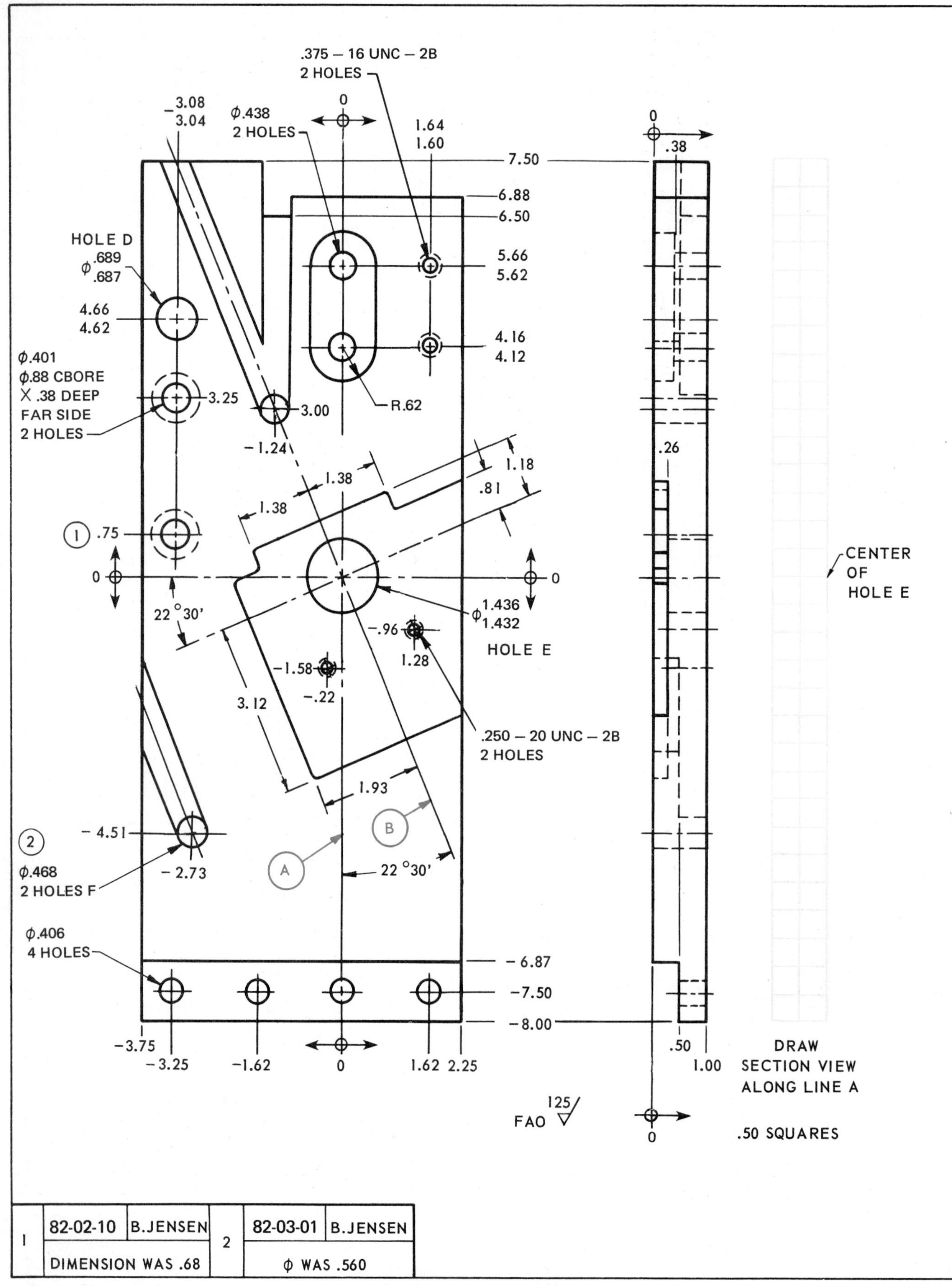
.375 – 16 UNC – 2B
2 HOLES
ϕ.438
2 HOLES
3.08
3.04
1.64
1.60
7.50
6.88
6.50
.38
HOLE D
ϕ .689
.687
5.66
5.62
4.66
4.62
4.16
4.12
ϕ.401
ϕ.88 CBORE
X .38 DEEP
FAR SIDE
2 HOLES
3.25
3.00
R.62
– 1.24
.26
1.18
1.38
.81
1.38
1 .75
0
0
22 °30'
ϕ 1.436
1.432
–.96
1.28
HOLE E
–1.58
–.22
3.12
.250 – 20 UNC – 2B
2 HOLES
CENTER
OF
HOLE E
1.93
B
A
2 – 4.51
– 2.73
ϕ.468
2 HOLES F
22 °30'
ϕ.406
4 HOLES
– 6.87
–7.50
–8.00
–3.75
– 3.25
–1.62
0
1.62
2.25
.50
1.00
DRAW
SECTION VIEW
ALONG LINE A
FAO 125
.50 SQUARES
1
82-02-10
B.JENSEN
DIMENSION WAS .68
2
82-03-01
B.JENSEN
ϕ WAS .560

CENTER OF HOLE E

DRAW SECTION VIEW ALONG LINE B

QUESTIONS

1. What are the overall (A) width, (B) height, (C) depth of the parts?
2. What is the width of the slots that are cut out from the **F** holes?
3. What is their depth?
4. What class of surface texture is required?
5. What is the depth of the recess adjacent to the (A) E hole, (B) Ø.438 holes?
6. How many degrees are there between line **B** and the horizontal?
7. What was the size of the **F** holes before they were changed?
8. What tolerance is required on the **E** hole?
9. What is the low limit on the **E** hole?
10. What tolerance is required on the **D** hole?
11. What is the high limit on the **D** hole?
12. Determine the maximum vertical distance from D hole to the four Ø.406 holes located at the bottom of the part.
13. Determine the (A) maximum, (B) minimum, distance between the two .375 tapped holes.
14. Determine the maximum horizontal distance between D hole and the .375 tapped holes.
15. What are the width, height, and depth of the cutout at the bottom of the part?
16. How deep are (A) the Ø.438 holes, (B) the Ø.401 holes?
17. How deep are the .250-20 UNC holes?
18. How many full threads do the .250-20 UNC tapped holes have?
19. What are the main alloys in the material?

ANSWERS

1. A. ______
   B. ______
   C. ______
2. ______
3. ______
4. ______
5. A. ______
   B. ______
6. ______
7. ______
8. .004
9. ______
10. ______
11. ______
12. ______
13. A. ______
    B. ______
14. ______
15. W ______
    H ______
    D ______
16. A. ______
    B. ______
17. ______
18. ______
19. ______

NOTE:

- ALL RADII R.06 UNLESS OTHERWISE SPECIFIED.
- DIMENSIONS ARE TAKEN FROM PLANES DESIGNATED 0–0, AND ARE PARALLEL TO THESE PLANES.
- UNLESS OTHERWISE STATED:
  ±.02 ON DIMENSIONS
  ±0.5° ON ANGLES.

MATERIAL	SAE 4020
SCALE	NOT TO SCALE
DRAWN	DATE
OIL CHUTE	A-57

UNIT 25

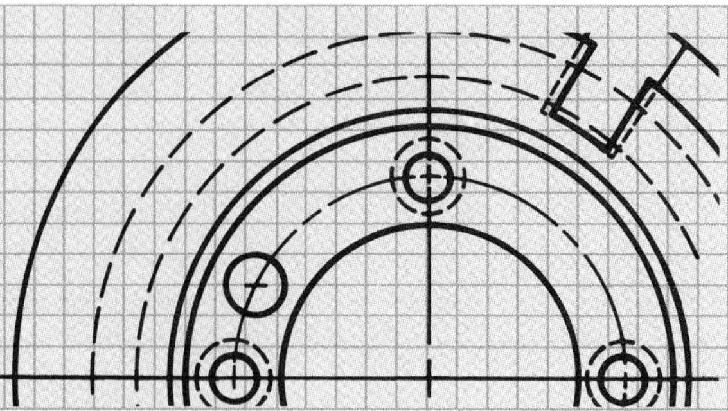

CASTINGS

Irregular or odd-shaped parts which are difficult to make from metal plate or bar stock may be cast to the desired shape. Casting processes for metals can be classified by either the type of mold or pattern or the pressure or force used to fill the mold. Conventional sand, shell, and plaster molds utilize a durable pattern, but the mold is used only once. Permanent molds and die-casting dies are machined in metal or graphite sections and are employed for a large number of castings. Investment casting and the full mold process involve both an expendable mold and pattern.

Sand Mold Casting

The most widely used casting process for metals uses a permanent pattern of metal or wood that shapes the mold cavity when loose molding material is compacted around the pattern. This material consists of a relatively fine sand plus a binder that serves as the adhesive.

Figures 25-1 and 25-2 show a typical sand mold, with the various provisions for pouring the molten metal and compensating for contraction of the solidifying metal, and a sand core for forming a cavity in the casting. Sand molds are prepared in flasks which consist of two or more sections: bottom (drag), top (cope), and intermediate sections (checks) when required.

The cope and drag are equipped with pins and lugs to insure the alignment of the flask. Molten metal is poured into the sprue and connecting runners provide flow channels for the metal to enter the mold cavity through gates. Riser cavities are located over the highest section of the casting.

The gating system, besides providing a way for the molten metal to enter the mold, functions as a venting system for the removal of gases from the mold and acts as a riser to furnish liquid metal to the casting during solidification.

In producing sand molds, a metal or wooden pattern must first be made. The pattern is slightly larger in every dimension than the part to be cast to allow for shrinkage when the casting cools. This is known as *shrinkage allowance,* and the patternmaker allows for it by using a shrink rule for each of the cast metals. Since shrinkage and draft are taken care of by the patternmaker, they are of no concern to the drafter.

Additional metal, known as machining or finish allowance, must be provided on the casting where a surface is to be finished. Depending on

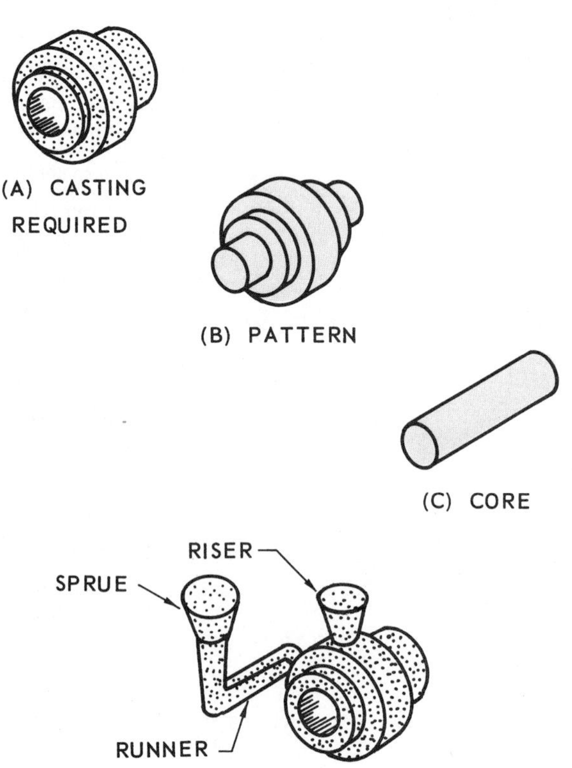

Fig. 25-1 Sand casting parts

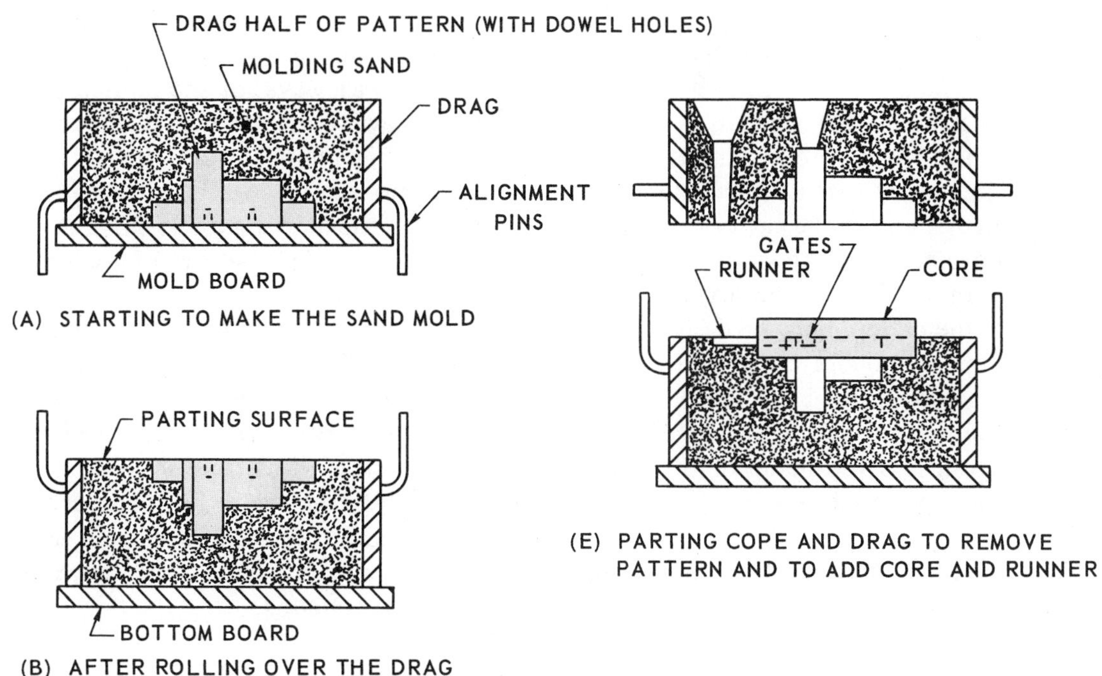

(A) STARTING TO MAKE THE SAND MOLD

(B) AFTER ROLLING OVER THE DRAG

(E) PARTING COPE AND DRAG TO REMOVE PATTERN AND TO ADD CORE AND RUNNER

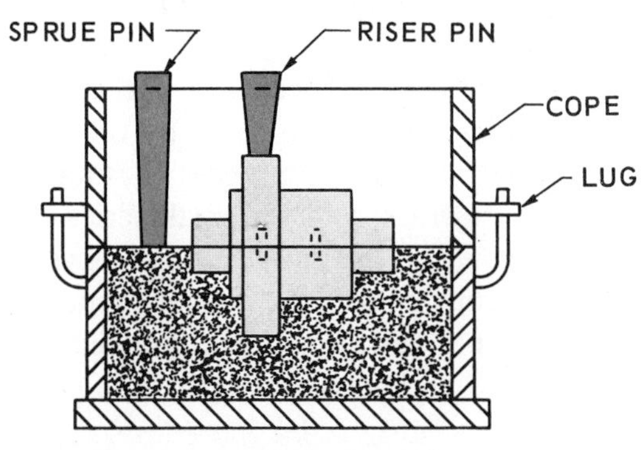

(C) PREPARING TO RAM MOLDING SAND IN COPE

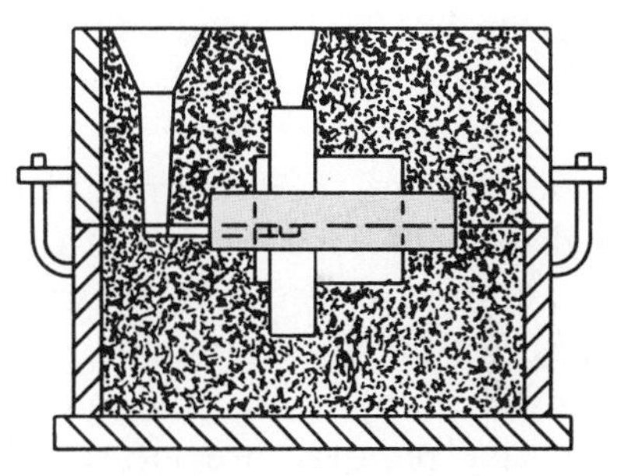

(F) SAND MOLD READY FOR POURING

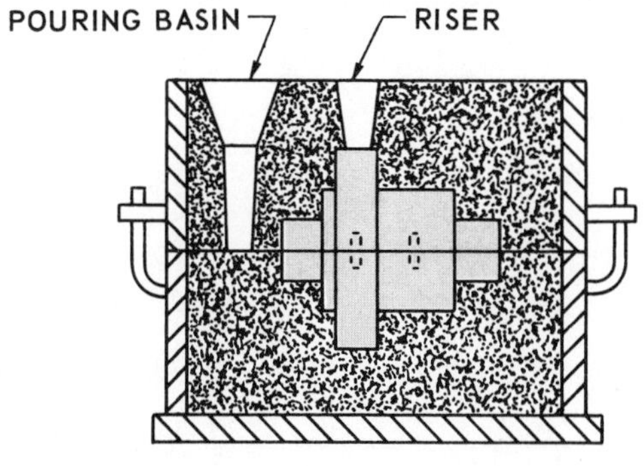

(D) REMOVING RISER AND GATE SPRUE PINS AND ADDING POURING BASIN

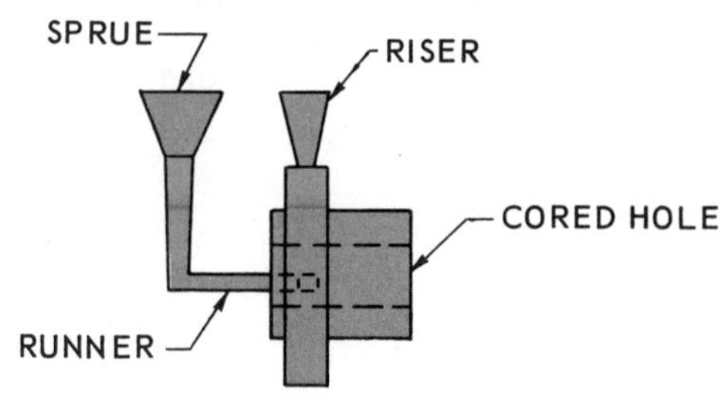

SPRUE, RISER, AND RUNNER TO BE REMOVED FROM CASTING

(G) CASTING AS REMOVED FROM THE MOLD

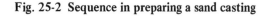

Fig. 25-2 Sequence in preparing a sand casting

the material being cast, from .06 to .12 inch is usually allowed on small castings for each surface requiring finishing.

When casting a hole or recess in a casting, a core is often used. A core is a mixture of sand and a bonding agent that is baked and hardened to the desired shape of the cavity in the casting, plus an allowance to support the core in the sand mold. In addition to the shape of the casting desired, the pattern must be designed to produce areas in the mold cavity to locate and hold the core. The core must be solidly supported in the mold, permitting only that part of the core that corresponds to the shape of the cavity in the casting to project into the mold.

Preparation of Sand Molds. The drag portion of the flask is first prepared in an upside-down position with the pins pointing down, figure 25-2(A). The drag half of the pattern is placed in position on the mold board and a light coating of parting compound is used as a release agent. The molding sand is then rammed or pressed into the drag flask. A bottom board is placed on the drag, the whole unit is rolled over, and the mold board is removed.

The cope half of the pattern is placed over the drag half and the cope portion of the flask is placed in position over the pins, figure 25-2(C). A light coating of parting compound is sprinkled throughout. Next the sprue pin and riser pin, tapered for easy removal, are located, and the molding sand is rammed into the cope flask. The sprue pin and riser pin are then removed, figure 25-2(D). A pouring basin may or may not be formed at the top of the gate sprue.

Now the cope is lifted carefully from the drag and the pattern is exposed. The runner and gate, passageways for the molten metal into the mold cavity, are formed in the drag sand. The pattern is removed and the core is placed in position, figure 25-2(E). The cope is then put back on the drag, figure 25-2(F).

The molten metal is poured into the pouring basin and runs down the sprue to a runner and through the gate and into the mold cavity. When the mold cavity is filled, the metal will begin to fill the sprue and the riser. Once the sprue and riser have been filled, the pouring should stop.

When the metal has hardened, the sand is broken and the casting removed, figure 25-2(G). The excess metal, gates, and risers, are removed and later remelted.

Full Mold Casting

The characteristic feature of the *full mold process* is the use of gasifiable patterns made of foamed plastic. These are not extracted from the mold, but are vaporized by the molten metal.

The full mold process is suitable for individual castings, and for small series of up to five castings. The full mold process is very economical, and it reduces the delivery time required for prototypes, articles urgently needed for repair jobs, and individual large machine parts.

REFERENCES AND SOURCE MATERIALS

1. Machine Design Materials Reference Issue, Mar. 1981.

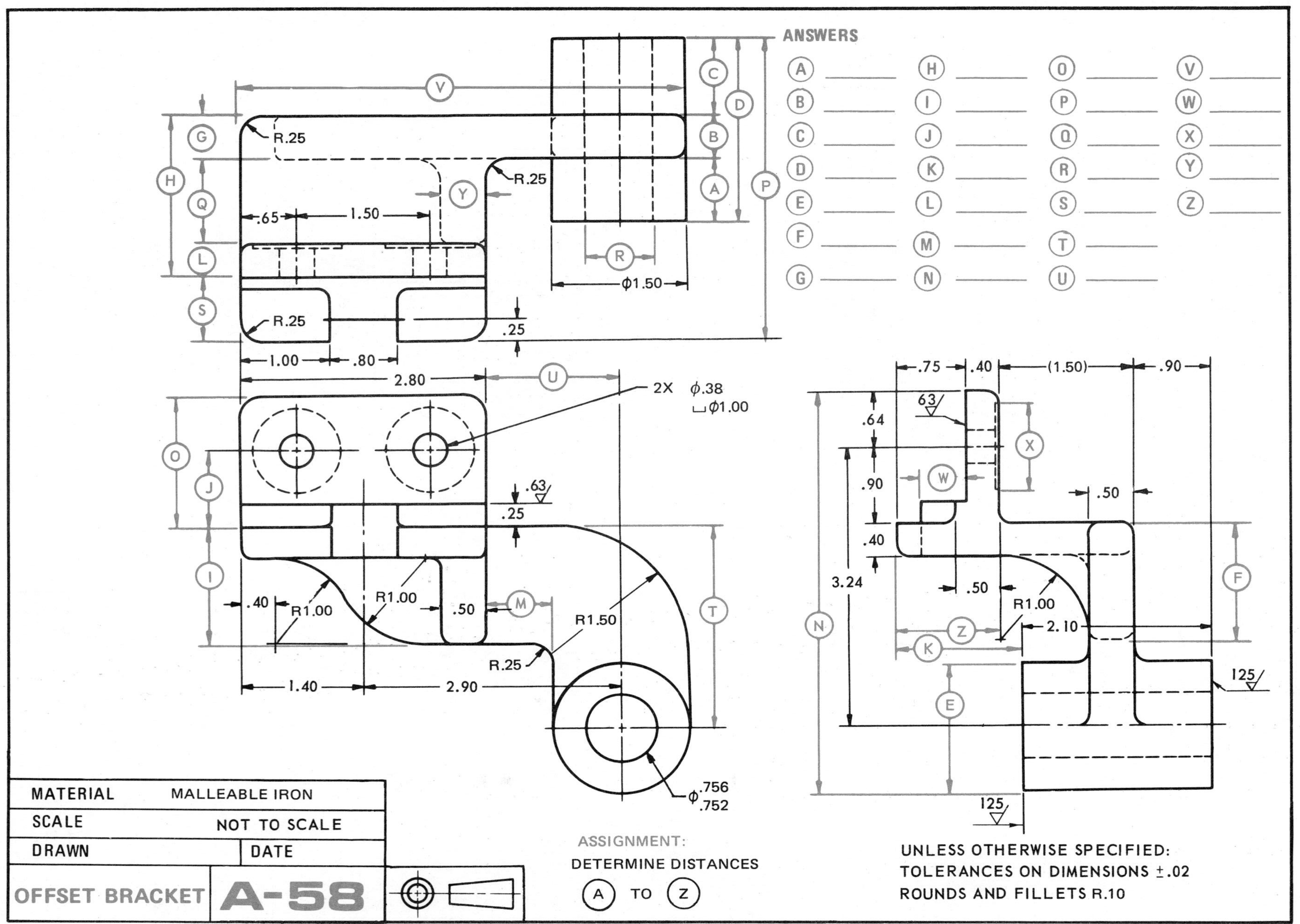
ANSWERS
A B C D E F G
H I J K L M N
O P Q R S T U
V W X Y Z
R.25
.65
1.50
ϕ1.50
.25
1.00
.80
2.80
2X ϕ.38
⌴ϕ1.00
.63/
.40
R1.00
.50
R1.50
1.40
2.90
ϕ.756
.752
.75
.40
(1.50)
.90
63/
.64
3.24
2.10
125/
MATERIAL
MALLEABLE IRON
SCALE
NOT TO SCALE
DRAWN
DATE
OFFSET BRACKET
A-58
ASSIGNMENT:
DETERMINE DISTANCES
A TO Z
UNLESS OTHERWISE SPECIFIED:
TOLERANCES ON DIMENSIONS ±.02
ROUNDS AND FILLETS R.10

QUESTIONS

1. Which line in the top view represents surface (1)?
2. Locate surface (A) in the left-side view and the front view.
3. Locate surface (8) in the front view.
4. How many surfaces are to be finished?
5. Which line in the left-side view represents surface (3)?
6. What is the center distance between holes (B) and (O) in the front view?
7. Determine distances at (4) (5) (6) and (11).
8. Locate surface (J) in the top view.
9. Which surface of the left-side view does line (14) represent?
10. What point in the front view is represented by line (15)?
11. What is the thickness of boss (E)?
12. Locate surface (G) in the left-side view.
13. Locate point (K) in the top view.
14. Locate surface (D) in the top view.
15. Determine distances at (M) (N) (S) (T).
16. What point or line in the top view is represented by point (16)?
17. What is the maximum horizontal distance between holes in the front view?
18. What would be the (A) width, (B) height, and (C) depth of the part before machining?

ANSWERS

1. ______
2. L.V. ______ F.V. ______
3. ______
4. ______
5. ______
6. ______
7. (4) ______ (5) ______ (6) ______ (11) ______
8. ______
9. ______
10. ______
11. ______
12. ______
13. ______
14. ______
15. (M) ______ (N) ______ (S) ______ (T) ______
16. ______
17. ______
18. A. ______ B. ______ C. ______

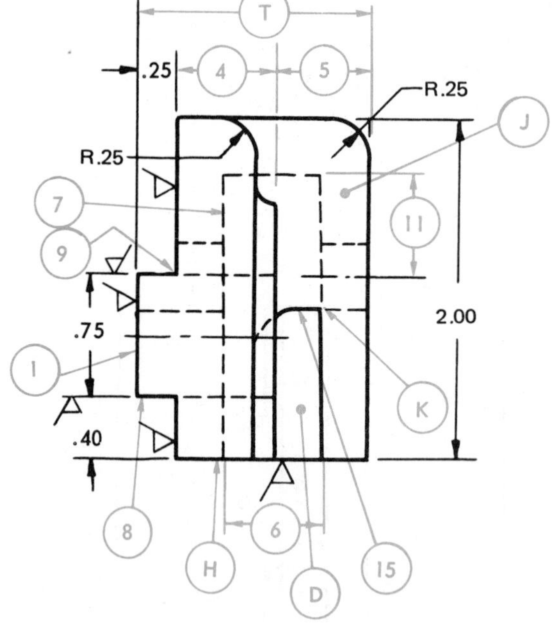

REVISIONS	I	82-03-04	C.J.
	1.15 WAS 1.20		

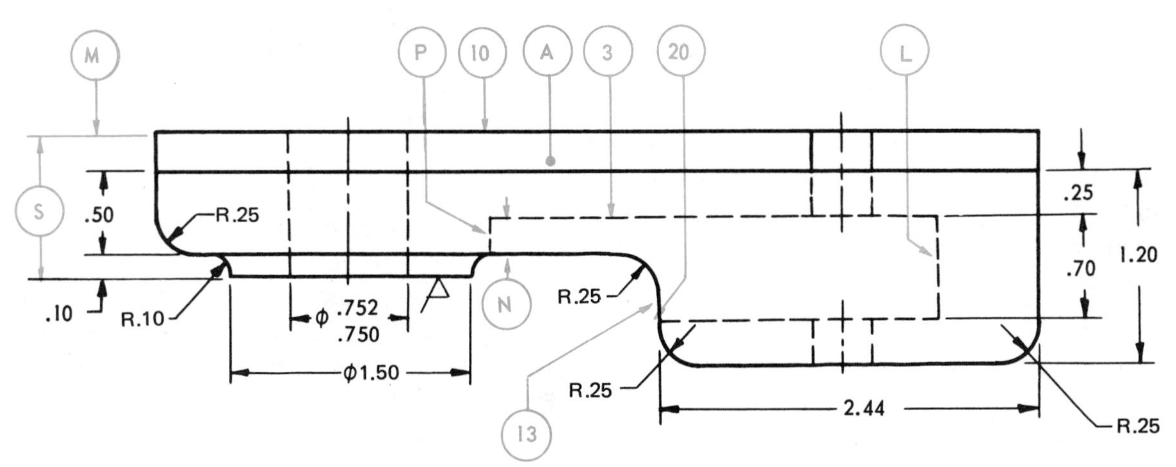

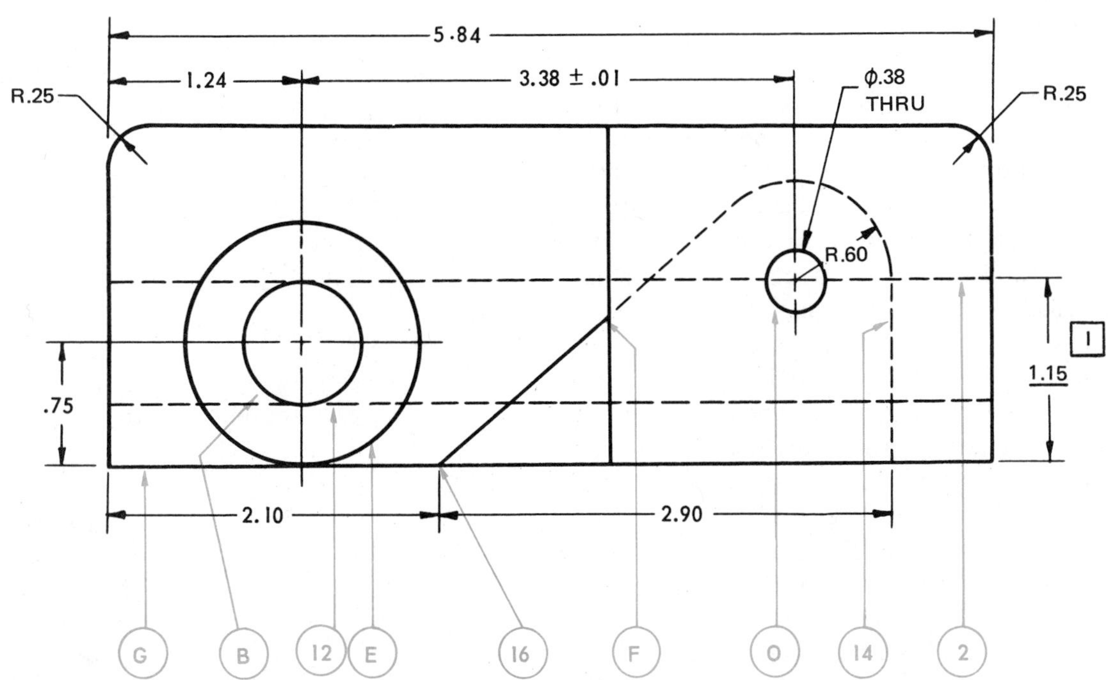

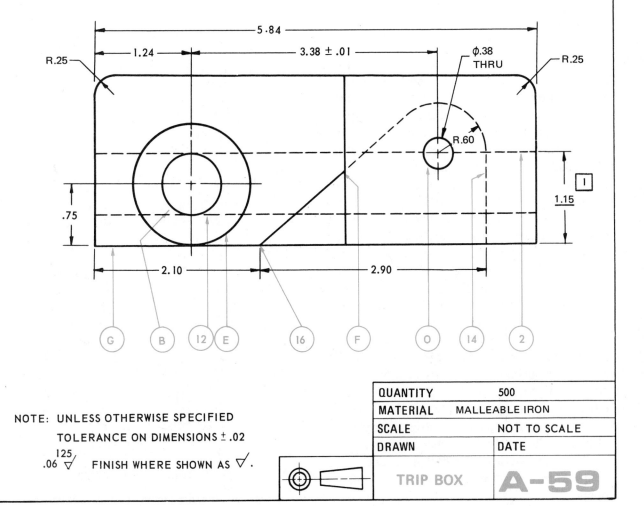

NOTE: UNLESS OTHERWISE SPECIFIED
TOLERANCE ON DIMENSIONS ± .02
.06 125 ∇ FINISH WHERE SHOWN AS ∇.

QUANTITY	500
MATERIAL	MALLEABLE IRON
SCALE	NOT TO SCALE
DRAWN	DATE
TRIP BOX	A-59

UNIT 26

CAST IRONS

Iron and the large family of iron alloys called steel are the most frequently specified metals. All commercial forms of iron and steel contain carbon, which is an integral part of the metallurgy of iron and steel.

Types of Cast Iron

Gray Iron. Gray iron is a supersaturated solution of carbon in an iron matrix. Generally, gray iron serves well in any machinery applications because of its fatigue resistance. Typical applications of gray iron include automotive blocks, flywheels, brake disks and drums, machine bases, and gears.

Ductile, or Nodular Iron. Ductile iron is not as available as gray iron, and is more difficult to control in production. However, ductile iron can be used where higher ductility or strength is required than is available in gray iron. Typical applications of ductile iron include crank shafts, heavy-duty gears, and automotive door hinges.

White Iron. White iron is produced by a process called chilling which prevents graphitic carbon from precipitating out. Because of their extreme hardness, white irons are used primarily for applications requiring wear and abrasion resistance such as mill liners and shot–blasting nozzles. Other uses include brick–making equipment, crushers, and pulverizers. The disadvantage of white iron is that it is very brittle.

Malleable Iron. Malleable iron is a white iron that has been converted to a malleable condition. It is commercial cast material which is similar to steel in many respects. It is strong and ductile, has good impact and fatigue properties, and has excellent machining characteristics.

The abbreviations for cast irons and steels found on drawings are shown in Table 1 of the Appendix.

CASTING DESIGN

Simplicity of Molding from Flat Back Patterns

Simple shapes such as the one shown in figure 26-1 are very easy to mold. In this case the flat face of the pattern is at the parting line and lies perfectly flat on the molding board. In this position no molding sand sifts under the flat surface to interfere with the drawing of the pattern. The simplicity with which flat back patterns of this type may be drawn from the mold is illustrated in figure 26-1.

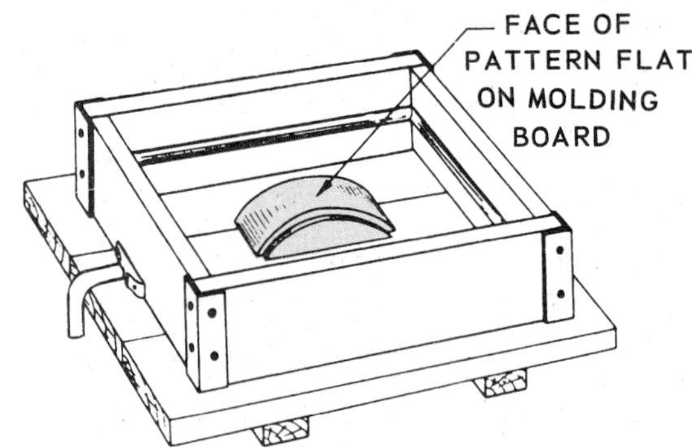

(A) PLACING THE PATTERN ON THE MOLDING BOARD

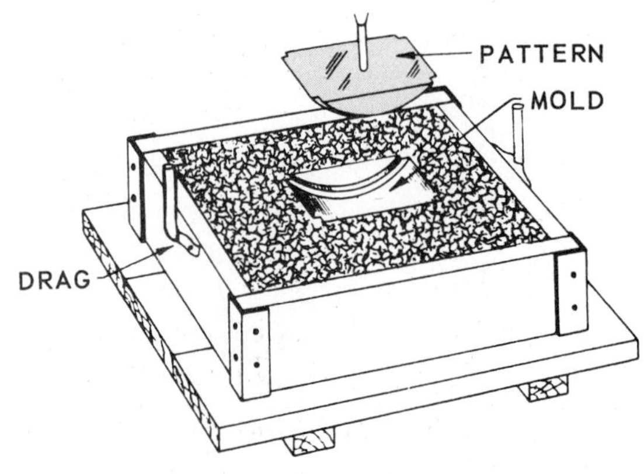

(B) DRAWING THE PATTERN

Fig. 26-1 Making a mold of a flat back pattern

Irregular or Odd-Shaped Castings

When a casting is to be made for an odd-shaped piece such as the offset bracket, figure 26-2, it is necessary to make the pattern for the bracket in one or more parts to facilitate the making of the mold. The difficulties with this are mostly due to the removal of the pattern from the sand mold.

The pattern for the bracket is made in two parts as shown in figure 26-3. The two adjacent flat surfaces of the divided pattern come together at the parting lines.

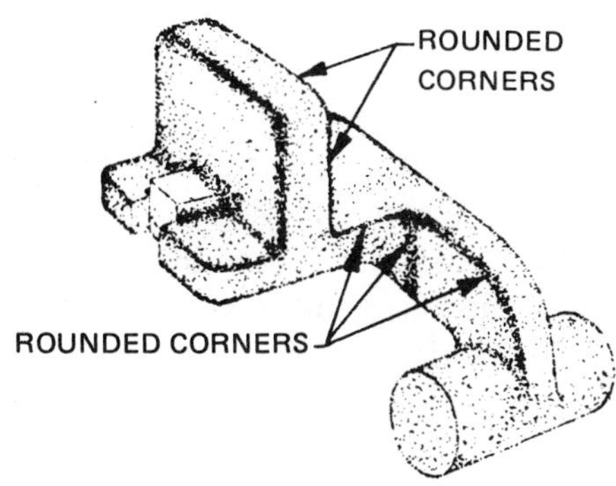

Fig 26-2 Casting of offset bracket shown in drawing A-58

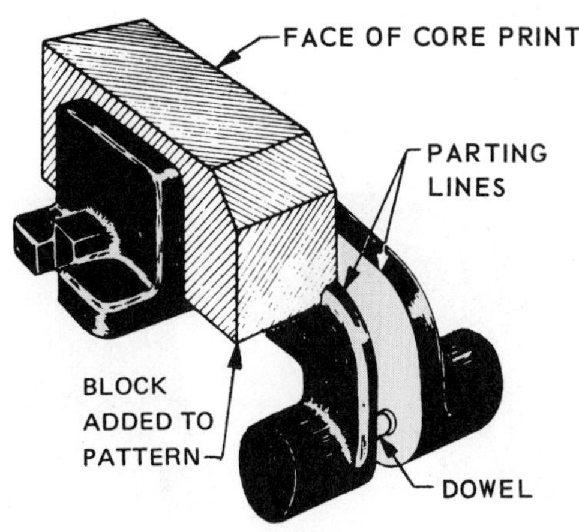

Fig. 26-3 Two-piece pattern for offset bracket

Set Cores

In examining the illustration of the offset bracket, it will be noted that there are rounded corners. In order to make it possible to mold these rounded corners, a block must be added to the pattern for ease in its removal from the mold.

This block, which becomes an integral part of the pattern, also acts as a core print for a *set core* (a core that has been baked hard), figure 26-4. It is made to conform to the shape of the faces of the casting, including the rounded corners.

The face of the core print also forms the parting line for one side of the two-part pattern. When making the mold for the bracket casting, this face, which corresponds to the flat face of a flat back pattern, is laid on the molding board with the drag of the flask in position. The sand is then rammed around the pattern.

When the drag is reversed, the cope, or upper part of the flask, is placed in position. The other half of the pattern is then joined with the first part, and the sand is rammed into the cope to flow around that part of the pattern which projects into it.

After the pattern is removed, the set core, which is formed in a core box and baked hard, is set in the impression in the mold made by the core print of the pattern. When poured into the mold, the molten metal fills the cavity made by the pattern and the faces of the set core as shown in figure 26-4 at **A** and **B** to form the casting.

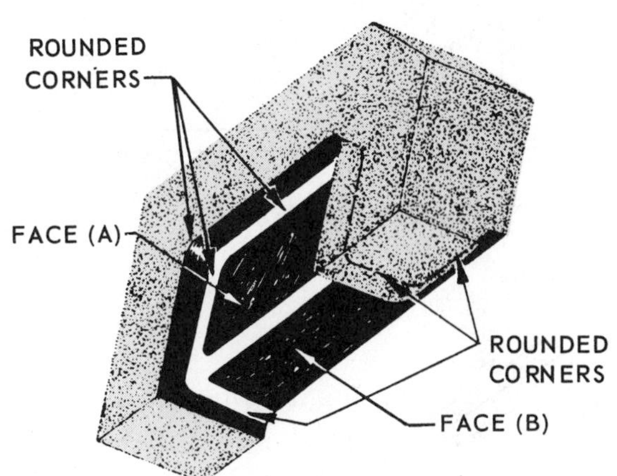

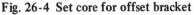

Fig. 26-4 Set core for offset bracket

Coping Down

The core print is made as part of the pattern to avoid removing molding sand in the drag which would correspond to the shape of the core print.

If this sand were dug out or *coped out,* as shown in figure 26-5, the remaining cavity would be again filled with molding sand when the cope was rammed. The sand would then hang below the parting line of the cope down into the drag.

When the mold is made by coping down, the hanging portion of the cope is supported by soldiers or gaggers embedded within the sand to hold the projecting part in position for subsequent operations, figure 26-6.

Coping down requires skill and takes time. The set core principle is at times preferred to coping down to avoid delay and assure a more even parting line on the casting.

Split Patterns

Irregularly shaped patterns which cannot be drawn from the sand are sometimes split so that one half of the pattern may be rammed in the drag as a simple flat back pattern while the cope half, when placed in position on the drag half, forms the mold in the cope. Patterns of this type are called *split patterns* and do not require coping down to the parting line which would be necessary if the pattern were made solid.

The pattern for the casting shown in figure 26-7, when made without a print for a core, can be drawn from the sand only by splitting the pattern on the parting line as illustrated.

The drawing must be examined to determine how the pattern should be constructed. This is important because the parting line must be located in a position which permits the halves of the pattern to be drawn from the sand without interference.

CORED CASTINGS

Cored castings have certain advantages over solid castings. Where practical, castings are designed with cored holes or openings for economy, appearance, and accessibility to interior surfaces.

Cored openings often improve the appearance of a casting. In most instances, cored cast-

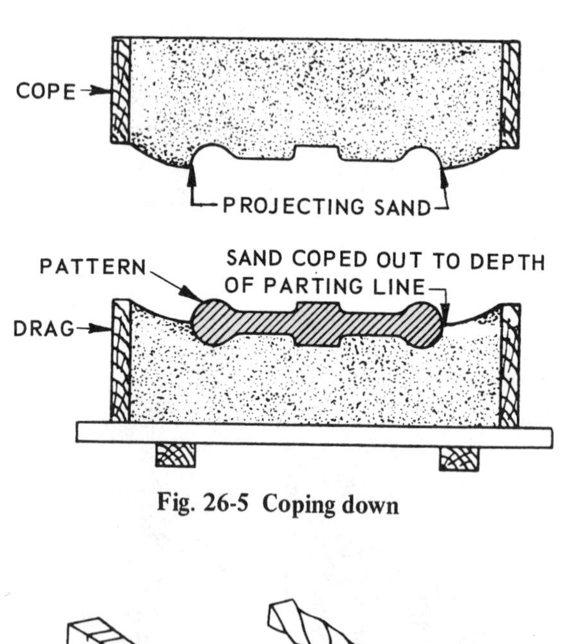

Fig. 26-5 Coping down

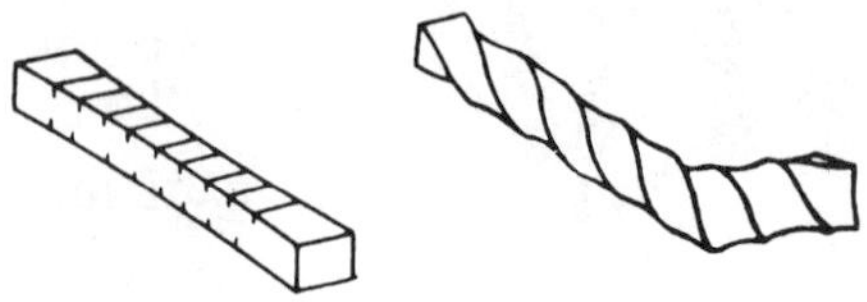

Fig 26-6 Soldiers and gaggers

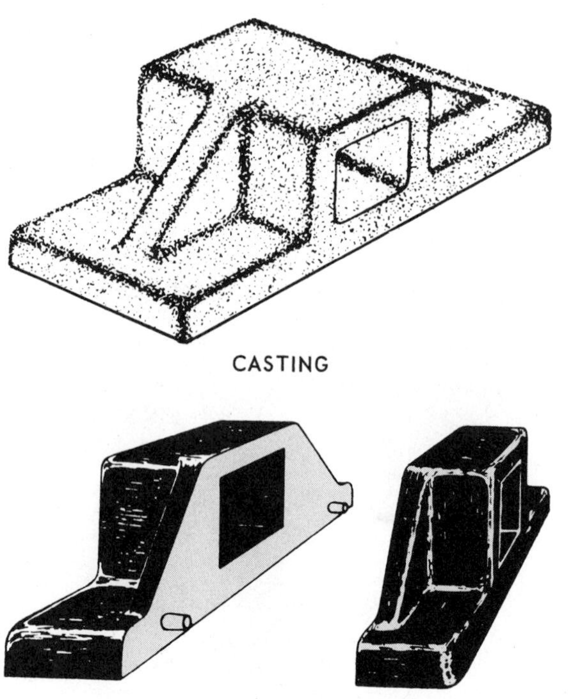

Fig. 26-7 Application of split pattern

ings are more economical than solid castings because of the savings in metal. While cored castings are lighter, they are designed without sacrificing strength. The openings cast in the part eliminate unnecessary machining.

Hand holes may also be formed by coring in order to provide an opening through which the interior of the casting may be reached. These openings also permit machining an otherwise inaccessible surface of the part as shown in figure 26-8.

MACHINING LUGS

It is difficult to hold and machine certain parts without using lugs. This is because of the design and nature of certain parts. The lugs are an integral part of the casting. They are sometimes removed to avoid interference with the functioning of the part. Lugs are usually represented in phantom outline.

An example of a machining lug is shown in figure 26-9. This lug is used to provide a flat surface on the end of the casting for centering and also to give a uniform center bearing for other machining operations. Both the function of the lug and its appearance determine whether or not it is removed after machining.

SURFACE COATINGS

Machined parts are frequently finished either to protect the surfaces from oxidation or for appearance. The type of finish depends on the use of the part. The finish commonly applied may be a protective coat of paint, lacquer, or a metallic plating. In some cases only a surface finish such as polishing or buffing may be specified.

The finish may be applied before any machining is done, between the various stages of machining, or after the piece has been completed. The type of finish is usually specified on the drawing in a notation similar to the one indicated on the auxiliary pump base, drawing A-60, which reads CASTING TO BE PAINTED WITH ALUMINUM BEFORE MACHINING.

The decorative finish on the drive housing should be added before machining so that there will be no accidental deposit of paint on the machined surfaces. This will insure the desired accuracy when assembled with other parts.

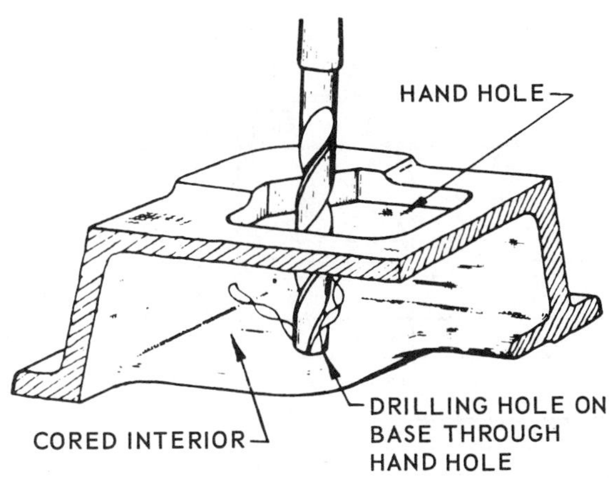

Fig. 26-8 Section of cored casting

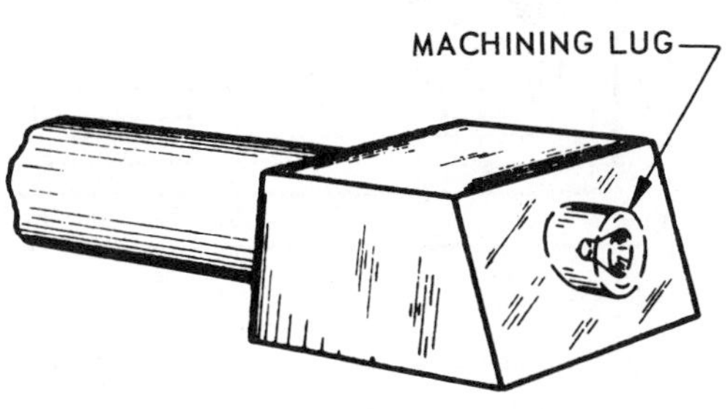

Fig. 26-9 Application of machining lug

QUESTIONS

1. Of what material is the base made?
2. How many finished surfaces are indicated?
3. Which circled letters on the drawing indicate the spaces that were cored when the casting was made?
4. Locate the surface in the top view that is represented by line (6).
5. What is the height of the cored area (X)?
6. What surface finish is used on the top pad?
7. What is the horizontal length of the pad (5)?
8. What reason might be given for openings (Q) (S) (Y) (Z)?
9. Determine distance (W).
10. What radius fillet would be used at (R)?
11. What is the total allowance added to the height for machining?
12. Determine distances (C) through (N).

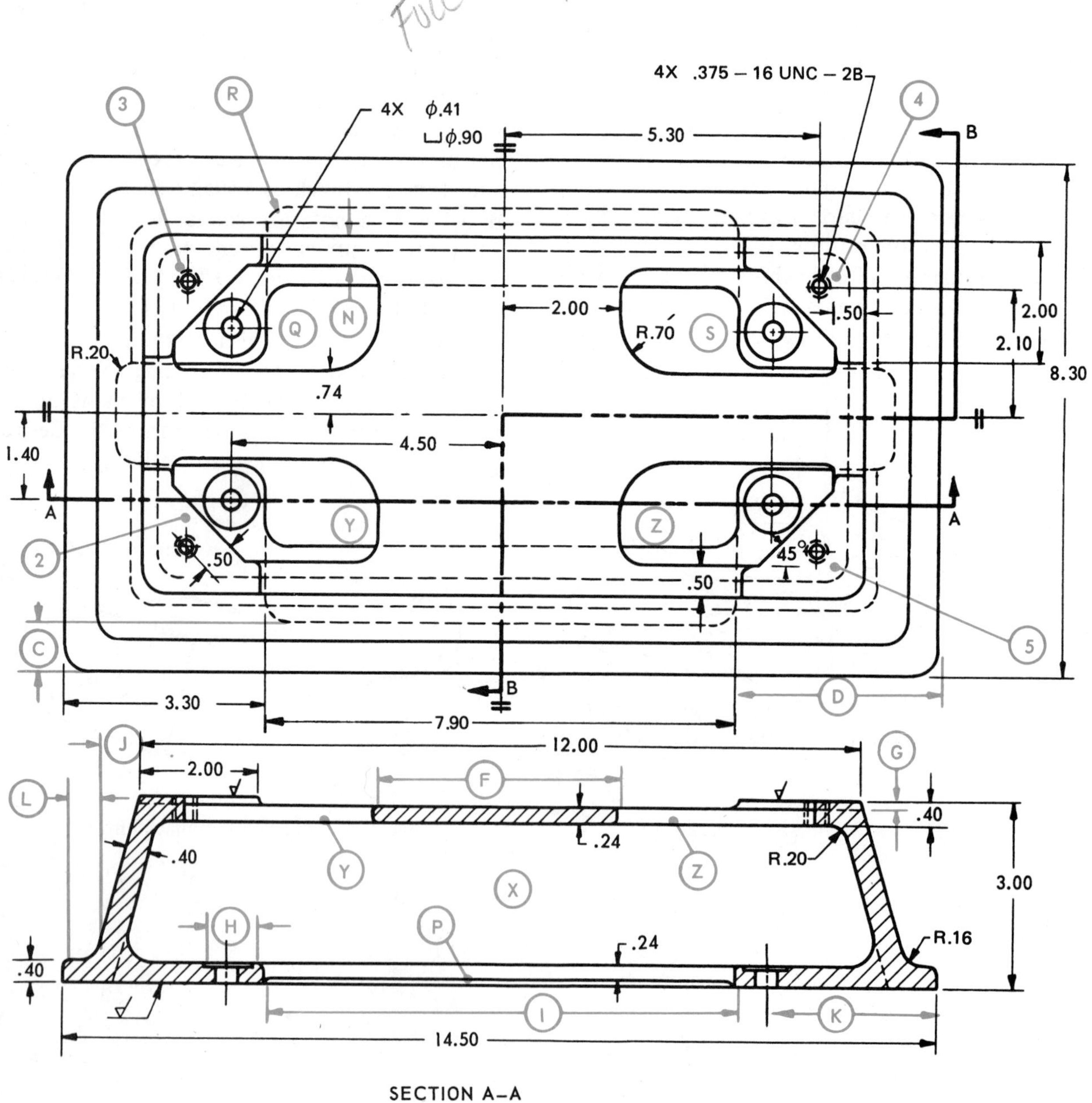

ANSWERS

1.	______	12. (C)	______
2.	______	(D)	______
3.	______	(E)	______
4.	______	(F)	______
5.	______	(G)	______
6.	______	(H)	______
7.	______	(I)	______
8.	______	(J)	______
9.	______	(K)	______
10.	______	(L)	______
11.	______	(M)	______
		(N)	______

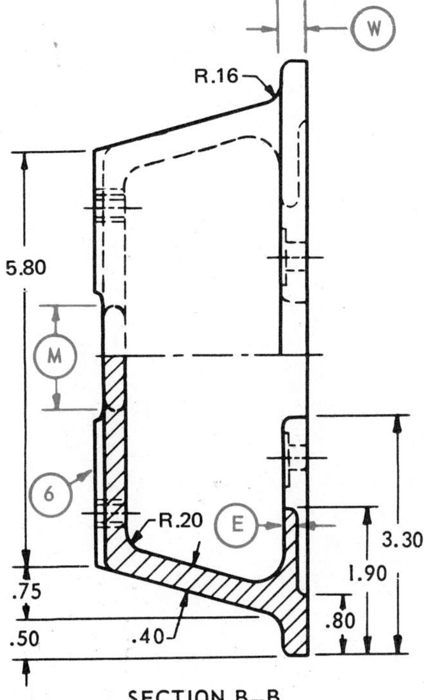

NOTE:

UNLESS OTHERWISE SPECIFIED:

- TOLERANCES ON DIMENSIONS ± .02
- ROUNDS AND FILLETS R.25
- FINISHES .06 125 ∇
- CASTING TO BE PAINTED ALUMINUM BEFORE MACHINING

MATERIAL	GRAY IRON	
SCALE	NOT TO SCALE	
DRAWN		DATE
AUXILIARY PUMP BASE		A-60

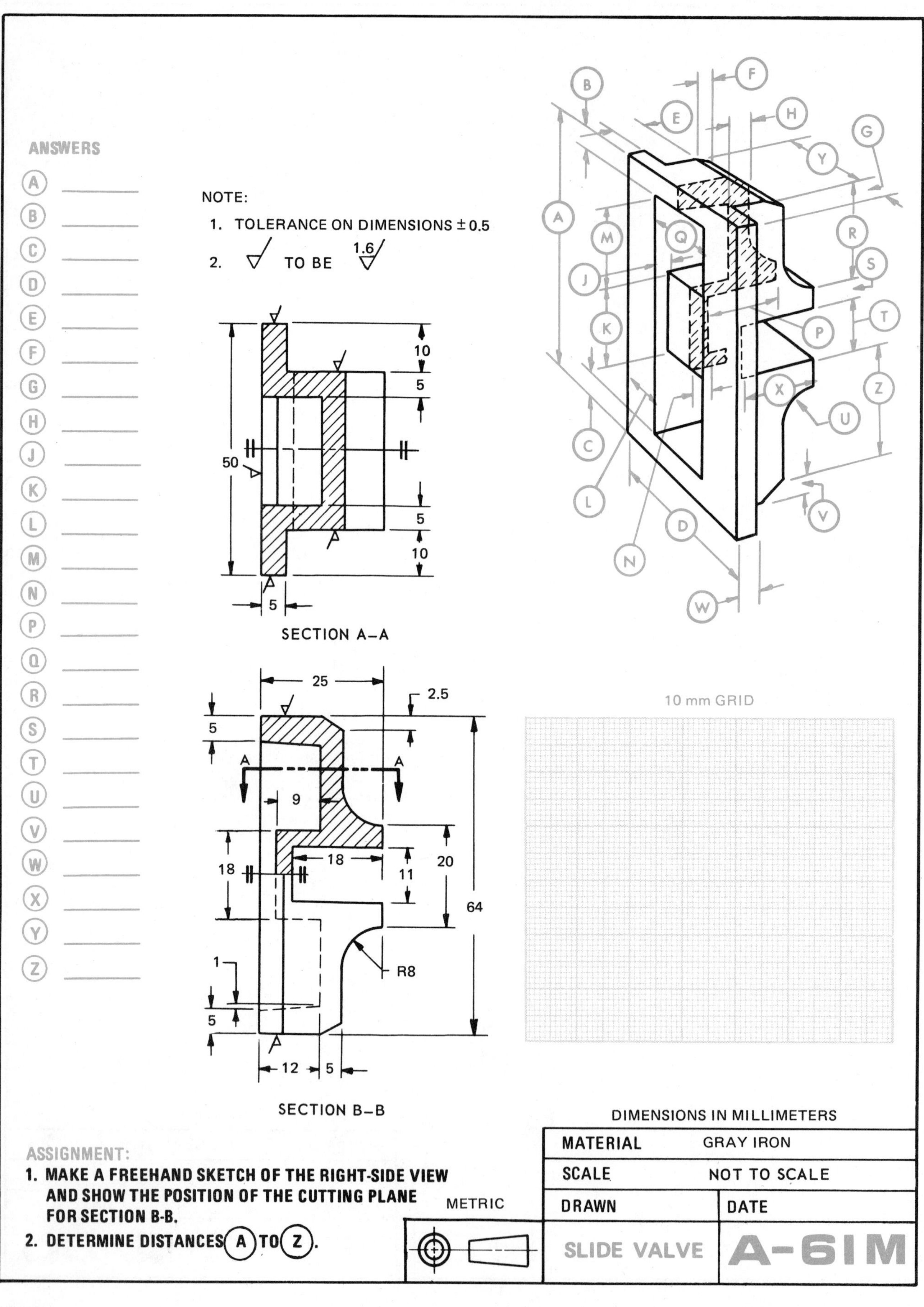
ANSWERS
A
B
C
D
E
F
G
H
J
K
L
M
N
P
Q
R
S
T
U
V
W
X
Y
Z
NOTE:
1. TOLERANCE ON DIMENSIONS ±0.5
2. TO BE 1.6
10
5
50
5
10
5
SECTION A–A
25
2.5
5
A
A
9
18
18
20
11
64
R8
1
5
12
5
SECTION B–B
10 mm GRID
DIMENSIONS IN MILLIMETERS
MATERIAL GRAY IRON
SCALE NOT TO SCALE
DRAWN
DATE
METRIC
SLIDE VALVE
A-61M
ASSIGNMENT:
1. MAKE A FREEHAND SKETCH OF THE RIGHT-SIDE VIEW AND SHOW THE POSITION OF THE CUTTING PLANE FOR SECTION B-B.
2. DETERMINE DISTANCES A TO Z.

UNIT 27

ALIGNMENT OF PARTS AND HOLES

Two important factors which must be considered when drawing an object are the number of views to be drawn and the time required to draw them. If possible, use time-saving devices, such as templates, for drawing standard features.

To simplify the representation of common features, a number of conventional drawing practices are used, figure 27-1. Many conventions deviate from true projection for the purpose of clarity; others are used to save drafting time. These conventions must be executed carefully; clarity is even more important than speed.

Foreshortened Projection

When the true projection of ribs or arms results in confusing foreshortening, these parts should be rotated until parallel to the line of the section or projection.

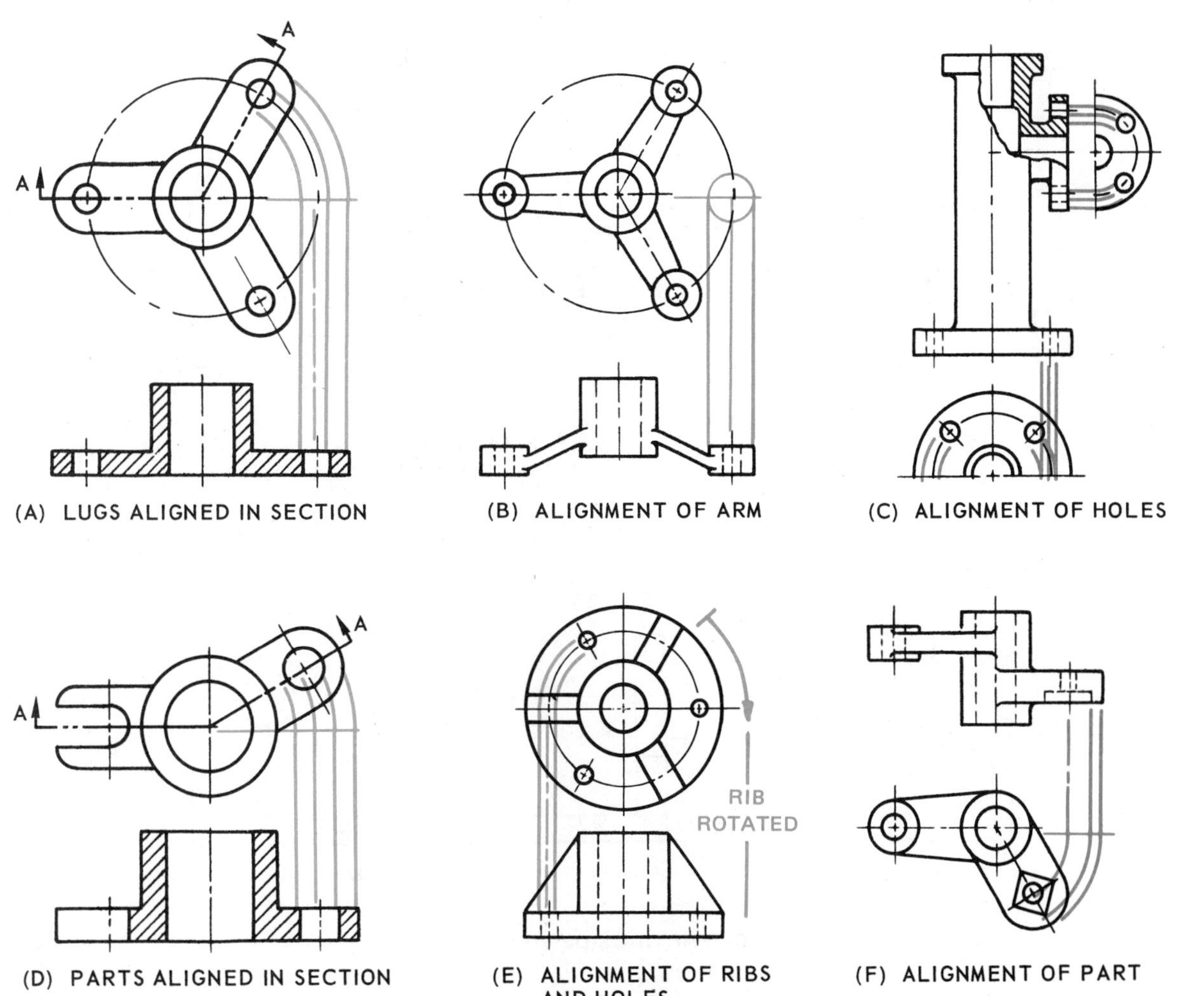

Fig. 27-1 Alignment of holes and parts to show their true relationship

Holes Revolved to Show True Center Distance

Drilled flanges in elevation or section should show the holes at their true distance from center, rather than the true projection.

PARTIAL VIEWS

Partial views, which show only a limited portion of the object with remote details omitted, should be used when necessary, to clarify the meaning of the drawing, figure 27-2. Such views are used to avoid the necessity of drawing many hidden features.

On drawings of objects where two side views can be used to better advantage than one, each need not be complete if together they depict the shape. Show only the hidden lines of features immediately behind the view, figure 27-2(C).

Another type of partial view is shown, figure 27-3. However, sufficient information is given in the partial view to complete the description of the object. The partial view is limited by a break line. Partial views are used because:

- They save time in drawing.
- They conserve space which might otherwise be needed for drawing the object.
- They sometimes permit the drawing to be made to scale large enough to bring out all details clearly, whereas if the whole view were drawn, lack of space might make it necessary to draw a smaller scale, resulting in the loss of detail clarity.
- If the part is symmetrical, a partial view (referred to as a half view) may be drawn on one side of the center line as shown in figure 27-4. In the case of the coil frame (drawing A-64) a partial view was used so that the object could be drawn to a larger scale for clarity saving time and space.

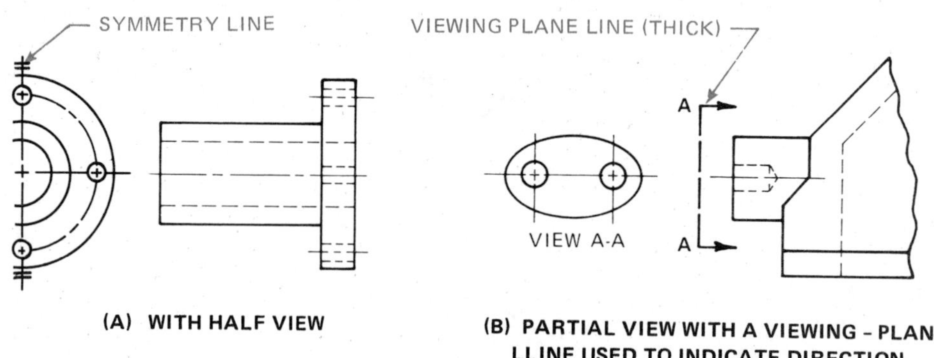

(A) WITH HALF VIEW

(B) PARTIAL VIEW WITH A VIEWING - PLANE LLINE USED TO INDICATE DIRECTION

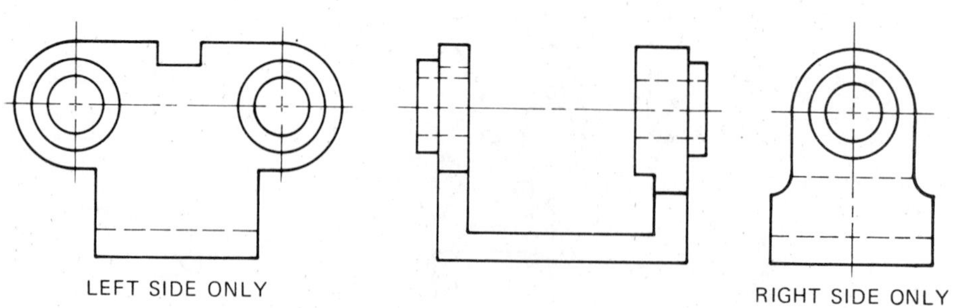

(C) PARTIAL SIDE VIEWS

Fig. 27-2 Partial views

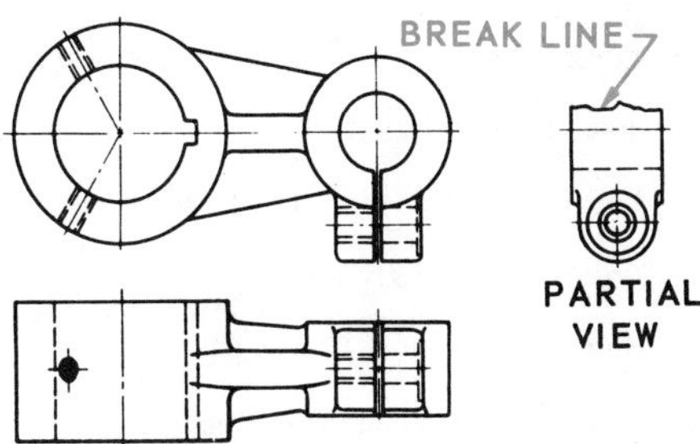

Fig. 27-3 Partial view

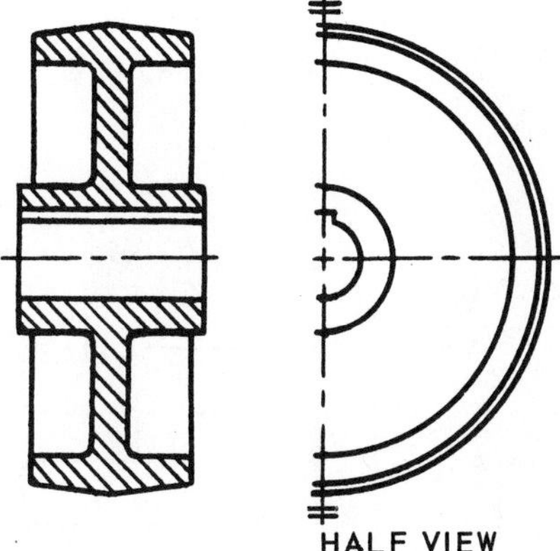

Fig. 27-4 Half views

NAMING OF VIEWS FOR SPARK ADJUSTER

The drawing of the spark adjuster, drawing A-62, illustrates several violations of true projection. The names of the views of the spark adjuster could be questionable. The importance lies not in the names, however, but in the relationship of the views to each other. This means that the right view must be on the right side of the front, the left view must be on the left side of the front view, etc. Any combination in figure 27-5 may be used for naming the views of the spark adjuster.

DRILL SIZES

Twist drills are the most common tools used in drilling. They are made in many sizes. Inch size twist drills are grouped according to decimal inch sizes; by number sizes, from 1 to 80 which correspond to the Stubbs steel wire gauge; by letter sizes A to Z; and by fractional sizes from 1/64th up. Twist drill sizes are listed in Table 3 of the Appendix.

Metric twist drill sizes are in millimeters and are classified as *preferred* and *available.* These sizes will eventually replace the fractional-inch, letter and number size drills which are presently in existence. Metric twist drill sizes are listed in Table 4 of the Appendix.

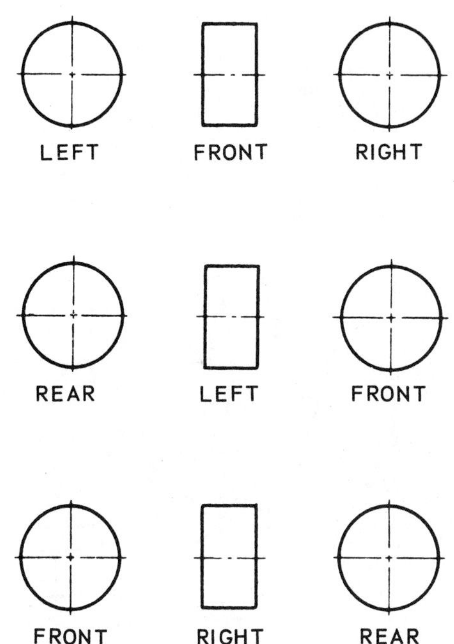

Fig. 27-5 Naming of views for spark adjuster, drawing A-62

QUESTIONS

1. What is radius (A)?
2. What surface is line (2) in the rear view?
3. Locate surface (B) in section A-A.
4. Locate surface (C) in section A-A.
5. Locate line (D) in the rear view.
6. Locate a surface or line in the front view that represents line (F).
7. Which point or line in the front view does line (G) represent?
8. What size cap screw would be used in (N) hole?
9. What is the diameter of (K) hole?
10. Locate the point or line in the rear view from which line (I) is projected.
11. Locate surface (H) in the rear view.
12. Locate (Z) in section A-A.
13. How thick is lug (7)?
14. What are the diameters of holes (L), (M), and (N)?
15. Determine angles (O), and (P).
16. Determine distances (Q) through (U).

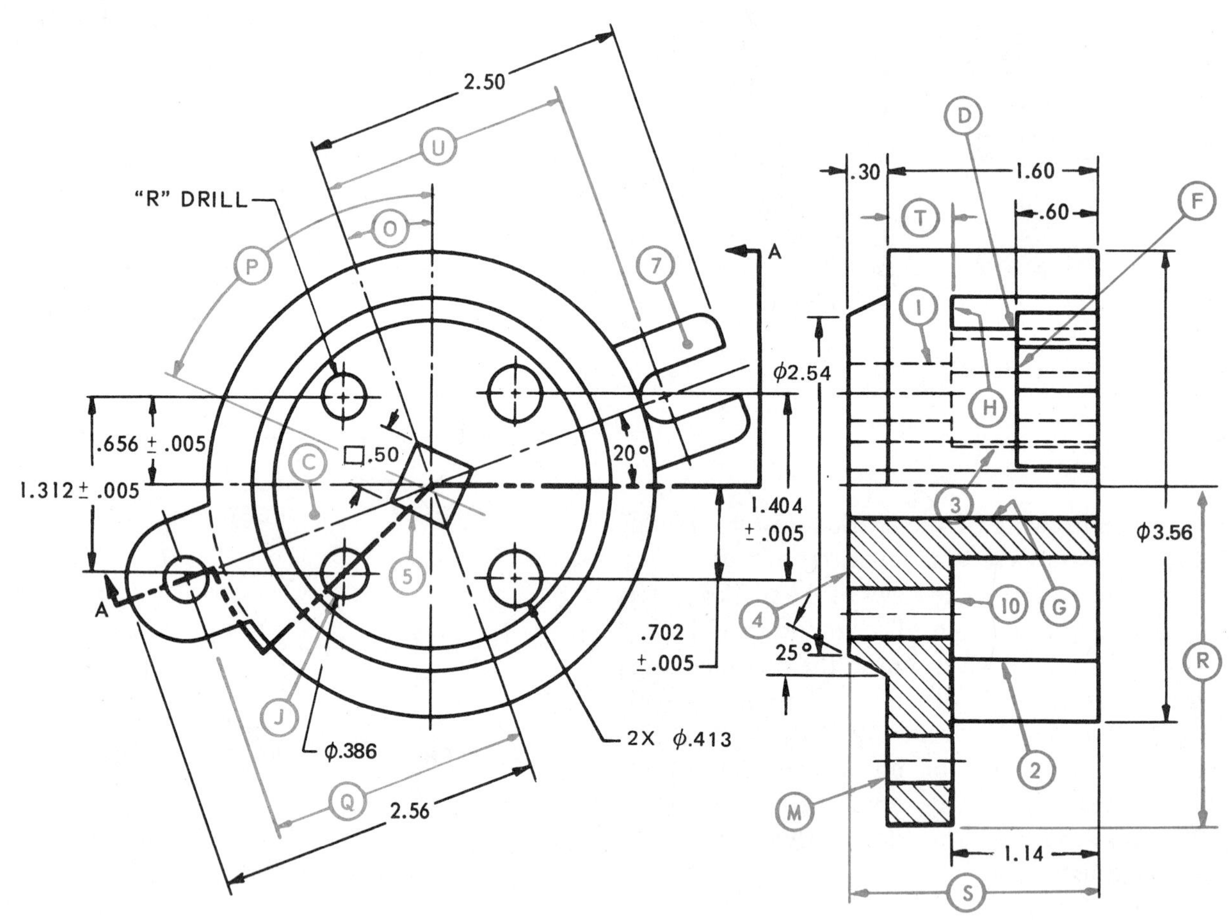

SECTION A–A

FRONT VIEW

RIGHT–SIDE VIEW

ANSWERS

1 ______
2 ______
3 ______
4 ______
5 ______
6 ______
7 ______
8 ______
9 ______
10 ______
11 ______
12 ______
13 ______
14 (L) ______
(M) ______
(N) ______
15 (O) ______
(P) ______
16 (Q) ______
(R) ______
(S) ______
(T) ______
(U) ______

NOTE: UNLESS OTHERWISE STATED
TOLERANCE ON TWO-PLACE DIMENSIONS ±.02
TOLERANCE ON THREE-PLACE DIMENSIONS ±.010
TOLERANCE ON ANGLES ±0.5°

REAR VIEW

MATERIAL	BAKELITE
SCALE	1:1
DRAWN	DATE
SPARK ADJUSTER	A-62

UNIT 28

BROKEN-OUT AND PARTIAL SECTIONS

When certain internal and external features of an object should be shown without drawing another view, broken-out and partial sections are used, figure 28-1. A cutting-plane line or a break line is used to indicate where the section is taken. In the front view of the following assignment, the raise block, drawing A-63M, two partial sections are used. Although this method of showing a partial section is not commonly used, it is an accepted practice.

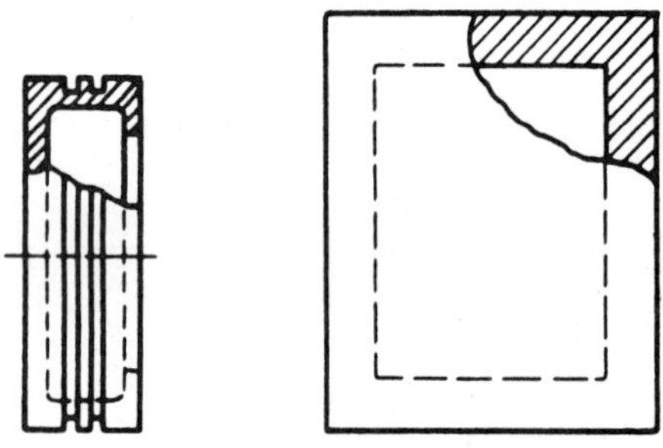

(A) BROKEN–OUT SECTIONS

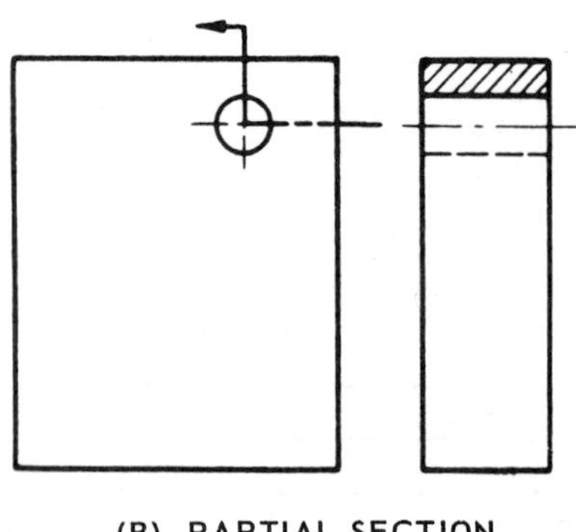

(B) PARTIAL SECTION

Fig. 28-1 Broken-out and partial sections

WEBS IN SECTION

The conventional (preferred) methods of representing a section of a part having webs or partitions are shown in figure 28-2. These methods are preferred to drawing the section in true projection. While the conventional methods are a violation of true projection, they are preferred over true projection because of clarity and ease in drawing.

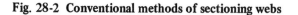

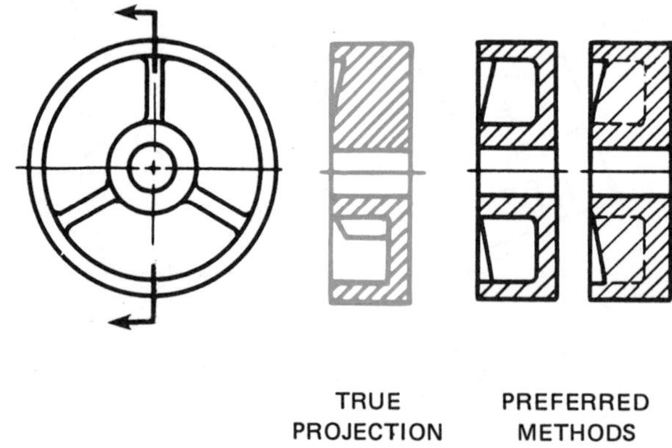

Fig. 28-2 Conventional methods of sectioning webs

RIBS IN SECTION

A true projection section view, figure 28-3(A), would be misleading when the cutting plane passes longitudinally through the center of a rib. To avoid this impression of solidity, a preferred section not showing the ribs section-lined or crosshatched is used. When there is an odd number of ribs, figure 28-3(B), the top rib is aligned with the bottom rib to show its true relationship with the hub and flange. If the rib is not aligned or revolved, it appears distorted on the section view and is misleading.

An alternate method of identifying ribs in a section view is shown in figure 28-3(C). If rib A

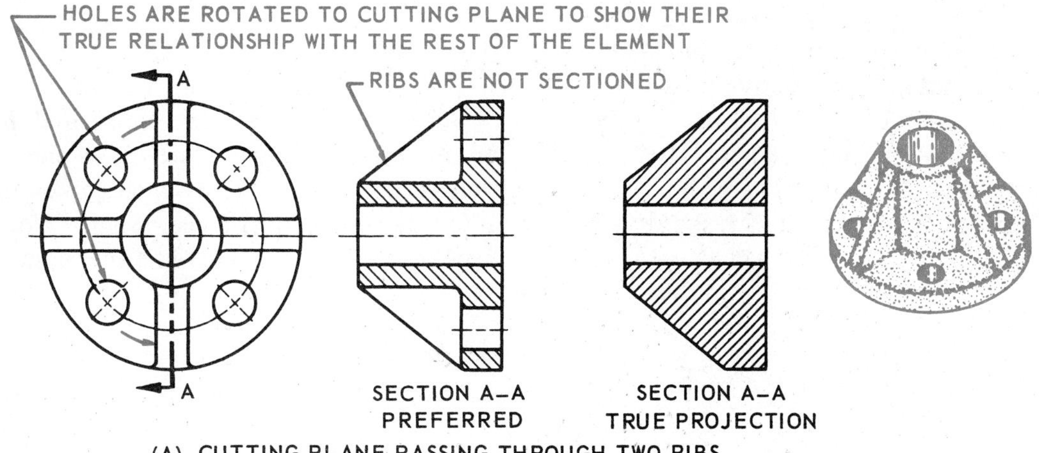

(A) CUTTING PLANE PASSING THROUGH TWO RIBS

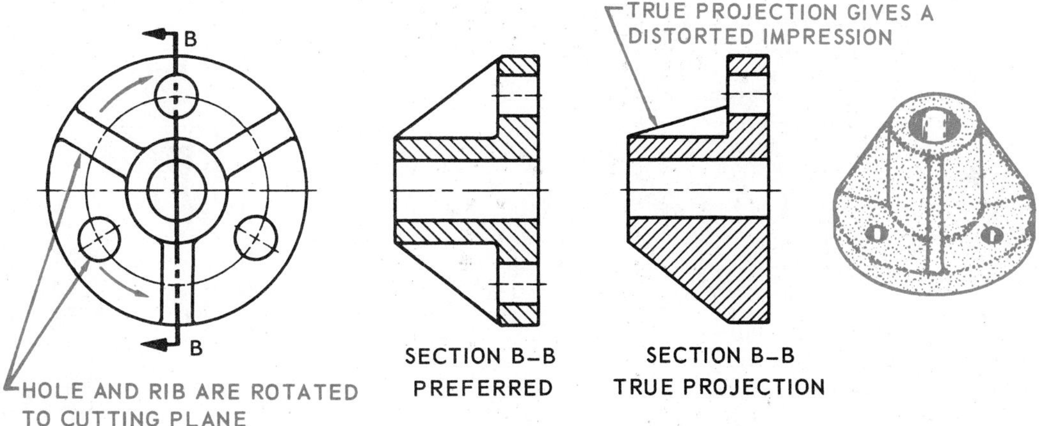

(B) CUTTING PLANE PASSING THROUGH ONE RIB AND ONE HOLE

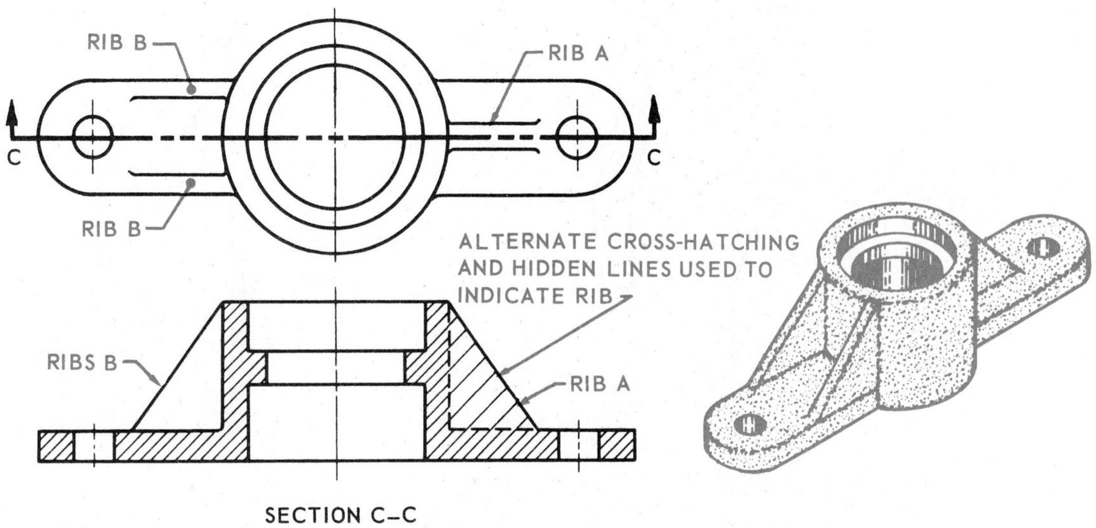

(C) ALTERNATE METHOD OF SHOWING RIBS IN SECTION

Fig. 28-3 Ribs in section

of the base were not sectioned as previously mentioned, it would appear exactiy like **B** in the section view and would be misleading. To distinguish between the ribs on the base, alternate section lining on the ribs is used. The line between the rib and solid portions is shown as a broken line.

SPOKES IN SECTION

Spokes in section are represented in the same manner as ribs. Figure 25-6 shows the preferred method of identifying spokes in section for aligned and unaligned designs. Note that the spokes are not sectioned in either case.

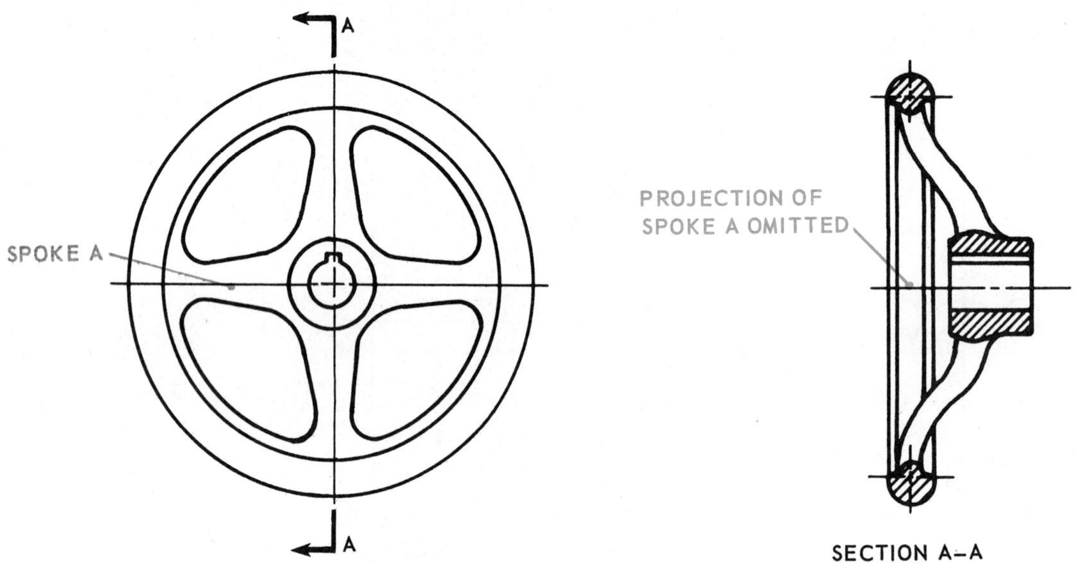

(A) CUTTING PLANE PASSING THROUGH TWO SPOKES

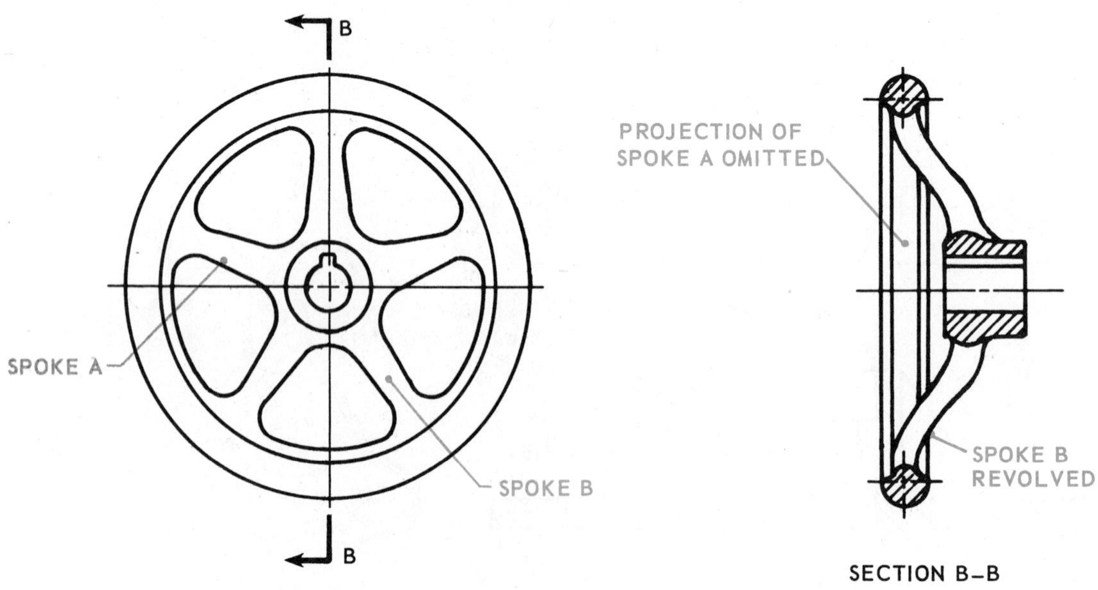

(B) CUTTING PLANE PASSING THROUGH ONE SPOKE

Fig. 28-4 Spokes in section

PAGE 201 IS INTENTIONALLY BLANK. ASSIGNMENT DRAWING A-63M BEGINS ON PAGE 202.

QUESTIONS

1. What is the diameter of the largest unthreaded hole?
2. What size is the smallest threaded hole?
3. What size are the smallest unthreaded holes?
4. Which surface does (7) represent in the top view?
5. Which surface does (1) represent in the top view?
6. Which line or surface does (6) represent in the left view?
7. Which line or surface does (V) represent in the front view?
8. What was the original width of the part?
9. By which line or surface is (H) represented in the left view?
10. Locate in the left view the line or surface that is represented by line (G).
11. Which line or surface represents (F) in the top view?
12. Which line represents surface (J) in the left view?
13. Determine the overall depth of the raise block.
14. Determine distances (A) through (E).
15. Determine distances (8) through (19).

	TOP VIEW
LEFT SIDE VIEW	FRONT VIEW

ARRANGEMENT OF VIEWS

NOTE: UNLESS OTHERWISE STATED

TOLERANCE ON DIMENSIONS ± 0.5

TOLERANCE ON ANGLES $\pm 0.5°$

ROUNDS AND FILLETS R3

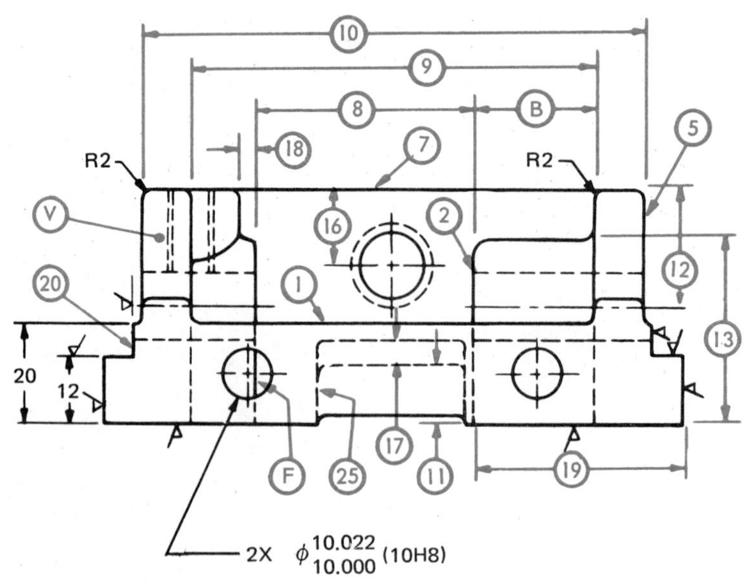

REVISIONS	I	82-03-04	R. KERR
	166 WAS 170		

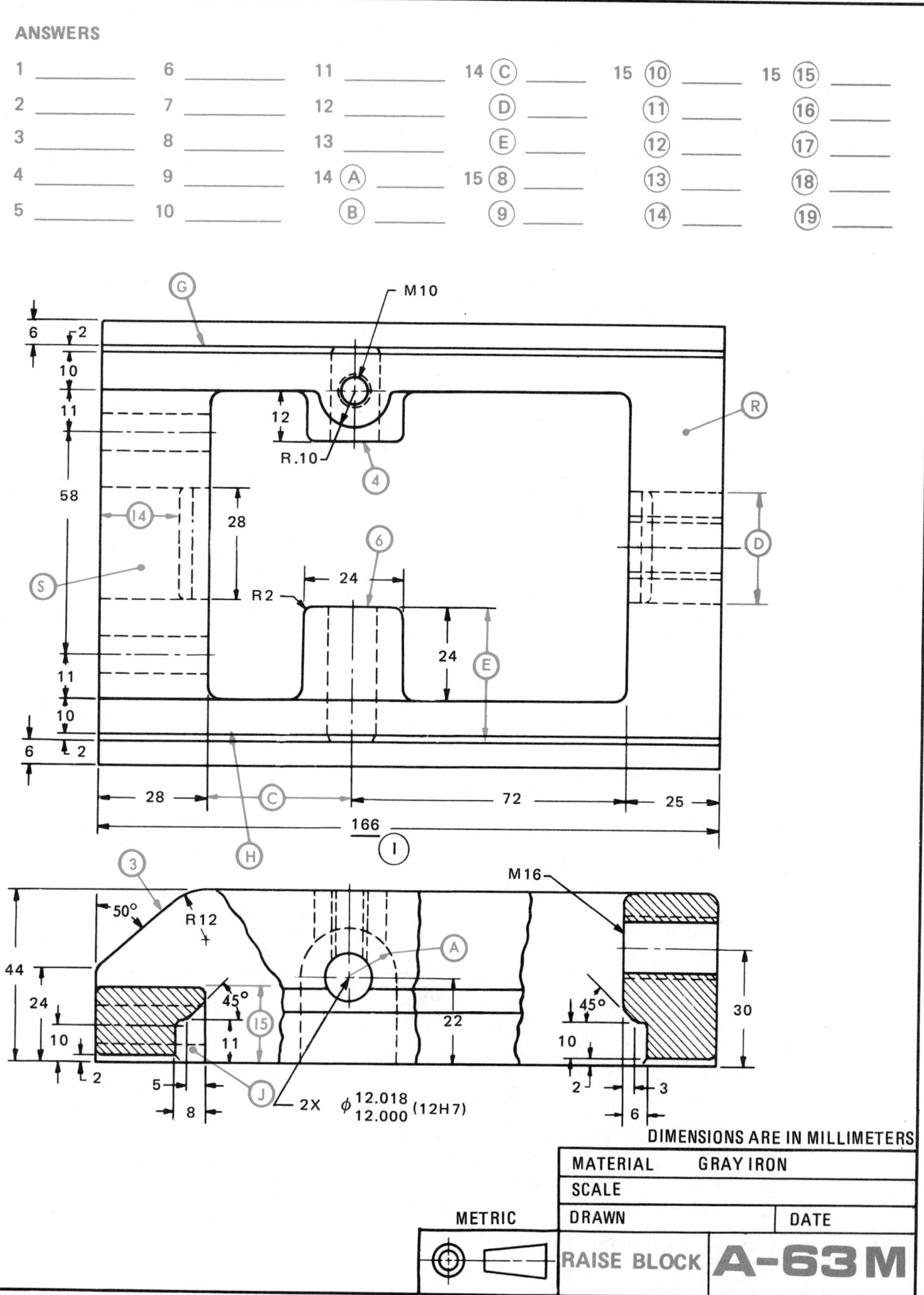
ANSWERS
1
2
3
4
5
6
7
8
9
10
11
12
13
14 A
B
C
D
E
15 8
9
10
11
12
13
14
15
16
17
18
19
M10
R.10
12
28
24
R2
58
11
10
6
2
72
25
28
166
M16
R12
50°
44
24
10
45°
11
22
30
5
8
3
6
2X ϕ 12.018 / 12.000 (12H7)
DIMENSIONS ARE IN MILLIMETERS
MATERIAL GRAY IRON
SCALE
DRAWN
DATE
METRIC
RAISE BLOCK
A-63M

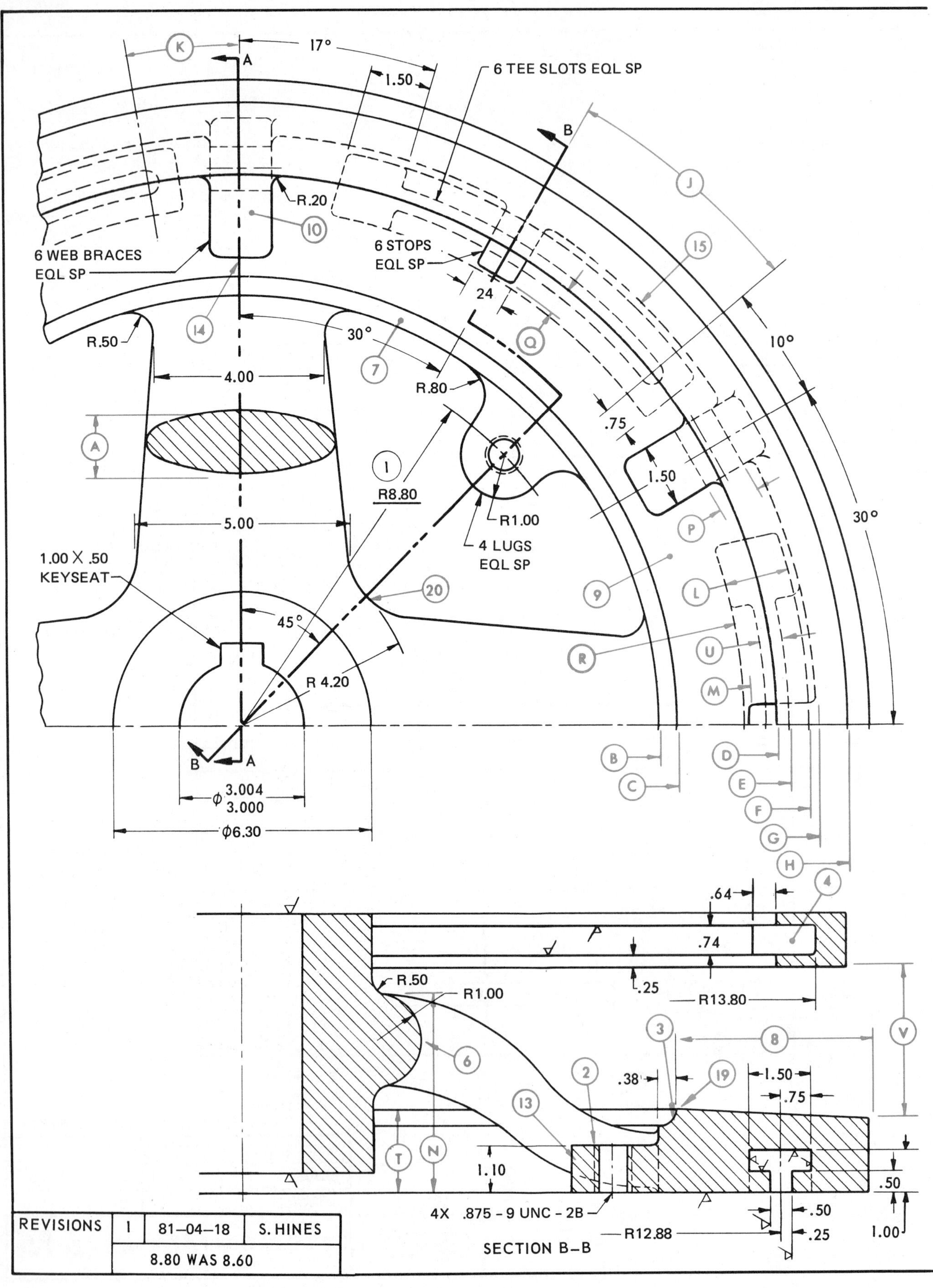

17°
A
1.50
6 TEE SLOTS EQL SP
B
R.20
6 WEB BRACES
EQL SP
6 STOPS
EQL SP
24
R.50
30°
10°
4.00
R.80
.75
1.50
R8.80
R1.00
4 LUGS
EQL SP
30°
5.00
1.00 X .50
KEYSEAT
45°
R 4.20
B
A
ϕ 3.004
3.000
ϕ6.30
.64
.74
.25
R13.80
R.50
R1.00
.38
1.50
.75
1.10
.50
4X .875 - 9 UNC - 2B
.50
R12.88
.25
1.00
SECTION B–B
REVISIONS
1
81–04–18
S. HINES
8.80 WAS 8.60

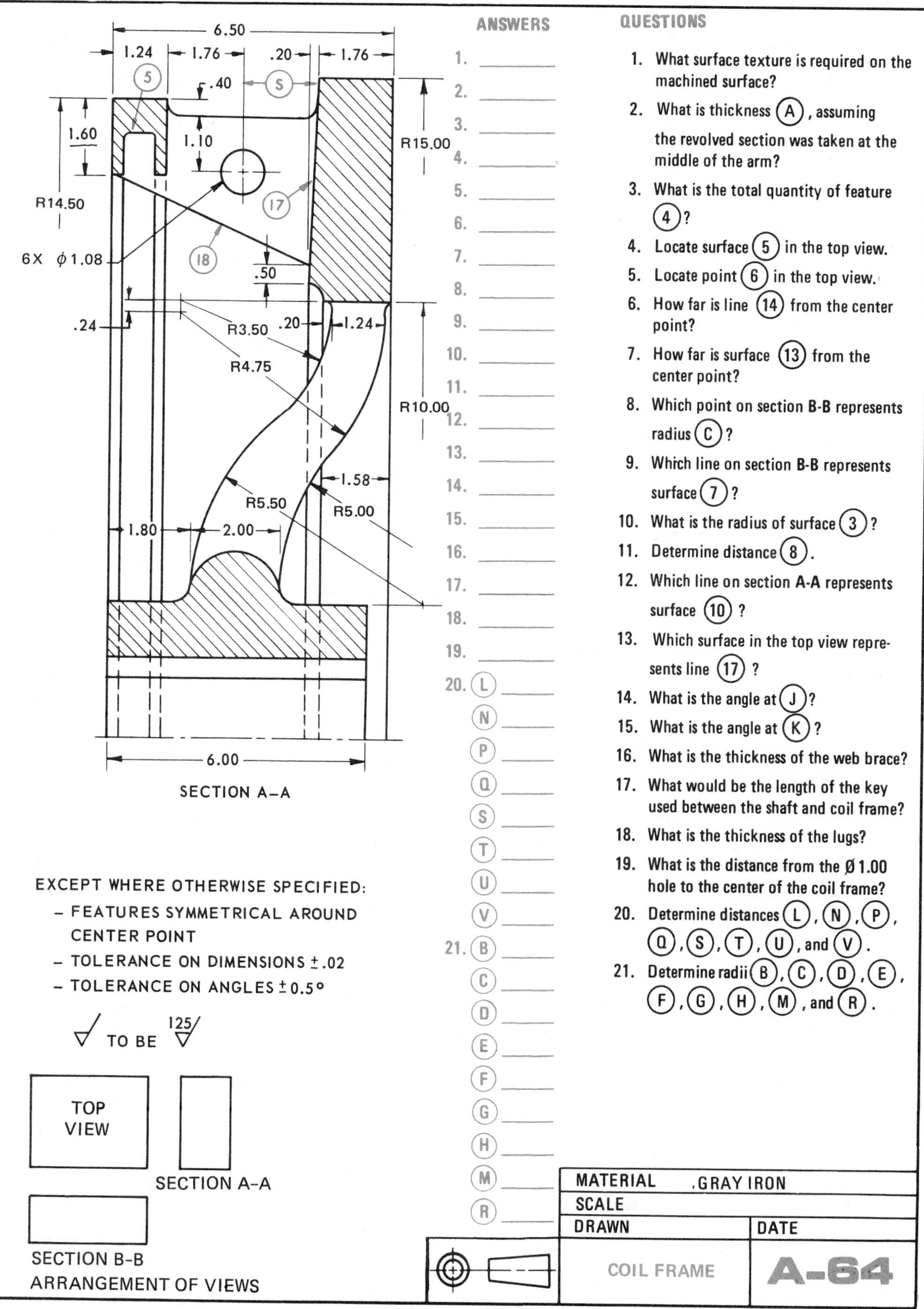
6.50
1.24
1.76
.20
1.76
5
.40
S
1.60
1.10
R15.00
R14.50
17
6X ϕ1.08
18
.50
.24
R3.50
.20
1.24
R4.75
R10.00
1.58
R5.50
R5.00
1.80
2.00
6.00
SECTION A–A
EXCEPT WHERE OTHERWISE SPECIFIED:
– FEATURES SYMMETRICAL AROUND CENTER POINT
– TOLERANCE ON DIMENSIONS ±.02
– TOLERANCE ON ANGLES ±0.5°
TO BE 125
TOP VIEW
SECTION A-A
SECTION B-B
ARRANGEMENT OF VIEWS
ANSWERS
1.
2.
3.
4.
5.
6.
7.
8.
9.
10.
11.
12.
13.
14.
15.
16.
17.
18.
19.
20. L
N
P
Q
S
T
U
V
21. B
C
D
E
F
G
H
M
R
QUESTIONS
1. What surface texture is required on the machined surface?
2. What is thickness (A), assuming the revolved section was taken at the middle of the arm?
3. What is the total quantity of feature (4)?
4. Locate surface (5) in the top view.
5. Locate point (6) in the top view.
6. How far is line (14) from the center point?
7. How far is surface (13) from the center point?
8. Which point on section B-B represents radius (C)?
9. Which line on section B-B represents surface (7)?
10. What is the radius of surface (3)?
11. Determine distance (8).
12. Which line on section A-A represents surface (10)?
13. Which surface in the top view represents line (17)?
14. What is the angle at (J)?
15. What is the angle at (K)?
16. What is the thickness of the web brace?
17. What would be the length of the key used between the shaft and coil frame?
18. What is the thickness of the lugs?
19. What is the distance from the Ø1.00 hole to the center of the coil frame?
20. Determine distances (L), (N), (P), (Q), (S), (T), (U), and (V).
21. Determine radii (B), (C), (D), (E), (F), (G), (H), (M), and (R).
MATERIAL .GRAY IRON
SCALE
DRAWN
DATE
COIL FRAME
A-64

UNIT 29

PIN FASTENERS

Pin fasteners offer an inexpensive and effective approach to assembly where loading is primarily in shear. They can be divided into two groups: semipermanent and quick release.

Semipermanent pin fasteners require application of pressure or the aid of tools for installation or removal. Representative types include machine pins (dowel, straight, taper, clevis, and cotter pins) and radial locking pins (grooved surface and spring).

Quick release fasteners are more elaborate self-contained pins which are used for rapid manual assembly or disassembly. They use a form of spring loaded mechanism to provide a locking action in assembly.

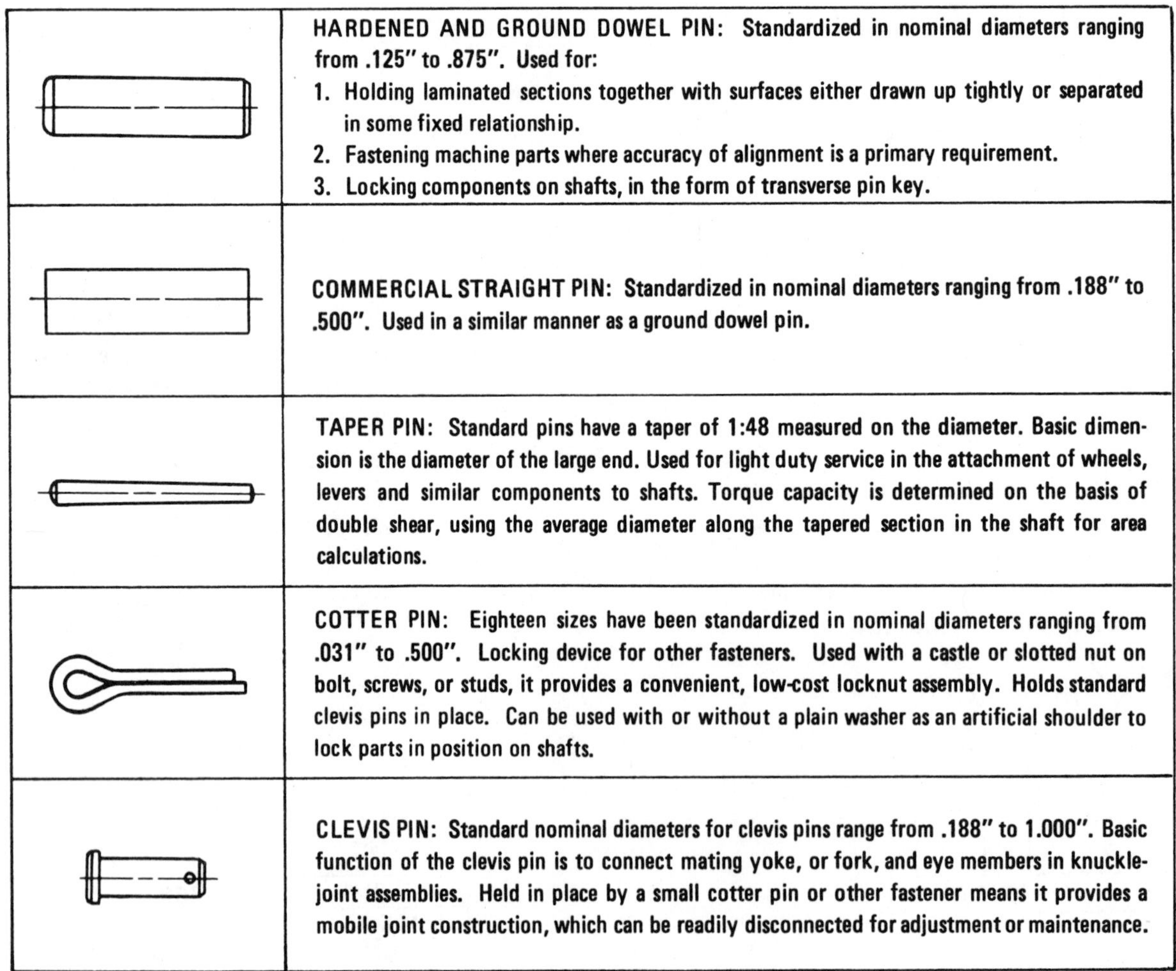

	HARDENED AND GROUND DOWEL PIN: Standardized in nominal diameters ranging from .125" to .875". Used for: 1. Holding laminated sections together with surfaces either drawn up tightly or separated in some fixed relationship. 2. Fastening machine parts where accuracy of alignment is a primary requirement. 3. Locking components on shafts, in the form of transverse pin key.
	COMMERCIAL STRAIGHT PIN: Standardized in nominal diameters ranging from .188" to .500". Used in a similar manner as a ground dowel pin.
	TAPER PIN: Standard pins have a taper of 1:48 measured on the diameter. Basic dimension is the diameter of the large end. Used for light duty service in the attachment of wheels, levers and similar components to shafts. Torque capacity is determined on the basis of double shear, using the average diameter along the tapered section in the shaft for area calculations.
	COTTER PIN: Eighteen sizes have been standardized in nominal diameters ranging from .031" to .500". Locking device for other fasteners. Used with a castle or slotted nut on bolt, screws, or studs, it provides a convenient, low-cost locknut assembly. Holds standard clevis pins in place. Can be used with or without a plain washer as an artificial shoulder to lock parts in position on shafts.
	CLEVIS PIN: Standard nominal diameters for clevis pins range from .188" to 1.000". Basic function of the clevis pin is to connect mating yoke, or fork, and eye members in knuckle-joint assemblies. Held in place by a small cotter pin or other fastener means it provides a mobile joint construction, which can be readily disconnected for adjustment or maintenance.

Fig. 29-1 Machine pins

Machine Pins

Five types of machine pins are commonly used today: ground dowel pins; commercial straight pins; taper pins; clevis pins; and standard cotter pins, figure 29-1.

Dowel Pins. Dowel pins or small straight pins have many uses. They are used to hold parts in alignment and to guide parts into desired positions. Dowel pins are most commonly used for the alignment of parts which are fastened with screws or bolts and must be accurately fitted together.

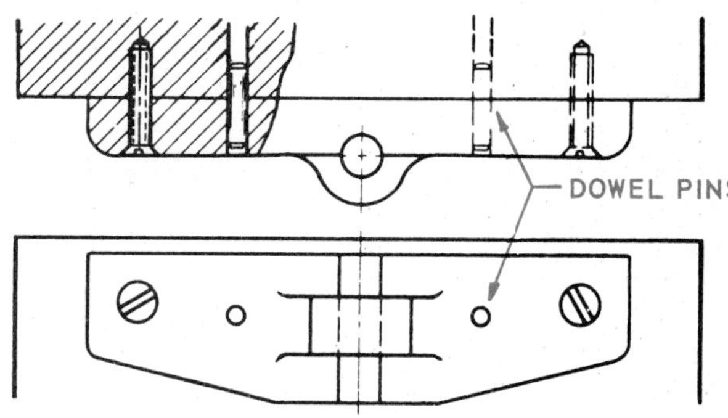

Fig. 29-2 Aligning parts with dowel pins

When two pieces are to be fitted together, as in the case of the part in figure 29-2, one method of alignment is to clamp the two pieces in the desired location, drill and ream the dowel holes, insert the dowel pins, and then drill and tap for the screw holes.

Drill jigs are frequently used when production in the interchangeability of parts is required or when the nature of the piece does not permit the transfer of the doweled holes from one piece to the other. A drill jig, figure 29-3, was used in drilling the dowel holes for the spider (drawing A-65M).

Taper Pins. Holes for taper pins are usually sized by reaming. A through hole is formed by step drills and straight fluted reamers. The present trend is toward the use of helically fluted taper reamers which provide more accurate sizing and require only a pilot hole the size of the small end of the taper pin. The pin is usually driven into the hole until it is fully seated. The taper of the pin aids hole alignment in assembly.

A tapered hole in a hub and a shaft is shown in figure 29-4. If the hub and shaft are drilled and reamed separately, a misalignment might

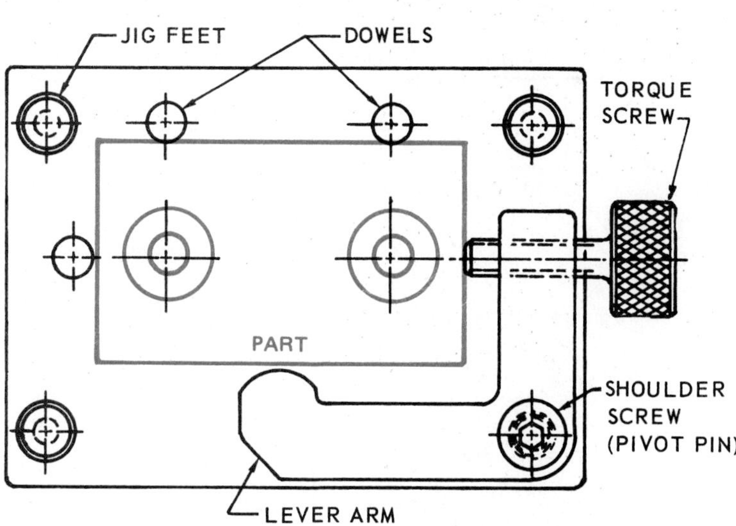

Fig. 29-3 Dowel pins used to align part during drillings

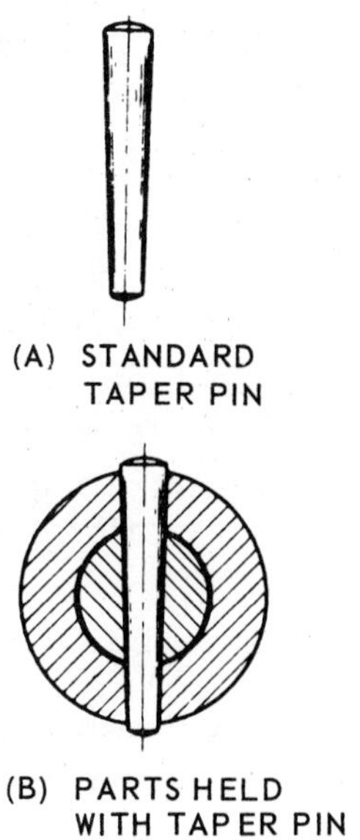

Fig. 29-4 Taper pin application

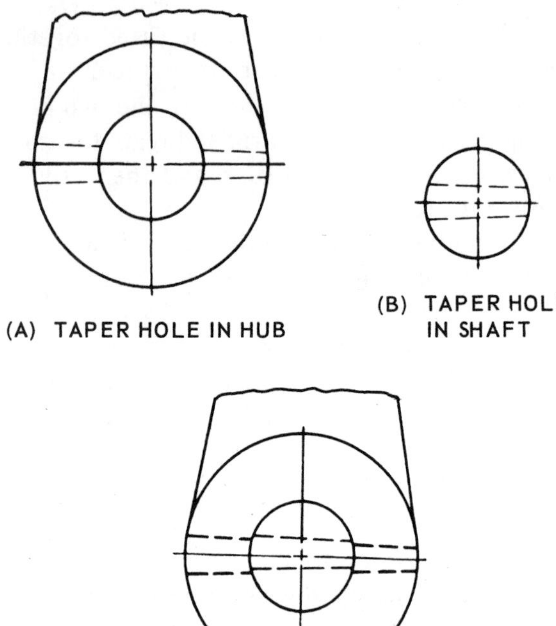

Fig. 29-5 Possibility of hole misalignment if holes are not drilled at assembly

occur as shown in figure 29-5. To prevent misalignment, the hub and shaft should be drilled at the same time as the parts are assembled. Each of the detailed parts should carry a note similar to the following: DRILL AND REAM FOR NO. 1 TAPER PIN AT ASSEMBLY.

Cotter Pins. The cotter pin is a standard machine pin commonly used as a fastener in the assembly of machine parts where great accuracy is not required, figure 29-6. There is no standard way to represent cotter pins in assembly drawings. The method of representation shown in figure 29-7 is, however, commonly used to indicate cotter pins.

Radial-Locking Pins

Low cost, ease of assembly, and high resistance to vibration and impact loads are common attributes of this group of commercial pin devices designed primarily for semipermanent fastening service. Two basic pin forms are used:

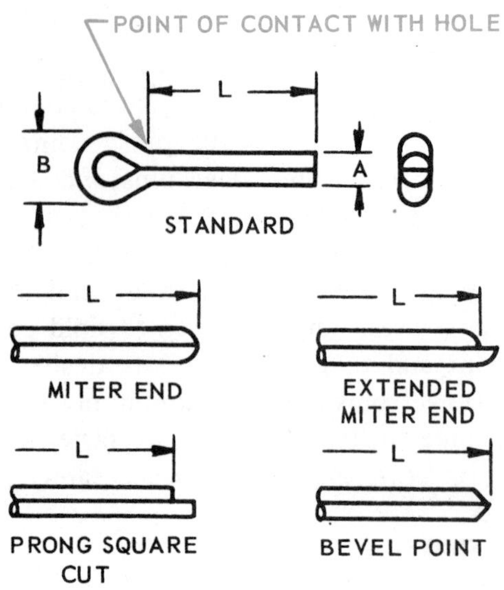

NOMINAL THREAD SIZE in.	(mm)	NOMINAL COTTER PIN SIZE in.	(mm)	COTTER PIN HOLE in.	(mm)	END CLEARANCE * in.	(mm)
.250	(6)	.062	(1.5)	.078	(1.9)	.12	(3)
.312	(8)	.078	(2)	.094	(2.4)	.12	(3)
.375	(10)	.094	(2.5)	.109	(2.8)	.14	(4)
.500	(12)	.125	(3)	.141	(3.4)	.18	(5)
.625	(14)	.156	(3)	.172	(3.4)	.25	(5)
.750	(20)	.156	(4)	.172	(4.5)	.25	(7)
1.000	(24)	.188	(5)	.203	(5.6)	.31	(8)
1.125	(27)	.188	(5)	.203	(5.6)	.39	(8)
1.250	(30)	.219	(6)	.234	(6.3)	.44	(10)
1.375	(36)	.219	(6)	.234	(6.3)	.44	(11)
1.500	(42)	.250	(6)	.266	(6.3)	.50	(12)
1.750	(48)	.312	(8)	.312	(8.5)	.55	(14)

* DISTANCE FROM EXTREME POINT OF BOLT OR SCREW TO CENTER OF COTTER PIN HOLE.

Fig. 29-6 Cotter pin data

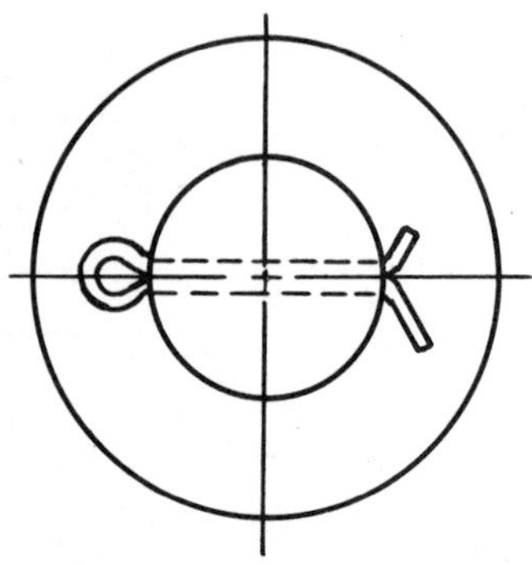

Fig. 29-7 Cotter pin in an assembly drawing

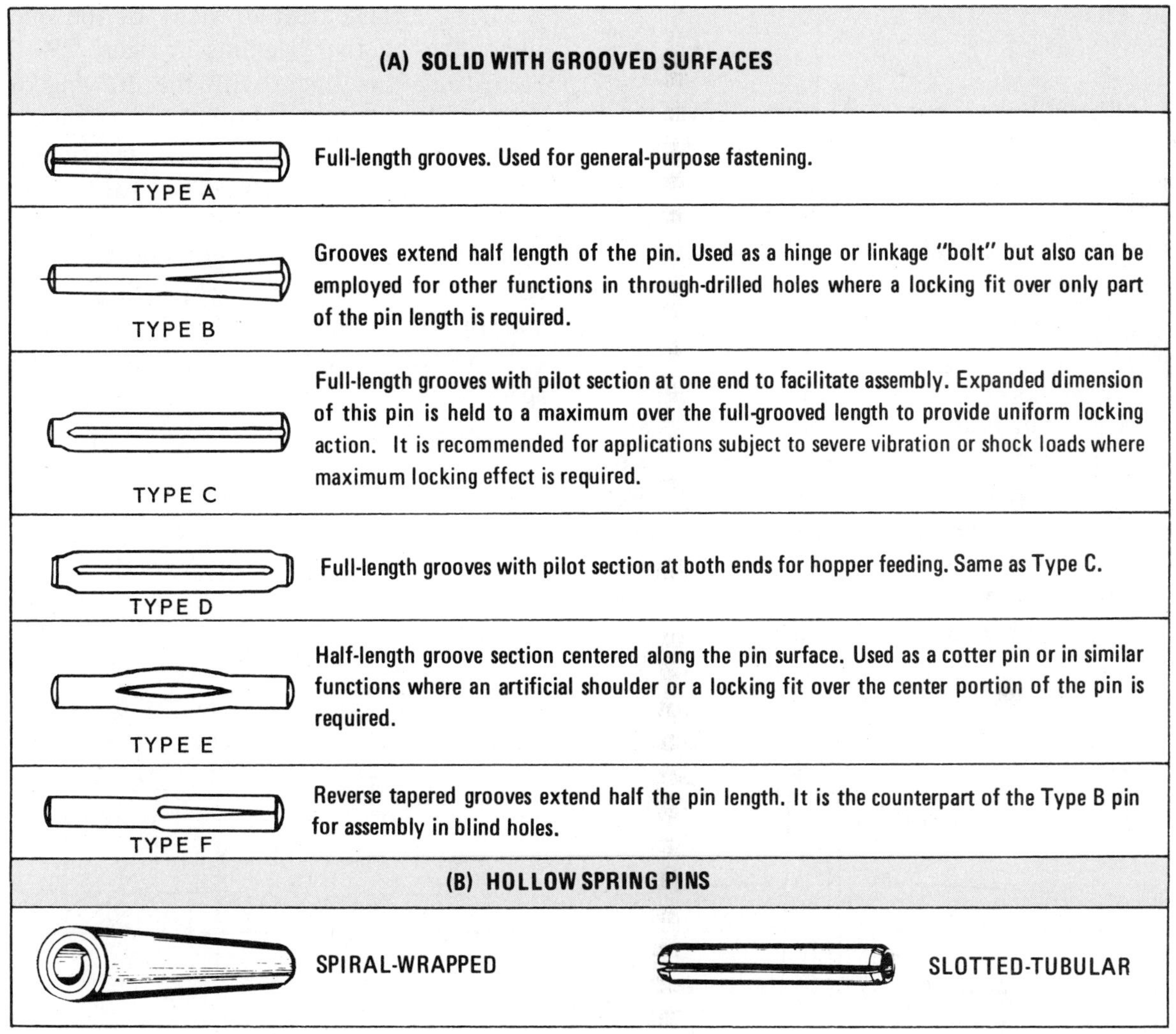

(A) SOLID WITH GROOVED SURFACES
TYPE A
Full-length grooves. Used for general-purpose fastening.
TYPE B
Grooves extend half length of the pin. Used as a hinge or linkage "bolt" but also can be employed for other functions in through-drilled holes where a locking fit over only part of the pin length is required.
TYPE C
Full-length grooves with pilot section at one end to facilitate assembly. Expanded dimension of this pin is held to a maximum over the full-grooved length to provide uniform locking action. It is recommended for applications subject to severe vibration or shock loads where maximum locking effect is required.
TYPE D
Full-length grooves with pilot section at both ends for hopper feeding. Same as Type C.
TYPE E
Half-length groove section centered along the pin surface. Used as a cotter pin or in similar functions where an artificial shoulder or a locking fit over the center portion of the pin is required.
TYPE F
Reverse tapered grooves extend half the pin length. It is the counterpart of the Type B pin for assembly in blind holes.
(B) HOLLOW SPRING PINS
SPIRAL-WRAPPED
SLOTTED-TUBULAR

Fig. 29-8 Radial locking pins

solid with grooved surfaces, and hollow spring pins which may be either slotted or spiral wrapped, figure 29-8. In assembly, radial forces produced by elastic action at the pin surface develop a secure, frictional-locking grip against the hole wall. These pins are reusable and can be removed and reassembled many times without appreciable loss of fastening effectiveness. Live spring action at the pin surface also prevents loosening under shock and vibration loads. The need for accurate sizing of holes is reduced since the pins accommodate variations.

Solid Pins with Grooved Surfaces. The locking action of groove pins is provided by parallel, longitudinal grooves uniformly spaced around the pin surface. Rolled or pressed into solid pin stock, the grooves expand the effective diameter of the pin. When the pin is driven into a drilled hole corresponding in size to the nominal pin diameter, elastic deformation of the raised groove edges produces a secure interference fit with the hole wall. Figure 29-8 shows the six standardized constructions of grooved pins.

Hollow Spring Pins. Spiral-wrapped and slotted-tubular pin forms are made to controlled diameters greater than the holes into which they are pressed. Compressed when driven into the hole, the pins exert spring pressure against the hole wall along their entire engaged length to develop a strong locking action.

SECTION THROUGH SHAFTS, PINS, AND KEYS

Shafts, bolts, nuts, rods, rivets, keys, pins, and similar solid parts, the axes of which lie on the cutting plane, are sectioned only when a broken-out section of the shaft is used to clearly indicate the key, keyseat, of pin, figure 29-9.

ARRANGEMENT OF VIEWS OF DRAWING A-65M

Parts which are to be fitted over shafts as a single unit are sometimes made in two or more pieces. This is done for ease in assembly and replacement on the main structure of a machine rather than for ease in manufacture. Drawing A-65M shows two parts which are bolted and doweled together to form one unit.

The arrangement of views of the spider is illustrated by the diagrams in figure 29-10. By comparing this figure with the drawing of the spider, note that the two halves together represent the top view. The right view is a full section of each half (**A** and **B**). The front view is a drawing of the front of part **A** only.

Although the front and side views are incomplete, the manner in which they are drawn and the arrangement of the views satisfies the demand for clearness and economy in time and space.

REFERENCES AND SOURCE MATERIALS

1. Machine Design-Fasteners Reference Issue Nov. 1981.

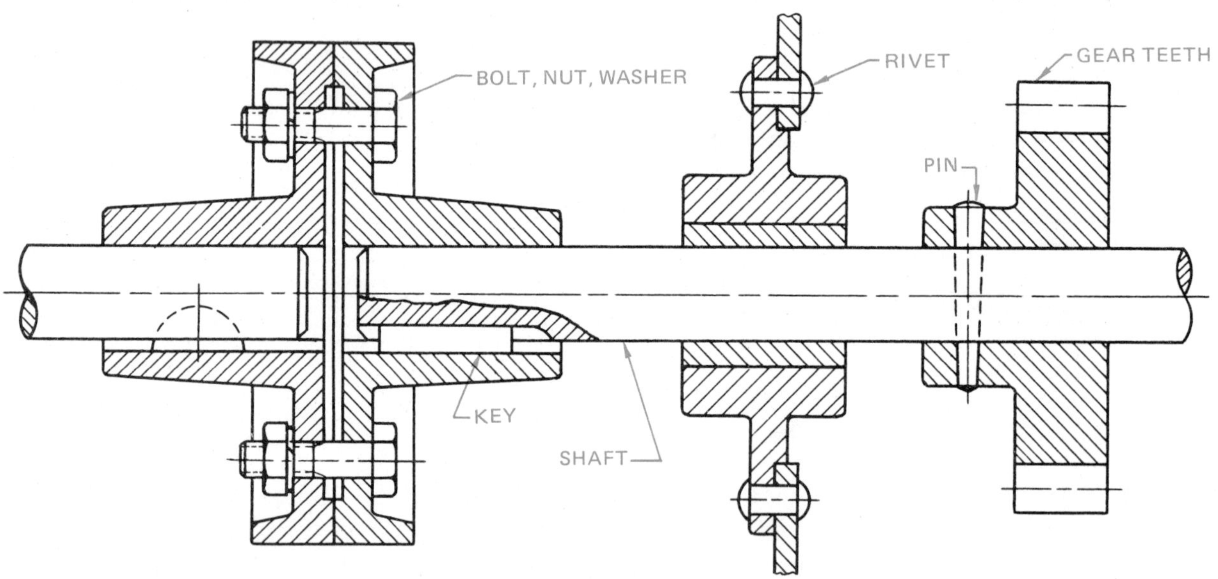

Fig. 29-9 Parts that are not section lined in section drawings

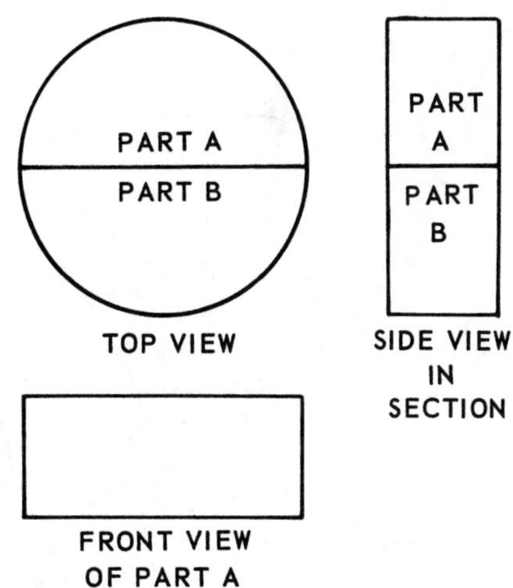

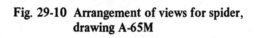

Fig. 29-10 Arrangement of views for spider, drawing A-65M

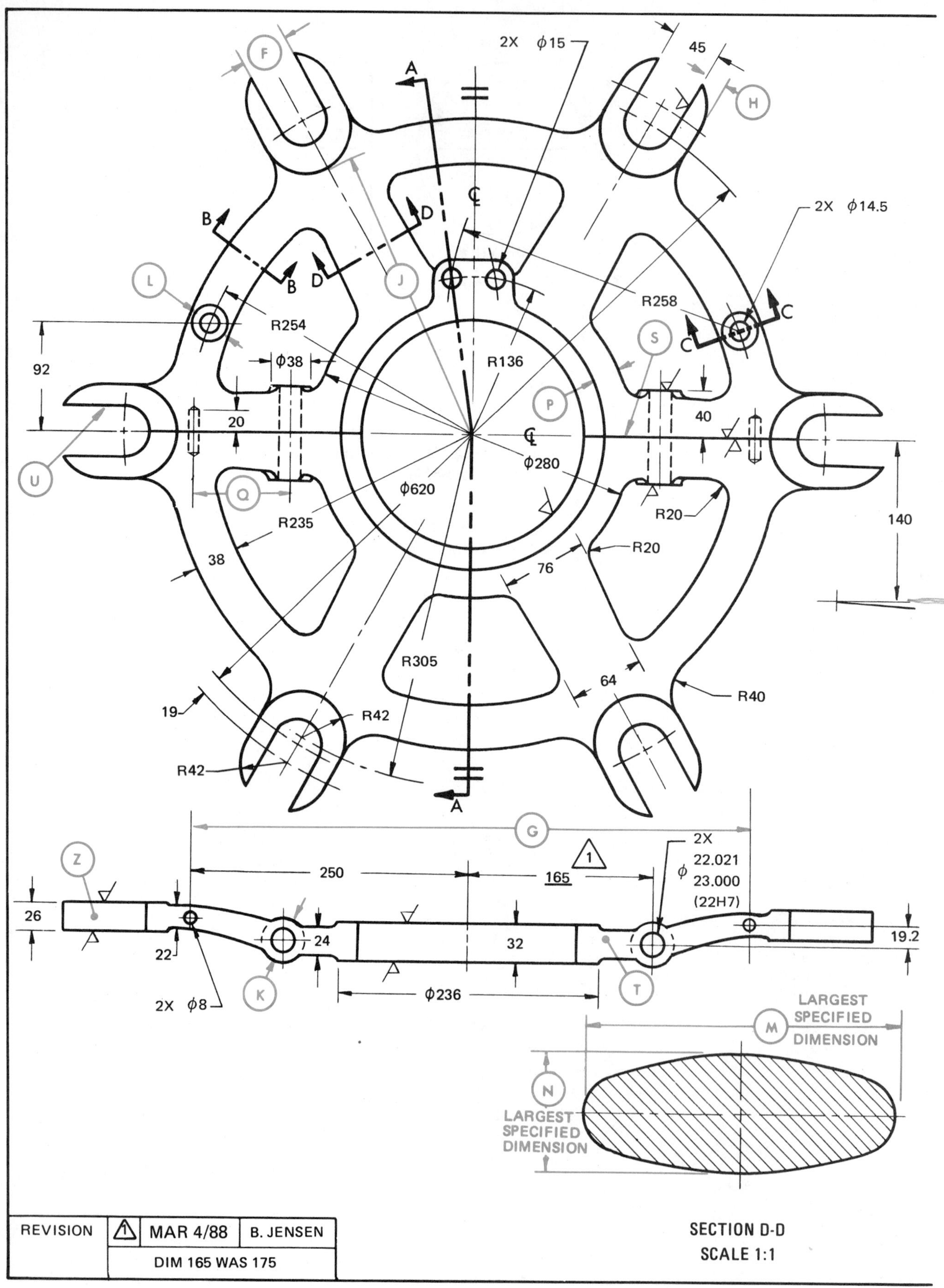

2X ϕ15
45
2X ϕ14.5
R258
R254
ϕ38
92
20
40
R136
ϕ280
ϕ620
R235
38
R20
R20
76
140
R305
64
R40
19
R42
R42
A
B
C
D
250
165
2X
ϕ 22.021
23.000
(22H7)
26
24
32
19.2
22
2X ϕ8
ϕ236
LARGEST SPECIFIED DIMENSION
LARGEST SPECIFIED DIMENSION
REVISION
MAR 4/88
B. JENSEN
DIM 165 WAS 175
SECTION D-D
SCALE 1:1

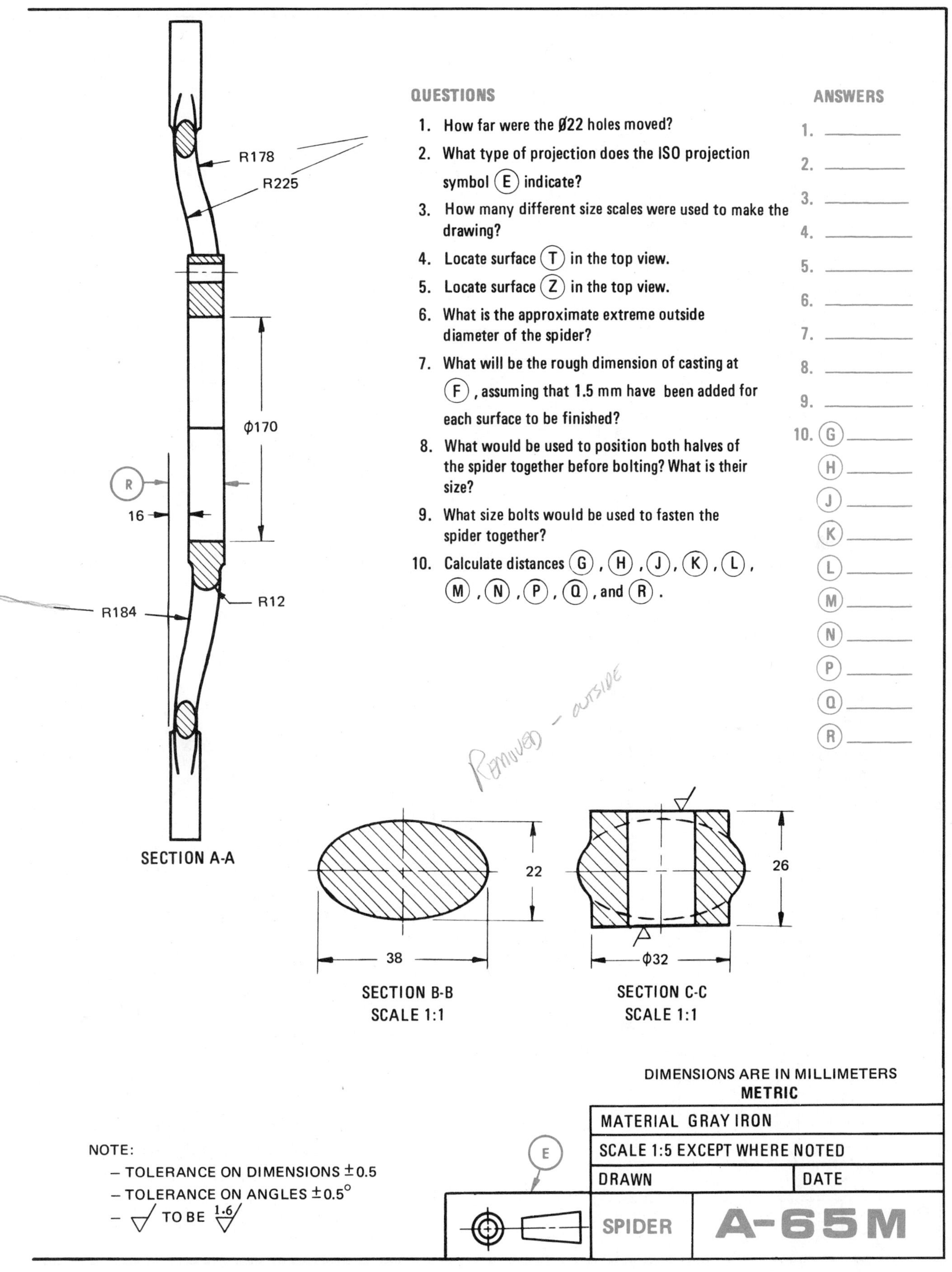
QUESTIONS
1. How far were the Ø22 holes moved?
2. What type of projection does the ISO projection symbol (E) indicate?
3. How many different size scales were used to make the drawing?
4. Locate surface (T) in the top view.
5. Locate surface (Z) in the top view.
6. What is the approximate extreme outside diameter of the spider?
7. What will be the rough dimension of casting at (F), assuming that 1.5 mm have been added for each surface to be finished?
8. What would be used to position both halves of the spider together before bolting? What is their size?
9. What size bolts would be used to fasten the spider together?
10. Calculate distances (G), (H), (J), (K), (L), (M), (N), (P), (Q), and (R).
ANSWERS
1.
2.
3.
4.
5.
6.
7.
8.
9.
10. G
H
J
K
L
M
N
P
Q
R
R178
R225
ϕ170
R
16
R184
R12
R12
SECTION A-A
22
38
SECTION B-B
SCALE 1:1
26
ϕ32
SECTION C-C
SCALE 1:1
DIMENSIONS ARE IN MILLIMETERS
METRIC
MATERIAL GRAY IRON
SCALE 1:5 EXCEPT WHERE NOTED
DRAWN
DATE
E
SPIDER
A-65M
NOTE:
– TOLERANCE ON DIMENSIONS ±0.5
– TOLERANCE ON ANGLES ±0.5°
– √ TO BE 1.6√

QUESTIONS

1. Which views are shown?
2. How many surfaces are to be finished?
3. How many scraped surfaces are indicated?
4. How many holes are to be tapped?
5. What is the purpose of tapped hole (R) ?
6. Which surface is (3) in the left-side view?
7. Which surface is (2) in the left-side view?
8. Which surface is (4) in the front view?
9. Which surface is (14) in the left-side view?
10. Which surface in the top view and front view is (9) ?
11. Which surface in the top view and front view is (8) ?
12. Which surface is (12) in the top view?
13. What is the name of part (V) ?
14. What is the purpose of part (V) ?
15. What do dotted lines at (W) represent?
16. Which surface is line (6) in the front view?
17. Which top view line indicates point (Z) ?
18. What is the depth of the tapped hole at (X) ?
19. Which edges or surfaces in the left and front views does line (T) represent?
20. What is the diameter of tap drill (Y) ? (See Appendix.)
21. Determine dimensions or operations at (A) to (Q) ; (20) to (52)

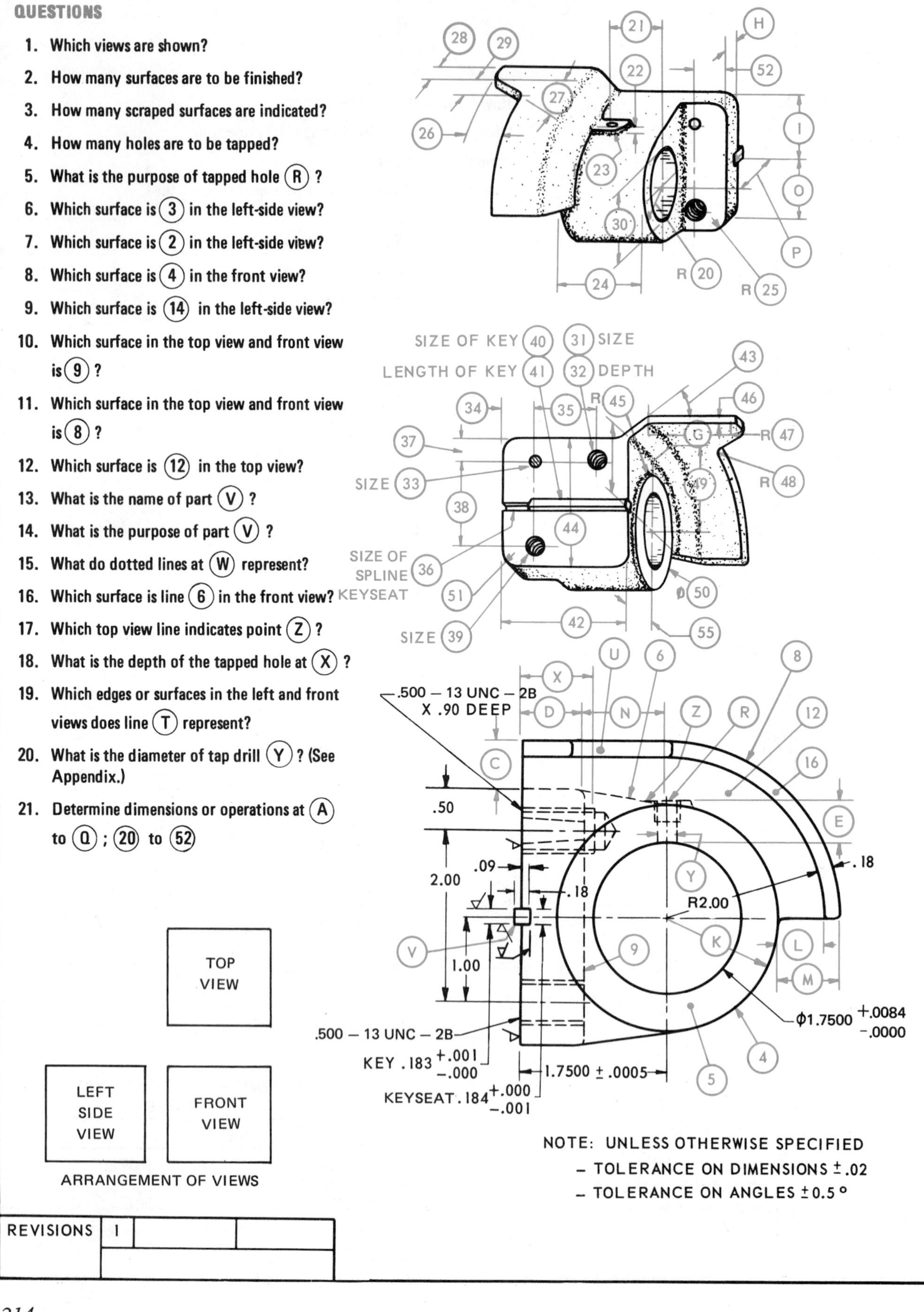

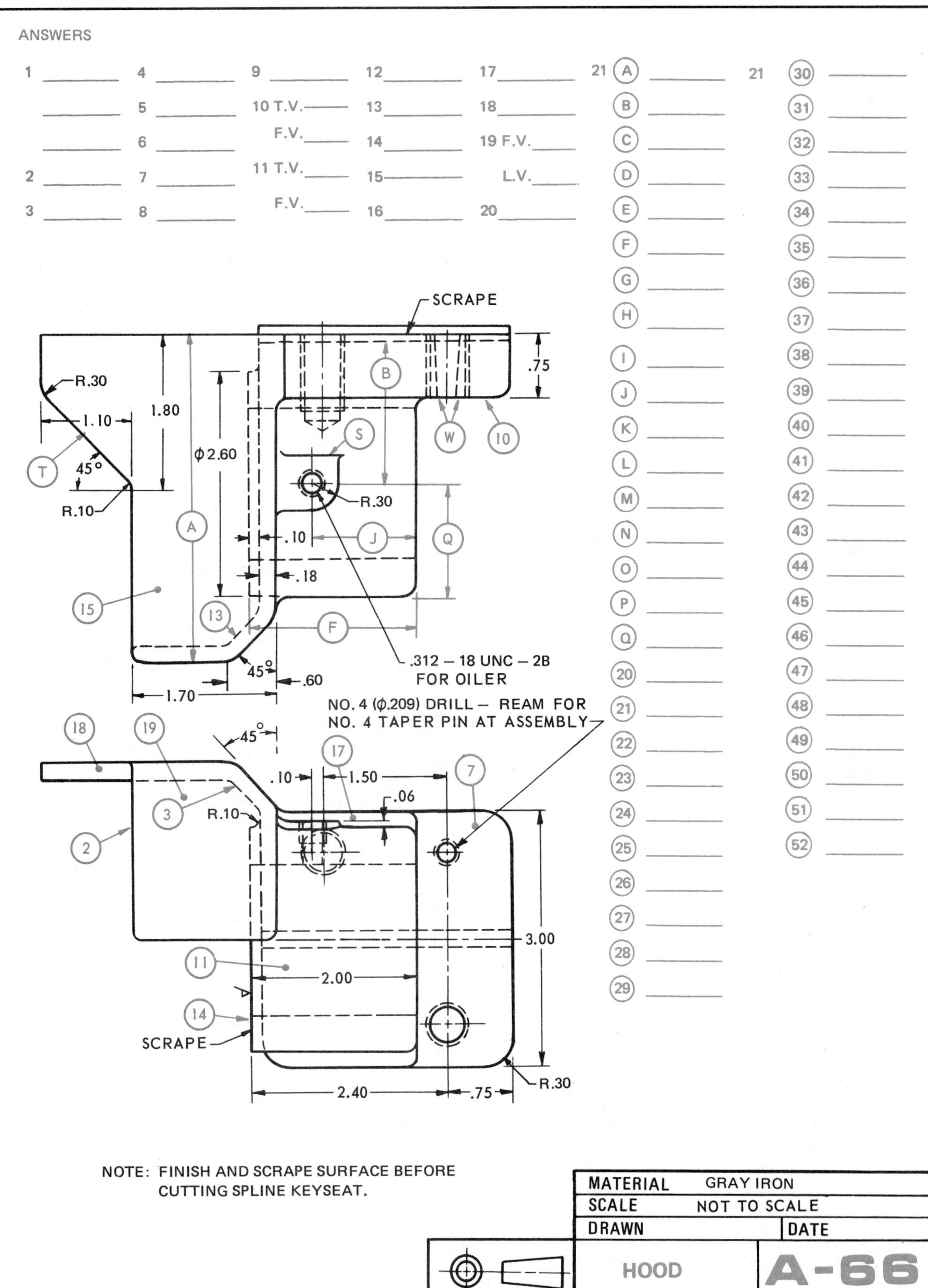
ANSWERS
1
2
3
4
5
6
7
8
9
10 T.V.
F.V.
11 T.V.
F.V.
12
13
14
15
16
17
18
19 F.V.
L.V.
20
21
SCRAPE
R.30
1.10
1.80
ϕ2.60
45°
R.10
.75
.10
.18
R.30
.312 – 18 UNC – 2B
FOR OILER
45°
.60
1.70
NO. 4 (ϕ.209) DRILL – REAM FOR
NO. 4 TAPER PIN AT ASSEMBLY
45°
.10
1.50
.06
R.10
3.00
2.00
SCRAPE
2.40
.75
R.30
NOTE: FINISH AND SCRAPE SURFACE BEFORE
CUTTING SPLINE KEYSEAT.
MATERIAL GRAY IRON
SCALE NOT TO SCALE
DRAWN
DATE
HOOD
A-66

UNIT 30

CHAIN DIMENSIONING

Most linear dimensions are intended to apply on a point-to-point basis. *Chain* dimensioning is applied directly from one feature to another, as shown in figure 30-1(A). Such dimensions locate surfaces and features directly between the points indicated, or between corresponding points on the indicated surfaces.

For example, a diameter applies to all diameters of a cylindrical surface (not merely to the diameter at the end where the dimension is shown), a thickness applies to all opposing points on the surfaces, and a hole-locating dimension applies from the hole axis perpendicular to the edge of the part on the same center line.

BASE LINE DIMENSIONING

When several dimensions extend from a common data point or points on a line or surface, figure 30-1(B), this is called *base line* or *datum* dimensioning. This form of dimensioning is preferred for parts to be manufactured by numerical control machines.

A *datum* is a feature (point, line, or surface) from which other features are located. A datum may be the center of a circle, the axis of a cylinder, or the axis of symmetry.

The location of a series of holes or step features as shown in figure 30-1(B) is an example of the use of base line dimensioning. Figure 30-2 illustrates more complex uses of base line dimensioning. Without this type of dimensioning the distances between the first and last steps could vary considerably because of the build-up of tolerances permitted between the adjacent holes. Dimensioning from a datum line controls the tolerances between the holes to the basic general tolerance of ±.02 inch.

In this system, the tolerance from the common point to each of the features must be

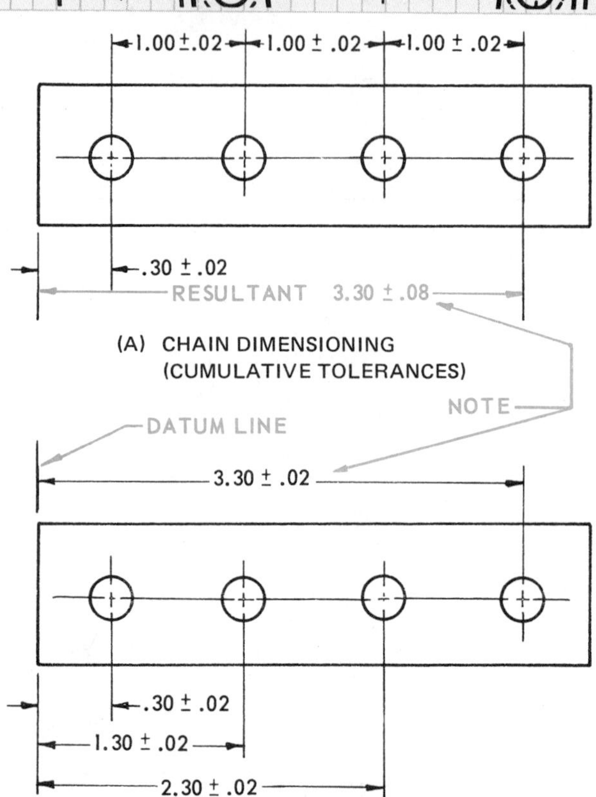

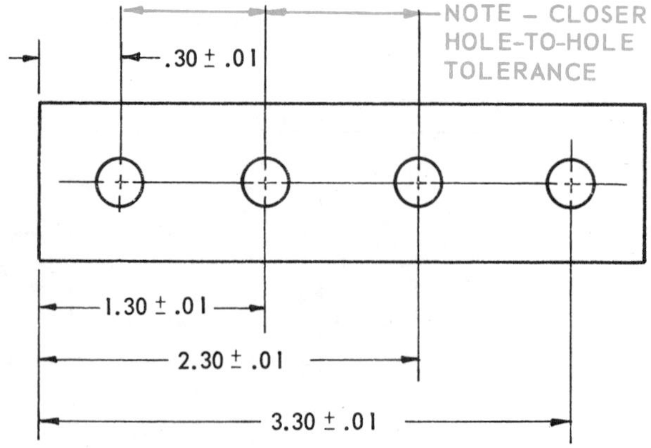

Fig. 30-1 Comparison between chain dimensioning and base line dimensioning

held to half the tolerance acceptable between individual features. For example, in figure 30-1(C), if a tolerance between two individual holes of ±.02 inch were desired, each of the dimensions shown would have to be held to ±.01 inch.

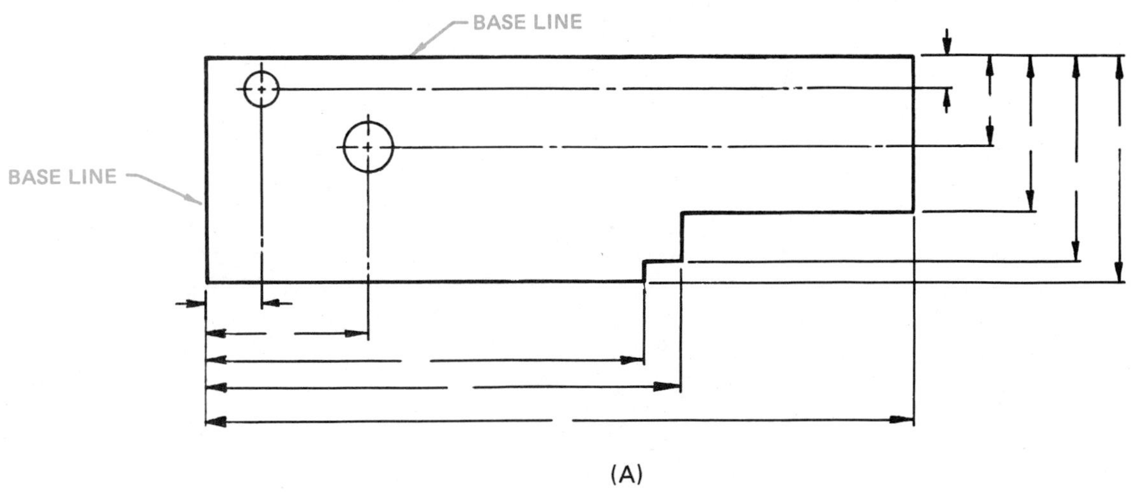

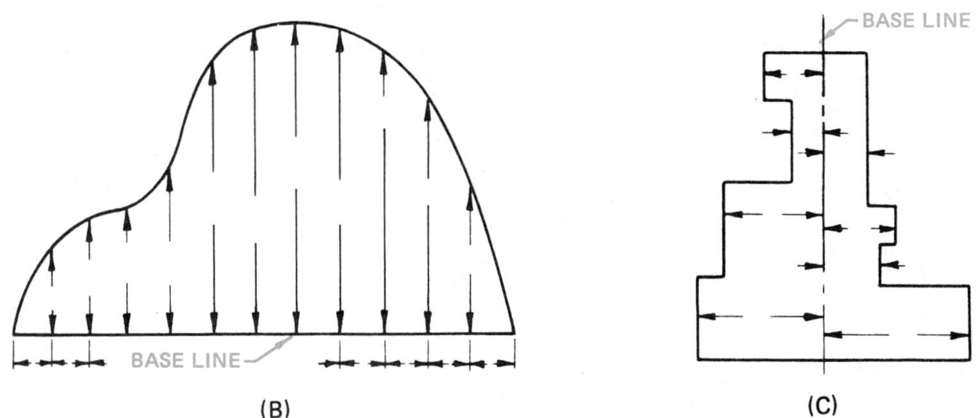

Fig. 30-2 Applications of base line dimensioning

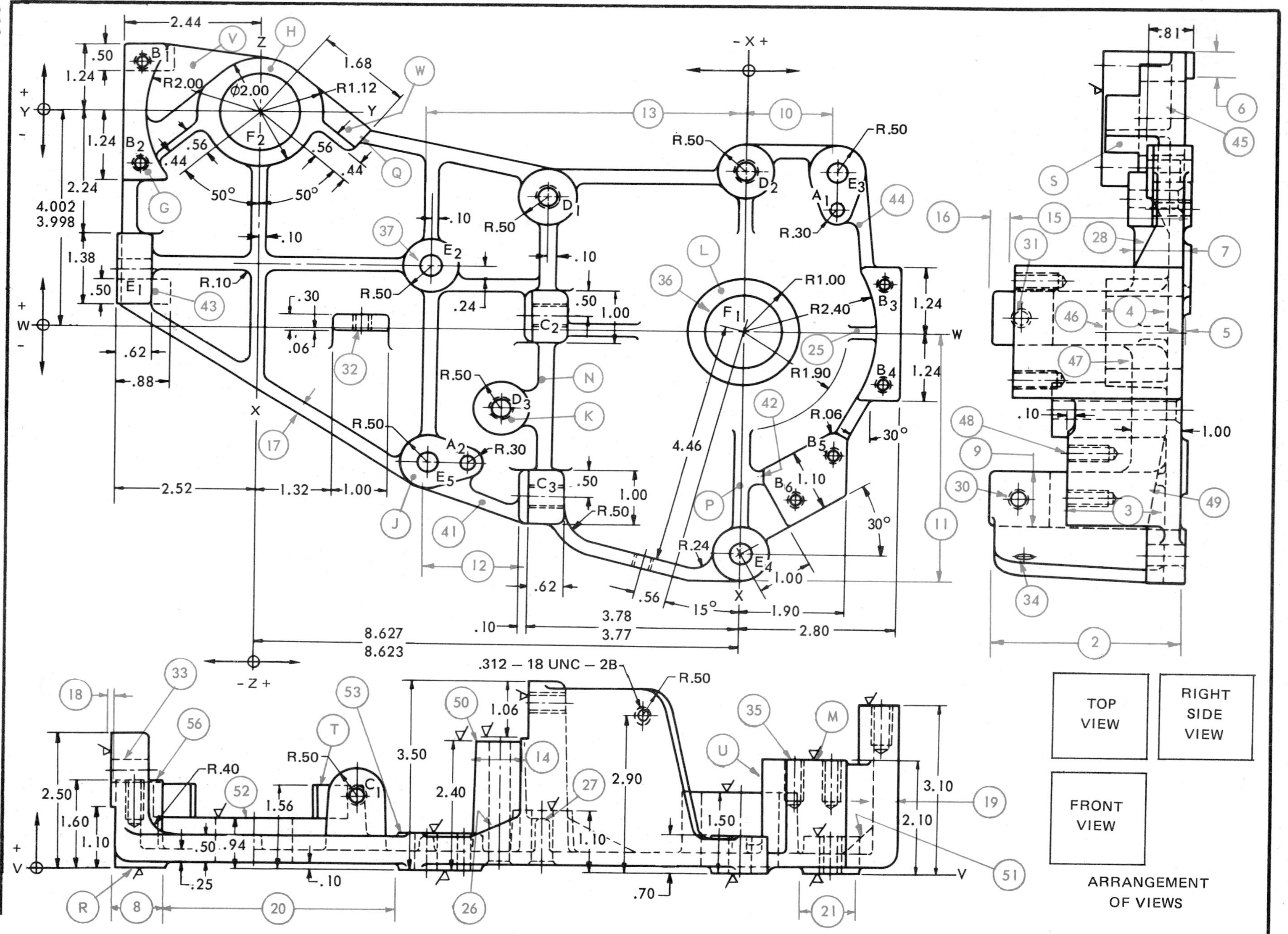
.312 — 18 UNC — 2B
TOP VIEW
RIGHT SIDE VIEW
FRONT VIEW
ARRANGEMENT OF VIEWS

QUESTIONS

NOTE: Where limit dimensions are given use larger limit.

1. What are the overall dimensions of the casting?
2. Show a roughness symbol meaning the same as the one given on the drawing.
3. Identify surfaces (G) to (U) on one of the other views.
4. Locate rib (25) on the front view.
5. Locate rib (26) on the top view.
6. Locate rib (27) on the right side view.
7. Locate rib (28) on the top view.
8. Give the sizes of the following holes: (30), (31), (32), (33), and (34).
9. How deep is hole (35)?
10. What is the tolerance on hole (36)?
11. How deep is hole (37)?
12. Determine distances (2) to (21). (Where limit dimensions are shown, use maximum size.)

ANSWERS

1. W ______
   D ______
   H ______
2. ______
3. (G) ______
   (H) ______
   (J) ______
   (K) ______
   (L) ______
   (M) ______
   (N) ______
   (P) ______
   (Q) ______
   (R) ______
   (S) ______
   (T) ______
   (U) ______
4. ______
5. ______
6. ______
7. ______
8. (30) ______
   (31) ______
   (32) ______
   (33) ______
   (34) ______
9. ______
10. ______
11. ______
12. (2) ______ (3) ______
    (4) ______ (5) ______
    (6) ______ (7) ______
    (8) ______ (9) ______
    (10) ______ (11) ______
    (12) ______ (13) ______
    (14) ______ (15) ______
    (16) ______ (17) ______
    (18) ______ (19) ______
    (20) ______ (21) ______

HOLE SIZE AND LOCATION						
HOLE	DISTANCE FROM					SIZE
	V–V	W–W	X–X	Y–Y	Z–Z	
A_1		2.32	1.62			ϕ .252 / .250
A_2		-2.50	-4.88			
B_1				.94	-2.12	.312 – 18 UNC–2B X .75 DEEP
B_2				-.94	-2.12	
B_3		.94	2.50			
B_4		-.94	2.50			
B_5		-2.28	1.62			
B_6		-3.08	.96			
C_1	1.38				-1.82	.375 – 16 UNC–2B
C_2	3.00	.28				
C_3	3.00	-3.12				
D_1		2.50	-3.50			ϕ .3752 / .3750 CSK .06 X 95°
D_2		3.00	.00			
D_3		-1.48	-4.28			
E_1	1.80			-2.94		ϕ.406
E_2		1.18	-5.56			
E_3		3.00	1.62			
E_4		-4.12	.00			
E_5		-2.50	-5.56			
F_1		.00	.00			ϕ 1.3765 / 1.3745
F_2				.00	.00	

– ALL FILLETS R.06 UNLESS OTHERWISE SHOWN
– ALL RIBS AND WEBS .24 THICK
– TOLERANCE ON DIMENSIONS ±.02
– TOLERANCE ON ANGLES ±0.5°
– SURFACES MARKED ✓ TO BE 250✓

MATERIAL	GRAY IRON
SCALE	NOT TO SCALE
DRAWN	DATE
INTERLOCK BASE	A-67

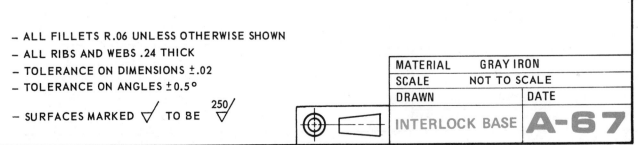

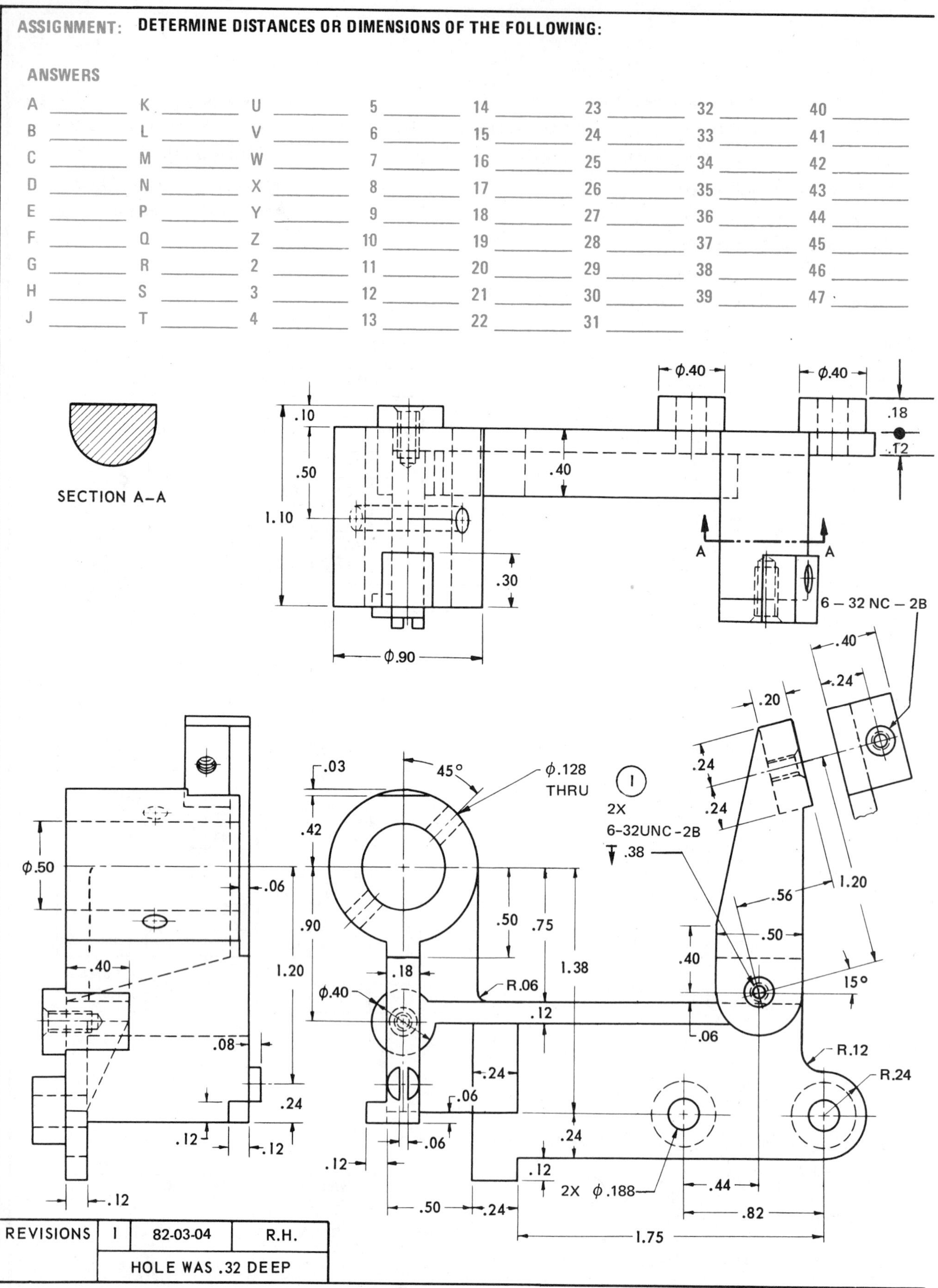
ASSIGNMENT: DETERMINE DISTANCES OR DIMENSIONS OF THE FOLLOWING:
ANSWERS
A B C D E F G H J
K L M N P Q R S T
U V W X Y Z 2 3 4
5 6 7 8 9 10 11 12 13
14 15 16 17 18 19 20 21 22
23 24 25 26 27 28 29 30 31
32 33 34 35 36 37 38 39
40 41 42 43 44 45 46 47
SECTION A–A
ϕ.40
ϕ.40
.18
.12
.10
.50
.40
1.10
.30
ϕ.90
A
A
6 – 32 NC – 2B
.40
.24
.20
.24
.24
.03
45°
ϕ.128
THRU
1
2X
6-32UNC-2B
.38
.42
ϕ.50
.06
.90
.50
.75
1.38
1.20
.56
1.20
.50
.40
15°
.40
.18
R.06
ϕ.40
.12
.06
.08
.24
R.12
R.24
.24
.06
.24
.12
.12
.06
.24
.12
.12
.12
2X ϕ.188
.44
.50
.24
.82
1.75
.12
REVISIONS
1
82-03-04
R.H.
HOLE WAS .32 DEEP

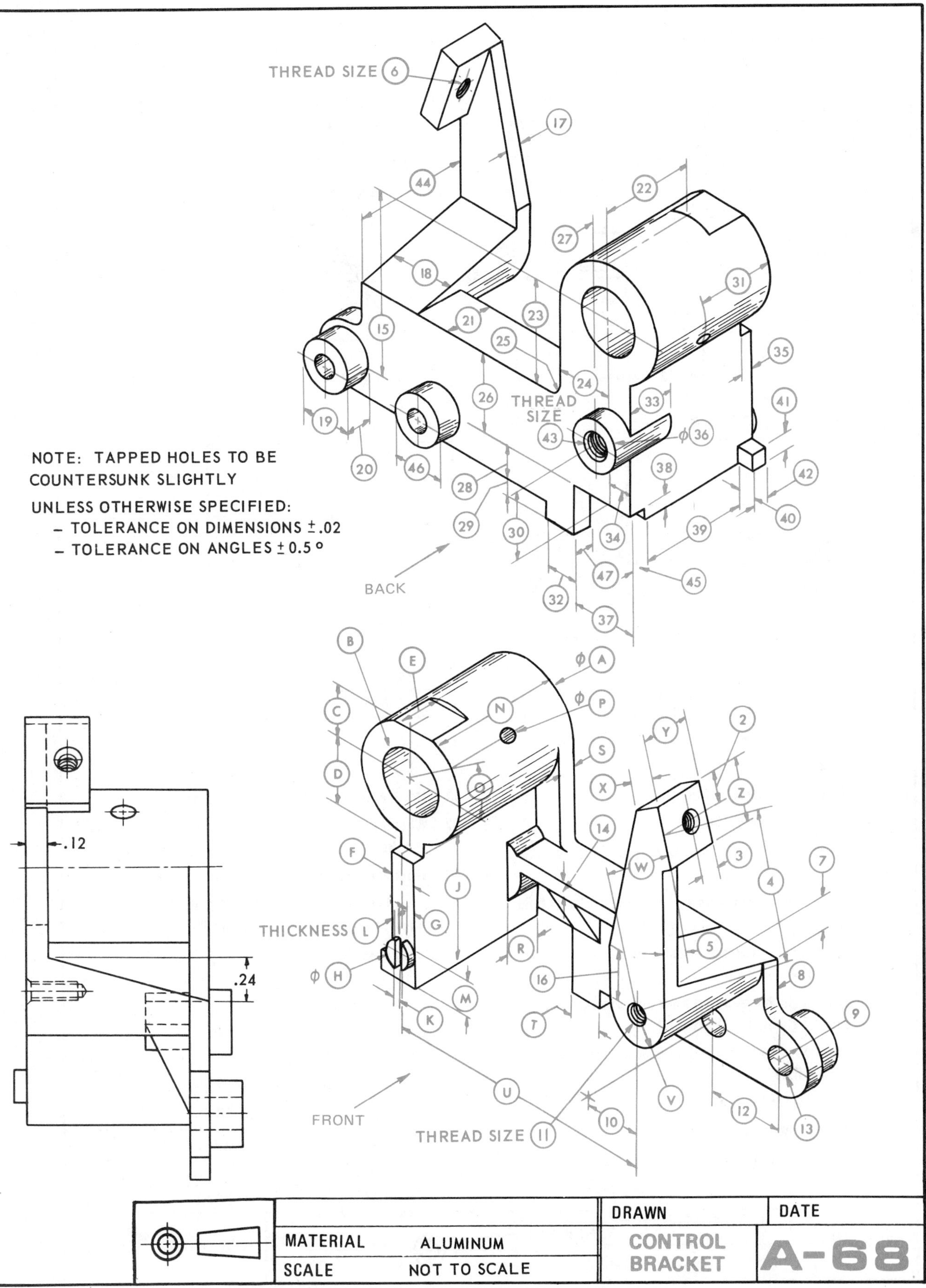
THREAD SIZE 6
NOTE: TAPPED HOLES TO BE COUNTERSUNK SLIGHTLY
UNLESS OTHERWISE SPECIFIED:
– TOLERANCE ON DIMENSIONS ±.02
– TOLERANCE ON ANGLES ±0.5°
THREAD SIZE 43
BACK
FRONT
THICKNESS L
THREAD SIZE 11
.12
.24
MATERIAL ALUMINUM
SCALE NOT TO SCALE
DRAWN
DATE
CONTROL BRACKET
A-68

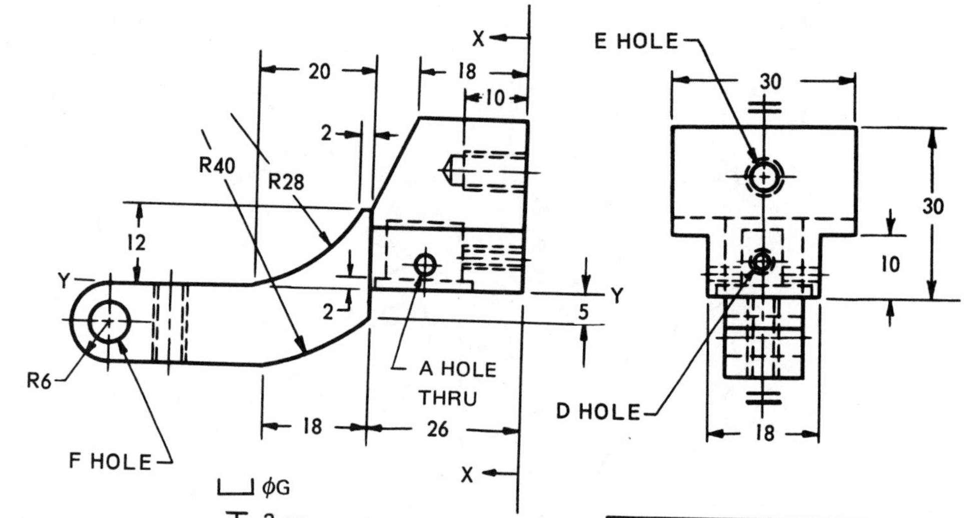

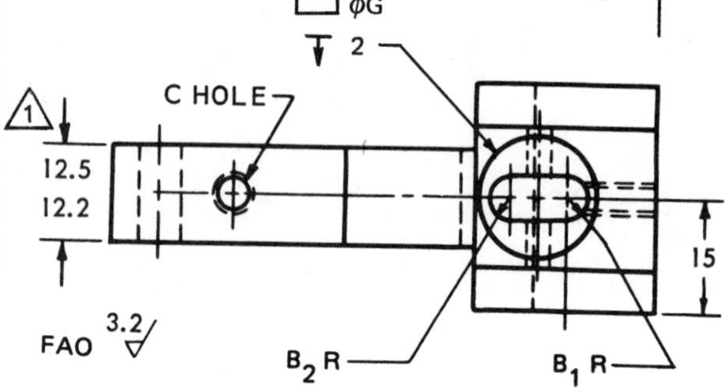

NOTE: UNLESS OTHERWISE SPECIFIED TOLERANCE ON DIMENSIONS ±0.5

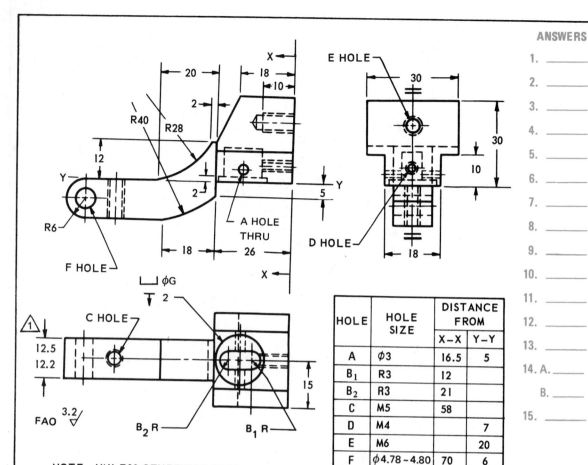

HOLE	HOLE SIZE	DISTANCE FROM	
		X–X	Y–Y
A	ϕ3	16.5	5
B_1	R3	12	
B_2	R3	21	
C	M5	58	
D	M4		7
E	M6		20
F	ϕ4.78–4.80	70	6
G	ϕ16	16.5	

ANSWERS

1. ______
2. ______
3. ______
4. ______
5. ______
6. ______
7. ______
8. ______
9. ______
10. ______
11. ______
12. ______
13. ______
14. A. ______
    B. ______
15. ______

QUESTIONS

1. What is the overall width?
2. What is the overall height?
3. What is the distance from X-X to the center of hole F?
4. What is the distance from Y-Y to the center of hole E?
5. What is the horizontal distance between the center lines of A and F holes?
6. What is the horizontal distance between the center lines of C hole and B_1 radius?
7. How many full threads does E hole have? (See Appendix.)
8. How deep is the spotface?
9. What is the tolerance of F hole?
10. What type of projection is used on this drawing?
11. How far apart are the D and E holes?
12. What is the tolerance on the thickness of the material at F hole?
13. What is the length of the C hole?
14. What is the distance from line X-X to the termination of both ends of the 28 radius?
15. What does FAO mean?

DIMENSIONS ARE IN MILLIMETERS

METRIC

MATERIAL	ALUMINUM
SCALE	NOT TO SCALE
DRAWN	DATE

REVISION	1	F. NEWMAN 6/8/88	CH. A. HEINEN
		DIMENSION WAS 12.4–12.7	

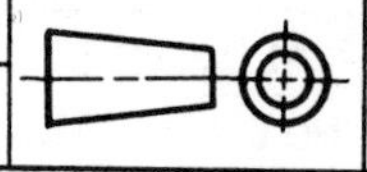

CONTACTOR **A-69M**

UNIT 31

ASSEMBLY DRAWINGS

The term *assembly drawing* refers to the type of drawing in which the various parts of a machine or structure are drawn in their relative positions in the completed unit.

In addition to showing how the parts fit together, the assembly drawing is used mainly:

- To represent the proper working relationships of the mating parts of a machine or structure and the function of each.
- To give a general idea of how the finished product should look.
- To aid in securing overall dimensions and center distances in assembly.
- To give the detailer data needed to design the smaller units of a larger assembly.
- To supply illustrations which may be used for catalogs, maintenance manuals, or other illustrative purposes.

In order to show the working relationship of interior parts, the principles of projection may be violated and details omitted for clarity. Assembly drawings should not be overly detailed because precise information describing part shapes is provided on detail drawings.

Detail dimensions which would confuse the assembly drawing should be omitted. Only such dimensions as center distances, overall dimensions, and dimensions showing the relationship of the parts as they apply to the mechanism as a whole should be included. There are times when a simple assembly drawing may be dimensioned so that no other detail drawings are needed. In such a case the assembly drawing becomes a working assembly drawing.

Sectioning is used more extensively on assembly drawings than on detail drawings. The conventional method of section lining is used on assembly drawings to show the relationship of the various parts, figure 31-1.

Subassembly Drawings

Subassembly drawings are often made of smaller mechanical units. When combined in final assembly, they make a single machine. For a lathe, subassembly drawings would be furnished for the headstock, the apron, and other units of the carriage. These units might be machined and assembled in different departments by following the subassembly drawings. The individual units would later be combined in final assembly according to the assembly drawing.

Identifying Parts of an Assembly Drawing

When a machine is designed, an assembly drawing or design layout is first drawn to visualize clearly the method of operation, shape, and clearances of the various parts. From this assembly drawing, the detail drawings are made and each part is given a part number.

To assist in the assembly of the machine, item numbers corresponding with the part numbers of various details are placed on the assembly drawing attached to the corresponding part with a leader. The part number is often enclosed in a small circle, called a balloon, which helps distinguish part numbers from dimensions.

BILL OF MATERIAL (ITEMS LIST)

A *bill of material* or *items list* is an itemized list of all the components shown on an assembly drawing or a detail drawing, figure 31-2. Often, a bill of material is placed on a separate sheet of paper for handling and duplicating. For castings, a pattern number would appear in the size column instead of the physical size of the part.

Standard components, which are purchased rather than fabricated, including bolts, nuts, and bearings, should have a part number and appear

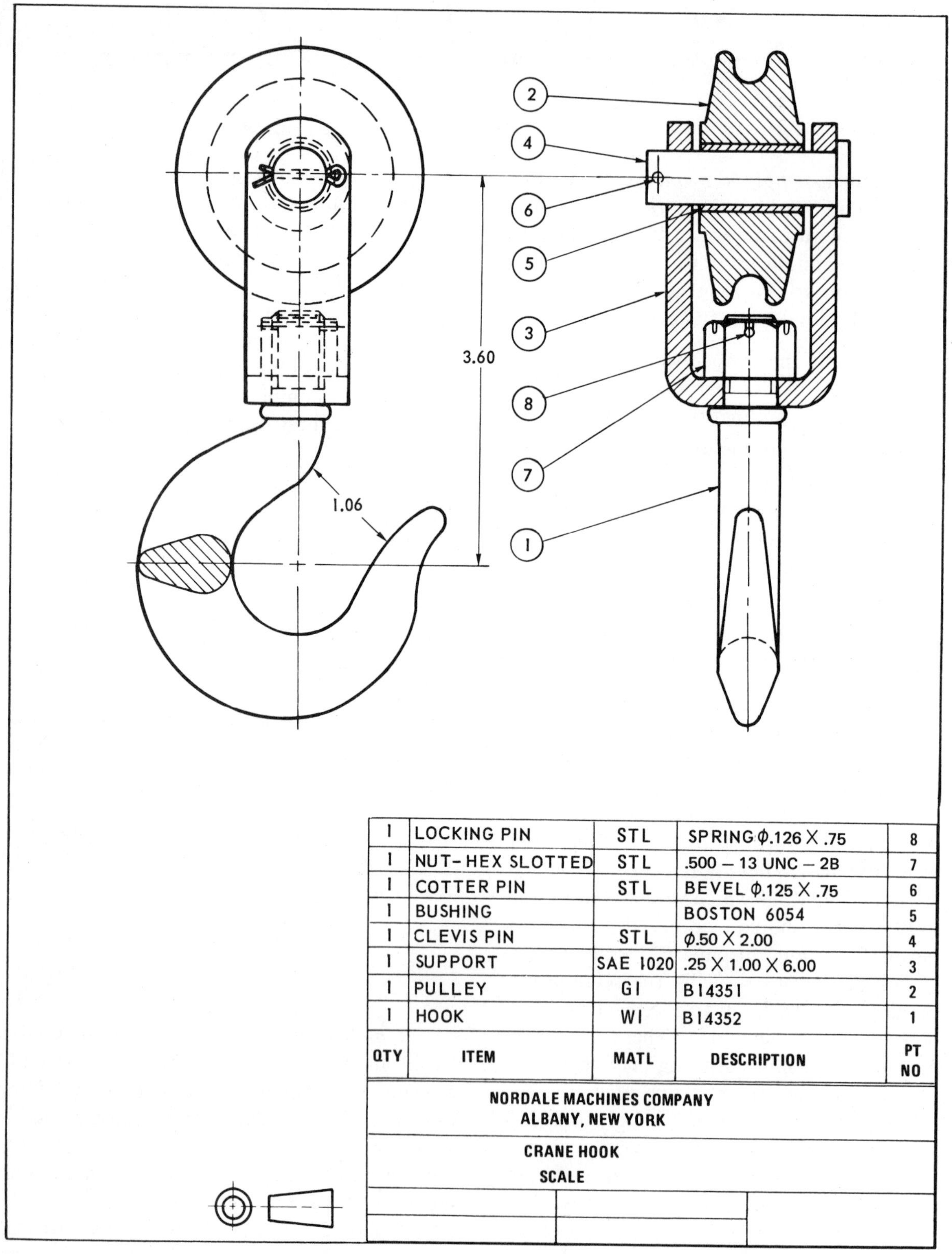

2
4
6
5
3
8
7
1
3.60
1.06
1 LOCKING PIN STL SPRING ϕ.126 X .75 8
1 NUT-HEX SLOTTED STL .500 – 13 UNC – 2B 7
1 COTTER PIN STL BEVEL ϕ.125 X .75 6
1 BUSHING BOSTON 6054 5
1 CLEVIS PIN STL ϕ.50 X 2.00 4
1 SUPPORT SAE 1020 .25 X 1.00 X 6.00 3
1 PULLEY GI B14351 2
1 HOOK WI B14352 1
QTY ITEM MATL DESCRIPTION PT NO
NORDALE MACHINES COMPANY
ALBANY, NEW YORK
CRANE HOOK
SCALE

Fig. 31-1 A typical assembly drawing

on the bill of material. There should be sufficient information in the descriptive column to enable the purchasing agent to order these parts.

Standard components are incorporated in the design of machine parts for economical production. These parts are specified on the drawing according to the manufacturer's specification. The use of manufacturers' catalogs is essential for determining detailing standards, characteristics of a special part, methods of representation, etc. However, it should be pointed out that manufacturer's catalogs are very unreliable for specifying parts; they should be used as a guide only. To protect the integrity of a design, a purchase part drawing must be made. This overcomes the frequent problem whereby the component supplier makes changes, unknown to the user, which frequently affect the design.

The four-wheel trolley (drawing A-71) includes many standard parts. Grease cups, lockwashers, Hyatt roller bearings, rivets, and nuts, all of which are standard purchased items, are used. These parts are not detailed but are described in the bill of material. However, the special countersunk head bolts and the taper washers, commonly called Dutchmen, are not standard parts and must therefore be made especially for this particular assembly.

SWIVELS AND UNIVERSAL JOINTS

A *swivel* is composed of two or more pieces constructed so that each part rotates in relation to the other around a common axis, figure 31-3(A).

A *universal joint* is composed of three or more pieces designed to permit the free rotation of two shafts whose axes deviate from a straight line, figure 31-3(B).

In practice, universal joints are constructed in many different forms. The design of the universal joints is influenced by the requirements of the working mechanism and the cost of the parts.

Universal joints are used by the designer where a rotating or swinging motion is wanted or where power must be delivered along shafts which are not in a straight line.

4	NUT–HEX REG	STL	.375 – 16 UNC	7
4	BOLT–HEX REG	STL	.375 – 16 UNC X 1.50	6
1	KEY	MS	WOODRUFF 608	5
2	BEARINGS	SKF	RADIAL BALL 620	4
1	SHAFT	CRS	ϕ1.00 X 6.50 LG	3
1	SUPPORT	MST	.375 X 2.00 X 5.50	2
1	BASE	GI	PATTERN – A3154	1
QTY	**ITEM**	**MATL**	**DESCRIPTION**	**PT NO**

Fig. 31-2 Bill of material for assembly drawing

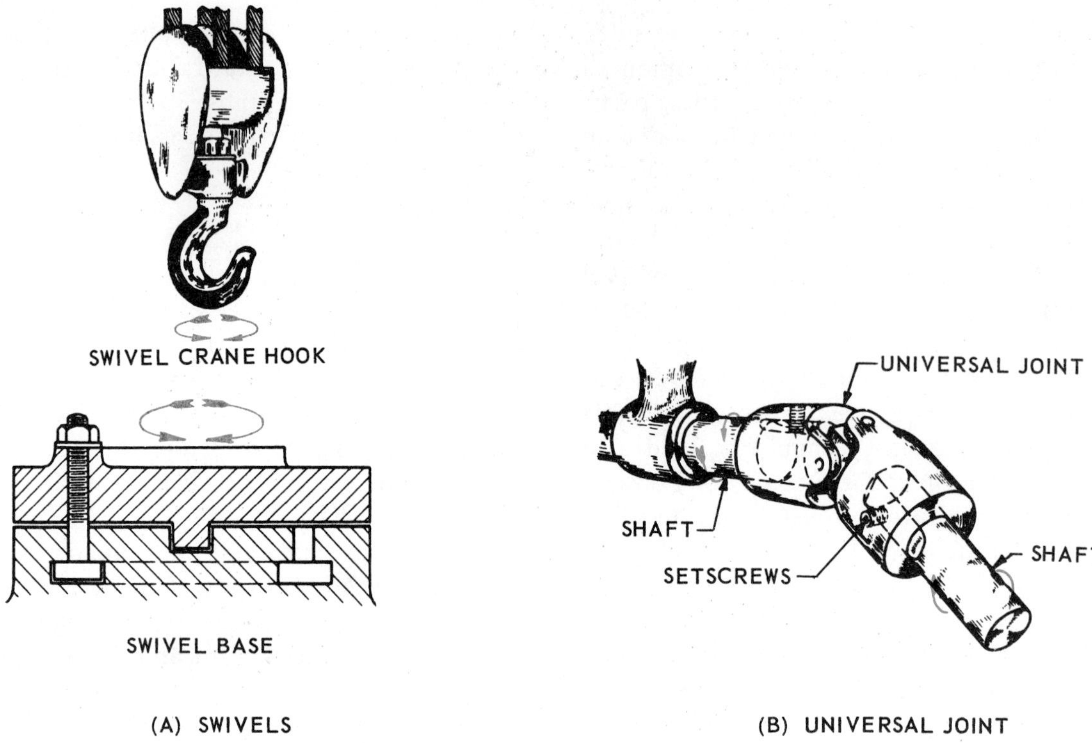

Fig. 31-3 Swivels and universal joints

PAGE 227 IS INTENTIONALLY BLANK. ASSIGNMENT DRAWING A-70 IS ON THE NEXT PAGE.

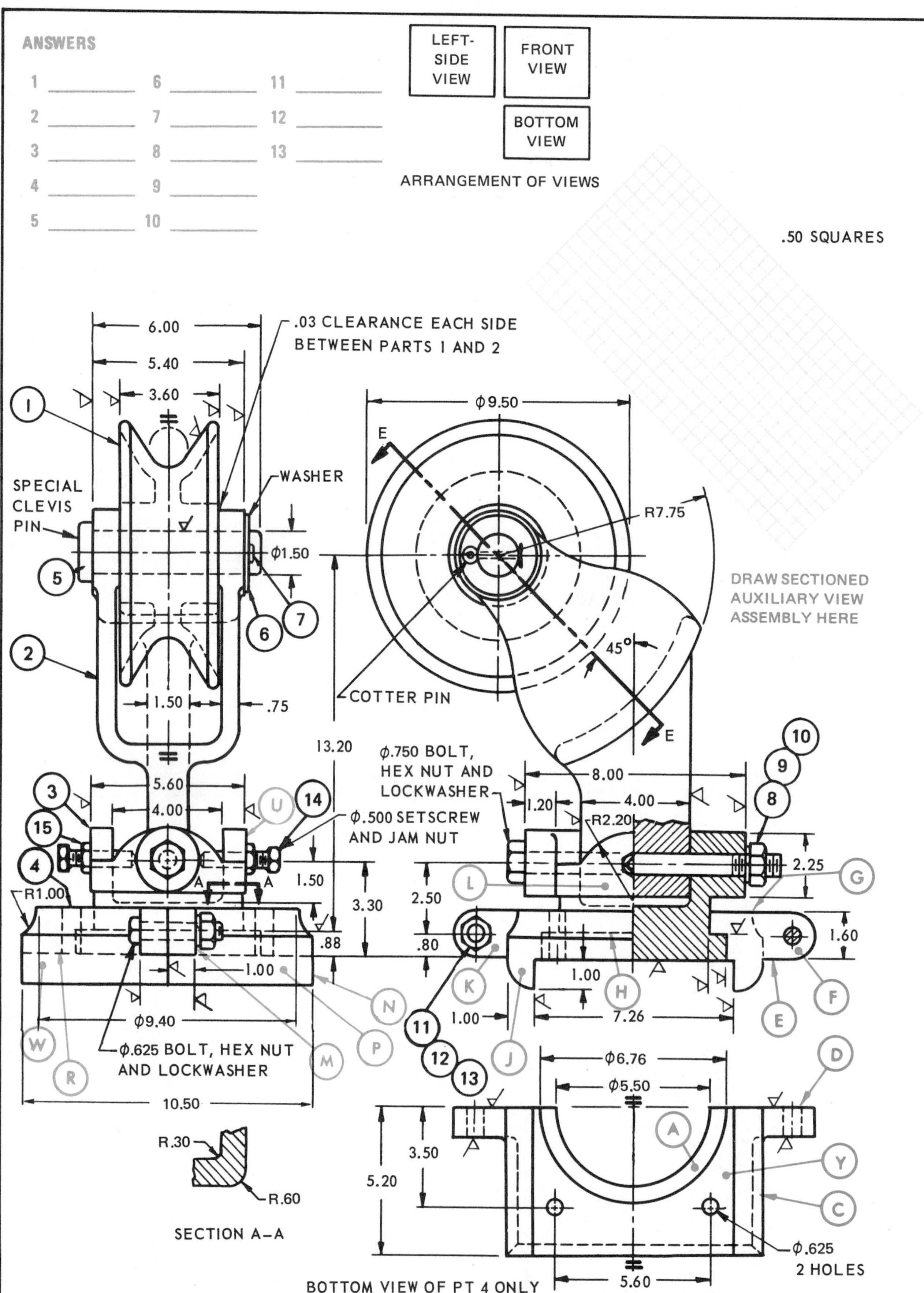
ANSWERS
1
2
3
4
5
6
7
8
9
10
11
12
13
LEFT-SIDE VIEW
FRONT VIEW
BOTTOM VIEW
ARRANGEMENT OF VIEWS
.50 SQUARES
DRAW SECTIONED AUXILIARY VIEW ASSEMBLY HERE
6.00
5.40
3.60
.03 CLEARANCE EACH SIDE BETWEEN PARTS 1 AND 2
SPECIAL CLEVIS PIN
WASHER
ϕ1.50
1.50
.75
5.60
4.00
ϕ.500 SETSCREW AND JAM NUT
R1.00
1.50
3.30
.88
13.20
1.00
ϕ9.40
ϕ.625 BOLT, HEX NUT AND LOCKWASHER
10.50
R.30
R.60
SECTION A–A
ϕ9.50
E
R7.75
COTTER PIN
45°
ϕ.750 BOLT, HEX NUT AND LOCKWASHER
8.00
1.20
4.00
R2.20
2.25
2.50
.80
1.60
1.00
7.26
ϕ6.76
ϕ5.50
3.50
5.20
5.60
ϕ.625 2 HOLES
BOTTOM VIEW OF PT 4 ONLY

SKETCHING ASSIGNMENT

1. Make a sectioned auxiliary view assembly taken on cutting-plane line E-E.
2. Make a three-view, sketch of part 3 in the space provided. Do not include dimensions.

BILL OF MATERIAL

Complete the bill of material for the universal trolley assembly and place the part numbers on the assembly drawing. Refer to manufacturer's catalogs and drafting manuals for standard components. Two lines in the bill of material may be used for one part if required. Sizes for cast iron parts need not be shown.

QUESTIONS

1. Locate surface (F) on the bottom view.
2. Is surface (J) shown in the bottom view?
3. Which line or surface in the left-side view represents surface (J)?
4. How many different parts are used to make the trolley?
5. What is the total number of parts used to make the trolley?
6. How many tapped holes (excluding the nuts) are there?
7. What is the total number of holes in parts (1) to (4)? Note: Count through holes as 2 holes.
8. Which line or surface in the bottom view represents point (G)?
9. Locate surface (A) in the front view.
10. Locate surface (Y) in the left-side view.
11. What is the overall height of the assembly?
12. Locate surface (L) in the left-side view.
13. What is the total number of surfaces that are to be finished for parts (1) to (4)? Note: One or more surfaces can be on the same plane. Where one finish mark is shown on matched surfaces of castings, this implies that both surfaces are to be finished.

SKETCH PT 3 HERE
1.00 SQUARES

	STAND	GI		4
	SWIVEL	GI		3
	SUPPORT	MI		2
	TROLLEY WHEEL	MI		1
QTY	ITEM	MATL	DESCRIPTION	PT NO

NOTE: UNLESS OTHERWISE SPECIFIED
- TOLERANCE ON DIMENSIONS ±.02
- TOLERANCE ON ANGLES ±0.5°
- SURFACES ✓ TO BE 125✓

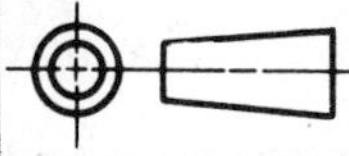

SCALE	
DRAWN	DATE
UNIVERSAL TROLLEY	A-70

UNIT 32

STRUCTURAL STEEL SHAPES

Structural steel is widely used in the metal trades for the fabrication of machine parts because the many standard shapes lend themselves to many different types of construction.

Remember that the steel produced at the rolling mills and shipped to the fabricating shop comes in a wide variety of shapes and forms (approximately 600). At this stage it is called *plain material.*

The great bulk of this material can be designated as one of the following and shown in figure 32-1.

Abbreviations

When structural steel shapes are designated on drawings, a standard method of abbreviating should be followed that will identify the group of shapes without reference to the manufacturer and without the use of inches and pounds per foot.

Therefore, it is recommended that structural steel be abbreviated as listed in figure 32-2.

The abbreviations shown are intended only for use on design drawings. When lists of materials are being prepared for ordering from the mills, the requirements of the respective mills from which the material is to be ordered should be observed.

S-shaped beams and all standard and miscellaneous channel have a slope on the inside flange of 16.67 percent (16.67 percent slope is equivalent to 9 degrees 28′ or a bevel of 1:6). All other beams have parallel face flanges.

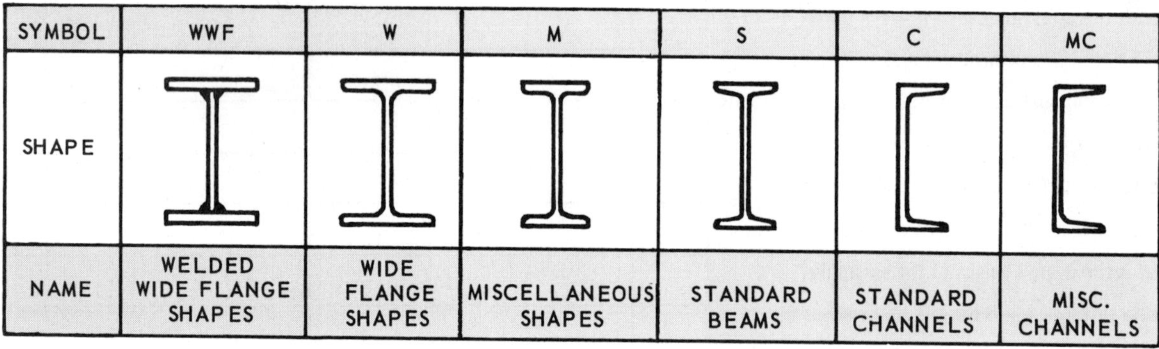

SYMBOL	WWF	W	M	S	C	MC
SHAPE						
NAME	WELDED WIDE FLANGE SHAPES	WIDE FLANGE SHAPES	MISCELLANEOUS SHAPES	STANDARD BEAMS	STANDARD CHANNELS	MISC. CHANNELS

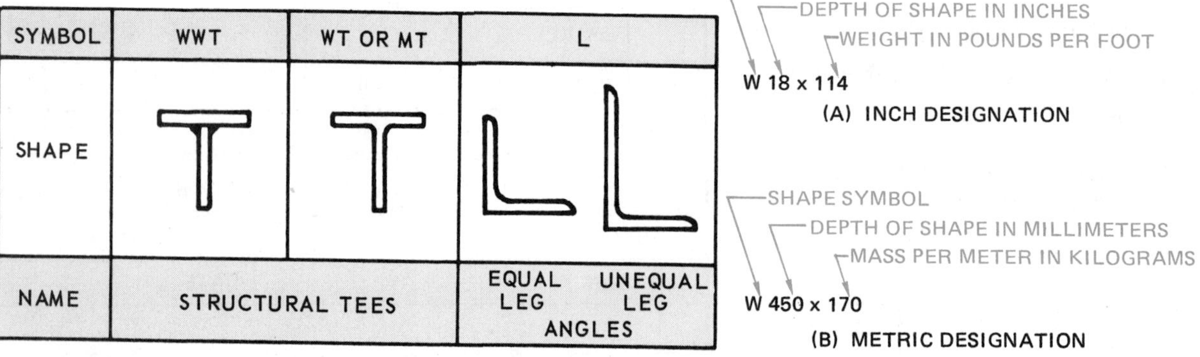

SYMBOL	WWT	WT OR MT	L	
SHAPE				
NAME	STRUCTURAL TEES		EQUAL LEG ANGLES	UNEQUAL LEG ANGLES

Fig. 32-1 Common structural steel shapes and drawing callout

Shape	U.S. Customary Examples See Note 1		Metric Size Examples See Note 2
	New Designation	Old Designation	
Welded Wide Flange Shapes (WWF Shapes)			
– Beam	WWF48 X 320	48WWF320	WWF1000 X 244
– Columns			WWF350 X 315
Wide Flange Shapes (W Shapes)	W24 X 76	24WF76	W600 X 114
	W14 X 26	14B26	W160 X 18
Miscellaneous Shapes (M Shapes)	M8 X 18.5	8M18.5	M200 X 56
	M10 X 9	10JR9.0	M160 X 30
Standard Beams (S Shapes)	S24 X 100	24I100	S380 X 64
Standard Channels (C Shapes)	C12 X 20.7	12C20.7	C250 X 23
Structural Tees			
– cut from WWF Shapes	WWT24 X 160	ST24WWF160	WWT280 X 210
– cut from W Shapes	WT12 X 38	ST12WF38	WT130 X 16
– cut from M Shapes	MT4 X 9.25	ST4M9.25	MT100 X 14
Bearing Piles (HP Shapes)	HP14 X 73	14BP73	HP350 X 109
Angles (L Shapes)	L6 X 6 X .75	L6 X 6 X 3/4	L75 X 75 X 6
(leg dimensions X thickness)	L6 X 4 X .62	16 X 4 X 5/8	L150 X 100 X 13
Plates (width X thickness)	20 X .50	20 X 1/2	500 X 12
Square Bar (side)	⎅ 1.00	Bar 1 ⎅	⎅ 25
Round Bar (diameter)	∅1.25	Bar 1-1/4 ∅	∅30
Flat Bar (width X thickness)	250 X .25	Bar 2-1/2 X 1/4	60 X 6
Round Pipe (type of pipe X OD X wall thickness)	12.75 OD X .375	12-3/4 X 3/8	XS 102 OD X 8
Square and Rectangular Hollow Structural Sections (outside dimensions X wall thickness)	HSS4 X 4 X .375	4 X 4RT X 3/8	HSS102 X 102 X 8
	HSS8 X 4 X .375	8 X 4RT X 3/8	
Steel Pipe Piles (OD X wall thickness)			320 OD X 6

Note 1–Values shown are nominal depth (inches) X weight per foot length (pounds).
Note 2–Values shown are nominal depth (millimeters) X mass per meter length (kilograms).
Note 3–Metric size examples shown are not necessarily the equivalents of the inch size examples shown.

Fig. 32-2 Abbreviations for shapes, plates, bars and tubes

PHANTOM OUTLINES

At times, a part or mechanism not included in the actual detail or assembly drawing is shown to clarify how the mechanism will connect with or operate from an adjacent part. This part is shown by drawing light dash lines (one long line and two short dashes) in the operating position. Such a drawing of the extra part is known as a *phantom drawing* or view drawn in *phantom,* figure 32-3.

On the drawing of the four-wheel trolley (drawing A-71), the track the wheels run on is an S beam. The wheels are set at an angle to the vertical plane in order to ride upon the sloping bottom flange of the S beam. The outline of the beam is shown by dash lines and, while not an integral part of the trolley, the outline or phantom view of the S beam shows clearly how the trolley operates.

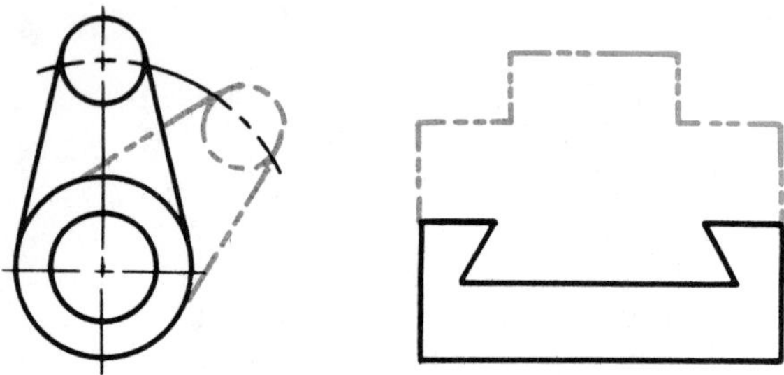

Fig. 32-3 Phantom lines

CONICAL WASHERS

Conical washers, figure 32-4, are available in a variety of sizes to accommodate the slopes found on structural steel shapes. A typical application can be found on the four-wheel trolley, parts D and W (drawing A-71).

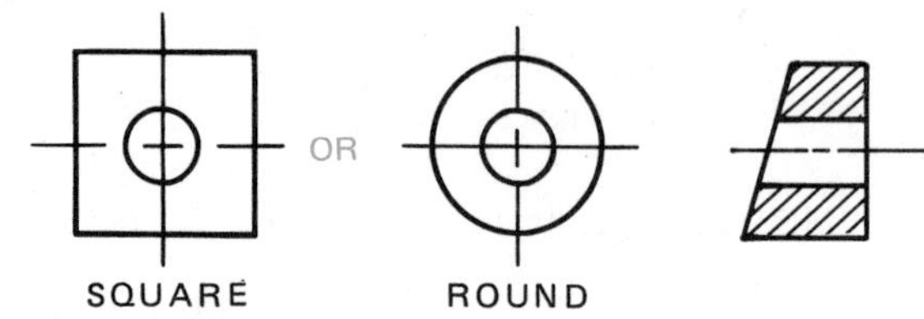

Fig. 32-4 Conical Washers

PAGE 233 IS INTENTIONALLY BLANK. ASSIGNMENT DRAWING A-71 IS ON THE NEXT PAGE.

SKETCHING ASSIGNMENT

1. Sketch the secondary auxiliary view of part (A) in the space provided. (Project from primary auxiliary view.)
2. In the space provided, make a detail drawing with dimensions of parts (C) and (D).

QUESTIONS

1. What does hidden line (E) indicate?
2. Which cutting plane in the primary auxiliary view indicates (A) where the section to the left of line **N-N** is taken, (B) where the section to the right of line **N-N** is taken?
3. What is the slope of angle (J)?
4. Locate part (K) in the section view.
5. What is the wheel diameter of the trolley?
6. Locate parts (2), (3), (5), (A), (C), (V), and (Z) in the primary auxiliary view.
7. What are the names of parts (T), (U), (V), (W), (X), and (Y)?
8. What is the diameter of the bearing rollers?
9. Determine distance (L).

ANSWERS

1. ______	(3) ______	(U) ______
2. (A) ______	(5) ______	(V) ______
(B) ______	(A) ______	(W) ______
3. ______	(C) ______	(X) ______
4. ______	(V) ______	(Y) ______
5. ______	(Z) ______	8. ______
6. (2) ______	7. (T) ______	9. ______

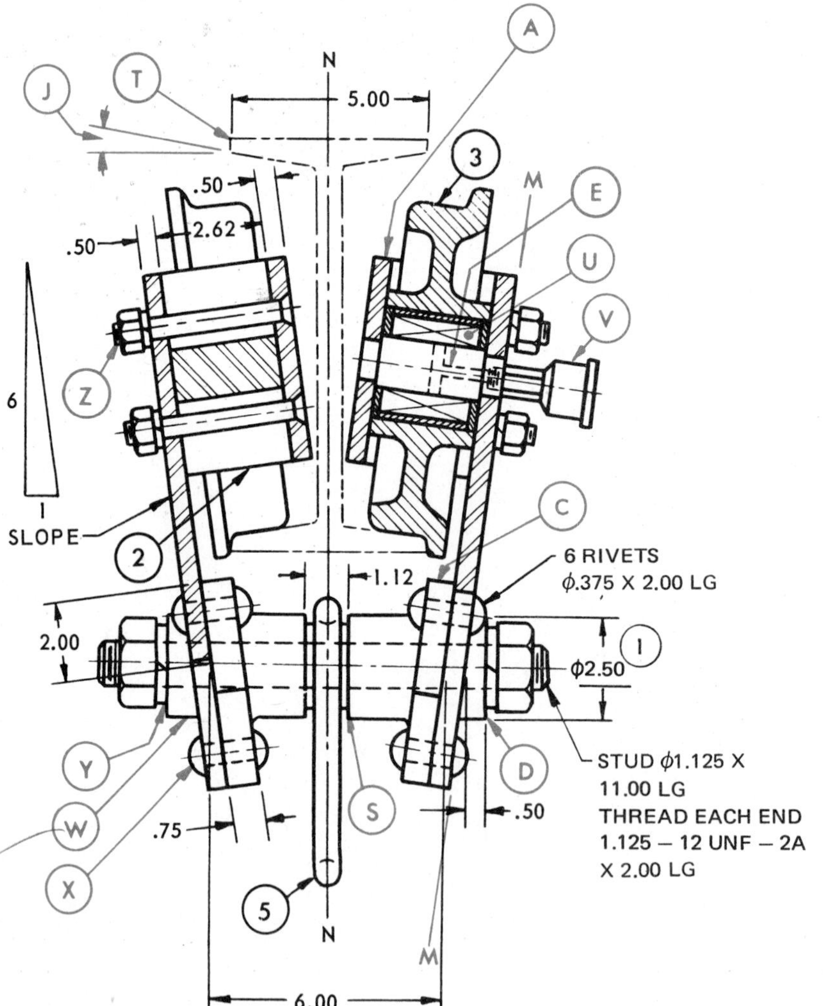

REVISIONS	I	82-04-21	F. NEWMAN
	2.50 WAS 2.60		

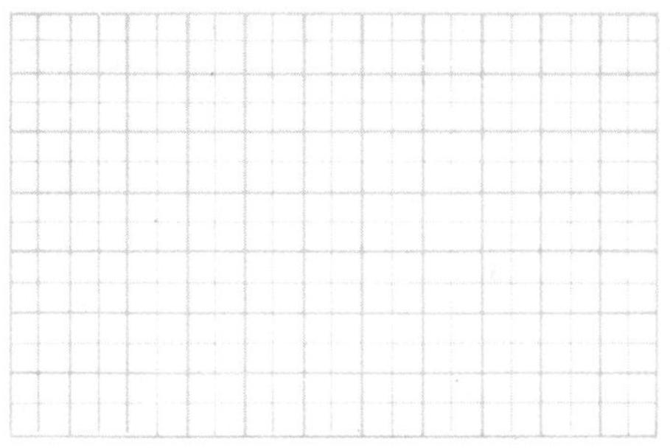

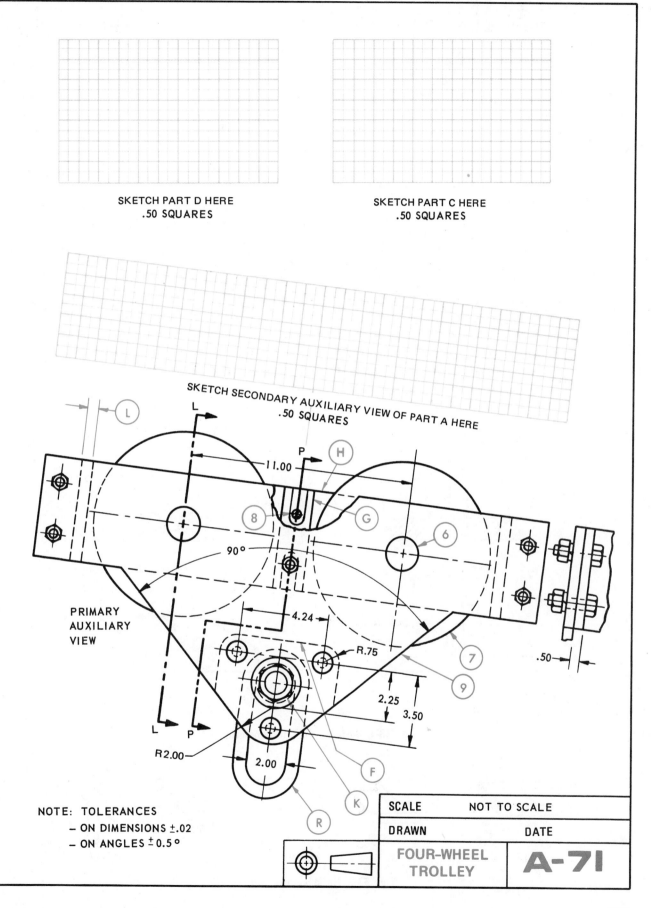

SKETCH PART D HERE
.50 SQUARES

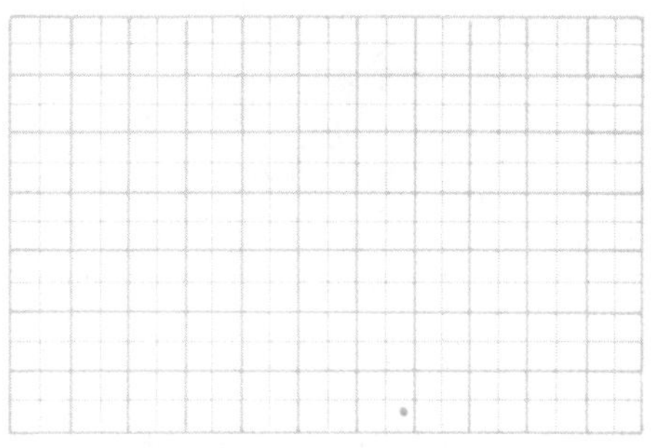

SKETCH PART C HERE
.50 SQUARES

SKETCH SECONDARY AUXILIARY VIEW OF PART A HERE
.50 SQUARES

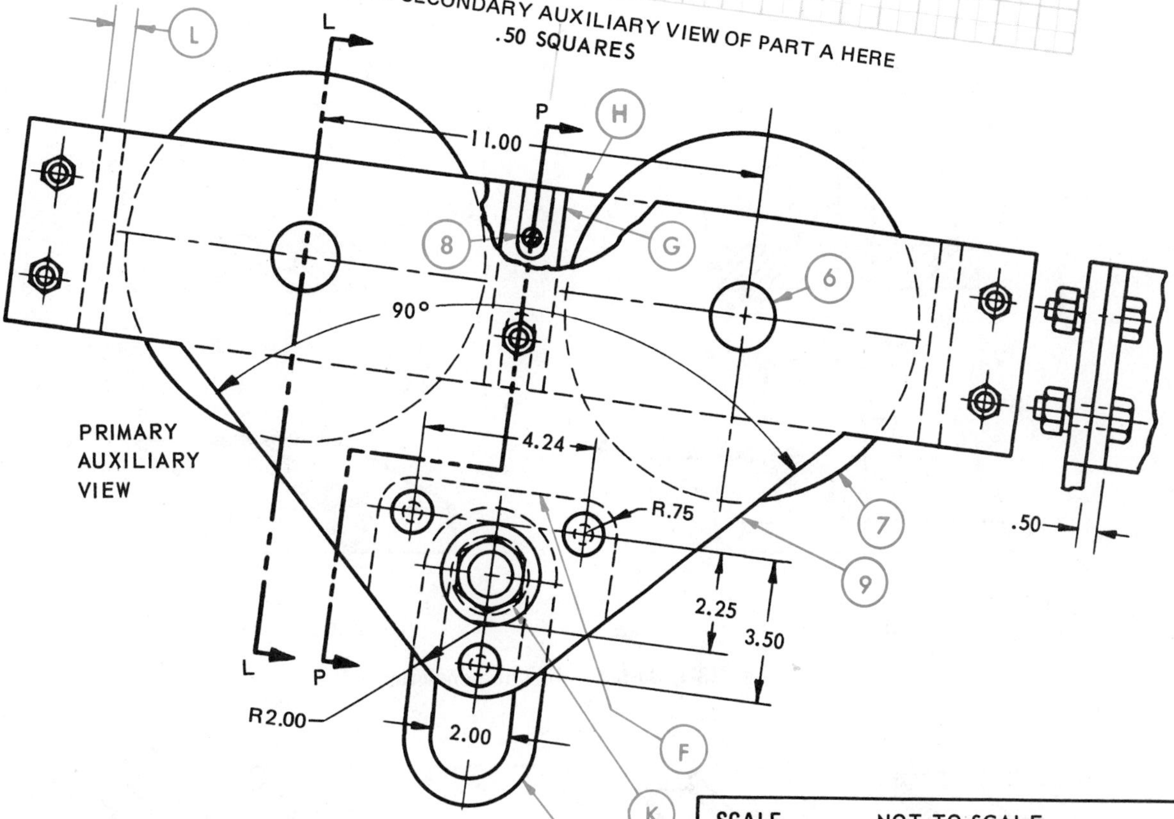

L
L
P
H
11.00
8
G
6
90°
PRIMARY
AUXILIARY
VIEW
4.24
R.75
7
.50
9
2.25
3.50
L
P
R2.00
2.00
F
K
R

NOTE: TOLERANCES
- ON DIMENSIONS ±.02
- ON ANGLES ± 0.5°

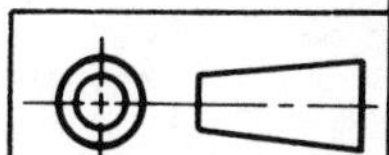

SCALE	NOT TO SCALE
DRAWN	DATE
FOUR-WHEEL TROLLEY	A-71

UNIT 33

WELDING DRAWINGS

The primary importance of welding is the uniting of pieces of metal so they will operate properly as a unit to support the loads to be carried. In order to design and build such a structure, to be economical and efficient, a basic knowledge of welding is essential. Figure 33-1 illustrates many basic welding terms.

The introduction of welding symbols on a drawing enables the designer to indicate clearly the type and size of weld required to meet the design requirements. It is becoming increasingly important for the designer to indicate the required type of weld correctly. Basic welding joints are shown in figure 33-2. Points which must be made clear are the type of weld, the joint penetration, the weld size, and the root opening (if any). These points can be clearly indicated on the drawing by the welding symbol.

WELDING SYMBOLS

Welding symbols are a shorthand language. They save time and money and if used correctly, insure understanding and accuracy. Welding symbols should be a universal language; for this reason the symbols of the American Welding Society have been adopted.

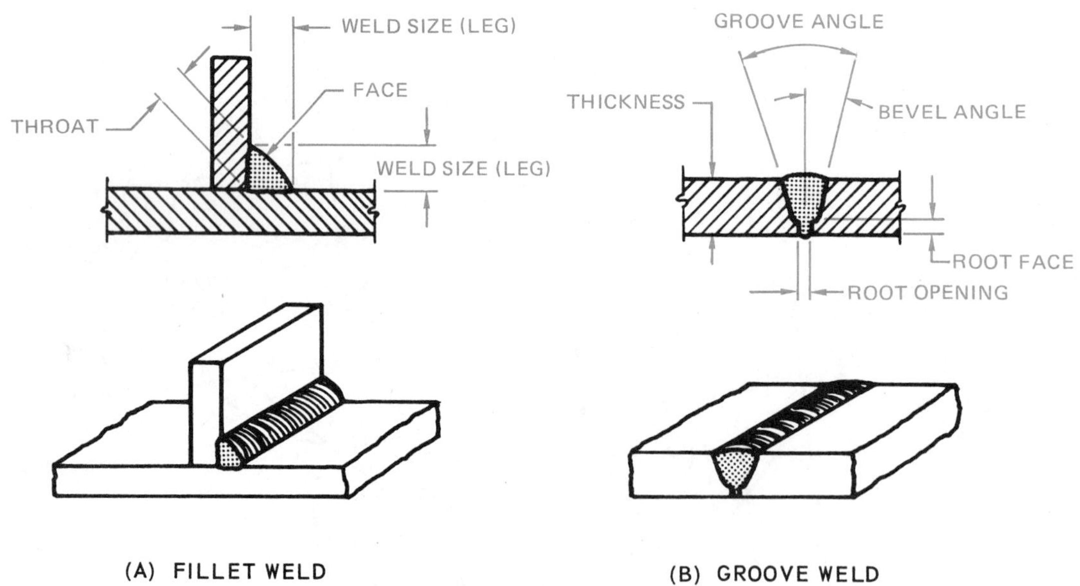

Fig. 33-1 Basic welding nomenclature

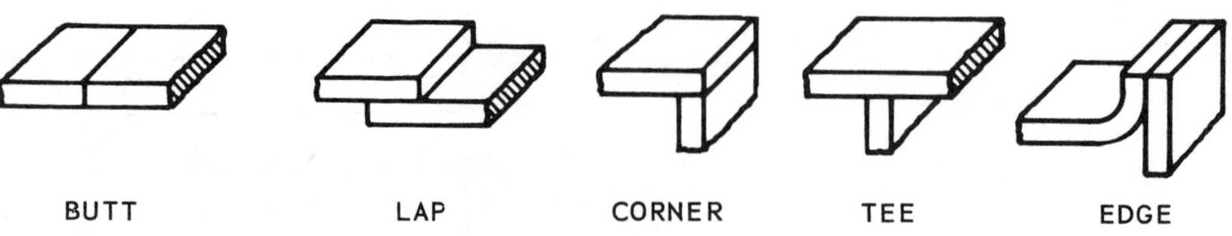

Fig. 33-2 Basic welding joints

A distinction between the terms *weld symbol* and *welding symbol* should be understood. The weld symbol indicates the type of weld. The welding symbol is a method of representing the weld on drawings. It includes supplementary information and consists of the following eight elements. Not all elements need be used unless required for clarity.

1. Reference line
2. Arrow
3. Basic weld symbol
4. Dimensions and other data
5. Supplementary symbols
6. Finish symbols
7. Tail
8. Specification, process, or other reference.

Figure 33-3 illustrates the position of the weld symbols and other information in relation to the basic welding symbol. The various weld symbols which may be applied to the basic welding symbol are shown in figure 33-4. Figure 33-5 shows the actual shape of many of the weld types symbolized in figure 33-4.

Supplementary symbols may also be added to the basic welding symbol. The supplementary symbols are illustrated in figure 33-6.

Any welding joint which is indicated by a symbol will always have an arrow side and an other side. The words arrow side, other side,

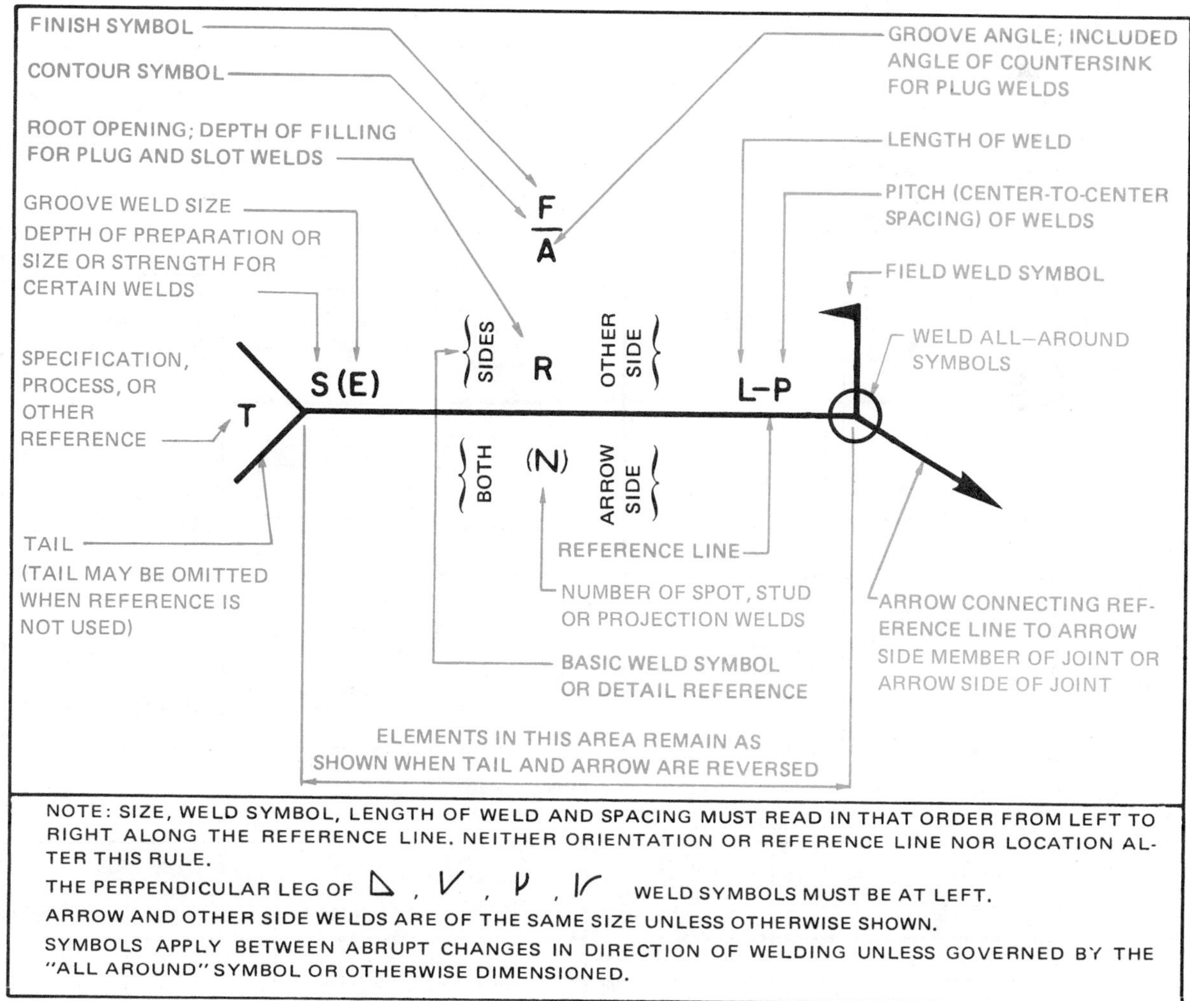

Fig. 33-3 Standard location of elements of a welding symbol

and both sides are used accordingly to locate the weld with respect to the joint.

Tail of Welding Symbol

The welding and allied process to be used may be specified by placing the appropriate letter designations from figure 33-7 in the tail of the welding symbol, figure 33-8.

Codes, specifications, or any other applicable documents may be specified by placing the reference in the tail of the welding symbol. Information contained in the referenced document need not be repeated in the welding symbol.

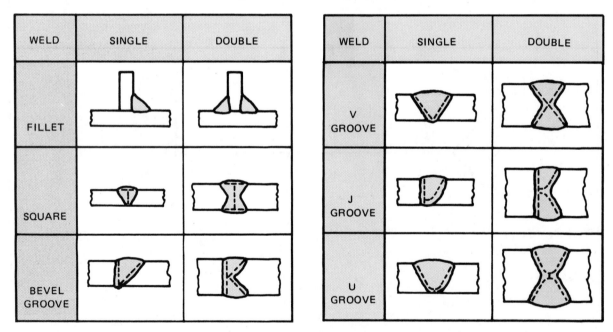

FILLET	PLUG OR SLOT	SPOT OR PROJECTION	STUD	SEAM	BACK OR BACKING	SURFACING	FLANGE	
							EDGE	CORNER

GROOVE							
SQUARE	SCARF	V	BEVEL	U	J	FLARE-V	FLARE-BEVEL

Fig. 33-4 Basic weld symbols

WELD	SINGLE	DOUBLE
FILLET		
SQUARE		
BEVEL GROOVE		

WELD	SINGLE	DOUBLE
V GROOVE		
J GROOVE		
U GROOVE		

Fig. 33-5 Types of welds

WELD ALL AROUND	FIELD WELD	MELT-THRU	BACKING OR SPACER MATERIAL	CONSUMABLE INSERT	CONTOUR		
					FLUSH	CONVEX	CONCAVE
			BACKING SPACER				

Fig. 33-6 Supplementary symbols

Welding Process	Welding Process (Specific)	Letter Designation
Brazing (B)	Infrared Brazing Torch Brazing Furnace Brazing Induction Brazing Resistance Brazing Dip Brazing	IRB TB FB IB RB DB
Oxyfuel Gas Welding (OFW)	Oxyacetylene Welding Oxyhydrogen Welding Pressure Gas Welding	OAW OHW PGW
Resistance Welding (RW)	Resistance-Spot Welding Resistance-Seam Welding Projection Welding Flash Welding Upset Welding Percussion Welding	RSW RSEW PW FW UW PEW
Arc Welding (AW)	Stud Arc Welding Plasma-Arc Welding Submerged Arc Welding Gas Tungsten-Arc Welding Gas Metal-Arc Welding Flux Cored Arc Welding Shielded Metal-Arc Welding Carbon-Arc Welding	SW PAW SAW GTAW GMAW FCAW SMAW CAW
Other Processes	Thermit Welding Laser Beam Welding Induction Welding Electroslag Welding Electron Beam Welding	TW LBW IW ESW EBW
Solid State Welding (SSW)	Ultrasonic Welding Friction Welding Forge Welding Explosion Welding Diffusion Welding Cold Welding	USW FRW FOW EXW DFW CW

Cutting Method	Letter Designation
Arc Cutting	AC
Air Carbon-Arc Cutting	AAC
Carbon-Arc Cutting	CAC
Metal-Arc Cutting	MAC
Plasma-Arc Cutting	PAC
Oxygen Cutting	OC
Chemical Flux Cutting	FOC
Metal Powder Cutting	POC
Oxygen-Arc Cutting	AOC

Fig. 33-7 Designation of welding process by letters

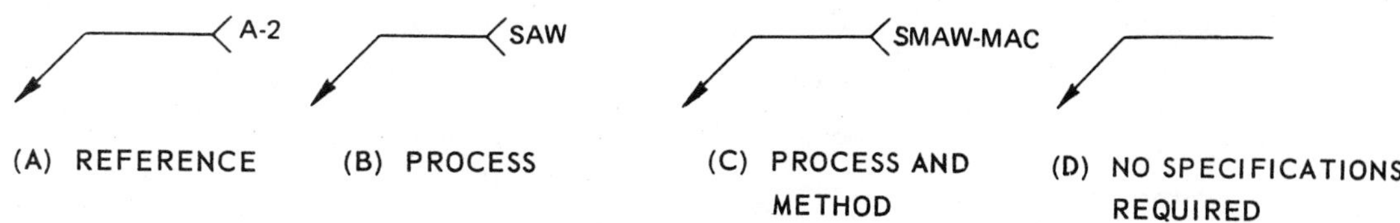

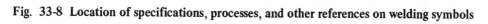

Fig. 33-8 Location of specifications, processes, and other references on welding symbols

Multiple Reference Lines

Two or more reference lines may be used to indicate a sequence of operations. The first operation is specified on the reference line nearest the arrow. Subsequent operations are specified sequentially on other reference lines, figure 33-9.

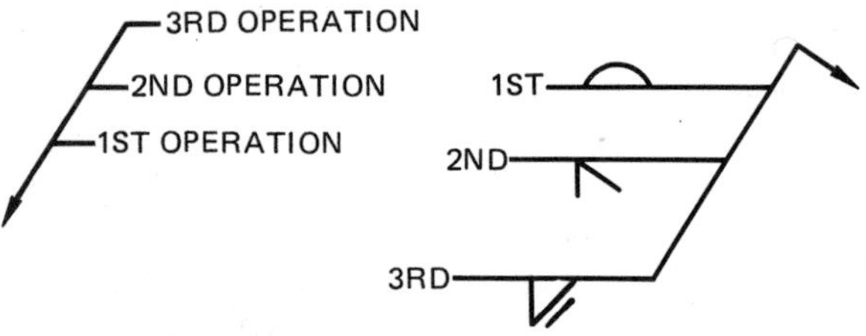

Fig. 33-9 Multiple reference lines

Welds on the arrow side of the joint are shown by placing the weld symbol on the bottom side of the reference line. Welds on the other side of the joint are shown by placing the weld symbol on the top side of the reference line. Welds on both sides of the joint are shown by placing the weld symbol on both sides of the reference line. A weld extending completely around a joint is indicated by means of a weld-all-around symbol placed at the intersection of the reference line and the arrow.

Field welds (welds not made in the shop or at the initial place of construction) are indicated by means of the field weld symbol placed at the intersection of the reference line and the arrow.

All weld dimensions on a drawing may be subject to a general note. Such a note might state: ALL FILLET WELDS .25 UNLESS OTHERWISE NOTED.

Only the basic fillet welds will be discussed in this Unit. Figure 33-10 illustrates several typical fillet welding symbols and the resulting welds.

FILLET WELDS

1. Fillet weld symbols are drawn with the perpendicular leg always to the left.

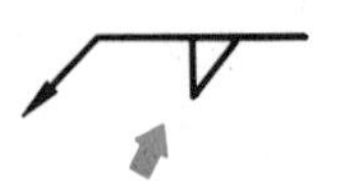
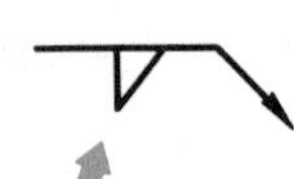
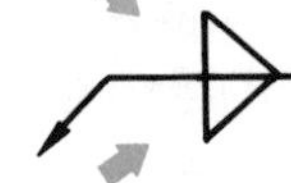

2. Dimensions of fillet welds are shown on the same side of the reference line and to the left of the weld symbol.

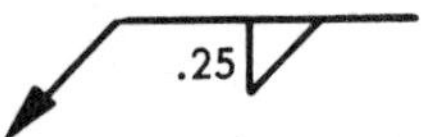

3. The dimensions of fillet welds on both sides of a joint are shown whether the dimensions are identical or different.

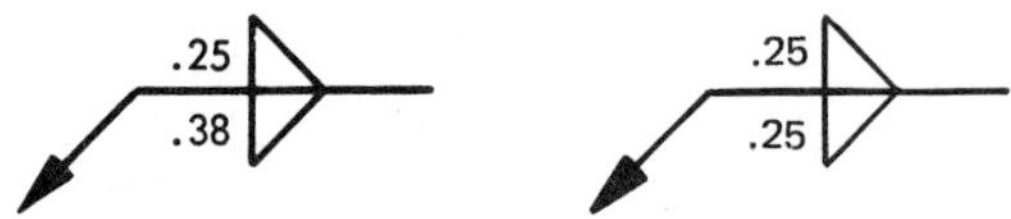

4. The dimension does not need to be shown when a general note is placed on the drawing to specify the dimension of fillet welds.

NOTE: SIZE OF FILLET WELDS .25 UNLESS OTHERWISE SPECIFIED.

5. The *length* of a fillet weld, when indicated on the welding symbol, is shown to the right of the weld symbol.

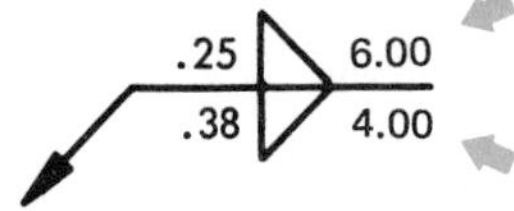

6. The *pitch* (center-to-center spacing) of an intermittent fillet weld is shown as the distance between centers of increments on one side of the joint. It is shown to the right of the length dimension following a hyphen.

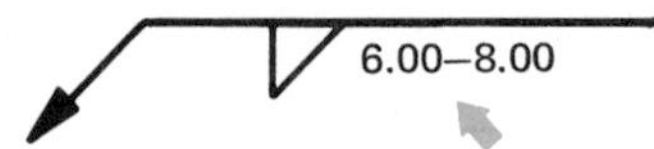

7. Staggered intermittent fillet welds are illustrated by staggering the weld symbols.

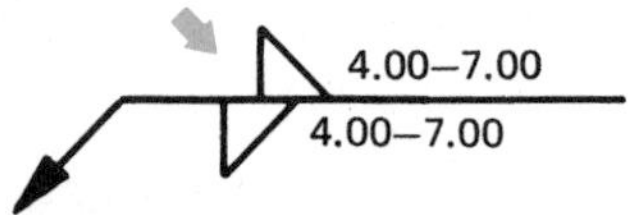

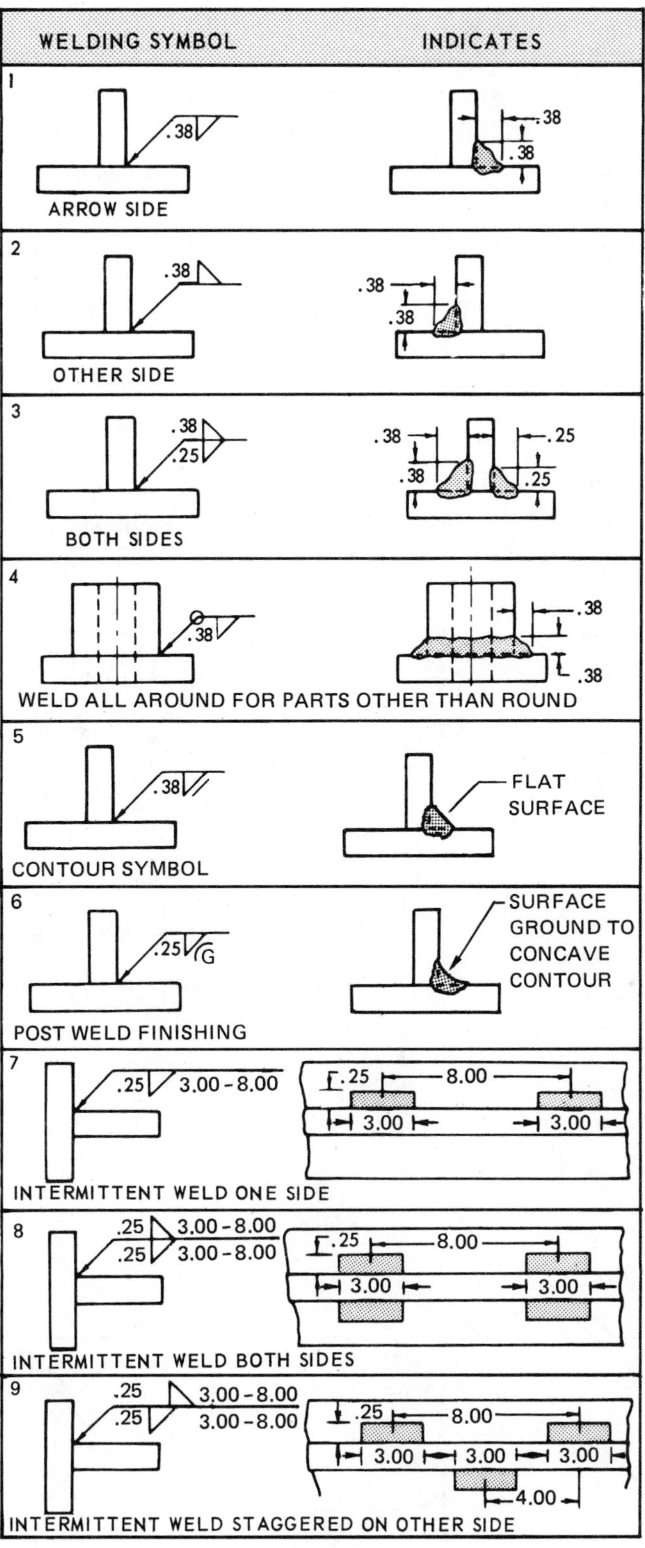

Fig. 33-10 Typical fillet welds

8. Fillet welds that are to be welded with approximately flat, convex, or concave faces without postweld finishing are specified by adding the flat, convex, or concave contour symbol to the weld symbol.

9. Fillet welds whose faces are to be finished approximately flat, convex, or concave by postweld finishing are specified by adding both the appropriate contour and finishing symbol to the weld symbol.

The following finishing symbols may be used to specify the method of finishing, but not the degree of finish:

C – Chipping
G – Grinding
H – Hammering
M – Machining
R – Rolling

10. A weld, with a length less than the available joint length whose location is significant, is specified on the drawing in a manner similar to that shown in figure 33-11.

11. Weld-All-Around Symbol. A continuous weld extending around a series of connected joints may be specified by the addition of the weld-all-around symbol at the junction of the arrow and reference line. The series of joints may involve different directions and may lie on more than one plane, figure 33-12(A).

Welds extending around a pipe or circular or oval holes do not require the weld-all-around symbol, figure 33-12(B).

REFERENCES AND SOURCE MATERIALS

1. ANSI/AWS A2.4-86

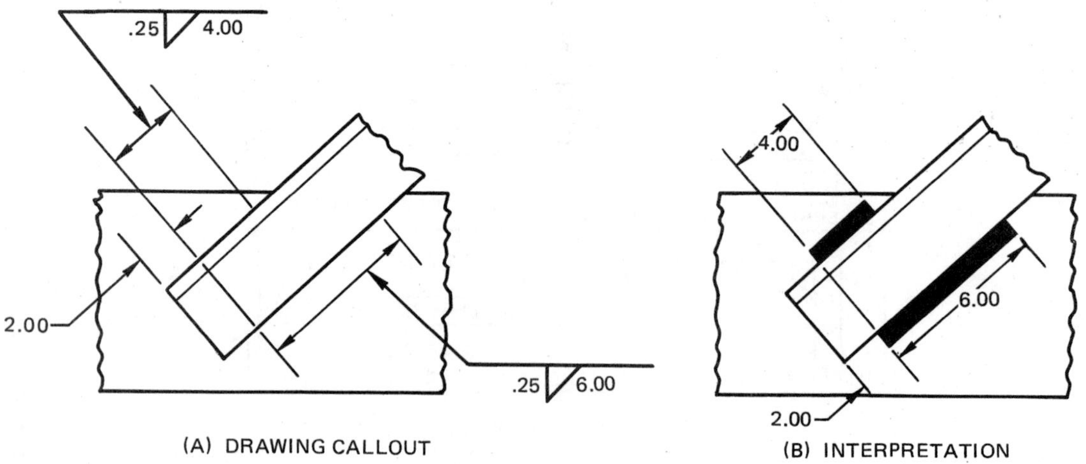

Fig. 33-11 Welds definitely located

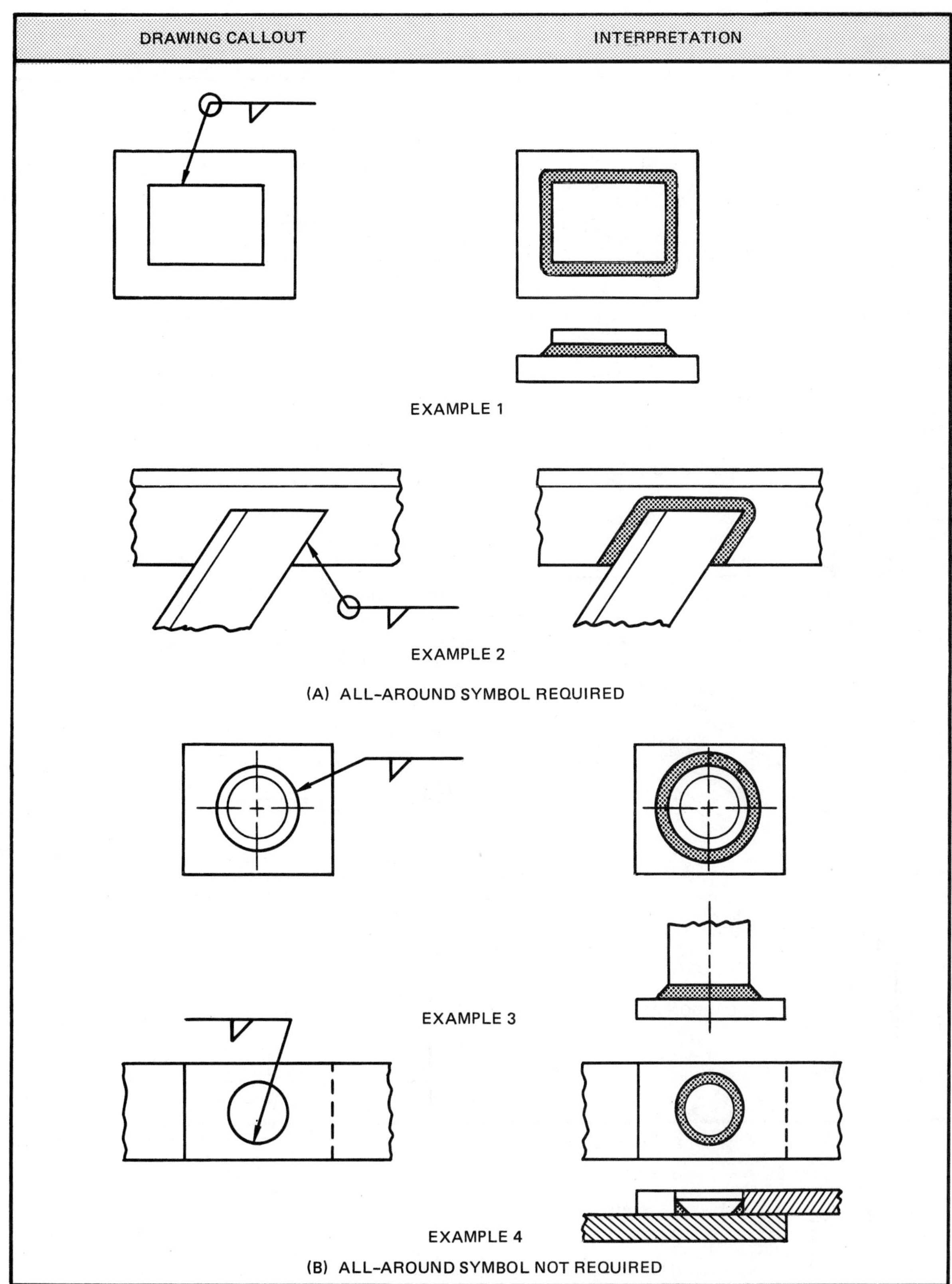

Fig. 33-12 The use of All-Around Symbols

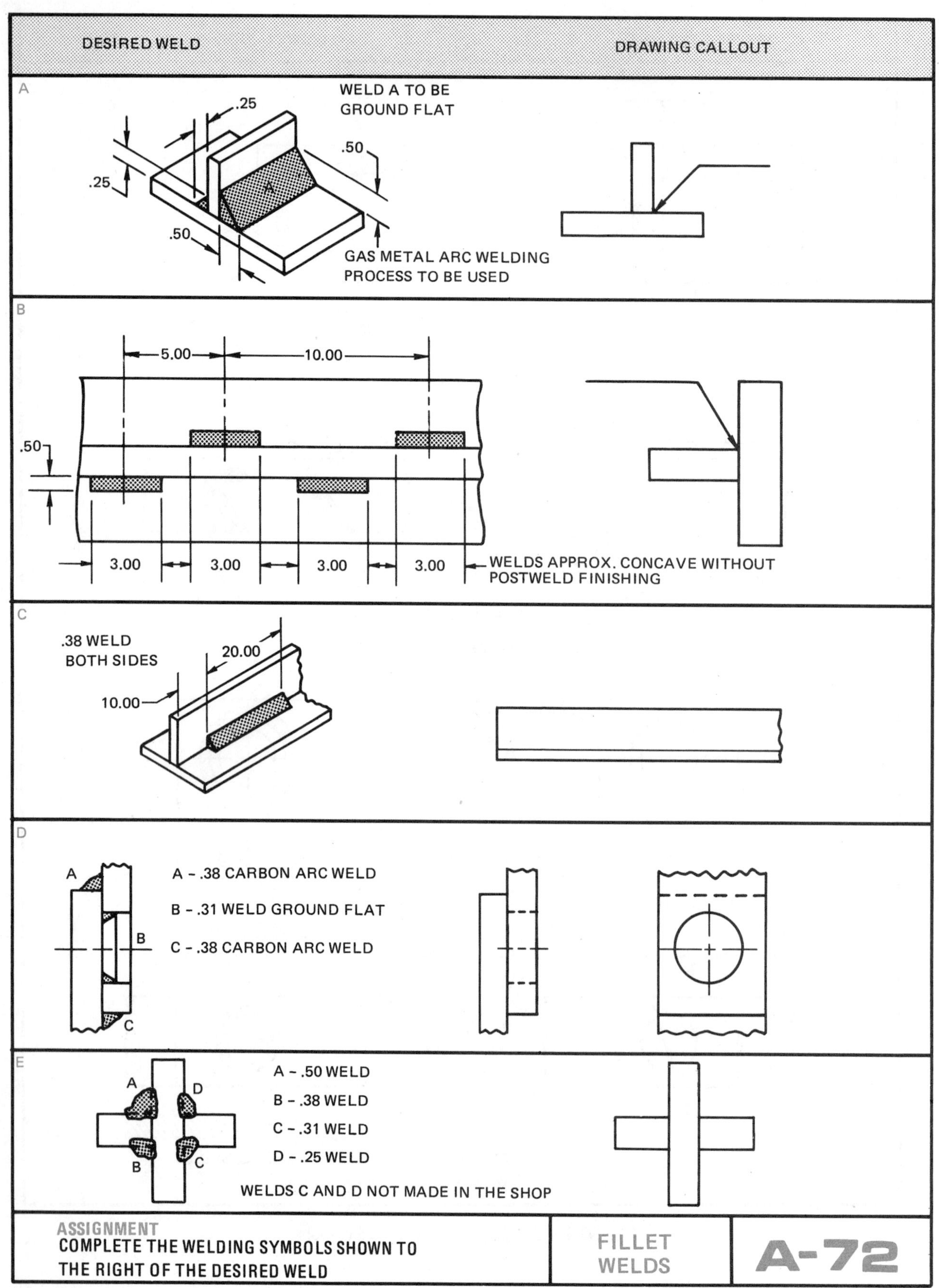
DESIRED WELD
DRAWING CALLOUT
A
WELD A TO BE
GROUND FLAT
.25
.50
.25
.50
GAS METAL ARC WELDING
PROCESS TO BE USED
B
5.00
10.00
.50
3.00
3.00
3.00
3.00
WELDS APPROX. CONCAVE WITHOUT
POSTWELD FINISHING
C
.38 WELD
BOTH SIDES
20.00
10.00
D
A - .38 CARBON ARC WELD
B - .31 WELD GROUND FLAT
C - .38 CARBON ARC WELD
E
A - .50 WELD
B - .38 WELD
C - .31 WELD
D - .25 WELD
WELDS C AND D NOT MADE IN THE SHOP
ASSIGNMENT
COMPLETE THE WELDING SYMBOLS SHOWN TO
THE RIGHT OF THE DESIRED WELD
FILLET
WELDS
A-72

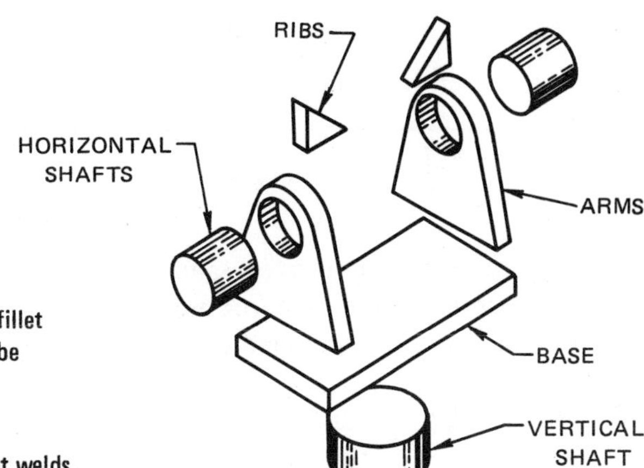

ASSIGNMENT

Add the following welds to the assembly drawing:

- Horizontal shafts welded both sides to arms with .25 fillet welds on outside and .19 fillet welds inside. Welds to be flat without postweld finishing.
- Arms welded both sides to base with .19 fillet welds.
- Ribs welded both sides to base and arms with .12 fillet welds.
- Vertical shaft welded to base with .25 fillet weld ground concave.

Process – Carbon Arc welding.

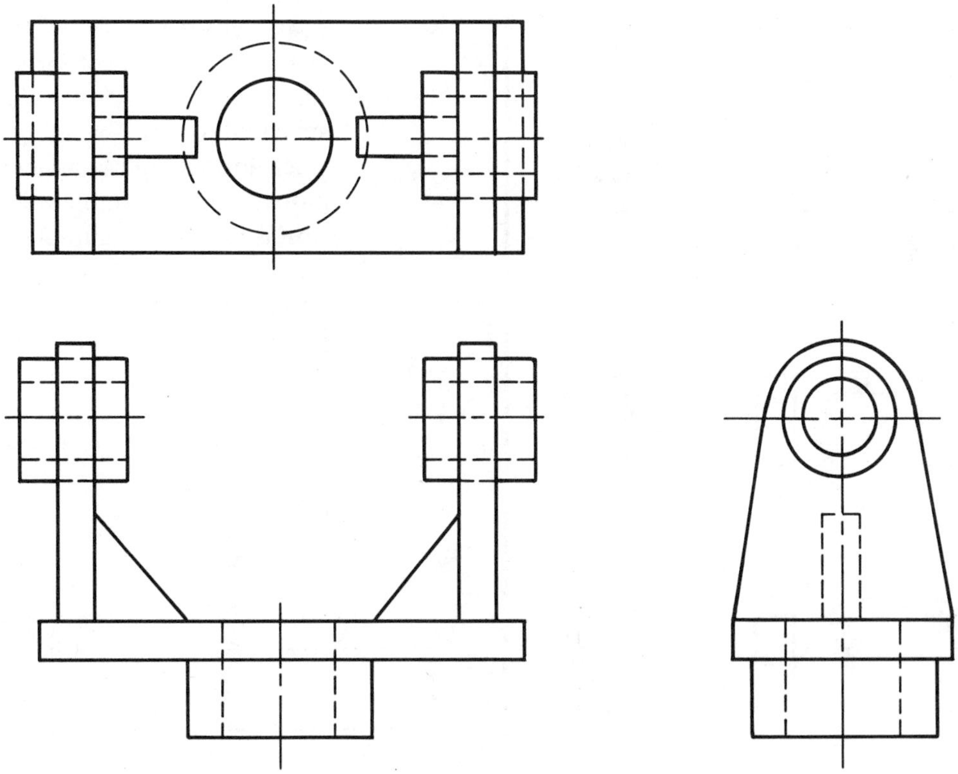

SHAFT SUPPORT	A-73

UNIT 34

GROOVE WELDS

1. Bevel-groove, J-groove, and flare-bevel-groove weld symbols are always drawn with the perpendicular leg to the left.
2. Dimensions of single groove welds are shown on the same side of the reference line as the weld symbol.
3. Each groove of a double-groove joint is dimensioned, however, the root opening need appear only once.
4. For bevel-groove and J-groove welds, a broken arrow is used, when necessary, to identify the member to be prepared, figure 34-1.
5. The depth of groove preparation "S" and size (E) of a groove weld when specified, is placed to the left of the weld symbol. Either or both may be shown. Except for square-groove welds, the groove weld size (E) in relation to the depth of the groove preparation "S" is shown as "S(E)," figure 34-2.
6. Only the groove weld size is shown for square-groove welds.
7. When no depth of groove preparation and no groove weld size are specified on the welding symbol for single-groove and symmetrical double-groove welds, complete joint preparation is required, figure 34-3.
8. When the groove welds extend only partly through the member being joined, the size of the weld is shown

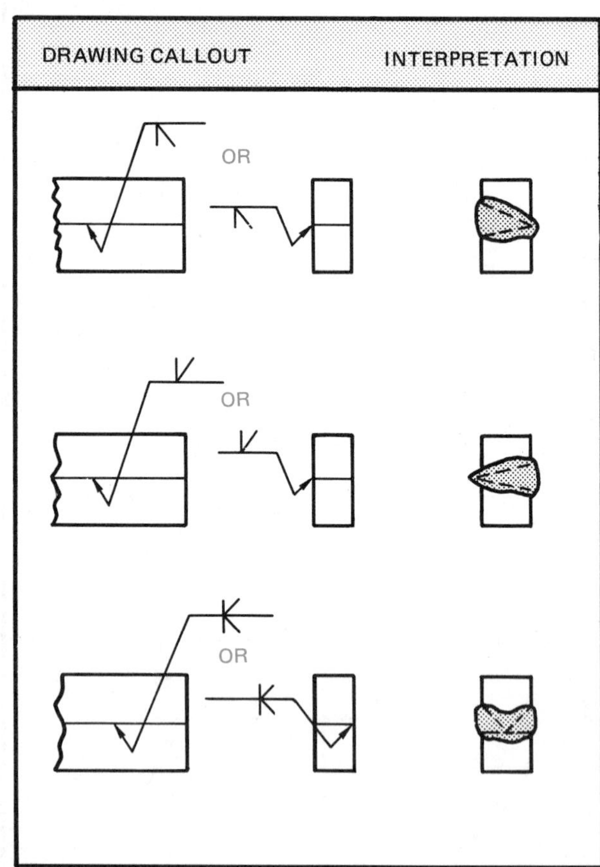

Fig. 34-1 Application of break in arrow on welding symbol

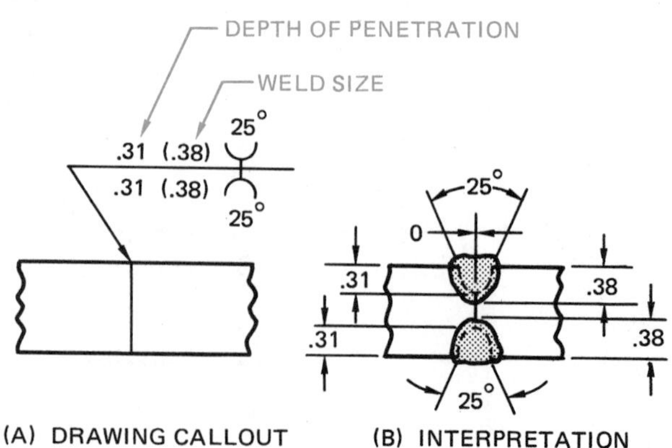

Fig. 34-2 Groove weld symbol showing use of combined dimensions

on the weld symbol, figures 34-4, 34-5 and 34-6.

9. A dimension not in parenthesis placed to the left of bevel, V-, J-, or U-groove weld symbol indicates only the depth of preparation.
10. Groove welds that are to be welded with approximately flush or convex faces without post-weld finishing are specified by adding the flush or convex contour symbol to the welding symbol.
11. Groove welds whose faces are to be finished flush or convex by postweld finishing are specified by adding both the appropriate contour and finishing symbol to the weld symbol. Standard finishing symbols are:
    C – Chipping
    G – Grinding

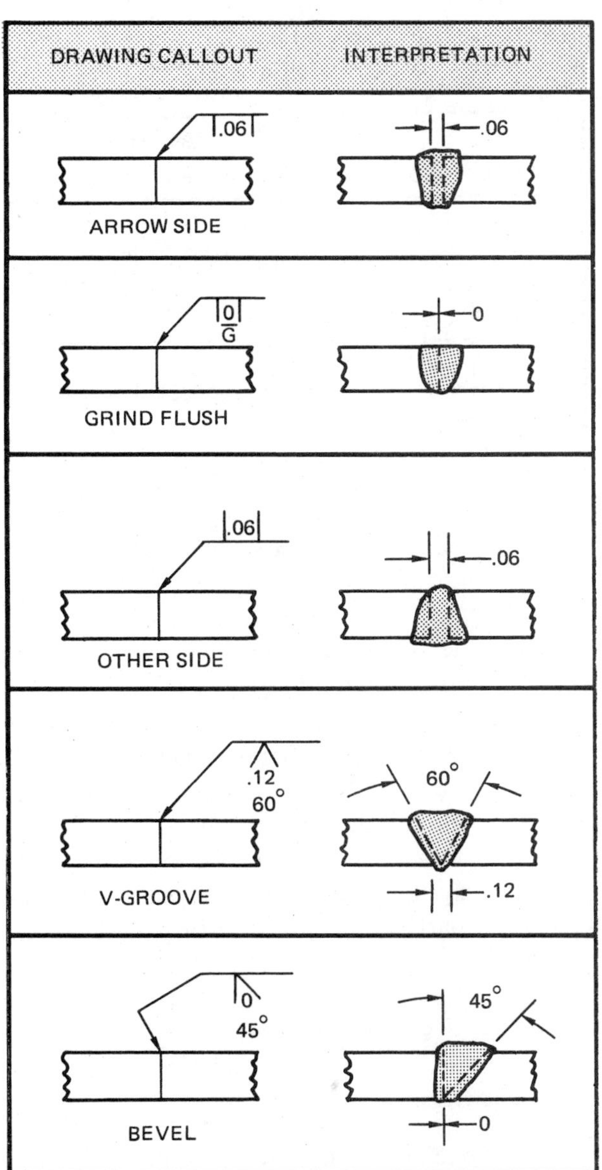

Fig. 34-3 Single-groove welds–complete joint penetration

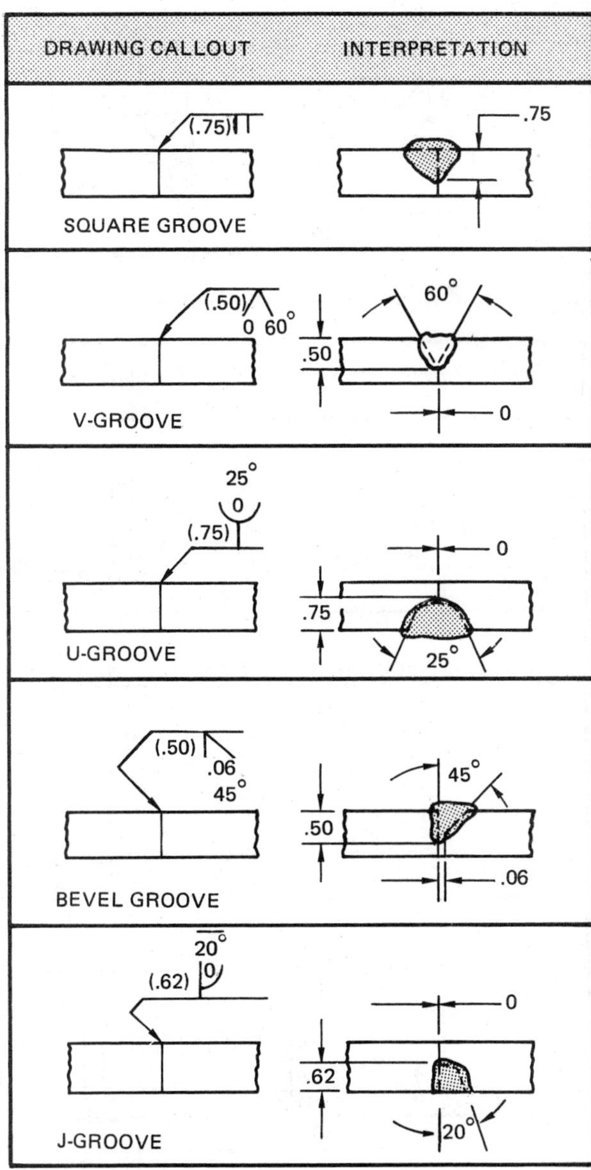

Fig. 34-4 Single-groove welds–partial penetration

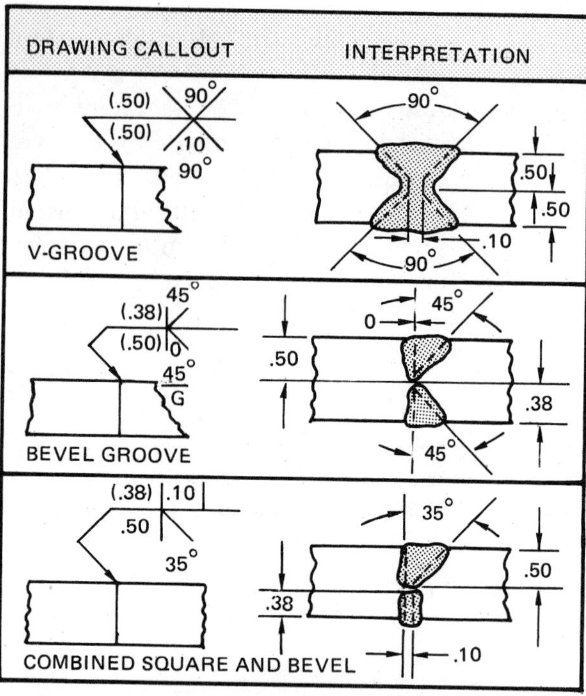

Fig. 34-5 Double-groove welds

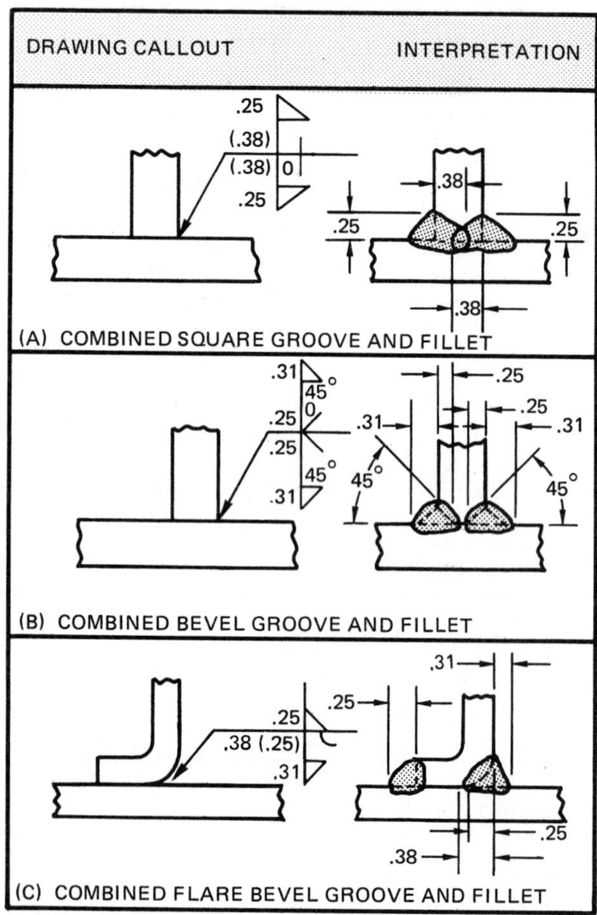

Fig. 34-6 Combined groove and fillet welds

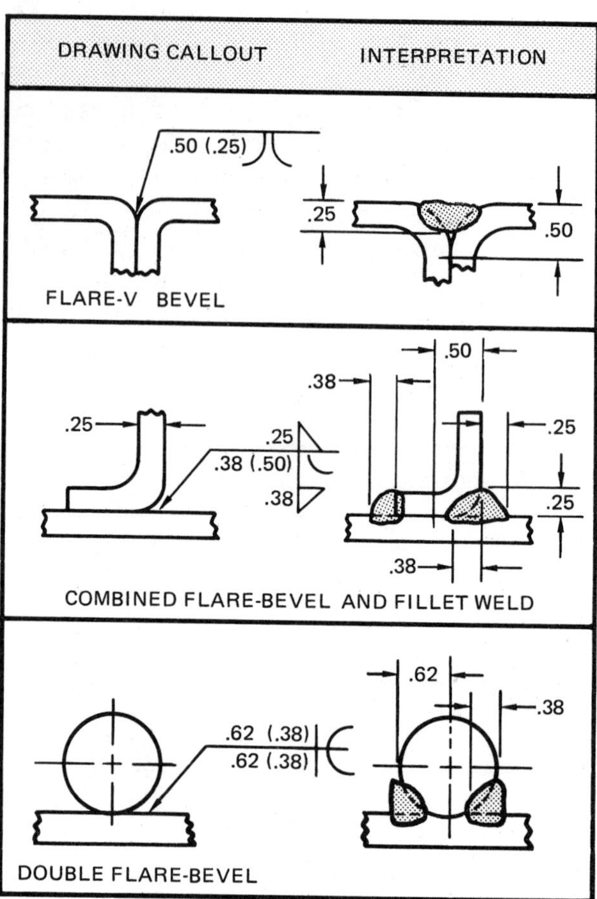

Fig. 34-7 Flare-V and flare bevel groove welds with partial joint preparation

H – Hammering
M – Machining
R – Rolling

12. The size of flare-groove welds when no weld size is given is considered as extending only to the tangent points indicated by dimension "S." For application of flare-groove welds with partial joint preparation, figure 34-7.

SUPPLEMENTARY SYMBOLS

Back and Backing Welds

The back or backing weld symbol is used to indicate bead-type back or backing welds of single groove welds.

The back and backing weld symbols are identical. The sequence of welding determines

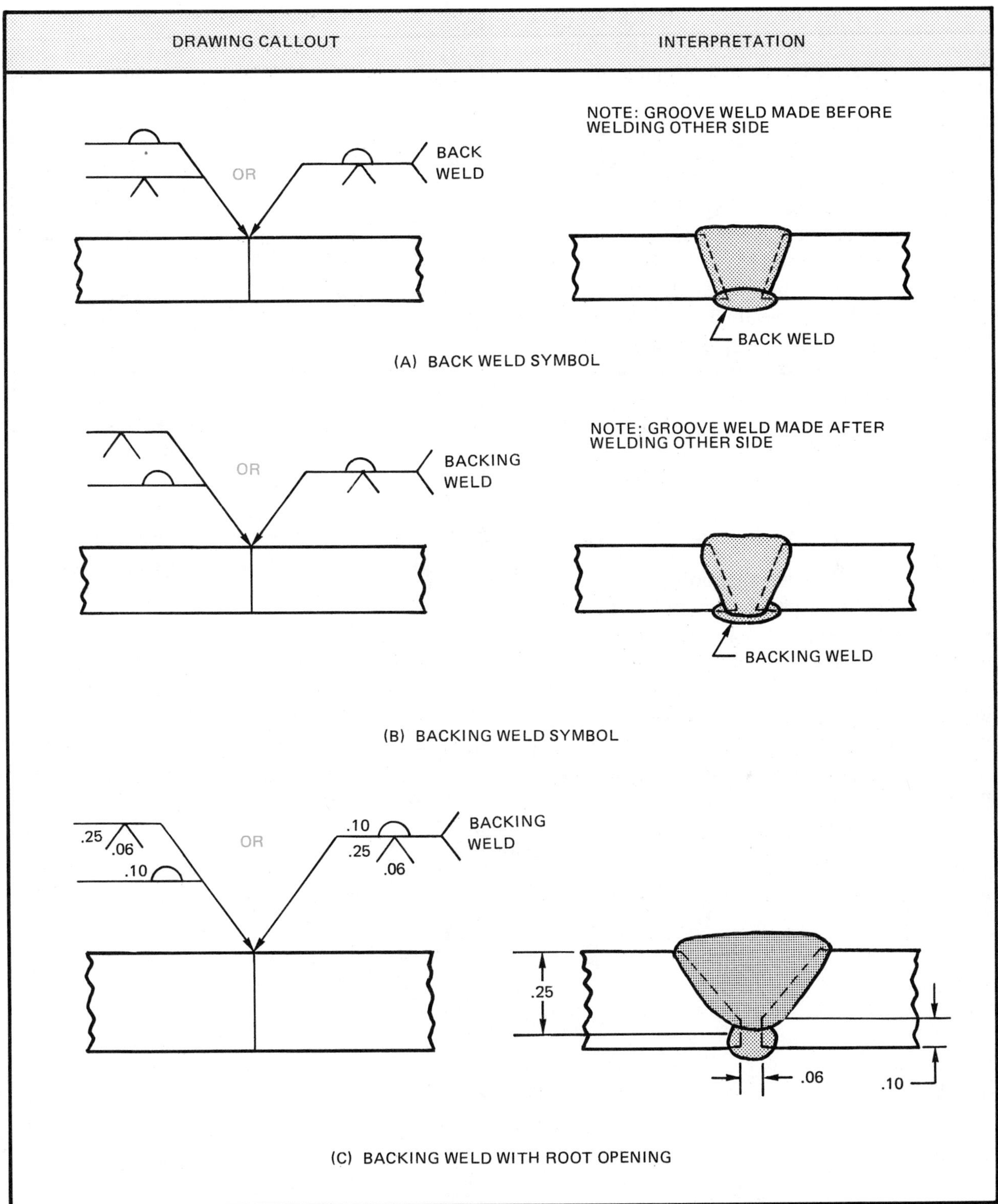

Fig. 34-8 Application of back and backing weld symbol

which designation applies. The back weld is made after the groove weld and the backing weld is made before the groove weld.

1. The back weld symbol is placed on the side of the reference line opposite a groove weld symbol. When a single

reference line is used, "BACK WELD" is specified in the tail of the symbol. Alternately, if a multiple reference line is used, the back weld symbol is placed on a reference line subsequent to the reference line specifying the groove weld, figure 34-8(A).

2. The backing weld symbol is placed on the side of the reference line opposite the groove weld symbol. When a single reference line is used, "BACKING WELD" is specified in the tail of the arrow. If a multiple reference line is used, the backing weld symbol is placed on a reference line prior to that specifying the groove weld, figures 34-8(B) and (C).

Melt-Through Symbol

The melt-through symbol is used only when complete root penetration plus visible root reinforcement is required in welds made from one side.

The melt-through symbol is placed on the side of the reference line opposite the weld symbol, figure 34-9.

The height of root reinforcement may be specified by placing the required dimension to the left of the melt-through symbol. The height of root reinforcement may be unspecified.

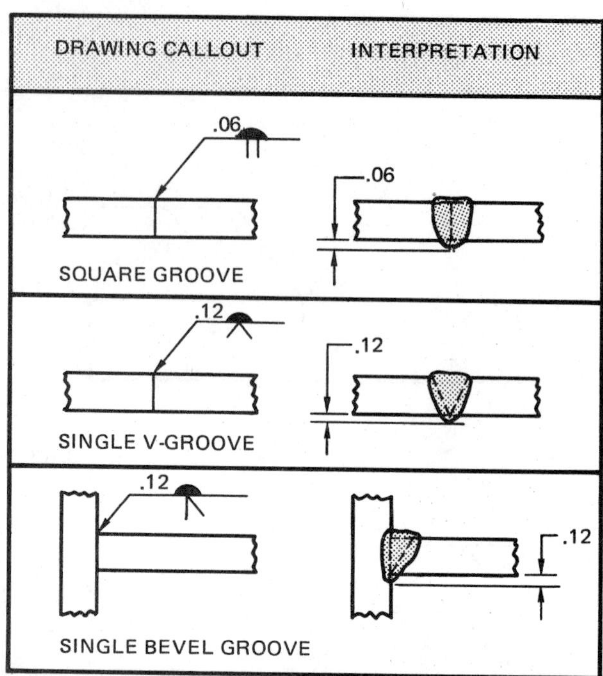

Fig. 34-9 Application of melt-through groove weld symbols

REFERENCES AND SOURCE MATERIALS

1. ANSI/AWS A2.4-86

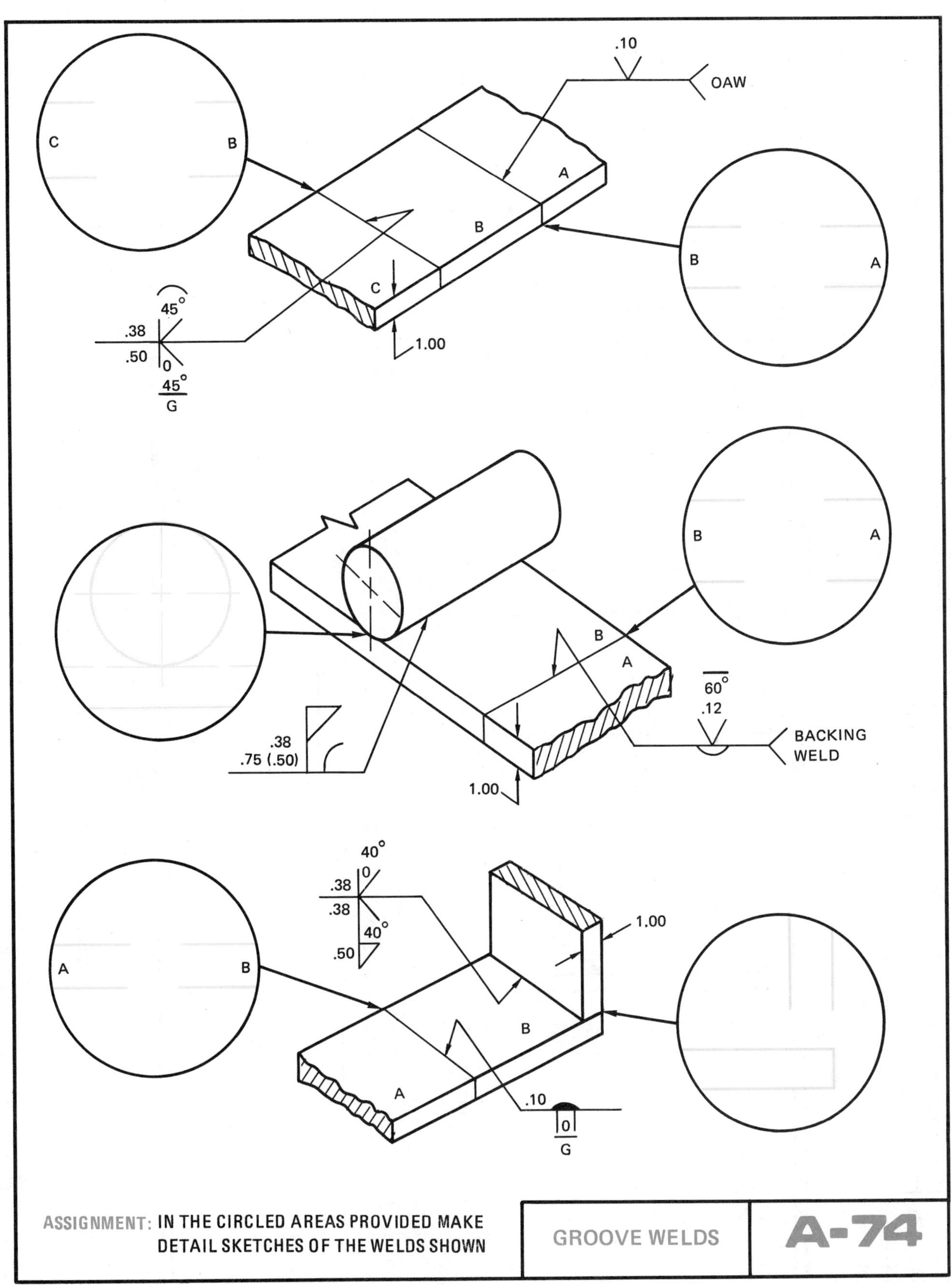

ASSIGNMENT: IN THE CIRCLED AREAS PROVIDED MAKE DETAIL SKETCHES OF THE WELDS SHOWN

GROOVE WELDS

A-74

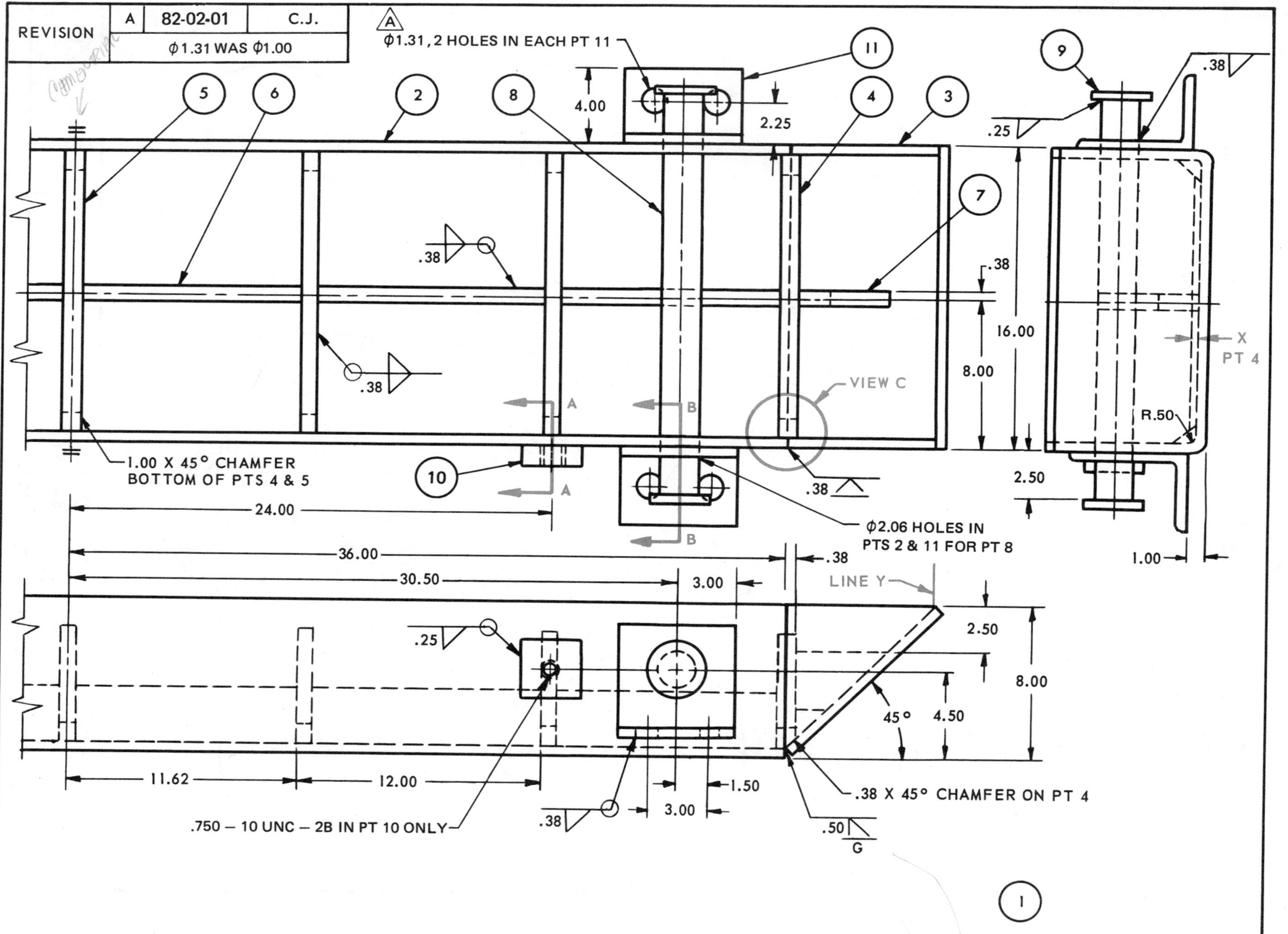
REVISION
A
82-02-01
C.J.
ϕ1.31 WAS ϕ1.00
ϕ1.31, 2 HOLES IN EACH PT 11
1.00 X 45° CHAMFER BOTTOM OF PTS 4 & 5
ϕ2.06 HOLES IN PTS 2 & 11 FOR PT 8
.38 X 45° CHAMFER ON PT 4
.750 — 10 UNC — 2B IN PT 10 ONLY
VIEW C
LINE Y
X PT 4
R.50

SKETCHING ASSIGNMENT

On the graph sections, complete the sketches and darken in welds.

QUESTIONS

1. How many Ø1.31 holes are there in the complete assembly?
2. How deep is the .750 tapped hole?
3. What was the original size of the Ø1.31 hole?
4. What is the overall height of the assembly?
5. Determine the distance from line Y to the center line of the assembly.
6. Determine distance X.
7. What is the developed width of part 2? Use inside dimensions of channel.
8. What is the clearance for fitting on the length of pt. 5?
9. What is the length of (A) pt. 2, (B) pt. 4, allow .25 for clearance, (C) pt. 7, (D) pt. 8?
10. What is the difference between pt. 4 and pt. 5?
11. How many 1.00 chamfers are needed?
12. What does G mean on .50 bevel weld symbols?
13. What type of weld is used to fasten pt. 11 to pt. 2?
14. What type of weld is used to join pt. 3 to pt. 2 at the sides?
15. How many parts make up the assembly?
16. Complete the missing sizes in the Bill of Material. Use inside travel for calculating part 2.

UNLESS OTHERWISE SPECIFIED

- TOLERANCE ON LINEAR DIMENSIONS ±.04
- TOLERANCE ON ANGLES ±0.5°
- TOLERANCE ON HOLES ±.004

1.00 SQUARES

2

SECTION A–A

1.00 SQUARES

2

SECTION B–B

.50 SQUARES

2

3

ENLARGED VIEW C

TOP VIEW

RIGHT SIDE VIEW

FRONT VIEW

ARRANGEMENT OF VIEWS

ANSWERS

1. ________
2. ________
3. ________
4. ________
5. ________
6. ________
7. ________
8. A. ________
   B. ________
   C. ________
   D. ________
9. ________
10. ________
11. ________
12. GRIND
13. ________
14. ________
15. ________

QTY	ITEM	MATL	DESCRIPTION	PT NO
4	LOCATING ANGLE	STL	6.00 X 4.00 X .50 X 6.00 LG	11
2	GROUND BAR	STL BAR	1.00 X 3.00 X 3.00	10
4	RETAINER	STL PL	.50 X ϕ3.00	9
2	DRAW BAR	STL RD	ϕ2.00 X	8
2	GUSSET	STL BAR	.75 X 3.00 X	7
6	GUSSET	STL BAR	.75 X 3.00 X 11.25	6
5	GUSSET	STL BAR	.75 X 6.00 X 14.75	5
2	GUSSET	STL BAR	.75 X 6.00 X	4
2	END PLATE	STL PL	.50 X 10.62 X 25.90	3
1	BASE	STL PL	.50 X X	2
1	SKID ASS'Y			1

BASE SKID

A-75

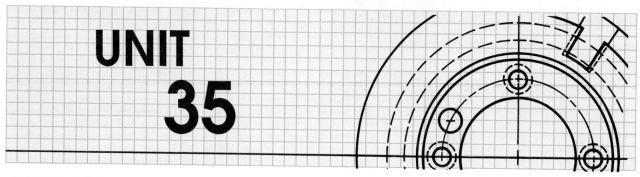

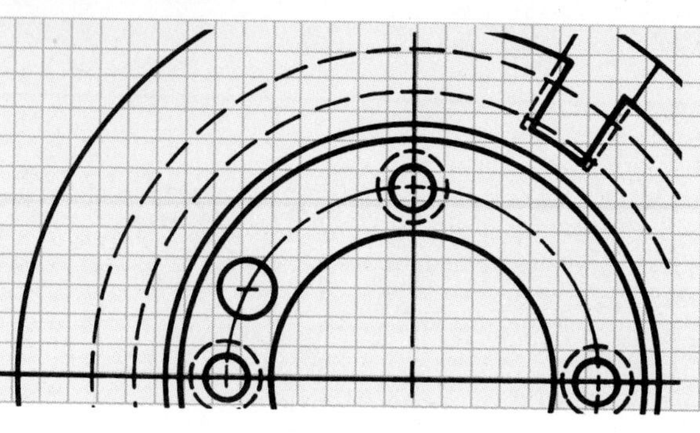

OTHER BASIC WELDS

Plug Welds (Figure 35-1)

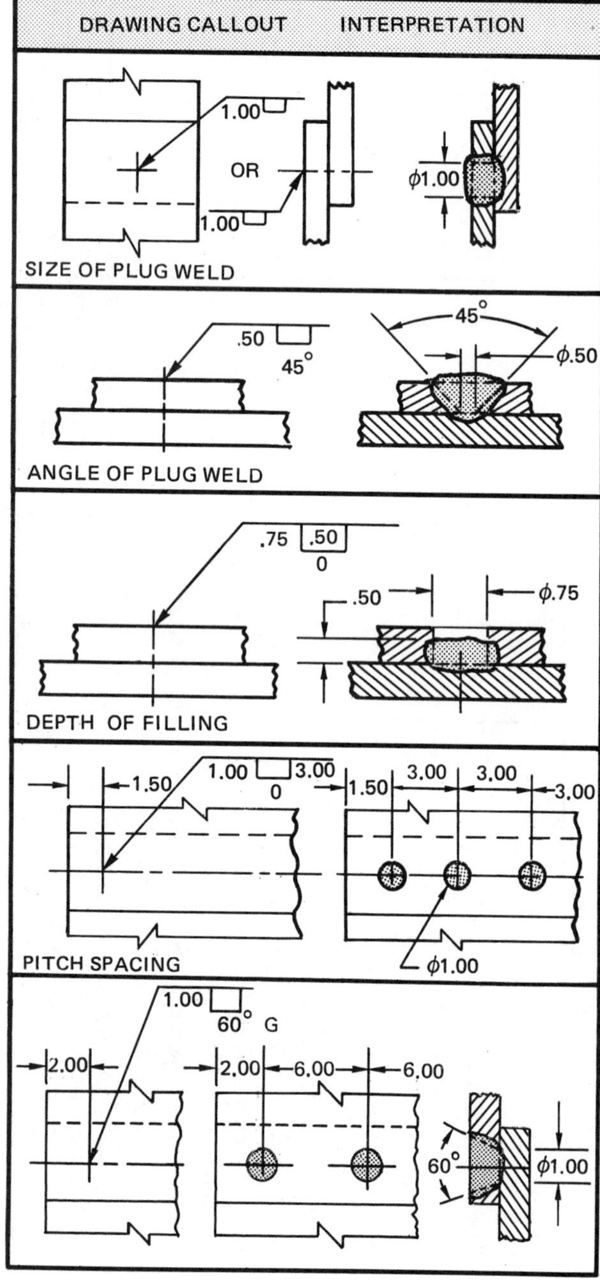

Fig. 35-1 Plug welds

1. Holes in the arrow-side member of a joint for plug welding are specified by placing the weld symbol below the reference line.

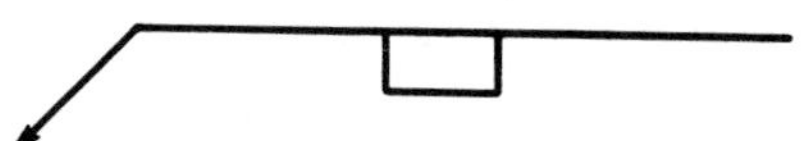

2. Holes in the other-side member of a joint for plug welding are indicated by placing the weld symbol above the reference line.

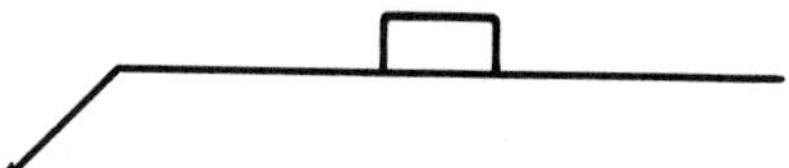

3. The size of a plug weld is shown on the same side and to the left of the weld symbol.

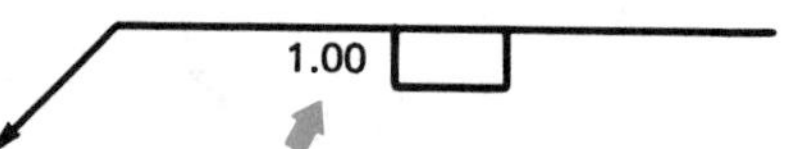

4. The included angle of countersink of plug welds is the user's standard, unless otherwise indicated. Included angle, when not the user's standard, is shown.

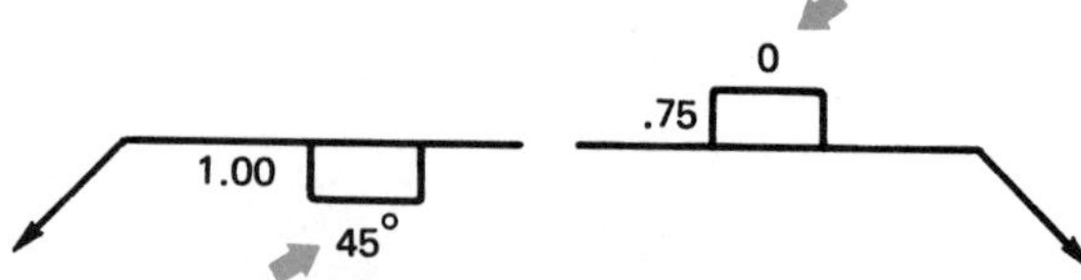

5. The depth of filling of plug welds is complete unless otherwise indicated. When the depth of filling is less than complete, the depth of filling, in inches or millimeters, is shown inside the weld symbol.

6. Pitch (center-to-center spacing) of plug welds is shown to the right of the weld symbol.

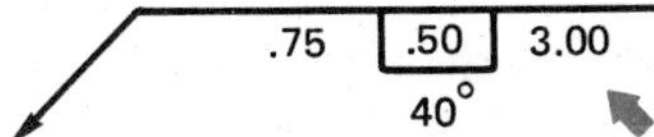

7. Plug welds that are to be welded with approximately flush or convex faces without postweld finishing are specified by adding the flush or convex contour symbol to the weld symbol.

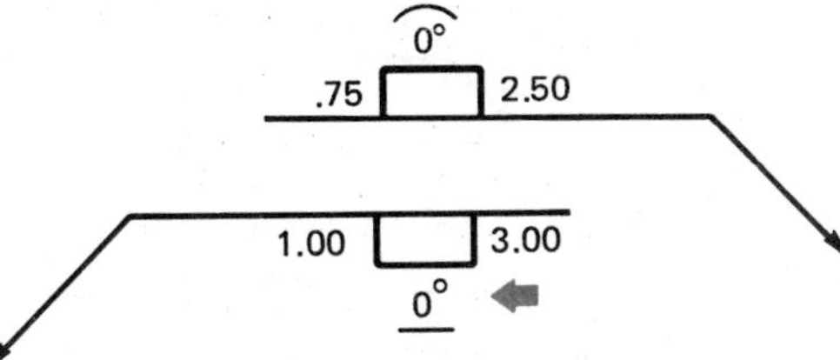

8. Plug welds whose faces are to be finished approximately flush or convex by postweld finishing are specified by adding both the appropriate contour and finishing symbol to the welding symbol. Welds that require a flat but not flush surface require an explanatory note in the tail of the symbol.

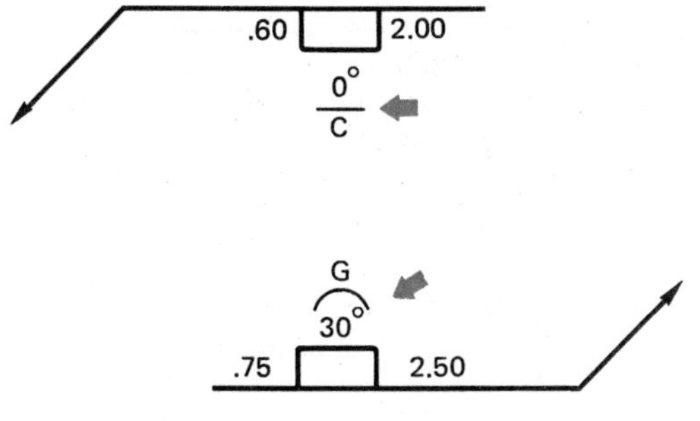

Slot Welds (Figure 35-2)

1. Slots in the arrow-side member of a joint for slot welding are indicated by placing the weld symbol below the reference line. Slot orientation must be shown on the drawing.

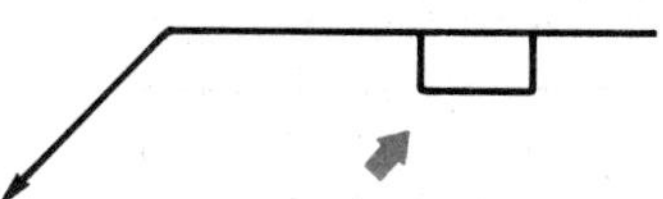

2. Slots in the other-side member of a joint for slot welding are indicated by placing the weld symbol above the reference line.

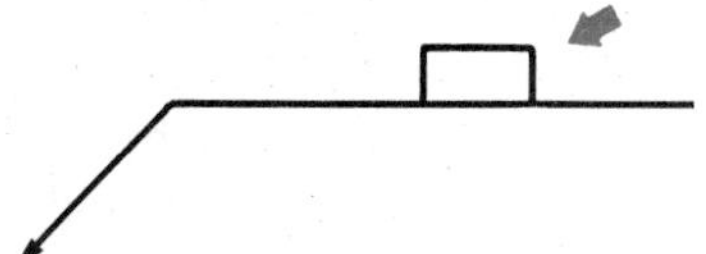

3. Depth of filling of slot welds is complete unless otherwise indicated. When the depth of filling is less than complete, the depth of filling, in inches or millimeters, is shown inside the welding symbol.

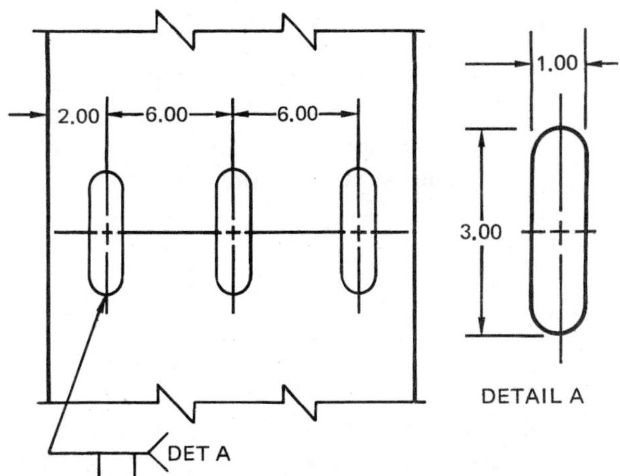

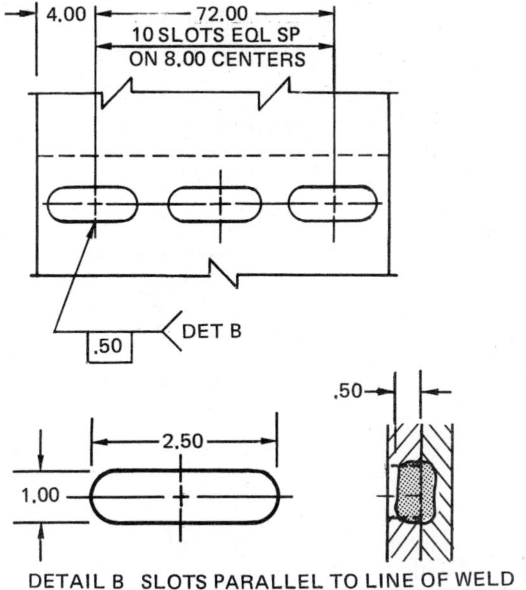

Fig. 35-2 Slot welds

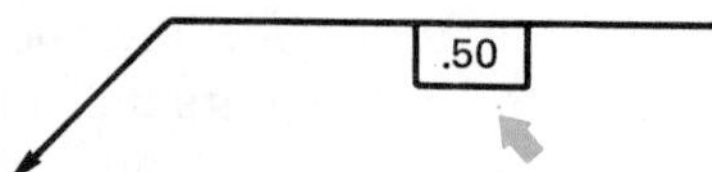

4. Length, width, spacing, included angle of countersink, orientation, and location of slot welds cannot be specified on the welding symbol. These data are to be specified on the drawing or by a detail with reference to it on the welding symbol.

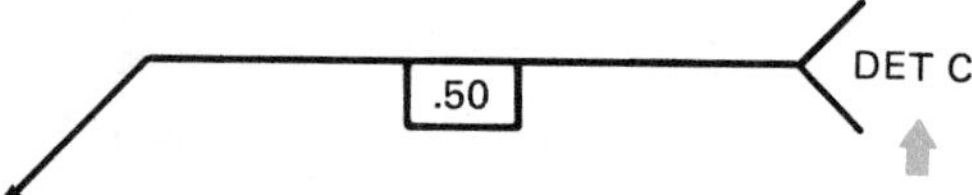

5. Slot welds that are to be welded with approximately flush or convex faces without postweld finishing are specified by adding the flush or convex contour symbol to the weld symbol.

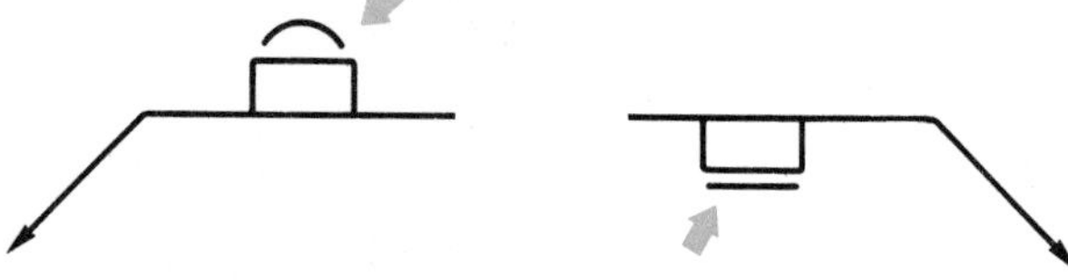

6. Slot welds whose faces are to be finished approximately flush or convex by postweld finishing are specified by adding both the appropriate contour and finishing symbol to the welding symbol. Welds that require a flat but not flush surface require an explanation note in the tail of the symbol.

Spot Welds (Figure 35-3)

1. The symbol for all spot or projection welds is a circle, regardless of the welding process used. There is no attempt to provide symbols for different ways of making a spot weld, such as resistance, arc, and electron beam welding. The symbol for a spot weld is a circle placed:
   - Below the reference line, indicating arrow side.
   - Above the reference line, indicating other side.
   - On the reference line, indicating that there is no arrow or other side.

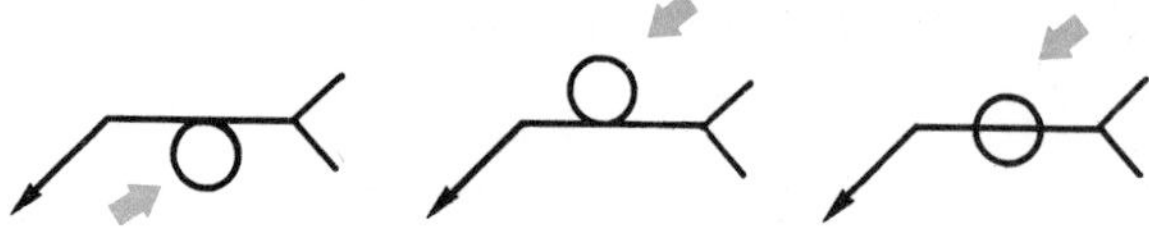

2. Dimensions of spot welds are shown on the same side of the reference line as the weld symbol, or on either side when the symbol is located astride the reference line and has no arrow-side or other-side significance. They are dimensioned by either the size or the strength. The size is designated as the diameter of the weld at the faying surfaces and is shown to the left of the weld symbol. The strength of the spot weld is designated in pounds (or new-

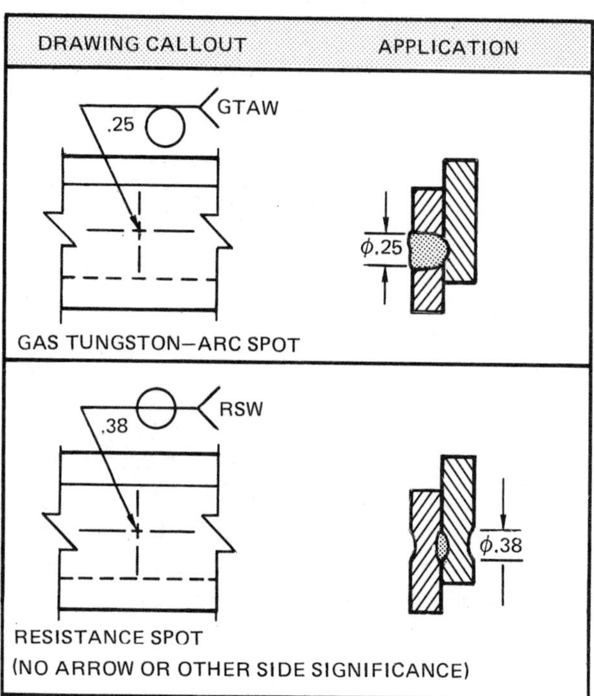

Fig. 35-3 Spot welds

tons) per spot and is shown to the left of the weld symbol.

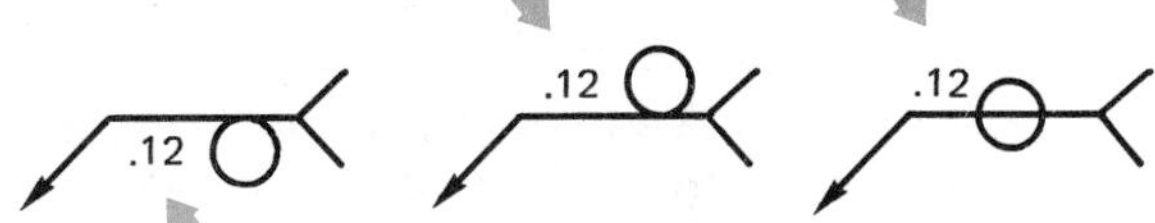

SPECIFYING DIAMETER OF SPOT

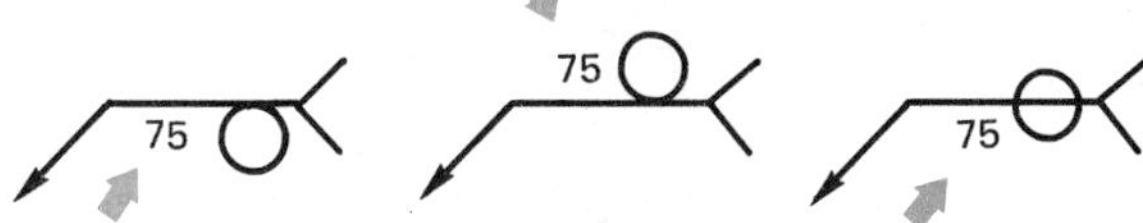

SPECIFYING STRENGTH OF SPOT

3. The process reference is to be indicated in the tail of the welding symbol.

4. When projection welding is used, the spot weld symbol is used and the projection welding process is referenced in the tail of the symbol. The spot weld symbol is located above or below (not on) the reference line to designate on which member the embossment is placed.

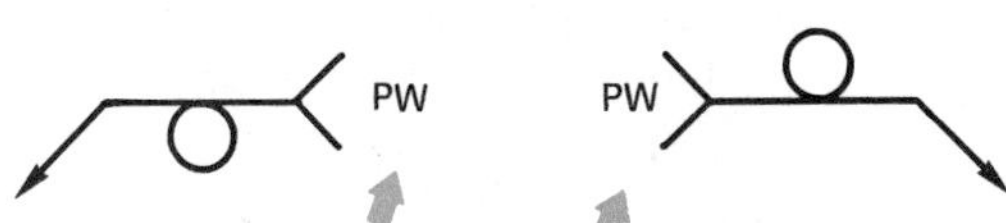

5. The pitch (center-to-center spacing) is shown to the right of the weld symbol.

6. When spot welding extends less than the distance between abrupt changes in the direction of the welding or less than the full length of the joint, the extent is dimensioned.

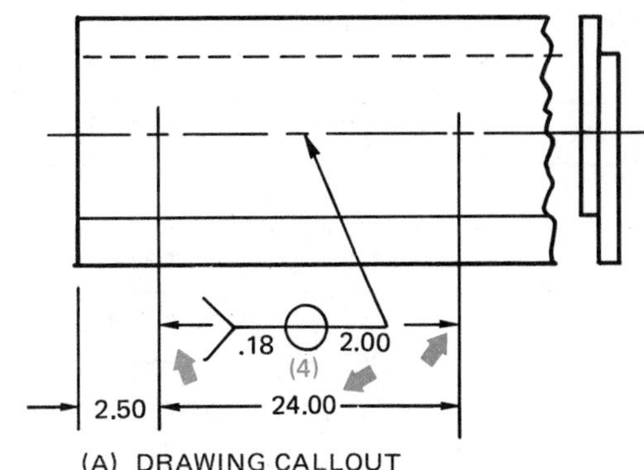

(A) DRAWING CALLOUT

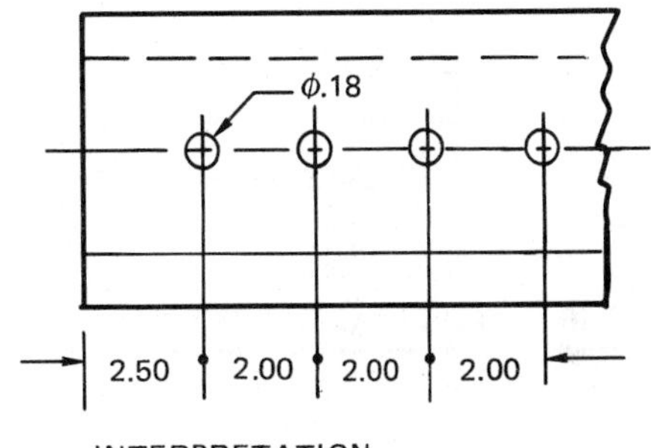

INTERPRETATION

7. Where the exposed surface of either member of a spot welded joint is to be welded with approximately flush or convex faces without postweld finishing, that surface is specified by adding the flush or convex contour symbol to the weld symbol.

8. Spot welds whose faces are to be finished approximately flush, or convex by postweld finishing are specified by adding both the appropriate contour and finishing symbol to the welding symbol. Welds that require a flat but not flush surface require an explanatory note in the tail of the symbol.

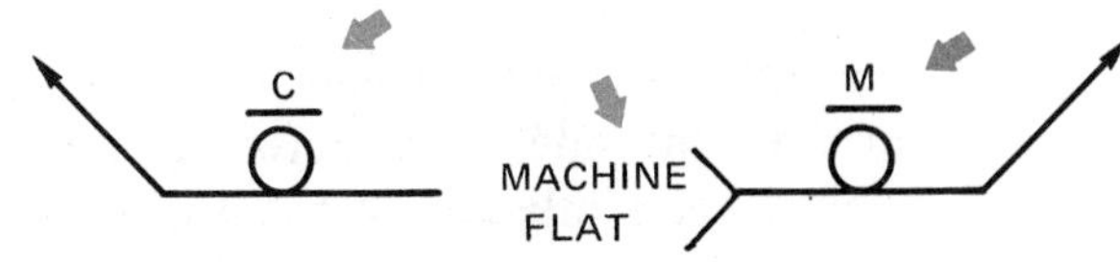

Seam Welds (Figure 35-4)

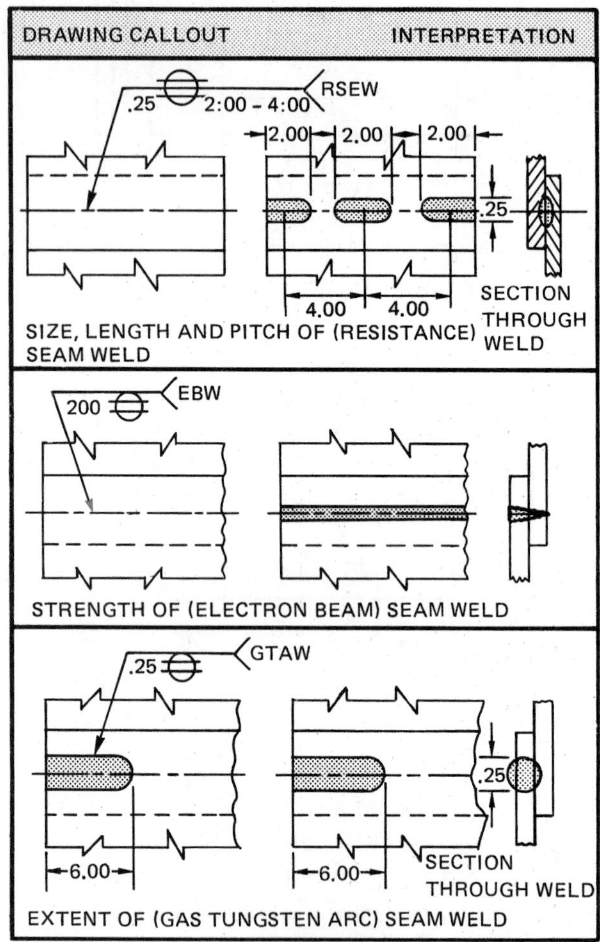

Fig. 35-4 Seam welds

1. The symbol for all seam welds is a circle traversed by two horizontal parallel lines. This symbol is used for all seam welds regardless of the way they are made. The seam weld symbol is placed (1) below the reference line to indicate arrow side, (2) above the reference line to indicate other side, and (3) on the reference line to indicate that there is no arrow or other side significance.

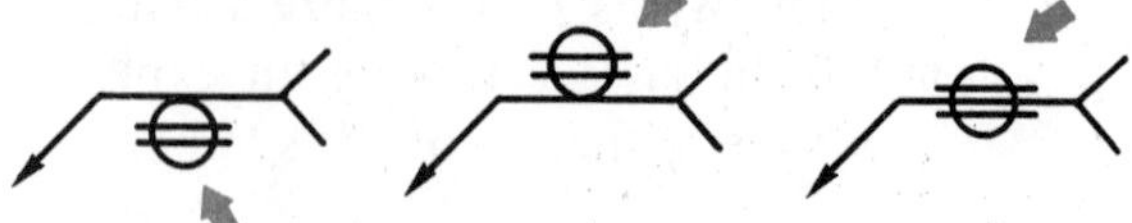

2. Dimensions of seam welds are shown on the same side of the reference line as the weld symbol or all on either side when the symbol is centered on the reference line. They are dimensioned by either size or strength. The size of seam welds is designated as the width of the weld at the faying surfaces and is shown to the left of the weld symbol. The strength of seam welds is designated in pounds per linear inch (lb/in.) or newtons per millimeter (N/mm) and is shown to the left of the weld symbols.

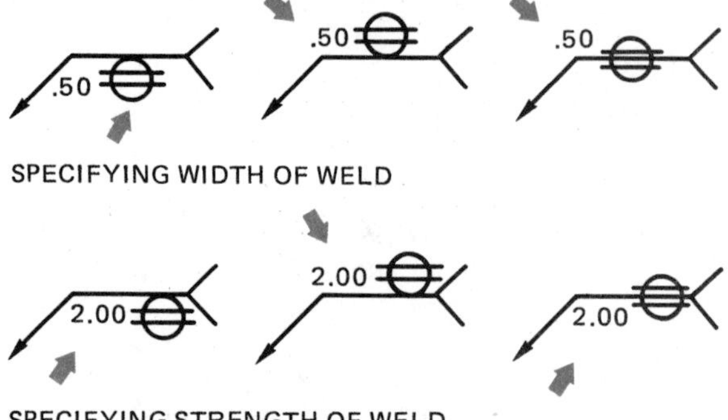

3. The process reference is indicated in the tail of the welding symbol.

4. The length of a seam weld, when indicated on the welding symbol, is shown to the right of the weld symbol. When seam welding extends for the full distance between abrupt changes in the

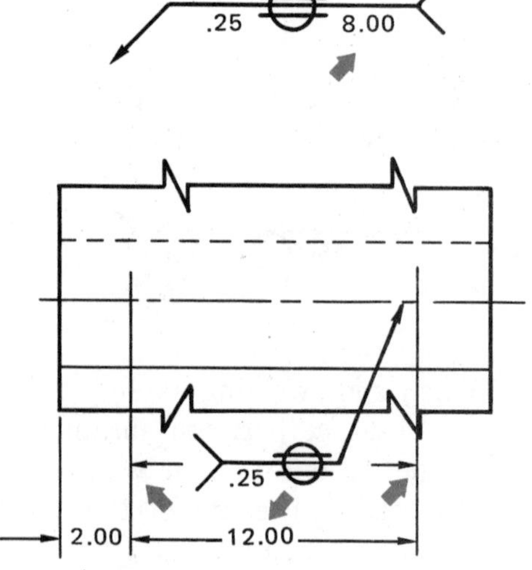

direction of the welding, no length dimension needs to be shown on the welding symbol. When a seam weld extends less than the full length of the joint, the extent of the weld should be shown.

5. The pitch of intermittent seam weld is shown as the distance between centers of the weld increments. The pitch is shown to the right of the length dimension.

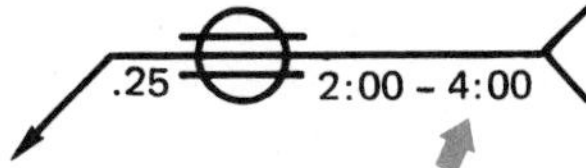

6. When the exposed surface of either member of a seamwelded joint is to be welded with approximately flush or convex faces without postweld finishing, that surface is specified by adding the flush or convex contour symbol to the weld symbol.

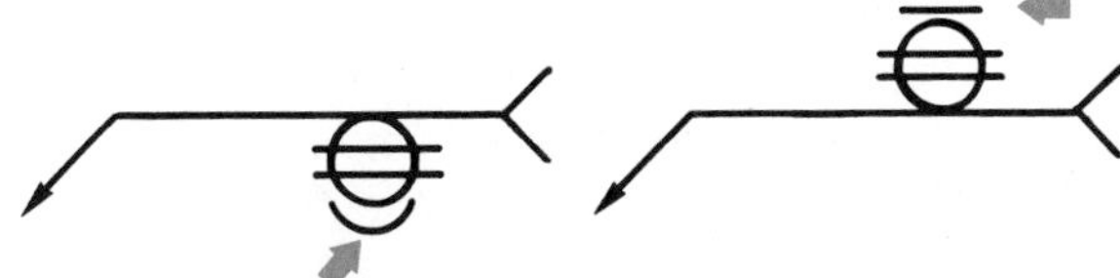

7. Seam welds whose faces are to be finished approximately flush or convex are specified by adding both the appropriate contour and finish symbol to the welding symbol.

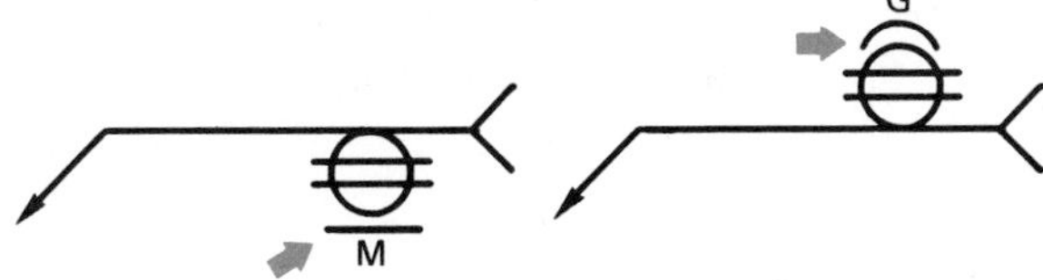

Flange Welds (Figure 35-5)

The following welding symbols are intended to be used for light-gage metal joints involving the flaring or flanging of the edges to be joined.

1. Edge-flange welds are shown by the edge-flange-weld symbol.

2. Corner-flange welds on joints detailed on the drawing are specified by the corner-flange weld symbol. Weld symbols are always drawn with the perpendicular leg to the left.

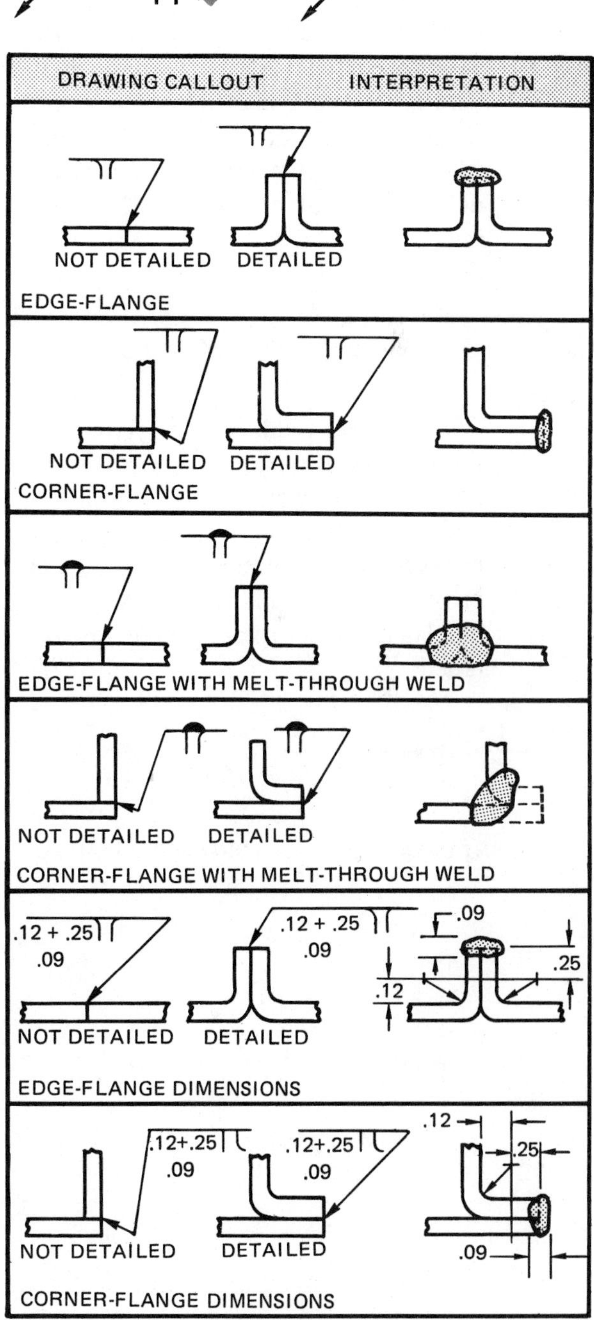

Fig. 35-5 Flange welds

3. Corner-flange welds on joints not detailed on the drawing are specified by the corner-flange weld symbol. A broken arrow points to the member being flanged.

4. Edge-flange welds requiring complete joint penetration are specified by the edge-flange weld symbol with the melt-through symbol placed on the opposite side of the reference line. The same welding symbol is used for joints either detailed or not detailed on the drawing.

5. Corner-flange welds requiring complete joint penetration are specified by the corner-flange weld symbol with the melt-through symbol placed on the opposite side of the reference line. A broken arrow points to the member to be flanged where the joint is not detailed.

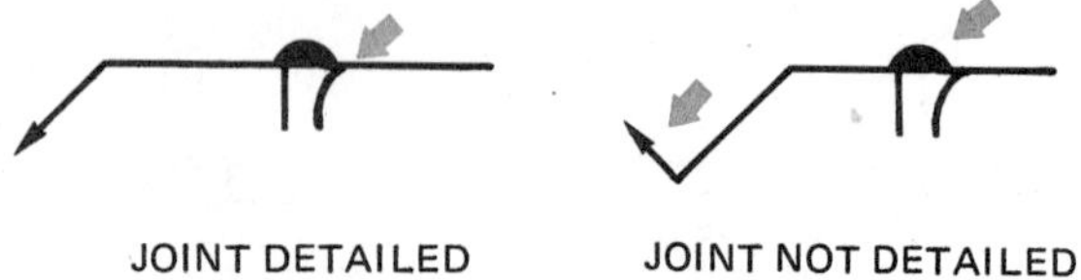

JOINT DETAILED JOINT NOT DETAILED

6. Dimensions of flange welds are shown on the same side of the reference line as the weld symbol. The radius and the height, separated by a plus (+) and placed to the left of the weld symbol. The radius and the height read in that order from left to right along the reference line.

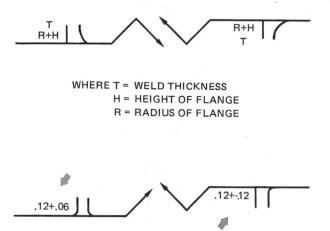

WHERE T = WELD THICKNESS
H = HEIGHT OF FLANGE
R = RADIUS OF FLANGE

7. The size (thickness) of flange welds is specified by a dimension placed above or below the flange dimensions.

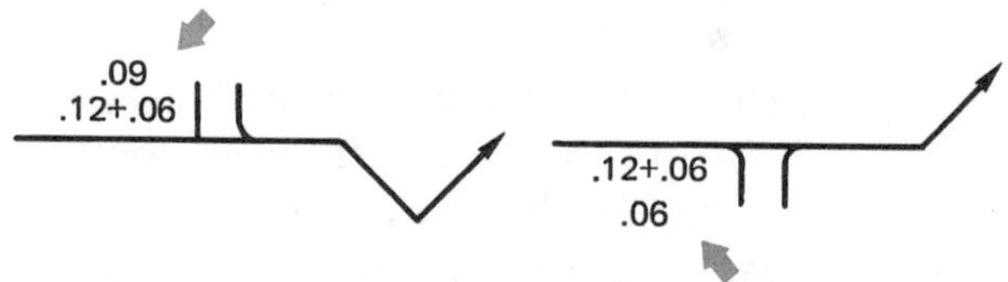

REFERENCES AND SOURCE MATERIALS

1. ANSI/AWS A2.4-86

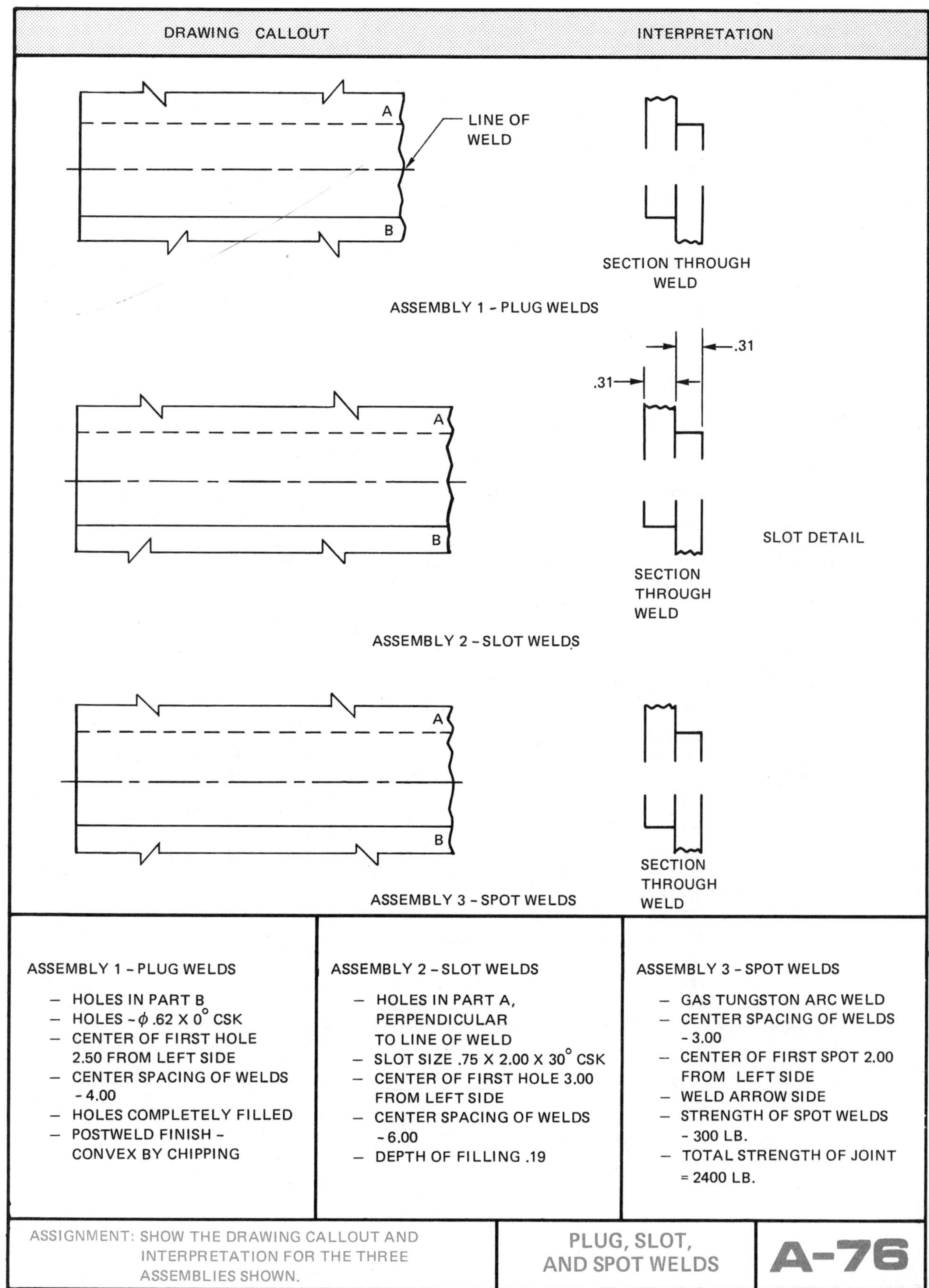

ASSEMBLY 1 - PLUG WELDS

- HOLES IN PART B
- HOLES - ϕ .62 X 0° CSK
- CENTER OF FIRST HOLE 2.50 FROM LEFT SIDE
- CENTER SPACING OF WELDS - 4.00
- HOLES COMPLETELY FILLED
- POSTWELD FINISH - CONVEX BY CHIPPING

ASSEMBLY 2 - SLOT WELDS

- HOLES IN PART A, PERPENDICULAR TO LINE OF WELD
- SLOT SIZE .75 X 2.00 X 30° CSK
- CENTER OF FIRST HOLE 3.00 FROM LEFT SIDE
- CENTER SPACING OF WELDS - 6.00
- DEPTH OF FILLING .19

ASSEMBLY 3 - SPOT WELDS

- GAS TUNGSTON ARC WELD
- CENTER SPACING OF WELDS - 3.00
- CENTER OF FIRST SPOT 2.00 FROM LEFT SIDE
- WELD ARROW SIDE
- STRENGTH OF SPOT WELDS - 300 LB.
- TOTAL STRENGTH OF JOINT = 2400 LB.

ASSIGNMENT: SHOW THE DRAWING CALLOUT AND INTERPRETATION FOR THE THREE ASSEMBLIES SHOWN.

PLUG, SLOT, AND SPOT WELDS

A-76

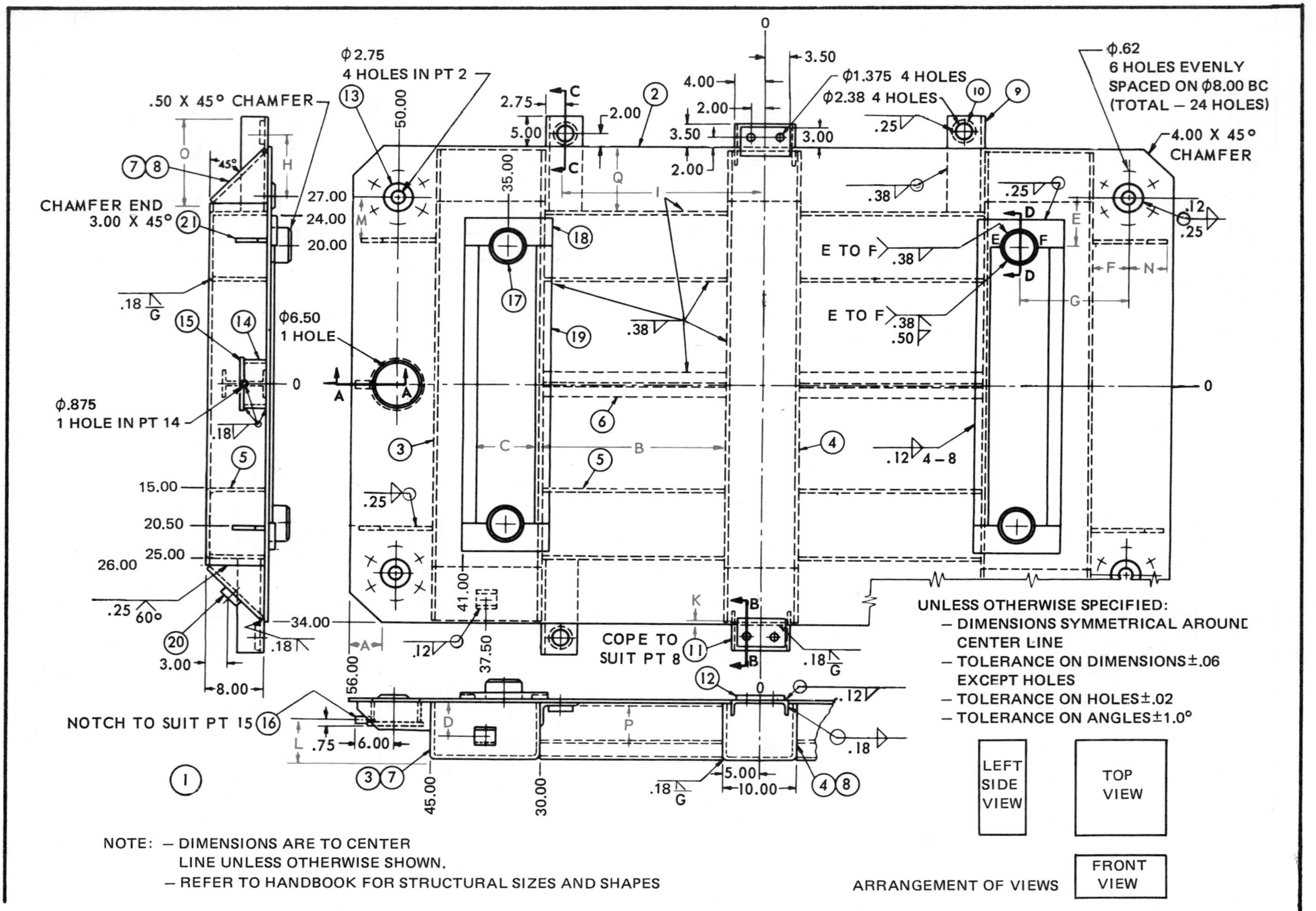

ϕ2.75
4 HOLES IN PT 2
.50 X 45° CHAMFER
CHAMFER END
3.00 X 45°
ϕ6.50
1 HOLE
ϕ.875
1 HOLE IN PT 14
ϕ1.375 4 HOLES
ϕ2.38 4 HOLES
ϕ.62
6 HOLES EVENLY
SPACED ON ϕ8.00 BC
(TOTAL – 24 HOLES)
4.00 X 45°
CHAMFER
E TO F
E TO F
COPE TO
SUIT PT 8
NOTCH TO SUIT PT 15
UNLESS OTHERWISE SPECIFIED:
– DIMENSIONS SYMMETRICAL AROUND
CENTER LINE
– TOLERANCE ON DIMENSIONS±.06
EXCEPT HOLES
– TOLERANCE ON HOLES±.02
– TOLERANCE ON ANGLES±1.0°
LEFT
SIDE
VIEW
TOP
VIEW
FRONT
VIEW
ARRANGEMENT OF VIEWS
NOTE: – DIMENSIONS ARE TO CENTER
LINE UNLESS OTHERWISE SHOWN.
– REFER TO HANDBOOK FOR STRUCTURAL SIZES AND SHAPES

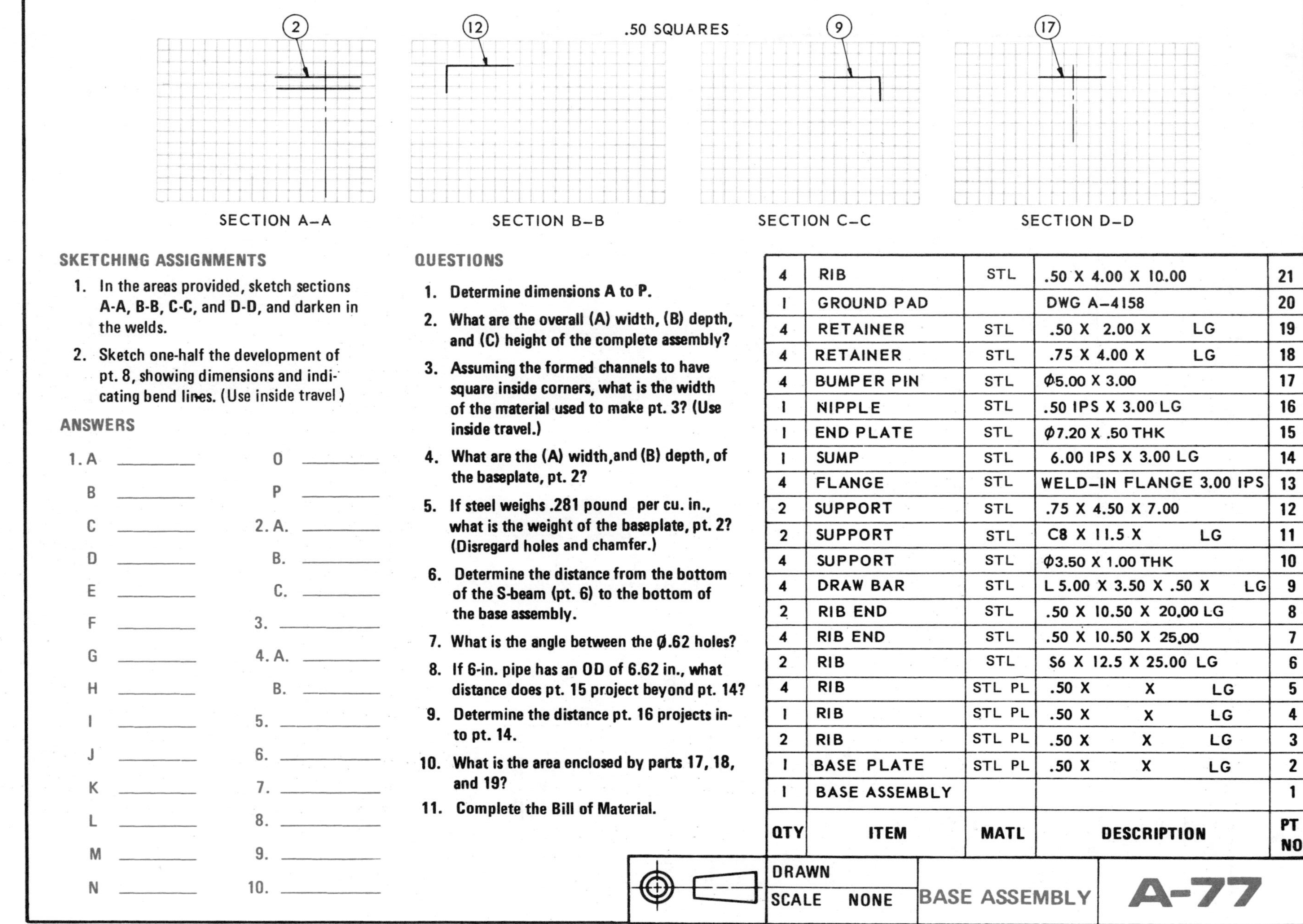

SKETCHING ASSIGNMENTS

1. In the areas provided, sketch sections A-A, B-B, C-C, and D-D, and darken in the welds.
2. Sketch one-half the development of pt. 8, showing dimensions and indicating bend lines. (Use inside travel.)

ANSWERS

1. A	______	O	______
B	______	P	______
C	______	2. A.	______
D	______	B.	______
E	______	C.	______
F	______	3.	______
G	______	4. A.	______
H	______	B.	______
I	______	5.	______
J	______	6.	______
K	______	7.	______
L	______	8.	______
M	______	9.	______
N	______	10.	______

QUESTIONS

1. Determine dimensions A to P.
2. What are the overall (A) width, (B) depth, and (C) height of the complete assembly?
3. Assuming the formed channels to have square inside corners, what is the width of the material used to make pt. 3? (Use inside travel.)
4. What are the (A) width, and (B) depth, of the baseplate, pt. 2?
5. If steel weighs .281 pound per cu. in., what is the weight of the baseplate, pt. 2? (Disregard holes and chamfer.)
6. Determine the distance from the bottom of the S-beam (pt. 6) to the bottom of the base assembly.
7. What is the angle between the Ø.62 holes?
8. If 6-in. pipe has an OD of 6.62 in., what distance does pt. 15 project beyond pt. 14?
9. Determine the distance pt. 16 projects into pt. 14.
10. What is the area enclosed by parts 17, 18, and 19?
11. Complete the Bill of Material.

QTY	ITEM	MATL	DESCRIPTION	PT NO
4	RIB	STL	.50 X 4.00 X 10.00	21
1	GROUND PAD		DWG A–4158	20
4	RETAINER	STL	.50 X 2.00 X LG	19
4	RETAINER	STL	.75 X 4.00 X LG	18
4	BUMPER PIN	STL	Ø5.00 X 3.00	17
1	NIPPLE	STL	.50 IPS X 3.00 LG	16
1	END PLATE	STL	Ø7.20 X .50 THK	15
1	SUMP	STL	6.00 IPS X 3.00 LG	14
4	FLANGE	STL	WELD–IN FLANGE 3.00 IPS	13
2	SUPPORT	STL	.75 X 4.50 X 7.00	12
2	SUPPORT	STL	C8 X 11.5 X LG	11
4	SUPPORT	STL	Ø3.50 X 1.00 THK	10
4	DRAW BAR	STL	L 5.00 X 3.50 X .50 X LG	9
2	RIB END	STL	.50 X 10.50 X 20.00 LG	8
4	RIB END	STL	.50 X 10.50 X 25.00	7
2	RIB	STL	S6 X 12.5 X 25.00 LG	6
4	RIB	STL PL	.50 X X LG	5
1	RIB	STL PL	.50 X X LG	4
2	RIB	STL PL	.50 X X LG	3
1	BASE PLATE	STL PL	.50 X X LG	2
1	BASE ASSEMBLY			1

DRAWN		BASE ASSEMBLY	A-77
SCALE	NONE		

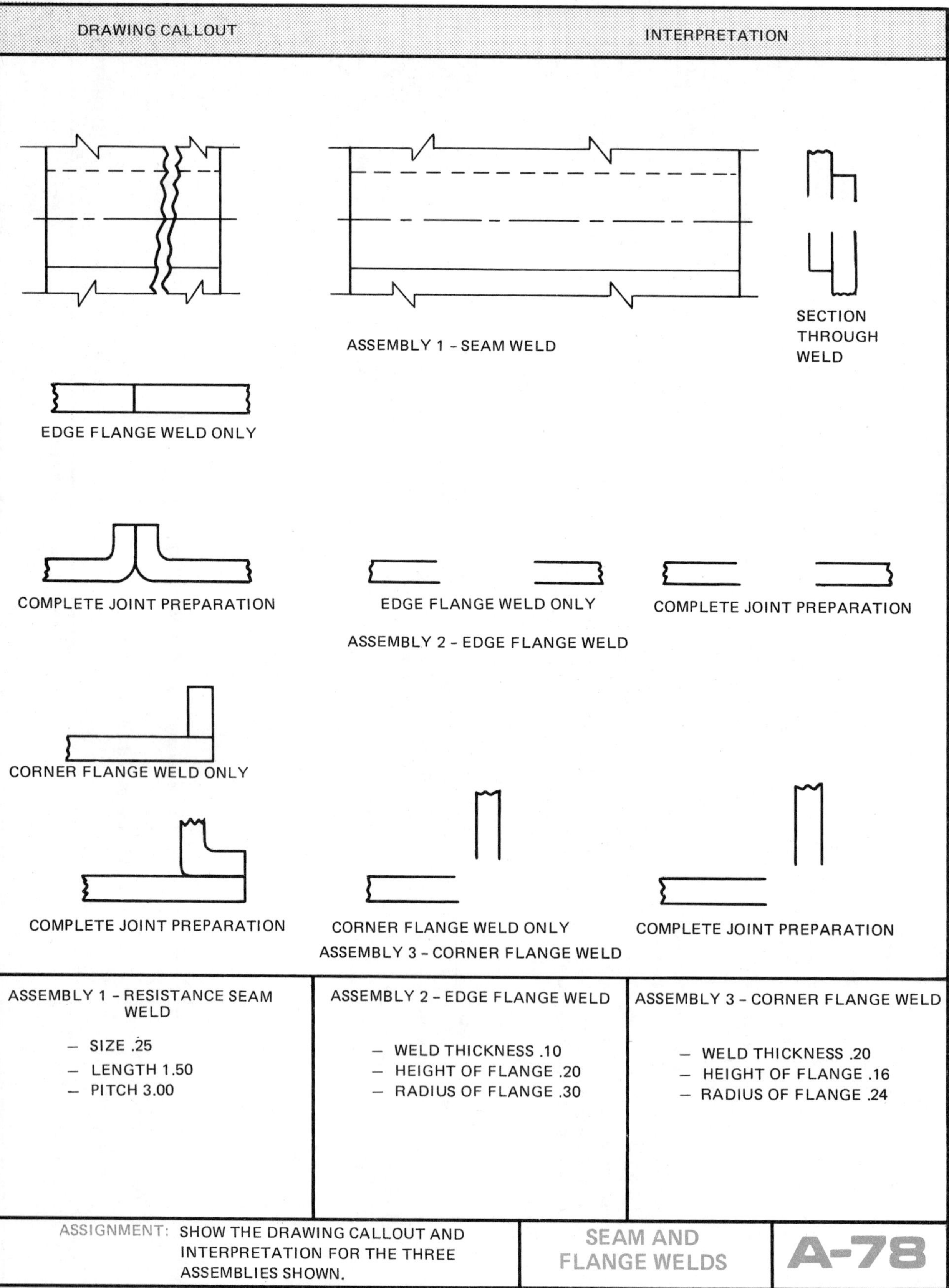

ASSEMBLY 1 - RESISTANCE SEAM WELD	ASSEMBLY 2 - EDGE FLANGE WELD	ASSEMBLY 3 - CORNER FLANGE WELD
– SIZE .25 – LENGTH 1.50 – PITCH 3.00	– WELD THICKNESS .10 – HEIGHT OF FLANGE .20 – RADIUS OF FLANGE .30	– WELD THICKNESS .20 – HEIGHT OF FLANGE .16 – RADIUS OF FLANGE .24

ASSIGNMENT: SHOW THE DRAWING CALLOUT AND INTERPRETATION FOR THE THREE ASSEMBLIES SHOWN.	SEAM AND FLANGE WELDS	A-78

UNIT 36

GEARS

The function of a gear is to transmit rotary or reciprocating motion from one machine part to another. Gears are often used to reduce or increase the r/min of a shaft. *Gears* are rolling cylinders or cones. They have teeth on their contact surfaces to insure the transfer of motion, figure 36-1.

There are many kinds of gears; they may be grouped according to the position of the shafts they connect. *Spur gears* connect parallel shafts; *bevel gears* connect shafts having intersecting axes; and *worm gears* connect shafts having axes which do not intersect. A spur gear with a rack converts rotary motion to reciprocating or linear motion. The smaller of two gears is the *pinion.*

A simple gear drive consists of a toothed driving wheel meshing with a similar driving wheel. Tooth forms are designed to insure uniform angular rotation of the driven wheel during tooth engagement.

SPUR GEARS

Spur gears, which are used for drives between parallel shafts, have teeth on the rim of the wheel, figure 36-2. A pair of spur gears operates as though it consists of two cylindrical surfaces with formed teeth which maintain constant speed ratio between the driving and the driven wheel.

Fig. 36-1 Gears

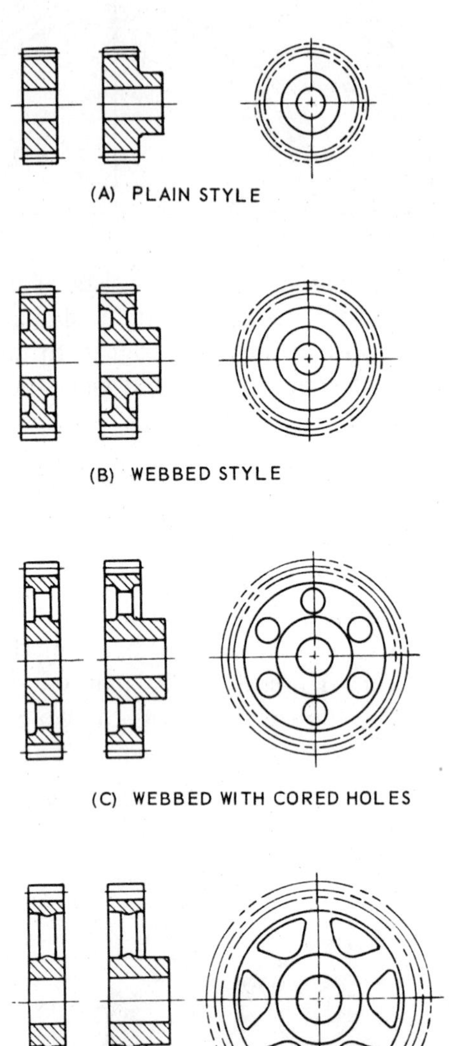

Fig. 36-2 Stock spur gear styles

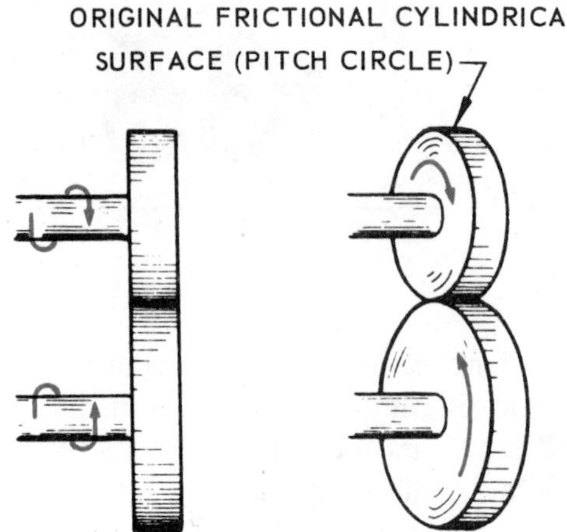

Fig. 36-3 The pitch circle of spur gears

Gear design is complex, dealing with such problems as strength, wear, noise, and material selection. Usually, a designer selects a gear from a catalog. Most gears are made of cast iron or steel. However, brass, bronze, and fiber are used when factors such as wear or noise must be considered.

Theoretically, the teeth of a spur gear are built around the original frictional cylindrical surface called the *pitch circle.*

The angle between the direction of pressure between contacting teeth and a line tangent to the pitch circle is the *pressure angle.*

The 14.5-degree pressure angle has been used for many years and remains useful for duplicate or replacement gearing.

The 20- and 25-degree pressure angles have become the standard for new gearing because of its smoother and quieter operation and greater load-carrying ability.

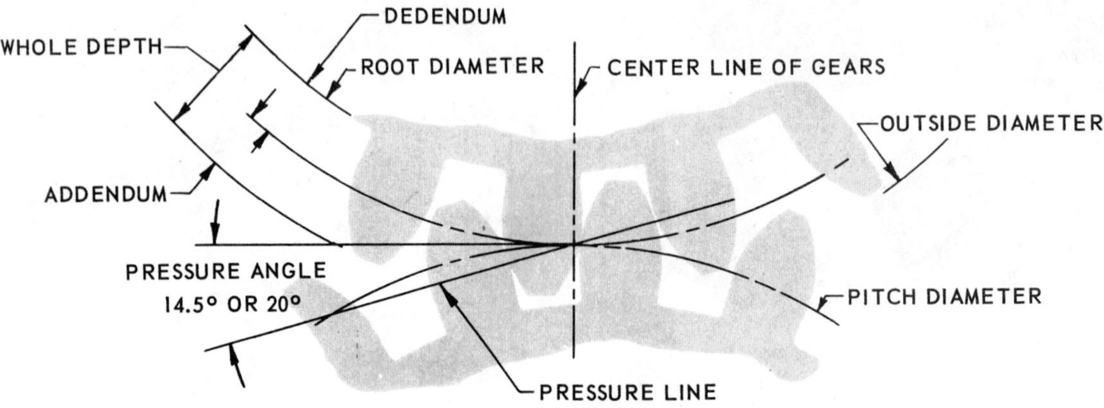

Fig. 36-4 Meshing of gear teeth showing symbols

Term and Symbol	Formula	
	Inch Gears	Metric Gears
Pitch Diameter–PD	PD = N ÷ DP	PD = MDL × N
Number of Teeth–N	N = PD × DP	N = PD ÷ MDL
Module–MDL		MDL = PD ÷ N
Diametral Pitch–DP	DP = N ÷ PD	
Addendum–ADD	14.5° or 20° ADD = 1 ÷ DP 20° stub ADD = 0.8 ÷ DP	14.5° or 20° ADD = MDL 20° stub ADD = 0.8 × MDL
Dedendum–DED	14.5° or 20° DED = 1.157 ÷ DP 20° stub DED = 1 ÷ DP	14.5° or 20° DED = 1.157 × MDL 20° stub DED = MDL
Whole Depth–WD	14.5° or 20° WD = 2.157 ÷ DP 20° stub WD = 1.8 ÷ DP	14.5° or 20° WD = 2.157 × MDL 20° stub WD = 1.8 × MDL
Clearance–CL	14.5° or 20° CL = 0.157 ÷ DP 20° stub CL = 0.2 ÷ DP	14.5° or 20° CL = 0.157 × MDL 20° stub CL = 0.2 × MDL
Outside Diameter–OD	14.5° or 20° OD = PD + 2ADD = (N + 2) ÷ DP	14.5° or 20° OD = PD + 2ADD = PD + 2MDL
	20° stub OD = PD + 2ADD = (N + 1.6) ÷ DP	20° stub OD = PD + 2ADD = PD + 1.6 MDL
Root Diameter–RD	14.5° or 20° RD = PD -2DED = (N -2.314) ÷ DP	14.5° or 20° RD = PD -2DED = PD -2.314 MDL
	20° stub RD = PD -2DED = (N -2) ÷ DP	20° stub RD = PD -2DED = PD -2 MDL
Base Circle–BC	BC = PD Cos PA	BC = PD Cos PA
Pressure Angle–PA	14.5° or 20°	14.5° or 20°
Circular Pitch–CP	CP = 3.1416 PD ÷ N = 3.1416 ÷ DP	CP = 3.1416 PD ÷ N = 3.1416 MDL
Circular Thickness–T	T = 3.1416 PD ÷ 2N = 1.5708 ÷ DP	T = 3.1416 PD ÷ 2N = 1.5708 PD ÷ N = 1.5708 MDL
Chordal Thickness–Tc	Tc = PD sin (90° ÷ N)	Tc = PD sin (90° ÷ N)
Chordal Addendum–ADDc	ADDc = ADD + T^2 ÷ 4PD	ADDc = ADD + T^2 ÷ 4PD
Working Depth–WKG DP	WKG DP = 2ADD	WKG DP = 2ADD

Fig. 36-5 Spur gear symbols and formulas

Gear Terms

The following terms, shown in figures 36-4 and 36-5, are used in spur gear train calculations.

Pitch Diameter (PD). The diameter of an imaginary circle on which the gear tooth is designed.

Number of Teeth (N). The number of teeth on the gear.

Diametral Pitch (DP). The diametral pitch is a ratio of the number of teeth (N) to a unit length of pitch diameter, DP = N/PD.

Outside Diameter (OD). The overall gear diameter.

Root Diameter (RD). The diameter at the bottom of the tooth.

Addendum (ADD). The radial distance from the pitch circle to the top of the tooth.

Dedendum (DED). The radial distance from the pitch circle to the bottom of the tooth.

Whole Depth (WD). The overall height of the tooth.

Clearance. The radial distance between the bottom of one tooth and the top of the mating tooth.

Circular Pitch. The distance measured from the point of one tooth to the corresponding point on the adjacent tooth on the circumference of the pitch diameter.

Circular Thickness. The thickness of a tooth or space measured on the circumference of the pitch diameter.

Chordal Thickness. The thickness of a tooth or space measured along a chord on the circumference of the pitch diameter.

Chordal Addendum. Chordal addendum, also known as corrected addendum, is the perpendicular distance from the chord to outside circumference of the gear.

Chordal Thickness and Corrected Addendum

After the gear teeth have been milled or generated, the width of the tooth space and the thickness of the tooth, measured on the pitch circle, should be equal.

Instead of measuring the curved length of line known as *circular thickness of tooth*, it is more convenient to measure the length of the straight line (*chordal thickness*) which connects the ends of that arc.

The *corrected* or *chordal addendum* is the radial distance extending from the addendum circle to the chord.

A gear tooth vernier caliper may be used to measure accurately the thickness of a gear tooth at the pitch line. To use the gear tooth vernier, which measures only a straight line or chordal distance, it is necessary to set the tongue to the computed chordal addendum, and then measure the chordal thickness.

Working Drawings of Spur Gears

The working drawings of gears which are normally cut from blanks are not complicated. A sectional view is sufficient unless a front view is required to show web or arm details. Since the teeth are cut to shape by cutters, they need not be shown in the front view, drawing A-80.

ANSI recommends the use of phantom lines for the outside and root circles. In the section view, the root and outside circles are shown as solid lines.

The dimensioning for the gear is divided into two groups because the finishing of the gear blank and the cutting of the teeth are separate operations in the shop. The gear blank dimensions are shown on the drawing while the gear tooth information is given in a table.

The only differences in terminology between inch-size and metric-size gear drawings are the terms *diametral pitch* and *module*.

For inch-size gears, the term *diametral pitch* is used instead of the term *module*. The diametral pitch is a ratio of the number of teeth to unit length of pitch diameter.

$$\text{Diametral pitch} = DP = \frac{N}{PD}$$

Module is the term used on metric gears. It is the length of pitch diameter per tooth measure in millimeters.

$$\text{Module} = MDL = \frac{PD}{N}$$

From these definitions it can be seen that the module is equal to the reciprocal of the diametral pitch and thus is not its metric dimensional equivalent. If the diametral pitch is known, the module can be obtained.

$$\text{Module} = 25.4 \div \text{diametral pitch}$$

Gears presently being used in North America are designed in the inch system and have a standard diametral pitch instead of a preferred standard module. Therefore, it is recommended that the diametral pitch be referenced beneath the module when gears designed with standard inch pitches are used. For gears designed with standard modules, the diametral pitch need not be referenced on the gear drawing. The standard modules for metric gears are 0.8, 1, 1.25, 2.25, 3, 4, 6, 7, 8, 9, 10, 12, and 16.

Examples of Spur Gear Calculations

The pitch diameter of a gear can easily be found if the number of teeth and diametral pitch are known. The outside diameter is equal to the pitch diameter plus two addendums. The addendum for a 14.5- or 20-degree spur gear tooth is equal to 1 ÷ DP.

Examples:

1. A 14.5-degree spur gear has a DP of 4 and 34 teeth.

$$\text{Pitch diameter} = \frac{N}{DP} = 34 \div 4 = 8.500 \text{ in.}$$

$$OD = PD + 2ADD = 8.500 + 2(1/4) = 9.000 \text{ in.}$$

2. The outside diameter of a 14.5-degree spur gear is 6.500 in. The gear has 24 teeth.

$$OD = \frac{N+2}{DP} = \frac{24+2}{DP}$$

$$DP = \frac{26}{6.500} = 4$$

$$\text{Addendum} = \frac{1}{DP} = 1/4 = .250 \text{ in.}$$

$$\begin{aligned}\text{Pitch diameter} &= OD - 2ADD \\ &= 6.500 - 2\,(.250) \\ &= 6.000 \text{ in.}\end{aligned}$$

3. A 14.5-degree spur gear has a module of 6.35 and 38 teeth.

$$\begin{aligned}\text{Pitch diameter} &= N \times MDL \\ &= 38 \times 6.35 = \\ &\quad 241.3 \text{ mm}\end{aligned}$$

$$\begin{aligned}OD = PD + 2ADD &= 241.3 + 2\,(6.35) \\ &= 254 \text{ mm}\end{aligned}$$

MODULE FOR METRIC SIZE GEARS	DIAMETRAL PITCH FOR INCH SIZE GEARS	PRESSURE ANGLE 14.5°	20°
6.35	4		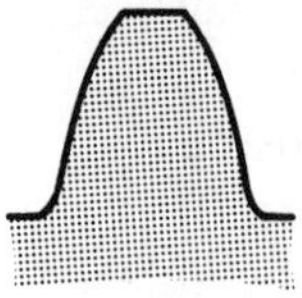
5.08	5		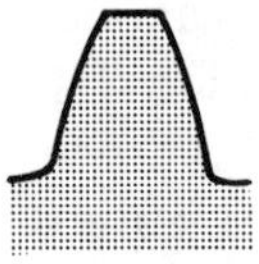
4.23	6		
3.18	8		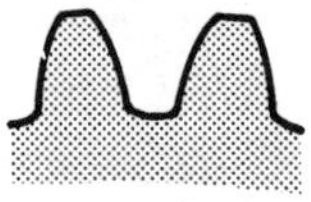
2.54	10		
2.17	12		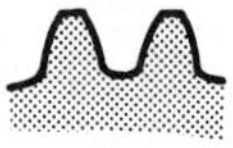
1.59	16		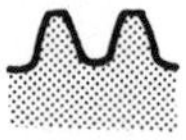
1.27	20		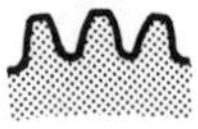
1.06	24		

NOTE: MODULE SIZES SHOWN ARE CONVERTED INCH SIZES

Fig. 36-6 Gear teeth sizes

Inch Gears				
Term and Symbol	Gear 1	Gear 2	Gear 3	Gear 4
Pitch Diameter–PD		5.000	3.000	2.250
Number of Teeth–N	24	40		36
Diametral Pitch–DP	6		10	
Addendum–ADD				
Dedendum–DED				
Whole Depth–WD				
Clearance–CL				
Outside Diameter–OD				
Root Diameter–RD				
Pressure Angle–PA	20°	14.5°	20°	14.5°
Circular Pitch–CP				
Circular Thickness–T				
Chordal Thickness–Tc				
Chordal Addendum–ADDc				

Metric Gears				
Term and Symbol	Gear 5	Gear 6	Gear 7	Gear 8
Pitch Diameter–PD		89.04		
Number of Teeth–N	40			30
Module–MDL	5.08	3.18	6	4
Addendum–ADD				
Dedendum–DED				
Whole Depth–WD				
Clearance–CL				
Outside Diameter–OD			228	
Root Diameter–RD				
Pressure Angle–PA	14.5°	20°	20°	14.5°
Circular Pitch–CP				
Circular Thickness–T				
Chordal Thickness–Tc				
Chordal Addendum–ADDc				

ASSIGNMENT

Complete the missing information for the gears shown.

SPUR GEAR CALCULATIONS

A-79

QUESTIONS

1. What is the hub diameter?
2. What is the maximum thickness of the spokes?
3. What is the average width of the spokes?
4. How many surfaces indicate that allowance must be added to pattern for finishing?
5. Determine distance (J) for the pattern. Assume .10 is allowed on pattern for each surface to be finished.
6. Determine distance (K) for the pattern.
7. What is the outside diameter of the pattern?
8. Determine distance (L) for the pattern.
9. What is the width of the pattern?
10. What is the diametral pitch of the gear?
11. Calculate the center-to-center distance if this gear were to mesh with a pinion having (A) 24 teeth, (B) 36 teeth, (C) 32 teeth.
12. Calculate distances (E) through (H) .
13. Calculate the following: addendum, dedendum, circular pitch, and root diameter.
14. Complete the missing information in the cutting data table.

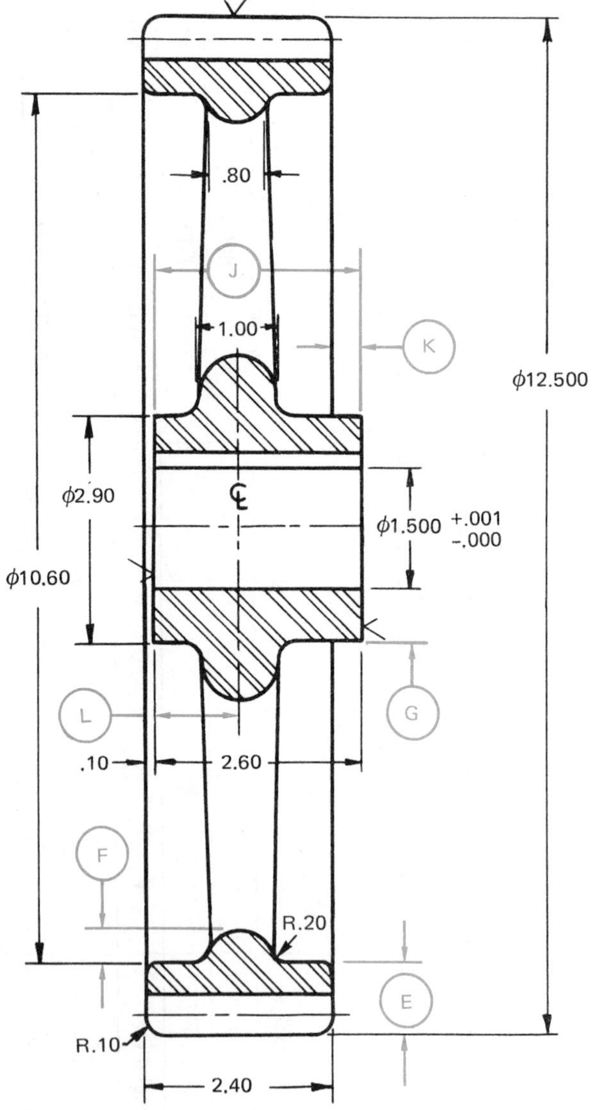

ANSWERS

1 ______	11 A ______	14 PD ______
2 ______	B ______	WD ______
3 ______	C ______	ADDc ______
4 ______	12 E ______	Tc ______
5 ______	F ______	
6 ______	G ______	
7 ______	H ______	
8 ______	13 ADD ______	
9 ______	DED ______	
10 ______	CP ______	
	RD ______	

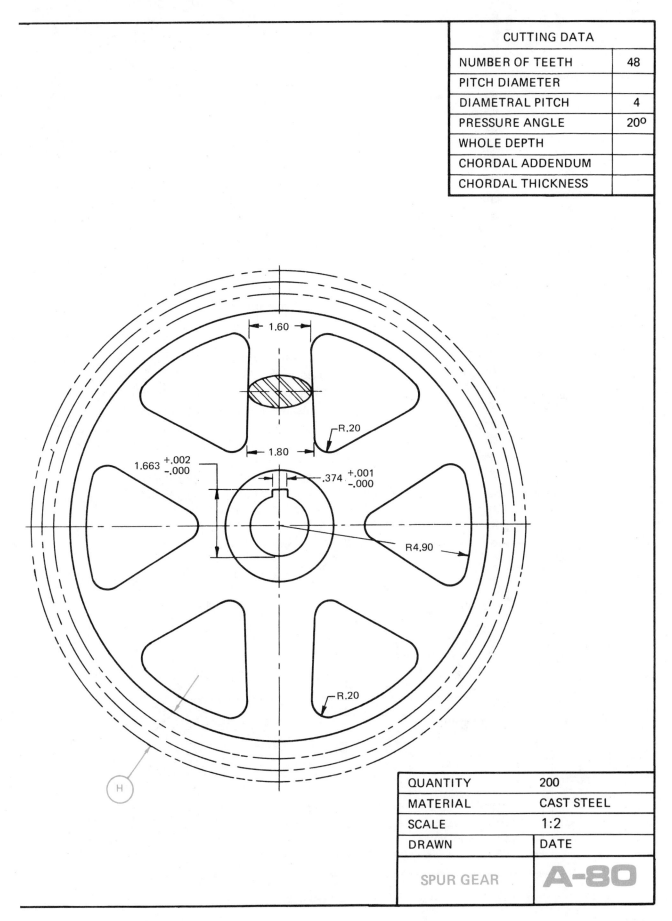

CUTTING DATA	
NUMBER OF TEETH	48
PITCH DIAMETER	
DIAMETRAL PITCH	4
PRESSURE ANGLE	20°
WHOLE DEPTH	
CHORDAL ADDENDUM	
CHORDAL THICKNESS	

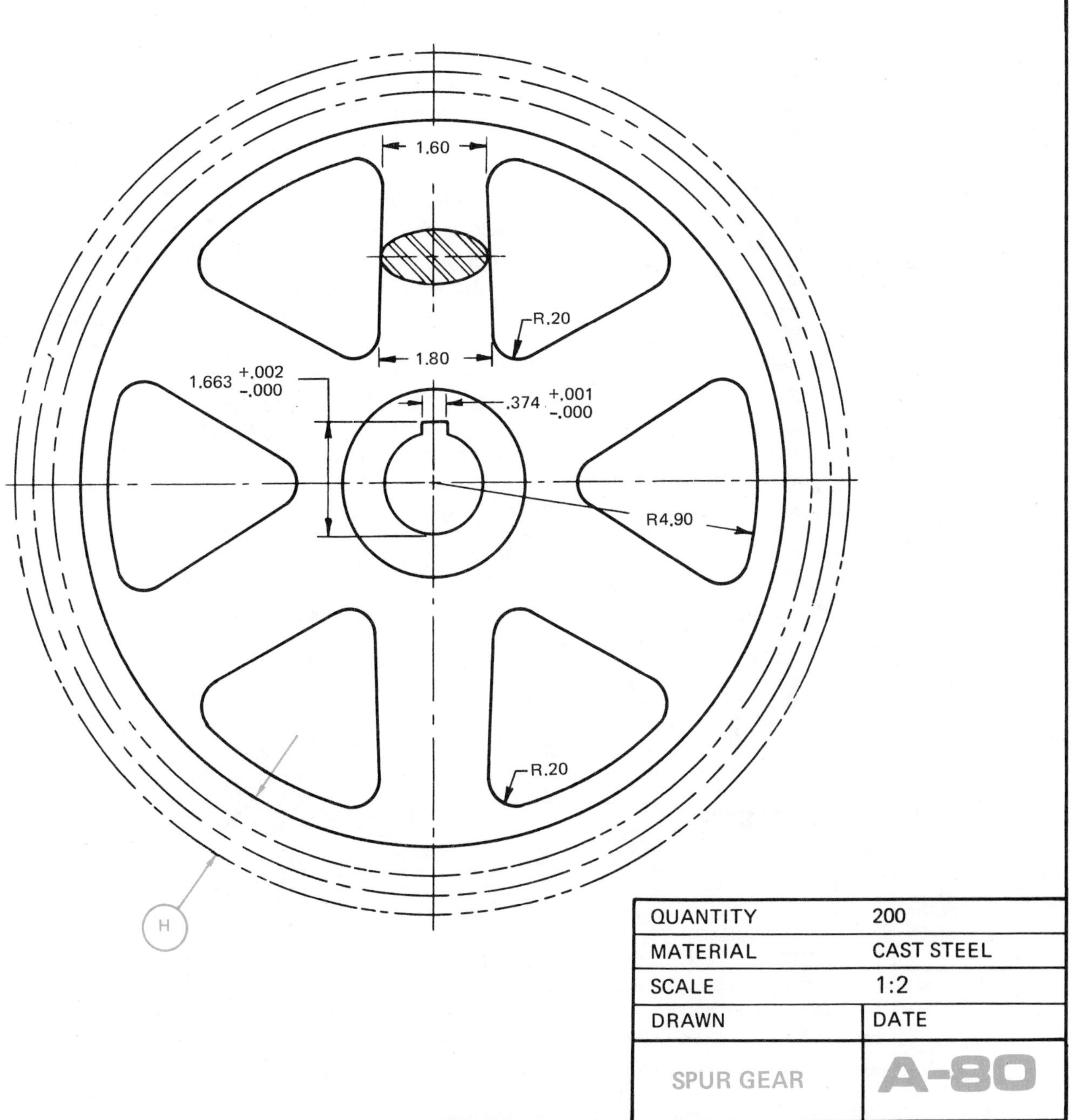

UNIT 37

BEVEL GEARS

Drawings of bevel gears may be more easily interpreted and understood as a result of having a working knowledge of the parts, principles, and formulas underlying spur gears.

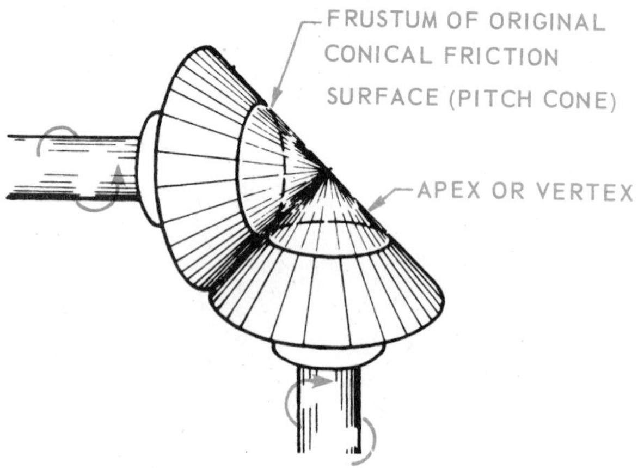

Fig. 37-1 Principle as applied to bevel gears

Spur gears transmit motion by or through shafts that are parallel and in the same plane, while bevel gears transmit motion between shafts which are in the same plane but whose axes would meet if extended. Theoretically, the teeth of a spur gear may be said to be built about the original frictional cylindrical surface known as *pitch circle,* while the teeth of a bevel gear are formed about the frustum of the original conical surface called *pitch cone,* figure 37-1.

One type of bevel gear which is commonly used is the miter gear. The term miter gear refers to a pair of bevel gears of the same size that transmit motion at right angles.

While any two spur gears of the same diametral pitch will mesh, this is not true of bevel gears except for miter gears. On each pair of mating bevel gears, the diameters of the gears determine the angles at which the teeth are cut.

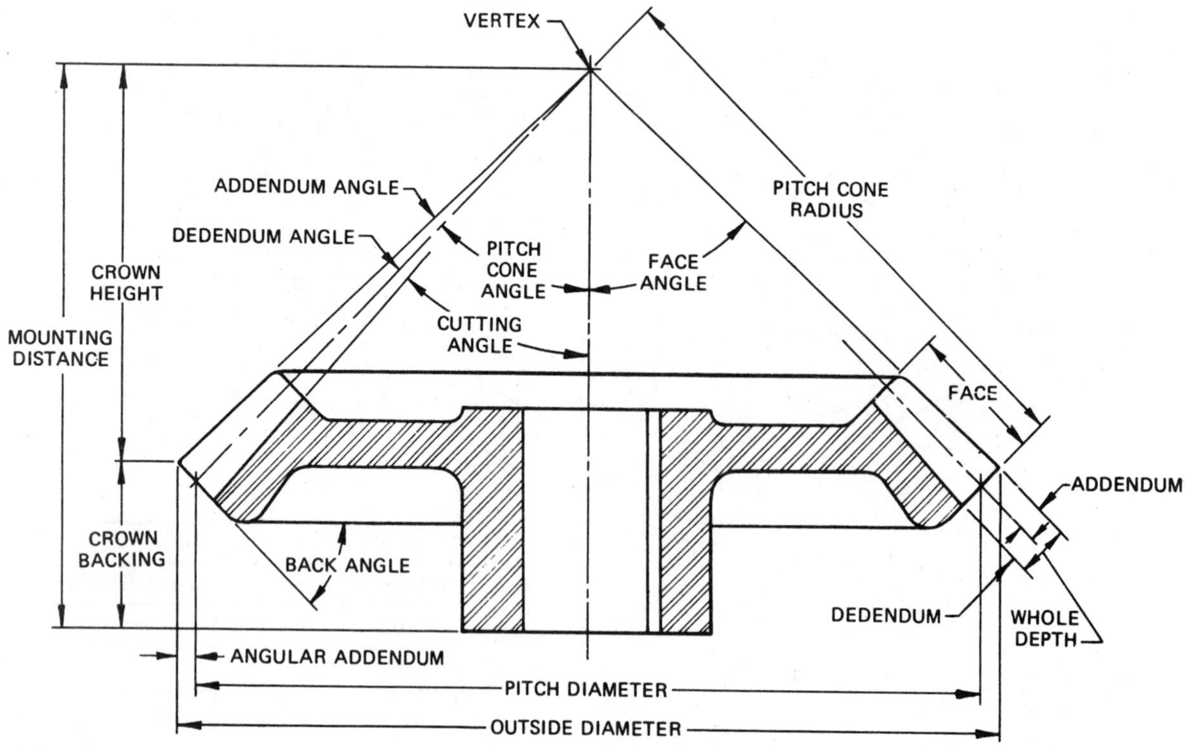

Fig. 37-2 Bevel gear nomenclature

Working drawings of bevel gears, like spur gears, give only the dimensions of the bevel gear blank. Cutting data for the teeth is given in a note or table. A single section view is used unless a second view is required to show such details as spokes. Sometimes both the bevel gear and pinion are drawn together. Dimensions and cutting data depend on the method used in cutting the teeth, but the information in figure 37-2 is commonly used.

Term	Formula
Addendum, dedendum, whole depth, pitch diameter, diametral pitch, number of teeth, circular pitch, chordal thickness, circular thickness	Same as for spur gears. Refer to Figure 36-5
Pitch cone angle (Pitch angle)	$\text{Tan pitch angle} = \dfrac{\text{PD of gear}}{\text{PD of pinion}} = \dfrac{\text{N of gear}}{\text{N of pinion}}$
Pitch cone radius	$\dfrac{\text{PD}}{2 \times \text{sin of pitch angle}}$
Addendum angle	$\text{Tan addendum angle} = \dfrac{\text{Addendum}}{\text{Pitch cone radius}}$
Dedendum angle	$\text{Tan dedendum angle} = \dfrac{\text{Dedendum}}{\text{Pitch cone radius}}$
Face angle	Pitch cone angle plus addendum angle
Cutting angle	Pitch cone angle minus dedendum angle
Back angle	Same as pitch cone angle
Angular addendum	Cosine of pitch cone angle X addendum
Outside diameter	Pitch diameter plus two angular addendums
Crown height	Divide 1/2 the outside diameter by the tangent of the face angle
Face width	1½ to 2½ times the circular pitch
Chordal addendum	$\text{Addendum} + \dfrac{\text{circular thickness}^2 \times \text{cos pitch cone angle}}{\text{4PD}}$

Fig. 37-3 Bevel gear formulas

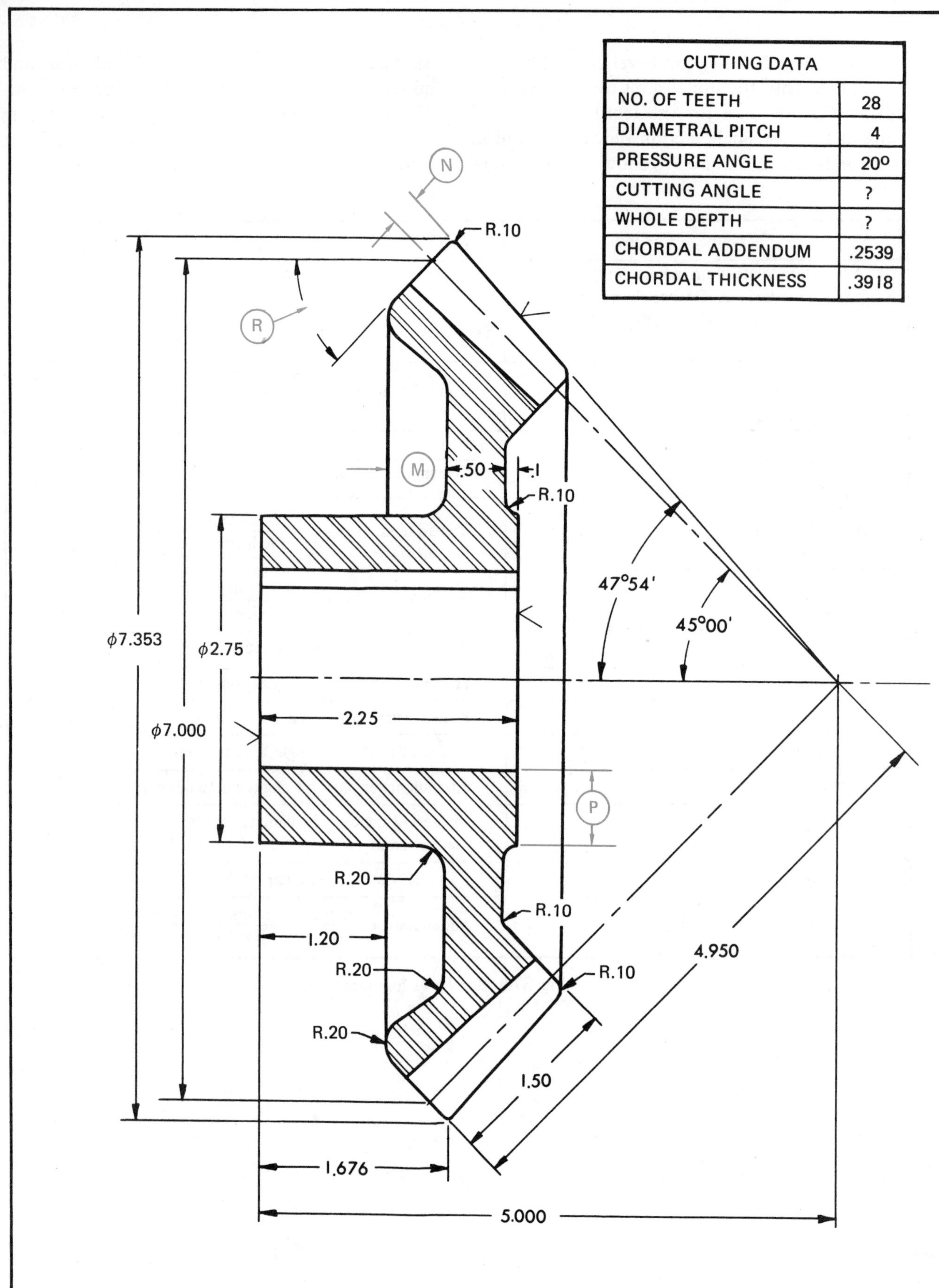

CUTTING DATA
NO. OF TEETH 28
DIAMETRAL PITCH 4
PRESSURE ANGLE 20°
CUTTING ANGLE ?
WHOLE DEPTH ?
CHORDAL ADDENDUM .2539
CHORDAL THICKNESS .3918
N
R
M
P
R.10
.50
.1
R.10
47°54'
45°00'
ϕ7.353
ϕ2.75
ϕ7.000
2.25
R.20
R.10
1.20
R.20
R.10
R.20
4.950
1.50
1.676
5.000

QUESTIONS

1. The drafter neglected to put on angle dimension (R) What should it be?
2. How many finished surfaces are indicated?
3. List those dimensions shown on the drawing which are not used by the patternmaker. Assume that the hole will not be cored.
4. What is the pitch cone angle?
5. What is the pitch diameter?
6. Indicate those dimensions on the drawing which the machinist would use to machine the blank before the teeth are cut.
7. What is the depth of teeth at the large end?
8. What is the pitch cone radius?
9. What is the face angle?
10. What is the addendum angle?
11. What is the cutting angle?
12. What is the mounting distance?
13. What is the crown height?
14. What is the angular addendum?
15. What is the face width?
16. Determine dimensions (M), (N), (P) .

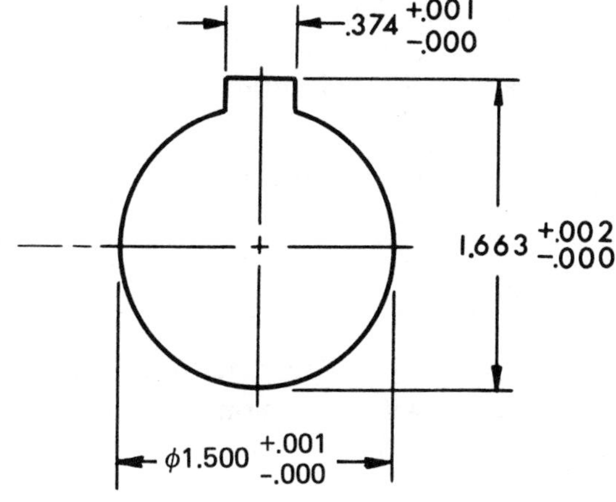

ANSWERS

1 ______
2 ______
3 ______
______
______
______
______
______
4 ______
5 ______
6 ______
______
______
______
______
7 ______
8 ______
9 ______
10 ______
11 ______
12 ______
13 ______
14 ______
15 ______
16M ______
N ______
P ______

QUANTITY	50
MATERIAL	CAST STEEL
SCALE	1:1
DRAWN	DATE
MITER GEAR	A-81

UNIT 38

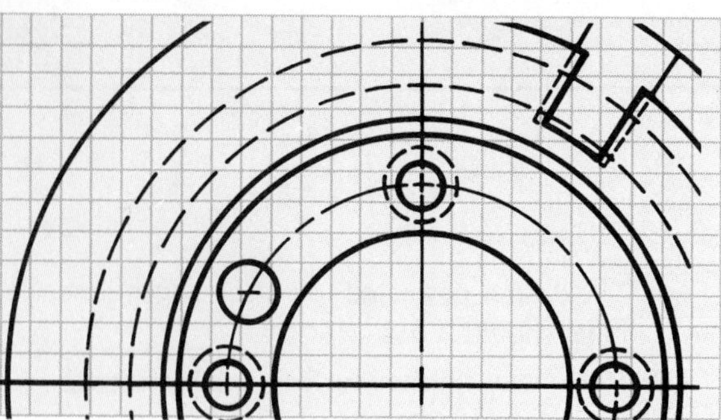

GEAR TRAINS

Center Distance

The center distance between the two shaft centers is determined by adding the pitch diameter of the two gears together and dividing the sum by 2.

Examples

1. An 8DP, 24-tooth pinion mates with a 90-tooth gear. Find the center distance.

   Pitch diameter of pinion = N ÷ DP = 24 ÷ 8 = 3.000 in.

   Pitch diameter of gear = N ÷ DP = 96 ÷ 8 = 12.000 in.

   Sum of the two pitch diameters = 3.000 + 12.000 = 15.000 in.

   Center distance = 1/2 sum of the two pitch diameters = $\frac{15.000}{2}$ = 7.500 in.

2. A 2.54 module, 28-tooth pinion mates with a 84-tooth gear. Find the center distance.

   Pitch diameter (PD)
   = number of teeth × module
   = 28 × 2.54 = ϕ71.12 (pinion)
   = 84 × 2.54 = ϕ213.36 (gear)

   Sum of the two pitch diameters
   = 71.12 + 213.36 = 284.48 mm

   Center distance = 1/2 sum of the two pitch diameters = 284.48 ÷ 2 = 142.24 mm

Ratio. The ratio of gears is a relationship between any of the following:

- The r/min (revolutions per minute) of the gears
- The number of teeth on the gears
- The pitch diameter of the gears

The ratio is obtained by dividing the larger value of any of the three by the corresponding smaller value.

3. A gear rotates at 90 r/min and the pinion at 360 r/min

   $$\text{Ratio} = \frac{360}{90} = 4 \text{ or ratio} = 4{:}1$$

4. A gear has 72 teeth; the pinion, 18 teeth.

   $$\text{Ratio} = \frac{72}{18} = 4 \text{ or ratio} = 4{:}1$$

5. A gear with a pitch diameter of 8.500 in. meshes with a pinion having a pitch diameter of 2.125 in.

   $$\text{Ratio} = \frac{\text{PD of gear}}{\text{PD of pinion}} = \frac{8.500}{2.125} = 4$$

   or ratio = 4:1

Figure 38-1 illustrates how this type of information would be shown on an engineering sketch.

Motor Drive

Drawing A-82 shows a motor drive similar to the type used to operate a load-ratio control switch on a power transformer.

The load-ratio control switch is operated by a small motor with a speed of 1080 r/min. The shaft speed at the switch is reduced to 9 r/min by a series of spur and miter gears.

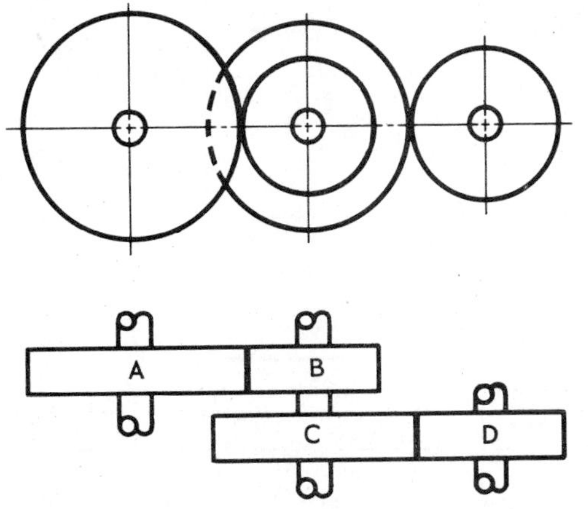

GEAR	PD	N	DP	R/MIN	CENTER DISTANCE
A	6.000	24	4	300	4.500
B	3.000	12	4	600	
C	6.000	48	8	600	4.000
D	2.000	16	8	1800	

Fig. 38-1 Gear train data

When the operator pushes a button, the motor is activated until the circuit breaker pointer rotates 90 degrees and one of the arms depresses the roller and breaks the circuit. During this time the load-ratio control switch shaft will rotate 360 degrees, moving the contactor in the load-ratio control switch one position. This will be shown on the dial by the position indicator.

To simplify the assembly, only the pitch diameters of the gears are shown and much of the hardware has been omitted.

When referring to the direction in which the gears rotate, the terms clockwise (CWISE) or (CW) and anticlockwise (AWISE) or counter clockwise (CCW) are used.

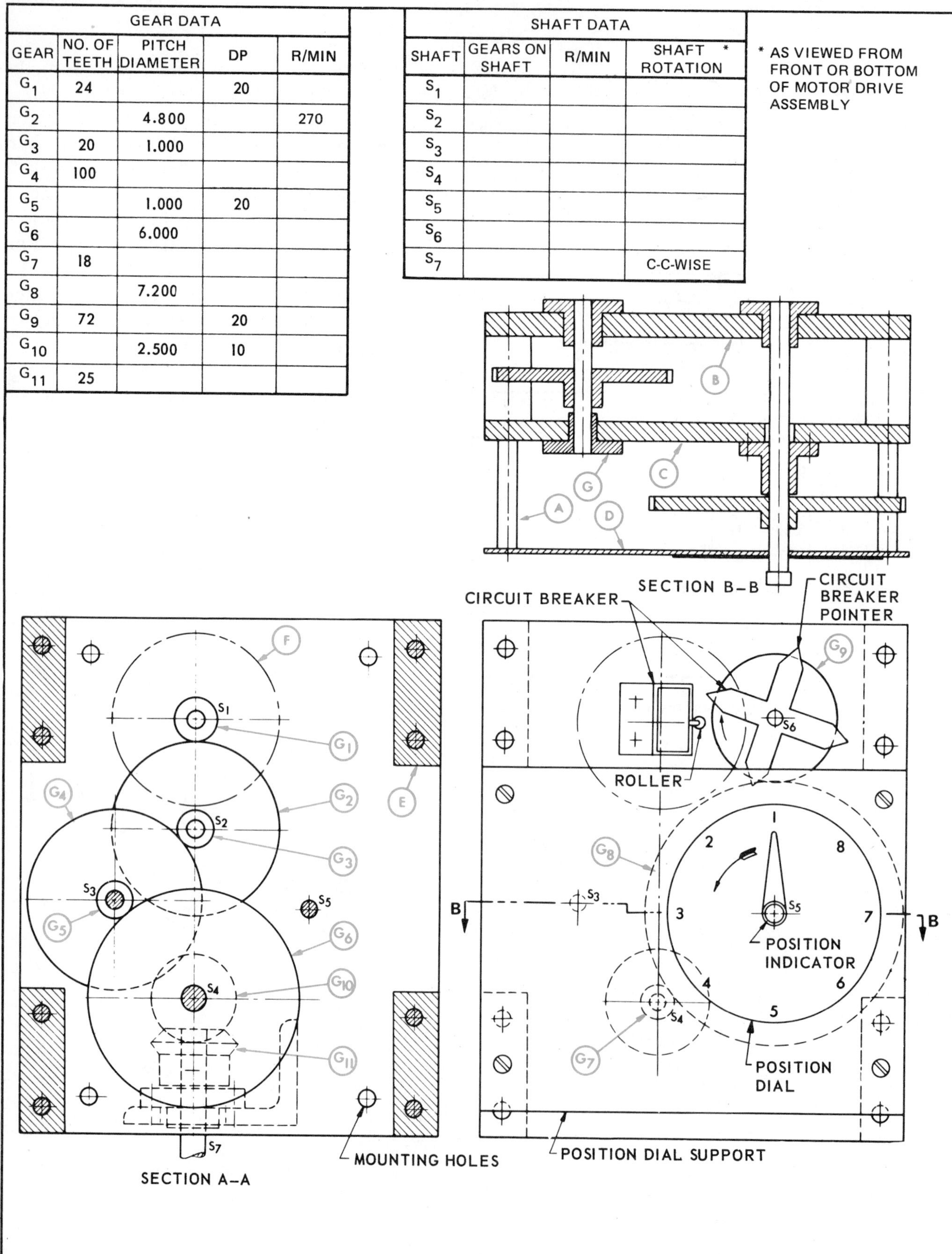

GEAR DATA				
GEAR	NO. OF TEETH	PITCH DIAMETER	DP	R/MIN
G_1	24		20	
G_2		4.800		270
G_3	20	1.000		
G_4	100			
G_5		1.000	20	
G_6		6.000		
G_7	18			
G_8		7.200		
G_9	72		20	
G_{10}		2.500	10	
G_{11}	25			

SHAFT DATA			
SHAFT	GEARS ON SHAFT	R/MIN	SHAFT ROTATION *
S_1			
S_2			
S_3			
S_4			
S_5			
S_6			
S_7			C-C-WISE

* AS VIEWED FROM FRONT OR BOTTOM OF MOTOR DRIVE ASSEMBLY

ASSIGNMENT

Complete the information shown in the gear and shaft tables.

QUESTIONS

1. What are the names of parts (A) to (G) ?
2. How many spur gears are shown?
3. How many miter gears are shown?
4. How many gear shafts are there?
5. What is the ratio between the following gears? (A) G_1 and G_2, (B) G_3 and G_4, (C) G_5 and G_6, (D) G_7 and G_8, (E) G_8 and G_9, (F) G_{10} and G_{11}.
6. What is the center-to-center distance between the following shafts? (A) S_1. and S_2, (B) S_2 and S_3, (C) S_3 and S_4, (D) S_4 and S_5, (E) S_5 and S_6.
7. How many seconds does it take to turn the load ratio control switch one position?
8. How many seconds does it take the position indicator to move continuously from position 4 to position 7?
9. What is the r/min ratio between the motor and the switch?
10. If the switch shaft S_7 rotates 1800 degrees, how many degrees does the position indicator rotate?

ANSWERS

1. (A) ______
(B) ______
(C) ______
(D) ______
(E) ______
(F) ______
(G) ______
2. ______
3. ______
4. ______
5. A. ______
B. ______
C. ______
D. ______
E. ______
F. ______
6. A. ______
B. ______
C. ______
D. ______
E. ______
7. ______
8. ______
9. ______
10. ______

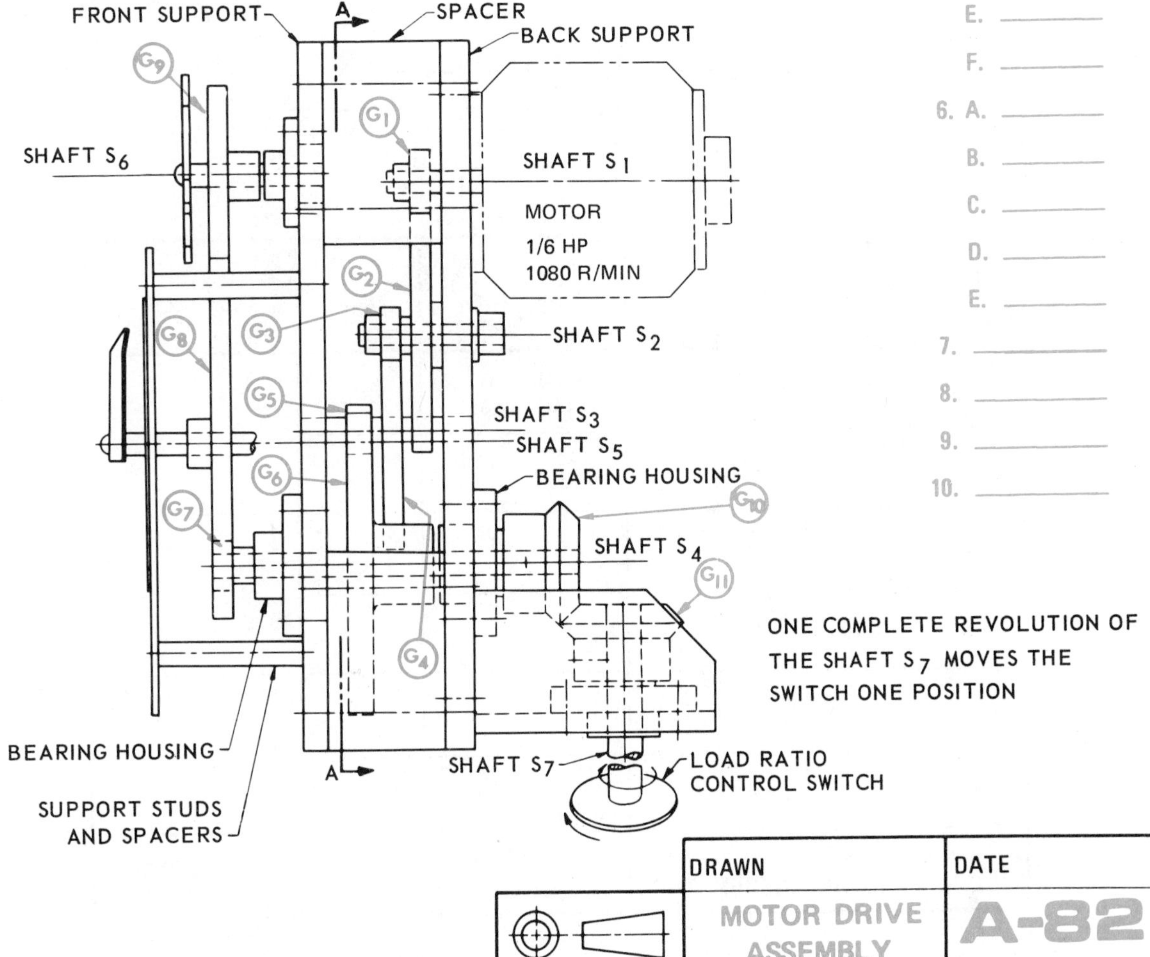

DRAWN	DATE
MOTOR DRIVE ASSEMBLY	A-82

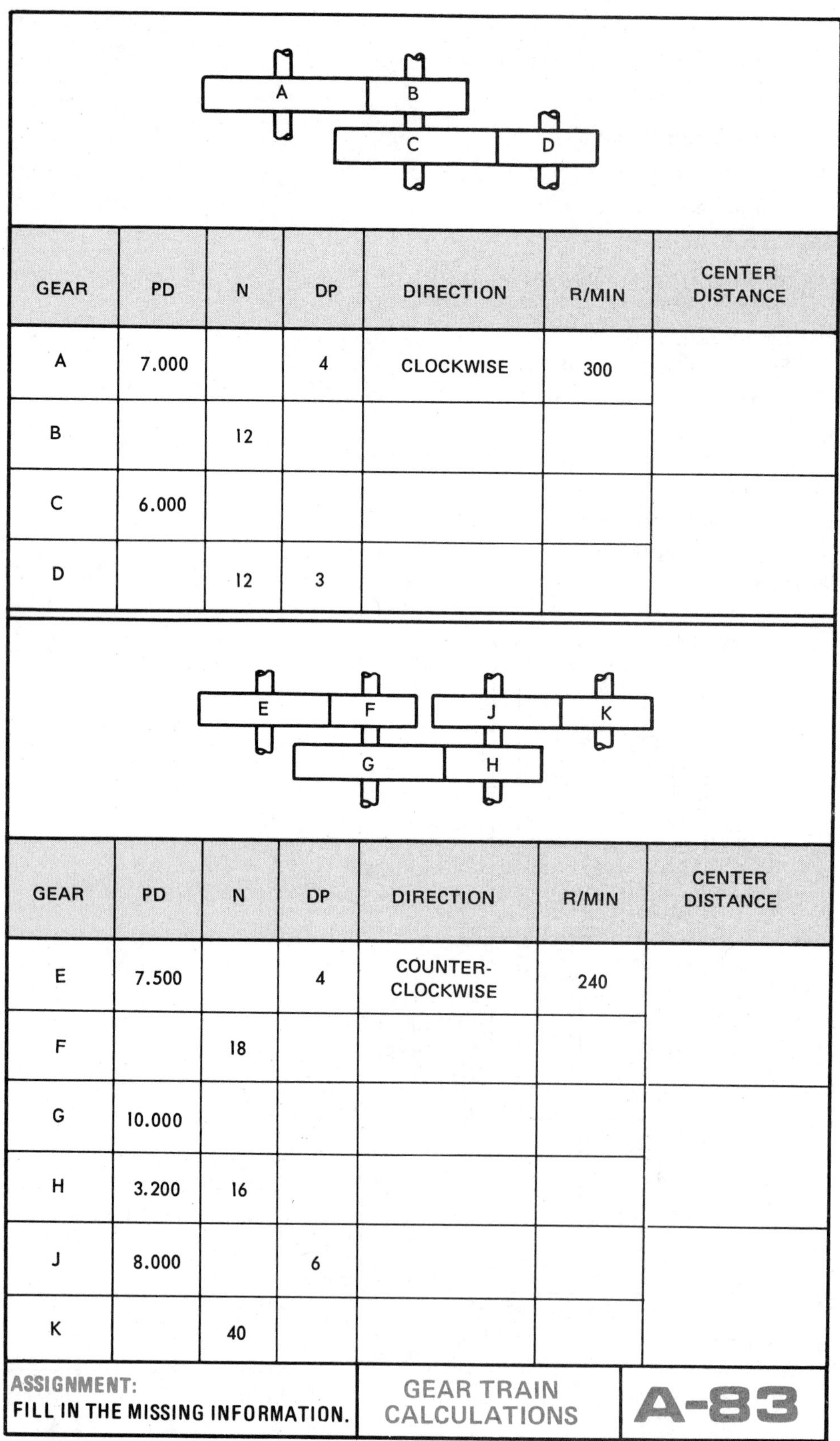

GEAR	PD	N	DP	DIRECTION	R/MIN	CENTER DISTANCE
A	7.000		4	CLOCKWISE	300	
B		12				
C	6.000					
D		12	3			

GEAR	PD	N	DP	DIRECTION	R/MIN	CENTER DISTANCE
E	7.500		4	COUNTER-CLOCKWISE	240	
F		18				
G	10.000					
H	3.200	16				
J	8.000		6			
K		40				

ASSIGNMENT:
FILL IN THE MISSING INFORMATION.

GEAR TRAIN CALCULATIONS

A-83

UNIT 39

CAMS

The cam is invaluable in the design of automatic machinery. Cams make it possible to impart any desired motion to another mechanism.

A *cam* is a rotating, oscillating, or reciprocating machine element which has a surface or groove formed to impart special or irregular motion to a second part called a *follower.* The follower rides against the curved surface of the cam. The distance that the follower rises and falls in a definite period of time is determined by the shape of the cam profile.

Types of Cams

The type and shape of cam used is dictated by the required relationship of the parts and the motions of both, figure 39-1. The cams which are generally used are either radial or cylindrical. The follower of a radial or face cam moves in a plane perpendicular to the axis of the cam, while in the cylindrical type of cam the movement of the follower is parallel to the cam axis.

A simple OD (outside diameter) or plate cam is shown in figure 39-2. The hole in the plate is bored off-center, causing the follower to

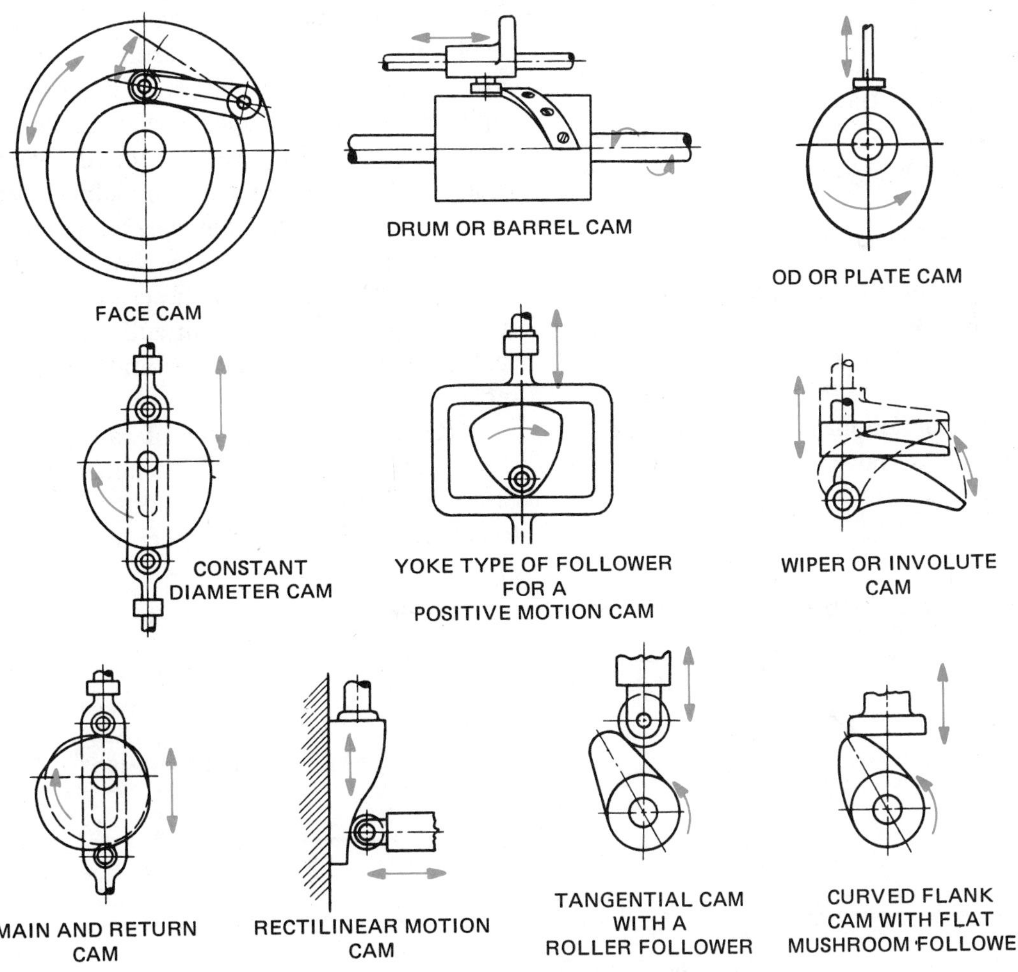

Fig. 39-1 Common types of cams

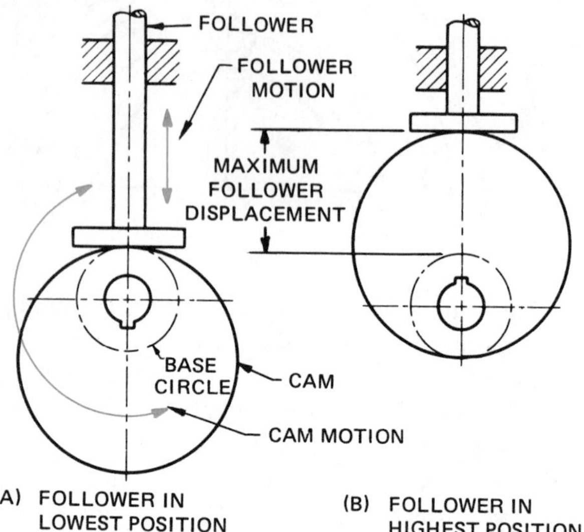

Fig. 39-2 Eccentric plate cam

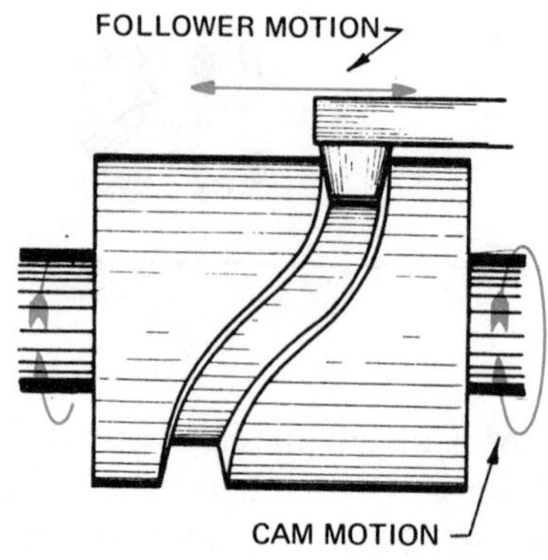

Fig. 39-3 Drum or barrel-type cam

move up and down as it revolves. The follower can be any type that will roll or slide on the surface of the cam. The follower used with this cam is called a flat face follower.

The cam shown in figure 39-3 is a drum or barrel-type cam that transmits motion transversely to a lever connected to a conical follower which rides in the groove as the cam revolves.

Cam Displacement Diagrams

In preparing cam drawings, a cam displacement diagram is drawn first to plot the motion of the follower. The curve on the drawing represents the path of the follower, not the face of the cam. The diagram may be any convenient length, but often it is drawn equal to the circumference of the base circle of the cam and the height is drawn equal to the follower displacement. The lines drawn on the motion diagram are shown as radial lines on the cam drawing, the sizes are transferred from the motion diagram to the cam drawing.

Figure 39-4 shows a cam displacement diagram having a modified uniform type of motion plus two dwell periods. Most cam displacement diagrams have 360-degree cam displacement angles. For drum or cylindrical grooved cams, the displacement diagram is often replaced by the developed surface of the cam.

The cylindrical feeder cam (drawing A-85 is a drum or barrel cam. In addition to the working views of the cam, a development of the contour of the grooves is shown. This development aids the machinist in scribing and laying out the contour of the cam action lobes on the surface of the cam blank preparatory to machining grooves.

Regardless of the type of cam or follower, the purpose of all cams is to impart motion to other mechanisms in various directions in order to actuate machines to do specific jobs.

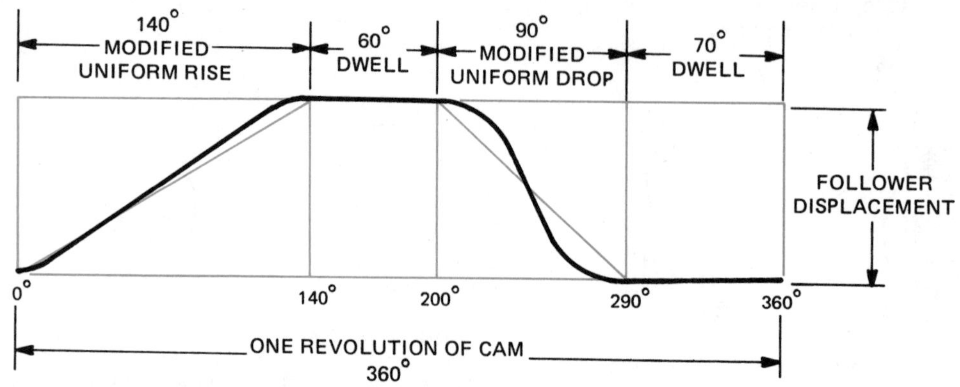

Fig. 39-4 Cam displacement diagram

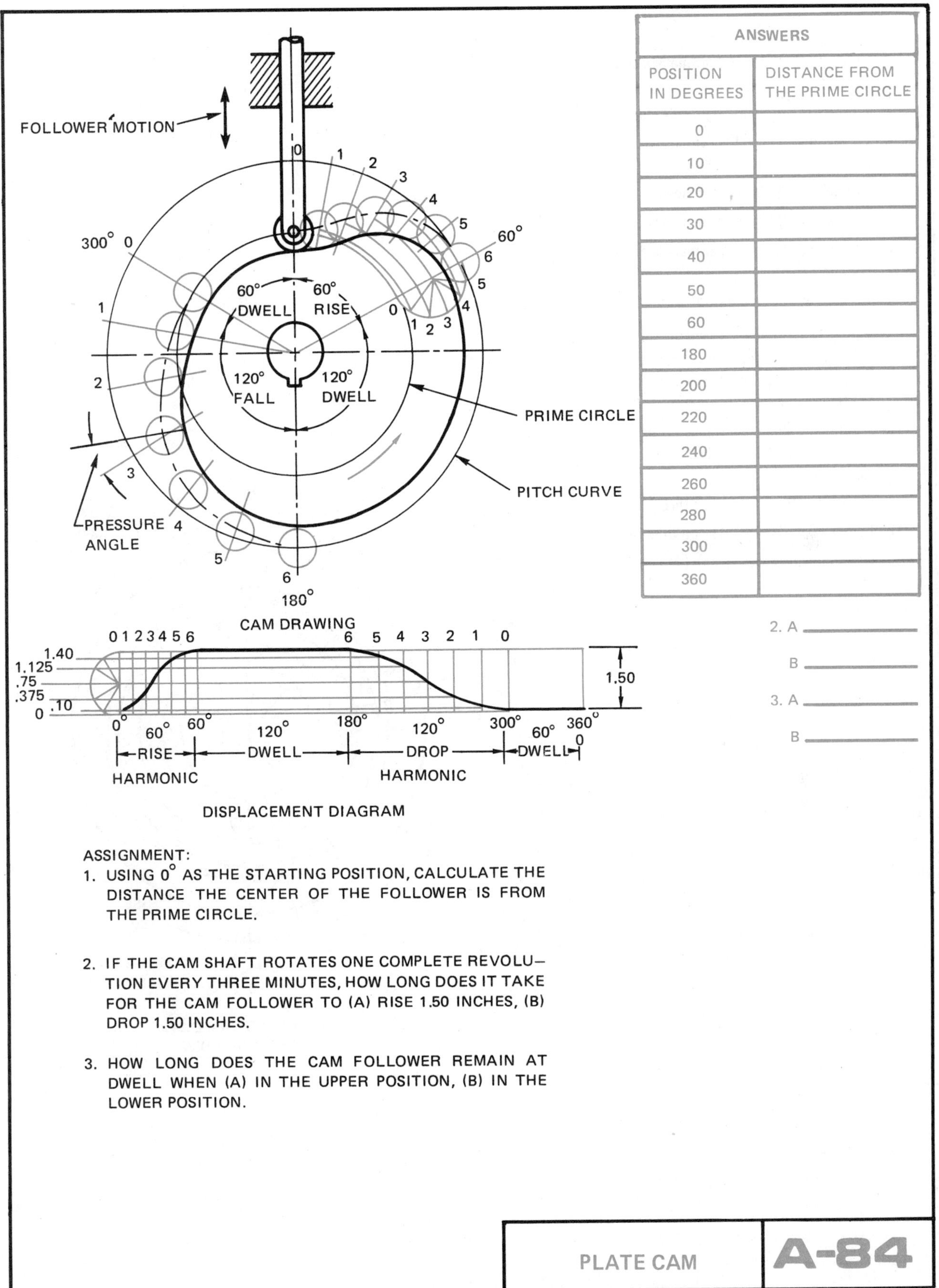

ANSWERS	
POSITION IN DEGREES	DISTANCE FROM THE PRIME CIRCLE
0	
10	
20	
30	
40	
50	
60	
180	
200	
220	
240	
260	
280	
300	
360	

2. A ______

B ______

3. A ______

B ______

ASSIGNMENT:

1. USING 0° AS THE STARTING POSITION, CALCULATE THE DISTANCE THE CENTER OF THE FOLLOWER IS FROM THE PRIME CIRCLE.

2. IF THE CAM SHAFT ROTATES ONE COMPLETE REVOLUTION EVERY THREE MINUTES, HOW LONG DOES IT TAKE FOR THE CAM FOLLOWER TO (A) RISE 1.50 INCHES, (B) DROP 1.50 INCHES.

3. HOW LONG DOES THE CAM FOLLOWER REMAIN AT DWELL WHEN (A) IN THE UPPER POSITION, (B) IN THE LOWER POSITION.

PLATE CAM	A-84

QUESTIONS

1. Through what thickness of metal is hole (2) drilled?
2. Locate line (4) in the right-side view.
3. Locate line (11) in the cam displacement diagram.
4. Locate line (6) in another view.
5. Locate line (9) in the right-side view.
6. What is the maximum permissible diameter of hole (2)?
7. Locate line (8) in the front view.
8. Locate line (43) in the right-side view.
9. What is the total follower displacement of the cam follower for (A) the finishing cut, (B) the roughing cut?
10. Assuming a .06 in. allowance for machining, what would be the outside diameter of the cam before finishing?
11. What is the total number of through holes?
12. Locate lines (30) to (41) on other view.
13. Determine distances (A) to (M).

ANSWERS

1.	______	12. (30)	______	13. (A)	______
2.	______	(31)	______	(B)	______
3.	______	(32)	______	(C)	______
4.	______	(33)	______	(D)	______
5.	______	(34)	______	(E)	______
6.	______	(35)	______	(F)	______
7.	______	(36)	______	(G)	______
8.	______	(37)	______	(H)	______
9. A.	______	(38)	______	(J)	______
B.	______	(39)	______	(K)	______
10.	______	(40)	______	(L)	______
11.	______	(41)	______	(M)	______

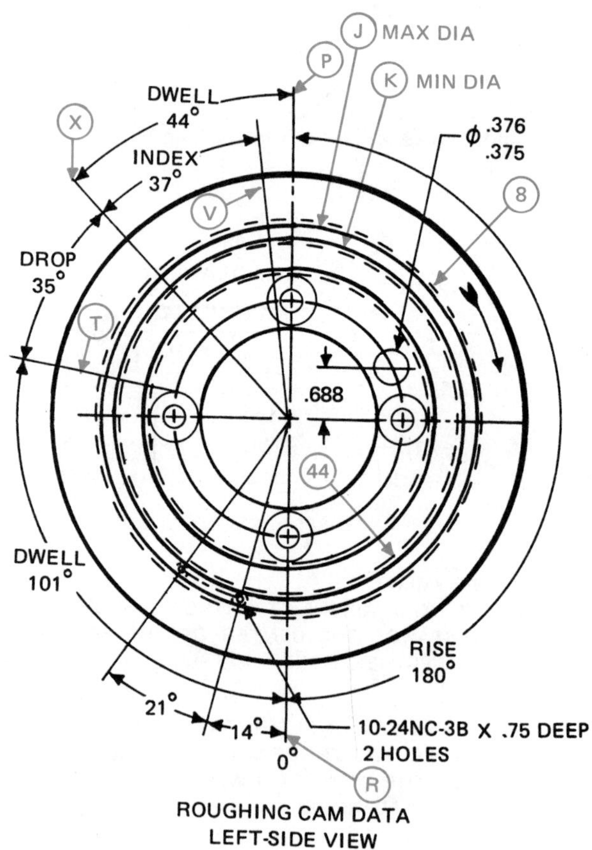

ROUGHING CAM DATA
LEFT-SIDE VIEW

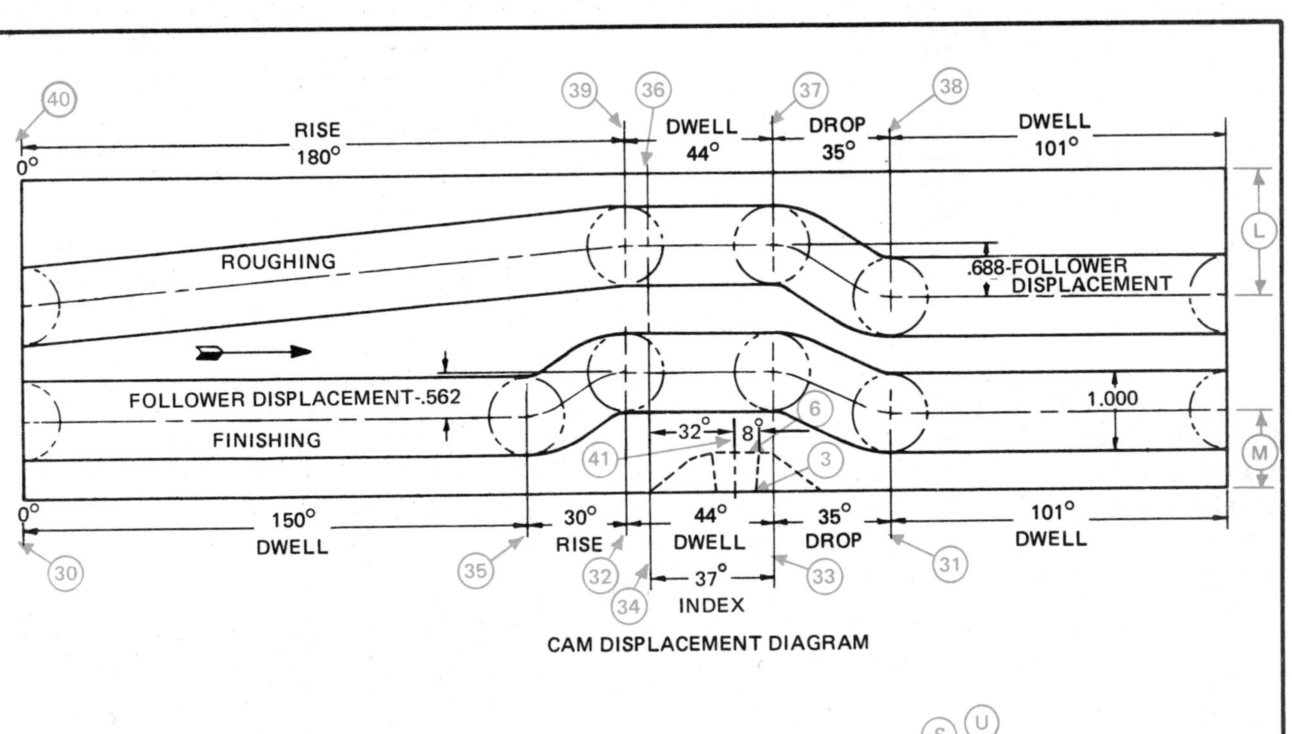

CAM DISPLACEMENT DIAGRAM

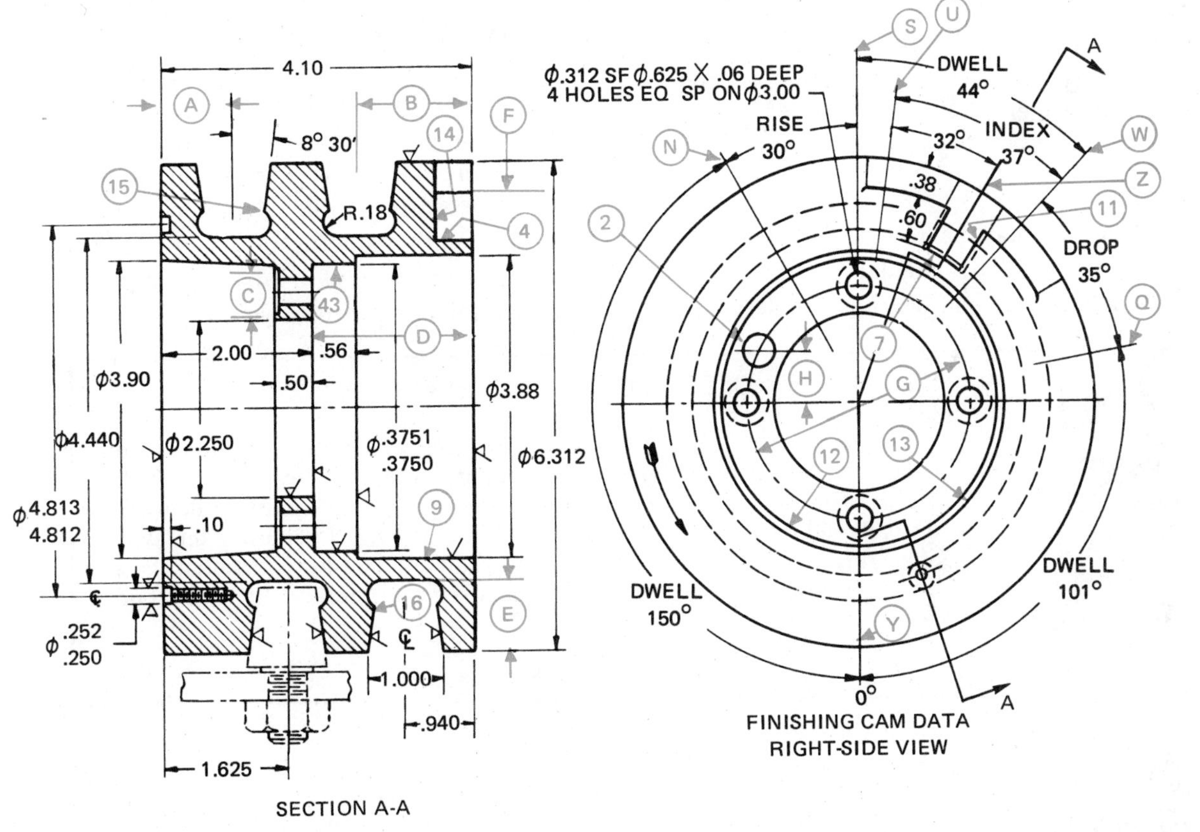

SECTION A-A

FINISHING CAM DATA
RIGHT-SIDE VIEW

NOTE: UNLESS OTHERWISE SHOWN:
- TOLERANCE ON TWO PLACE DECIMAL DIMENSIONS ± .02
- TOLERANCE ON THREE PLACE DECIMAL DIMENSIONS ± .005
- TOLERANCE ON ANGLES ± 30′
- SURFACES ▽ TO BE 63▽

MATERIAL	
SCALE	NOT TO SCALE
DRAWN	DATE
CYLINDRICAL FEEDER CAM	A-85

UNIT 40

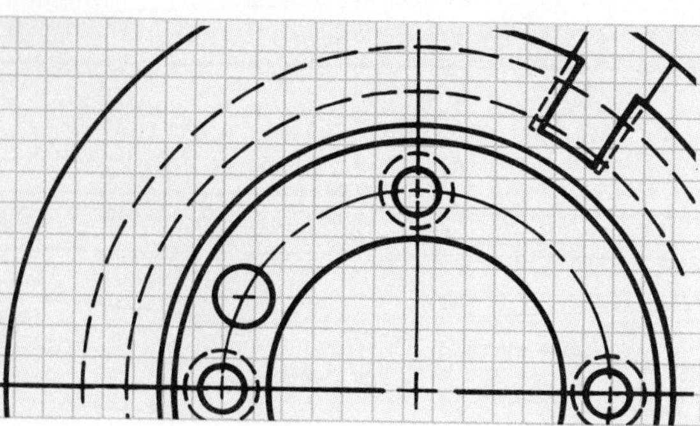

ANTIFRICTION BEARINGS

Antifriction bearings, also known as *roller-element* bearings, use a type of rolling element between the loaded members. Relative motion is accommodated by rotation of the elements. Roller-element bearings are usually housed in bearing races conforming to the element shapes. In addition, a cage or separator is often used to locate the elements within the bearings. These bearings are usually categorized by the form of the rolling element and in some instances by the load type they carry, figures 40-1 and 40-2. Roller-element bearings are generally classified as either ball or roller.

Ball Bearings

Ball bearings may be roughly divided into three categories: radial, angular contact, and thrust. Radial-contact ball bearings are designed for applications in which the load is primarily radial with only low-magnitude thrust loads. Angular-contact bearings are used where loads are combined radial and high thrust, and where precise shaft location is required. Thrust bearings handle loads which are primarily thrust.

Roller Bearings

Roller bearings have higher load capacities than ball bearings for a given envelope size. They are widely used in moderate-speed, heavy-duty applications. The four principal types of roller bearings are: cylindrical, needle, tapered, and spherical. Cylindrical roller bearings utilize cylinders with approximate length-diameter ratios ranging from 1:1 to 1:3 as rolling elements. Needle roller bearings utilize cylinders or needles of greater length-diameter ratios. Tapered and spherical roller bearings are capable of supporting combined radial and thrust loads.

SINGLE ROW, DEEP GROOVE BALL BEARINGS

The *Single Row, Deep Groove Ball Bearing* will sustain, in addition to radial load, a substantial thrust load in either direction . . . even at very high speeds. This advantage results from the intimate contact existing between the balls and the deep, continuous groove in each ring. When using this type of bearing, careful alignment between the shaft and housing is essential. This bearing is also available with seals, which serve to exclude dirt and retain lubricant.

ANGULAR CONTACT BALL BEARINGS

The *Angular Contact Ball Bearing* supports a heavy thrust load in one direction . . . sometimes combined with a moderate radial load. A steep contact angle, assuring the highest thrust capacity and axial rigidity, is obtained by a high thrust supporting shoulder on the inner ring and a similar high shoulder on the opposite side of the outer ring. These bearings can be mounted singly or, when the sides are flush ground, in tandem for constant thrust in one direction; mounted in pairs, also when sides are flush ground, for a combined load . . . either face-to-face or back-to-back.

Fig. 40-1A Roller-element bearings

CYLINDRICAL ROLLER BEARINGS

The *Cylindrical Roller Bearing* has high radial capacity and provides accurate guiding of the rollers, resulting in a close approach to true rolling. Consequent low friction permits operation at high speed. Those types which have flanges on one ring only allow a limited free axial movement of the shaft in relation to the housing. They are easy to dismount even when both rings are mounted with a tight fit. The double row type assures maximum radial rigidity and is particularly suitable for machine tool spindles.

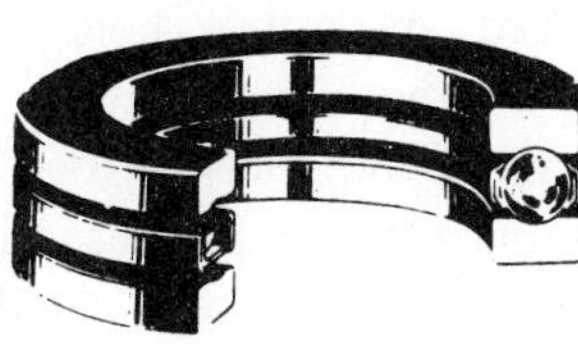

BALL THRUST BEARINGS

The *Ball Thrust Bearing* is designed for thrust load in one direction only. The load line through the balls is parallel to the axis of the shaft . . . resulting in high thrust capacity and minimum axial deflection. Flat seats are preferred . . . particularly where the load is heavy . . . or where close axial positioning of the shaft is essential; as for example, in machine tool spindles.

SPHERICAL ROLLER THRUST BEARINGS

The *Spherical Roller Thrust Bearing* is designed to carry heavy thrust loads, or combined loads which are predominantly thrust. This bearing has a single row of rollers which roll on a spherical outer race with full self-alignment. The cage, centered by an inner ring sleeve, is constructed so that lubricant is pumped directly against the inner ring's unusually high guide flange. This ensures good lubrication between the roller ends and the guide flange. The spherical roller thrust bearing operates best with relatively heavy oil lubrication.

Fig. 40-1B Roller-element bearings, continued

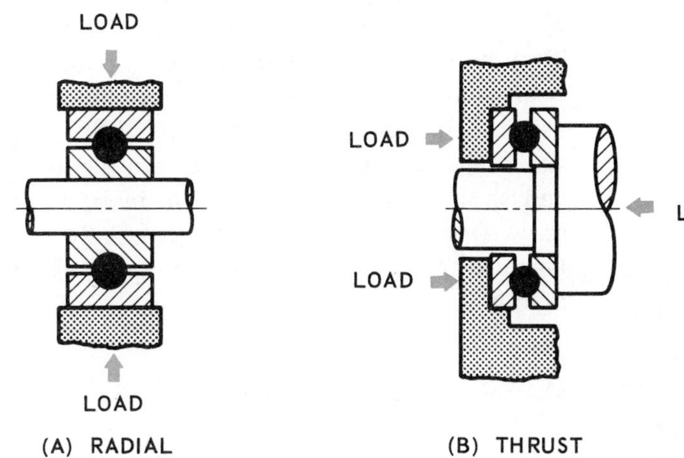

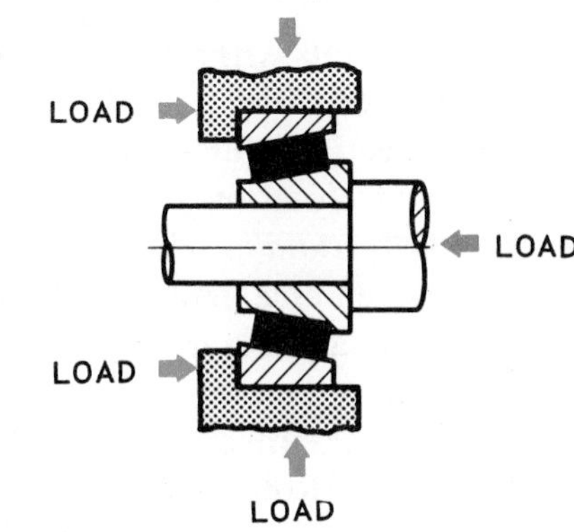

Fig. 40-2 Types of bearing loads

The rolling elements of tapered roller bearings are truncated cones. Spherical roller bearings are available with both barrel and hourglass roller shapes. The primary advantage of spherical roller bearings is their self-aligning capability. Plain bearings are covered in Unit 23.

RETAINING RINGS

Retaining rings, or snap rings, are designed to provide a removable shoulder to locate, retain, or lock components accurately on shafts and in bores and housings, figures 40-3 and 40-4. They are easily installed and removed. Since they are usually made of spring steel, retaining rings have a high shear strength and impact capacity. In addition to fastening and positioning, a number of rings are designed for taking up end play caused by accumulated tolerances or wear in the parts being retained.

O-RING SEALS

O-rings are used as an axial mechanical seal (a seal which forms a running seal between a moving shaft and a housing) or a static seal (no moving parts). The advantage of using an O-ring as a gasket-type seal, figure 40-5, over conventional gaskets is that the nuts need not be tightened uniformly and sealing compounds are not required. A rectangular groove is the most common type of groove used for O-rings.

CLUTCHES

Clutches are used to start and stop machines or rotating elements without starting or stopping the prime mover. They are also used for automatic disconnection, quick starts and stops, and to permit shaft rotation in one direction only such as the *overrunning* clutch shown in figure 40-6. A full complement of sprags between concentric inner and outer races transmits power from one race to the other by wedging action of the sprags when either race is rotated in the driving direction. Rotation in the opposite direction frees the sprags and the clutch is disengaged or overruns, figure 40-6. This type of clutch is used in the power drive, drawing A-86.

BELT DRIVES

A *belt drive* consists of an endless flexible belt connecting two wheels or pulleys. Belt drives depend on friction between belt and pulley surfaces for transmission of power.

In a V-belt drive, the belt has a trapezoidal cross section, and runs in V-shaped grooves on the pulleys. These belts are made of cords or cables, impregnated and covered with rubber or other organic compound. The covering is formed to produce the required cross section. V-belts are usually manufactured as endless belts, although open-end and link types are available.

In the case of V-belts, the friction for the transmission of the driving force is increased by the wedging of the belt into the grooves on the pulley.

V-Belt Sizes

To facilitate interchangeability and to insure uniformity. V-belt manufacturers have developed industrial standards for the various types of V-belts, figure 40-7. Industrial V-belts are made in two types: heavy duty (conventional and narrow) and light duty. Conventional belts are available in A, B, C, D, and E sections. Narrow belts are made in 3V, 5V, and 8V sections. Light-duty belts come in 3L, 4L, and 5L sections.

Sheaves and Bushings

Sheaves, the grooved wheels of pulleys, are sometimes equipped with tapered bushings for ease of installation and removal, figure 40-8. They have extreme holding power, providing the equivalent of a shrink fit. The sheave and bushing used in the power drive, drawing A-86, have a six-hole drilling arrangement in both the bushing and sheave making it possible to insert the cap screw from either side. This is especially advantageous for applications where space is at a premium.

REFERENCES AND SOURCE MATERIAL

1. A.O. Dehayt, "Basic Bearing Types," *Machine Design* 40, No. 14

AXIAL ASSEMBLY RINGS

INTERNAL **EXTERNAL**

BASIC TYPES: Designed for axial assembly. Internal ring is compressed for insertion into bore or housing, external ring expanded for assembly over shaft. Both rings seat in deep grooves and are secure against heavy thrust loads and high rotational speeds.

INTERNAL **EXTERNAL**

INVERTED RINGS: Same tapered construction as basic types, with lugs inverted to abut bottom of groove. Section height increased to provide higher shoulder, uniformly concentric with housing or shaft. Rings provide better clearance, more attractive appearance than basic types.

END PLAY RINGS

INTERNAL **EXTERNAL**

BOWED RINGS: For assemblies in which accumulated tolerances cause objectionable end play between ring and retained part. Bowed construction permits rings to provide resilient end-play takeup in axial direction while maintaining tight grip against groove bottom.

EXTERNAL **INTERNAL**

RADIAL RINGS: Bowed E-rings are used for providing resilient end-play takeup in an assembly.

SELF-LOCKING RINGS

EXTERNAL **EXTERNAL**

CIRCULAR EXTERNAL RINGS: The push-on type of fastener with inclined prongs which bend from their initial position to grip the shaft. Ring at left has arched rim for increased strength and thrust load capacity; extra-long prongs accommodate wide shaft tolerances. Ring at right has flat rim, shorter locking prongs, smaller OD.

INTERNAL

CIRCULAR INTERNAL RINGS: Designed for use in bores and housings. Functions in same manner as external types except that locking prongs are on the outside rim.

RADIAL LOCKING RINGS

EXTERNAL

CRESCENT RING: Has a tapered section similar to the basic axial types. Remains circular after installation on a shaft and provides a tight grip against the groove bottom.

EXTERNAL

E-RINGS: Provide a large bearing shoulder on small-diameter shafts and is often used as a spring retainer. Three heavy prongs, spaced approximately 120 degrees apart, provide contact surface with groove bottom.

Fig. 40-3 Stamped retaining rings

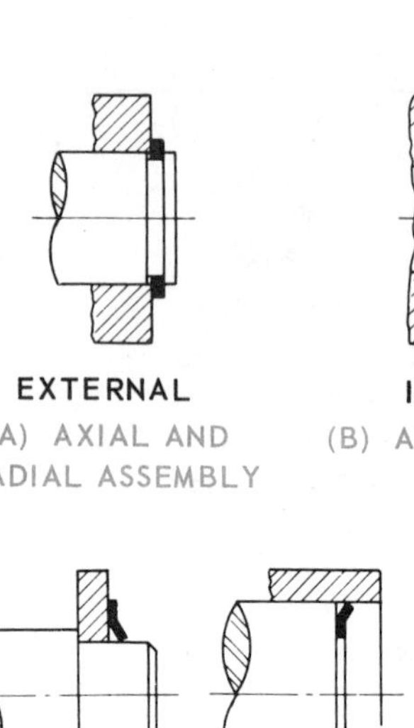

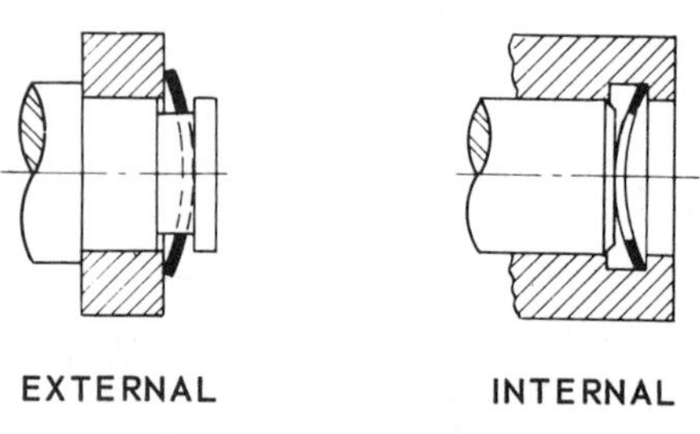

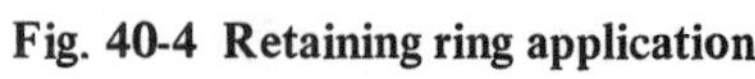

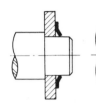

Fig. 40-4 Retaining ring application

Fig. 40-5 O-ring seal

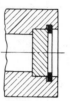

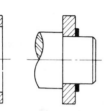

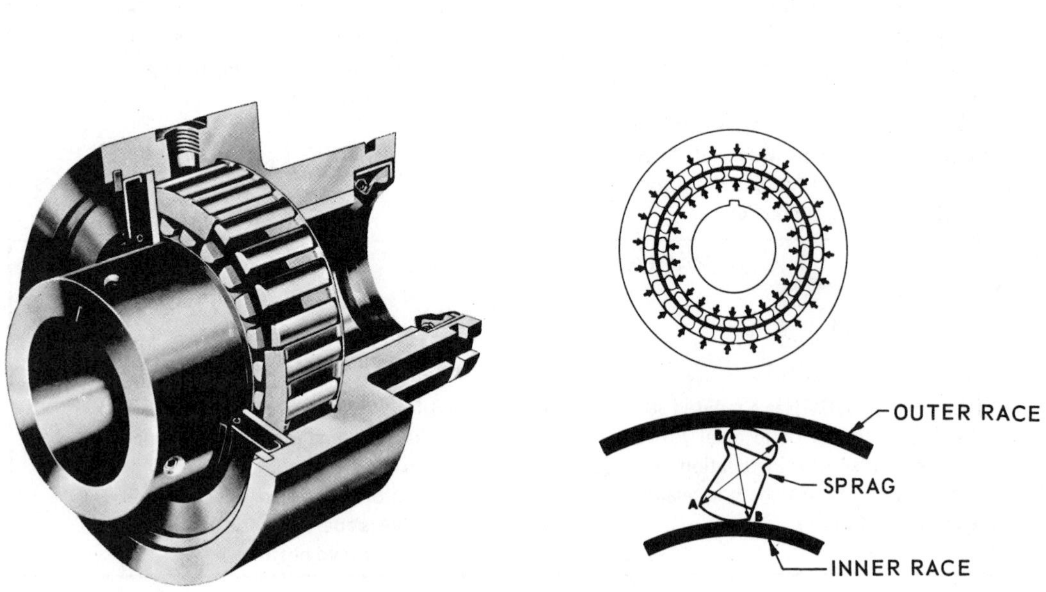

Fig. 40-6 Overrunning clutch

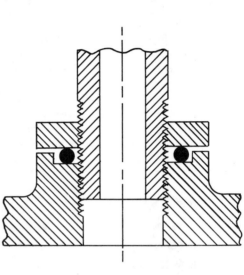

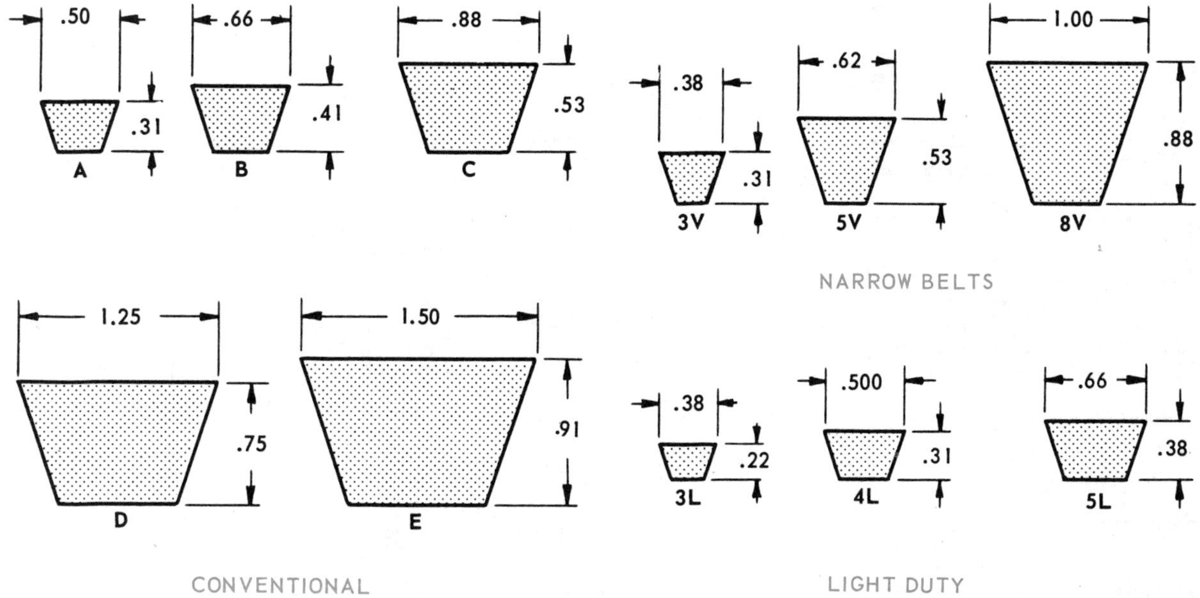

Fig. 40-7 V-belt sizes

Fig. 40-8 V-belt sheave and bushing

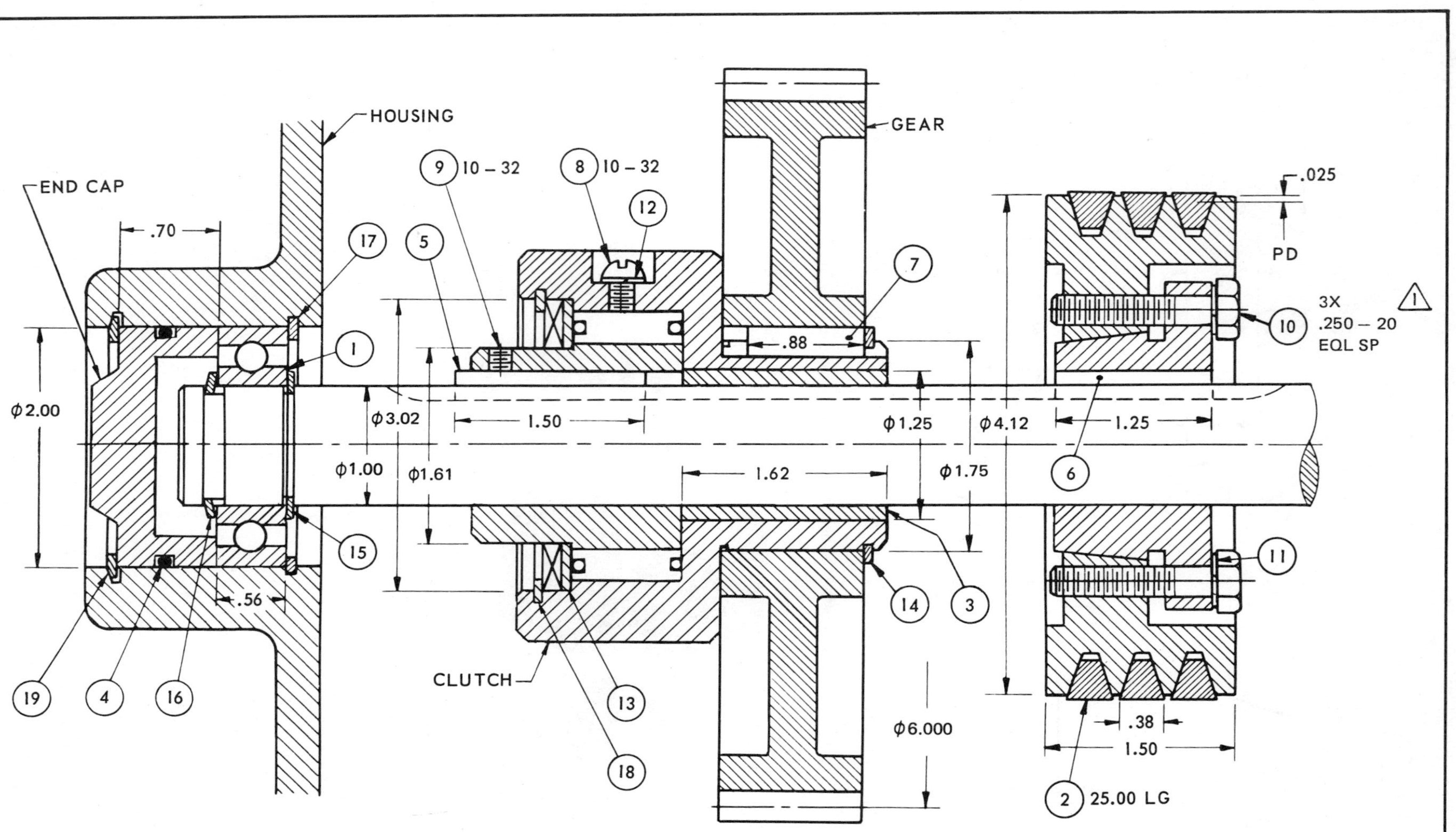

NOTE: ALL DIMENSIONS SHOWN ARE NOMINAL SIZE.

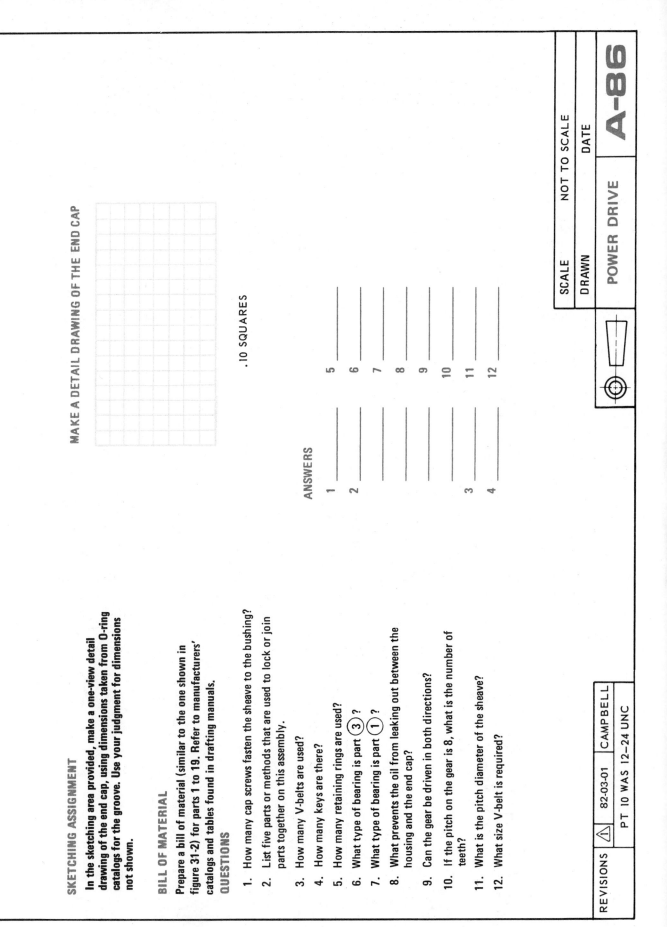

SKETCHING ASSIGNMENT

In the sketching area provided, make a one-view detail drawing of the end cap, using dimensions taken from O-ring catalogs for the groove. Use your judgment for dimensions not shown.

BILL OF MATERIAL

Prepare a bill of material (similar to the one shown in figure 31-2) for parts 1 to 19. Refer to manufacturers' catalogs and tables found in drafting manuals.

QUESTIONS

1. How many cap screws fasten the sheave to the bushing?
2. List five parts or methods that are used to lock or join parts together on this assembly.
3. How many V-belts are used?
4. How many keys are there?
5. How many retaining rings are used?
6. What type of bearing is part (3) ?
7. What type of bearing is part (1) ?
8. What prevents the oil from leaking out between the housing and the end cap?
9. Can the gear be driven in both directions?
10. If the pitch on the gear is 8, what is the number of teeth?
11. What is the pitch diameter of the sheave?
12. What size V-belt is required?

MAKE A DETAIL DRAWING OF THE END CAP

.10 SQUARES

ANSWERS

1 ________ 5 ________

2 ________ 6 ________

________ 7 ________

________ 8 ________

________ 9 ________

________ 10 ________

3 ________ 11 ________

4 ________ 12 ________

REVISIONS	△	82-03-01	CAMPBELL
	PT 10 WAS 12–24 UNC		

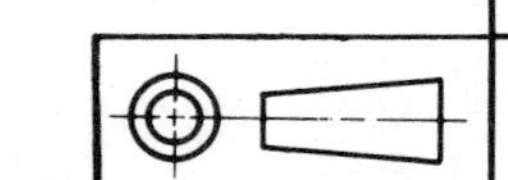

SCALE	NOT TO SCALE
DRAWN	DATE
POWER DRIVE	A-86

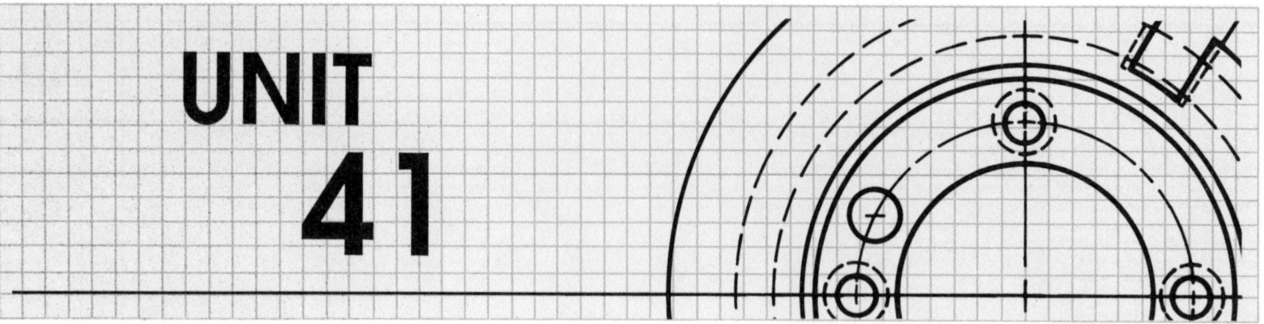

RATCHET WHEELS

Ratchet wheels are used to transform reciprocating or oscillatory motion into intermittent motion, to transmit motion in one direction only, or to serve as an indexing device.

Common forms of ratchets and pawls are shown in figure 41-1. The teeth in the ratchet engage with the teeth in the pawl, permitting rotation in one direction only.

When a ratchet wheel and pawl are designed, points A, B, and C, as shown in figure 41-1(A) are positioned on the same circle to insure that the smallest forces are acting on the system.

Mechanical Advantage

Mechanical advantage occurs when a weight in one place lifts a heavier weight in another place, or a force applied at one point on a lever

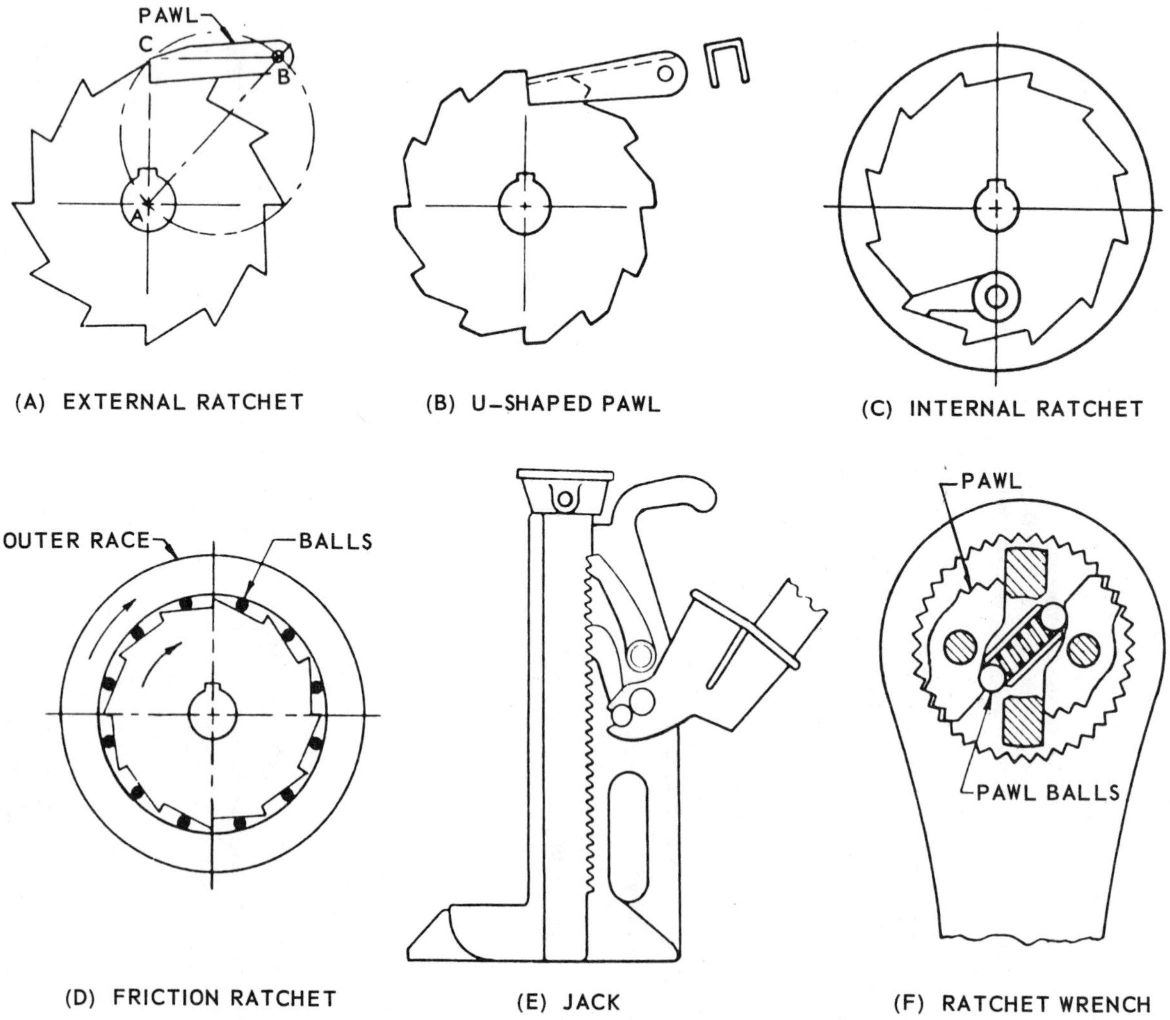

PAWL
C
B
A
(A) EXTERNAL RATCHET

(B) U–SHAPED PAWL

(C) INTERNAL RATCHET

OUTER RACE
BALLS
(D) FRICTION RATCHET

(E) JACK

PAWL
PAWL BALLS
(F) RATCHET WRENCH

Fig. 41-1 Rachets and pawls

produces a greater force at another point. Examples of mechanical advantage are the teeter-totter, the winch, and the gears shown in figure 41-2.

The teeter-totter shows how mechanical advantage can be applied. If a 10 pound (lb.) weight is placed on one end of a teeter-totter and a 20 lb. weight is placed on the other end, the 10 lb. weight goes up and the 20 lb. weight goes down when placed equidistant from the fulcrum. If the fulcrum is moved to make the distance from the 10 lb. weight to the fulcrum twice that of the distance between the fulcrum and the 20 lb. weight, the weights become balanced. If the fulcrum is moved still closer to the 20 lb. weight, the 10 lb. weight goes down and the 20 lb. weight goes up.

As the distance between the fulcrum and the 10 lb. weight increases, the distance that the 10 lb. weight moves must increase proportionately to maintain the same movement of the 20 lb. weight. Thus, a light mass can move a heavier weight, but in doing so, the smaller weight must travel farther than the larger one.

Mechanical advantage also occurs when a winch or different size gears are used. With the winch design shown in figure 41-2(B), a mechanical advantage of 10 is obtained because the handle is 10 times the distance from the fulcrum as compared to the center of the rope.

The mechanical advantage of gears can easily be determined by obtaining the ratio between the number of teeth on the gears or the ratio between the pitch diameters.

In the winch, drawing A-87, the center of the handle bar is 10 in. from the center of the shaft to which the pinion gear is attached. Half the pitch diameter of the pinion is .625 in. This produces a mechanical advantage of 10:.625 or 16:1. Further mechanical advantage is gained through the gear attached to the Ø1.00 shaft on which the rope revolves.

The hand will move a distance of approximately 63 in. when turning the handle one complete revolution. This will turn the rope drum one fifth of a revolution (50:10 teeth ratio), winding a Ø.25 rope up approximately 4 in. A greater force can be exerted using the winch than by simply pulling the rope.

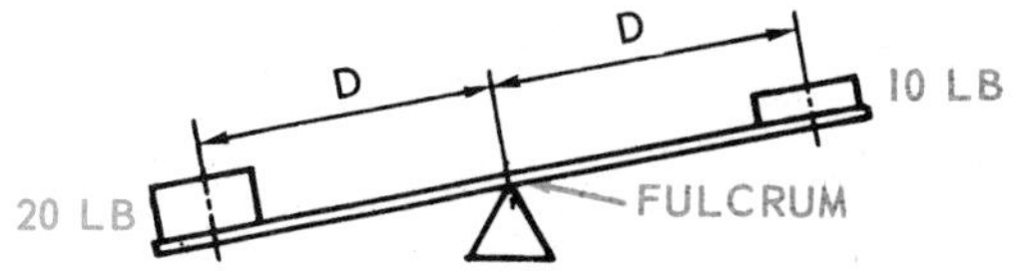

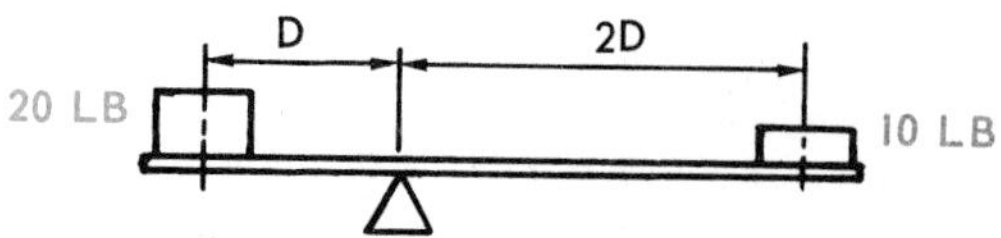

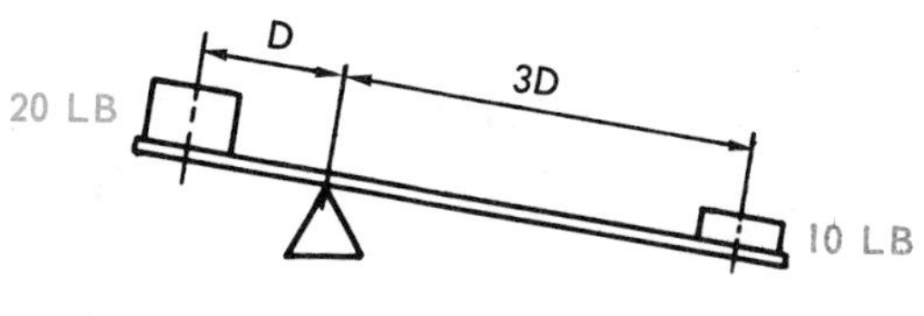

(A) TEETER–TOTTER

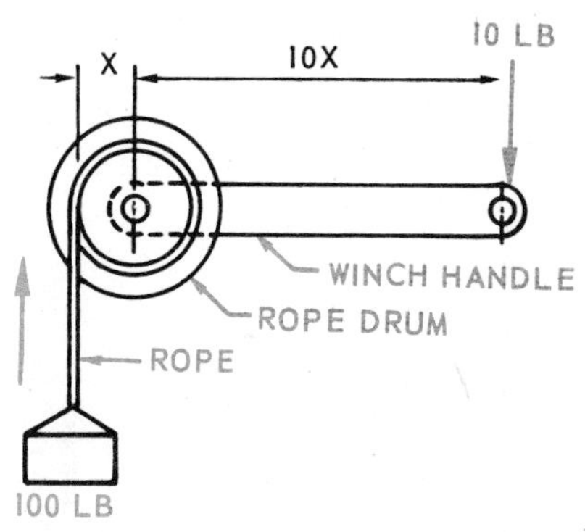

(B) WINCH

MECHANICAL ADVANTAGE=3:1

MECHANICAL ADVANTAGE=5:1

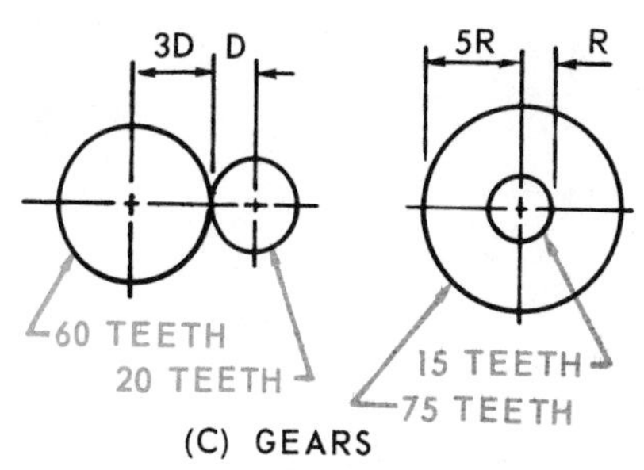

(C) GEARS

Fig. 41-2 Mechanical advantage

ASSIGNMENT: MAKE DETAIL DRAWINGS OF PARTS. NOTE, ALL DIMENSIONS SHOWN ARE NOMINAL SIZES. ALLOWANCES AND TOLERANCES ARE TO BE DETERMINED.

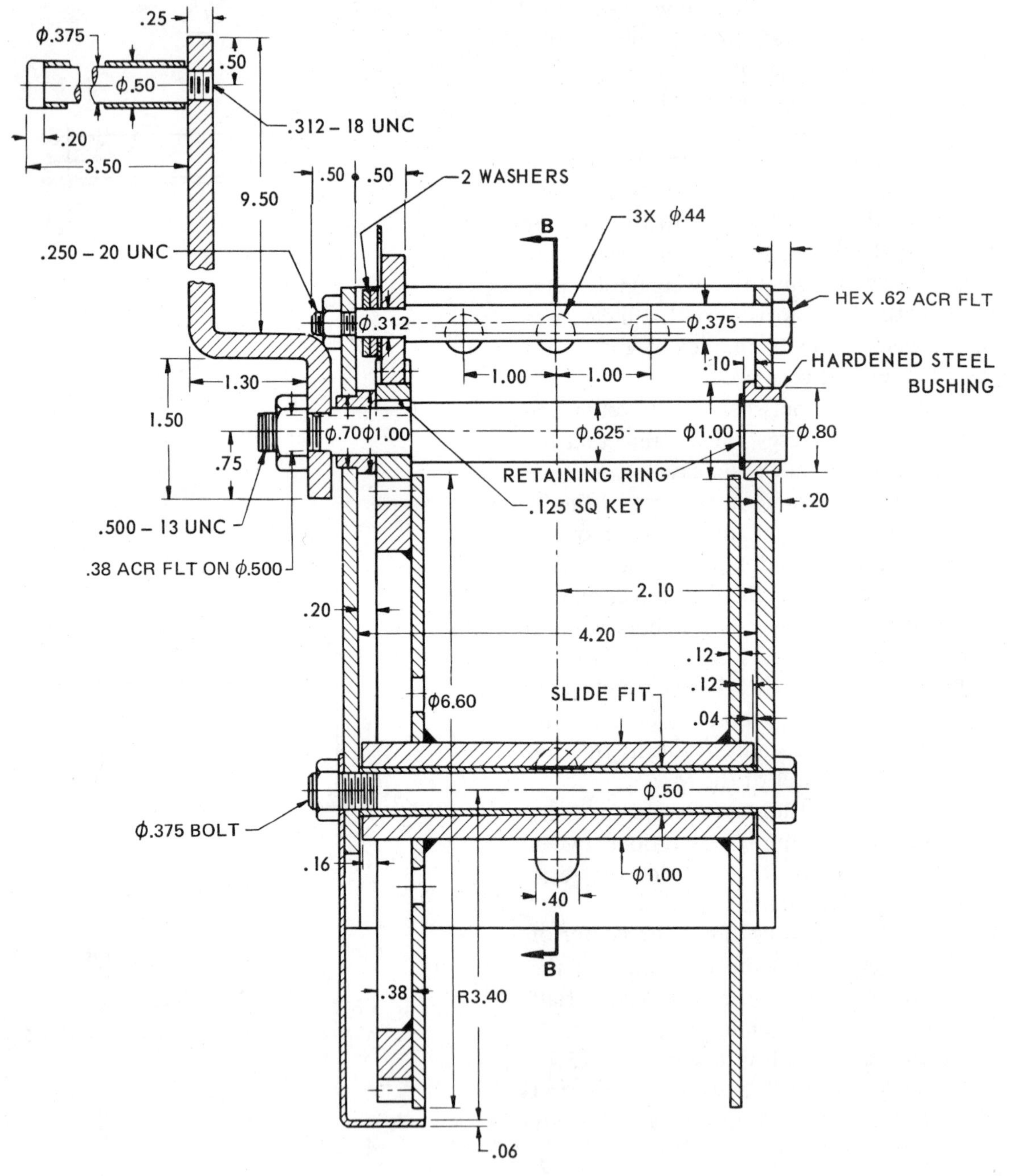

SECTION A–A

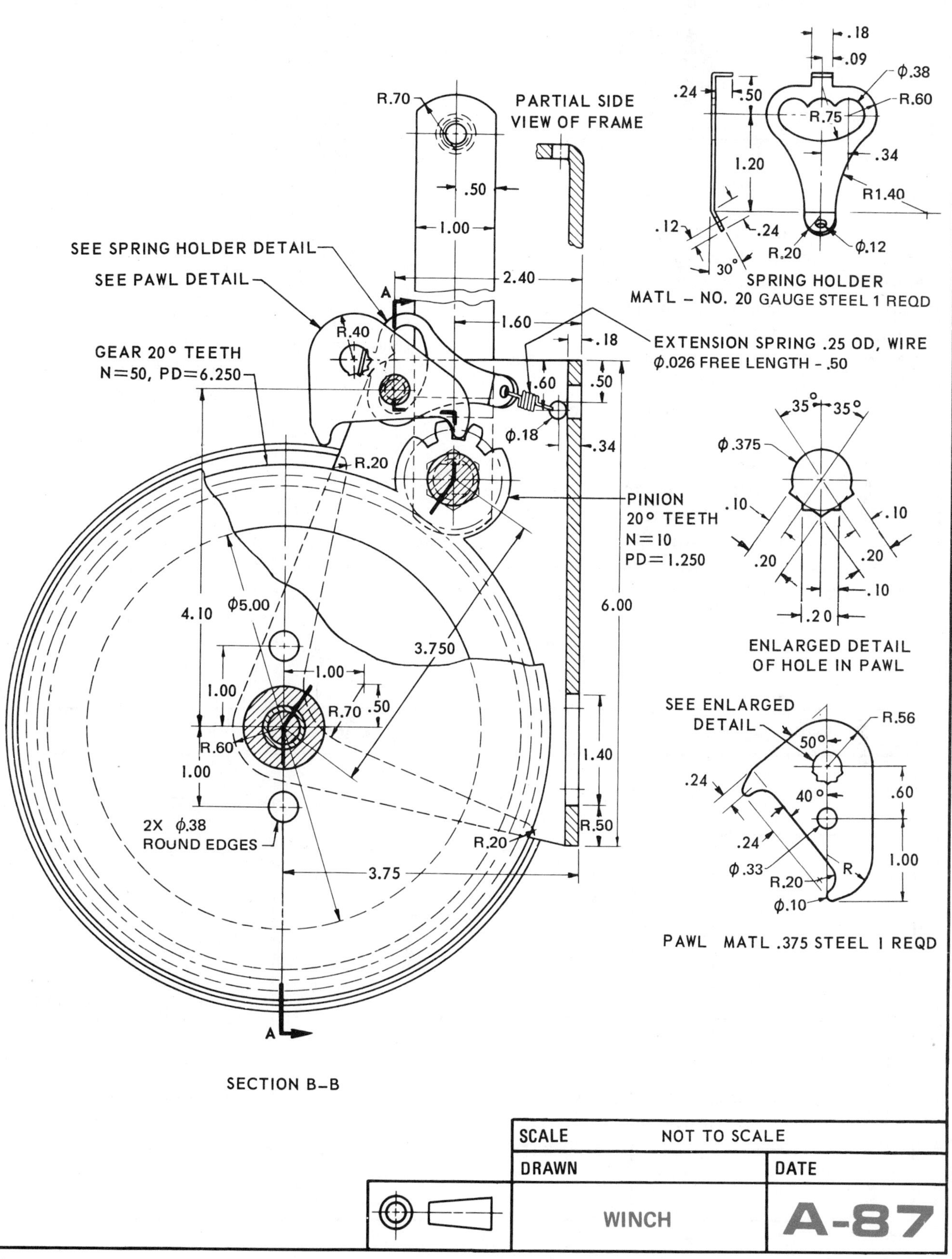
R.70
PARTIAL SIDE VIEW OF FRAME
.50
1.00
SEE SPRING HOLDER DETAIL
SEE PAWL DETAIL
2.40
1.60
.18
GEAR 20° TEETH N=50, PD=6.250
R.40
.60
.50
ϕ.18
.34
R.20
4.10
ϕ5.00
3.750
PINION 20° TEETH N=10 PD=1.250
6.00
1.00
1.00
R.70
.50
R.60
1.00
1.40
2X ϕ.38 ROUND EDGES
R.20
R.50
3.75
SECTION B–B
.18
.09
ϕ.38
.24
.50
R.60
R.75
.34
1.20
R1.40
.12
.24
R.20
ϕ.12
30°
SPRING HOLDER
MATL – NO. 20 GAUGE STEEL 1 REQD
EXTENSION SPRING .25 OD, WIRE ϕ.026 FREE LENGTH - .50
35°
35°
ϕ.375
.10
.10
.20
.20
.10
.20
ENLARGED DETAIL OF HOLE IN PAWL
SEE ENLARGED DETAIL
R.56
50°
.24
40°
.60
.24
ϕ.33
R.20
R
1.00
ϕ.10
PAWL MATL .375 STEEL 1 REQD
SCALE NOT TO SCALE
DRAWN
DATE
WINCH
A-87

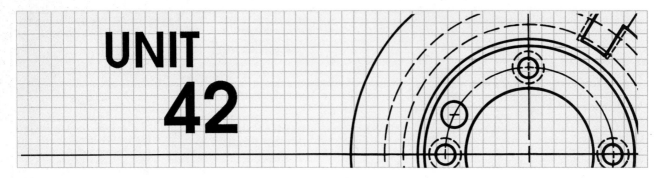

UNIT 42

MODERN ENGINEERING TOLERANCING

An engineering drawing of a manufactured part conveys information from the designer to the manufacturer and inspector. It must contain all information necessary for the part to be correctly manufactured. It must also enable the inspector to determine precisely whether the finished parts are acceptable.

Therefore, each drawing must convey three essential items of information: the material to be used, the size or dimensions of the part, and the shape or geometrical characteristics of the part. The drawing must also indicate the permissible variation of size and form.

The actual size of a feature must be within the size limits specified on the drawing. Each measurement made at any cross section of the feature must not be greater than the maximum limit of size, nor smaller than the minimum limit of size, figure 42-1. Although each part is within the prescribed tolerance zones, the parts may not be usable because of their deviation from their form.

Formerly, tolerances for which there were no precise interpretations were often shown. While tolerancing of geometrical characteristics was sometimes limited to notes such as,

PARALLEL WITH SURFACE B WITHIN .001

or

STRAIGHT WITHIN .005

it did not precisely express permissible variations.

In order to meet functional requirements, it is often necessary to control errors of form including: squareness, roundness, and flatness, as well as deviation from true size. In the case

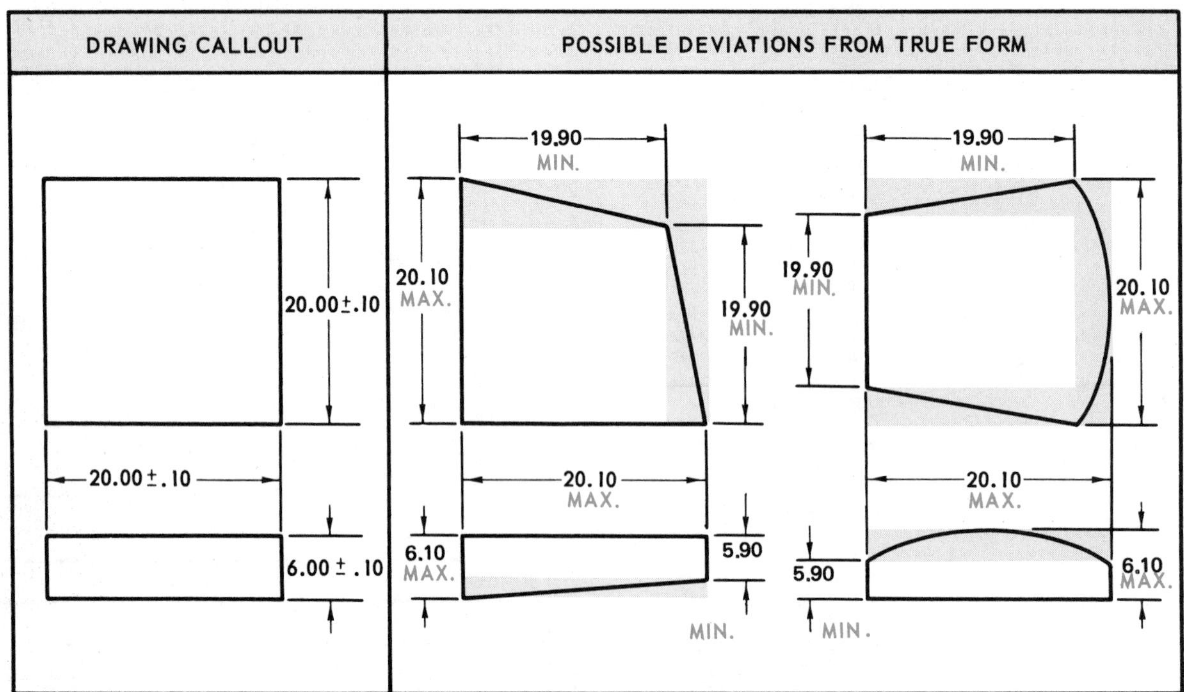

Fig. 42-1 Deviations permitted by toleranced dimensions

of mating parts, such as holes and shafts, it is usually necessary to insure that they do not cross the boundary of perfect form at their maximum material size (the smallest hole or the largest shaft) because of being bent or otherwise deformed. This condition is shown in figure 42-2, where features are not permitted to cross the boundary of perfect form at the least material size (the largest hole or the smallest shaft).

The system of *geometric tolerancing* offers a precise interpretation of drawing requirements. Geometric tolerancing controls geometrical characteristics of parts. These characteristics include: flatness, roundness, angularity, profile, and position. Other techniques, such as datum systems, datum targets, and projected tolerance zones were developed in order to facilitate this precise interpretation.

Geometric tolerances need not be used for every feature of a part. Generally, if each feature meets all dimensional tolerances, form variations will be adequately controlled by the accuracy of the manufacturing process and equipment used. A geometric tolerance is used when geometric errors must be limited more closely than might ordinarily be expected from the manufacturing process. A geometric tolerance is also used to state functional or interchangeability requirements.

The symbols, figure 42-3, and descriptions used in this text reflect the recommendations of ANSI Y14.5M.

GEOMETRIC TOLERANCING

A geometric tolerance is the maximum permissible variation of form, orientation, or location of a feature from that indicated or specified on a drawing. The tolerance value represents the width or diameter of the tolerance zone, within which the point, line, or surface of the feature should lie.

Feature Control Frame

The geometric tolerance is shown on the drawing by a *feature control frame*, figure 42-4.

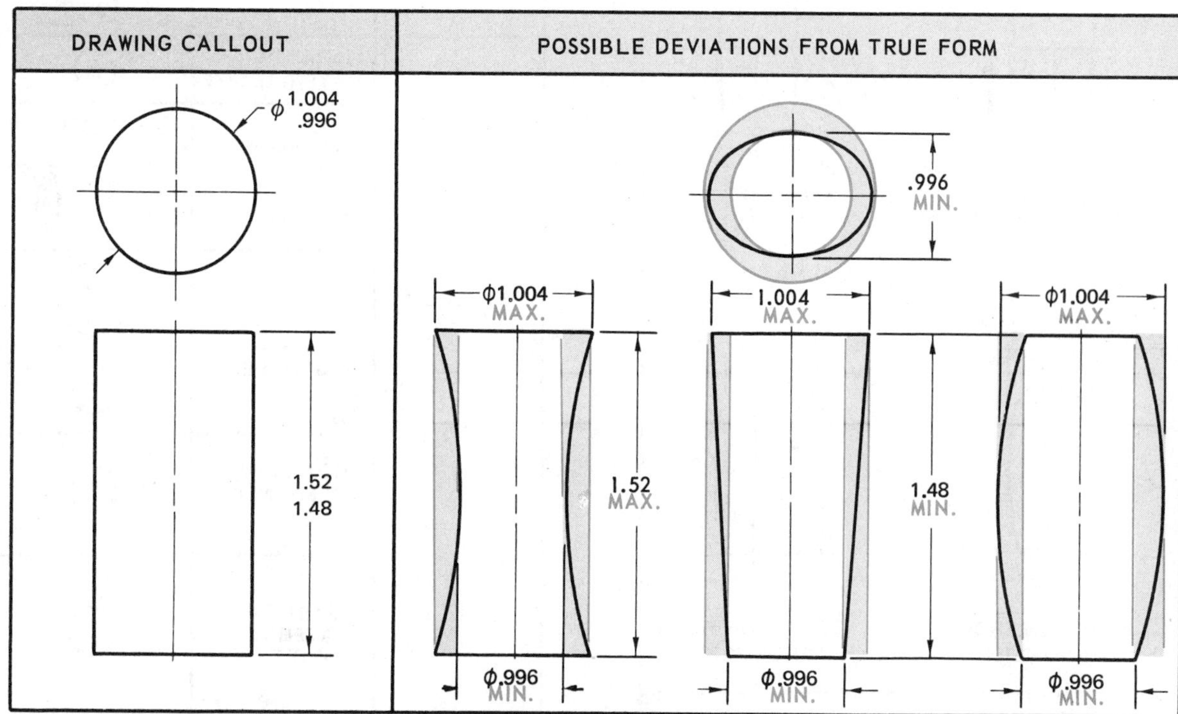

(B) CYLINDRICAL FEATURES

Fig. 42-1 Deviations permitted by toleranced dimensions, continued

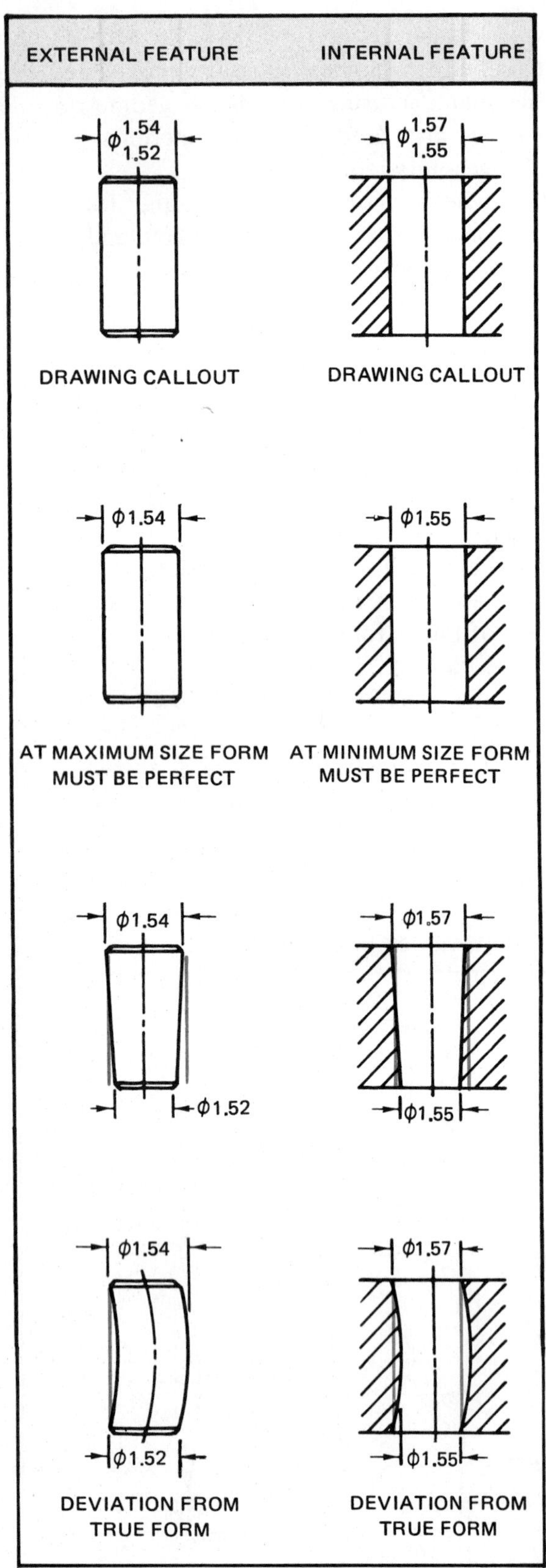

Fig. 42-2 Examples of deviation of form when perfect form at the maximum material size is required

GEOMETRIC CHARACTERISTIC SYMBOLS			
CHARACTERISTIC			SYMBOL
FORM TOLERANCES	FORM OF A LINE	STRAIGHTNESS	—
		CIRCULARITY (ROUNDNESS)	○
		PROFILE OF A LINE	⌒
	FORM OF A SURFACE	FLATNESS	▱
		CYLINDRICITY	⌭
		PROFILE OF A SURFACE	⌓
	ORIENTATION OF RELATED FEATURES	ANGULARITY	∠
		PARALLELISM	//
		PERPENDICULARITY	⊥
TOLERANCE FOR LOCATION		POSITION	⌖
		CONCENTRICITY	◎
		SYMMETRY	SEE UNIT 50
CIRCULAR		RUNOUT	↗ OR ➚
		TOTAL	⌰ OR ⌰
SUPPLEMENTARY SYMBOLS		MAXIMUM MATERIAL SIZE	Ⓜ
		BASIC DIMENSION	[XX]
		DATUM FEATURE	ANSI [-A-] ISO [A]
		DATUM TARGET	ϕ.50 / A2
		REGARDLESS OF FEATURE SIZE	Ⓢ
		PROJECTED TOLERANCE ZONE	Ⓟ
		LEAST MATERIAL CONDITION	Ⓛ

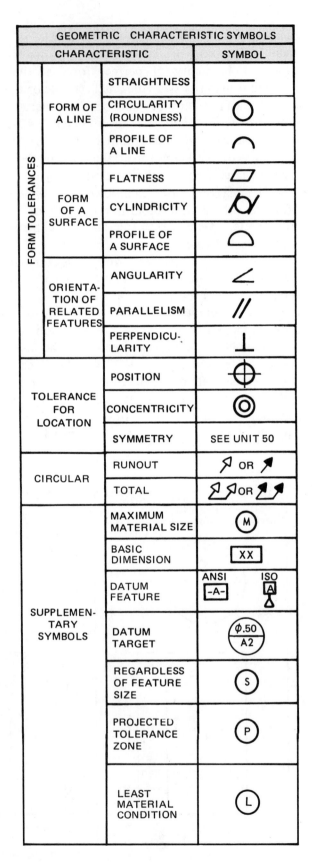

Fig. 42-3 Geometric characteristic symbols

A feature control frame consists of a rectangular frame divided into two or more compartments. The first compartment (starting from the left) contains the geometric characteristic. The second compartment contains the allowable tolerance. Where applicable, the tolerance is preceded by the diameter symbol and followed by a modifying symbol. Other compartments are added when datums must be specified.

The feature control frame is related to the feature by one of the following methods.

(A) Running a leader from the frame to the feature, figure 42-5(A). This method is used when the control of the surface elements is required.
(B) Running a leader from the frame to an extension line of the surface but not in line with the dimension, figure 42-5(B).
(C) Locating the frame below the size dimension to control the center line, axis, or center plane of the feature, figure 42-5(C).
(D) Attaching a side or end of the frame to an extension line extending from a plane surface feature, figure 42-5(D).
(E) Locating the frame below or attached to the leader directed callout or dimension pertaining to the feature, figure 42-5(E).

Figure 42-6 illustrates the preferred location of the feature control frame.

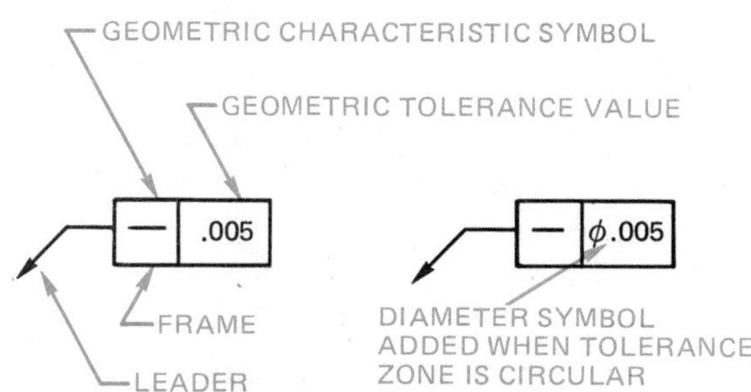

(A) SINGLE FEATURE CONTROL

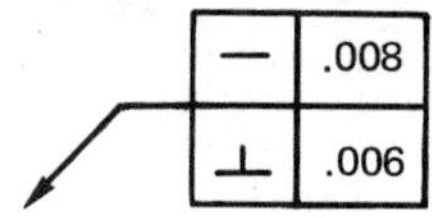

(B) COMBINED FEATURE CONTROL FRAMES

Fig. 42-4 Feature control frame

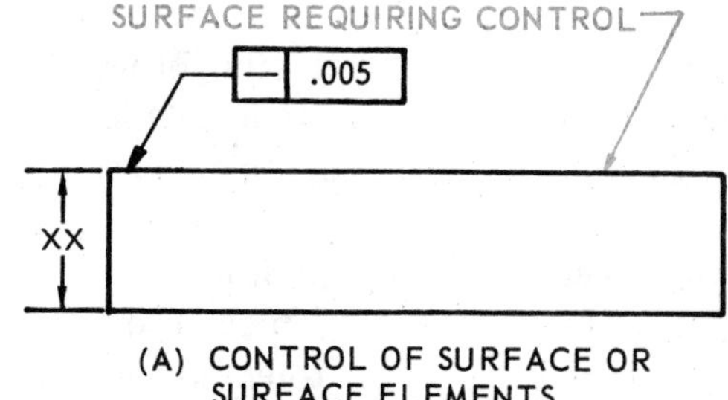

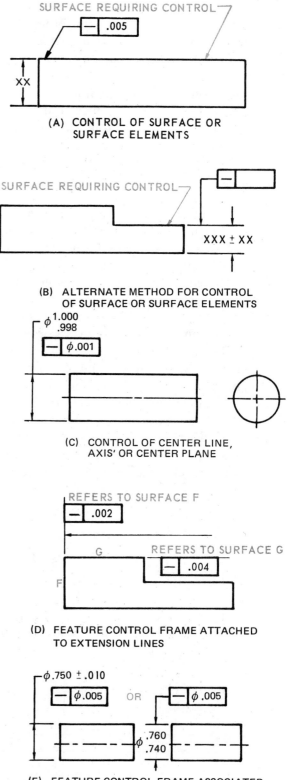

(A) CONTROL OF SURFACE OR SURFACE ELEMENTS

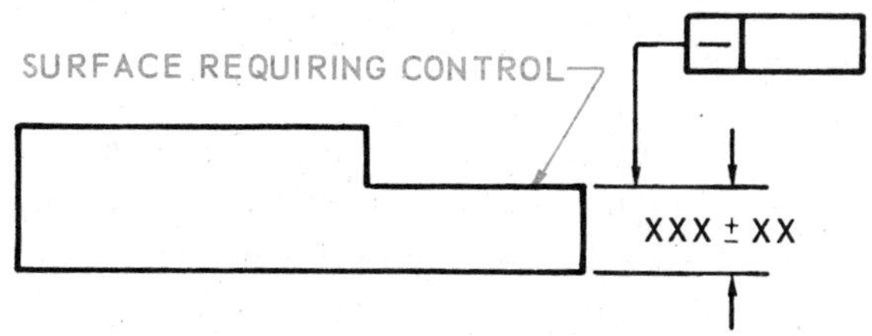

(B) ALTERNATE METHOD FOR CONTROL OF SURFACE OR SURFACE ELEMENTS

ϕ 1.000 / .998

ϕ.001

(C) CONTROL OF CENTER LINE, AXIS' OR CENTER PLANE

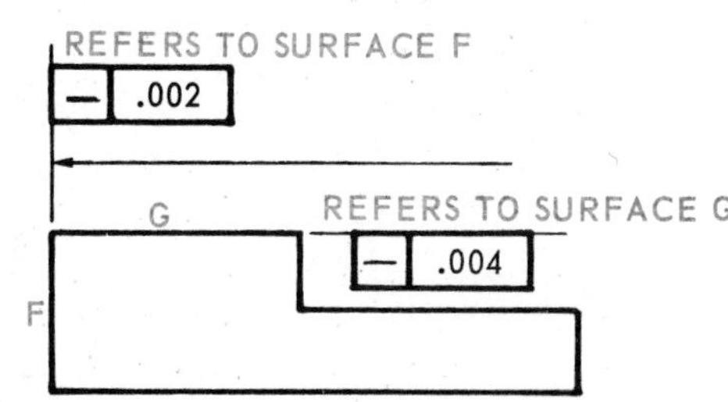

(D) FEATURE CONTROL FRAME ATTACHED TO EXTENSION LINES

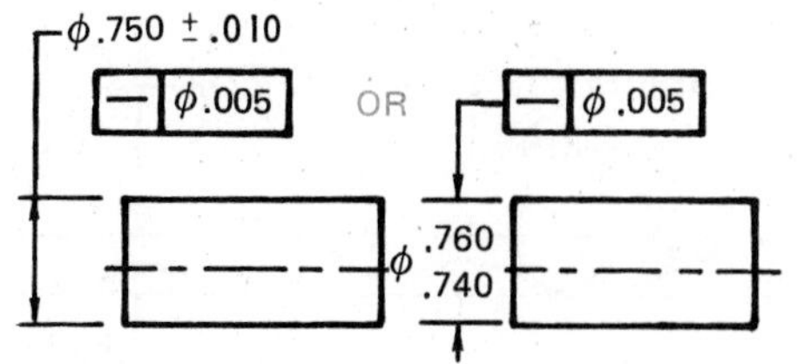

(E) FEATURE CONTROL FRAME ASSOCIATED WITH SIZE DIMENSIONS

Fig. 42-5 Application of feature control frame

FORM TOLERANCES

Form tolerances control straightness, flatness, circularity, and cylindricity. Orientation tolerances control angularity, parallelism, and perpendicularity.

Form tolerances are applicable to single (individual) features or elements of single features; therefore, form tolerances are not related to datums.

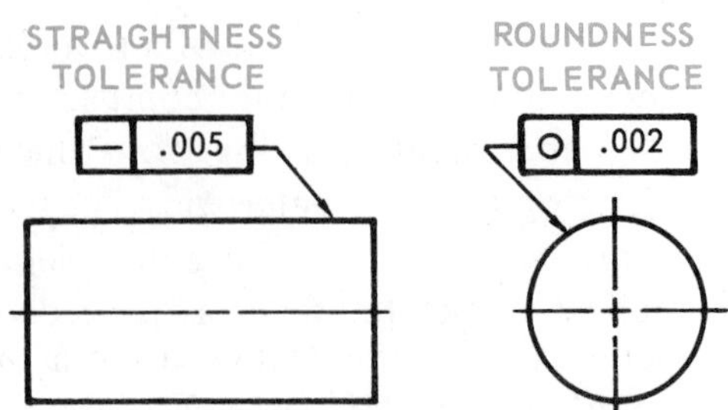

Fig. 42-6 Preferred location of feature control symbol

Straightness

Lines and Surfaces. Straightness is fundamentally a characteristic of a line, such as the edge of a part or a line scribed on a surface. Figure 42-7(A) states in symbolic form that the line shall be straight within .005 in. This means the line shall be contained within a tolerance zone consisting of an area between two parallel straight lines in the same plane, separated by the specified tolerance.

When a feature control symbol is intended to apply to a surface, the leader touches the surface or an extension line from the surface.

Figure 42-8 shows an example of a cylindrical part where all circular elements of the surface are to be within the specified size tolerances. Each longitudinal element of the surface must lie between two parallel lines separated by the amount of the prescribed straightness tolerance and in a plane common with the nominal axis of the feature. The feature control frame is attached to a leader directed to the surface or extension line of the surface but not to the size dimension. The straightness tolerance must be less than the size tolerance. Since the limits of size must be adhered to, the full straightness tolerance may not be permitted in the case of tapering or barreling of the surface.

Application to Center Lines. When a feature control symbol is intended to apply to an axis, center line, or median plane of a feature, the leader is directed to the feature size dimension, figure 42-9(A).

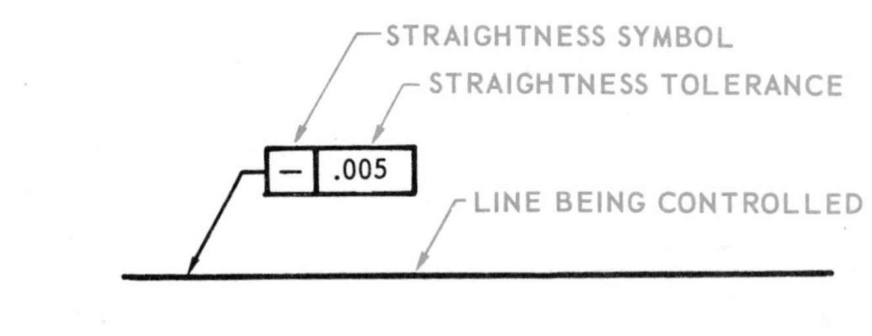

(A) STRAIGHTNESS SYMBOL APPLIED TO A LINE

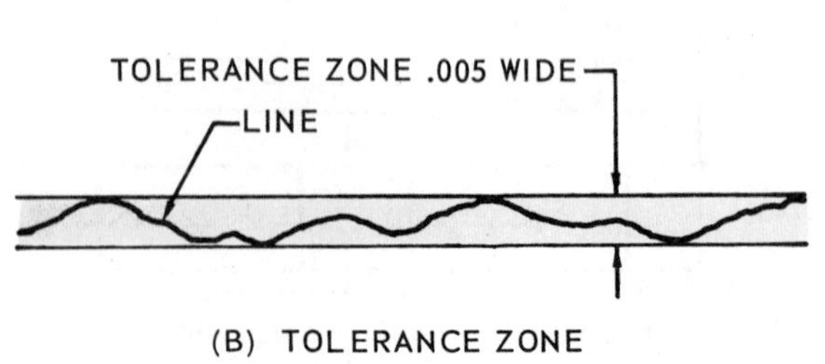

(B) TOLERANCE ZONE

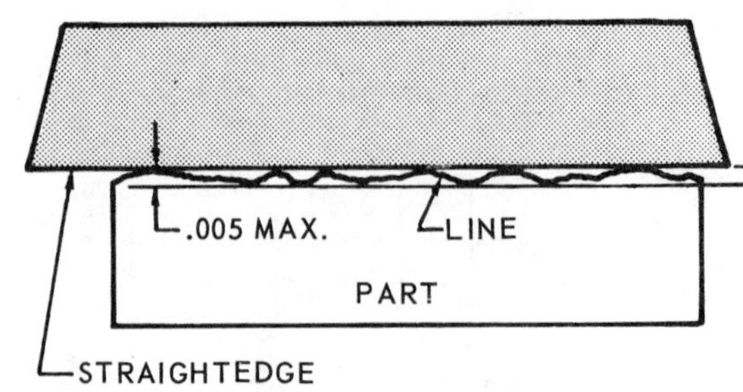

(C) CHECKING WITH STRAIGHTEDGE

Fig. 42-7 Straightness symbol and application (applying to a surface)

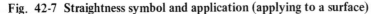

When controlling one of the features sharing the same center line, the method shown in figure 42-9(B) is used.

When controlling all of the features sharing the same center line, the method shown in figure 42-9(C) is used. This method is not shown in ANSI Y14.5M-1982 but is covered in ISO standards.

When the resulting tolerance zone is circular or cylindrical, such as when controlling the straightness of a center line, figure 42-9, a diameter symbol precedes the tolerance value in the feature control frame.

Straightness in a Specified Length. It is often desirable on long parts to specify a straightness tolerance over a specified length, either with or without a maximum overall tolerance. This requirement is specified on a drawing by including the specified length with the tolerance as shown in Figure 42-10.

A straightness tolerance applied to a flat surface indicates straightness control in one direction only and must be directed to the line on the drawing representing the surface to be controlled and the direction in which the control is required, as shown in figure 42-11.

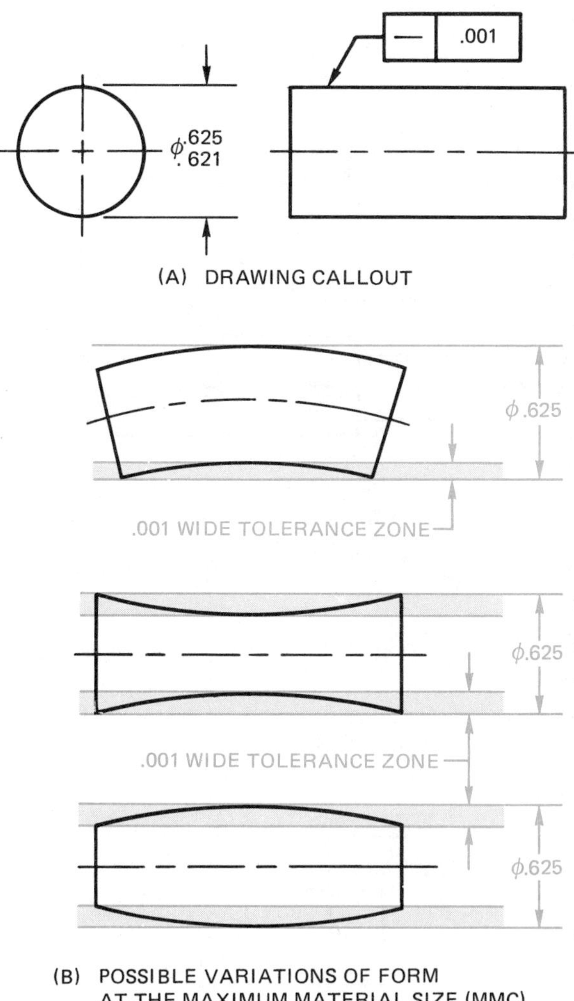

Fig. 42-8 Specifying straightness of surface elements

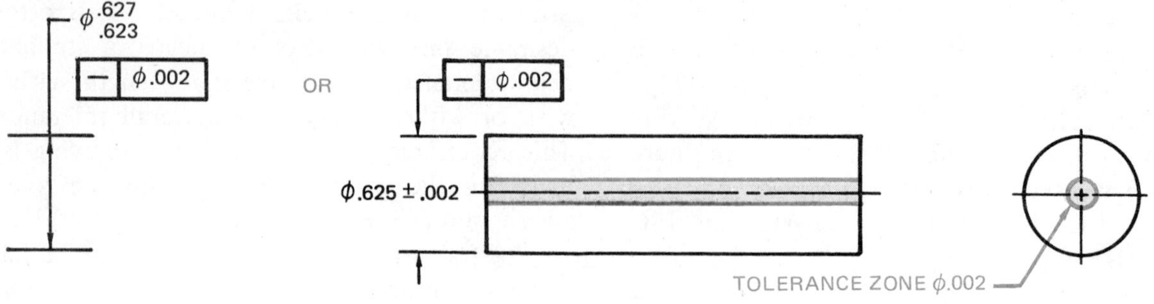

(A) SINGLE CYLINDRICAL FEATURE

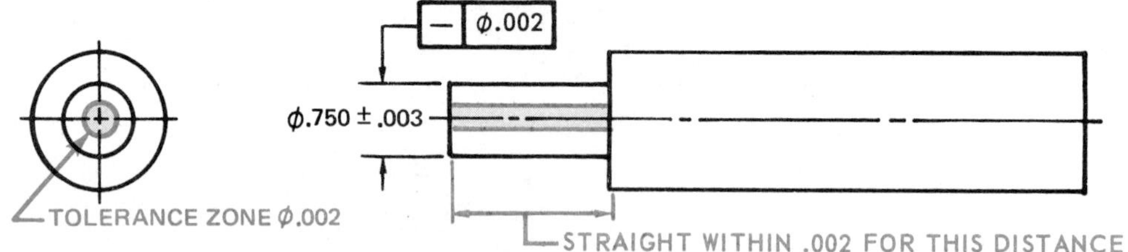

(B) CONTROLLING ONE FEATURE SHARING A COMMON CENTER LINE

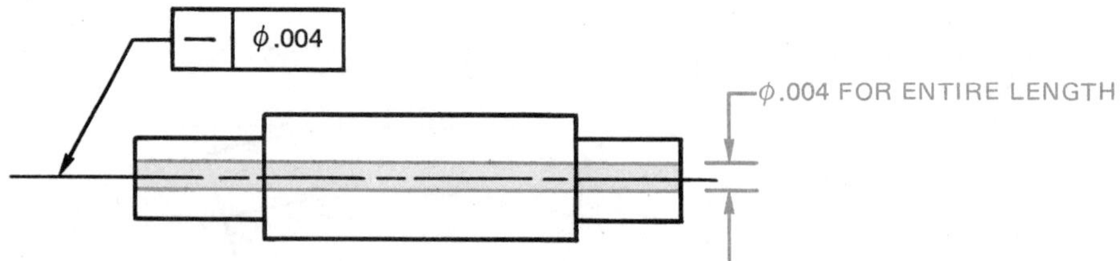

(C) CONTROLLING ALL FEATURES SHARING A COMMON CENTER LINE (150 ONLY)

Fig. 42-9 Application of straightness symbol to center lines

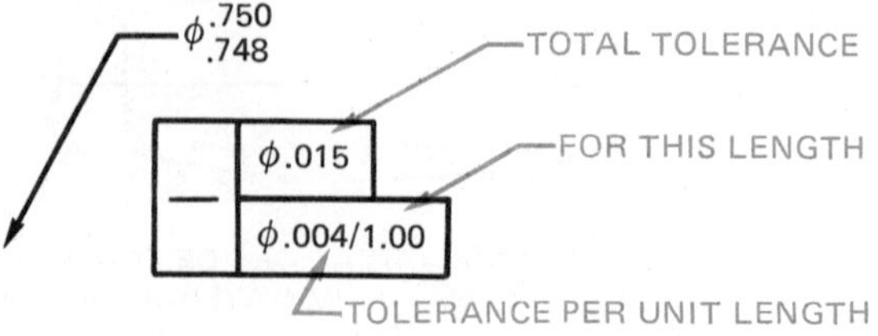

Fig. 42-10 Tolerance in a specified length

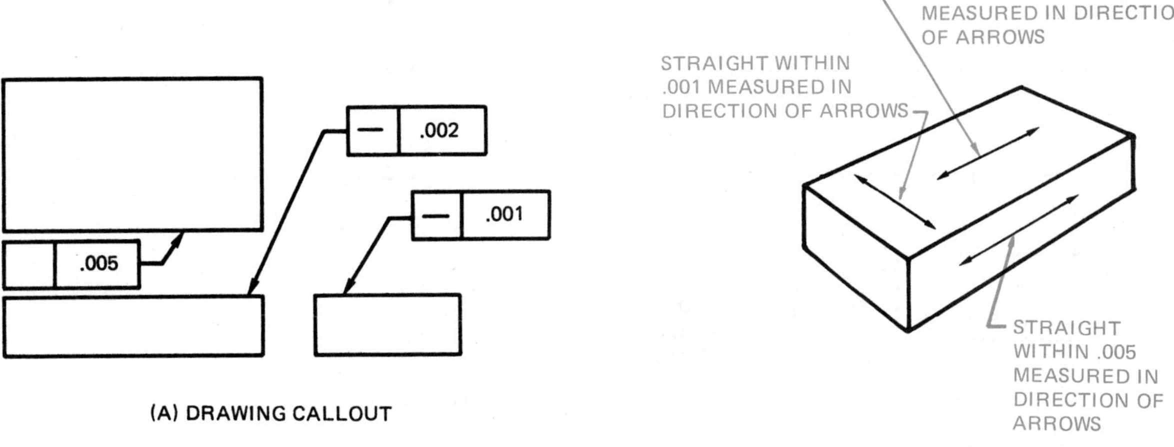

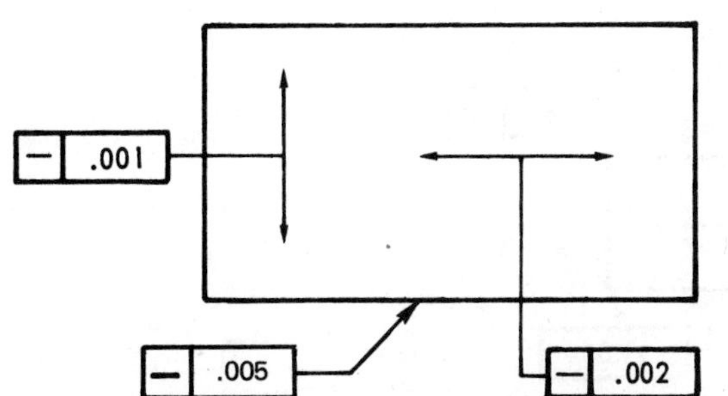

(C) THREE STRAIGHTNESS TOLERANCES ON ONE VIEW

Fig. 42-11 Straightness in several directions

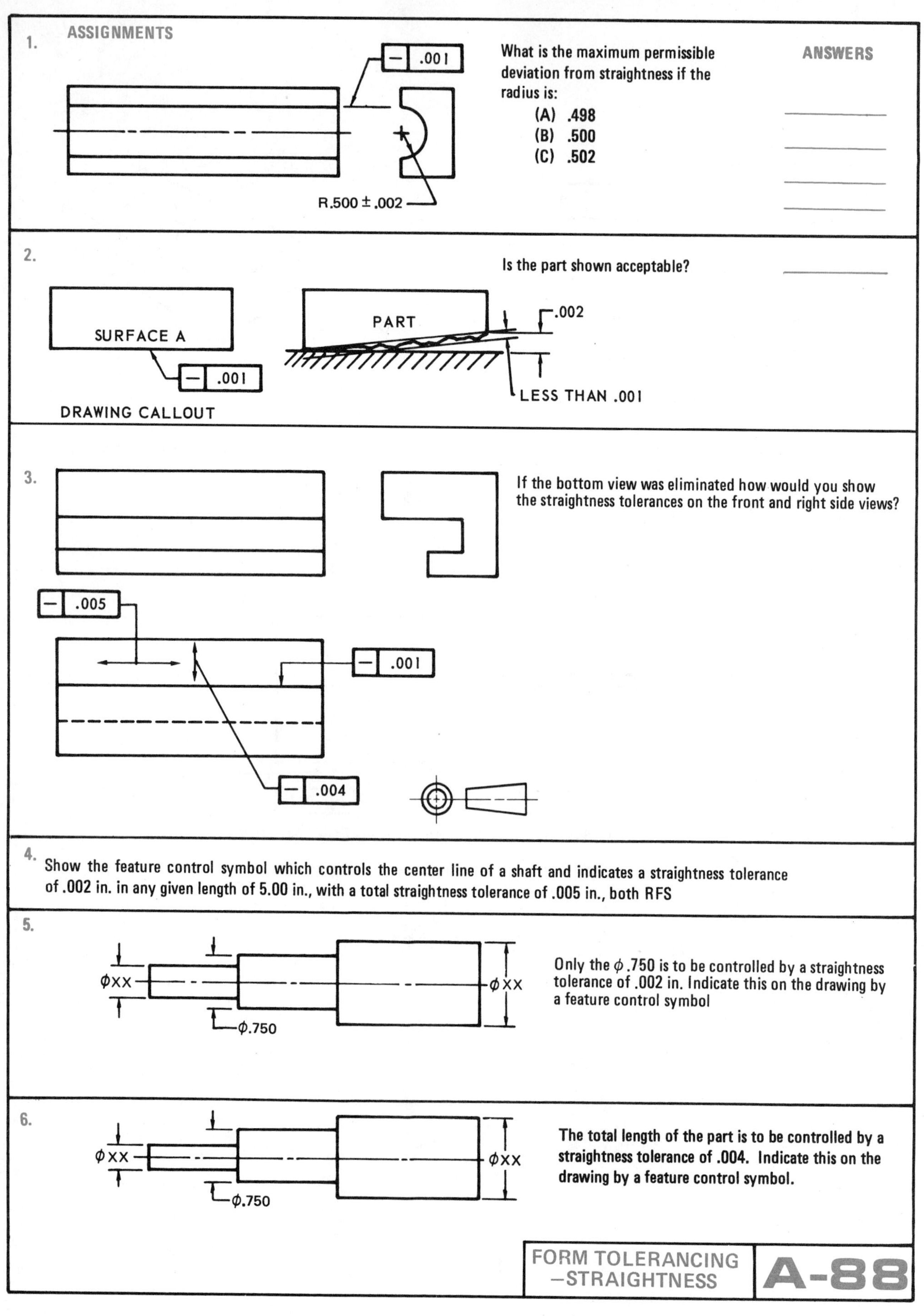

ASSIGNMENTS
1.
.001
R.500 ± .002
What is the maximum permissible deviation from straightness if the radius is:
(A) .498
(B) .500
(C) .502
ANSWERS
2.
Is the part shown acceptable?
SURFACE A
.001
DRAWING CALLOUT
PART
.002
LESS THAN .001
3.
If the bottom view was eliminated how would you show the straightness tolerances on the front and right side views?
.005
.001
.004
4.
Show the feature control symbol which controls the center line of a shaft and indicates a straightness tolerance of .002 in. in any given length of 5.00 in., with a total straightness tolerance of .005 in., both RFS
5.
ϕXX
ϕ.750
ϕXX
Only the ϕ .750 is to be controlled by a straightness tolerance of .002 in. Indicate this on the drawing by a feature control symbol
6.
ϕXX
ϕ.750
ϕXX
The total length of the part is to be controlled by a straightness tolerance of .004. Indicate this on the drawing by a feature control symbol.
FORM TOLERANCING
—STRAIGHTNESS
A-88

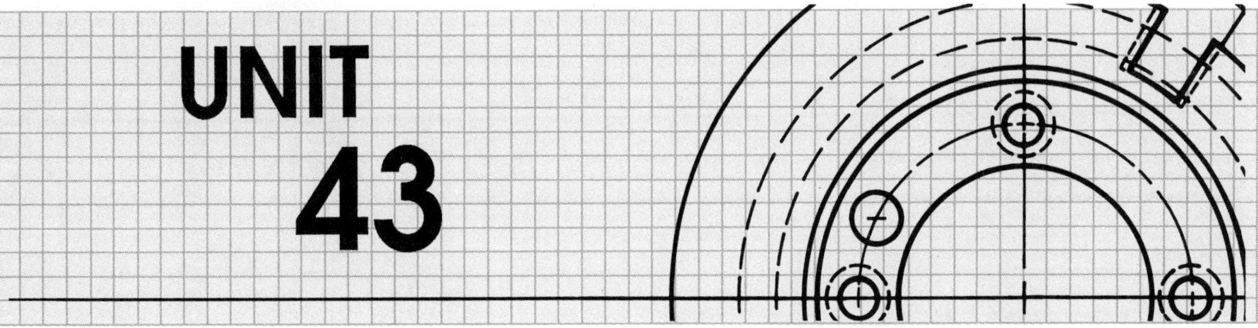

MODIFYING SYMBOLS

Definitions

Regardless of Feature Size (RFS). The term used to indicate that a geometric tolerance or datum reference applies at any increment of size of the feature and any applicable geometric tolerances. See Unit 47 for application.

Maximum Material Condition (MMC). The condition in which a feature of size contains the maximum amount of material within the stated limits of size; for example, minimum hole diameter, maximum shaft diameter, figure 43-1.

Least Material Condition (LMC). The condition in which a feature of size contains the least amount of material within the stated limits of size; for example, maximum hole diameter, minimum shaft diameter. See Unit 47 for application.

Virtual Condition. The boundary generated by the collective effects of the specified MMC limit of size and any applicable geometric tolerances, figure 43-2.

Parts are generally toleranced so they will assemble when mating features are at MMC. Additional tolerance on form or location is permitted when features depart from their MMC size.

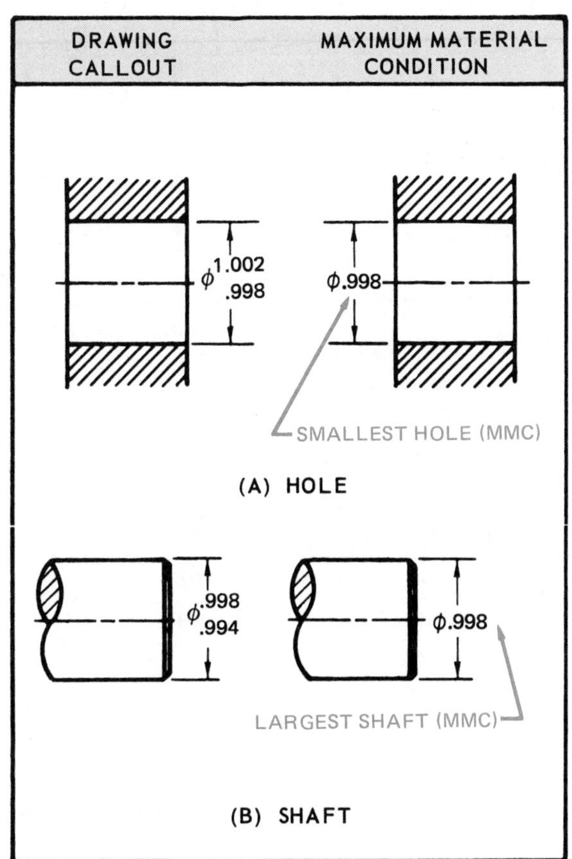

Fig. 43-1 Maximum material condition (MMC)

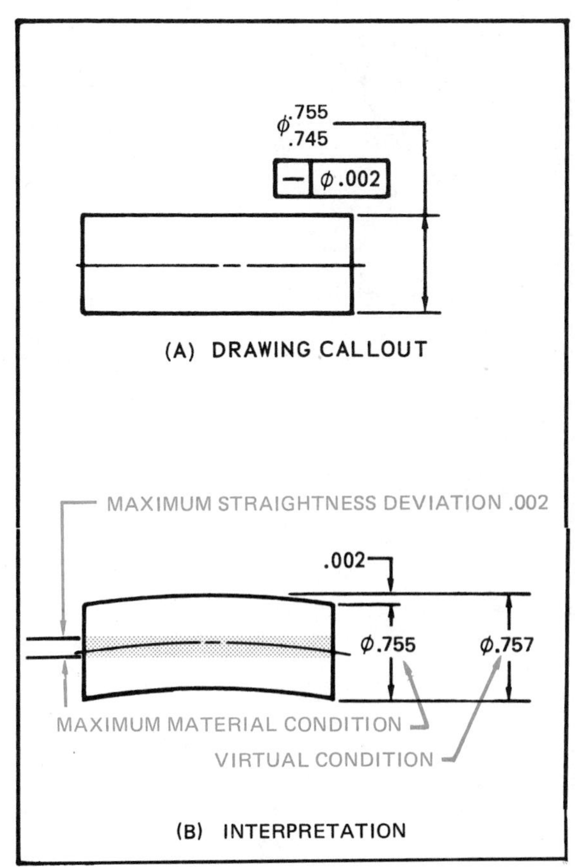

Fig 43-2 Virtual condition

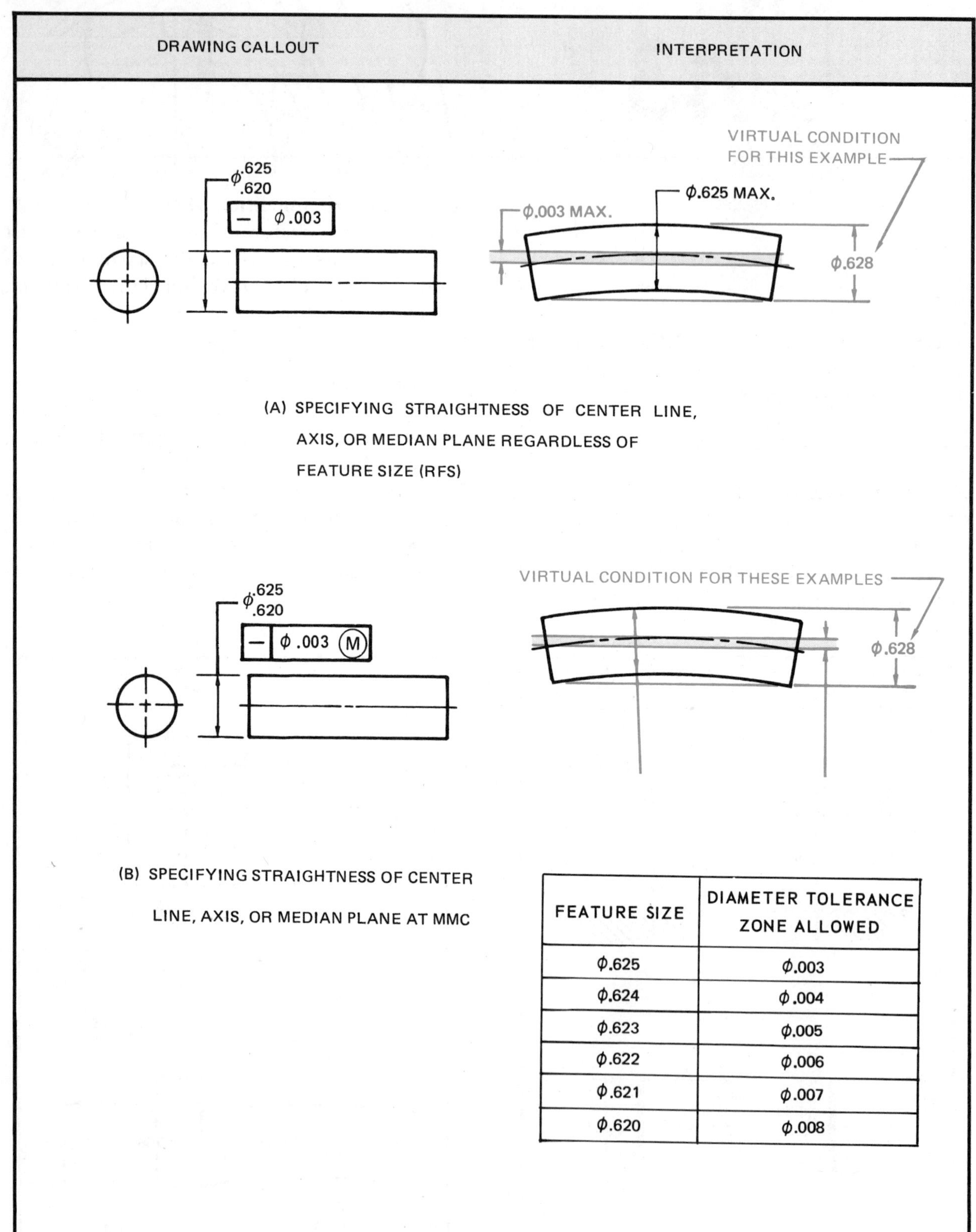

FEATURE SIZE	DIAMETER TOLERANCE ZONE ALLOWED
ϕ.625	ϕ.003
ϕ.624	ϕ.004
ϕ.623	ϕ.005
ϕ.622	ϕ.006
ϕ.621	ϕ.007
ϕ.620	ϕ.008

Fig. 43-3 A comparison between straightness–RFS and straightness–MMC.

STRAIGHTNESS–RFS AND MMC

Figure 43-3 shows examples of cylindrical parts where all circular elements of the surface are to be within the specified size tolerances; however, the boundary of perfect form at MMC may be violated. This violation is permissible when the feature control frame is associated with the size dimension, or attached to an extension of the dimension line. In this instance, a diameter symbol precedes the tolerance value and the tolerance is applied on either an RFS or MMC basis. Normally, the straightness tolerance is smaller than the size tolerance. The collective effect of size and form variation can produce a virtual condition equal to the MMC size plus the straightness tolerance, figure 43-2. The derived axis or center line of the feature must lie within a cylindrical tolerance zone as specified.

Straightness–RFS

When applied on an RFS basis, as in figure 43-3(A), the maximum straightness tolerance is .003 in. regardless of the feature size. Note, the absence of the modifying symbol indicates that RFS applies. The regardless-of-feature-size symbol (S), figure 43-8 (page 312), is shown only with a tolerance of position (see Unit 47). This symbol is used only in ANSI standards and has not been adopted by other countries.

Straightness–MMC

If the straightness tolerance of .003 in. is required only at MMC, further geometric variation can be permitted without jeopardizing assembly, as the features approach their least material size, figure 43-3(B).

If the tolerance can be modified on an MMC basis, this is specified on the drawing by including the symbol (M) immediately after the tolerance value in the feature control frame, figures 43-3(B) and 43-5.

If the virtual condition must be kept within the maximum material boundary of ϕ.625, the form tolerance must be specified as zero at MMC, as shown in figure 43-6.

It is sometimes necessary to insure that the geometrical tolerance does not vary over the full range permitted by size variations. For such applications a maximum limit may be applied to the geometrical tolerance, in addition to the tolerance permitted at the maximum material limit, as shown in figure 43-7.

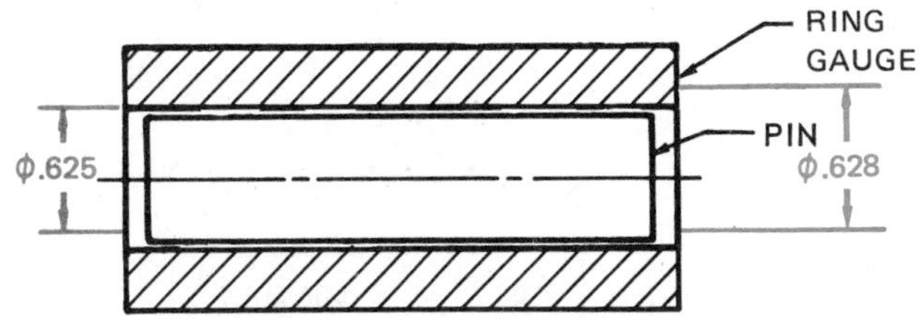

(A) MAXIMUM DIAMETER OF PIN WITH PERFECT FORM WITH GAUGE HOLE OF ϕ.628

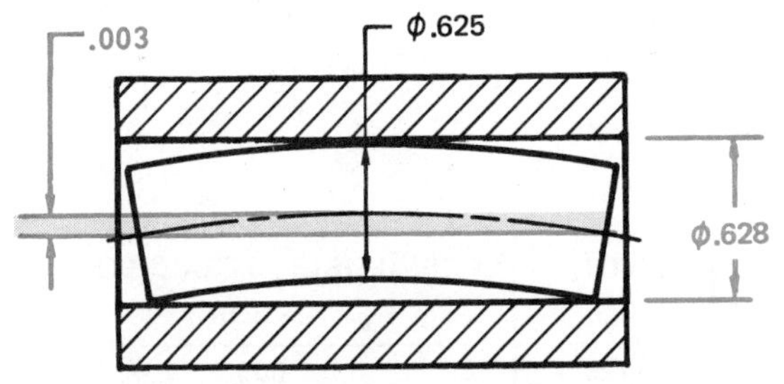

(B) WITH THE PIN AT MAXIMUM DIAMETER OF .625 THE GAUGE WILL ACCEPT THE PIN WITH UP TO A MAXIMUM OF .003 VARIATION IN STRAIGHTNESS.

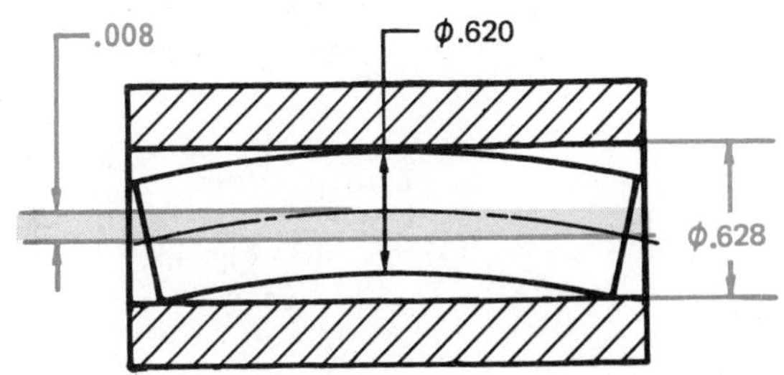

(C) WITH THE PIN AT MINIMUM DIAMETER OF .620 THE GAUGE WILL ACCEPT THE PIN WITH UP TO A MAXIMUM OF .008 VARIATION IN STRAIGHTNESS AND THE PIN IS ACCEPTABLE.

Fig. 43-4 Size and tolerance variations for pin shown in figure 43-3(B)

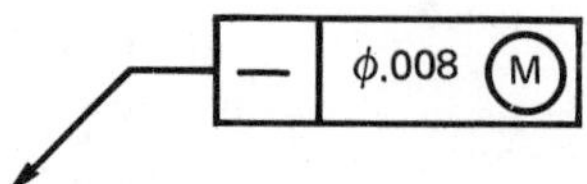

Fig. 43-5 Application of MMC symbol to tolerance

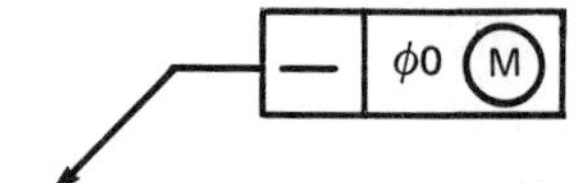

Fig. 43-6 MMC symbol with zero tolerance

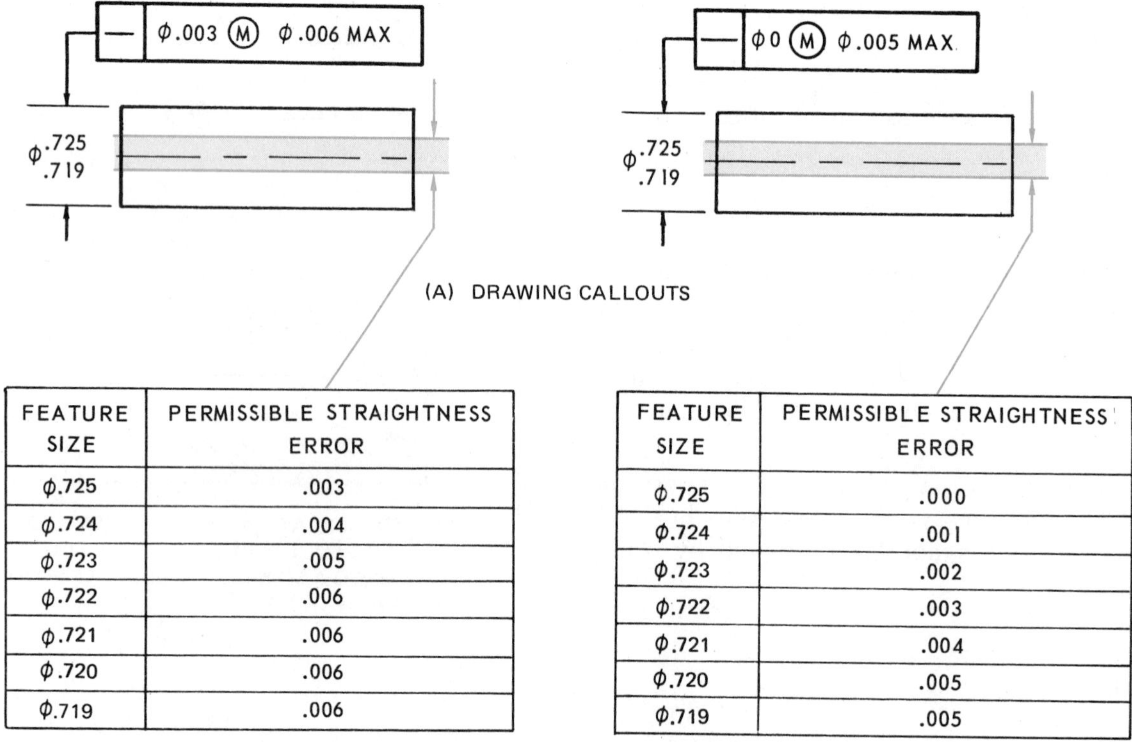

FEATURE SIZE	PERMISSIBLE STRAIGHTNESS ERROR
ϕ.725	.003
ϕ.724	.004
ϕ.723	.005
ϕ.722	.006
ϕ.721	.006
ϕ.720	.006
ϕ.719	.006

FEATURE SIZE	PERMISSIBLE STRAIGHTNESS ERROR
ϕ.725	.000
ϕ.724	.001
ϕ.723	.002
ϕ.722	.003
ϕ.721	.004
ϕ.720	.005
ϕ.719	.005

(B) PERMISSIBLE STRAIGHTNESS VARIATIONS

Fig. 43-7 Tolerance with a maximum value

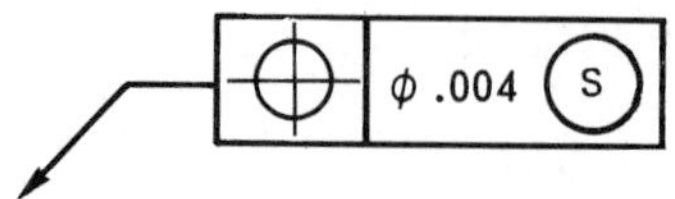

Fig. 43-8 Regardless-of-feature-size (RFS) symbol

1. What is the virtual condition for each of the features shown?

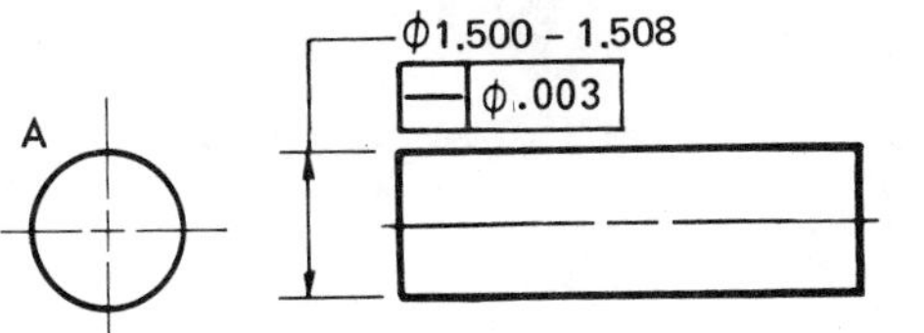

(A) ______

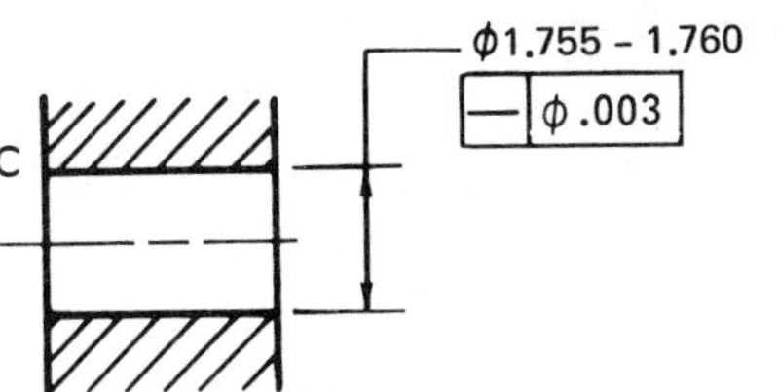

(C) ______

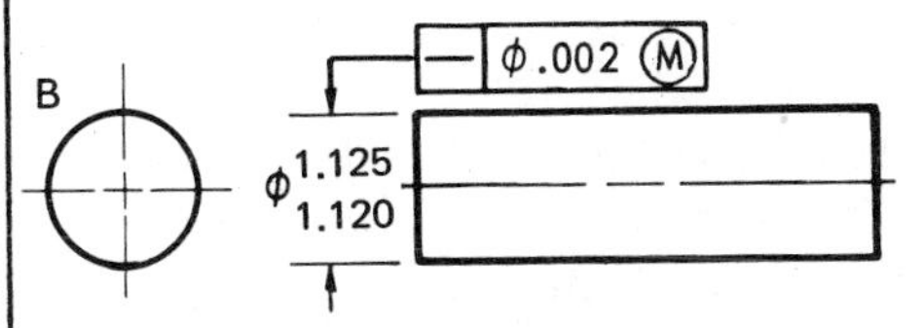

(B) ______

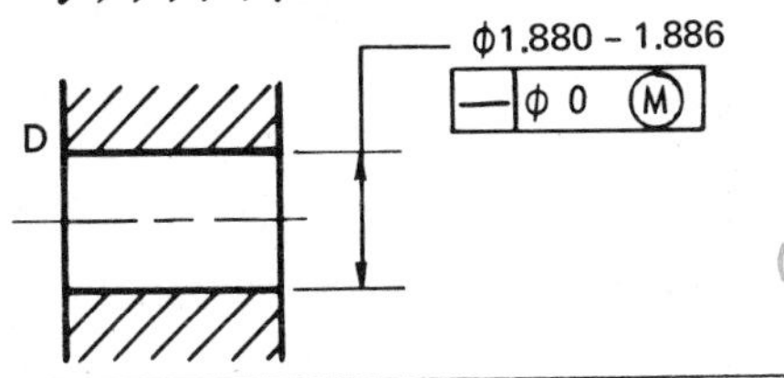

(D) ______

2. Complete the charts showing the largest permissible straightness error for the feature sizes shown.

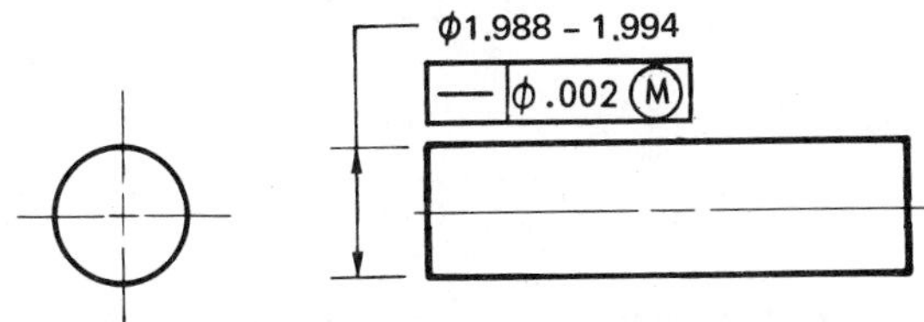

FEATURE SIZE	PERMISSIBLE STRAIGHTNESS TOLERANCE
ϕ1.994	
ϕ1.993	
ϕ1.992	
ϕ1.991	
ϕ1.990	
ϕ1.989	
ϕ1.988	

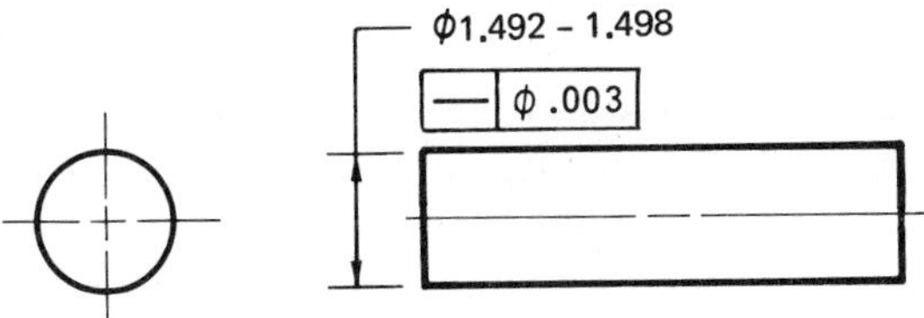

FEATURE SIZE	PERMISSIBLE STRAIGHTNESS TOLERANCE
ϕ1.498	
ϕ1.497	
ϕ1.496	
ϕ1.495	
ϕ1.494	
ϕ1.493	
ϕ1.492	

3. Design a ring gauge to check the pins shown below.

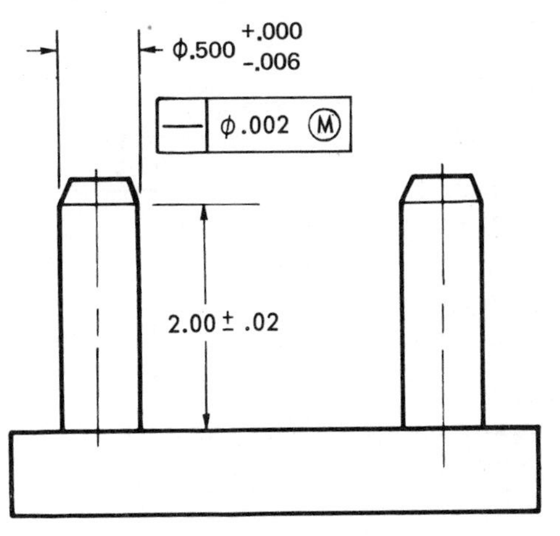

4. With reference to the drawing, are the following parts acceptable?

PART	FEATURE SIZE	STRAIGHTNESS DEVIATION
A	ϕ1.126	.002
B	ϕ1.123	.007
C	ϕ1.122	.007
D	ϕ1.121	.004
E	ϕ1.120	.009

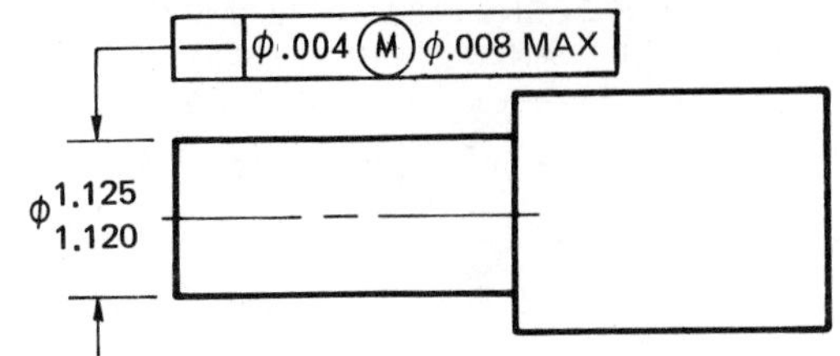

STRAIGHTNESS – RFS, MMC

A-89

UNIT 44

FORM TOLERANCES

Form tolerancing is applied to the surface of an object to control its shape in terms of flatness and circularity.

FLATNESS

Flatness is a condition in which all surface elements are in one plane. On such a surface, all line elements in two or more directions are straight. A flatness symbol and tolerance are applied to a line representing the surface of a part by a feature control frame, figure 44-1.

If the same control is wanted on two or more surfaces, a note may be added instead of repeating the symbol, figure 44-2.

Flatness on an MMC basis is very useful for controlling relatively thin parts which may be subject to bending or dishing. It is also preferred to a straightness tolerance for hexagons, squares, and other relatively long parts having opposing flat surfaces. The symbol is the same as for the flatness of a surface, except the modifier (M) is placed after the tolerance value, and the feature control frame is directed toward the thickness dimension, figure 44-3.

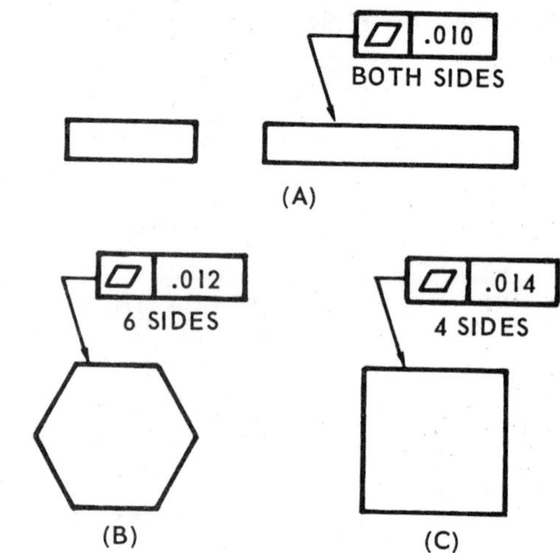

Fig. 44-2 Controlling flatness on two or more surfaces

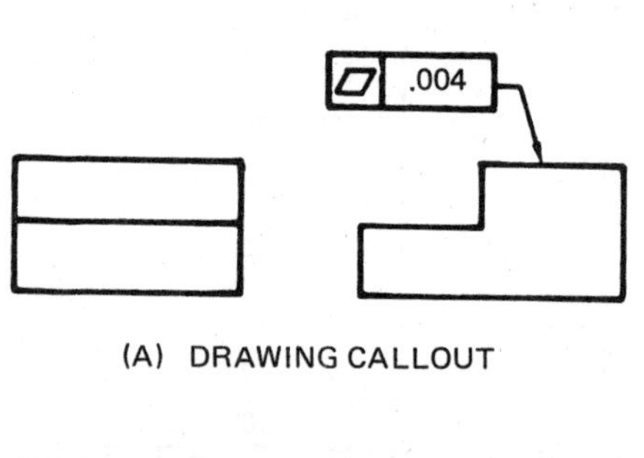

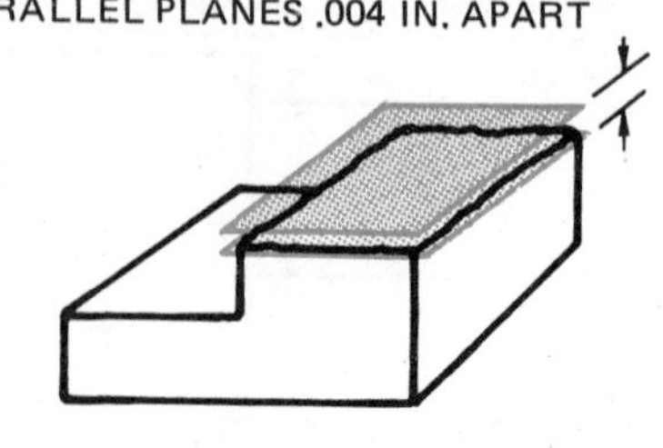

Fig. 44-1 Flatness of a surface

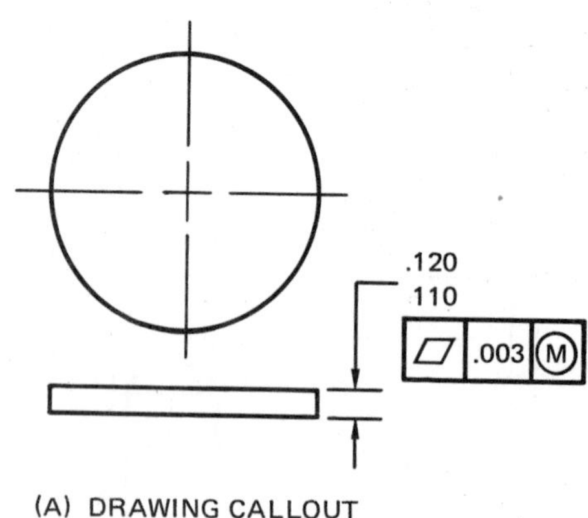

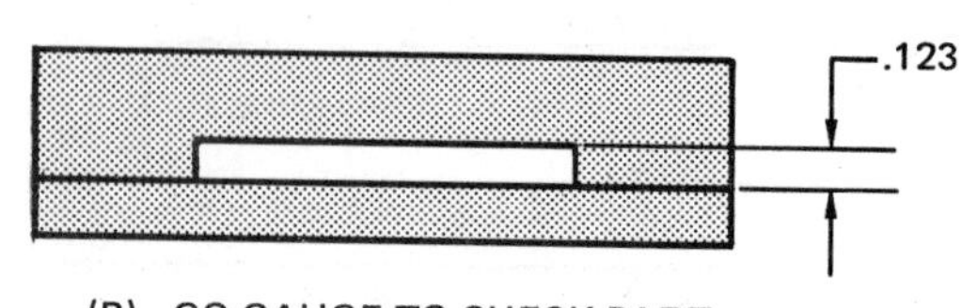

Fig. 44-3 Tolerance flatness on a MMC basis for thin parts

Where the considered surface is associated with a size dimension, the flatness tolerance must by less than the size tolerance.

To ensure that the flatness tolerance remains stable over the entire length, a maximum limit may be applied to the geometric tolerance, in addition to the tolerance permitted by the maximum material limit. The size of the limited area, e.g., 1.00 X 1.00 in. is specified to the right of the flatness tolerance, separated by a slash line, figure 44-4.

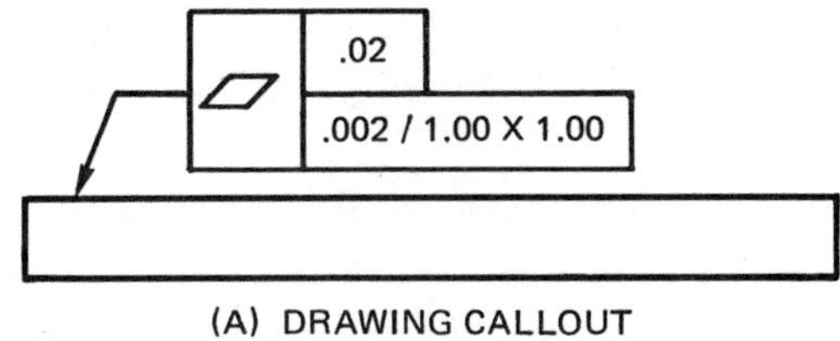

(A) DRAWING CALLOUT

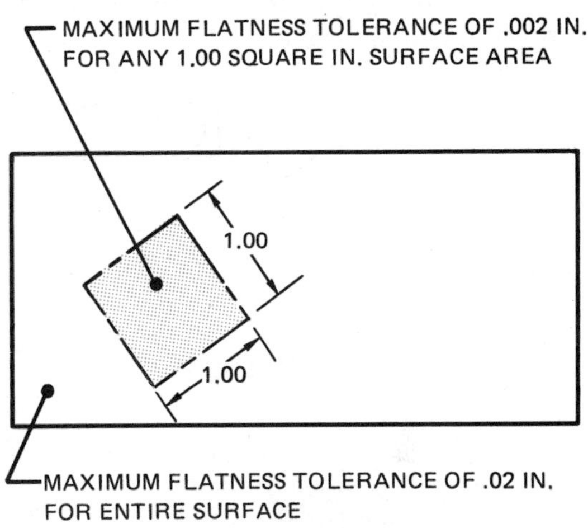

(B) INTERPRETATION

Fig. 44-4 Specifying flatness on a total and limited basis

CIRCULARITY

Circularity is a condition of a circular line or the surface of a circular feature in which all points on the line are equidistant from a common center point.

Errors of circularity (out-of-roundness) in a circular feature may occur as ovality, as lobing, or as random irregularities from a true circle. These errors are illustrated in figure 44-5.

The geometric characteristic symbol for circularity is a circle. A circularity tolerance may be specified by using this symbol in the feature control frame, figure 44-6(A). It may be expressed on an MMC basis, but when not specified as such, applies to regardless-of-feature-size.

Circularity Tolerance – RFS. A circularity tolerance specifies the width of an annular tolerance zone, bounded by two concentric circles in the same plane, within which the circular line of the feature in that plane should be, figures 44-6 and 44-7. The circularity must be less than half the size tolerance except for parts subject to a restrained specification.

Circularity on an MMC Basis. It is often advisable to ensure that errors of circularity do not cause the outline of the feature to cross the maximum material boundary or to control the amount it does cross, in order to guarantee that the part will join satisfactorily with its mating parts. This is accomplished by specifying the circularity on an MMC basis. A tolerance on this basis is generally directed to the diametral dimension, such as the zero MMC tolerance shown in

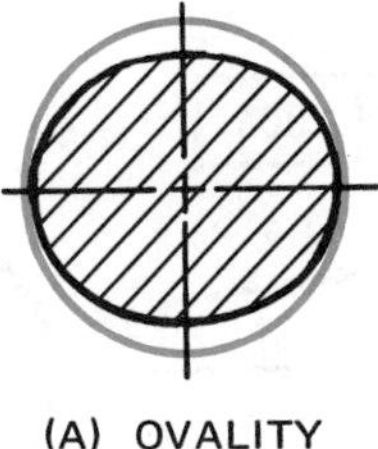

(A) OVALITY

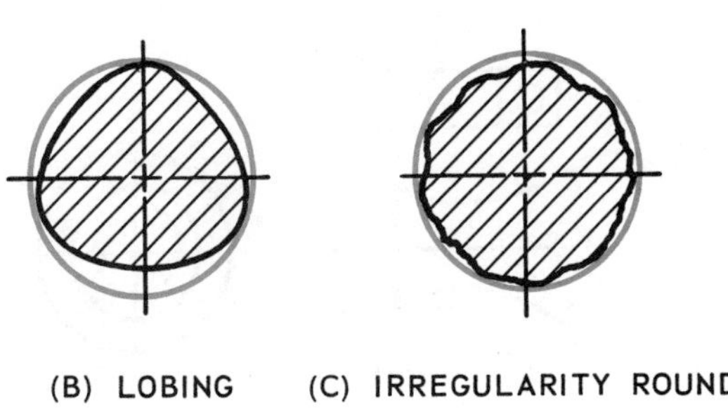

(B) LOBING (C) IRREGULARITY ROUND

Fig. 44-5 Common circularity errors

figure 44-8. This type of tolerancing does not prevent the part from crossing the least material boundary of ϕ1.996.

Although this type of tolerance controls circularity at individual cross sections, it does not ensure that the complete part is within the virtual condition, since this may be affected by lack of straightness. Therefore, to control cylindrical parts on an MMC basis, a cylindricity tolerance is usually preferred.

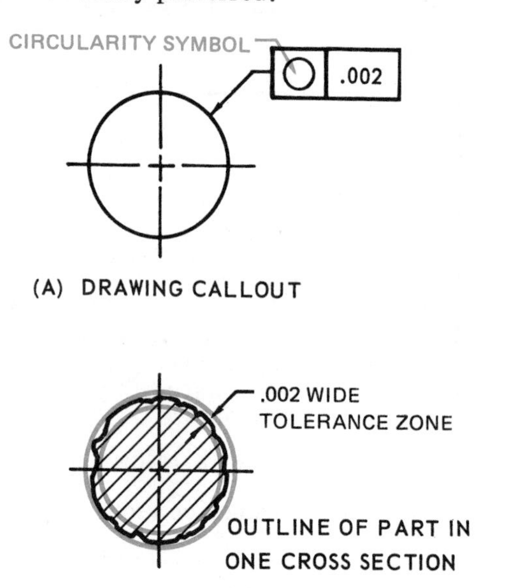

Fig. 44-6 Circularity tolerance

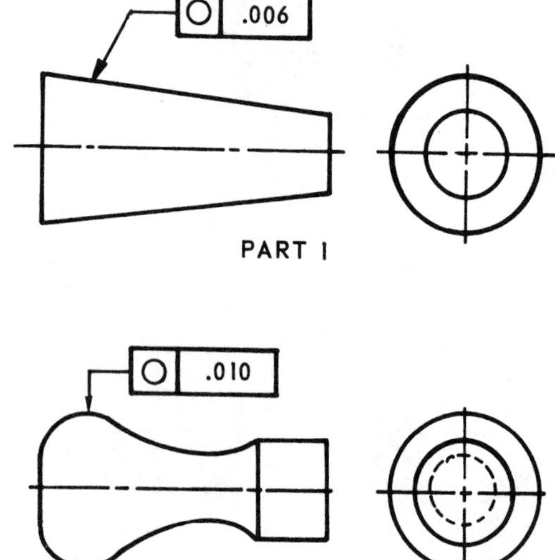

Fig. 44-7 Circularity tolerance of noncylindrical parts

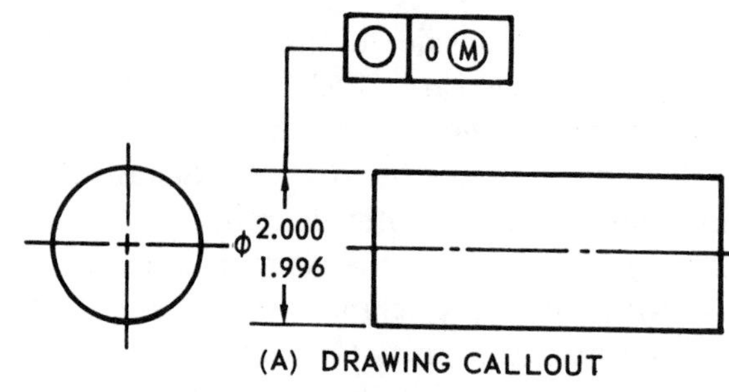

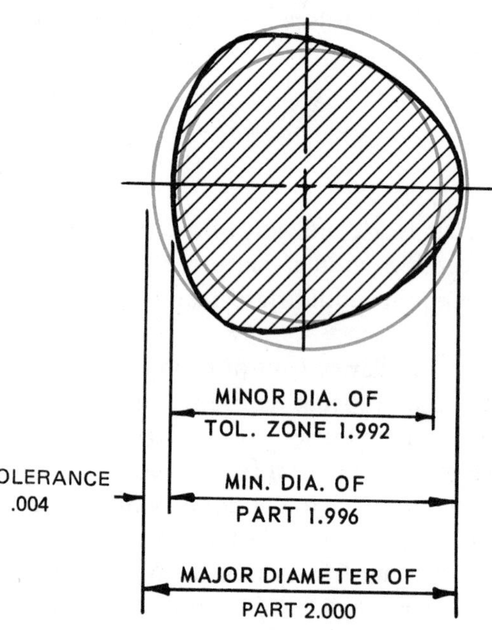

(B) FULL INTERPRETATION

Fig. 44-8 Circularity tolerance on an MMC basis

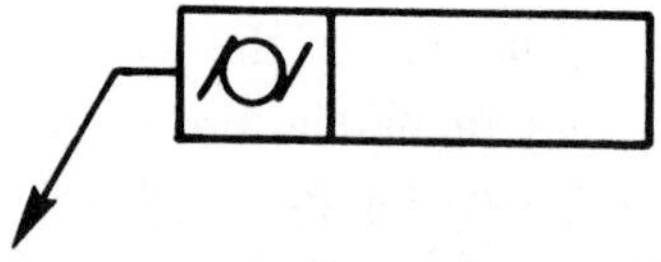

Fig. 44-9 Cylindricity symbol

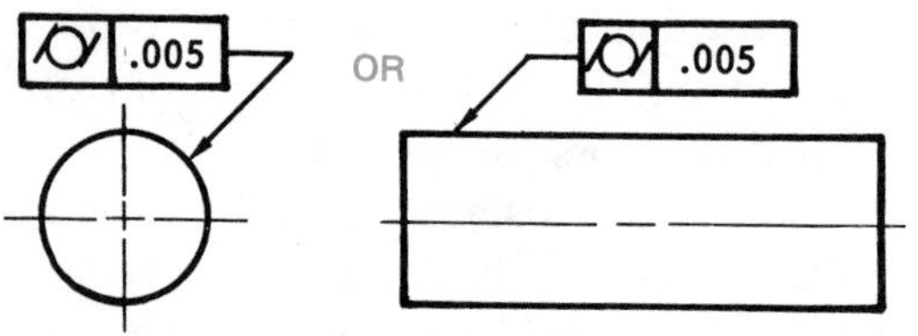

Fig. 44-10 Cylindricity tolerance directed toward either view

CYLINDRICITY

Cylindricity is the condition of a surface forming a cylinder where the surface elements in cross sections parallel to the axis are straight and parallel and in cross sections perpendicular to the axis are round. Cylindricity is a combination of geometrical form tolerances for circularity, straightness, and parallelism of the surface elements.

Cylindricity tolerances can only be applied to cylindrical surfaces, such as round holes and shafts. A conical surface must be controlled by a combination of tolerances for circularity, straightness, and angularity.

Errors of cylindricity may be caused by out-of-roundness, such as ovality or lobing; by errors of straightness caused by bending or diametral variation; by errors of parallelism such as conicity or taper; and by random irregularities from a true cylindrical form.

The geometric characteristic symbol for cylindricity consists of a circle with two tangent lines at 60 degrees, figure 44-9. It is used in a feature control frame and is directed toward the cylindrical surface, in either the side or end view, figures 44-10 and 44-11. When not modified on an MMC basis, a cylindricity tolerance applies regardless of feature size. Additionally, the surface must be within the specified limits of size.

The *tolerance zone* is the annular space between two coaxial cylinders having a difference in radii equal to the specified tolerance, within which the entire surface of the feature must lie. The axis of the annular tolerance zone does not always coincide with the center line of the part. The diameters of the tolerance zone do not necessarily fall within the diameter limits of the part.

Because the measurement of cylindricity on an RFS basis is a tedious and time-consuming procedure, a cylindricity tolerance should be on an MMC basis.

When cylindricity on an MMC basis is required, it is expressed as shown in figure 44-11. Such a tolerance on a cylindrical feature is easily checked by means of a ring or plug gauge to encompass the whole length. It should be noted, however, that the diameter of such a gauge will be equal to the maximum material size of the feature plus (or minus for internal diameters) twice the specified cylindricity tolerance. For the part shown in figure 44-11 it would be ϕ2.0030.

Whenever a fit between mating cylindrical features is required, a cylindricity tolerance of zero MMC is usually the preferred condition because it controls straightness and parallelism as well as circularity.

However, a cylindricity tolerance larger than zero may be specified. It is also satisfactory to specify a larger diameter tolerance with a zero geometrical tolerance, figure 44-13. Instead of specifying a maximum diameter of 1.247 in. with a cylindricity tolerance of .003 in., it would be better to specify a maximum diameter of ϕ1.250 with a cylindricity tolerance of zero MMC. The cylindricity tolerance must be less than half the size tolerance.

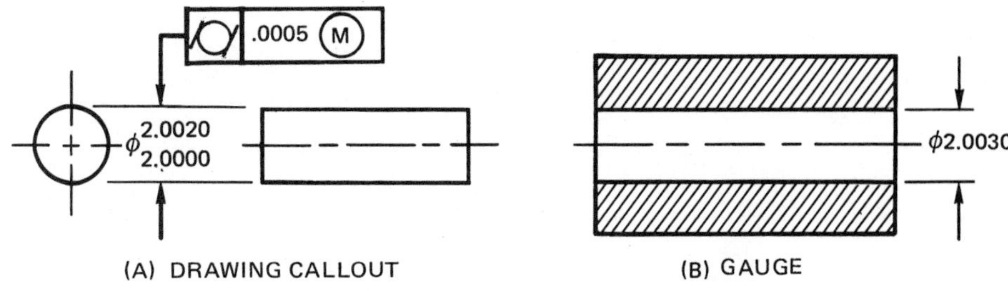

Fig. 44-11 Cylindricity tolerance on an MMC basis

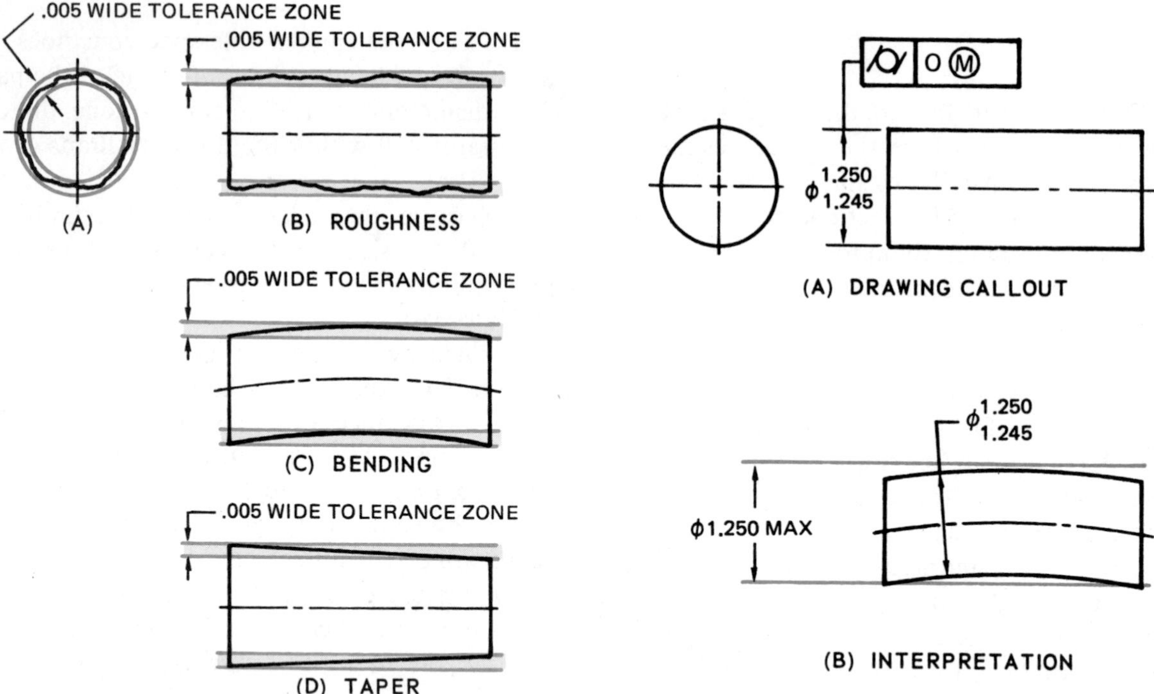

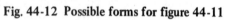

Fig. 44-12 Possible forms for figure 44-11

Fig. 44-13 Cylindricity tolerance on a zero MMC basis

ASSIGNMENTS

1. Add the following flatness tolerances to surface M of Part 1.
   – Maximum flatness tolerance of .01 for entire surface.
   – Limited area flatness tolerance of .002 for any .75 X .75 area.
2. Show the flatness and size tolerance zones for Part 2.
3. Sketch a GO-gauge to check Part 3.

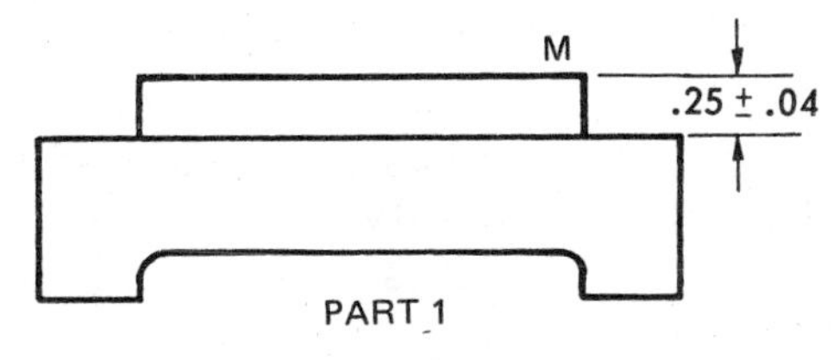

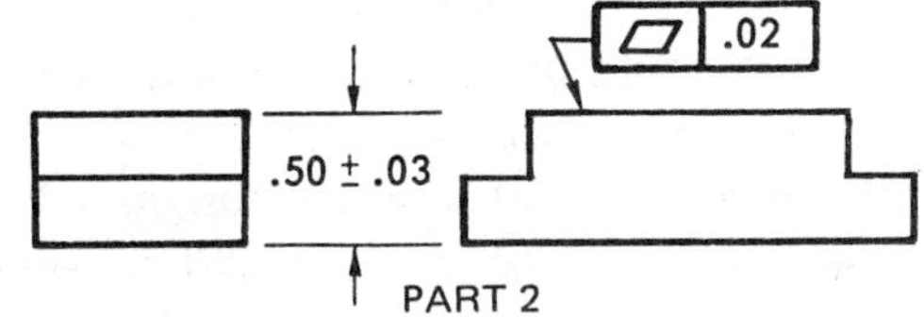

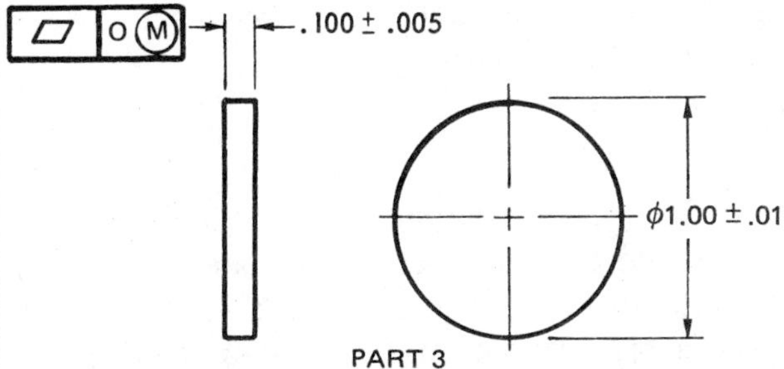

4. If measurements made at cross sections **A-A**, **B-B** and **C-C** indicate that all points on the circumference fall within the two annular rings shown, does the part meet the specified circularity tolerances? If not, explain why.

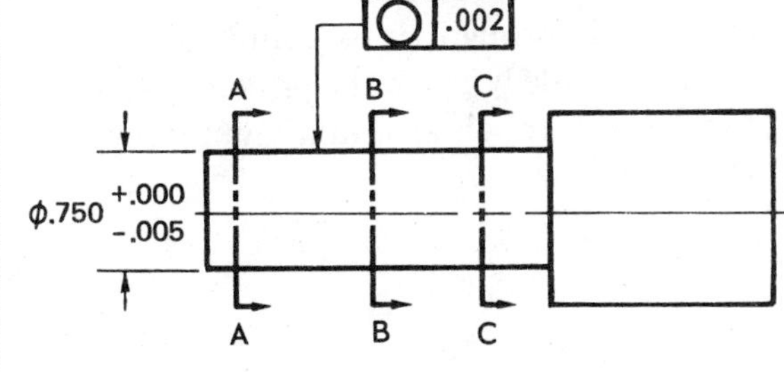

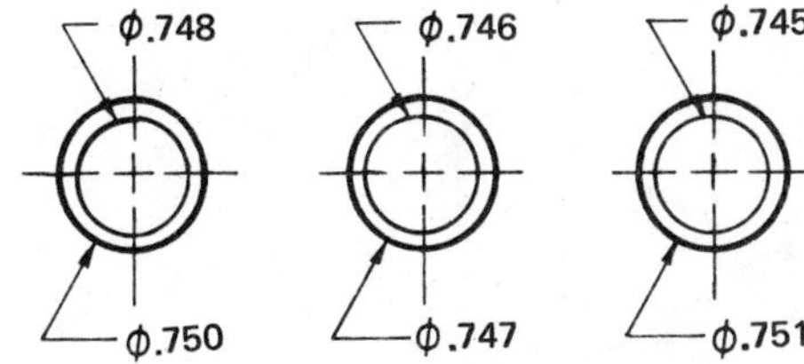

5. Sketch the tolerance zone for the cylindricity tolerance shown indicating its size and shape.

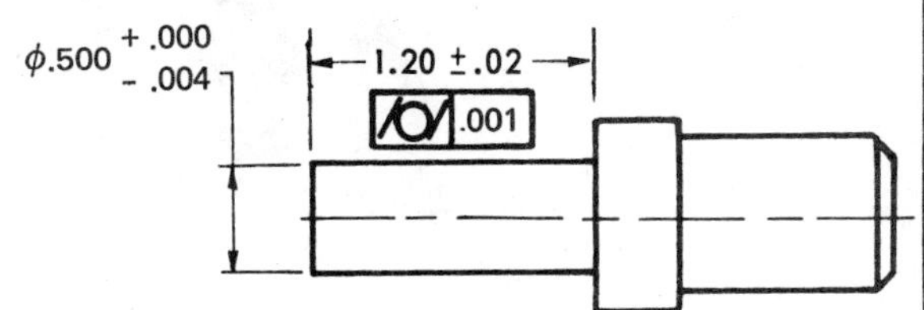

6. The parts shown below must assemble without interference. Add the largest cylindricity tolerances on an MMC basis to each part which will ensure this condition.

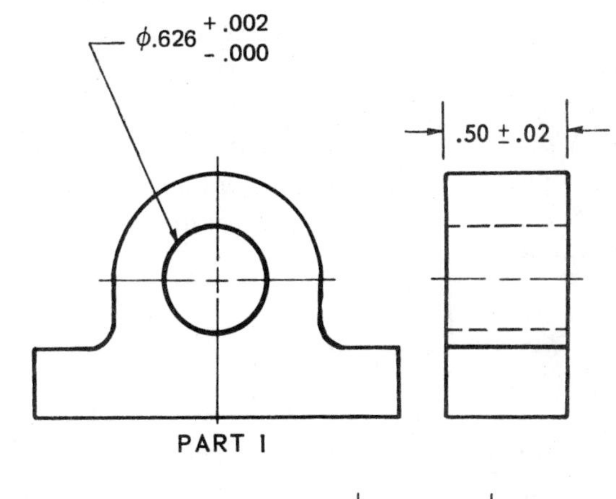

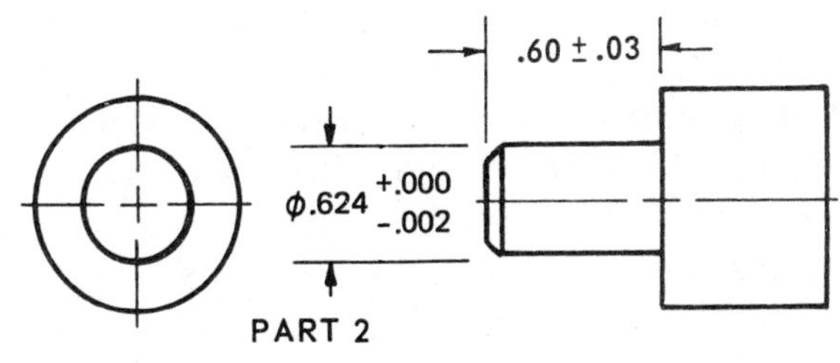

7. Add circularity tolerances to the .750 and .500 diameter features so that the features will not cross the boundary of perfect form at the maximum material size.

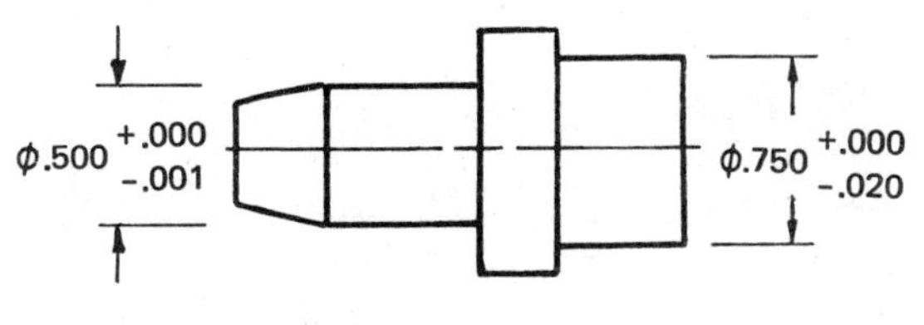

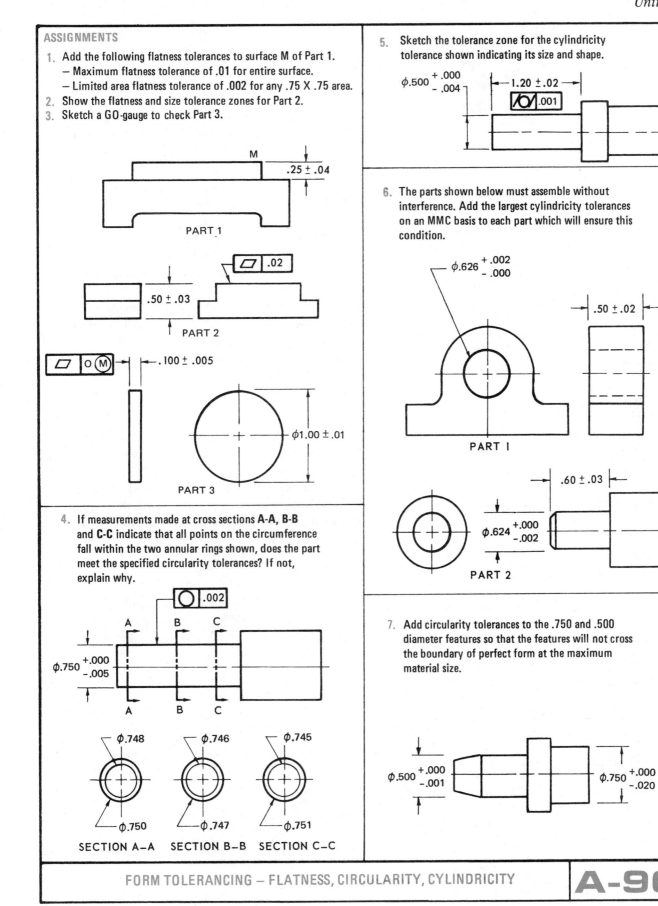

FORM TOLERANCING – FLATNESS, CIRCULARITY, CYLINDRICITY

A-90

UNIT 45

DATUMS AND THE THREE-PLANE METHOD OF TOLERANCING

A *datum* is a point, line, plane, or other geometrical surface from which dimensions are measured, or to which geometrical tolerances are referenced. A datum has an exact form and represents an exact or fixed location for purposes of manufacture or measurement.

A *datum feature* is a feature of a part, such as an edge, surface, or hole, which forms the basis for a datum or is used to establish the location of a datum.

DATUMS FOR GEOMETRICAL TOLERANCING

Datums are exact geometrical points, lines, or surfaces, each based on one or more datum features of the part. Surfaces are usually either flat or cylindrical, but other shapes are used when necessary. Since the datum features are physical surfaces of the part, they are subject to manufacturing errors and variations. For example, a flat surface of a part, if greatly magnified, will show some irregularity. If brought into contact with a perfect plane, this flat surface will touch only at the highest points, figure 45-1. The true datums exist only in theory but are considered to be in the form of locating surfaces of machines, fixtures, and gauging equipment on which the part rests or with which it makes contact during manufacture and measurement.

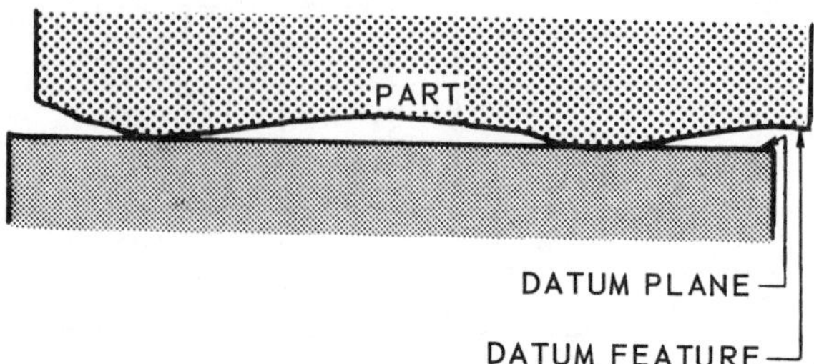

Fig. 45-1 Magnified section of a flat surface

Although these surfaces are not perfect geometrical surfaces, they are intended to be of a sufficiently high quality to render any errors they introduce insignificant in comparison with the great tolerances.

Usually only one datum is required for orientation purposes, but positional relationships may require a datum system consisting of two or three datums. These datums are designated as *primary, secondary,* and *tertiary.* When these datums are mutually perpendicular plane surfaces, they are referred to as a *three-plane system* or a *datum reference frame.*

Primary Datum

If the primary datum feature is a flat surface, it could be laid on a suitable plane surface, such as the surface of a gauge, which would then become a primary datum, figure 45-2. Theoretically, there will be a minimum of three high spots on the flat surface coming in contact with the gauge surface.

Secondary Datum

If the part is brought into contact with a secondary plane while lying on the primary plane, it will theoretically touch at a minimum of two points, figure 45-3.

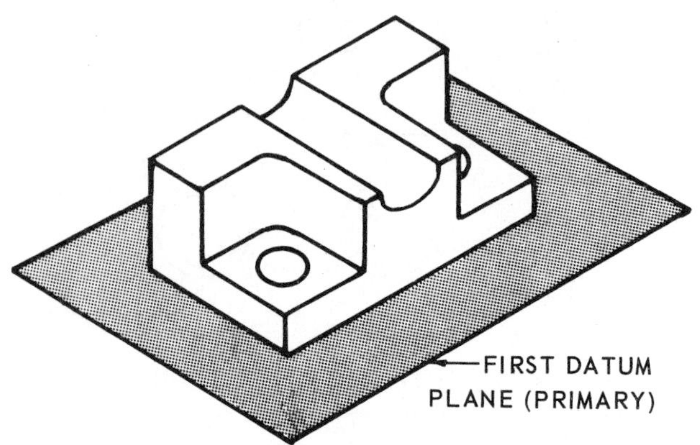

Fig. 45-2 Primary datum

Tertiary Datum

The part can be slid along while maintaining contact with both the primary and secondary planes until it contacts a third plane, figure 45-4. This plane then becomes the tertiary datum and the part will, in theory, touch it at only one point.

These three planes constitute a datum system from which measurements can be taken. They will appear on the drawing as shown in figure 45-5, except that the datum features should be identified in their correct sequence by the methods described later in the unit.

DATUM IDENTIFYING SYMBOL

Datum symbols have two functions. They indicate the datum surface or feature on the drawing and identify the datum feature so it can be easily referred to in other requirements.

There are two methods of datum symbolization for such purposes: the American method used in ANSI standards, and the ISO method which is used in most other countries.

ANSI Symbol

In the ANSI system every datum feature is identified by a capital letter enclosed in a rectangular box. A dash is placed before and after the letter to indicate it applies to a datum feature, see figure 45-6(A).

ISO Symbol

The ISO method is used in Canadian, British, and most other national standards. The datum symbol is a triangle, figure 45-6(B). When identification of a datum is necessary to permit reference to it in another requirement, each datum feature is identified by a capital letter placed in a square frame and connected to the datum indicator symbol.

This identifying symbol may be directed to the datum feature in any of the following methods: by attaching a side, end, or corner of the symbol frame to an extension line from the feature, figure 45-7; by running a leader with arrowheads from the symbol frame to the feature, figure 45-8; by adding the symbol to a dimension or a feature control frame pertaining to the feature, figures 45-8 and 45-9.

Multiple Datum Features

If a single datum is established by two datum features, such as two flat surfaces, figure 45-10, the features are identified by separate letters. Both letters are then placed in the same compartment of the feature control frame, separated by a dash. The datum in this case is the common line between the two datum features.

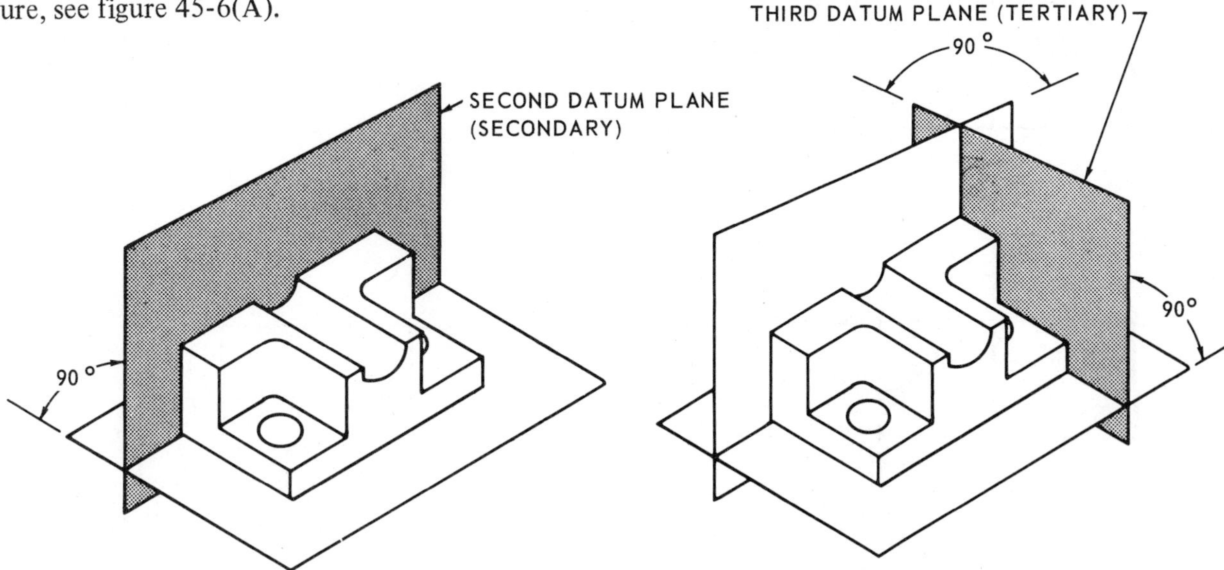

Fig. 45-3 Secondary datum

Fig. 45-4 Tertiary datum

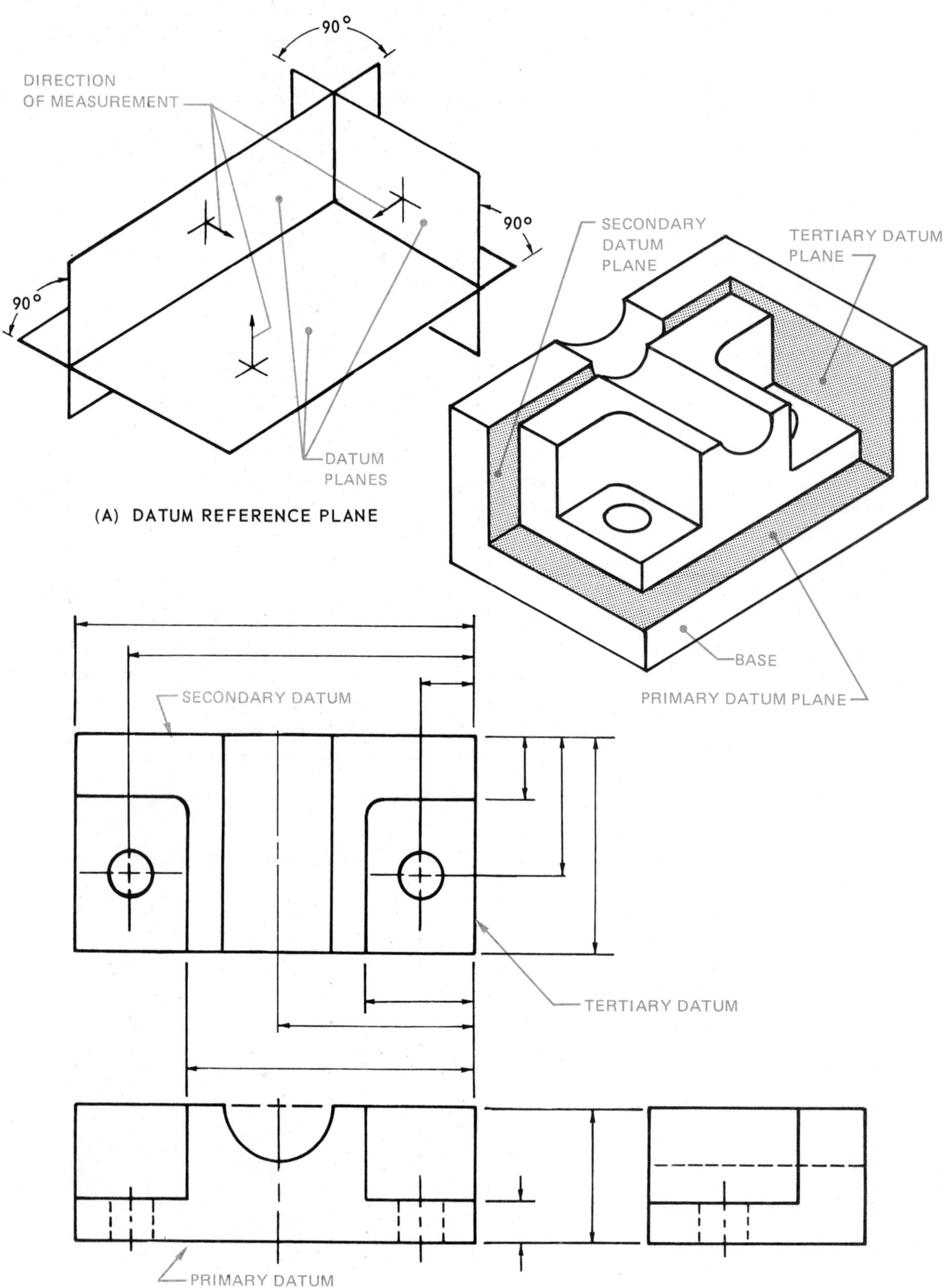

Fig. 45-5 Three-plane datum system

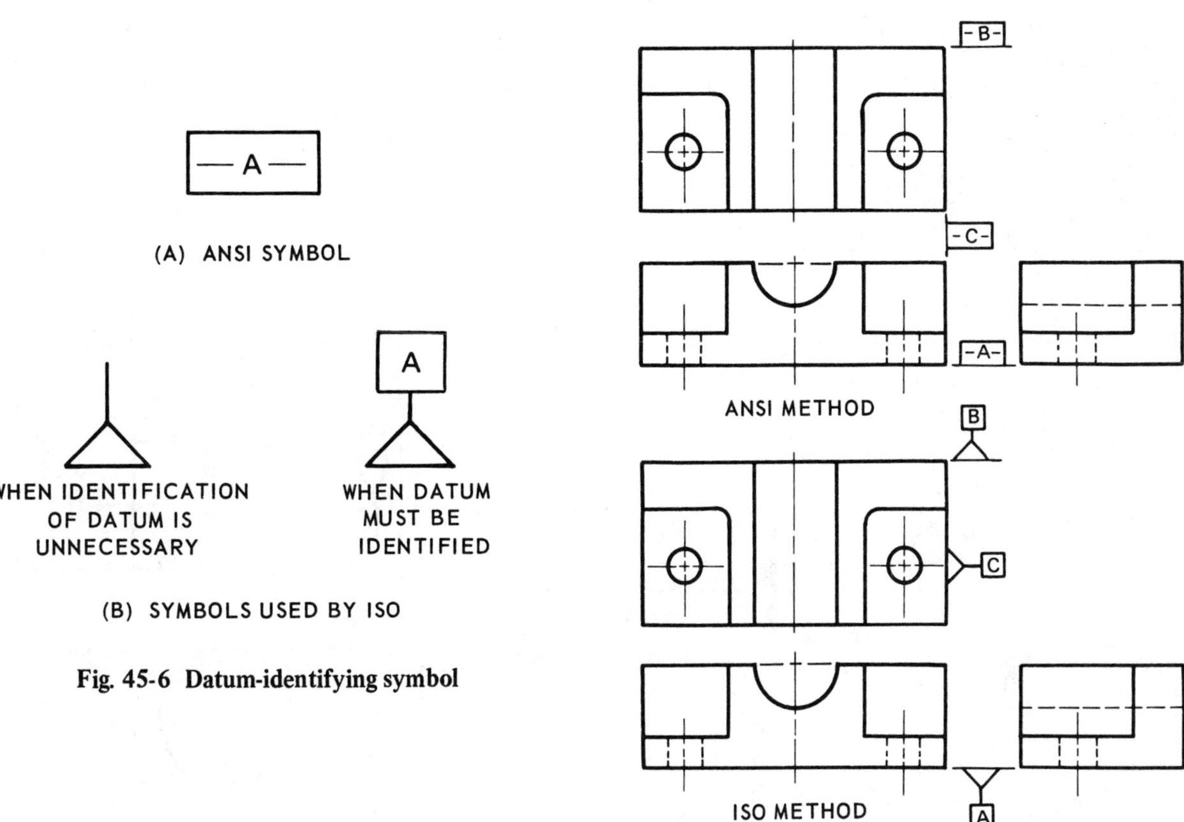

Fig. 45-6 Datum-identifying symbol

Fig. 45-7 Identifying datums for part shown in figure 45-5

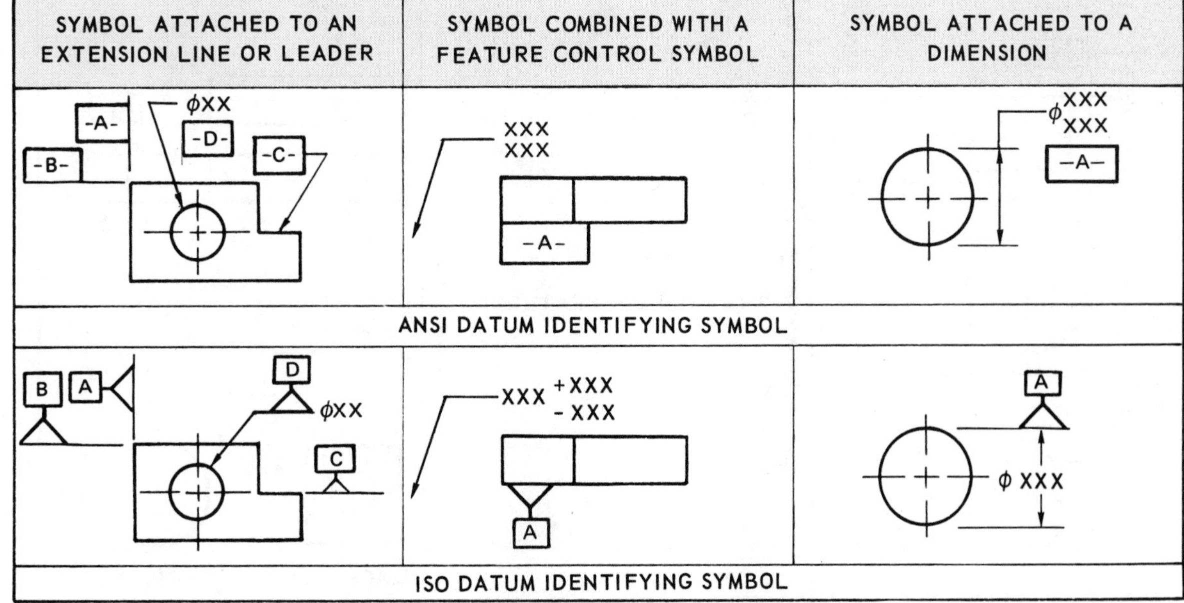

Fig. 45-8 Common methods of connecting datum-identifying symbol to datum

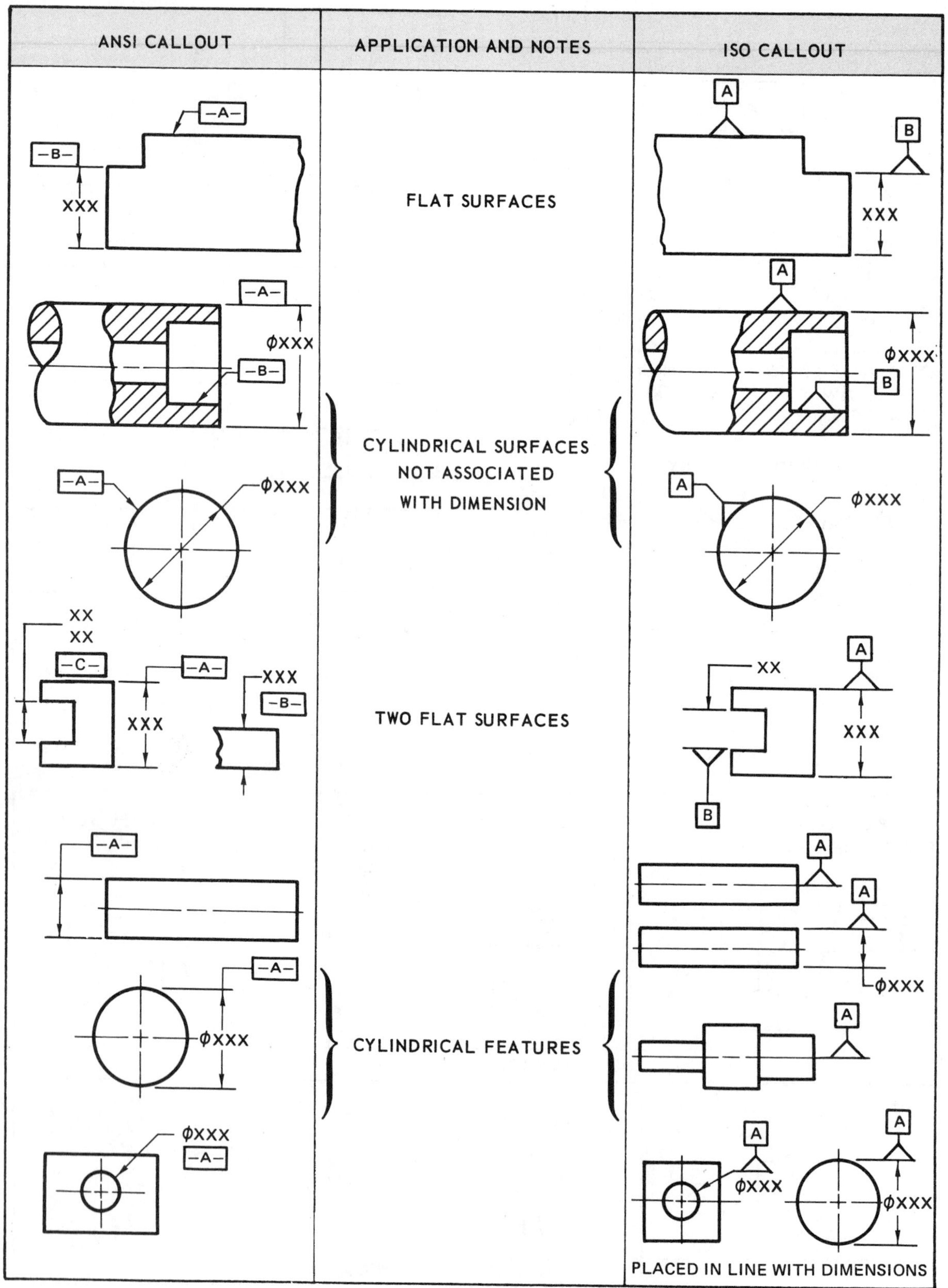

Fig. 45-9 Methods of connecting datum-identifying symbol to feature

Combined Feature Control and Datum Identifying Symbol

When a feature is controlled by a positional or form tolerance and serves as a datum, the feature control and datum identifying symbol are combined, figure 45-11 through 45-13. Applications of combined feature control and datum identifying symbols are shown in Units 46 to 50.

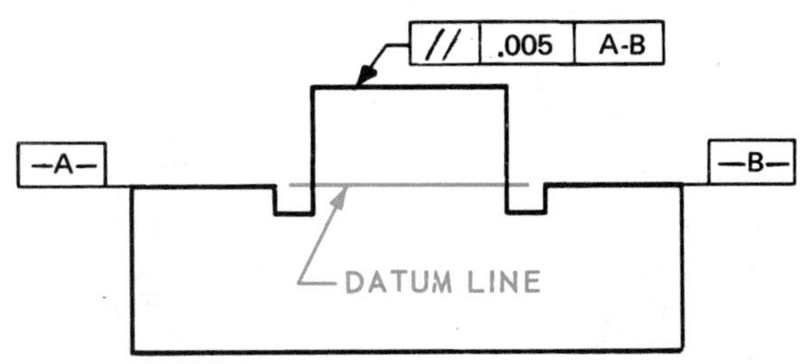

Fig. 45-10 Two datum features for one datum

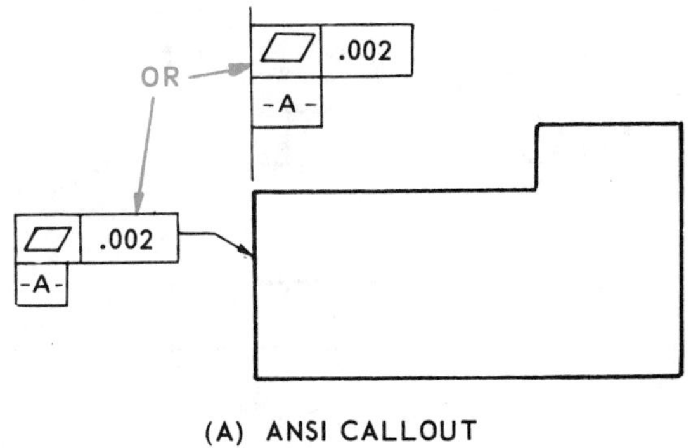

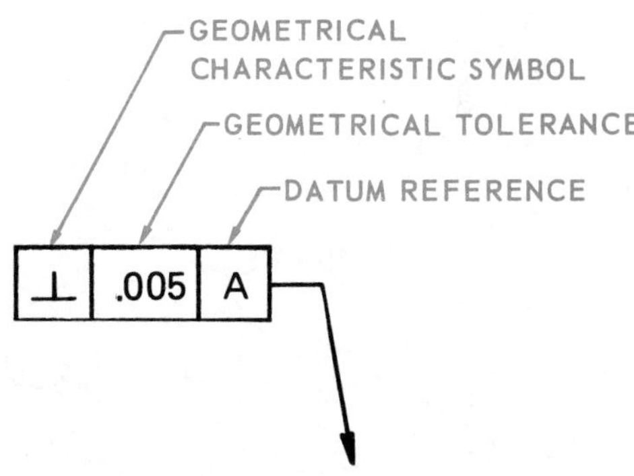

Fig. 45-12 Feature control frame referenced to a datum

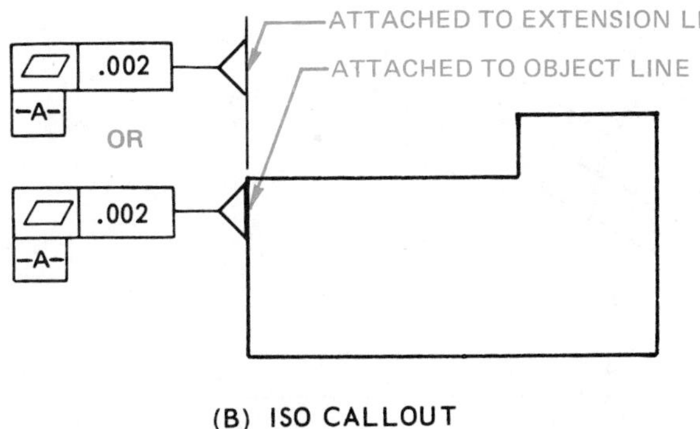

Fig. 45-11 Datum surface attached to the feature control frame

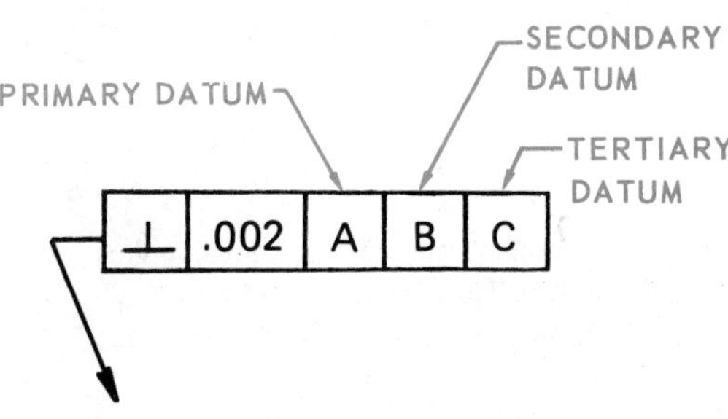

Fig. 45-13 Feature referenced to more than one datum

ASSIGNMENTS

1. Show the following information on the two-view drawing:
   – Surface **A** is datum **A** and is to be straight within 0.2 for the 100 mm length; the straightness error should not exceed 0.05 for any 25 mm length.
   – Surface **B** is datum **B** and is to be flat within 0.1. Surface **B** is parallel to datum **C-D** within 0.2.
   – Surface **C** and **D** are datum features **C** and **D** respectively which form a single datum. Surfaces **C** and **D** each to be flat within 0.05.

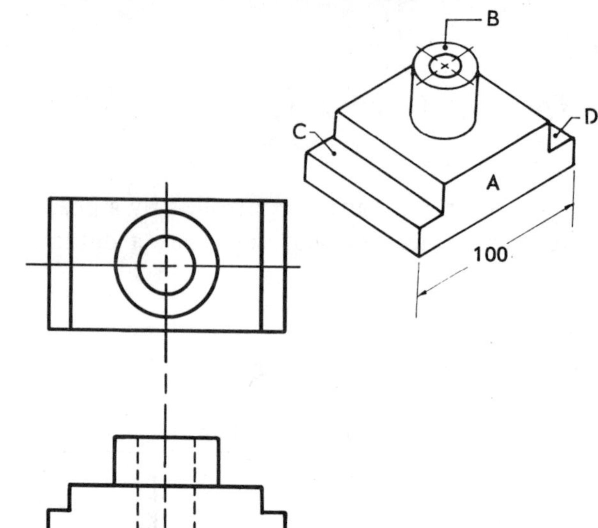

2. A. Pins 1, 2, and 3 are used to establish the secondary and tertiary datums for the part shown. What is used for the primary datum?
   B. On the two-view drawing of the part identify the primary, secondary and tertiary datum planes as **A, B,** and **C** respectively.
   C. How far is the center of the hole from (1) tertiary datum, (2) secondary datum?
   D. The back of the slot is to be flat within 0.3 mm. Place the form tolerance on the drawing.

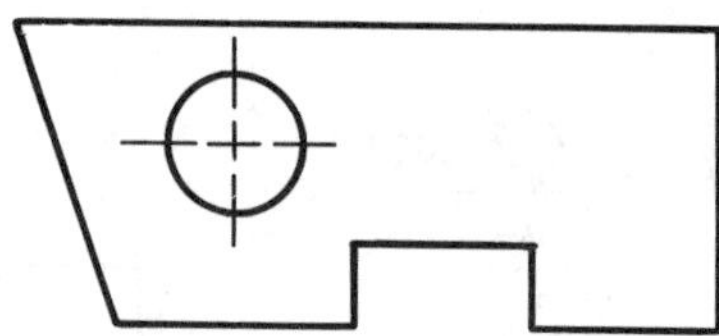

(A) ________

(C) ________

____

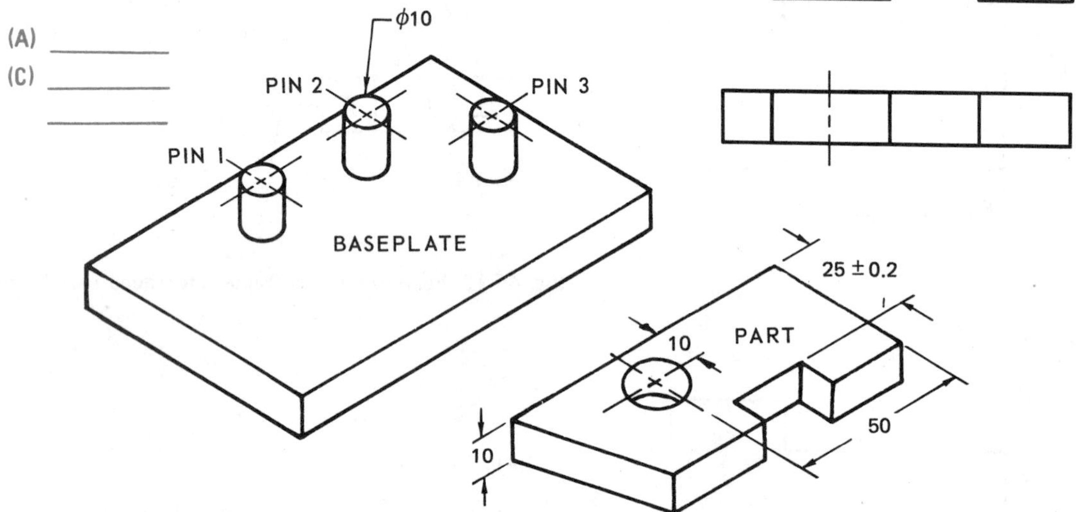

3. What is the minimum number of contact points in a three-plane datum system for (A) primary datum, B) secondary datum, (C) tertiary datum?

(A) ________ (B) ________ (C) ________

4. A. The bottom surface of the part shown is to be identified as datum **B** and should be flat within 0.1 mm. Specify both the datum and the tolerance on the drawing.
   B. What is the maximum height of the part?

(B) ________

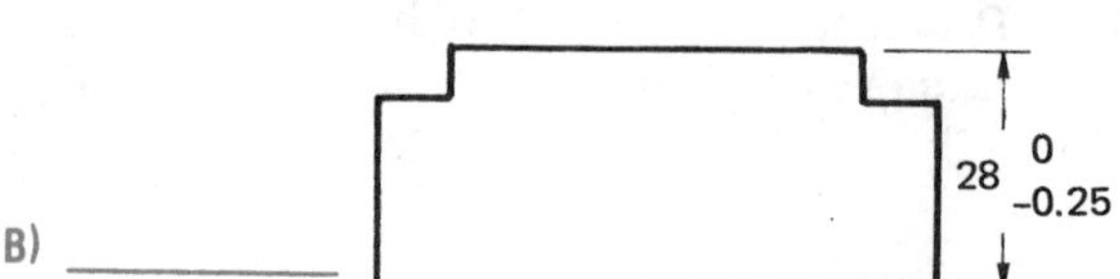

METRIC DIMENSIONS ARE IN MILLIMETERS	DATUMS AND DATUM DIMENSIONING	A-91M

UNIT 46

ORIENTATION TOLERANCING

Orientation is the angular relationship existing between two or more lines, surfaces, or other features. *Angularity* is the general geometric characteristic for orientation. This term describes relationships of any angle, between straight lines or surfaces such as flat or cylindrical surfaces. Special terms are used for two particular types of angularity. These are *perpendicularity,* or squareness, for features related to each other by a 90-degree angle; and *parallelism* for features related to one another by an angle of zero, figure 46-1.

An orientation tolerance indicates a relationship between two or more features. Whenever possible, the feature to which the controlled feature is related should be designated as a datum.

Tolerance of Flat Surfaces – Regardless of Feature Size

Figure 46-2 shows three simple parts in which one flat surface is designated as a datum

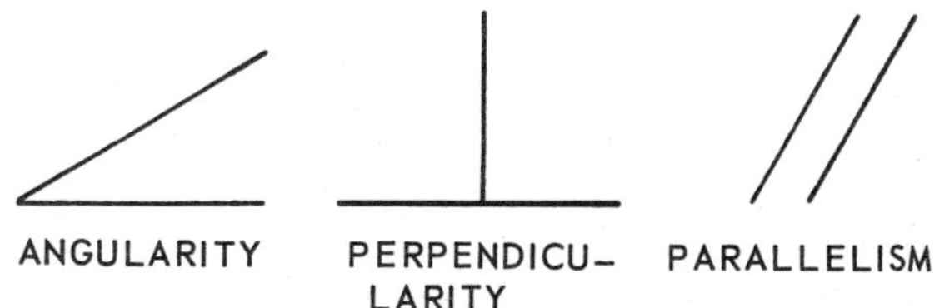

Fig. 46-1 Orientation symbols

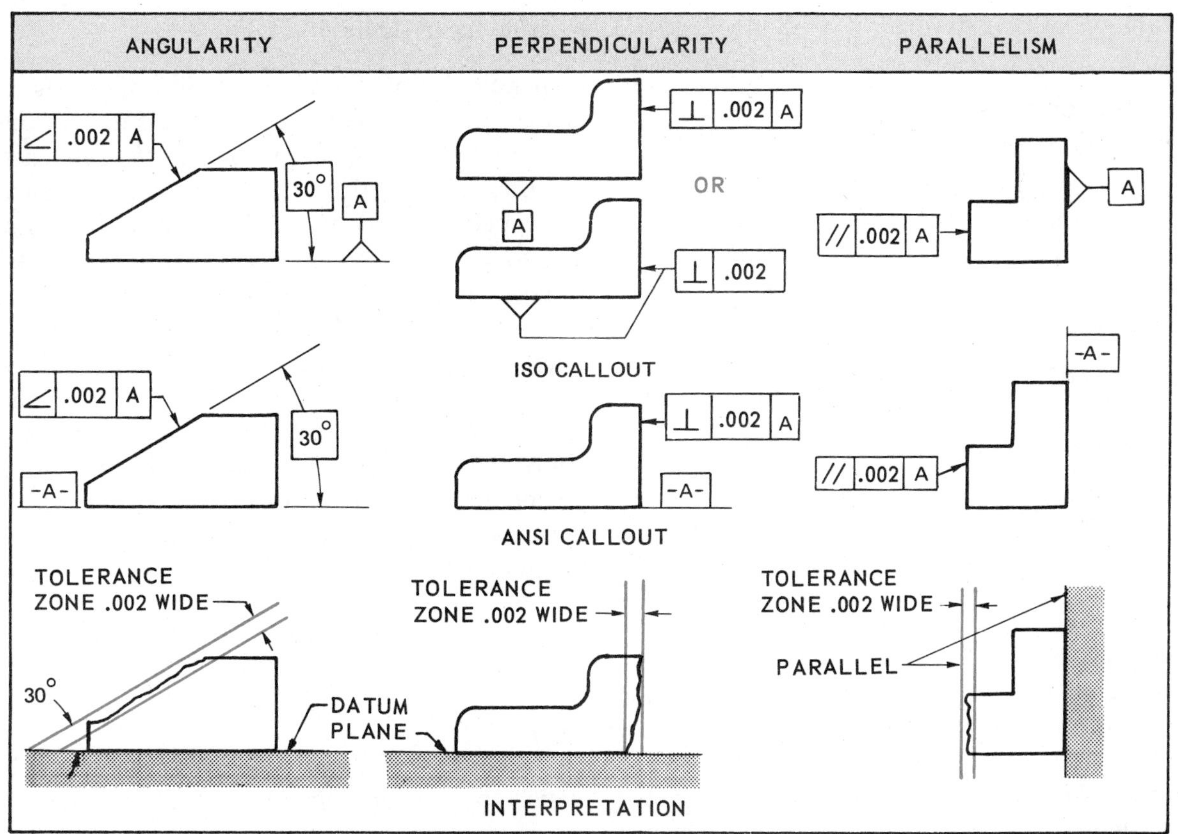

Fig. 46-2 Orientation tolerancing

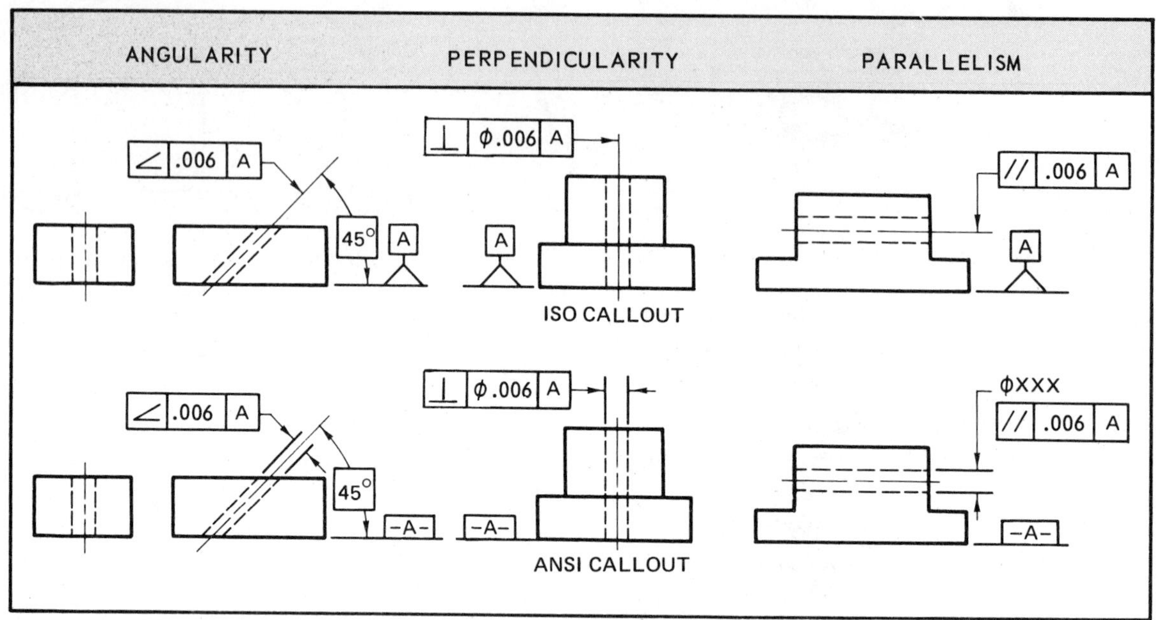

Fig. 46-3 Orientation of lines and surfaces

feature and another flat surface is related to it by one of the orientation tolerances.

Angularity. is the condition of a surface or axis at a specified angle (other than 30°) from a datum plane or axis, figures 46-2, 46-3, and 46-4.

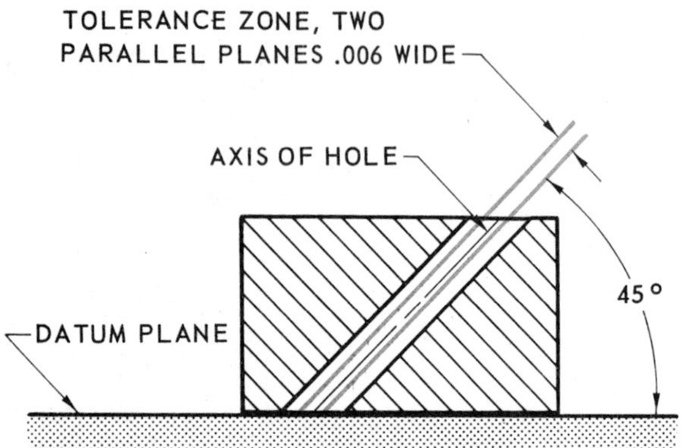

Fig. 46-4 Tolerance zone for angularity shown in figure 46-3

Parallelism. is the condition of a surface equidistant at all points from a datum plane or an axis equidistant along its length to a datum axis. A parallelism tolerance specifies: (a) a tolerance zone defined by two planes or lines parallel to a datum plane, or axis, within which the line elements of the surface (as in figure 46-2) or axis of the considered feature must lie (see figure 46-3); or (b) a cylindrical tolerance zone whose axis is parallel to a datum axis within which the axis of the considered feature must lie (as in figure 46-5).

Perpendicularity. is the condition of a surface, median plane, or axis at a right angle to a datum plane or axis. A perpendicularity tolerance specifies one of the following: (a) a tolerance zone specified by two parallel planes perpendicular to a datum plane, or axis, within which the surface or median plane of the considered feature must lie (see figures 46-2 and 46-3); (b) a tolerance zone defined by two parallel planes perpendicular to a datum axis within which the axis of the considered feature must lie (see figure 46-6); (c) a cylindrical tolerance zone perpendicular to a datum plane within which the axis of the considered feature must lie (see figure 46-3).

Tolerancing of Lines Related to Surfaces – Regardless of Feature Size

Figure 46-3 shows some simple parts in which the axis or center line of a hole is related by an orientation tolerance to a flat surface. The flat surface is designated as the datum feature.

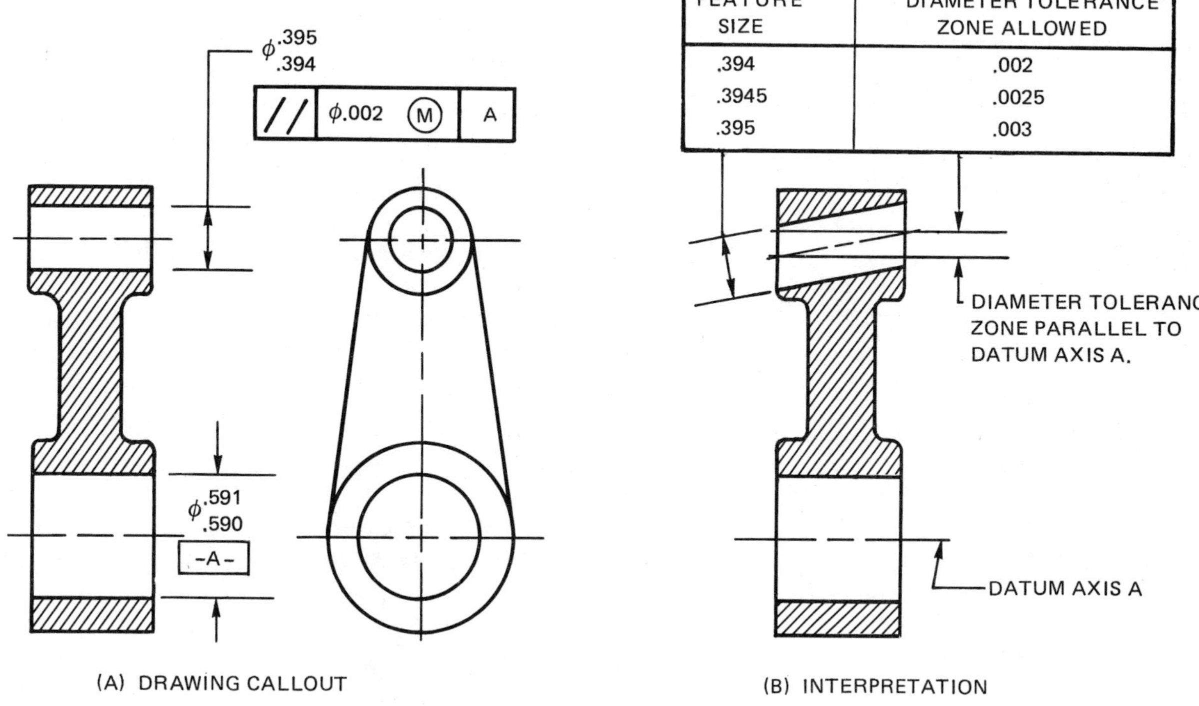

(A) DRAWING CALLOUT

(B) INTERPRETATION

Fig. 46-5 Specifying parallelism for an axis

The center line of the hole must be contained within a tolerance zone consisting of the space between two parallel planes. These planes are separated by a specified tolerance of .006 in. and are related to the datum by one of the basic angles, 45 degrees, 90 degrees, or 0 degrees. Figure 46- 4 clearly illustrates the tolerance zone for angularity.

When the tolerance is one of perpendicularity, the tolerance zone planes can be revolved around the feature axis without affecting the angle. The tolerance zone therefore becomes a cylinder. This cylindrical zone is perpendicular to the datum and has a diameter equal to the specified tolerance, figure 46-7.

Control in Two Directions

The feature control frame shown in figure 46-3 controls angularity and parallelism with the base (datum A) only. If control with a side is also required, the side should be designated as

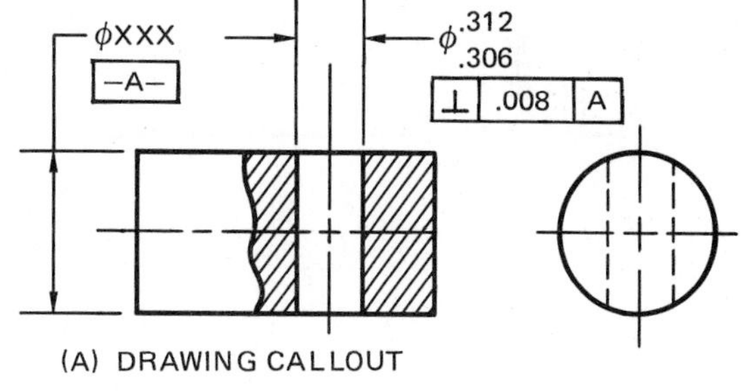

(A) DRAWING CALLOUT

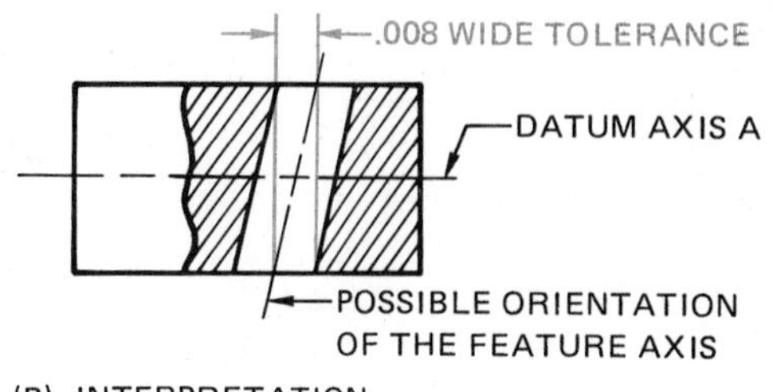

(B) INTERPRETATION

Fig. 46-6 Specifying perpendicularity for an axis

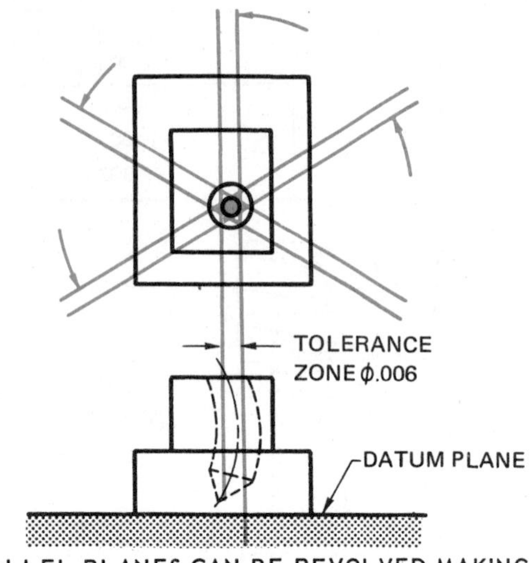

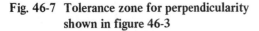

Fig. 46-7 Tolerance zone for perpendicularity shown in figure 46-3

the secondary datum, figure 46-8. The tolerance zone will be shaped like a prism whose bases are parallelograms (a parallelepiped).

Control of Center Lines

Tolerances intended to control orientation of the center line of a feature are applied to drawings as shown in figure 46-9. The resultant tolerance zone is illustrated in figure 46-10. Perpendicularity is considered to apply in all directions, unless a note such as THIS DIRECTION ONLY or DIRECTION OF ARROW ONLY is added.

In some cases it is more important to control elements of the cylindrical surface than of its axis or center line. To control elements of

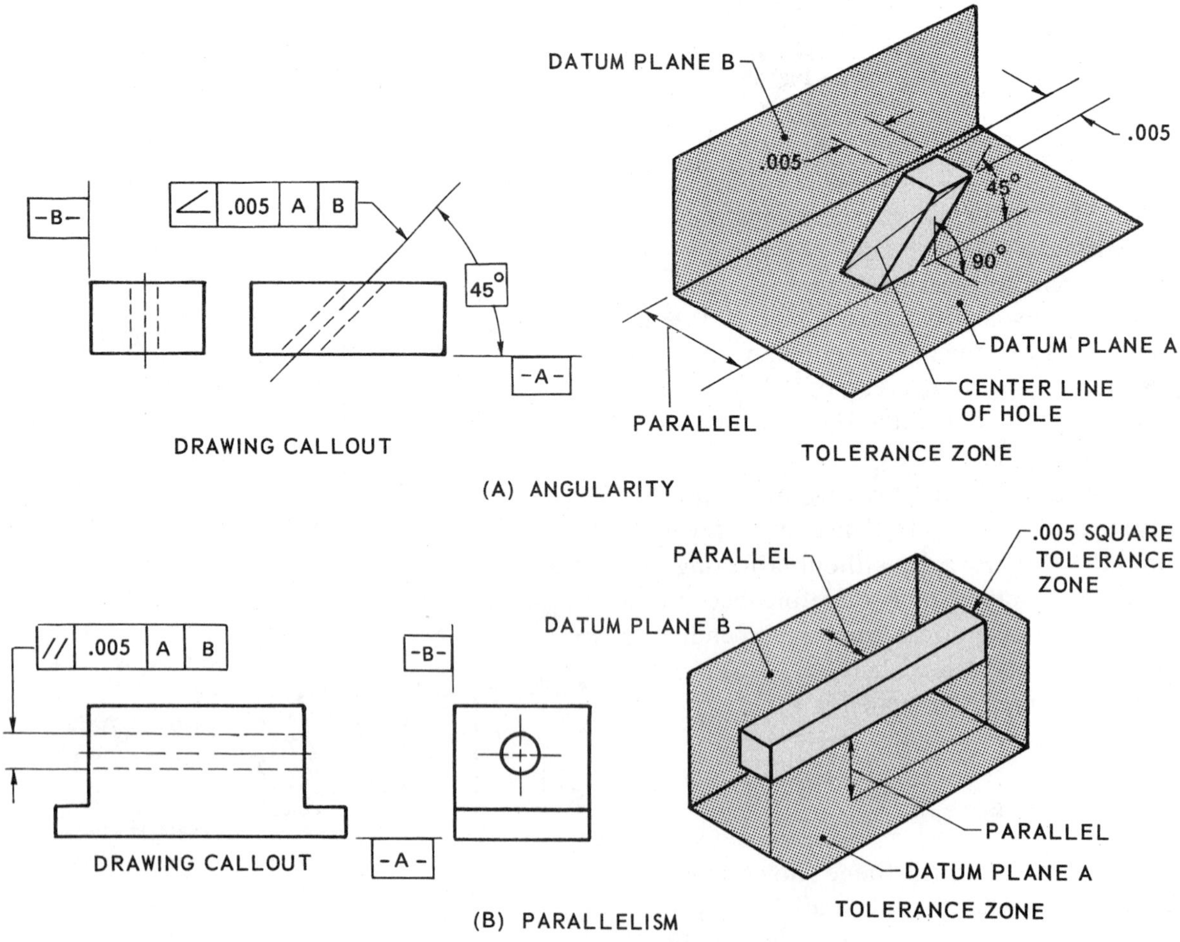

Fig. 46-8 Control of orientation in two directions

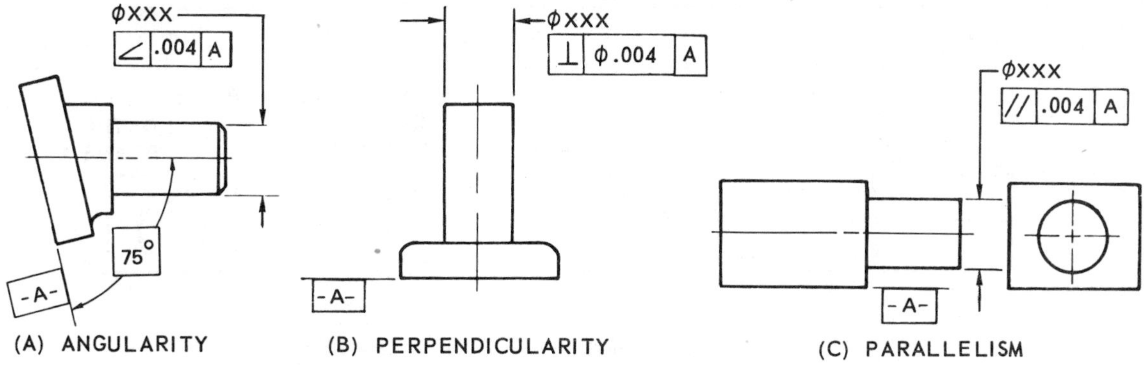

Fig. 46-9 Orientation of external cylindrical features

the cylindrical surface, the tolerance is directed to the surface with a single arrowhead, figure 46-9.

Because the cylindrical features represent size, orientation symbols may be applied on an MMC basis. This is indicated by adding the modifier symbol after the tolerance, figure 46-10. This method provides additional permissible variation in orientation as the size of the feature decreases to its least material size.

A comparison between the basic geometrical tolerancing methods is shown in figure 46-11.

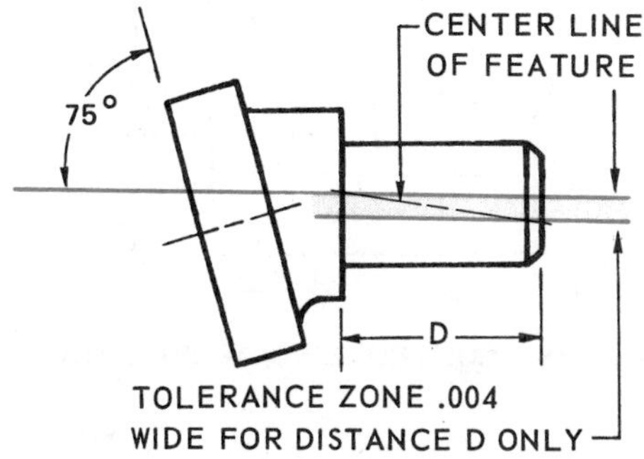

Fig. 46-10 Tolerance zone for angularity shown in figure 46-9

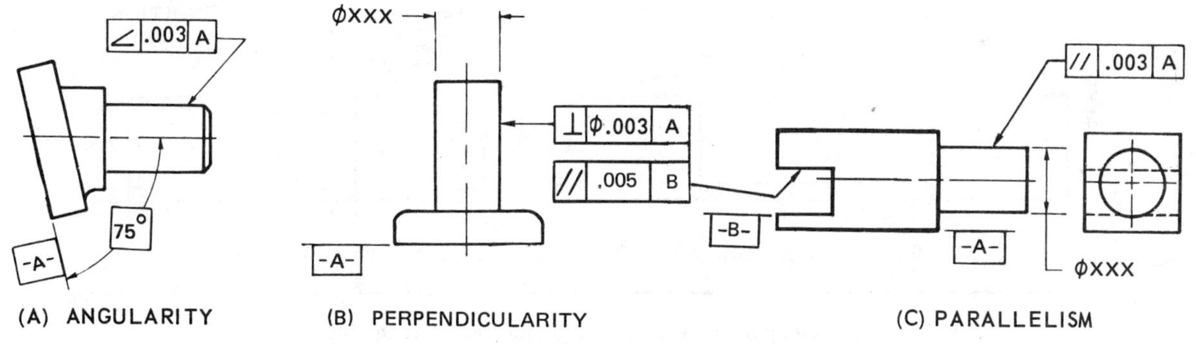

Fig. 46-11 Orientation of surface elements

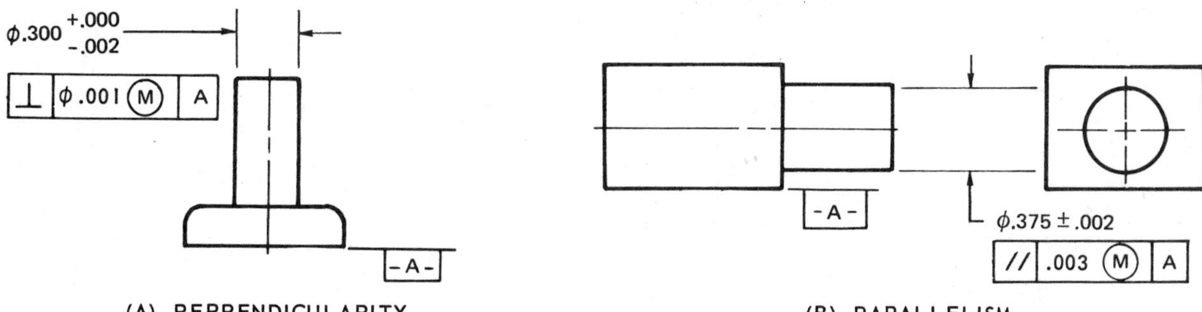

Fig. 46-12 Perpendicularity and parallelism on MMC basis

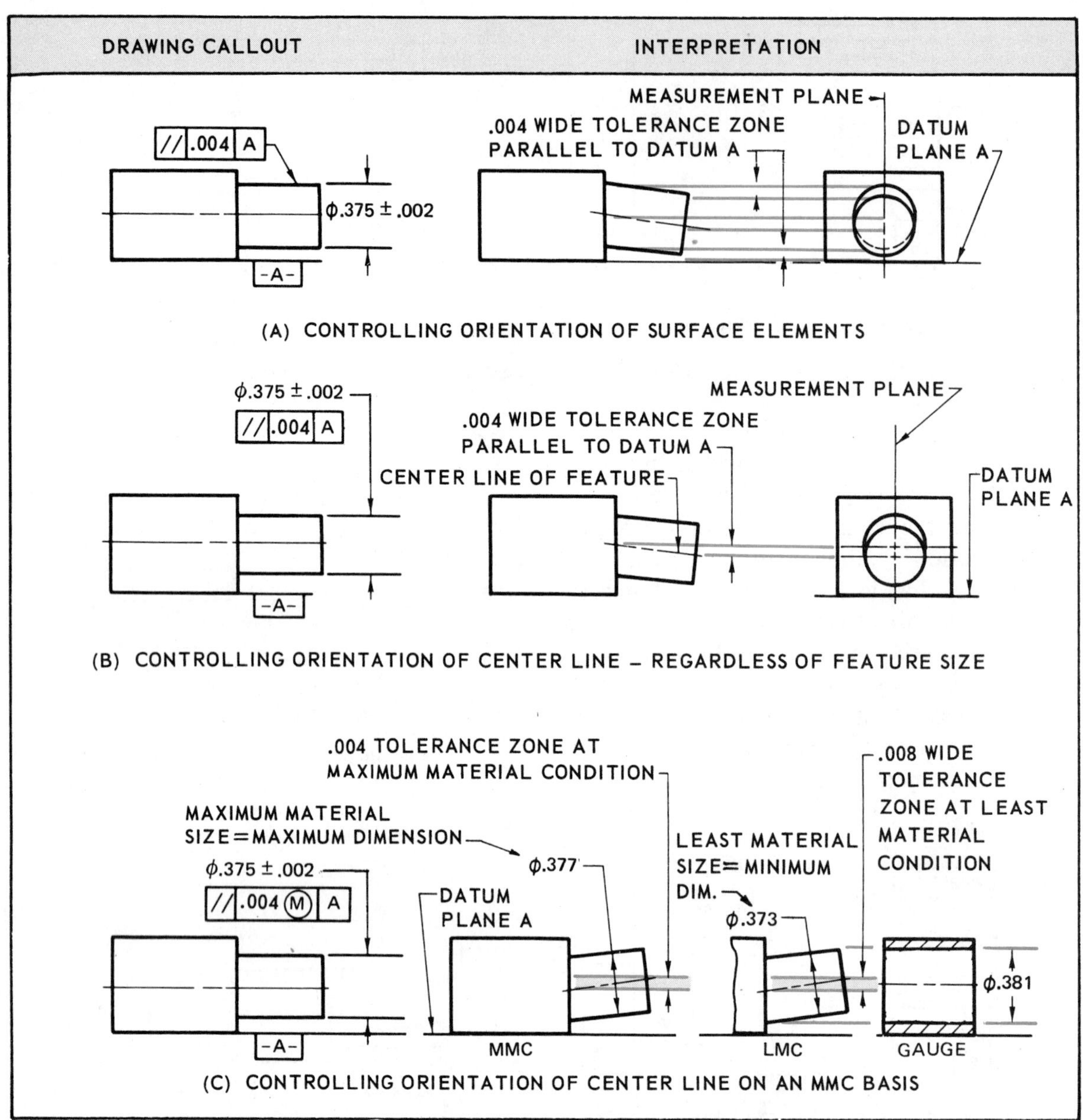
DRAWING CALLOUT
INTERPRETATION
MEASUREMENT PLANE
.004 WIDE TOLERANCE ZONE
PARALLEL TO DATUM A
DATUM
PLANE A
// .004 A
ϕ.375 ± .002
-A-
(A) CONTROLLING ORIENTATION OF SURFACE ELEMENTS
ϕ.375 ± .002
// .004 A
MEASUREMENT PLANE
.004 WIDE TOLERANCE ZONE
PARALLEL TO DATUM A
CENTER LINE OF FEATURE
DATUM
PLANE A
-A-
(B) CONTROLLING ORIENTATION OF CENTER LINE – REGARDLESS OF FEATURE SIZE
.004 TOLERANCE ZONE AT
MAXIMUM MATERIAL CONDITION
.008 WIDE
TOLERANCE
ZONE AT LEAST
MATERIAL
CONDITION
MAXIMUM MATERIAL
SIZE=MAXIMUM DIMENSION
LEAST MATERIAL
SIZE= MINIMUM
DIM.
ϕ.375 ± .002
ϕ.377
// .004 Ⓜ A
DATUM
PLANE A
ϕ.373
ϕ.381
-A-
MMC
LMC
GAUGE
(C) CONTROLLING ORIENTATION OF CENTER LINE ON AN MMC BASIS

Fig. 46-13 A comparison of the basic geometrical tolerancing methods

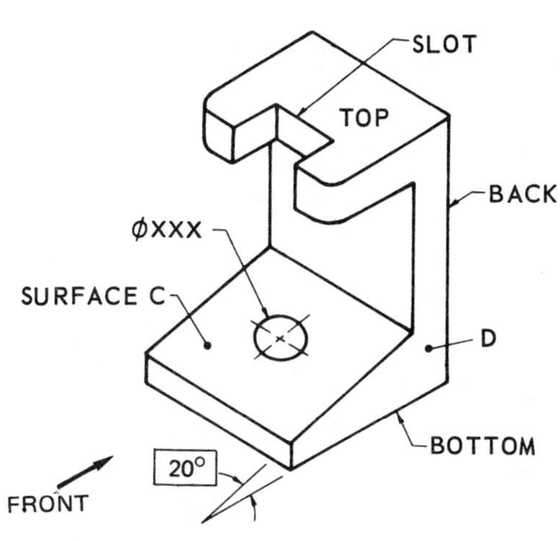

ASSIGNMENTS

1. From the information given, prepare a three-view sketch of the part showing the datums and geometric tolerances.
   - Bottom to be datum **A**
   - Back to be datum **B**
   - Hole to be perpendicular to bottom within .001
   - Back to be perpendicular to bottom within .01
   - Top to be parallel with bottom within .005
   - Surface **C** to have an angularity tolerance of .008 with the bottom. Surface **D** to be the secondary datum for this feature.
   - The sides of the slot to be parallel to each other within .002 and to be perpendicular to back within .004. One side of the slot to be Datum **E**.

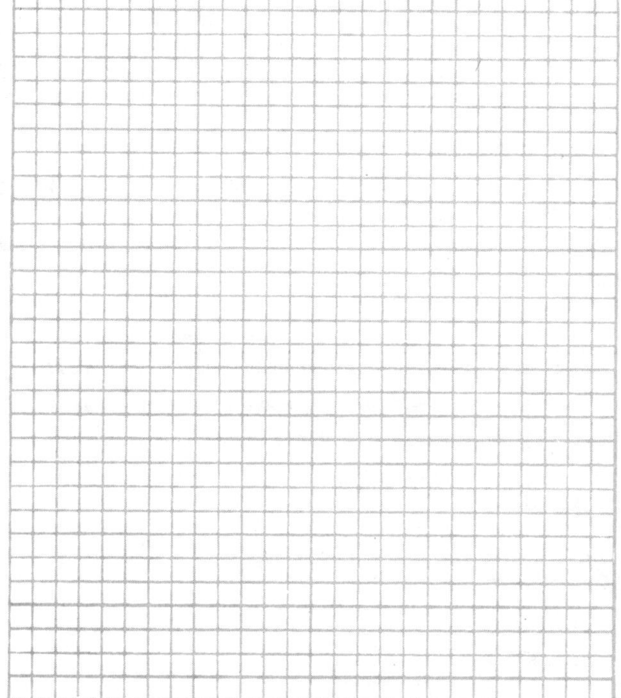

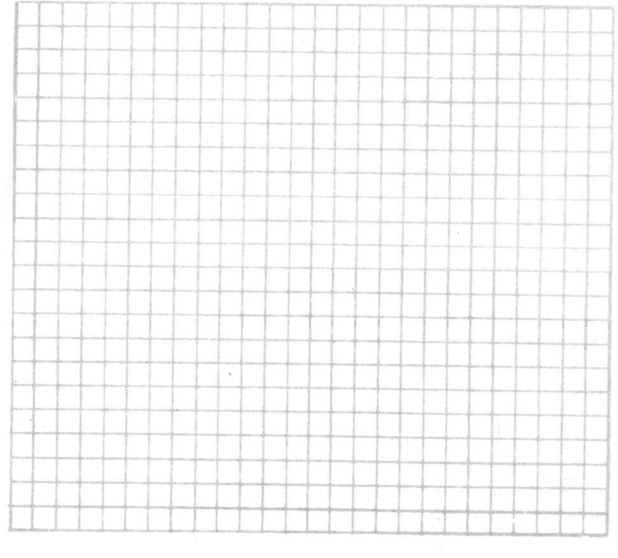

2. Prepare a three-view sketch (top, front, and left-side views) showing the datums and feature control symbols from the information supplied.
   - Surfaces A, B, C, D, and E are datums A, B, C, D, and E respectively.
   - Surface D of the dovetail must have an angularity tolerance of .003 with datums A and E.
   - Surface C should be perpendicular to datum B within .002.
   - Surface F must be perpendicular to datums A, E, and D within .001.

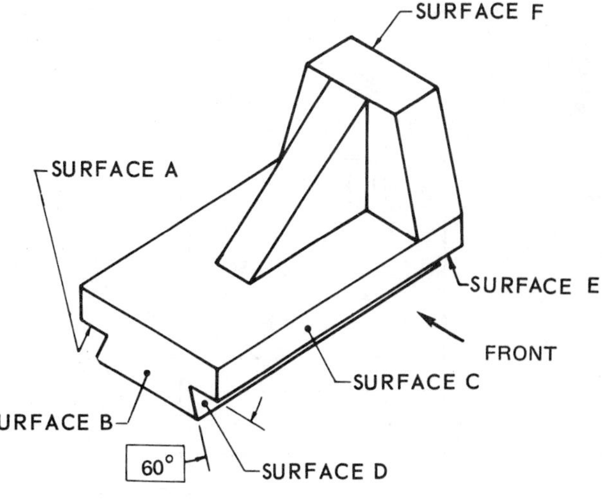

ORIENTATION TOLERANCING – ANGULARITY, PERPENDICULARITY, PARALLELISM

A-92

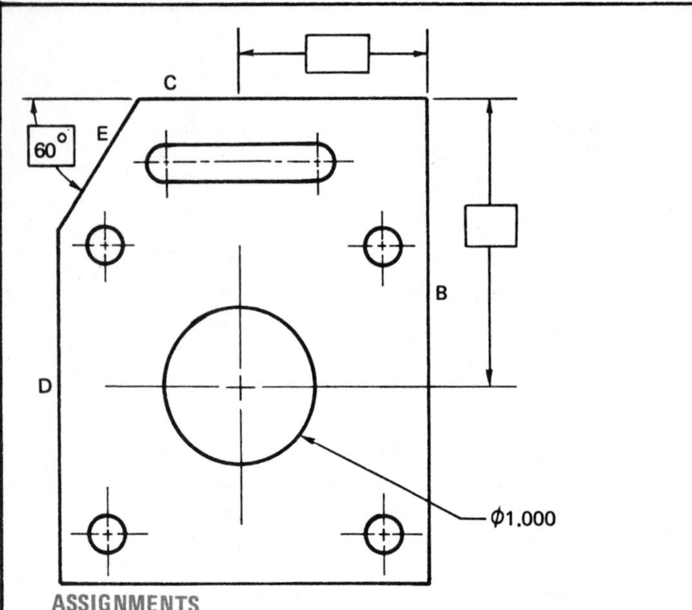

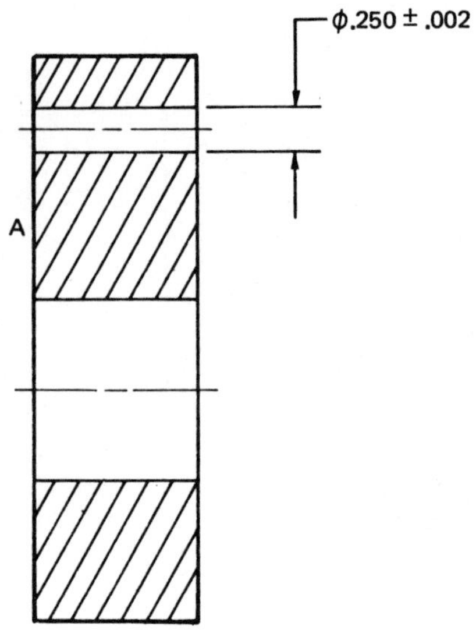

ASSIGNMENTS

1. Add the following information to the drawing:
   - Surfaces marked A, B, and C are datums A, B, and C, respectively.
   - Surface A is perpendicular to datums B and C within .01.
   - Surface D is parallel to datum B within .004.
   - The slot is parallel to datum C within .001 and perpendicular to datum A within .002 at MMC.
   - The Ø 1.000 hole is perpendicular to datum A within .001 at MMC.
   - Surface E has an angularity tolerance of .006. With Datum C.
   - Indicate which dimensions are basic.

2. Calculate the maximum diameter tolerance zone allowed for the diameters shown in the tables below. The feature control frame is associated with Ø $\frac{2.005}{2.000}$

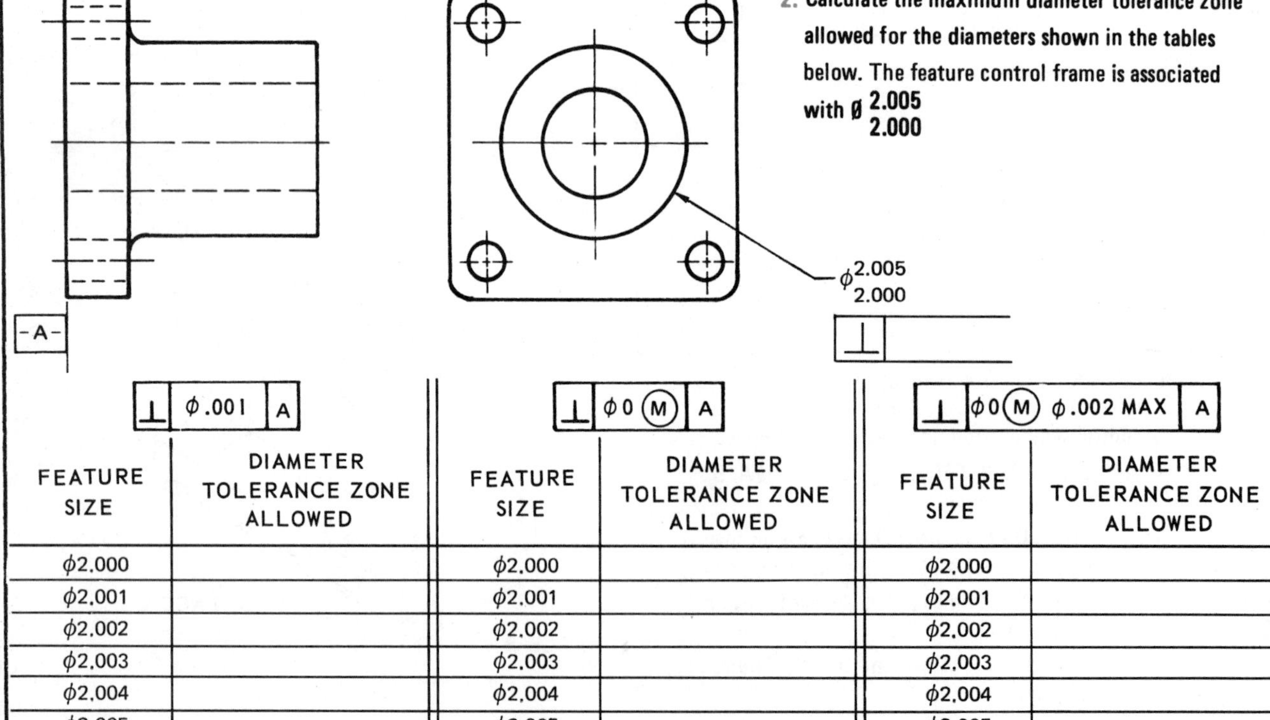

⊥ | ϕ .001 | A

FEATURE SIZE	DIAMETER TOLERANCE ZONE ALLOWED
ϕ2.000	
ϕ2.001	
ϕ2.002	
ϕ2.003	
ϕ2.004	
ϕ2.005	

⊥ | ϕ 0 Ⓜ | A

FEATURE SIZE	DIAMETER TOLERANCE ZONE ALLOWED
ϕ2.000	
ϕ2.001	
ϕ2.002	
ϕ2.003	
ϕ2.004	
ϕ2.005	

⊥ | ϕ 0 Ⓜ ϕ .002 MAX | A

FEATURE SIZE	DIAMETER TOLERANCE ZONE ALLOWED
ϕ2.000	
ϕ2.001	
ϕ2.002	
ϕ2.003	
ϕ2.004	
ϕ2.005	

ORIENTATION TOLERANCING – ANGULARITY, PERPENDICULARITY, PARALLELISM **A-93**

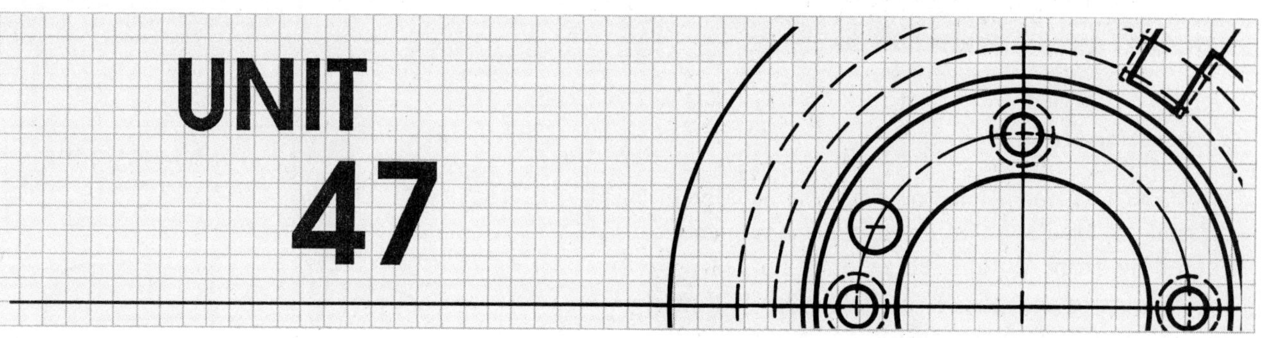

TOLERANCING OF FEATURES BY POSITION

The location of features is one of the most frequent applications of dimensions on technical drawings. Tolerancing is accomplished either by applying coordinate tolerances to the dimensions or by geometric (positional) tolerancing.

A positional tolerance defines a zone within which the center, axis, or center plane of a feature of size is permitted to vary from true (theoretically exact) position. Basic dimensions establish the true position from specified datum features and between interrelated features. A positional tolerance is indicated by the positional symbol, a tolerance, and appropriate datum references placed in a feature control frame.

Positional tolerancing is especially useful when applied on an MMC basis to groups or patterns of holes. This method meets functional requirements in most cases and at the same time permits inspection with simple gauge procedures.

Most examples in these units will be devoted to the principles involved in the location of small holes, because they represent the most commonly used applications. The same principles apply to the location of other features such as slots, tabs, bosses, and noncircular holes.

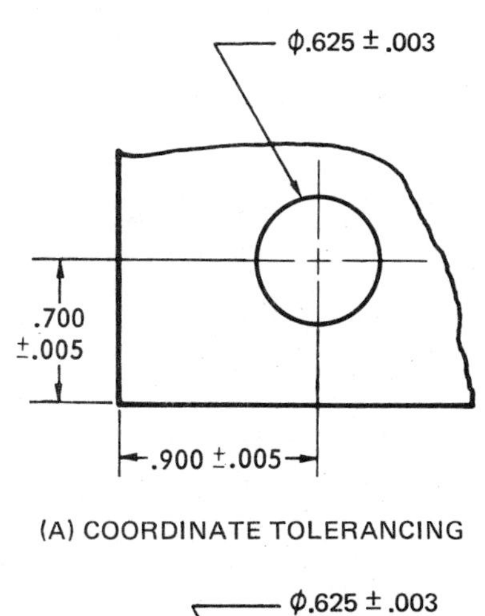

(A) COORDINATE TOLERANCING

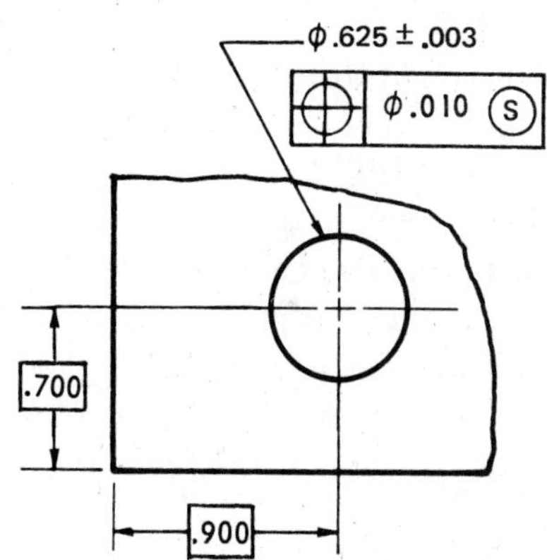

(B) POSITIONAL TOLERANCING RFS

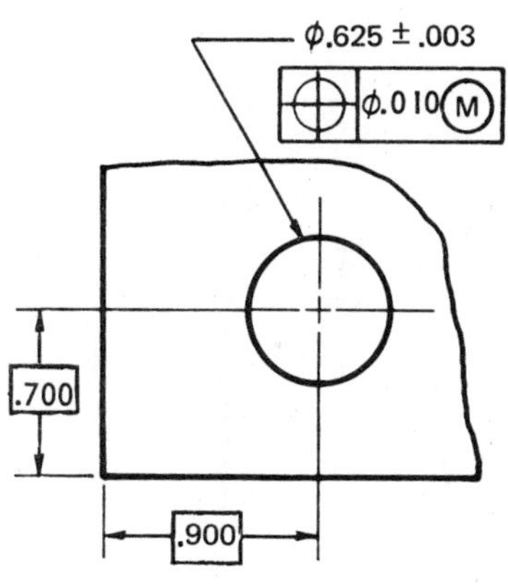

(C) POSITIONAL TOLERANCING MMC

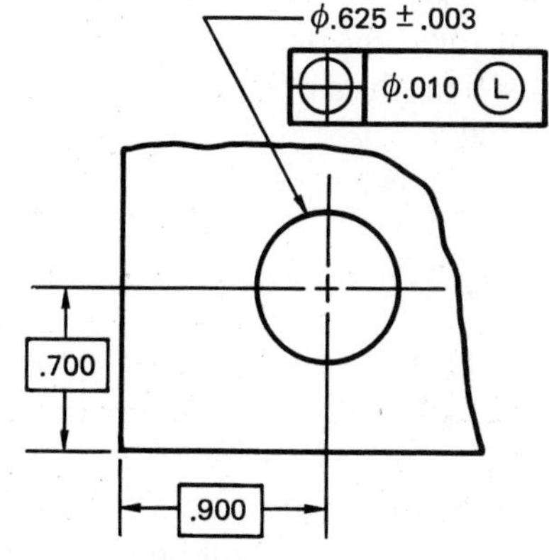

(D) POSITIONAL TOLERANCING LMC

Fig. 47-1 Comparison of tolerancing methods

Tolerancing Methods

The location of a single hole is usually indicated by rectangular coordinate dimensions, extending from suitable edges or other features of the part to the axis of the hole. Other dimensioning methods, such as that of polar coordinates, may be used when circumstances warrant. There are two standard methods of tolerancing the location of holes, coordinate tolerancing and positional tolerancing.

- Coordinate tolerancing refers to the tolerances applied directly to the coordinate dimensions or to applicable tolerances specified in a general tolerance note, figure 47-1(A).
- Positional tolerancing encompasses:
  - (A) Positional tolerancing, regardless of feature size (RFS), figure 47-1(B).
  - (B) Positional tolerancing, maximum material condition (MMC) basis, figure 47-1(C).
  - (C) Positional tolerancing, least material condition (LMC) basis, figure 47-1(D).

    The RFS and LMC symbols, figure 47-1, are used only with a tolerance of position. The United States is the only country to adopt these two symbols.

    When specified on an LMC basis, the specified tolerance is dependent upon the size of the feature.

These positional tolerancing methods are part of the system of geometric tolerancing. To explain and understand the advantages and disadvantages of the positional tolerancing methods, the widely used method of coordinate tolerancing must be analyzed.

COORDINATE TOLERANCING

Coordinate dimensions and tolerances may be applied to the location of a single hole, figure 47-2.

If the two coordinate tolerances are equal, the tolerance zone formed will be a square. Unequal tolerances result in a rectangular tolerance zone. Polar dimensioning where one of the locat-

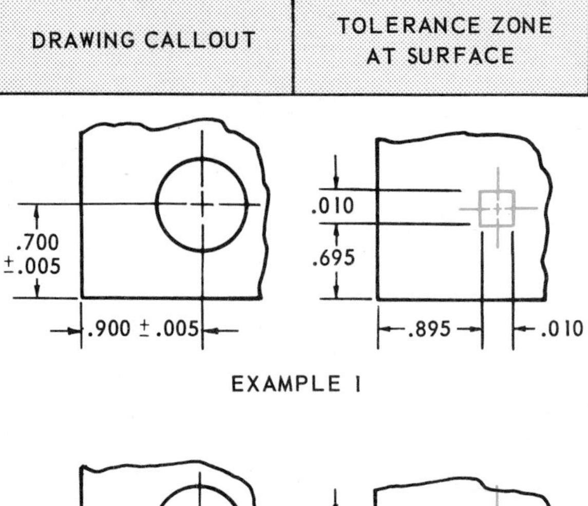

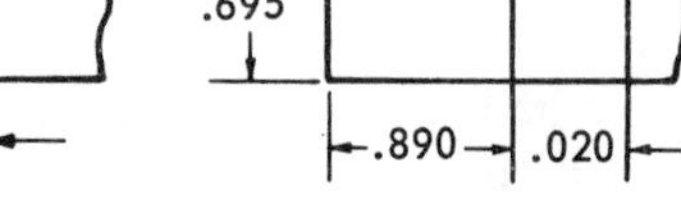

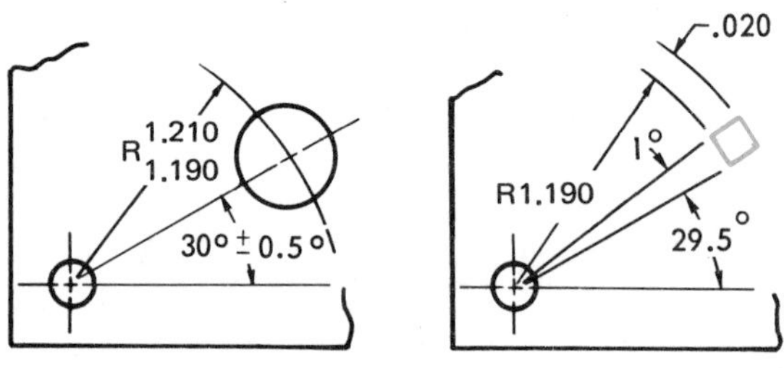

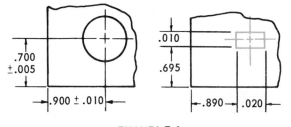

Fig. 47-2 Tolerance zones for coordinate tolerances

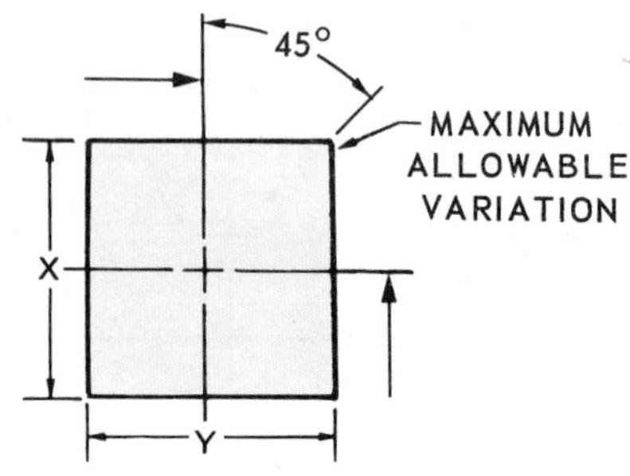

Fig. 47-3 Maximum permissible error for square tolerance zone

ing dimensions is a radius gives a circular ring section tolerance zone. For simplicity and because it is most frequently used, square tolerance zones are utilized in the analysis of most of the examples, figure 47-3.

The tolerance zone extends for the full depth of the hole. In most of the illustrations, tolerances will be analyzed as they apply at the surface of the part, where the axis is represented as a point.

The actual position of the center of the hole may be within the rectangular tolerance zone. For square tolerance zones, the maximum allowable variation from the theoretical exact position occurs in a direction of 45 degrees from the direction of the coordinate dimensions.

For the examples shown in figure 47-2, the tolerance zones are shown with their maximum tolerance values in figure 47-4.

A quick and easy method of finding the maximum positional error permitted with coordinate tolerancing is by using a chart similar to that shown in figure 47-5.

In the first example shown in figure 47-2, the tolerance in both directions is .010 in. The extension of the vertical and horizontal bars in the chart intersect at point A which lies between the radii of .014 and .015 in. When rounded off to three decimal places it indicates a maximum permissible variation of position of .014 in. In the second example shown in figure 47-2, the extension of the vertical and horizontal lines at .010 and .020 respectively in the chart intersect at point B indicating a maximum variation of position of .022 in.

POSITIONAL TOLERANCING

Positional tolerancing is part of the system of geometric tolerancing. In this system, the location of features is shown by coordinate dimensions or by polar and angular dimensions. The dimensions are shown, however, without direct tolerances. These dimensions called basic dimensions, establish the true position from specified datum features and between interrelated features. Basic dimension means a numerical value used to describe the theoretical exact size, and is shown enclosed in a rectangular frame, figure 47-6. Basic dimensions are exempt from any tolerance note. It is the basis from which permissible variations are established by tolerances on other dimensions, or in feature control frames.

Applications of basic dimensions include location of datum targets and establishing the true position of the center of a feature when using positional tolerancing. Each such dimension is enclosed in a rectangular frame to indicate that it represents an exact value to which tolerances shown in the general note do not apply, or whose tolerances are expressed in the feature control frame, figure 47-6.

Symbol for Position

The geometric characteristic symbol for position is a circle and two solid center lines shown in the feature control frame in the same way as for other geometric symbols. Positional

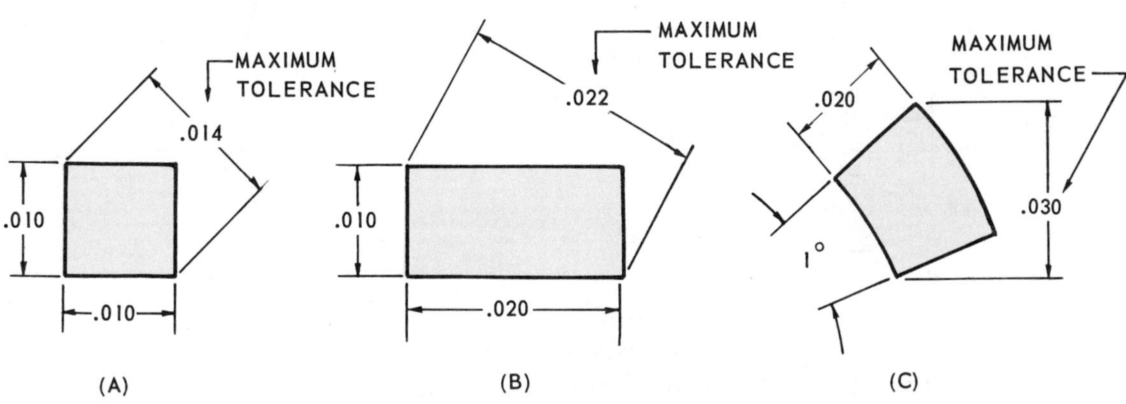

Fig. 47-4 Tolerance zones for examples in figure 47-2

tolerances are specified on a MMC or RFS basis, or LMC (least material condition).

When specified on an MMC basis, the specified tolerance is dependent on the size of the feature, figure 47-6.

When specified on an RFS basis, the specified tolerance is independent of the size of the feature, figure 47-7.

Positional Tolerancing—Regardless of Feature Size (RFS)

Positional tolerancing on this basis requires the positional tolerance or datum reference, or both, be maintained regardless of actual feature sizes, figure 47-8. RFS, where applied to the positional tolerance of circular features, requires the axis of each feature to be located within the specified positional tolerance regardless of the size of the feature. This positional control is more than the MMC principle.

It has already been shown that the rectangular coordinate tolerancing the maximum permissible error in location is not the value indicated by the horizontal and vertical tolerances, but is instead equivalent to the length of the diagonal between the two tolerances. For square tolerance zones this amount is 1.4 times the specified tolerance values. The specified tolerance can therefore be increased to an amount equal to the diagonal of the coordinate tolerance zone without affecting the clearance between the hole and its mating part.

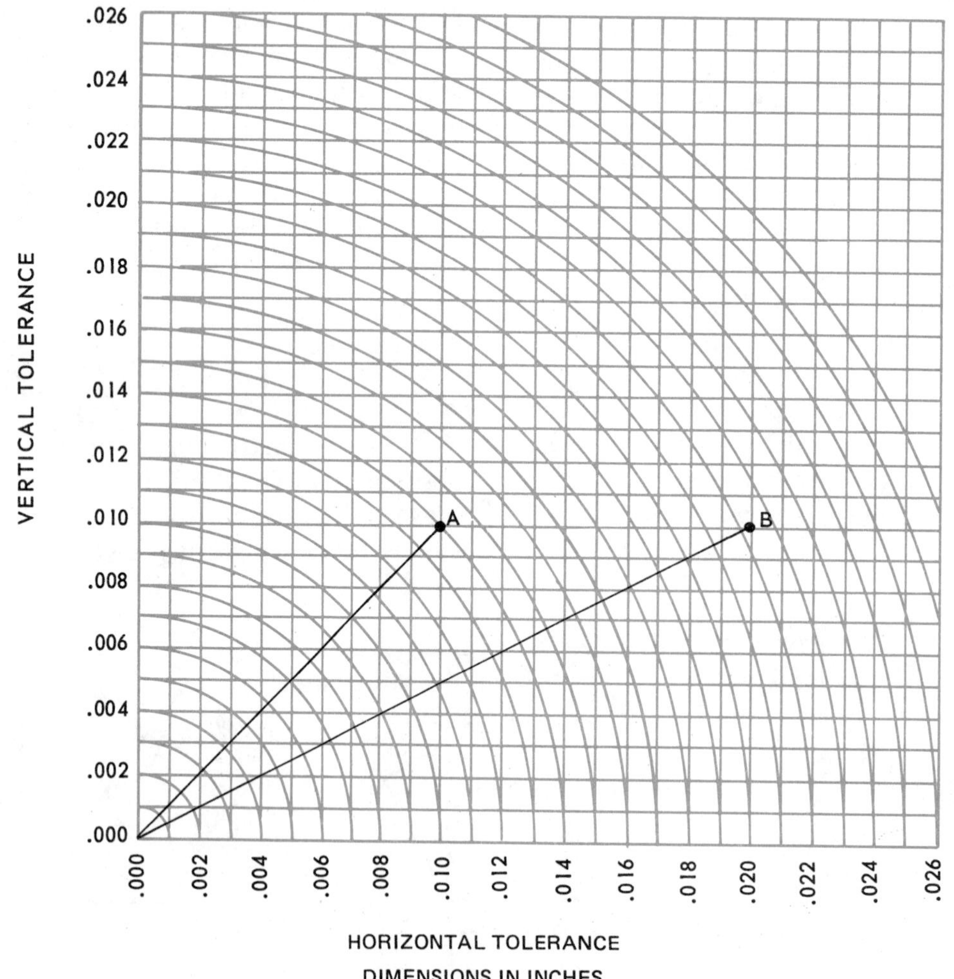

Fig. 47-5 Chart for calculating maximum tolerance using coordinate tolerancing

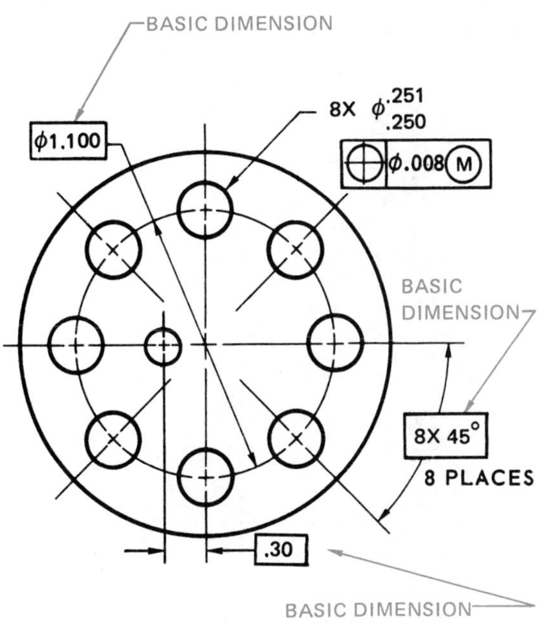

Fig. 47-6 Identifying basic dimensions

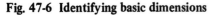

Fig. 47-7 Regardless-of-feature-size symbol shown with positional tolerance

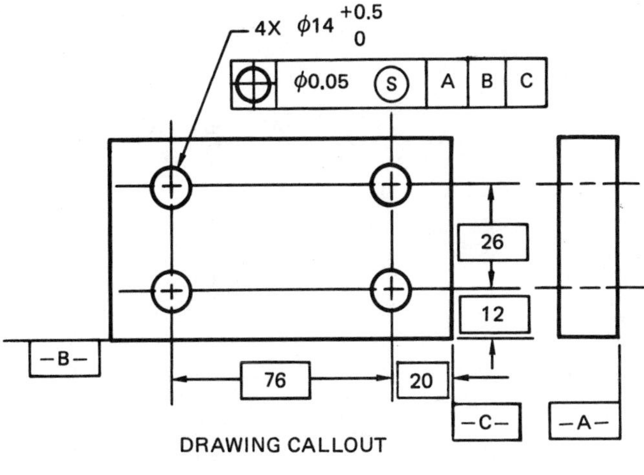

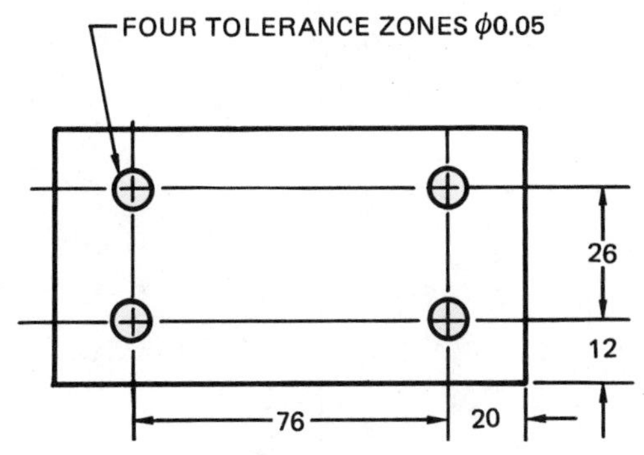

Fig. 47-8 Positional tolerancing–RFS

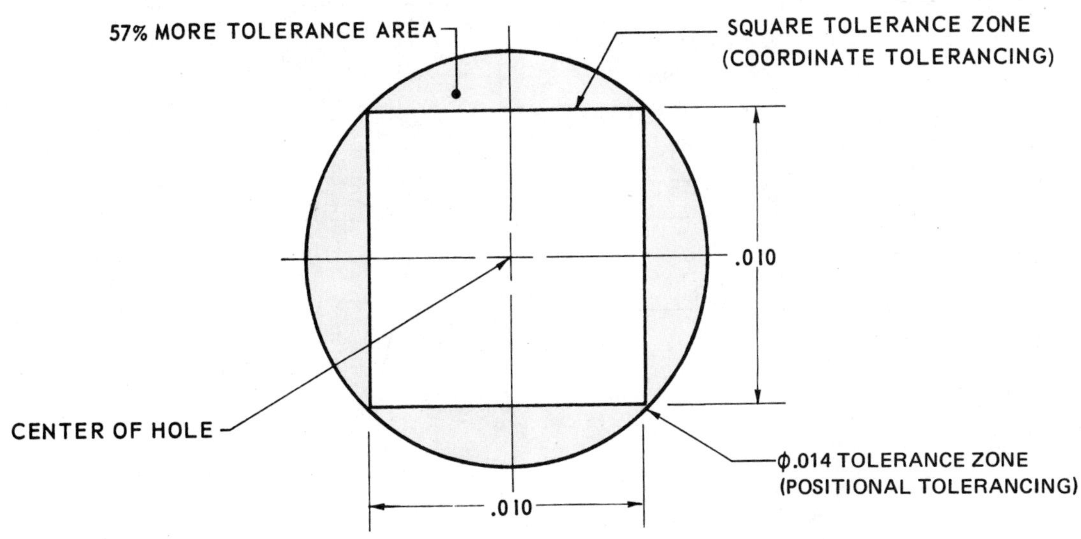

Fig. 47-9 Relationship of tolerance zones

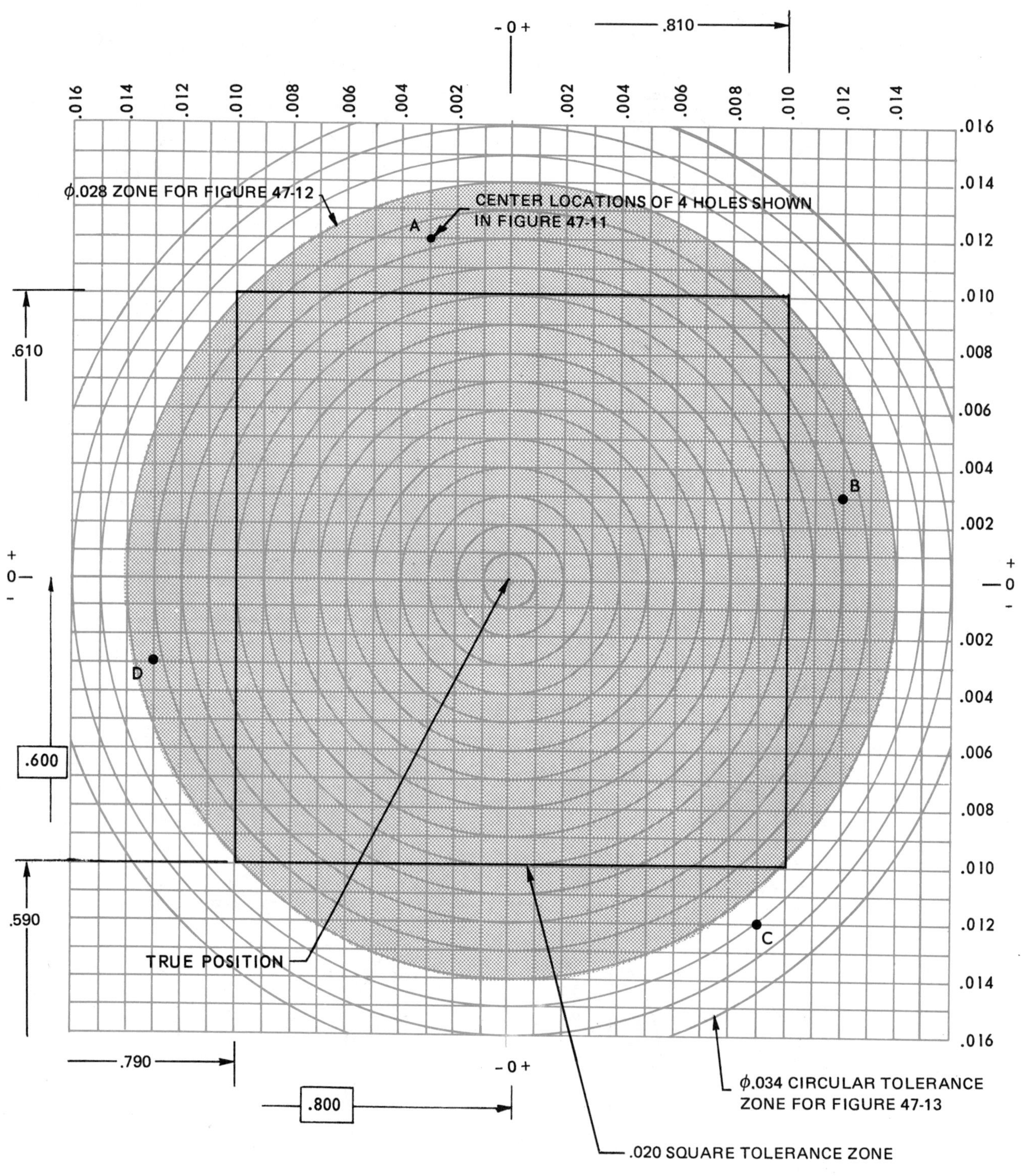

Fig. 47-10 Chart for evaluating positional tolerancing

It is practical to replace coordinate tolerances with a positional tolerance having a value equal to the diagonal of the coordinate tolerance zone. This provides 57 percent more tolerance area, figure 47-9, and would probably result in the rejection of fewer parts for positional errors.

A simple method for checking positional tolerance errors is to take coordinate measurements and evaluate them on a chart as shown in figure 47-10. For example, the four parts shown in figure 47-11 were rejected when the coordinate tolerances were applied to them.

If the part had been toleranced using the positional tolerance MMC method shown in figure 47-12, and given a tolerance of ∅.028 (equal to the diagonal of the coordinate tolerance zone), three of the parts, A, B, and D, would not have been rejected.

Positional Tolerancing – Maximum Material Condition

The problems of tolerancing for the position of holes are simplified when positional tolerancing is applied on an MMC basis. This permits an increase in positional variations as the size departs from the maximum material size and also permits the use of functional GO, NO-GO gauges.

To apply a positional tolerance on an MMC basis, the symbol Ⓜ is added in the feature control symbol after the tolerance and the feature control symbol is associated with the size dimension.

Consider the part shown in figure 47-13. A positional tolerance applied to the hole on an MMC basis means that the boundary of the hole

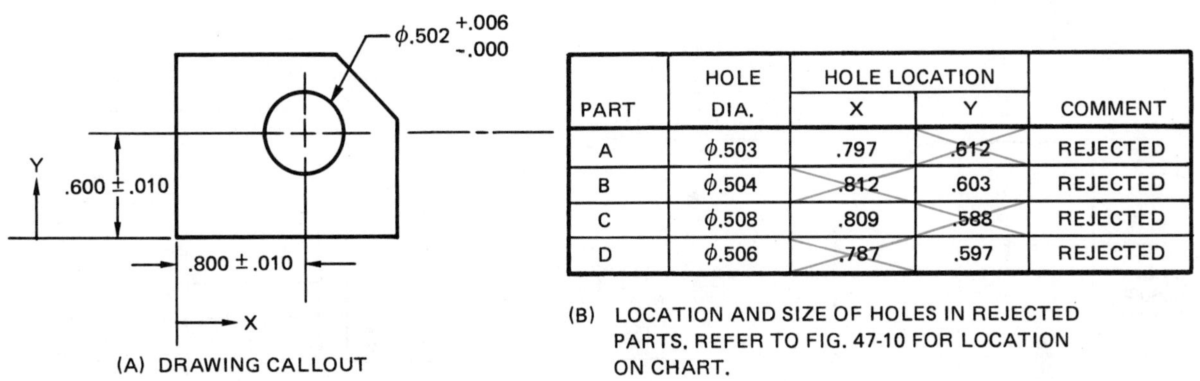

PART	HOLE DIA.	HOLE LOCATION X	HOLE LOCATION Y	COMMENT
A	ϕ.503	.797	.612	REJECTED
B	ϕ.504	.812	.603	REJECTED
C	ϕ.508	.809	.588	REJECTED
D	ϕ.506	.787	.597	REJECTED

(B) LOCATION AND SIZE OF HOLES IN REJECTED PARTS. REFER TO FIG. 47-10 FOR LOCATION ON CHART.

Fig. 47-11 Parts A to D rejected because hole centers do not lie within coordinate tolerance zone

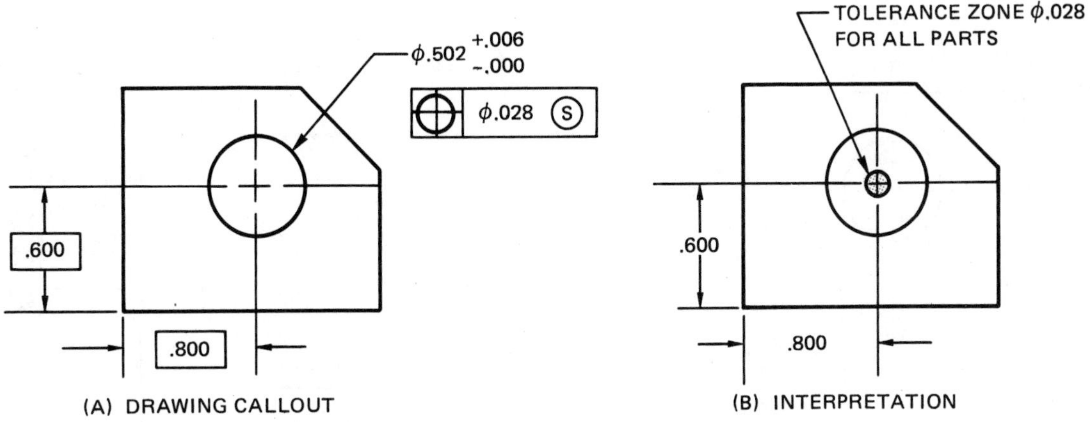

Fig. 47-12 Positional tolerancing–regardless of feature size

must fall outside a perfect cylinder having a diameter equal to the maximum material size of the feature (Ø.502) minus the positional tolerance (Ø.028). This cylinder (Ø.474) is located with its axis at true position. The hole must be within the limits of Ø.502 and Ø.508. A simple gauge to check the part is shown in figure 47-14.

If the part shown in figure 47-11 had been toleranced using the positional tolerance (MMC) method and given a tolerance of Ø.028 at MMC, figure 47-13, part C which was rejected using the RFS tolerancing method (figure 47-12) would not have been rejected if it had been straight. The positional tolerance can be increased to Ø.034 for a part having a diameter of .508 in. without jeopardizing the function of the part, figure 47-15.

Positional Tolerancing–Least Material Condition

Where positional tolerancing at LMC is specified, the stated positional tolerance applies when the feature contains the least amount of material permitted by its toleranced size dimension. Where the feature departs from its LMC size, an increase in positional tolerancing is allowed, which is equal to the amount of such departure. Specifying LMC is limited to positional tolerancing applications where MMC does

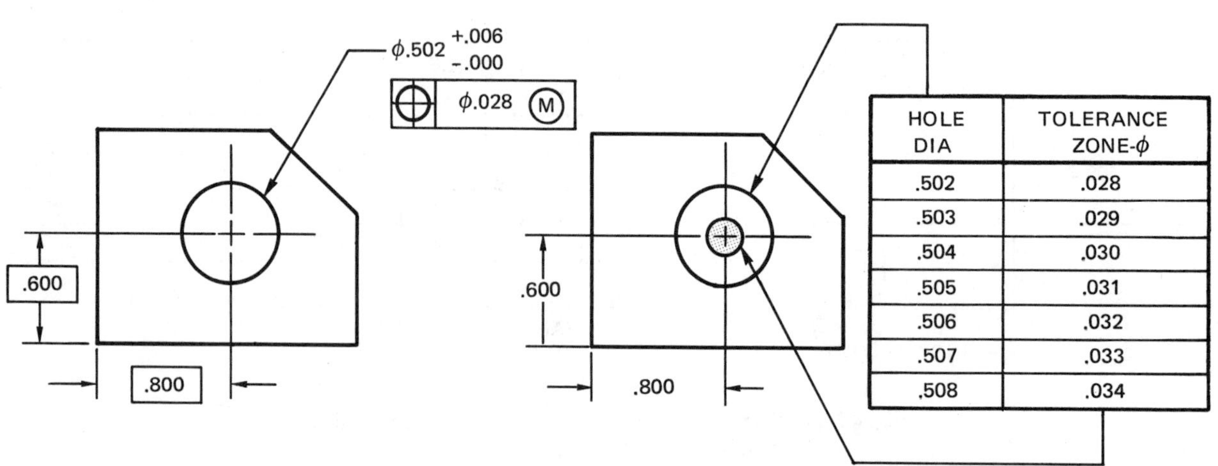

HOLE DIA	TOLERANCE ZONE-ϕ
.502	.028
.503	.029
.504	.030
.505	.031
.506	.032
.507	.033
.508	.034

Fig. 47-13 Positional tolerancing–maximum material condition

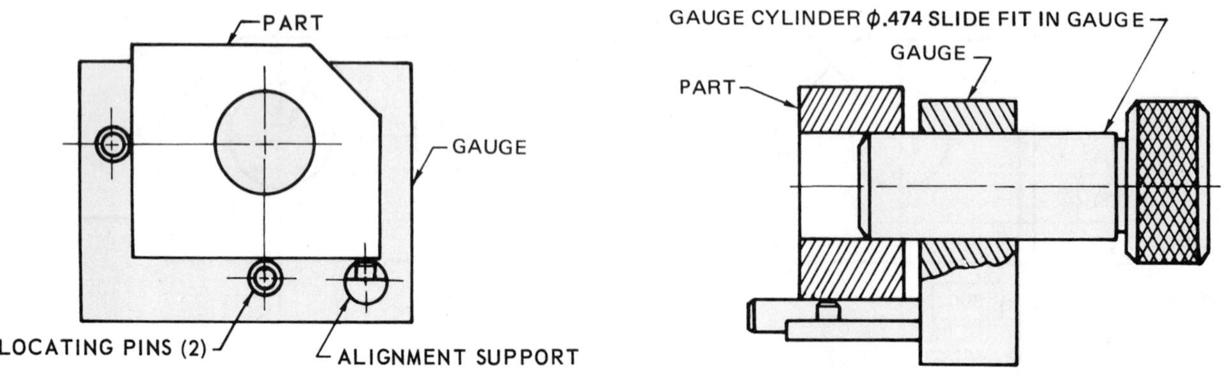

Fig. 47-14 Gauge for part shown in figure 47-13

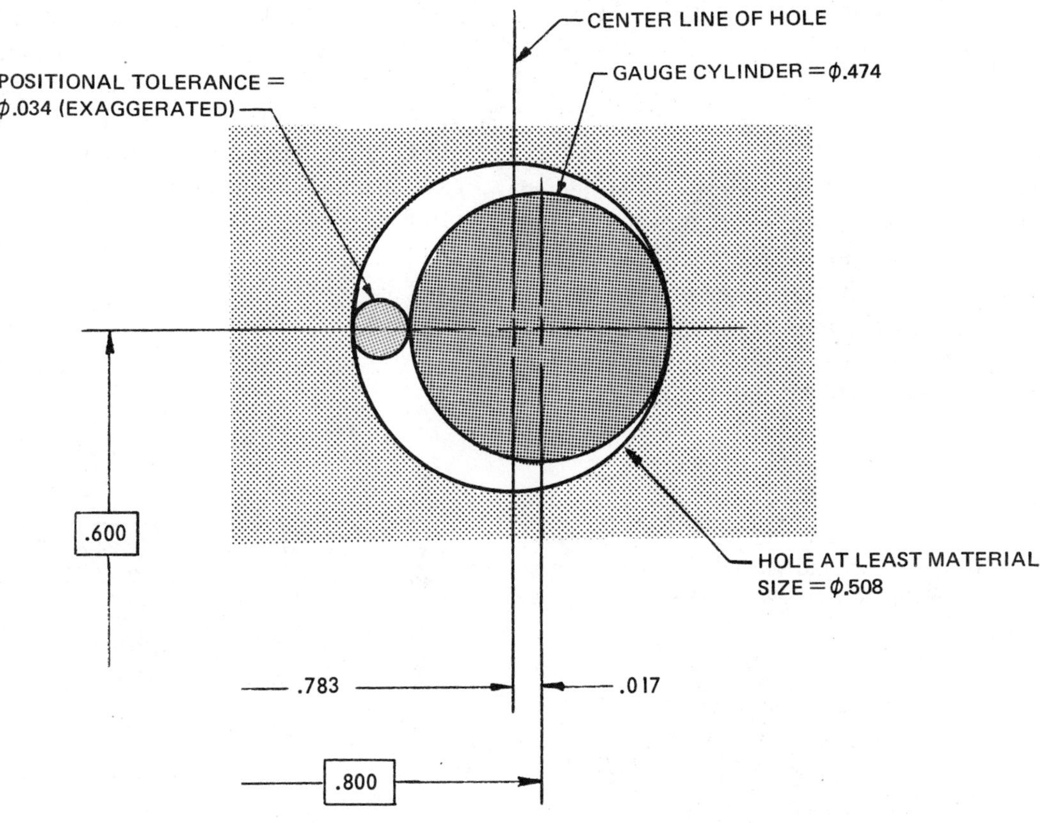

Fig. 47-15 Positional variations for tolerancing on an MMC basis

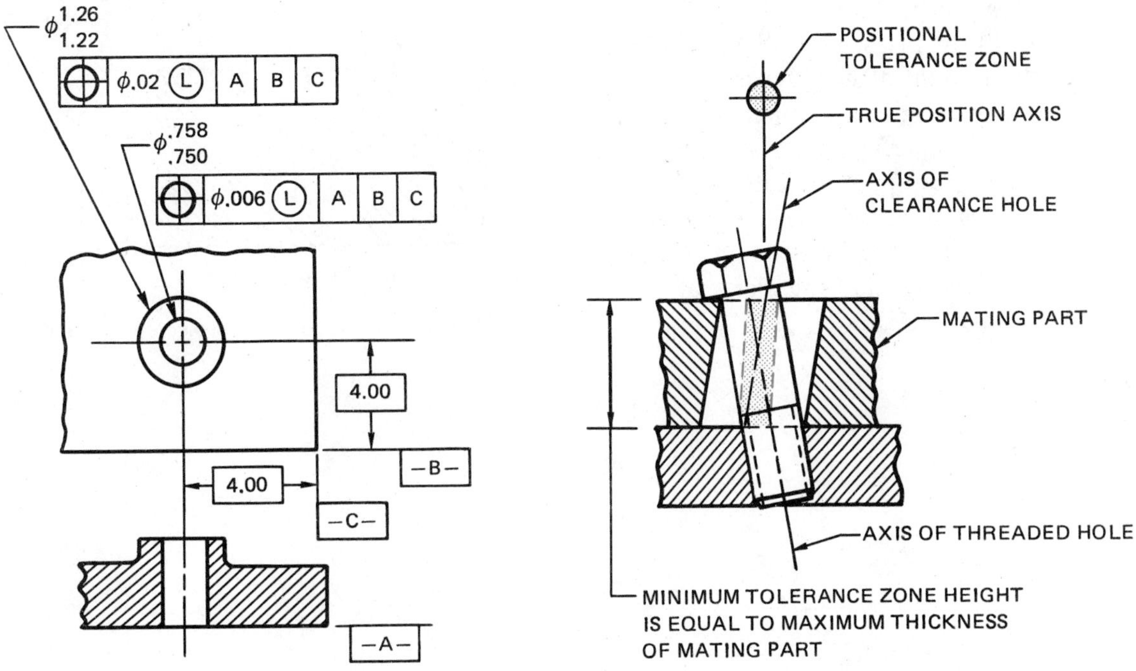

Fig. 47-16 LMC applied to a boss and hole

Fig. 47-17 Basis for projected tolerance zone

not provide the desired control and RFS is too restrictive, figure 47-16.

PROJECTED TOLERANCE ZONE

The symbol used to indicate a projected tolerance zone is shown in figure 47-3. The application of a projected tolerance zone is recommended where the variation in perpendicularity of threads or press-fit holes could cause fasteners such as screws, or pins to interfere with mating parts, figures 47-17 and 47-18.

Variations of Form (Envelope Principle)

The form of an individual feature is controlled by its limits of size to the extent prescribed by the following:

(A) The surface or surfaces of a feature shall not extend beyond a boundary (envelope) of perfect form at MMC. The boundary is the true geometric form represented by the drawing. No variation of form is permitted if the feature is produced at its MMC limit of size.

(B) Where the actual size of a feature has departed from MMC toward LMC, a variation in form is allowed equal to the amount of such departure.

(C) There is no requirement for a boundary of perfect form at LMC. Thus, a feature produced at its LMC limit of size is permitted to vary from true form to the maximum variation allowed by the boundary of perfect form at MMC.

Perfect Form at MMC Not Required. Where it is desired to permit a surface or surfaces of a feature to exceed the boundary of perfect form at MMC, a note such as PERFECT FORM AT MMC NOT REQD is specified on the drawing, exempting the pertinent size dimension from the provisions described above.

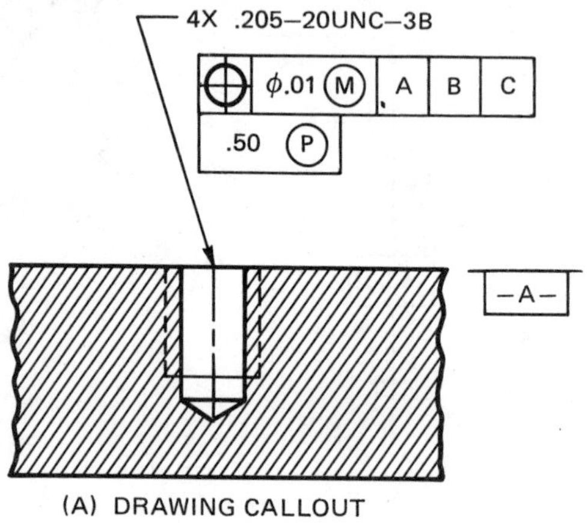

(A) DRAWING CALLOUT

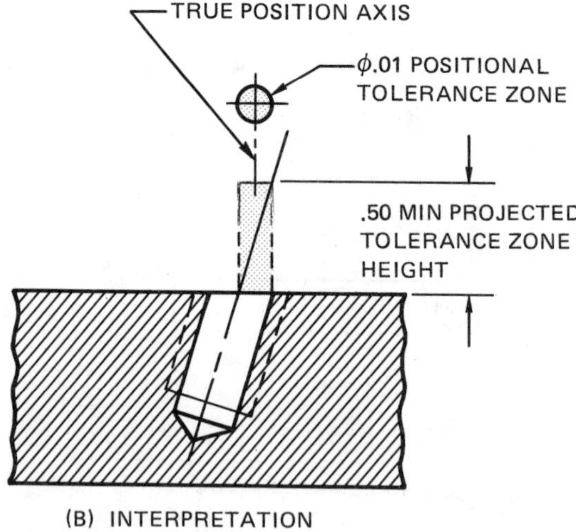

(B) INTERPRETATION

Fig. 47-18 Specifying a projected tolerance zone

PAGE 345 IS INTENTIONALLY BLANK. ASSIGNMENT DRAWING A-94 IS ON THE NEXT PAGE.

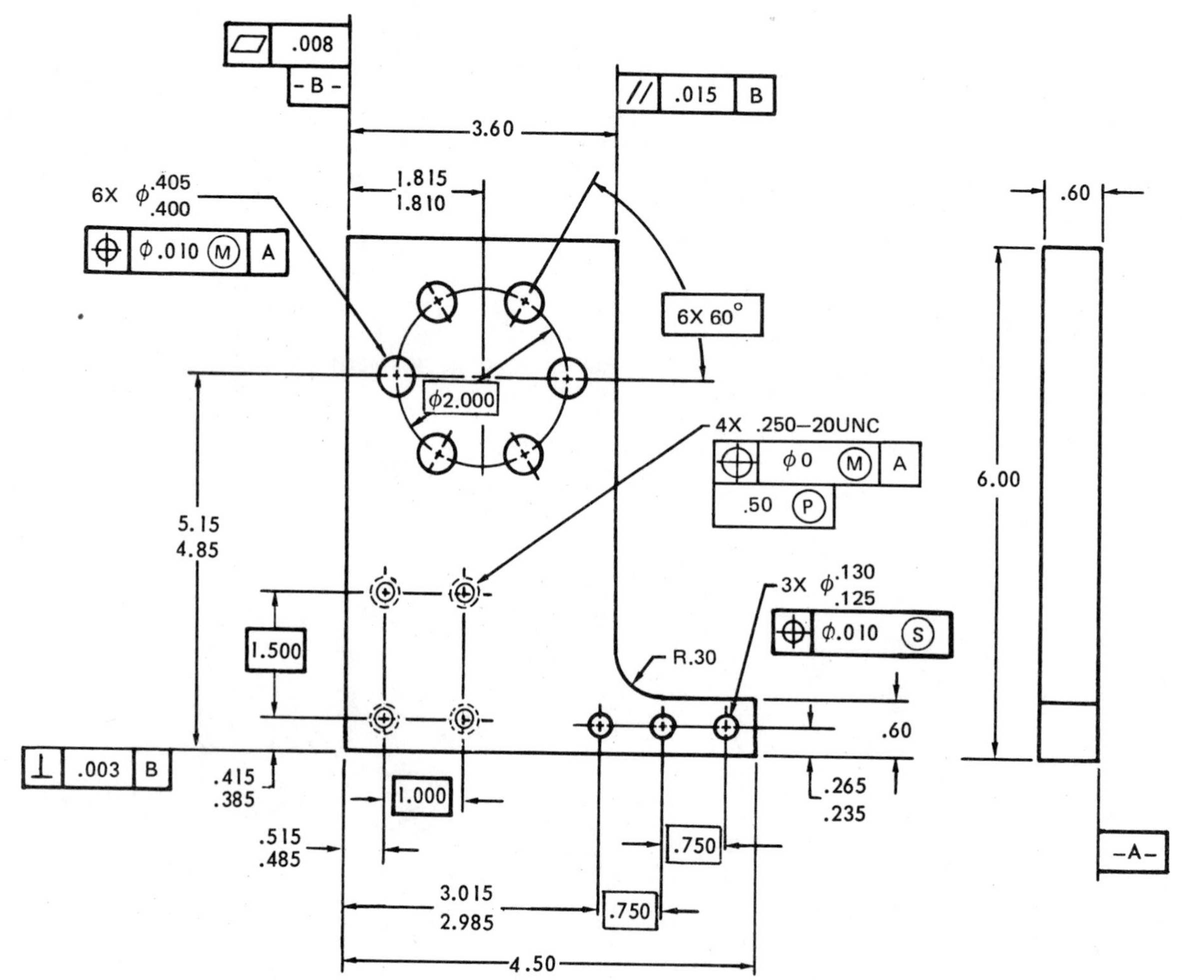

UNSPECIFIED TOLERANCE ±.02

QUESTIONS

1. How many datum surface are there on the drawing?
2. How many basic dimensions are shown on the drawing?
3. What is the maximum flatness deviation permitted on datum surface B?
4. What is the thickness tolerance permitted on the part?
5. How many form tolerances are shown?
6. How many locational tolerances are indicated?
7. What is the height of the projected tolerance zone above the surface of the part shown?
8. The part shown below was rejected because it did not meet the requirements specified on the drawing. List the reasons why it was not acceptable. Lines and surfaces are labeled for identification purposes.
9. Show the maximum positional tolerance allowed for each of the feature sizes in the table below.

ANSWERS

1. ______
2. ______
3. ______
4. ______
5. ______
6. ______
7. ______
8. ______
______
______
9. ______

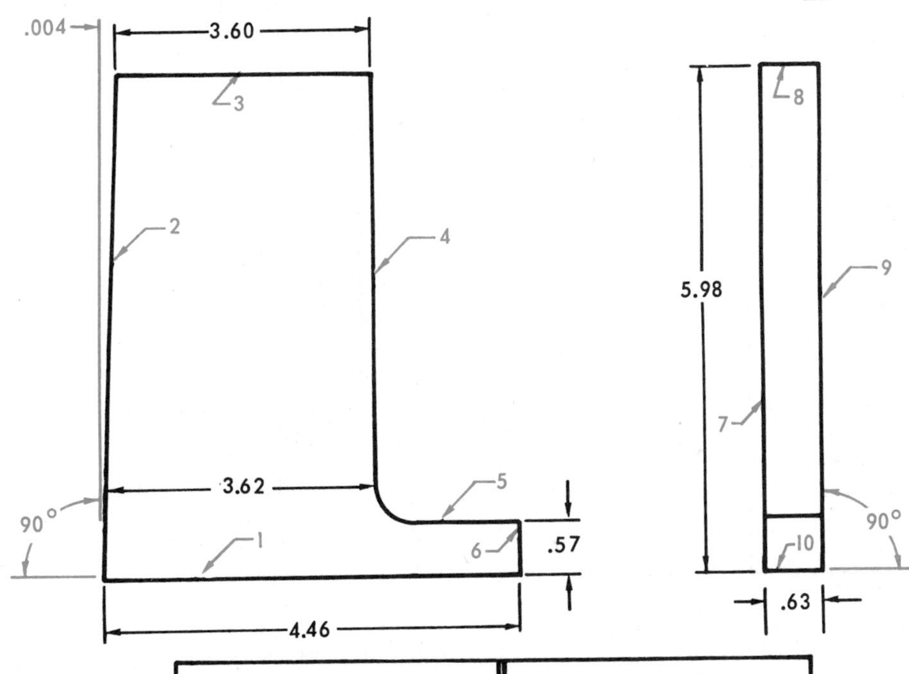

ϕ.125 – .130 HOLES		ϕ.400 – .405 HOLES	
FEATURE SIZE	MAX. POSN. TOLERANCE ALLOWED	FEATURE SIZE	MAX. POSN. TOLERANCE ALLOWED
ϕ.125		ϕ.400	
ϕ.126		ϕ.401	
ϕ.127		ϕ.402	
ϕ.128		ϕ.403	
ϕ.129		ϕ.404	
ϕ.130		ϕ.405	

UNIT 48

DATUMS FOR TOLERANCING BY POSITION

Occasionally it is preferable to have a hole related to surfaces or features (datums) other than the outside edges of the part. The first consideration in such applications is determination of the primary datum feature. Usually, the surface on which the hole is made is specified as the primary datum. This insures that the true position of the axis is either perpendicular to this surface or at the basic angle, if it is other than 90 degrees. This surface is rested on the gauging plane or surface plate for measuring purposes. Secondary and tertiary datum features are then selected and identified if required.

Figure 48-1 shows a part having three datums specified, while figure 48-2 illustrates a gauge which could be used to check such a part.

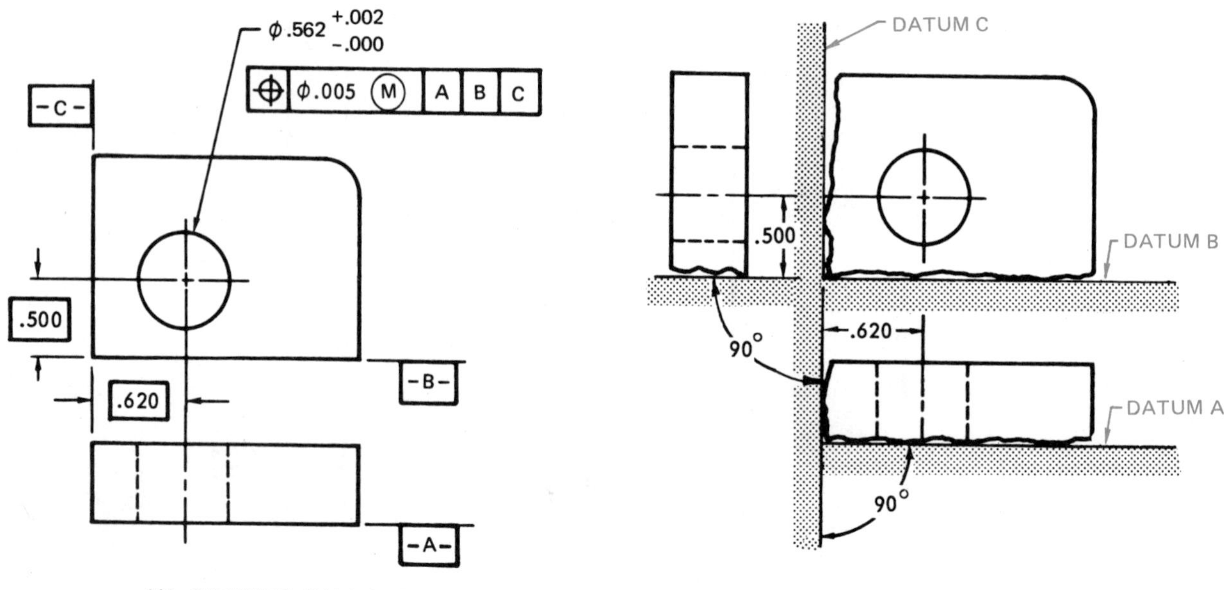

Fig. 48-1 Part with three datum features specified

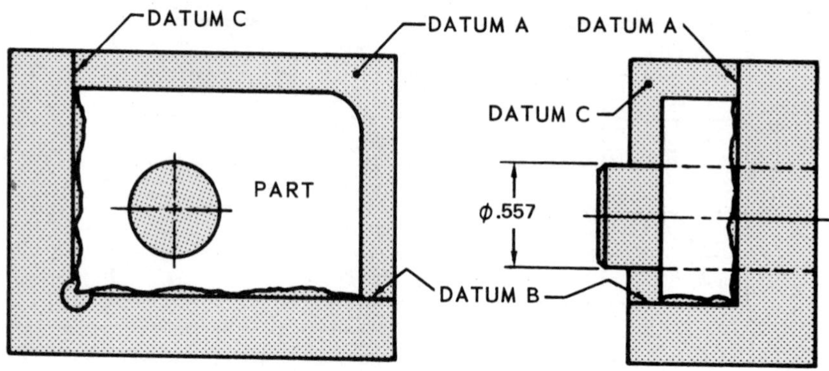

Fig. 48-2 Gauge for part in figure 48-1

DATUM TARGETS

It may be inadvisable to use a complete surface of a feature as a datum for the following reasons:

- The surface of a feature may be so large that a gauge designed to make contact with the full surface may be too expensive or too cumbersome to use.
- Functional requirements of a part may necessitate using only part of the surface as a datum feature.
- A surface selected as a datum feature may not be sufficiently accurate.

The datum target method is a useful technique for overcoming such problems. In this method, certain points, lines, or small areas on the surfaces are selected as the basis for establishing datums, figure 48-3. For flat surfaces, this usually requires three target points or areas for a primary datum; two for a secondary datum; and one for a tertiary datum. Often the use of such target areas eliminates the need for costly machining which might otherwise be required to produce surfaces suitable for use as datum features.

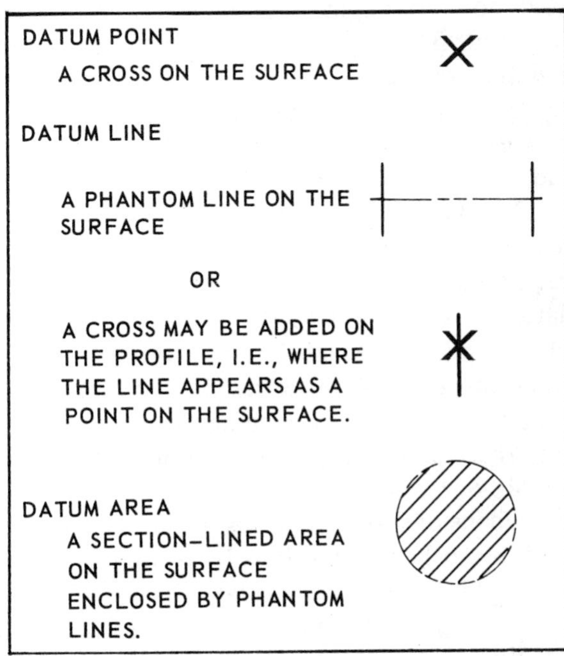

Fig. 48-3 Datum targets

Targets need not be used for all datums. It is logical, for example, to use targets for the primary datum and other features or surfaces for secondary and tertiary datums. Datum targets should be spaced as far apart as possible to provide maximum rigidity for making measurements.

Each datum target is shown on a view of the part in its desired location by a datum target symbol, figure 48-4. The datum target symbol is a circle divided into two parts. The upper part contains the size of the datum area. The lower part contains a letter and number which identifies

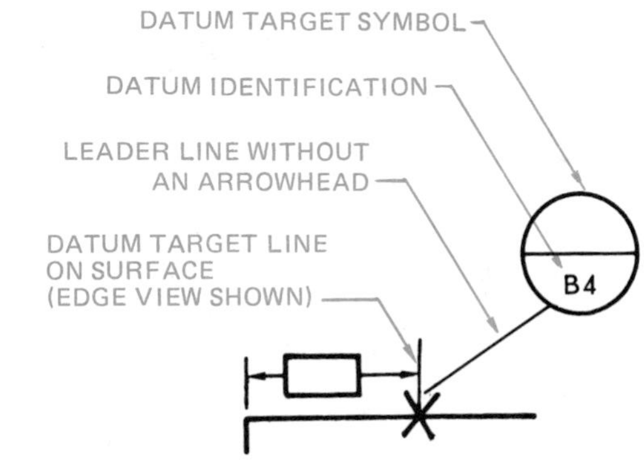

(A) A LINE AS A DATUM TARGET

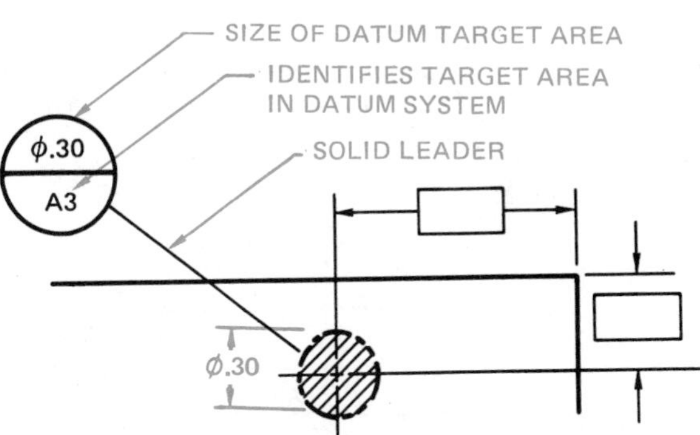

(B) NEAR SURFACE USED AS A DATUM TARGET

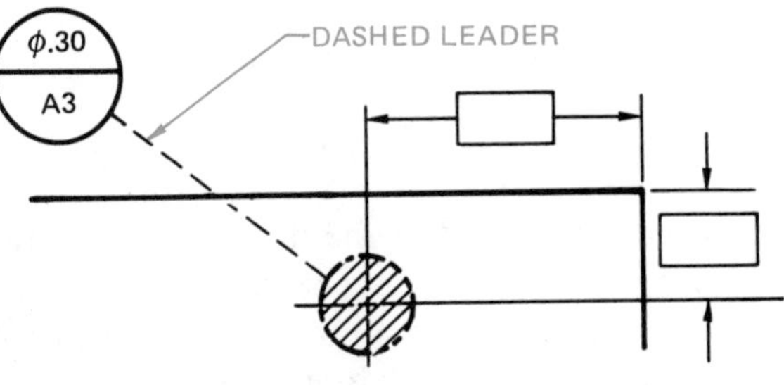

(C) FAR HIDDEN SURFACE USED AS A DATUM TARGET

Fig. 48-4 Datum target symbols

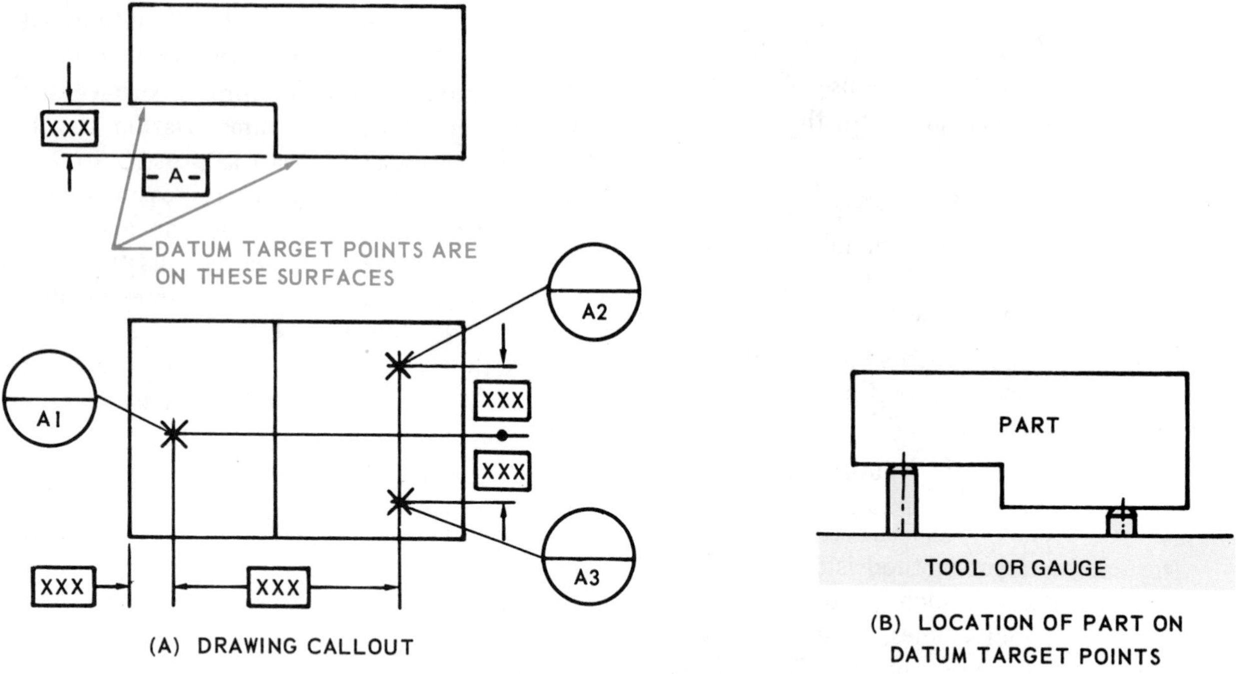

(A) DRAWING CALLOUT

(B) LOCATION OF PART ON DATUM TARGET POINTS

Fig. 48-5 Datum target points on different planes used as a datum

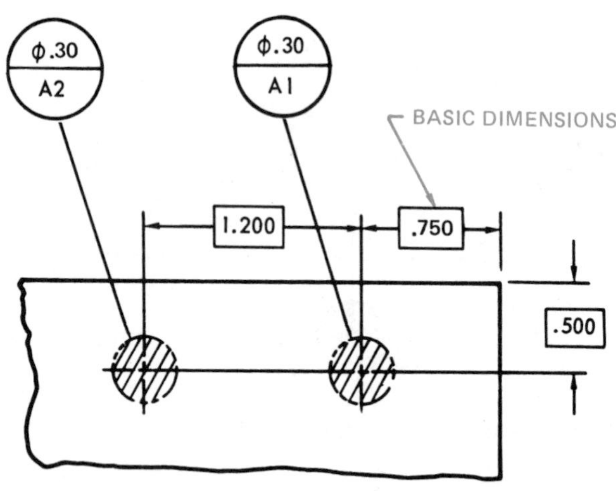

Fig. 48-6 Datum target areas

that particular target in the datum system. Targets are numbered consecutively, for each datum. For example, in a three-plane, six-point datum system, if the datums are A, B, and C, the datum targets would A1, A2, A3, B1, B2 and C1.

The datum target symbol is placed outside the part outline with a radial (leader) line directed to the target and ending without an arrowhead. The use of a solid leader indicates the datum target is on the near (visible) surface. The use of a dashed leader line, figure 48-4(C), indicates that the datum target is on the far (hidden) surface.

Targets need not be in the same plane. They must be located on different surfaces to meet functional requirements, figure 48-5. When a datum target area is used, the size of the area is shown inside the datum target symbol. Basic dimensions are used to locate the datum targets and are enclosed in a rectangular frame indicating that the general tolerance does not apply, figure 48-6.

Dimensions locating a set of datum targets should be either dimensionally related to one another or have a common origin.

The application and use of a surface and three lines as datum features are shown in figure 48-7.

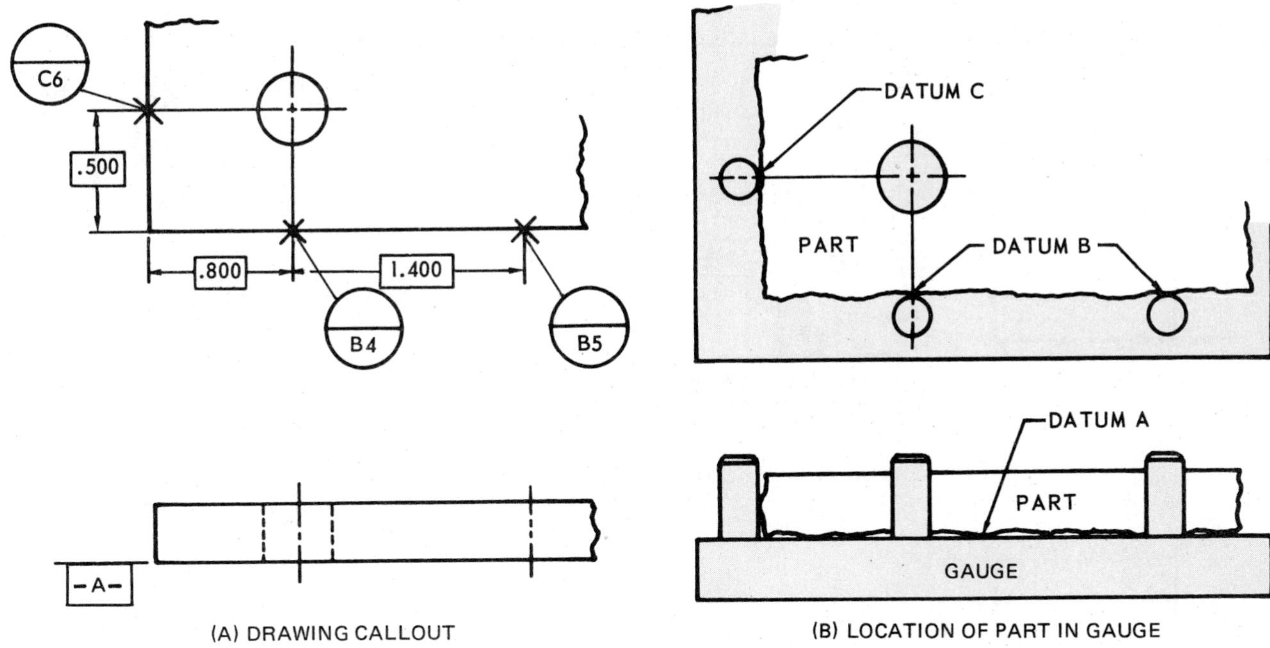

Fig. 48-7 Part with a surface and three lines used as datum features

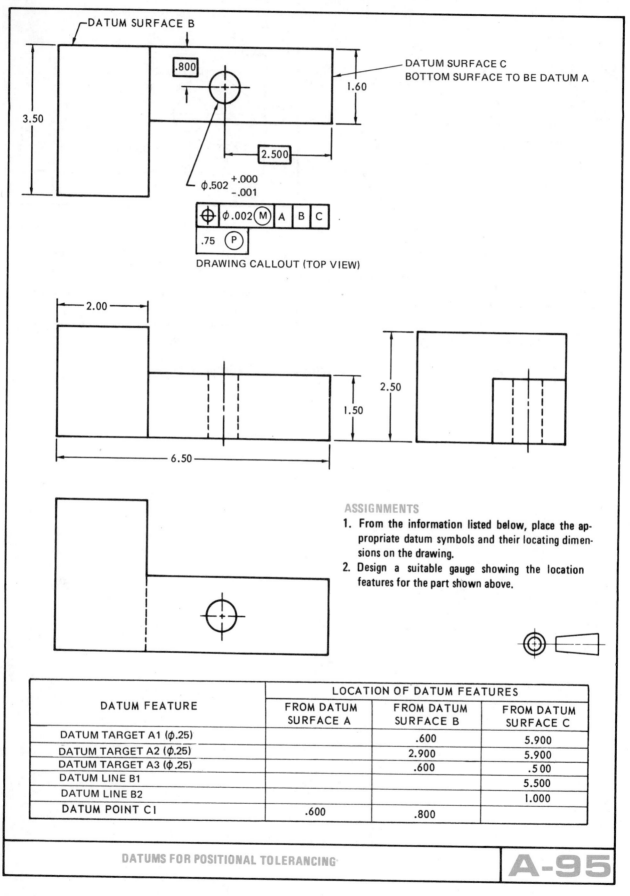

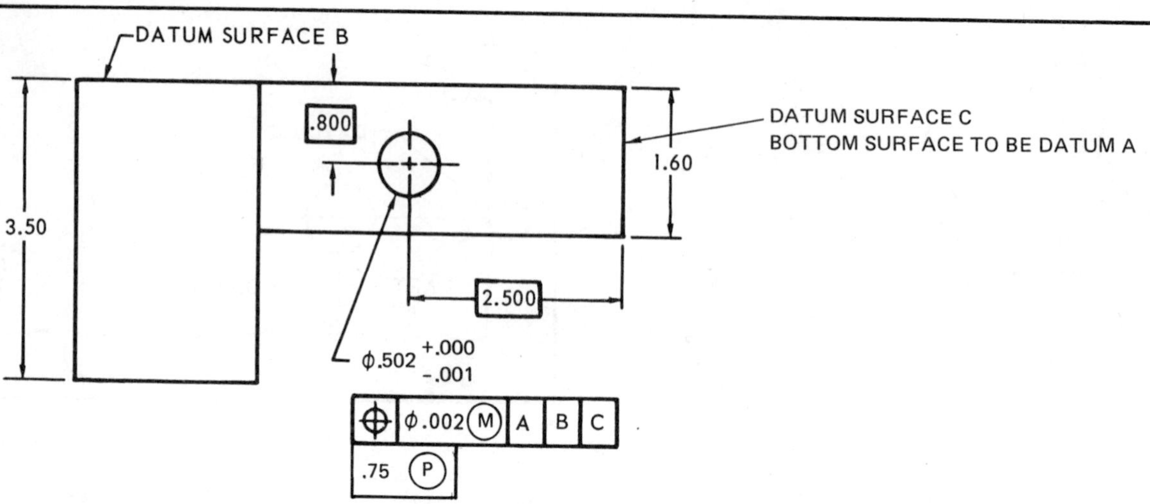

DATUM SURFACE B

DATUM SURFACE C
BOTTOM SURFACE TO BE DATUM A

.800

1.60

3.50

2.500

ϕ.502 +.000 -.001

⌖	ϕ.002 (M)	A	B	C
.75 (P)				

DRAWING CALLOUT (TOP VIEW)

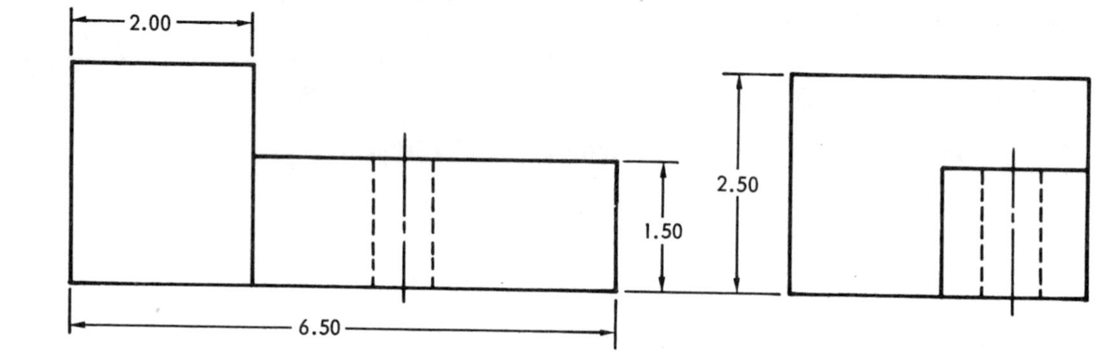

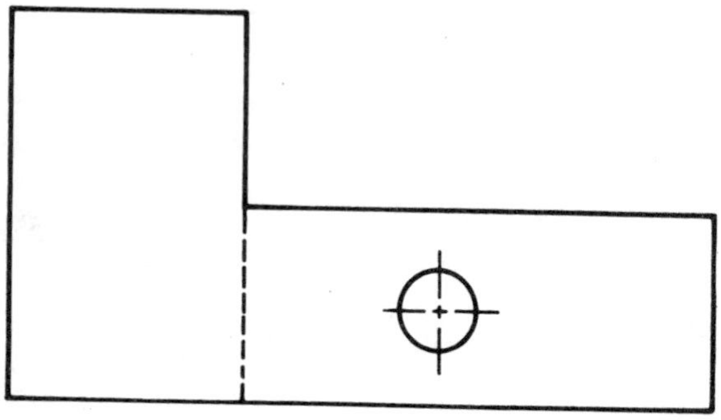

2.00

2.50

1.50

6.50

ASSIGNMENTS

1. From the information listed below, place the appropriate datum symbols and their locating dimensions on the drawing.
2. Design a suitable gauge showing the location features for the part shown above.

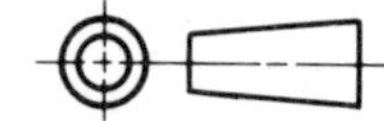

DATUM FEATURE	LOCATION OF DATUM FEATURES		
	FROM DATUM SURFACE A	FROM DATUM SURFACE B	FROM DATUM SURFACE C
DATUM TARGET A1 (ϕ.25)		.600	5.900
DATUM TARGET A2 (ϕ.25)		2.900	5.900
DATUM TARGET A3 (ϕ.25)		.600	.500
DATUM LINE B1			5.500
DATUM LINE B2			1.000
DATUM POINT C1	.600	.800	

DATUMS FOR POSITIONAL TOLERANCING

A-95

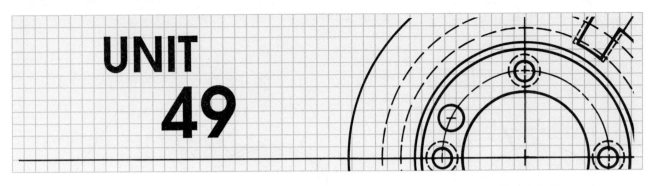

UNIT 49

PROFILE OF A LINE

A *profile* is the outline form or shape of a line or surface. A line profile may be the outline of a part or feature as depicted in a view on a drawing. A line profile may represent the edge of a part or it may refer to line elements of a surface in a single direction.

The symbol for profile of a line is a semicircle, figure 49-1. This is shown in the usual manner by including the symbol and tolerance in the feature control frame directed toward the line to be controlled, as shown in figure 49-2.

Bilateral and Unilateral Tolerances

The profile tolerance zone is usually equally arranged around the basic profile in a form known as a *bilateral tolerance zone.* The width of this zone is always measured perpendicular to the profile surface.

Occasionally, it is advisable to have the tolerance zone on one side of the basic profile instead of being equally divided on both sides. Such zones are called *unilateral tolerance zones* and are indicated with a phantom line close to the profile surface. The tolerance is directed to this line, as shown in figure 49-3.

Extent of Controlled Profile

The profile is generally intended to extend to the first abrupt change or sharp corner. If the extent is not clearly identified by sharp corners or by basic profile dimensions, it must be indicated by a note under the feature control frame, figure 49-4.

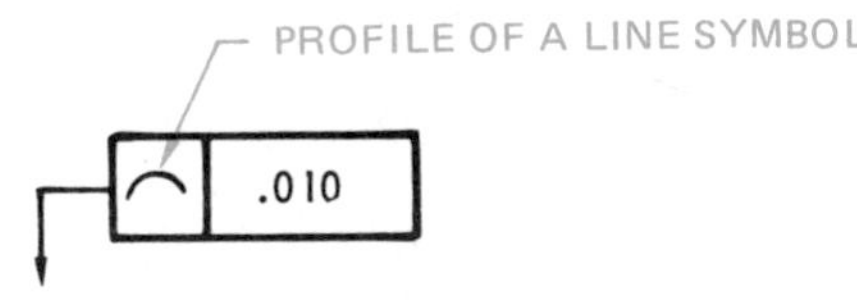

Fig. 49-1 Profile of a line tolerance

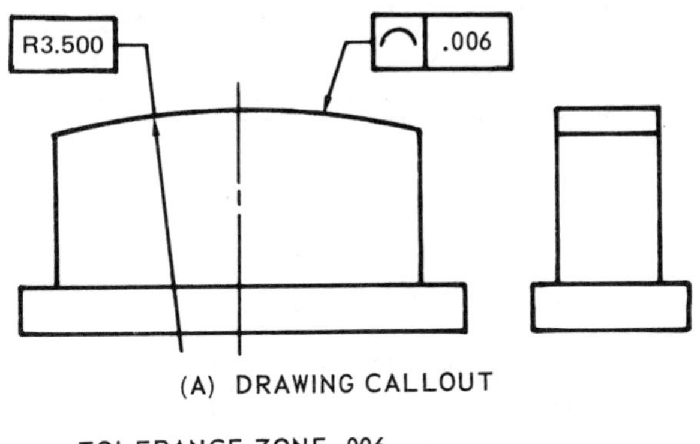

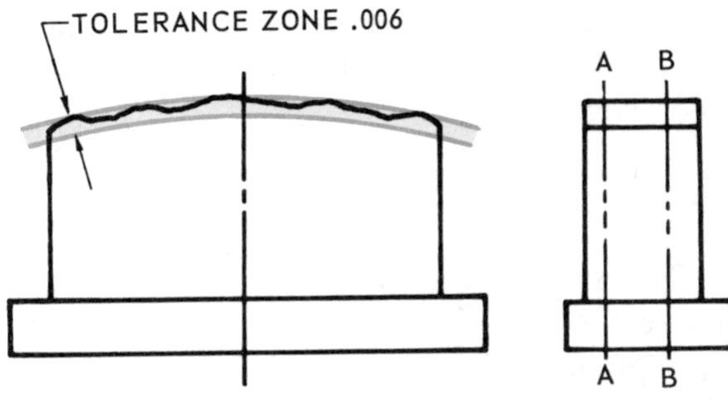

Fig. 49-2 Simple profile with profile of a line tolerance

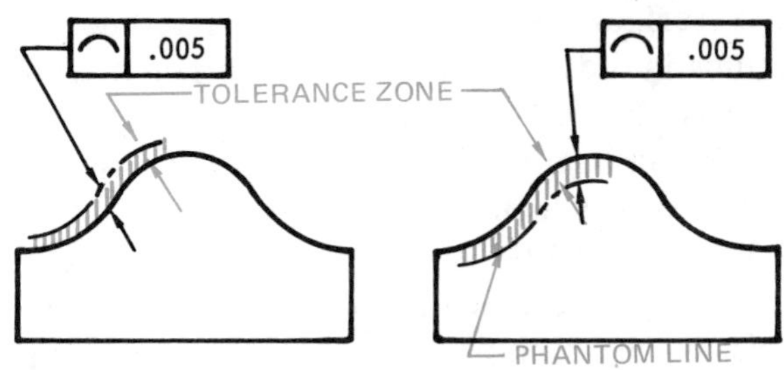

Fig. 49-3 Unilateral tolerance zones

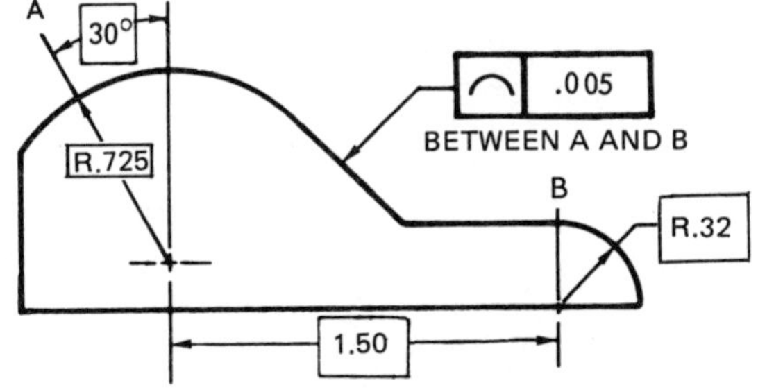

Fig. 43-4 Specifying extent of profile tolerance

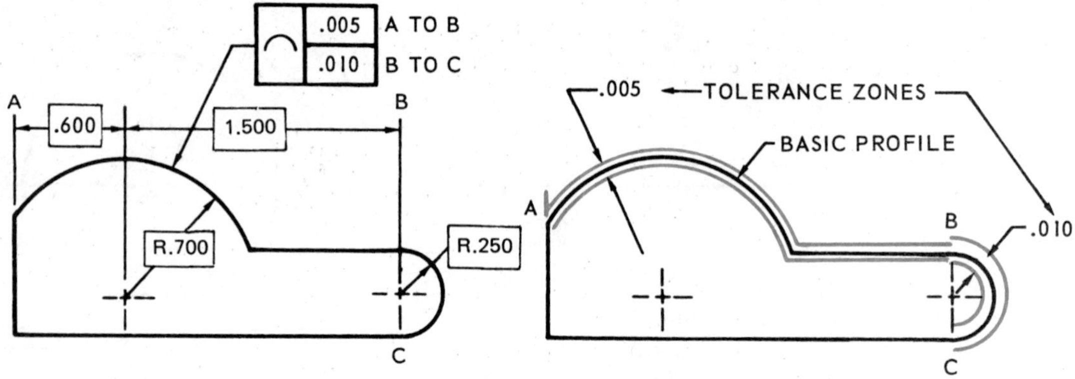

Fig. 49-5 Dual profile of a line tolerance zones

Dual Tolerances

If different profile tolerances must be applied to different parts of a profile, the tolerances are shown in a combined feature control frame. However, the extent of each should be clearly stated, figure 49-5.

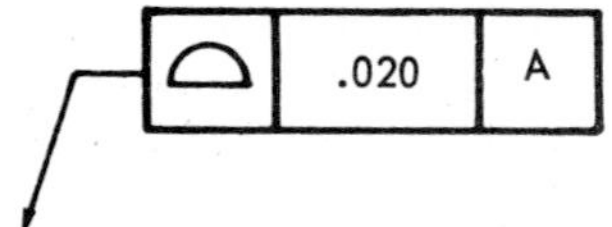

Fig. 49-6 Profile of a surface symbol

PROFILE OF A SURFACE

The symbol for profile of a surface is a semicircle enclosed by a straight line at the bottom, figure 49-6.

The tolerance zone established by the profile of a structure tolerance is a three dimensional zone extending the length and width (or circumference) of the feature. Usually, profile of a surface tolerancing requires reference to datums in order to provide proper orientation of a profile, figure 49-7.

Functional GO, NO-GO gauges are used to check parts which are toleranced on an MMC basis. Figure 49-8 shows two identical parts. One is toleranced with a profile of a line tolerance which applies to the circumference of the part. This is often referred to as *all-around tolerancing.* The symbol used to designate "all around" is placed on the leader. The second part has a profile of a surface tolerance.

For figure 49-8(A), a suitable gauge would be made to the maximum material size plus the profile tolerance. If the profile tolerance is zero, the profile gauge will also check the maximum limit of size. The part must be capable of passing through the gauge.

When a profile of a surface tolerance is applied on an MMC basis, as shown in figure 49-8(B), a functional GO, NO-GO gauge is required, similar to that shown in figure 49-8(A); except that it must be of such length as to completely encompass the part.

Profile tolerancing for copular surfaces in one place is covered in Unit 50.

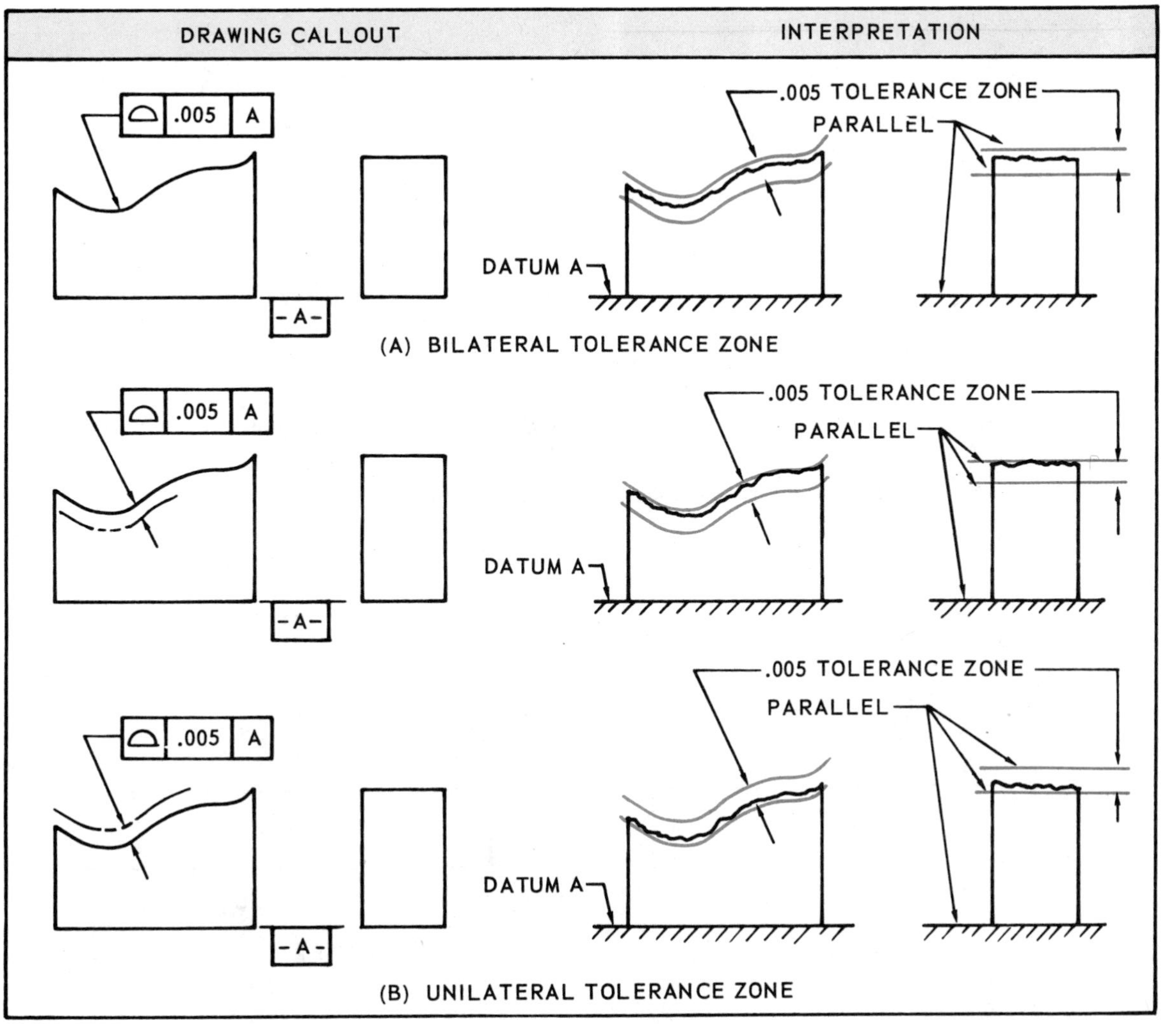

DRAWING CALLOUT	INTERPRETATION
.005 A; -A-	.005 TOLERANCE ZONE; PARALLEL; DATUM A
(A) BILATERAL TOLERANCE ZONE	
.005 A; -A-	.005 TOLERANCE ZONE; PARALLEL; DATUM A
.005 A; -A-	.005 TOLERANCE ZONE; PARALLEL; DATUM A
(B) UNILATERAL TOLERANCE ZONE	

Fig. 49-7 Profile of a surface referenced to a datum

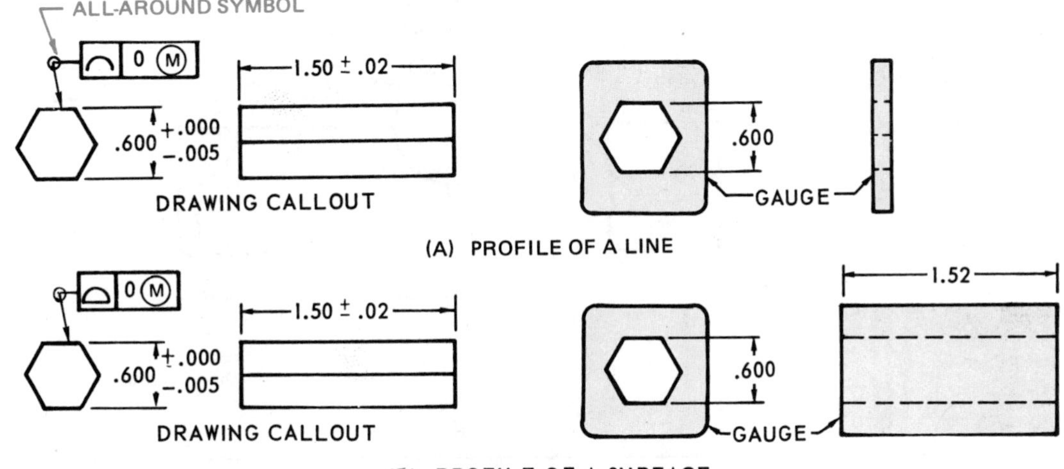

Fig. 49-8 A comparison of profile tolerances

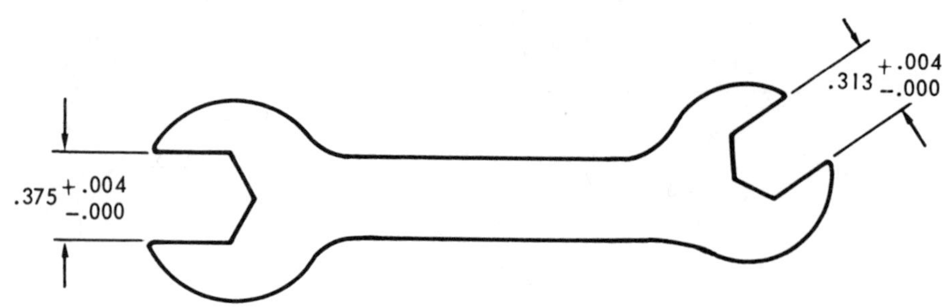

ASSIGNMENTS

1. Profile tolerances are required for the wrench. The wrench openings are to have unilateral profile of a surface tolerances of zero tolerance at MMC. The remainder of the wrench is to have a bilateral profile of a line tolerance of .030 in. Add the necessary geometric tolerances to the drawing.
2. Make a sketch showing the tolerance zone for the profile of the wrench.
3. Design suitable gauges to check the following parts.

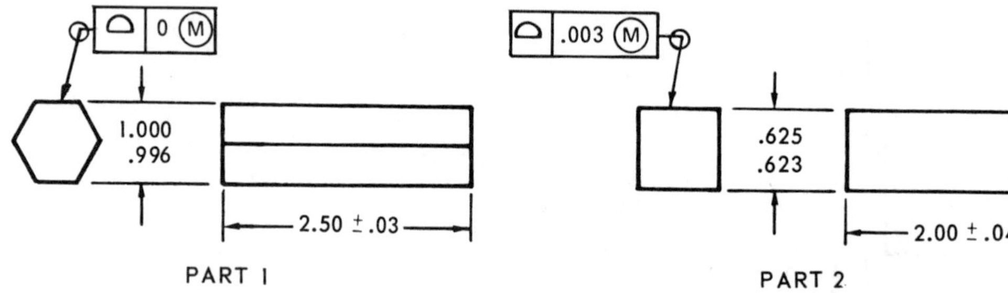

4. The cross-sectional shape of the part shown below must be controlled to ensure that it does not cross the boundary of perfect form at the maximum material size. If all dimensions are to be held to +.000, -.010, add a suitable profile of a line tolerance to the part.

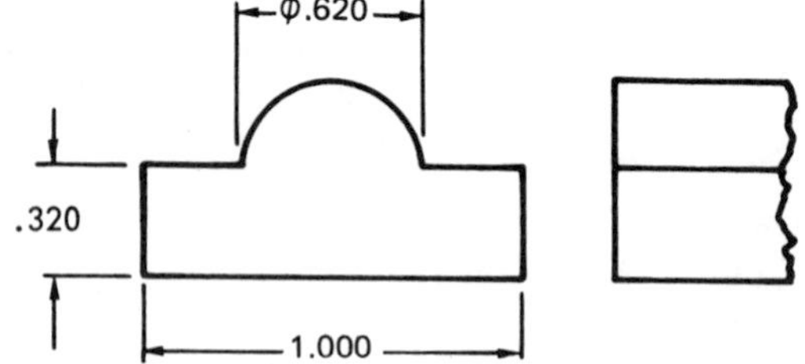

5. The profile of the part shown below must be controlled from A to B within a profile of a line bilateral tolerance of .004 except that the .562 in. portion can be permitted to vary vertically by ±.005. Show how this would be specified on the drawing and sketch the resulting tolerance zone.

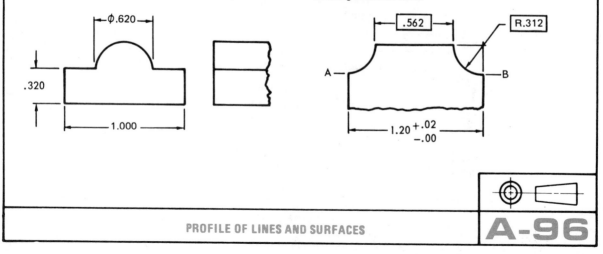

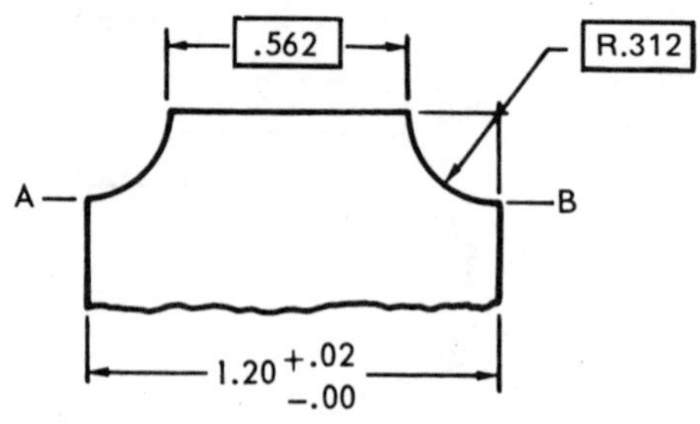

PROFILE OF LINES AND SURFACES

A-96

UNIT 50

CORRELATIVE TOLERANCES

Correlative geometric tolerancing refers to tolerancing for the control of two or more features which should be related in position or attitude. Examples of correlative tolerancing include: coplanarity, for controlling two or more flat surfaces; positional tolerance at MMC, for symmetrical relationships, such as for controlling features equally disposed around a center plane; concentricity and coaxiality, for controlling features having common axes or center lines; and runout, for controlling surfaces related to an axis.

Coplanarity

Coplanarity is the condition of two or more surfaces having all elements in the same plane. The profile of a surface tolerance may be used where it is desirable to treat two or more surfaces as a single interrupted or noncontinuous surface. In this case a control is provided similar to that achieved by a flatness tolerance applied to a single plane surface. As shown in figure 50-1, the profile of a surface tolerance establishes a tolerance zone defined by two parallel planes within which the considered surfaces must lie.

Coaxiality Controls

Coaxiality is the condition where the axes of two or more surfaces of revolution are coincident. The amount of permissible variation from coaxiality may be expressed by a positional tolerance, a runout tolerance, or a concentricity tolerance. The type of tolerance selected depends on the functional requirements of the design.

Positional Tolerance Control

Where the surfaces of revolution are cylindrical and the control of the axes can be applied on a material condition basis, positional tolerance is recommended. Coaxiality may be controlled by specifying a positional tolerance at MMC. The datum feature may be specified on either an MMC or RFS basis. In figure 50-2, the datum feature is applied on an MMC basis. In such cases, any departure of the datum feature from MMC may result in additional displacement between its axis and the axis of the feature.

Where it is necessary to control coaxiality of related features within their limits of size, a zero positional tolerance at MMC is specified. The datum feature is normally specified on an MMC basis, see figure 50-3. Boundaries of perfect form are thereby established that are truly coaxial, when both features are at MMC.

A positional tolerance may also be used to control the alignment of two or more holes located on a common axis. It is used where a tolerance of location alone does not provide the necessary control of alignment of these holes,

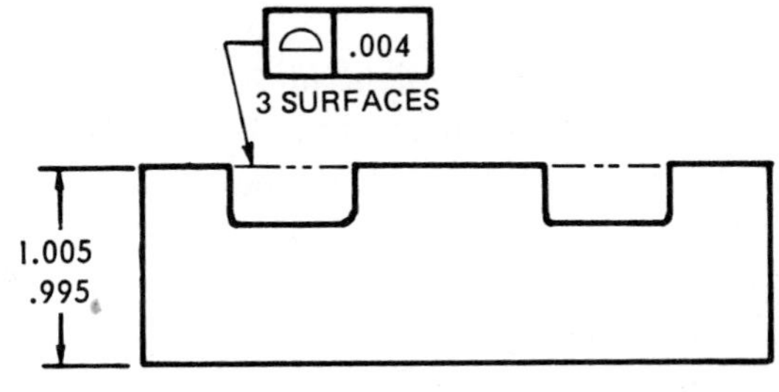

(A) DRAWING CALLOUT

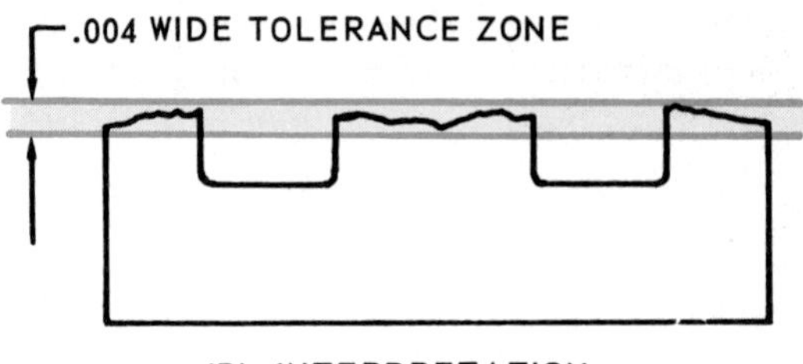

(B) INTERPRETATION

Fig. 50-1 Specifying profile of a surface for coplanar surfaces

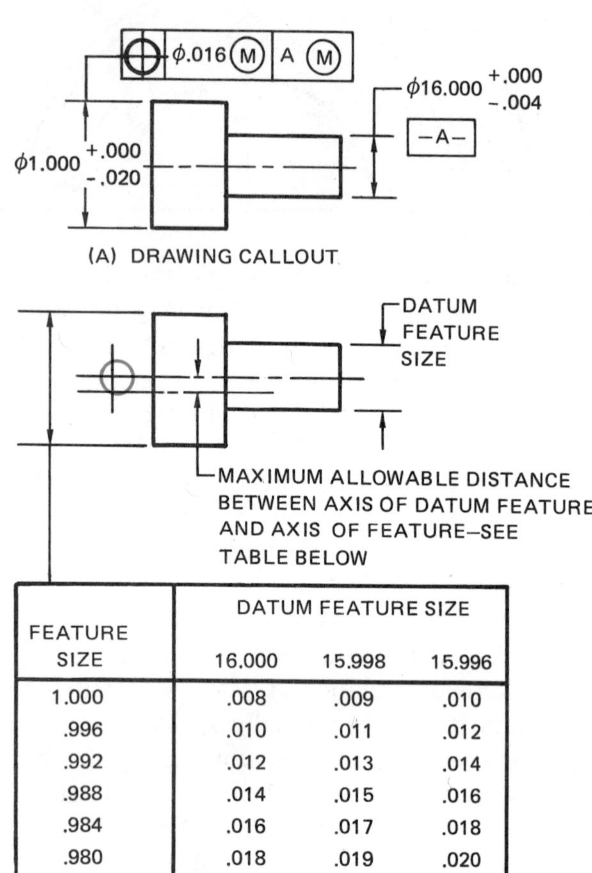

FEATURE SIZE	DATUM FEATURE SIZE		
	16.000	15.998	15.996
1.000	.008	.009	.010
.996	.010	.011	.012
.992	.012	.013	.014
.988	.014	.015	.016
.984	.016	.017	.018
.980	.018	.019	.020

(B) INTERPRETATION

Fig. 50-2 Positional tolerancing for coaxiality

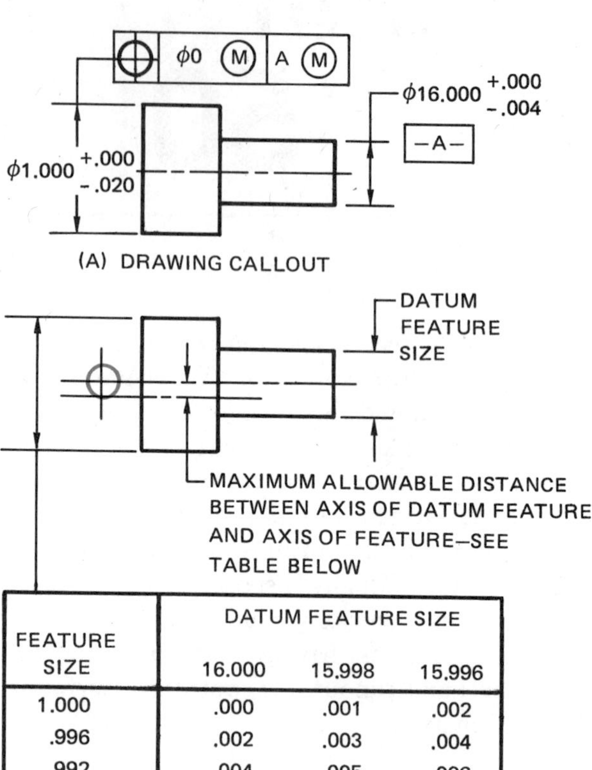

FEATURE SIZE	DATUM FEATURE SIZE		
	16.000	15.998	15.996
1.000	.000	.001	.002
.996	.002	.003	.004
.992	.004	.005	.006
.988	.006	.007	.008
.984	.008	.009	.010
.980	.010	.011	.012

(B) INTERPRETATION

Fig. 50-3 Zero positional tolerancing at MMC for coaxiality

and a separate requirement must be specified. Figure 50-4 shows an example of two coaxial holes of different sizes which require the same control of alignment. A single feature control symbol, supplemented by a note such as TWO COAXIAL HOLES, is used.

Runout Tolerance Control

Where a combination of surfaces of revolution are cylindrical, conical, or spherical relative to a common datum axis, a runout tolerance is recommended, see figure 50-5.

Runout

Runout is the deviation in position of a surface of revolution as a part is revolved around a datum axis.

A *runout tolerance* represents the maximum permissible variation of position of a surface, measured at a fixed point, when the part is revolved without axial movement through 360 degrees around the datum axis. The tolerance zone is the area, or space, of uniform thickness between two shapes coaxial with the datum axis and parallel to the true profile of the controlled surface.

For this type of control, MMC is not applicable because the tolerance controls surface elements.

There are two concepts in runout tolerancing. One is known as *circular runout* and concerns runout of each circular element or cross section. The other is called *total runout* and provides composite control of all surface elements simultaneously. Circular runout is simpler than total runout and is generally easier to measure.

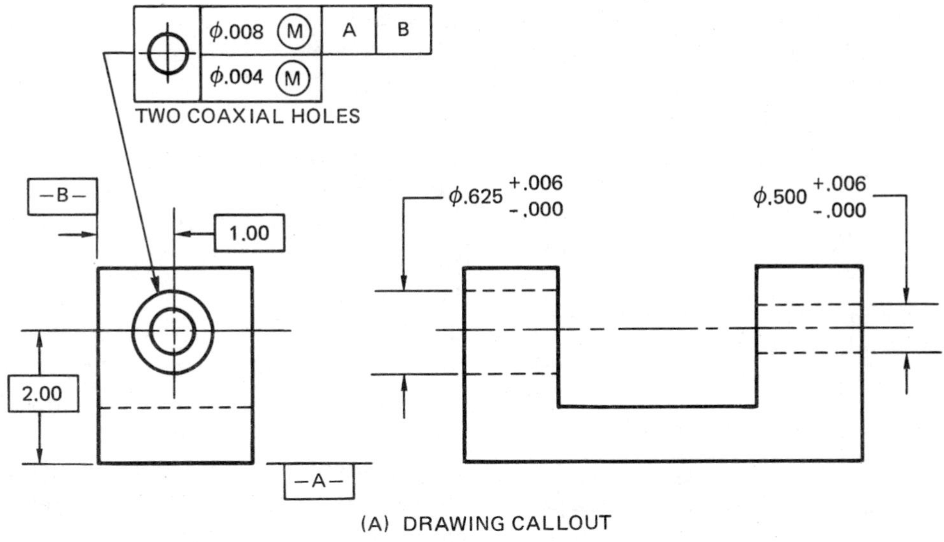

(A) DRAWING CALLOUT

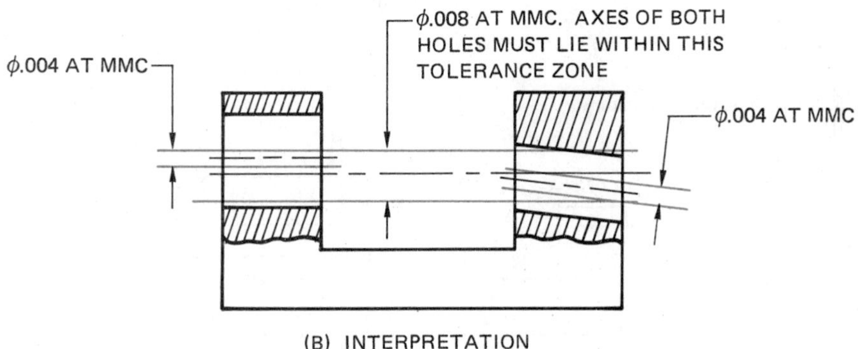

(B) INTERPRETATION

Fig. 50-4 Positional tolerancing for coaxial holes of different size

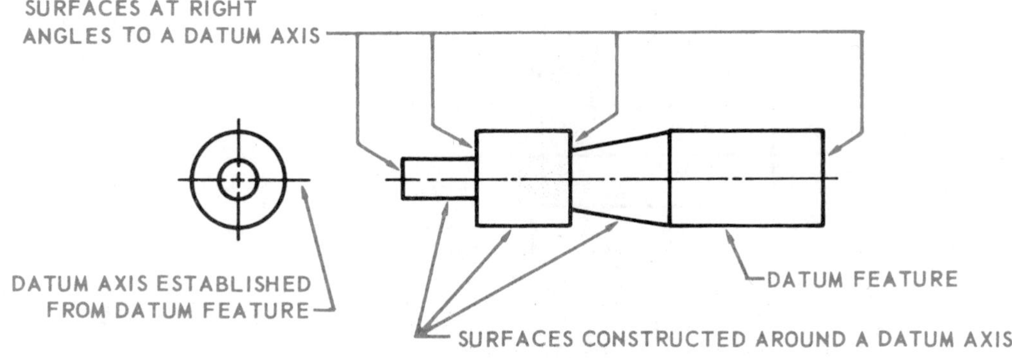

Fig. 50-5 Features applicable to runout tolerancing

There are two geometrical characteristic symbols for runout, one for circular and one for total runout, figure 50-6.

Circular Runout. Circular runout means that the runout tolerance must be met in each circular element, and a sufficient number of measurements must be made at different positions along the surface to guarantee that the tolerance is met at all cross sections of the part.

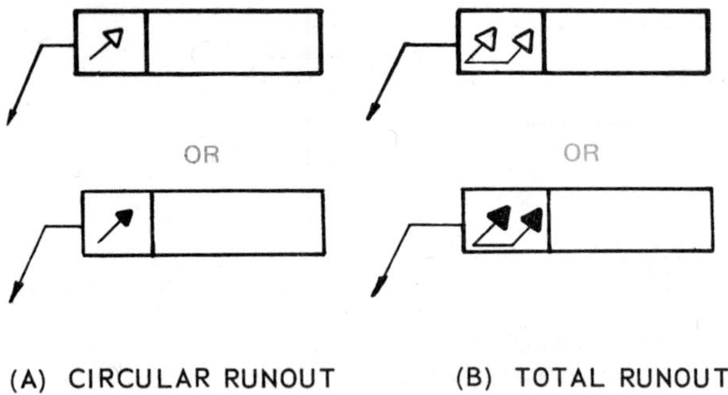

Fig. 50-6 Runout tolerance

Thus, in figure 50-7, the surface is measured at several positions along the surface. At each position the indicator movement during one revolution of the part must not exceed the specified tolerance, in this case .004 in. Circular runout does not control the profile elements of the surface.

When a runout tolerance applies to a specific portion of a surface, a chain line is drawn adjacent to the surface profile to show the desired length.

Total Runout. Total runout concerns the runout of a complete surface, not merely the runout of each circular element. For measurement purposes, the checking indicator must traverse the full length or extent of the surface while the part is revolved about its datum axis. Total runout is the difference between the lowest indicator

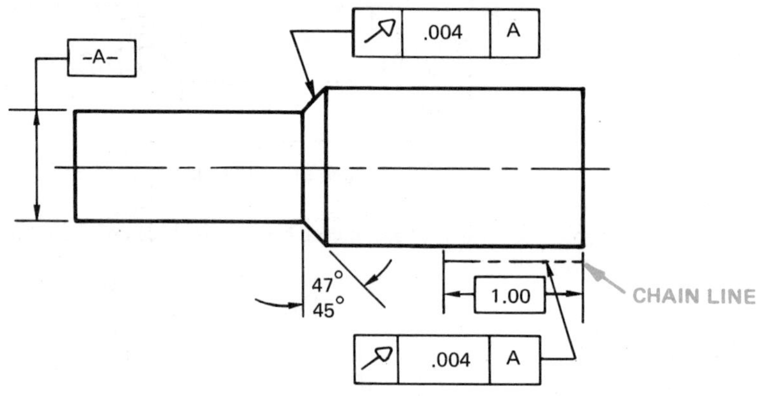

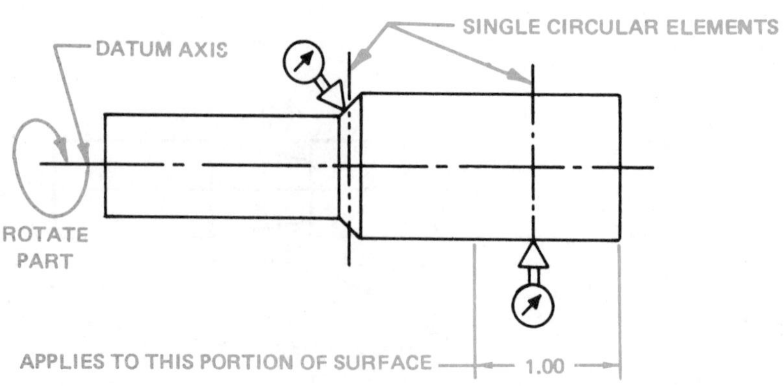

Fig. 50-7 Specifying circular runout relative to a datum diameter

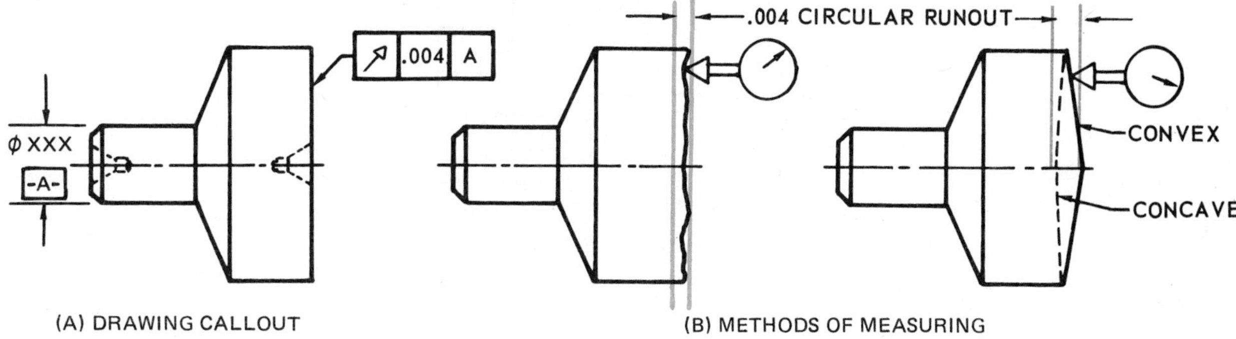

Fig. 50-8 Runout perpendicular to axis

reading in any position and the highest reading in that or any other position on the same surface, figures 50-8 and 50-9.

Concentricity Tolerance Control

Concentricity is the condition where the axes of all cross-sectional elements of a surface of revolution are common to the axis of a datum feature. A concentricity tolerance specifies a cylindrical tolerance zone whose axis coincides with a datum axis and within which all cross-sectional axes of the feature being controlled must lie. The geometric characteristic symbol used for concentricity consists of two concentric circles, figure 50-10.

The specified tolerance and the datum reference apply only on an RFS basis, see figure 50-11. The feature axis must lie within a cylindrical zone .005 in. in diameter, regardless of feature size, whose axis coincides with the datum axis. A concentricity tolerance requires the establishment and verification of axis irrespective of surface conditions.

Symmetry

Symmetry is a condition in which a feature(s) is symmetrically disposed about the center plane of a datum feature. Where it is required that a feature be located symmetrically with respect to the center plane of a datum feature, positional tolerancing is used. A symmetrical relationship may be controlled by specifying a positional tolerance at MMC, as in

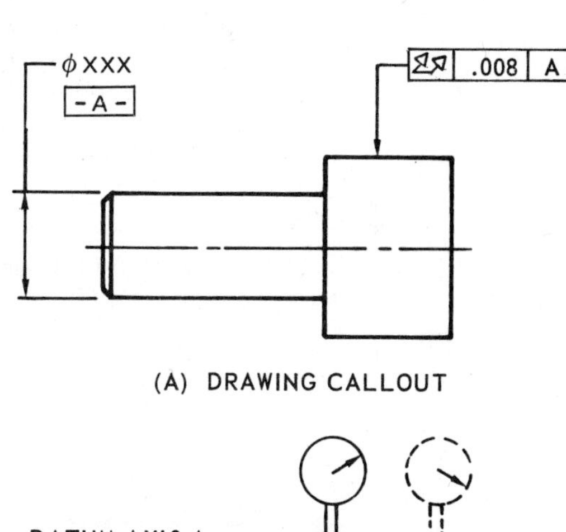

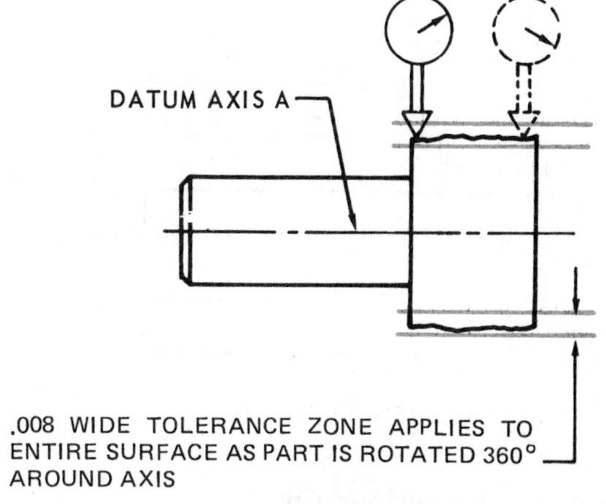

Fig. 50-9 Specifying total runout relative to a datum diameter

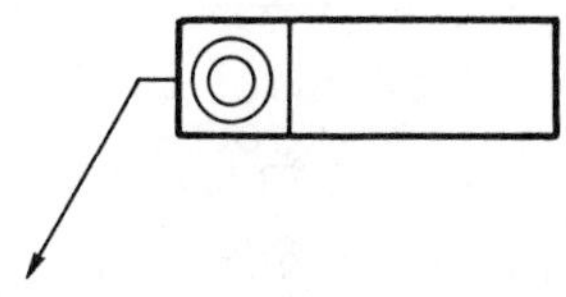

Fig. 50-10 Concentricity or coaxiality tolerancing

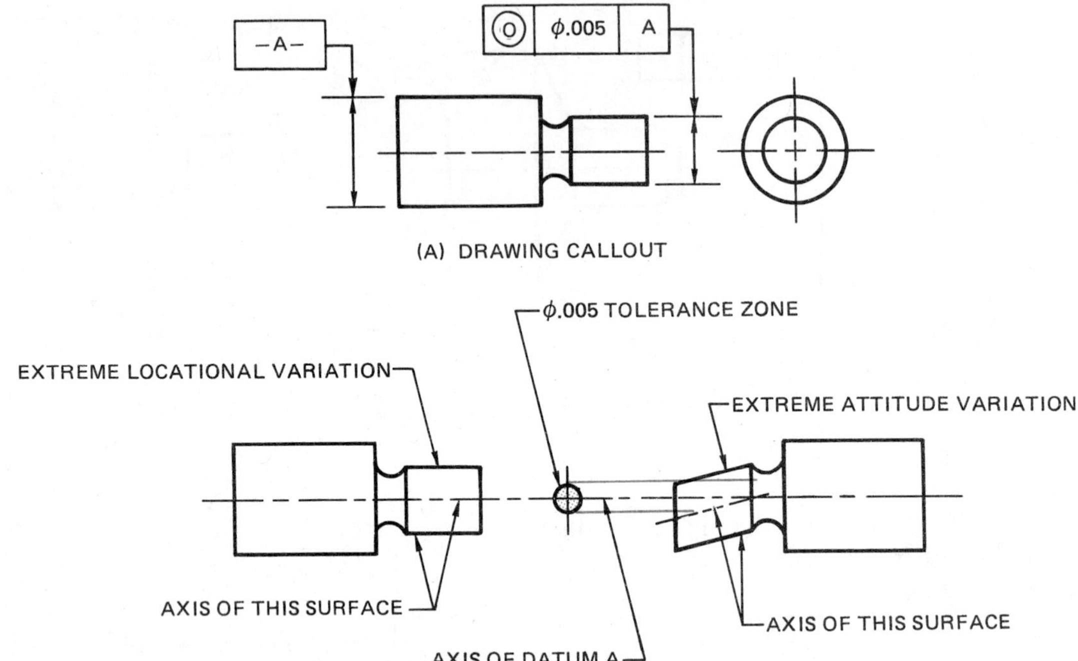

Fig. 50-11 Concentricity tolerancing for coaxiality

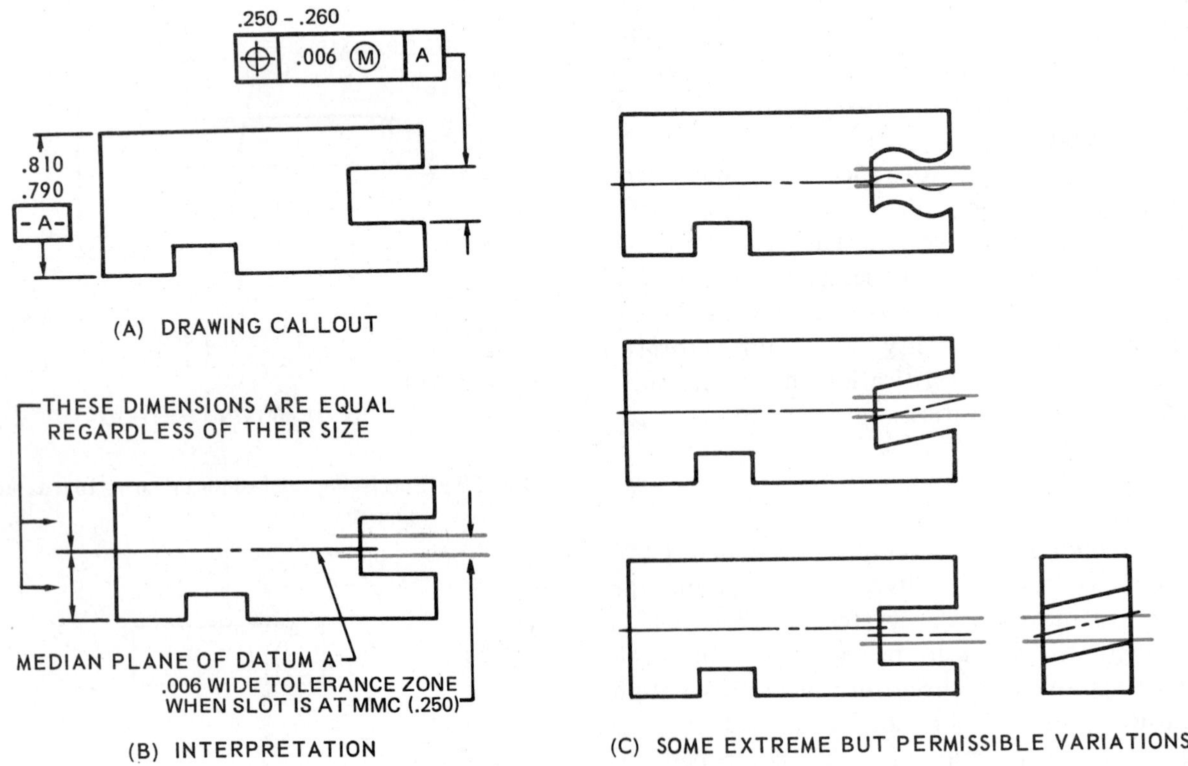

Fig. 50-12 Tolerancing of symmetrical features

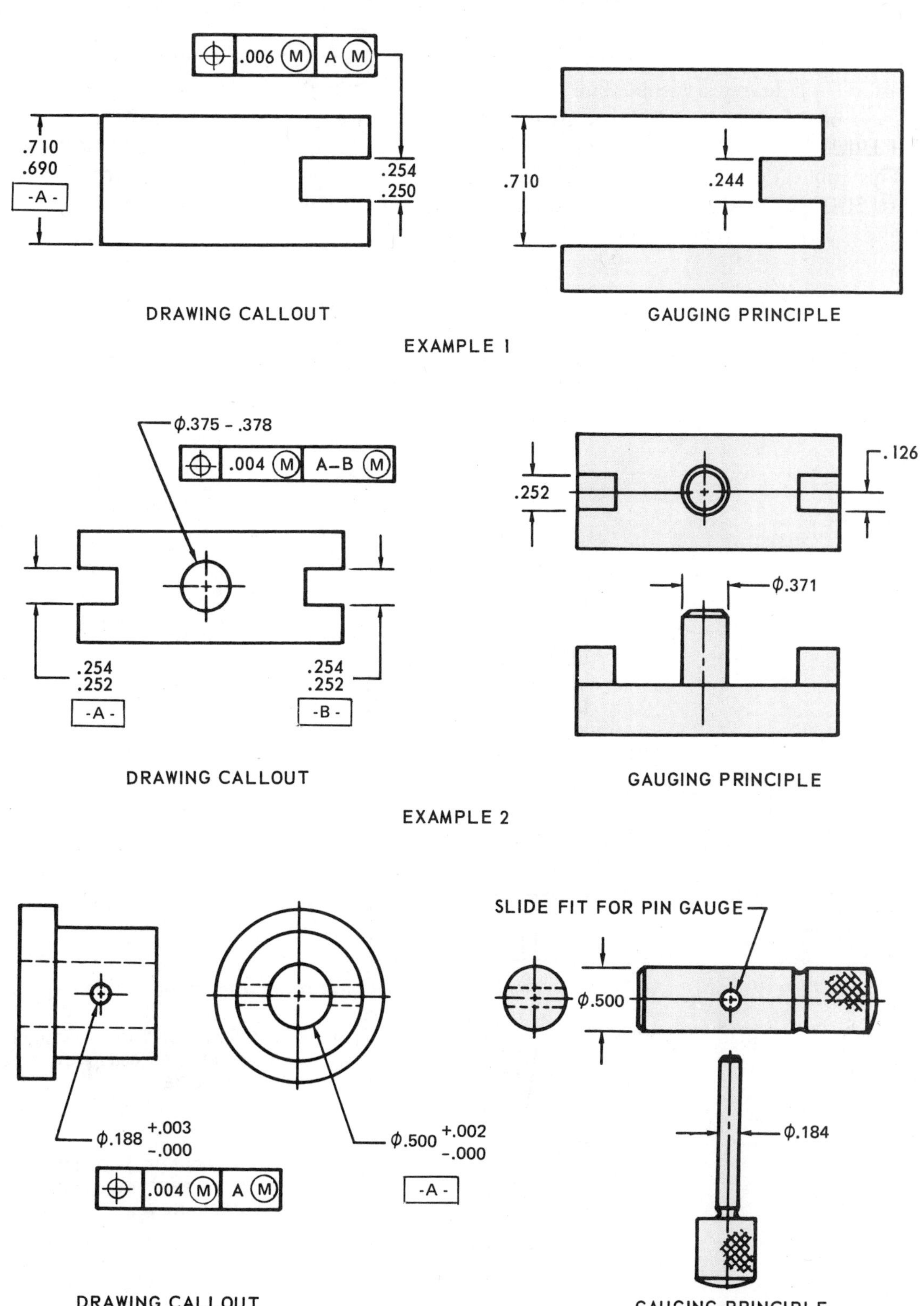

Fig. 50-13 Examples of symmetrical parts toleranced on an MMC basis

figure 50-12. The datum feature may be specified either on an MMC or an RFS basis.

Where it is necessary to control the symmetry of related features within their limits of size, a zero positional tolerance is specified. The datum feature is normally specified on an MMC basis. This application is the same as that shown in figure 50-2 except that it applies a tolerance to a center plane location. Figure 50-13 shows examples of symmetrical parts toleranced on an MMC basis.

Some designs may require a control of the symmetrical relationship between features regardless of their actual sizes. In such cases, both the specified positional tolerance and the datum reference apply on an RFS basis, see figure 50-14.

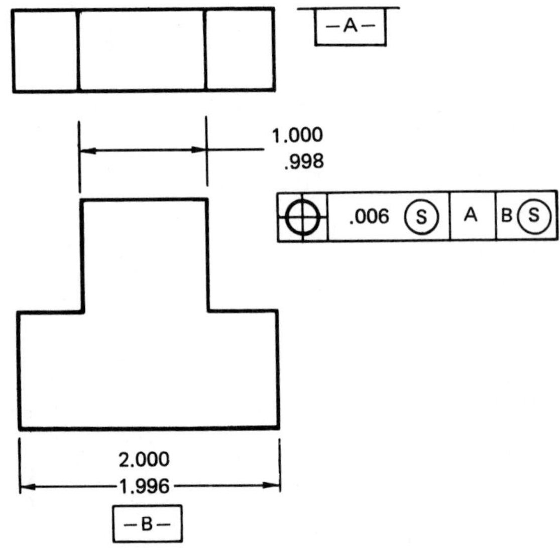

(A) DRAWING CALLOUT

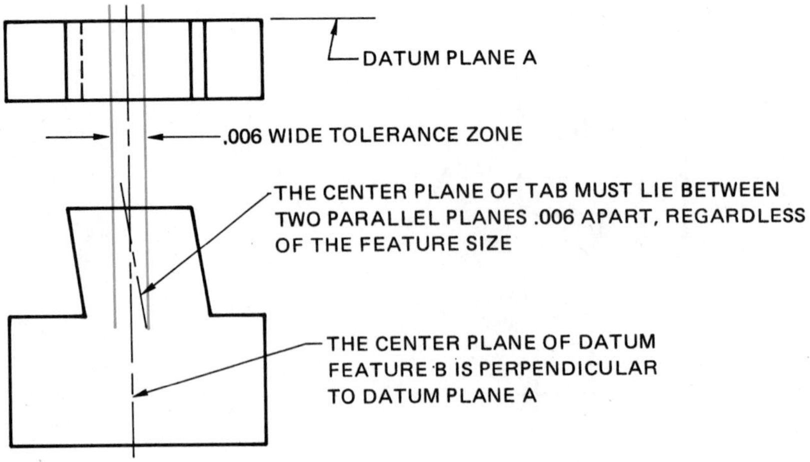

DATUM PLANE A

.006 WIDE TOLERANCE ZONE

THE CENTER PLANE OF TAB MUST LIE BETWEEN TWO PARALLEL PLANES .006 APART, REGARDLESS OF THE FEATURE SIZE

THE CENTER PLANE OF DATUM FEATURE B IS PERPENDICULAR TO DATUM PLANE A

Fig. 50-14 Positional tolerance RFS for symmetry

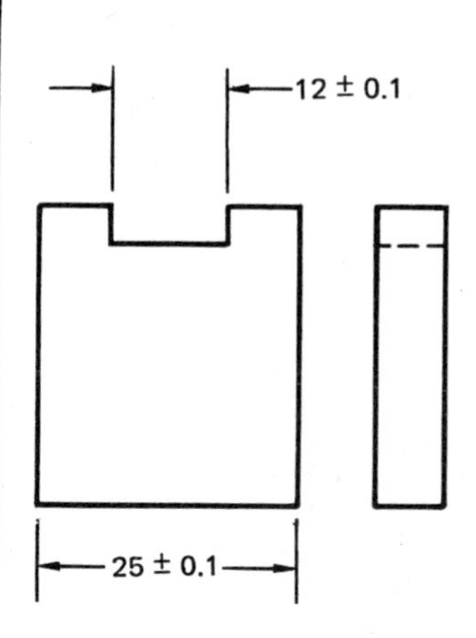

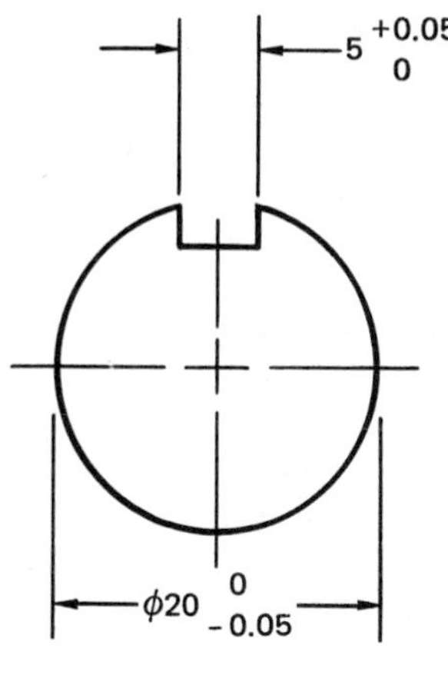

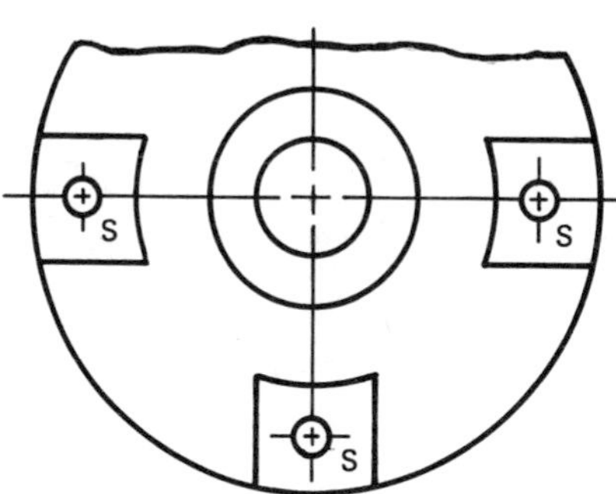

ASSIGNMENTS

1. The 12 mm slot is to be symmetrical within zero tolerance at MMC with the 25 mm width, regardless of the width size. Add a suitable geometric tolerance to the drawing.
2. The 5 mm keyseat is to be symmetrically located on the Ø20 mm shaft within 0.05 MMC when the diameter is at its maximum permissible size. Add a suitable geometric tolerance to the drawing.

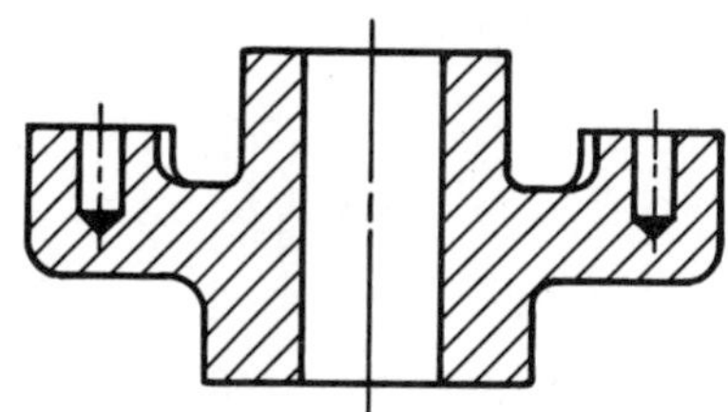

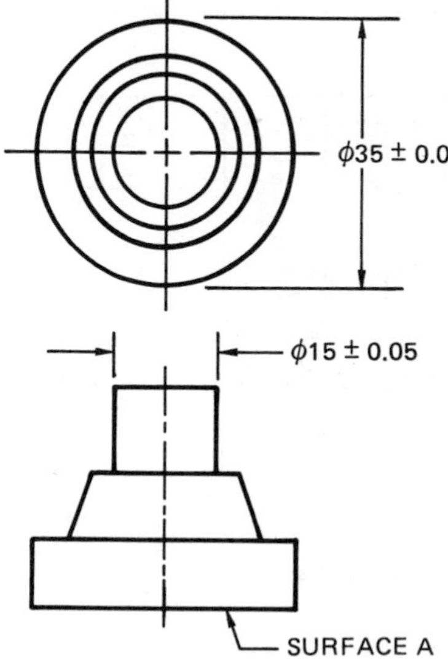

3. The Ø15 mm must be concentric with the Ø35 mm within 0.03 mm when resting on surface **A**. Specify this on the drawing.

4. The four surfaces marked **S** must be coplanar with one another within 0.01 mm and perpendicular with the axis of the center hole within 0.03 MMC. Add the geometric tolerances to the drawing.

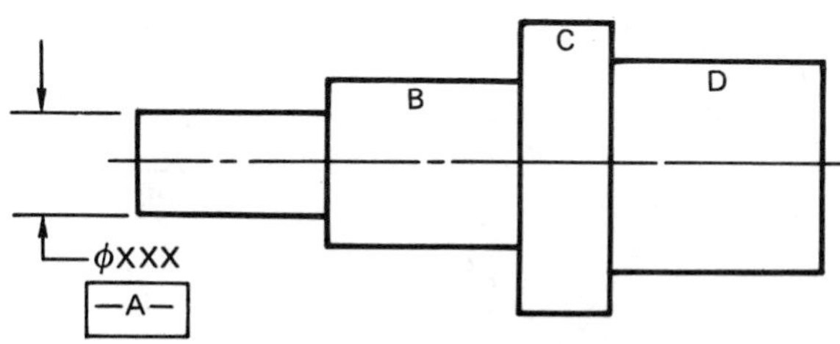

5. Diameters **B** and **C** are to have a circular runout of 0.1 mm with reference to datum **A**. Diameter **D** is to have a total runout of 0.05 mm with reference to datum **A**. Add the geometric tolerances to the drawing.

ALL DIMENSIONS ARE IN MILLIMETERS
METRIC

CORRELATIVE TOLERANCES – COPLANARITY, SYMMETRY, CONCENTRICITY, RUNOUT	A-97M

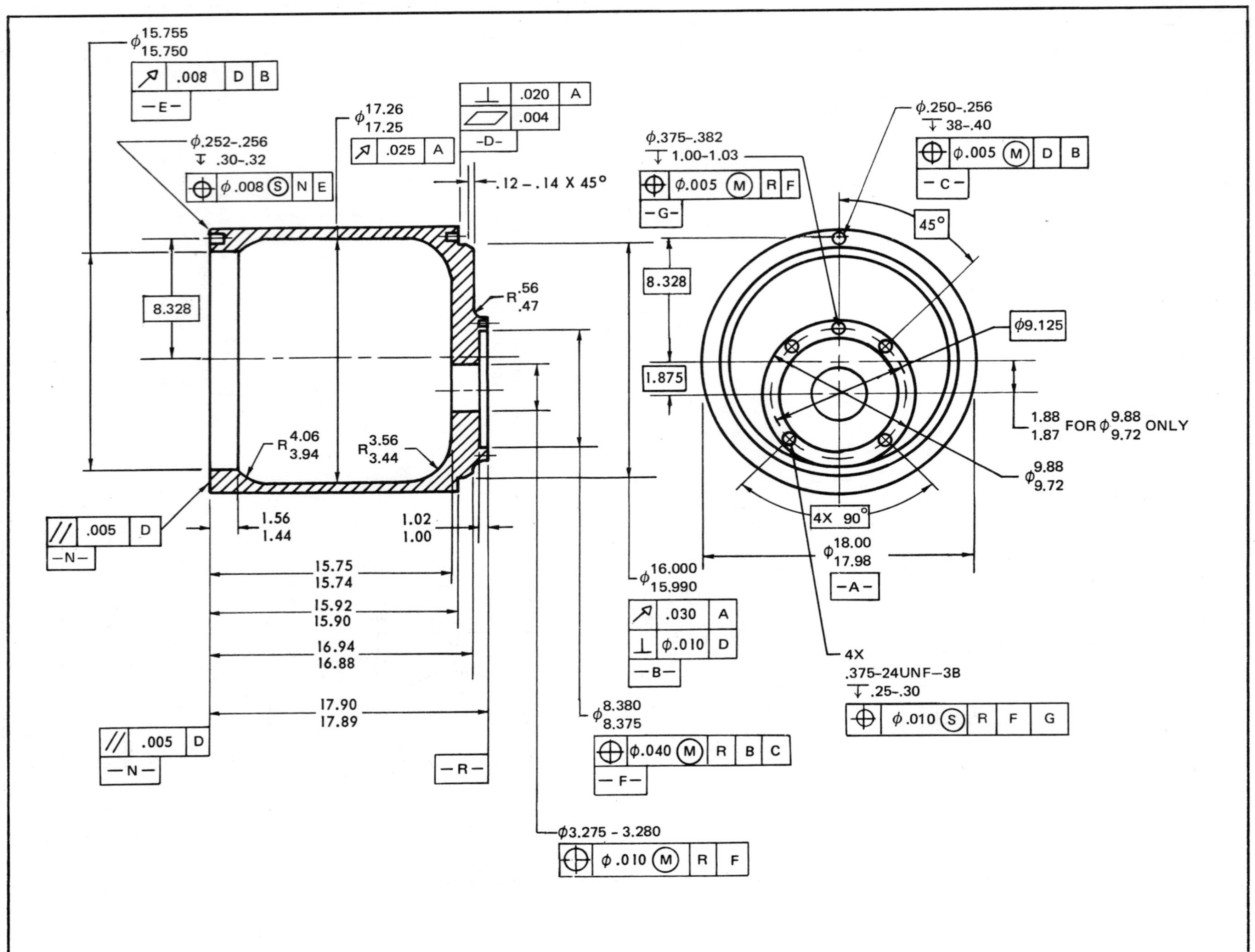

φ 15.755 15.750
.008 D B
– E –
φ.252-.256
⊤ .30-.32
φ.008 Ⓢ N E
φ 17.26 17.25
.025 A
.020 A
.004
-D-
.12 – .14 X 45°
φ.375-.382
⊤ 1.00-1.03
φ.005 Ⓜ R F
– G –
φ.250-.256
⊤ 38-.40
φ.005 Ⓜ D B
– C –
45°
8.328
8.328
R .56 .47
φ9.125
1.875
1.88
1.87
FOR φ 9.88 9.72 ONLY
φ 9.88 9.72
R 4.06 3.94
R 3.56 3.44
4X 90°
.005 D
– N –
1.56
1.44
1.02
1.00
φ 18.00 17.98
– A –
15.75
15.74
φ 16.000 15.990
15.92
15.90
.030 A
φ.010 D
– B –
16.94
16.88
4X
.375-24UNF–3B
⊤ .25-.30
φ.010 Ⓢ R F G
17.90
17.89
φ 8.380 8.375
.005 D
– N –
– R –
φ.040 Ⓜ R B C
– F –
φ3.275 - 3.280
φ.010 Ⓜ R F

SKETCHING ASSIGNMENTS

1. Sketch a suitable gauge to check the Ø3.275-3.280 hole.
2. Make a sketch of datum surface **D** showing the permissible tolerance zone.
3. Make a sketch of datum surface **N** showing the permissible tolerance zone.

QUESTIONS

1. How many datum surfaces or points are indicated?
2. How many basic dimensions are indicated?
3. How many datum surfaces are flat?
4. How many datum surfaces are circular?
5. How many dimensions show positional tolerancing?
6. How many form tolerances are required?
7. How many orientation tolerances are shown?
8. How many features use datum **A** as a reference?
9. If the diameter of datum **F** was Ø8.375, what would be the maximum permissible positional tolerance?
10. What is the tertiary datum for the positional tolerance of the diameter shown as datum **F**?
11. The geometric tolerances placed on datum surface **D** controls ______ and ______
12. With reference to the Ø3.275-3.280 hole, what variation from the basic dimension [1.875] is permitted when the hole is (A) Ø3.275, (B) Ø3.280?
13. With reference to the Ø.252-.256 hole, what is the maximum vertical variation from true position if the hole was (A) Ø.252, (B) Ø.256?
14. What is (A) maximum, (B) minimum wall thickness permitted at the Ø17.25-17.26?
15. With reference to the Ø.250-.256 hole, what vertical variation from the basic dimension [8.328] is permitted when the hole is (A) Ø.250, (B) Ø.254, (C) Ø.256?

ANSWERS

1. ______
2. ______
3. ______
4. ______
5. ______
6. ______
7. ______
8. ______
9. ______
10. ______
11. ______ ______

12\. A. ______ B. ______
13\. A. ______ B. ______
14\. A. ______ B. ______
15\. A. ______ B. ______ C. ______

HOUSING

A-98

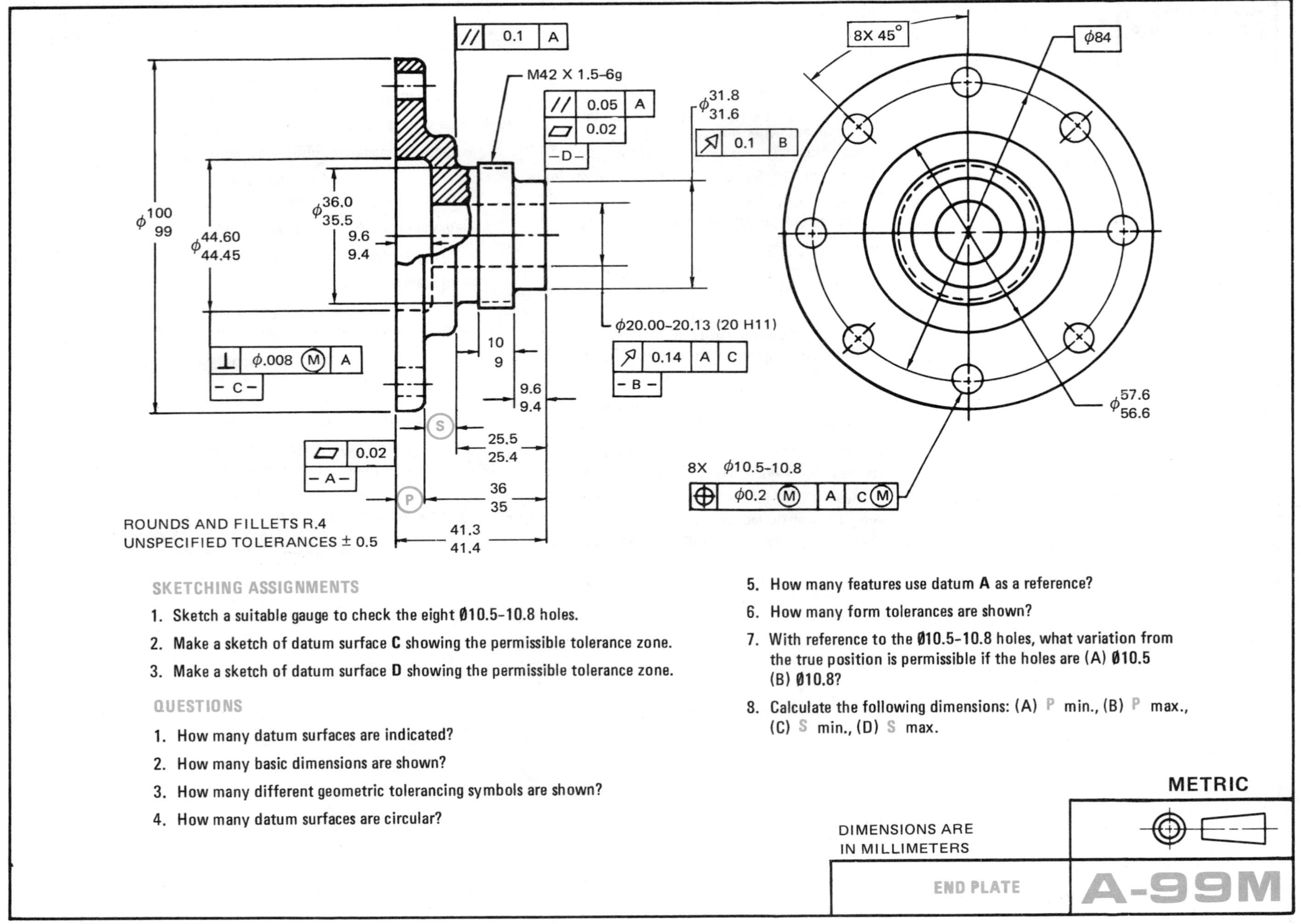

SKETCHING ASSIGNMENTS

1. Sketch a suitable gauge to check the eight Ø10.5-10.8 holes.
2. Make a sketch of datum surface **C** showing the permissible tolerance zone.
3. Make a sketch of datum surface **D** showing the permissible tolerance zone.

QUESTIONS

1. How many datum surfaces are indicated?
2. How many basic dimensions are shown?
3. How many different geometric tolerancing symbols are shown?
4. How many datum surfaces are circular?
5. How many features use datum **A** as a reference?
6. How many form tolerances are shown?
7. With reference to the Ø10.5-10.8 holes, what variation from the true position is permissible if the holes are (A) **Ø10.5** (B) **Ø10.8**?
8. Calculate the following dimensions: (A) P min., (B) P max., (C) S min., (D) S max.

APPENDIX

Term	Abbreviation / Symbol
And	&
Across Flats	ACR FLT
Angular	ANG
Approximate	APPROX
Assembly	ASSY
Basic	BSC
Bill of Material	B/M
Bolt Circle	BC
Brass	BR
Bronze	BRZ
Brown and Sharpe Gauge	B & S GA
Bushing	BUSH
Carbon Steel	CS
Casting	CSTG
Cast Iron	CI
Centimeter	cm
Center Line	℄ or CL
Center to Center	C to C
Chamfer	CHAM
Circularity	CIR
Cold Rolled Steel	CRS
Concentric	CONC
Copper	COP
Counterbore	CBORE or ⌴
Countersink	CSK or ⌵
Cubic Centimeter	cm^3
Cubic Meter	m^3
Datum	DAT .. [A] (ISO) .. [-A-] (ANSI)
Deep	↧
Degree (Angle)	°
Diameter	∅ or DIA
Diametral Pitch	DP
Dimension	DIM
Drawing	DWG
Eccentric	ECC
Equally Spaced	EQL SP
Figure	FIG
Finish All Over	FAO
Gauge	GA
Gray Iron	GI
Heat Treat	HT TR
Head	HD
Heavy	HVY
Hexagon	HEX
Hydraulic	HYD
Inside Diameter	ID
International Organization for Standardization	ISO
Iron Pipe Size	IPS
Kilogram	kg
Kilometer	km
Large End	LE
Least Material Condition	Ⓛ
Left Hand	LH
Long	LG
Machined	√
Machine Steel	MST
Malleable Iron	MI
Material	MATL
Maximum	MAX
Maximum Material Condition	Ⓜ or MMC
Meter	m
Metric Thread	M
Micrometer	μm
Mild Steel	MS
Millimeter	mm
Minimum	MIN
Minute (Angle)	(′)
Minute (arc)	′
Nominal	NOM
Not to Scale	___
Number	NO
Outside Diameter	OD
Parallel	PAR
Perpendicular	PERP
Pitch	P
Pitch Circle Diameter	PCD
Pitch Diameter	PD
Plate	PL
Projected Tolerance Zone	Ⓟ
Radian	rad
Radius	R
Reference or Reference Dimension	( )
Regardless of Feature Size	Ⓡ
Revolutions per Minute	R/MIN
Right Hand	RH
Second (Arc)	″
Second (Time)	s
Section	SECT
Slotted	SLOT
Socket	SOCK

Table 1 Abbreviations and Symbols used on Technical Drawings

Spherical Radius	SR
Spotface	SF or ⌴
Square	SQ or □
Square Centimeter	cm^2
Square Meter	m^2
Steel	STL
Straight	STR
Symmetrical	SYM or ╪
Taper Pipe Thread	NPT
Thick	THK
Thread	THD
Through	THRU
Tolerance	TOL
Undercut	UCUT
United States Gage	USG
Wrought Iron	WI
Wrought Steel	WS

Table 1 Abbreviations and Symbols used on Technical Drawings (cont'd)

Fraction	Decimals Inch		Millimeters	Fraction	Decimals Inch		Millimeters
	Two Place	Three Place			Two Place	Three Place	
1/64	.02	.016	.04	33/64	.52	.516	13.1
1/32	.03	.031	.08	17/32	.53	.531	13.5
3/64	.05	.047	1.2	35/64	.55	.547	13.9
1/16	.06	.062	1.6	9/16	.56	.562	14.3
5/64	.08	.078	2	37/64	.58	.578	14.7
3/32	.09	.094	2.4	19/32	.59	.594	15.1
7/64	.11	.109	2.8	39/64	.61	.609	15.5
1/8	.12	.125	3.2	5/8	.62	.625	15.9
9/64	.14	.141	3.6	41/64	.64	.641	16.3
5/32	.16	.156	4	21/32	.66	.656	16.7
11/64	.17	.172	4.4	43/64	.67	.672	17.1
3/16	.19	.188	4.8	11/16	.69	.688	17.5
13/64	.20	.203	5.2	45/64	.70	.703	17.9
7/32	.22	.219	5.6	23/32	.72	.719	18.3
15/64	.23	.234	6	47/64	.73	.734	18.7
1/4	.25	.250	6.4	3/4	.75	.750	19.1
17/64	.27	.266	6.8	49/64	.77	.766	19.5
9/32	.28	.281	7.1	25/32	.78	.781	19.9
19/64	.30	.297	7.5	51/64	.80	.797	20.2
5/16	.31	.312	7.9	13/16	.81	.812	20.6
21/64	.33	.328	8.3	53/64	.83	.828	21
11/32	.34	.344	8.7	27/32	.84	.844	21.4
23/64	.36	.359	9.1	55/64	.86	.859	21.8
3/8	.38	.375	9.5	7/8	.88	.875	22.2
25/64	.39	.391	9.9	57/64	.89	.891	22.6
13/32	.41	.406	10.3	29/32	.91	.906	23
27/64	.42	.422	10.7	59/64	.92	.922	23.4
7/16	.44	.438	11.1	15/16	.94	.938	23.8
29/64	.45	.453	11.5	61/64	.95	.953	24.2
15/32	.47	.469	11.9	31/32	.97	.969	24.6
31/64	.48	.484	12.3	63/64	.98	.984	25
1/2	.50	.500	12.7	1	1.00	1.000	25.4

Table 2 Chart for Converting Inch Dimensions to Millimeters

Decimal Inch and Millimeter Equivalents of Number Size Drills						Decimal Inch and Millimeter Equivalents of Letter Size Drills		
No.	Decimal Inch	mm	No.	Decimal Inch	mm	Letter	Decimal Inch	mm
1	.2280	5.8	31	.1200	3.0	A	.234	5.9
2	.2210	5.6	32	.1160	2.9	B	.238	6.0
3	.2130	5.4	33	.1130	2.9	C	.242	6.1
4	.2090	5.3	34	.1110	2.8	D	.246	6.2
5	.2055	5.2	35	.1100	2.8	E	.250	6.4
6	.2040	5.2	36	.1065	2.7	F	.257	6.5
7	.2010	5.1	37	.1040	2.6	G	.261	6.6
8	.1990	5.1	38	.1015	2.6	H	.266	6.8
9	.1960	5.0	39	.0095	2.5	I	.272	6.9
10	.1935	4.9	40	.0980	2.5	J	.277	7.0
11	.1910	4.9	41	.0960	2.4	K	.281	7.1
12	.1890	4.8	42	.0935	2.4	L	.290	7.4
13	.1850	4.7	43	.0890	2.3	M	.295	7.5
14	.1820	4.6	44	.0860	2.2	N	.302	7.7
15	.1800	4.6	45	.0820	2.1	O	.316	8.0
16	.1770	4.5	46	.0810	2.1	P	.323	8.2
17	.1730	4.4	47	.0785	2.0	Q	.332	8.4
18	.1695	4.3	48	.0760	1.9	R	.339	8.6
19	.1660	4.2	49	.0730	1.9	S	.348	8.8
20	.1610	4.1	50	.0700	1.8	T	.358	9.1
21	.1590	4.0	51	.0670	1.7	U	.368	9.3
22	.1570	4.0	52	.0635	1.6	V	.377	9.6
23	.1540	3.9	53	.0595	1.5	W	.386	9.8
24	.1520	3.9	54	.0550	1.4	X	.397	10.1
25	.1495	3.8	55	.0520	1.3	Y	.404	10.3
26	.1470	3.7	56	.0465	1.2	Z	.413	10.5
27	.1440	3.7	57	.0430	1.1			
28	.1405	3.6	58	.0420	1.1			
29	.1360	3.5	59	.0410	1.0			
30	.1285	3.3	60	.0400	1.0			

Table 3 Number and Letter-Size Drills

Metric Drill Sizes (mm)		Reference Decimal Equivalent (Inches)	Metric Drill Sizes (mm)		Reference Decimal Equivalent (Inches)	Metric Drill Sizes (mm)		Reference Decimal Equivalent (Inches)
Preferred	**Available**		**Preferred**	**Available**		**Preferred**	**Available**	
–	0.40	.0157	2.2	–	.0866	10	–	.3937
–	0.42	.0165	–	2.3	.0906	–	10.3	.4055
–	0.45	.0177	2.4	–	.0945	10.5	–	.4134
–	0.48	.0189	2.5	–	.0984	–	10.8	.4252
0.5	–	.0197	2.6	–	.1024	11	–	.4331
–	0.52	.0205	–	2.7	.1063	–	11.5	.4528
0.55	–	.0217	2.8	–	.1102	12	–	.4724
–	0.58	.0228	–	2.9	.1142	12.5	–	.4921
0.6	–	.0236	3	–	.1181	13	–	.5118
–	0.62	.0244	–	3.1	.1220	–	13.5	.5315
0.65	–	.0256	3.2	–	.1260	14	–	.5512
–	0.68	.0268	–	3.3	.1299	–	14.5	.5709
0.7	–	.0276	3.4	–	.1339	15	–	.5906
–	0.72	.0283	–	3.5	.1378	–	15.5	.6102
0.75	–	.0295	3.6	–	.1417	16	–	.6299
–	0.78	.0307	–	3.7	.1457	–	16.5	.6496
0.8	–	.0315	3.8	–	.1496	17	–	.6693
–	0.82	.0323	–	3.9	.1535	–	17.5	.6890
0.85	–	.0335	4	–	.1575	18	–	.7087
–	0.88	.0346	–	4.1	.1614	–	18.5	.7283
0.9	–	.0354	4.2	–	.1654	19	–	.7480
–	0.92	.0362	–	4.4	.1732	–	19.5	.7677
0.95	–	.0374	4.5	–	.1772	20	–	.7874
–	0.98	.0386	–	4.6	.1811	–	20.5	.8071
1	–	.0394	4.8	–	.1890	21	–	.8268
–	1.03	.0406	5	–	.1969	–	21.5	.8465
1.05	–	.0413	–	5.2	.2047	22	–	.8661
–	1.08	.0425	5.3	–	.2087	–	23	.9055
1.1	–	.0433	–	5.4	.2126	24	–	.9449
–	1.15	.0453	5.6	–	.2205	25	–	.9843
1.2	–	.0472	–	5.8	.2283	26	–	1.0236
1.25	–	.0492	6	–	.2362	–	27	1.0630
1.3	–	.0512	–	6.2	.2441	28	–	1.1024
–	1.35	.0531	6.3	–	.2480	–	29	1.1417
1.4	–	.0551	–	6.5	.2559	30	–	1.1811
–	1.45	.0571	6.7	–	.2638	–	31	1.2205
1.5	–	.0591	–	6.8	.2677	32	–	1.2598
–	1.55	.0610	–	6.9	.2717	–	33	1.2992
1.6	–	.0630	7.1	–	.2795	34	–	1.3386
–	1.65	.0650	–	7.3	.2874	–	35	1.3780
1.7	–	.0669	7.5	–	.2953	36	–	1.4173
–	1.75	.0689	–	7.8	.3071	–	37	1.4567
1.8	–	.0709	8	–	.3150	38	–	1.4961
–	1.85	.0728	–	8.2	.3228	–	39	1.5354
1.9	–	.0748	8.5	–	.3346	40	–	1.5748
–	1.95	.0768	–	8.8	.3465	–	41	1.6142
2	–	.0787	9	–	.3543	42	–	1.6535
–	2.05	.0807	–	9.2	.3622	–	43.5	1.7126
2.1	–	.0827	9.5	–	.3740	45	–	1.7717
–	2.15	.0846	–	9.8	.3858	–	46.5	1.8307

Table 4 Metric Twist Drill Sizes

Size		Coarse Thread Series UNC & NC		Fine Thread Series UNF & NF		Extra Fine Series UNEF & NEF		8-Pitch Thread Series 8 N		12-Pitch Thread Series 12 N		16-Pitch Thread Series 16 N	
Number or Fraction	Decimal	Threads Per Inch	Tap Drill	Threads Per Inch	Tap Drill	Threads Per Inch	Tap Drill	Threads Per Inch	Tap Drill	Threads Per Inch	Tap Drill	Threads Per Inch	Tap Drill
	.060			80	3/64								
1	.073	64	No. 53	72	No. 53								
2	.086	56	No. 50	64	No. 50								
3	.099	48	No. 47	56	No. 45								
4	.112	40	No. 43	48	No. 42								
5	.125	40	No. 38	44	No. 37								
6	.138	32	No. 36	40	No. 33								
8	.164	32	No. 29	36	No. 29								
10	.190	24	No. 25	32	No. 21								
12	.216	24	No. 16	28	No. 14	32	No. 13						
1/4	.250	20	No. 7	28	No. 3	32	7/32						
5/16	.312	18	F	24	I	32	9/32						
3/8	.375	16	5/16	24	Q	32	11/32						
7/16	.438	14	U	20	25/64	28	13/32						
1/2	.500	13	27/64										
1/2	.500	12		20	29/64	28	15/32			12	27/64		
3/16	.562	12	31/64	18	33/64	24	33/64			12	31/64		
5/8	.625	11	17/32	18	37/64	24	37/64			12	35/64		
3/4	.750	10	21/32	16	11/16	20	45/64			12	43/64	16	11/16
7/8	.875	9	49/64	14	13/16	20	53/64			12	51/64	16	13/16
1	1.000	8	7/8	12	59/64	20	61/64	8	7/8	12	59/64	16	15/16
1 1/8	1.125	7	63/64	12	1 3/64	18	1 5/64	8	1	12	1 3/64	16	1 1/16
1 1/4	1.250	7	1 7/64	12	1 11/64	18	1 3/16	8	1 1/8	12	1 11/64	16	1 3/16
1 3/8	1.375	6	1 7/32	12	1 19/64	18	1 5/16	8	1 1/4	12	1 19/64	16	1 5/16
1 1/2	1.500	6	1 11/32	12	1 27/64	18	1 7/16	8	1 3/8	12	1 27/64	16	1 7/16
1 3/4	1.750	5	1 9/16			16	1 11/16	8	1 5/8	12	1 43/64	16	1 11/16
2	2.000	4 1/2	1 25/32			16	1 15/16	8	1 7/8	12	1 51/64	16	1 15/16
2 1/4	2.250	4 1/2	2 1/32					8	2 1/8	12	2 11/64	16	2 3/16
2 1/2	2.500	4	2 1/4					8	2 3/8	12	2 27/64	16	2 7/16
2 3/4	2.750	4	2 1/2					8	2 5/8	12	2 43/64	16	2 11/16
3	3.000	4	2 3/4					8	2 7/8	12	2 59/64	16	2 15/16

Color shows unified thread

Table 5 Unified and American (Inch) Threads

Nominal Size DIA (mm)		Series with Graded Pitches				Series with Constant Pitches																	
		Coarse		Fine		4		3		2		1.5		1.25		1		0.75		0.5		0.35	
Preferred		Thread Pitch	Tap Drill Size	Thread Pitch	Tap Drill Size	Thread Pitch	Tap Drill Size	Thread Pitch	Tap Drill Size	Thread Pitch	Tap Drill Size	Thread Pitch	Tap Drill Size	Thread Pitch	Tap Drill Size	Thread Pitch	Tap Drill Size	Thread Pitch	Tap Drill Size	Thread Pitch	Tap Drill Size	Thread Pitch	Tap Drill Size
1.6		0.35	1.25																				
	1.8	0.35	1.45																				
2		0.4	1.6																				
	2.2	0.45	1.75																				
2.5		0.45	2.05																			0.35	2.15
3		0.5	2.5																			0.35	2.65
	3.5	0.6	2.9																			0.35	3.15
4		0.7	3.3																	0.5	3.5		
	4.5	0.75	3.7																	0.5	4		
5		0.8	4.2																	0.5	4.5		
*6		1	5															0.75	5.2				
**6.3		1	5.3																				
8		1.25	6.7	1	7											1	7	0.75	7.2				
10		1.5	8.5	1.25	8.7									1.25	8.7	1	9	0.75	9.2				
12		1.75	10.2	1.25	10.8							1.5	10.5	1.25	10.7	1	11						
	14	2	12	1.5	12.5							1.5	12.5	1.25	12.7	1	13						
16		2	14	1.5	14.5							1.5	14.5			1	15						
	18	2.5	15.5	1.5	16.5					2	16	1.5	16.5			1	17						
20		2.5	17.5	1.5	18.5					2	18	1.5	18.5			1	19						
	22	2.5	19.5	1.5	20.5					2	20	1.5	20.5			1	21						
24		3	21	2	22					2	22	1.5	22.5			1	23						
	27	3	24	2	25					2	25	1.5	25.5			1	26						
30		3.5	26.5	2	28					2	28	1.5	28.5			1	29						
	33	3.5	29.5	2	31					2	31	1.5	31.5										
36		4	32	3	33					2	34	1.5	34.5										
	39	4	35	3	36					2	37	1.5	37.5										
42		4.5	37.5	3	39	4	38	3	39	2	40	1.5	40.5										
	45	4.5	39	3	42	4	41	3	42	2	43	1.5	43.5										
48		5	43	3	45	4	44	3	45	2	46	1.5	46.5										

* ISO thread size
** ANSI thread size (to be discontinued)

Table 6 Metric Threads

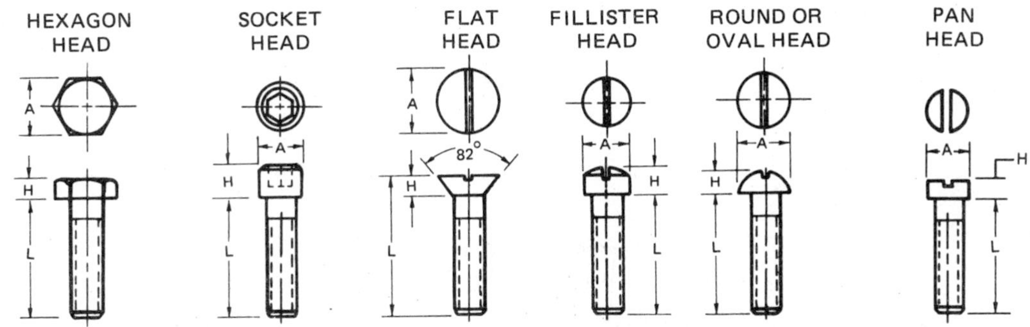

U.S. Customary (Inches)

Nominal Size		Hexagon Head		Socket Head		Flat Head		Fillister Head		Round or Oval Head	
Fraction	Decimal	A	H	A	H	A	H	A	H	A	H
1/4	.250	.44	.17	.38	.25	.50	.14	.38	.22	.44	.19
5/16	.312	.50	.22	.47	.31	.62	.18	.44	.25	.56	.25
3/8	.375	.56	.25	.56	.38	.75	.21	.56	.31	.62	.27
7/16	.438	.62	.30	.66	.44	.81	.21	.62	.36	.75	.33
1/2	.500	.75	.34	.75	.50	.88	.21	.75	.41	.81	.35
5/8	.625	.94	.42	.94	.62	1.12	.28	.88	.50	1.00	.44
3/4	.750	1.12	.50	1.12	.75	1.38	.35	1.00	.59	1.25	.55
7/8	.875	1.31	.58	1.31	.88	1.62	.42	1.12	.69		
1	1.000	1.50	.67	1.50	1.00	1.88	.49	1.31	.78		
1 1/8	1.125	1.69	.75	1.69	1.12	2.06	.53				
1 1/4	1.250	1.88	.84	1.88	1.25	2.31	.60				
1 1/2	1.500	2.25	1.00	2.25	1.50	2.81	.74				

Metric (Millimeters) Sizes

Nominal Size	Hexagon Head		Socket Head			Flat Head		Fillister Head		Pan Head	
	A	H	A	H	Key Size	A	H	A	H	A	H
M3	5.5	2	5.5	3	2.5	5.6	1.6	6	2.4	5.6	1.9
4	7	2.8	7	4	3	7.5	2.2	8	3.1	7.5	2.5
5	8.5	3.5	9	5	4	9.2	2.5	10	3.8	9.2	3.1
6	10	4	10	6	5	11	3	12	4.6	11	3.8
8	13	5.5	13	8	6	14.5	4	16	6	14.5	5
10	17	7	16	10	8	18	5	20	7.5	18	6.2
12	19	8	18	12	10						
14	22	9	22	14	12						
16	24	10	24	16	14						
18	27	12	27	18	14						
20	30	13	30	20	17						
22	36	15	33	22	17						
24	36	15	36	24	19						
27	41	17	40	27	19						
30	46	19	45	30	22						

Table 7 Common Cap Screws

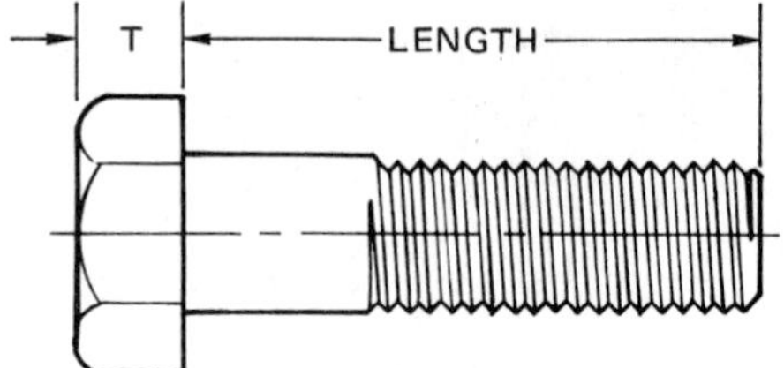

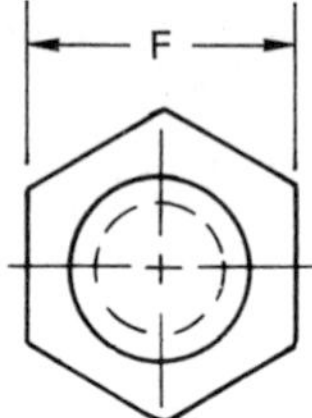

U.S. Customary (Inches)			
Nominal Size Fraction	Nominal Size Decimal	Width Across Flats	Thickness
1/4	.250	.44	.17
5/16	.312	.50	.22
3/8	.375	.56	.25
7/16	.438	.62	.30
1/2	.500	.75	.34
5/8	.625	.94	.42
3/4	.750	1.12	.50
7/8	.875	1.31	.58
1	1.000	1.50	.67
1 1/8	1.125	1.69	.75
1 1/4	1.250	1.88	.84
1 3/8	1.375	2.06	.91
1 1/2	1.500	2.25	1.00

Metric (Millimeters)		
Nominal Size (Millimeters)	Width Across Flats	Thickness
4	7	2.8
5	8	3.5
6	10	4
8	13	5.5
10	17	7
12	19	8
14	22	9
16	24	10
18	27	12
20	30	13
22	32	14
24	36	15
27	41	17
30	46	19
33	50	21
36	55	23

Table 8 Hexagon-Head Bolts and Cap Screws

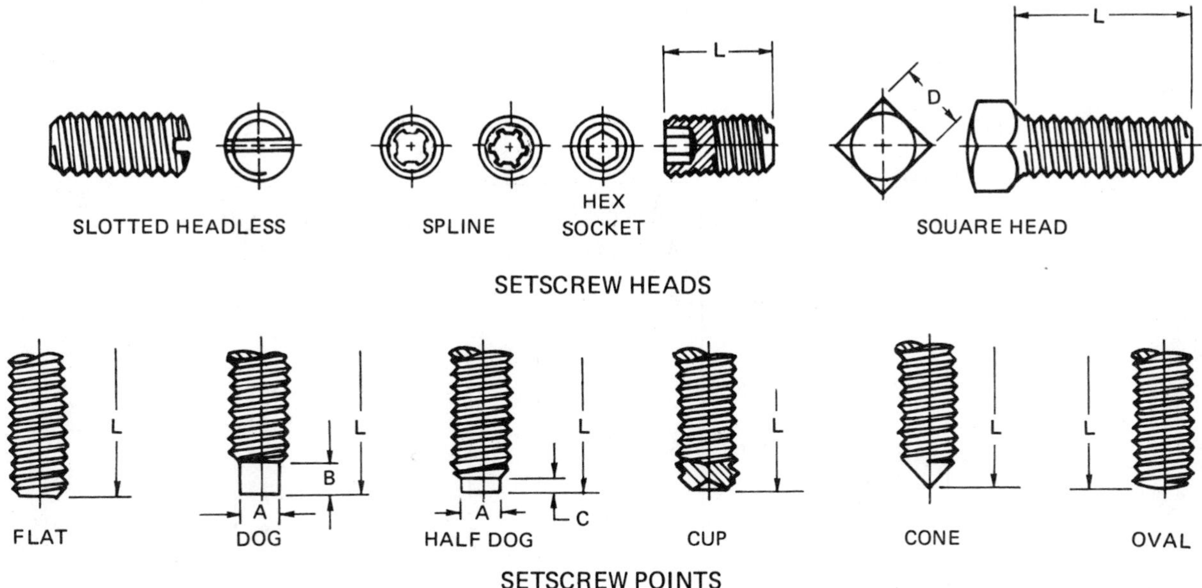

U.S. Customary (Inches)			Metric (Millimeters)	
Nominal Size		Key Size	Nominal Size	Key Size
Number	Decimal			
4	.112	.050	M 1.4	0.7
5	.125	.062	2	0.9
6	.138	.062	3	1.5
8	.164	.078	4	2
10	.190	.094	5	2.5
12	.216	.109	6	3
1/4	.250	.125	8	4
5/16	.312	.156	10	5
3/8	.375	.188	12	6
1/2	.500	.250	16	8

Table 9 Setscrews

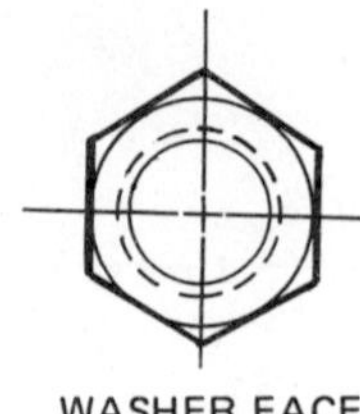

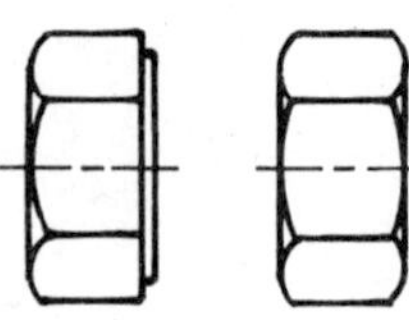

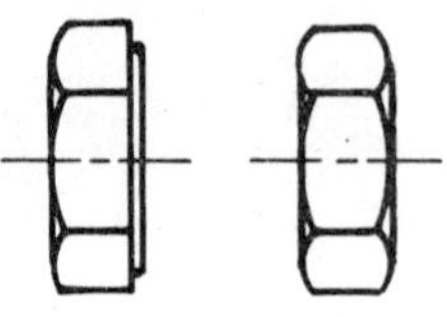

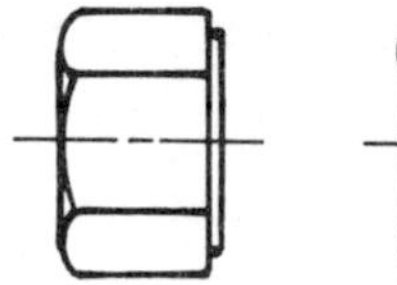
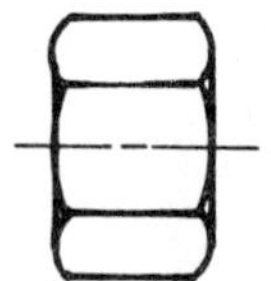

U.S. Customary (Inches)

Nominal Size		Distance	Thickness		
Fraction	Decimal	Across Flats	Regular	Jam	Thick
1/4	.250	.44	.22	.16	.28
5/16	.312	.50	.27	.19	.33
3/8	.375	.56	.33	.22	.41
7/16	.438	.69	.38	.25	.45
1/2	.500	.75	.44	.31	.56
9/16	.562	.88	.48	.31	.61
5/8	.625	.94	.55	.38	.72
3/4	.750	1.12	.64	.42	.81
7/8	.875	1.31	.75	.48	.91
1	1.000	1.50	.86	.55	1.00
1 1/8	1.125	1.69	.97	.61	1.16
1 1/4	1.250	1.88	1.06	.72	1.25
1 3/8	1.375	2.06	1.17	.78	1.38
1 1/2	1.500	2.25	1.28	.84	1.50

Metric (Millimeters)

Nominal Size (Millimeters)	Distance Across Flats	Thickness		
		Regular	Jam	Thick
4	7	3	2	5
5	8	4	2.5	5
6	10	5	3	6
8	13	6.5	5	8
10	17	8	6	10
12	19	10	7	12
14	22	11	8	14
16	24	13	8	16
18	27	15	9	18.5
20	30	16	9	20
22	32	18	10	22
24	36	19	10	24
27	41	22	12	27
30	46	24	12	30
33	50	26		
36	55	29		
39	60	31		

Table 10 Hexagon-Head Nuts

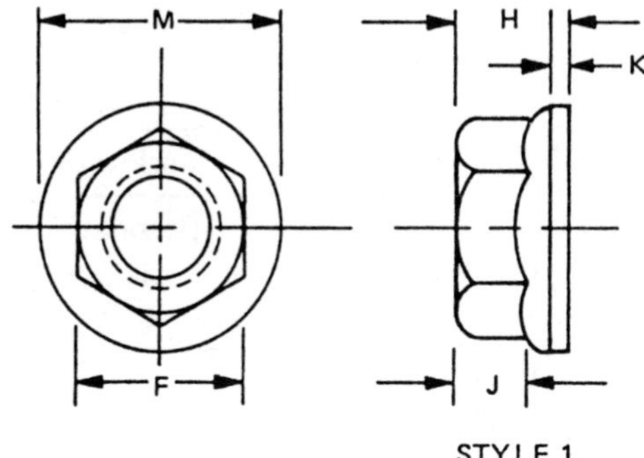

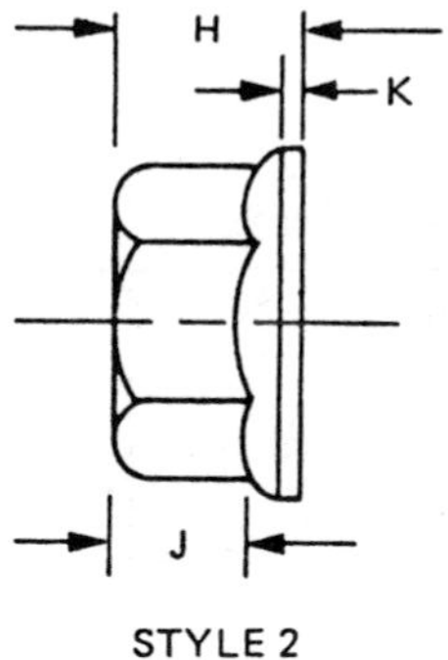

Metric (Millimeters)							
Nominal Nut Size and Thread Pitch	Width Across Flats	Style 1				Style 2	
	F	H	J	K	M	H	J
M6 X 1	10	5.8	3	1	14.2	6.7	3.7
M8 X 1.25	13	6.8	3.7	1.3	17.6	8	4.5
M10 X 1.5	15	9.6	5.5	1.5	21.5	11.2	6.7
M12 X 1.75	18	11.6	6.7	2	25.6	13.5	8.2
M14 X 2	21	13.4	7.8	2.3	29.6	15.7	9.6
M16 X 2	24	15.9	9.5	2.5	34.2	18.4	11.7
M20 X 2.5	30	19.2	11.1	2.8	42.3	22	12.6

Table 11 Hex Flange Nuts

FLAT WASHER

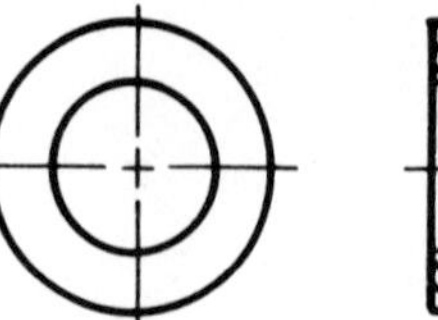

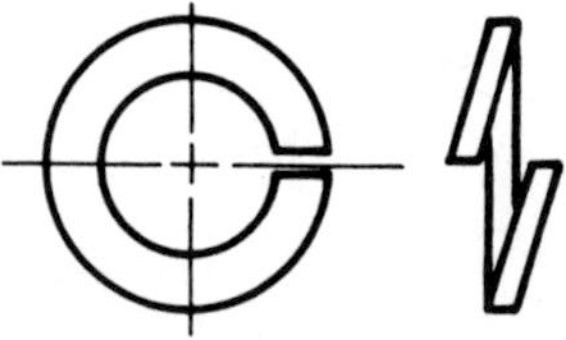

Nominal Screw Size		Flat Washer			Lockwasher		
Number or Fraction	Decimal	Inside Dia A	Outside Dia B	Thickness C	Inside Dia A	Outside Dia B	Thickness C
6	.138	.16	.38	.05	.14	.25	.03
8	.164	.19	.44	.05	.17	.29	.04
10	.190	.22	.50	.05	.19	.33	.05
12	.216	.25	.56	.07	.22	.38	.06
1/4	.250 N	.28	.63	.07	.26	.49	.06
1/4	.250 W	.31	.73	.07			
5/16	.312 N	.34	.69	.07	.32	.59	.08
5/16	.312 W	.38	.88	.08			
3/8	.375 N	.41	.81	.07	.38	.68	.09
3/8	.375 W	.44	1.00	.08			
7/16	.438 N	.47	.92	.07	.45	.78	.11
7/16	.438 W	.50	1.25	.08			
1/2	.500 N	.53	1.06	.10	.51	.87	.12
1/2	.500 W	.56	1.38	.11			
5/8	.625 N	.66	1.31	.10	.64	1.08	.16
5/8	.625 W	.69	1.75	.13			
3/4	.750 N	.81	1.47	.13	.76	1.27	.19
3/4	.750 W	.81	2.00	.15			
7/8	.875 N	.94	1.75	.13	.89	1.46	.22
7/8	.875 W	.94	2.25	.17			
1	1.000 N	1.06	2.00	.13	1.02	1.66	.25
1	1.000 W	1.06	2.50	.17			
1 1/8	1.125 N	1.25	2.25	.13	1.14	1.85	.28
1 1/8	1.125 W	1.25	2.75	.17			
1 1/4	1.250 N	1.38	2.50	.17	1.27	2.05	.31
1 1/4	1.250 W	1.38	3.00	.17			
1 3/8	1.375 N	1.50	2.75	.17	1.40	2.24	.34
1 3/8	1.375 W	1.50	3.25	.18			
1 1/2	1.500 N	1.62	3.00	.17	1.53	2.43	.38
1 1/2	1.500 W	1.62	3.50	.18			

U.S. Customary (Inch) Washers

N–SAE Sizes (Narrow)
W–Standard Plate (Wide)

Table 12 Common Washer Sizes

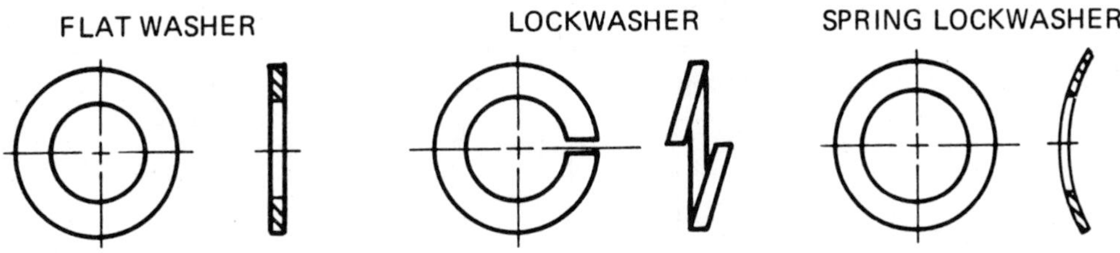

Bolt Size	Flat Washers			Lockwashers			Spring Lockwashers		
	ID	OD	Thickness	ID	OD	Thickness	ID	OD	Thickness
2	2.2	5.5	0.5	2.1	3.3	0.5			
3	3.2	7	0.5	3.1	5.7	0.8			
4	4.3	9	0.8	4.1	7.1	0.9	4.2	8	0.3 0.4
5	5.3	11	1	5.1	8.7	1.2	5.2	10	0.4 0.5
6	6.4	12	1.5	6.1	11.1	1.6	6.2	12.5	0.5 0.7
7	7.4	14	1.5	7.1	12.1	1.6	7.2	14	0.5 0.8
8	8.4	17	2	8.2	14.2	2	8.2	16	0.6 0.9
10	10.5	21	2.5	10.2	17.2	2.2	10.2	20	0.8 1.1
12	13	24	2.5	12.3	20.2	2.5	12.2	25	0.9 1.5
14	15	28	2.5	14.2	23.2	3	14.2	28	1 1.5
16	17	30	3	16.2	26.2	3.5	16.3	31.5	1.2 1.7
18	19	34	3	18.2	28.2	3.5	18.3	35.5	1.2 2
20	21	36	3	20.2	32.2	4	20.4	40	1.5 2.25
22	23	39	4	22.5	34.5	4	22.4	45	1.75 2.5
24	25	44	4	24.5	38.5	5			
27	28	50	4	27.5	41.5	5			
30	31	56	4	30.5	46.5	6			

Metric (Millimeter) Washers

Table 12 Common Washer Sizes (cont'd)

U.S. Customary (Inches)					Metric (Millimeters)					
Diameter of Shaft	Square Key		Flat Key		Diameter of Shaft (mm)		Square Key		Flat Key	
	Nominal Size		Nominal Size				Nominal Size		Nominal Size	
Inclusive	W	H	W	H	Over	Up To	W	H	W	H
.500- .562	.125	.125	.125	.094	12	17	5	5		
.625- .875	.188	.188	.188	.125	17	22	6	6		
.938-1.250	.250	.250	.250	.188	22	30	7	7	8	7
1.312-1.375	.312	.312	.312	.250	30	38	8	8	10	8
1.438-1.750	.375	.375	.375	.250	38	44	9	9	12	8
1.812-2.250	.500	.500	.500	.375	44	50	10	10	14	9
2.312-2.750	.625	.625	.625	.438	50	58	12	12	16	10

Table 13 Square and Flat Stock Keys

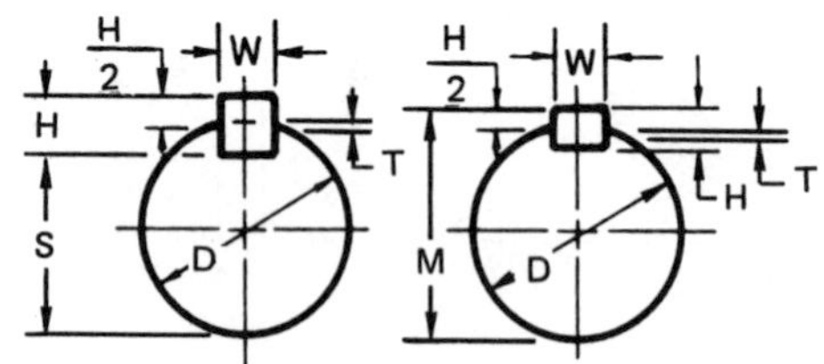

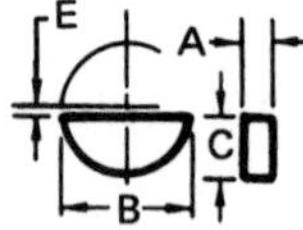

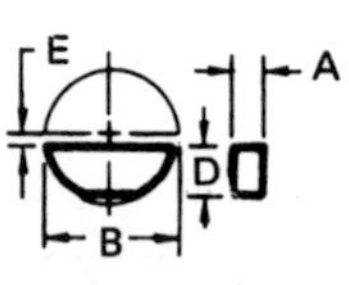

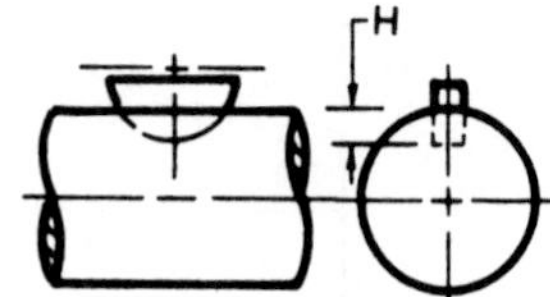

C = Allowance for parallel keys = .005 in. or 0.12 mm

W = Normal key width (inch or millimeters)

$$S = D - \frac{H}{2} - T = \frac{D - H + \sqrt{D^2 - W^2}}{2} \qquad T = \frac{D - \sqrt{D^2 - W^2}}{2}$$

$$M = D - T + \frac{H}{2} + C = \frac{D + H + \sqrt{D^2 - W^2} + C}{2}$$

Key No.	Nominal (A x B)		U.S. Customary (Inches)				Metric (Millimeters)			
			Key			Keyseat	Key			Key Seat
	Millimeters	Inches	E	C	D	H	E	C	D	H
204	1.6 x 6.4	0.062 x 0.250	.05	.20	.19	.10	0.5	2.8	2.8	4.3
304	2.4 x 12.7	0.094 x 0.500	.05	.20	.19	.15	1.3	5.1	4.8	3.8
305	2.4 x 15.9	0.094 x 0.625	.06	.25	.24	.20	1.5	6.4	6.1	5.1
404	3.2 x 12.7	0.125 x 0.500	.05	.20	.19	.14	1.3	5.1	4.8	3.6
405	3.2 x 15.9	0.125 x 0.625	.06	.25	.24	.18	1.5	6.4	6.1	4.6
406	3.2 x 19.1	0.125 x 0.750	.06	.31	.30	.25	1.5	7.9	7.6	6.4
505	4.0 x 15.9	0.156 x 0.625	.06	.25	.24	.17	1.5	6.4	6.1	4.3
506	4.0 x 19.1	0.156 x 0.750	.06	.31	.30	.23	1.5	7.9	7.6	5.8
507	4.0 x 22.2	0.156 x 0.875	.06	.38	.36	.29	1.5	9.7	9.1	7.4
606	4.8 x 19.1	0.188 x 0.750	.06	.31	.30	.21	1.5	7.9	7.6	5.3
607	4.8 x 22.2	0.188 x 0.875	.06	.38	.36	.28	1.5	9.7	9.1	7.1
608	4.8 x 25.4	0.188 x 1.000	.06	.44	.43	.34	1.5	11.2	10.9	8.6
609	4.8 x 28.6	0.188 x 1.250	.08	.48	.47	.39	2.0	12.2	11.9	9.9
807	6.4 x 22.2	0.250 x 0.875	.06	.38	.36	.25	1.5	9.7	9.1	6.4
808	6.4 x 25.4	0.250 x 1.000	.06	.44	.43	.31	1.5	11.2	10.9	7.9

Table 14 Woodruff Keys

	U.S. Customary (Inches)				Metric			
		Wall Thickness				Wall Thickness (Millimeters)		
Nominal Pipe Size	Outside Diameter	Schedule 40 Pipe *	Schedule 80 Pipe **	Schedule 160 Pipe	Outside Diameter (Millimeters)	Standard	Extra Strong	Double Extra Strong
.125 (1/8)	.405	.068	.095	–	10.29	1.75	2.44	—
.250 (1/4)	.540	.088	.119	–	13.72	2.29	3.15	—
.375 (3/8)	.675	.091	.126	–	17.15	2.36	3.28	—
.500 (1/2)	.840	.109	.147	.188	21.34	2.82	3.84	7.80
.750 (3/4)	1.050	.113	.154	.219	26.67	2.92	3.99	8.08
1.00	1.315	.133	.179	.250	33.4	3.45	4.65	9.37
1.25	1.660	.140	.191	.250	42.4	3.63	4.98	9.98
1.50	1.900	.145	.200	.281	48.3	3.76	5.18	10.44
2.00	2.375	.154	.218	.344	60.3	3.99	5.66	11.35
2.50	2.875	.203	.276	.375	73.0	5.26	7.16	14.40
3.00	3.500	.216	.300	.438	88.9	5.61	7.77	15.62
3.50	4.000	.226	.318	–	101.6	5.87	8.25	16.54
4.00	4.500	.237	.337	.531	114.3	6.15	8.74	17.53
5.00	5.563	.258	.375	.625	141.3	6.68	9.73	19.51
6.00	6.625	.280	.432	.719	168.3	7.26	11.20	22.45
8.00	8.625	.322	.500	.096	219.1	8.36	12.98	22.73
10.00	10.750	.365	.594	1.125	273.1	9.45	12.95	—
12.00	12.750	.406	.688	1.312	323.9	—	12.95	—
14.00	14.000	.438	.750	1.406	355.6	9.73	12.95	—
16.00	16.000	.500	.844	1.594	406.4	9.73	12.95	—

*Standard Pipe
**Extra Strong Pipe

Nominal pipe sizes are specified in inches.
Outside diameter and wall thicknesses are specified in millimeters.

Table 15 American Standard Wrought Steel Pipe

North American Gauges								European Gauges					
Ferrous Metals, Such as Galvanized Steel, Tin Plate				Galvanized Steel, Tin Plate, Copper, Strip Steel and Steel, Copper, and Aluminum Tubes				Nonferrous Metals, Such as Copper, Brass, Aluminum		Nonferrous			
U.S. Standard (USS)		U.S. Standard (Revised)		Birmingham (BWG)		New Birmingham (BG)		Browne And Sharpe (B & S)		Imperial Standard (SWG)		Electrical Steel	
Gauge	In.	Gauge	In.	Gauge	In.	Gauge	In.	Gauge	In.	Gauge	In.	Gauge	In.
		3	.239					3	.229				
4	.234	4	.224	4	.238	4	.250	4	.204	4	.232		
5	.219	5	.209	5	.220	5	.223	5	.182	5	.212		
6	.203	6	.194	6	.203	6	.198	6	.162	6	.192		
7	.188	7	.179	7	.180	7	.176	7	.144	7	.176		
8	.172	8	.164	8	.165	8	.157	8	.129	8	.160		
9	.156	9	.149	9	.148	9	.140	9	.114	9	.144		
10	.141	10	.135	10	.134	10	.125	10	.102	10	.128		
11	.125	11	.120	11	.120	11	.111	11	.091	11	.116	11	.125
12	.109	12	.105	12	.109	12	.099	12	.081	12	.104	12	.109
13	.094	13	.090	13	.095	13	.088	13	.072	13	.092	13	.094
14	.078	14	.075	14	.083	14	.079	14	.064	14	.080	14	.078
15	.070	15	.067	15	.072	15	.070	15	.057	15	.072	15	.070
16	.063	16	.060	16	.065	16	.063	16	.051	16	.064	16	.063
17	.056	17	.054	17	.058	17	.056	17	.045	17	.056	17	.056
18	.050	18	.048	18	.049	18	.050	18	.040	18	.048	18	.050
19	.044	19	.042	19	.042	19	.044	19	.036	19	.040	19	.044
20	.038	20	.036	20	.035	20	.039	20	.032	20	.036	20	.038
21	.034	21	.033	21	.032	21	.035	21	.029	21	.032		
22	.031	22	.030	22	.028	22	.031	22	.025	22	.028	22	.031

Table 16 Sheet Metal Gauges and Thicknesses

North American Gauges								European Gauges					
Ferrous Metals, Such as Galvanized Steel, Tin Plate				Galvanized Steel, Tin Plate, Copper, Strip Steel and Steel, Copper, and Aluminum Tubes				Nonferrous Metals, Such as Copper, Brass, Aluminum		Nonferrous			
U.S. Standard (USS)		U.S. Standard (Revised)		Birmingham (BWG)		New Birmingham (BG)		Browne And Sharpe (B & S)		Imperial Standard (SWG)		Electrical Steel	
Gauge	In.	Gauge	In.	Gauge	In.	Gauge	In.	Gauge	In.	Gauge	In.	Gauge	In.
23	.028	23	.027	23	.025	23	.028	23	.023	23	.024	23	.028
24	.025	24	.024	24	.022	24	.025	24	.020	24	.022	24	.025
25	.022	25	.021	25	.020	25	.022	25	.018	25	.020	25	.022
26	.019	26	.018	26	.018	26	.020	26	.016	26	.018	26	.019
27	.017	27	.016	27	.016	27	.017	27	.014	27	.016	27	.017
28	.016	28	.015	28	.014	28	.016	28	.013	28	.015	28	.016
29	.014	29	.014	29	.013	29	.014	29	.011	29	.014	29	.014
30	.012	30	.012	30	.012	30	.012	30	.010	30	.012	30	.013
31	.011	31	.011	31	.010	31	.011	31	.009				
32	.010	32	.010	32	.009			32	.008	32	.011	32	.010
33	.009	33	.009	33	.008	33	.009	33	.007	33	.010		
34	.008	34	.008	34	.007	34	.008	34	.006	34	.009		
				35	.005	35	.007			35	.008		
36	.007	36	.007	36	.004	36	.006	36	.005				
										37	.007		
38	.006	38	.006							38	.006		
						38	.005	38	.004				
						40	.004			40	.005		
										42	.004		

Fig. 16 Sheet Metal Gauges and Thicknesses (cont'd)

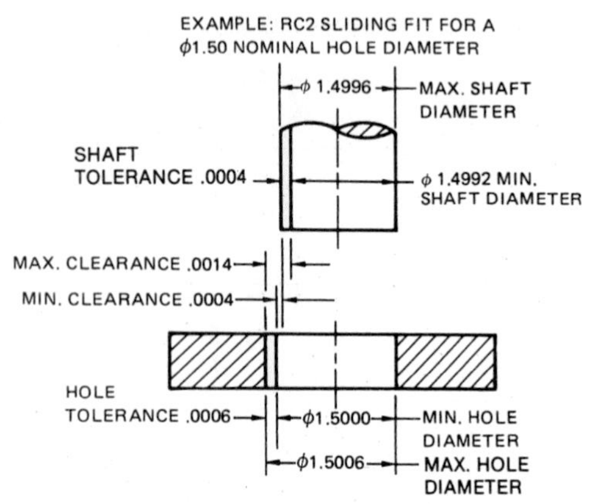

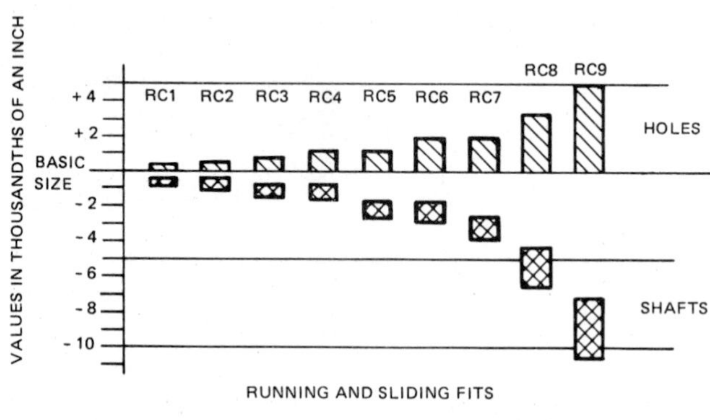

Nominal Size Range Inches		Class RC1 Precision Sliding			Class RC2 Sliding Fit			Class RC3 Precision Running			Class RC4 Close Running			Class RC5 Medium Running		
		Hole Tol. GR5	Minimum Clearance	Shaft Tol. GR4	Hole Tol. GR6	Minimum Clearance	Shaft Tol. GR5	Hole Tol. GR7	Minimum Clearance	Shaft Tol. GR6	Hole Tol. GR8	Minimum Clearance	Shaft Tol. GR7	Hole Tol. GR8	Minimum Clearance	Shaft Tol. GR7
Over	To	-0		+0	-0		+0	-0		+0	-0		+0	-0		+0
0	.12	+0.15	0.10	-0.12	+0.25	0.10	-0.15	+0.40	0.30	-0.25	+0.60	0.30	-0.40	+0.60	0.60	-0.40
.12	.24	+0.20	0.15	-0.15	+0.30	0.15	-0.20	+0.50	0.40	-0.30	+0.70	0.40	-0.50	+0.70	0.80	-0.50
.24	.40	+0.25	0.20	-0.15	+0.40	0.20	-0.25	+0.60	0.50	-0.40	+0.90	0.50	-0.60	+0.90	1.00	-0.60
.40	.71	+0.30	0.25	-0.20	+0.40	0.25	-0.30	+0.70	0.60	-0.40	+1.00	0.60	-0.70	+1.00	1.20	-0.70
.71	1.19	+0.40	0.30	-0.25	+0.50	0.30	-0.40	+0.80	0.80	-0.50	+1.20	0.80	-0.80	+1.20	1.60	-0.50
1.19	1.97	+0.40	0.40	-0.30	+0.60	0.40	-0.40	+1.00	1.00	-0.60	+1.60	1.00	-1.00	+1.60	2.00	-1.00
1.97	3.15	+0.50	0.40	-0.30	+0.70	0.40	-0.50	+1.20	1.20	-0.70	+1.80	1.20	-1.20	+1.80	2.50	-1.20
3.15	4.73	+0.60	0.50	-0.40	+0.90	0.50	-0.60	+1.40	1.40	-0.90	+2.20	1.40	-1.40	+2.20	3.00	-1.40
4.73	7.09	+0.70	0.60	-0.50	+1.00	0.60	-0.70	+1.60	1.60	-1.00	+2.50	1.60	-1.60	+2.50	3.50	-1.60
7.09	9.85	+0.80	0.60	-0.60	+1.20	0.60	-0.80	+1.80	2.00	-1.20	+2.80	2.00	-1.80	+2.80	4.50	-1.80
9.85	12.41	+0.90	0.80	-0.60	+1.20	0.80	-0.90	+2.00	2.50	-1.20	+3.00	2.50	-2.00	+3.00	5.00	-2.00
12.41	15.75	+1.00	1.00	-0.70	+1.40	1.00	-1.00	+2.20	3.00	-1.40	+3.50	3.00	-2.20	+3.50	6.00	-2.20

Nominal Size Range Inches		Class RC6 Medium Running			Class RC7 Free Running			Class RC8 Loose Running			Class RC9 Loose Running		
		Hole Tol. GR9	Minimum Clearance	Shaft Tol. GR8	Hole Tol. GR9	Minimum Clearance	Shaft Tol. GR8	Hole Tol. GR10	Minimum Clearance	Shaft Tol. GR9	Hole Tol. GR11	Minimum Clearance	Shaft Tol. GR10
Over	To	-0		+0	-0		+0	-0		+0	-0		+0
0	.12	+1.00	0.60	-0.60	+1.00	1.00	-0.60	+1.60	2.50	-1.00	+2.50	4.00	-1.60
.12	.24	+1.20	0.80	-0.70	+1.20	1.20	-0.70	+1.80	2.80	-1.20	+3.00	4.50	-1.80
.24	.40	+1.40	1.00	-0.90	+1.40	1.60	-0.90	+2.20	3.00	-1.40	+3.50	6.00	-2.20
.40	.71	+1.60	1.20	-1.00	+1.60	2.00	-1.00	+2.80	3.50	-1.60	+4.00	6.00	-2.80
.71	1.19	+2.00	1.60	-1.20	+2.00	2.50	-1.20	+3.50	4.50	-2.00	+5.00	7.00	-3.50
1.19	1.97	+2.50	2.00	-1.60	+2.50	3.00	-1.60	+4.00	5.00	-2.50	+6.00	8.00	-4.00
1.97	3.15	+3.00	2.50	-1.80	+3.00	4.00	-1.80	+4.50	6.00	-3.00	+7.00	9.00	-4.50
3.15	4.73	+3.50	3.00	-2.20	+3.50	5.00	-2.20	+5.00	7.00	-3.50	+9.00	10.00	-5.00
4.73	7.09	+4.00	3.50	-2.50	+4.00	6.00	-2.50	+6.00	8.00	-4.00	+10.00	12.00	-6.00
7.09	9.85	+4.50	4.00	-2.80	+4.50	7.00	-2.80	+7.00	10.00	-4.50	+12.00	15.00	-7.00
9.85	12.41	+5.00	5.00	-3.00	+5.00	8.00	-3.00	+8.00	12.00	-5.00	+12.00	18.00	-8.00
12.41	15.75	+6.00	6.00	-3.50	+6.00	10.00	-3.50	+9.00	14.00	-6.00	+14.00	22.00	-9.00

Table 17 Running and Sliding Fits (Values in Thousandths of an Inch)

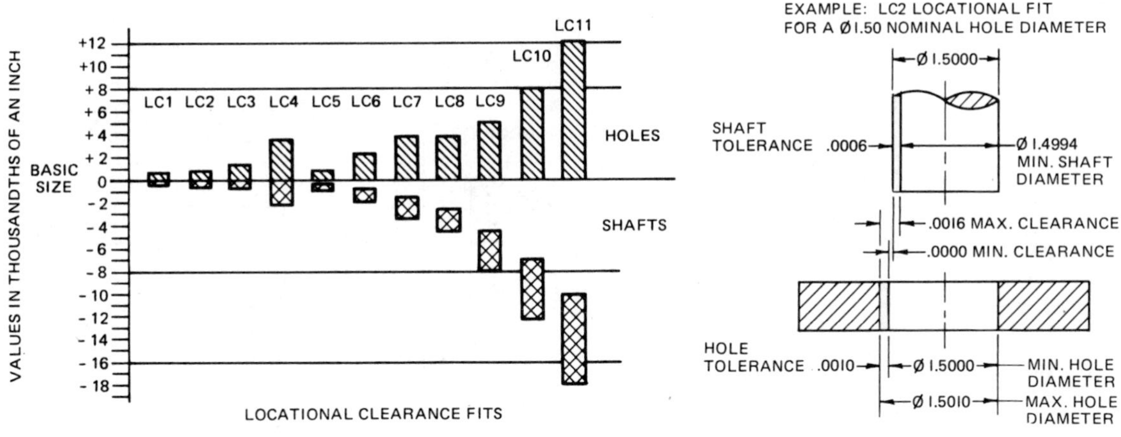

Nominal Size Range Inches		Class LC1			Class LC2			Class LC3			Class LC4			Class LC5			Class LC6		
		Hole Tol. GR6	Minimum Clearance	Shaft Tol. GR5	Hole Tol. GR8	Minimum Clearance	Shaft Tol. GR7	Hole Tol. GR10	Minimum Clearance	Shaft Tol. GR9	Hole Tol. GR7	Minimum Clearance	Shaft Tol. GR6	Hole Tol. GR9	Minimum Clearance	Shaft Tol. GR8	Hole Tol. GR9	Minimum Clearance	Shaft Tol. GR8
Over	To	-0		+0	-0		+0	-0		+0	-0		+0	-0		+0	-0		+0
0	.12	+0.25	0	-0.15	+0.4	0	-0.25	+0.6	0	-0.4	+1.6	0	-1.0	+0.4	0.10	-0.25	+1.0	0.3	-0.6
.12	.24	+0.30	0	-0.20	+0.5	0	-0.30	+0.7	0	-0.5	+1.8	0	-1.2	+0.5	0.15	-0.30	+1.2	0.4	-0.7
.24	.40	+0.40	0	-0.25	+0.6	0	-0.40	+0.9	0	-0.6	+2.2	0	-1.4	+0.6	0.20	-0.40	+1.4	0.5	-0.9
.40	.71	+0.40	0	-0.30	+0.7	0	-0.40	+1.0	0	-0.7	+2.8	0	-1.6	+0.7	0.25	-0.40	+1.6	0.6	-1.0
.71	1.19	+0.50	0	-0.40	+0.8	0	-0.50	+1.2	0	-0.8	+3.5	0	-2.0	+0.8	0.30	-0.50	+2.0	0.8	-1.2
1.19	1.97	+0.60	0	-0.40	+1.0	0	-0.60	+1.6	0	-1.0	+4.0	0	-2.5	+1.0	0.40	-0.60	+2.5	1.0	-1.6
1.97	3.15	+0.70	0	-0.50	+1.2	0	-0.70	+1.8	0	-1.2	+4.5	0	-3.0	+1.2	0.40	-0.70	+3.0	1.2	-1.8
3.15	4.73	+0.90	0	-0.60	+1.4	0	-0.90	+2.7	0	-1.4	+5.0	0	-3.5	+1.4	0.50	-0.90	+3.5	1.4	-2.2
4.73	7.09	+1.00	0	-0.70	+1.6	0	-1.00	+2.5	0	-1.6	+6.0	0	-4.0	+1.6	0.60	-1.00	+4.0	1.6	-2.5
7.09	9.85	+1.20	0	-0.80	+1.8	0	-1.20	+2.8	0	-1.8	+7.0	0	-4.5	+1.8	0.60	-1.20	+4.5	2.0	-2.8
9.85	12.41	+1.20	0	-0.90	+2.0	0	-1.20	+3.0	0	-2.0	+8.0	0	-5.0	+2.0	0.70	-1.20	+5.0	2.2	-3.0
12.41	15.75	+1.40	0	-1.00	+2.2	0	-1.40	+3.5	0	-2.2	+9.0	0	-6.0	+2.2	0.70	-1.40	+6.0	2.5	-3.5

Nominal Size Range Inches		Class LC7			Class LC8			Class LC9			Class LC10			Class LC11		
		Hole Tol. GR10	Minimum Clearance	Shaft Tol. GR9	Hole Tol. GR10	Minimum Clearance	Shaft Tol. GR9	Hole Tol. GR11	Minimum Clearance	Shaft Tol. GR10	Hole Tol. GR12	Minimum Clearance	Shaft Tol. GR11	Hole Tol. GR13	Minimum Clearance	Shaft Tol. GR12
Over	To	-0		+0	-0		+0	-0		+0	-0		+0	-0		+0
0	.12	+1.6	0.6	-1.0	+1.6	1.0	-1.0	+2.5	2.5	-1.6	+1.0	4.0	-2.5	+6.0	5.0	-4.0
.12	.24	+1.8	0.8	-1.2	+1.8	1.2	-1.2	+3.0	2.8	-1.8	+5.0	4.5	-3.0	+7.0	6.0	-5.0
.24	.40	+2.2	1.0	-1.4	+2.2	1.6	-1.4	+3.5	3.0	-2.2	+6.0	5.0	-3.5	+9.0	7.0	-6.0
.40	.71	+2.8	1.2	-1.6	+2.8	2.0	-1.6	+4.0	3.5	-2.8	+7.0	6.0	-4.0	+10.0	8.0	-7.0
.71	1.19	+3.5	1.6	-2.0	+3.5	2.5	-2.0	+5.0	4.5	-3.5	+8.0	7.0	-5.0	+12.0	10.0	-8.0
1.19	1.97	+4.0	2.0	-2.5	+4.0	3.6	-2.5	+6.0	5.0	-4.0	+10.0	8.0	-6.0	+16.0	12.0	-10.0
1.97	3.15	+4.5	2.5	-3.0	+4.5	4.0	-3.0	+7.0	6.0	-4.5	+12.0	10.0	-7.0	+18.0	14.0	-12.0
3.15	4.73	+5.0	3.0	-3.5	+5.0	5.0	-3.5	+9.0	7.0	-5.0	+14.0	11.0	-9.0	+22.0	16.0	-14.0
4.73	7.09	+6.0	3.5	-4.0	+6.0	6.0	-4.0	+10.0	8.0	-6.0	+16.0	12.0	-10.0	+25.0	18.0	-16.0
7.09	9.85	+7.0	4.0	-4.5	+7.0	7.0	-4.5	+12.0	10.0	-7.0	+18.0	16.0	-12.0	+28.0	22.0	-18.0
9.85	12.41	+8.0	4.5	-5.0	+8.0	7.0	-5.0	+12.0	12.0	-8.0	+20.0	20.0	-12.0	+30.0	28.0	-20.0
12.41	15.75	+9.0	5.0	-6.0	+9.0	8.0	-6.0	+14.0	14.0	-9.0	+22.0	22.0	-14.0	+35.0	30.0	-22.0

Table 18 Locational Clearance Fits (Values in Thousandths of an Inch)

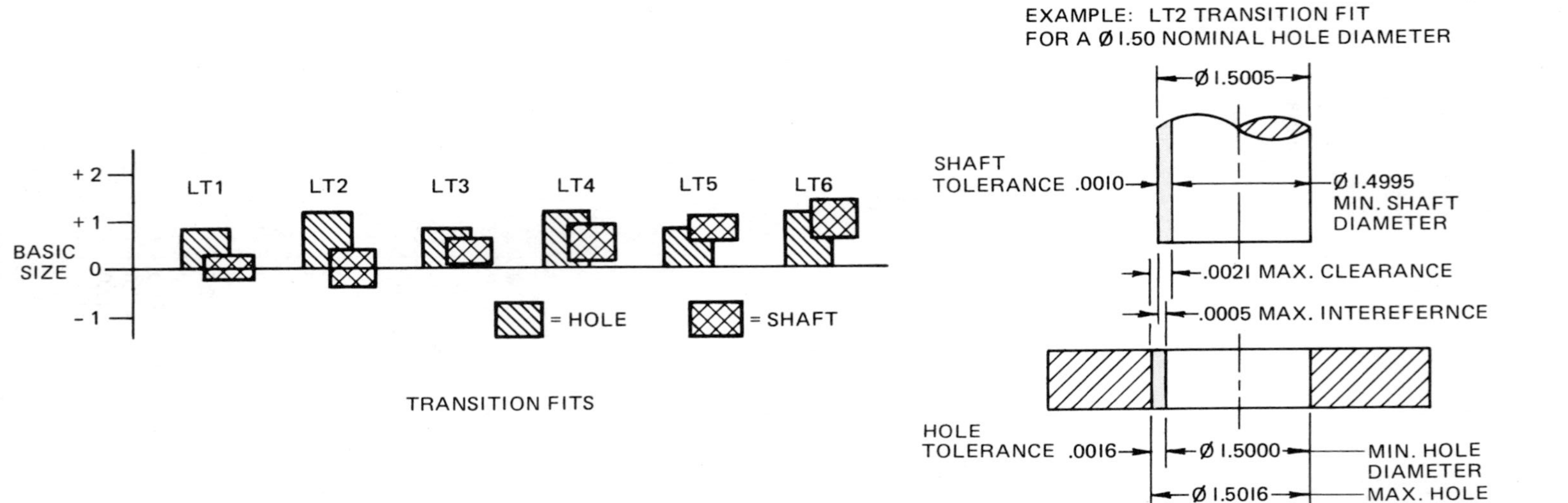

Nominal Size Range Inches		Class LT1			Class LT2			Class LT3			Class LT4			Class LT5			Class LT6		
		Hole Tol. GR7	Maximum Interference	Shaft Tol. GR6	Hole Tol. GR8	Maximum Interference	Shaft Tol. GR7	Hole Tol. GR7	Maximum Interference	Shaft Tol. GR6	Hole Tol. GR8	Maximum Interference	Shaft Tol. GR7	Hole Tol. GR7	Maximum Interference	Shaft Tol. GR6	Hole Tol. GR8	Maximum Interference	Shaft Tol. GR7
Over	To	-0		+0	-0		+0	-0		+0	-0		+0	-0		+0	-0		+0
0	.12	+0.4	0.10	-0.25	+0.6	0.20	-0.4	+0.4	0.25	-0.25	+0.6	0.4	-0.4	+0.4	0.5	-0.25	+0.6	0.65	-0.4
.12	.24	+0.5	0.15	-0.30	+0.7	0.25	-0.5	+0.5	0.40	-0.30	+0.7	0.6	-0.5	+0.5	0.6	-0.30	+0.7	0.80	-0.5
.24	.40	+0.6	0.20	-0.40	+0.9	0.30	-0.6	+0.6	0.50	-0.40	+0.9	0.7	-0.6	+0.6	0.8	-0.40	+0.9	1.00	-0.6
.40	.71	+0.7	0.20	-0.40	+1.0	0.30	-0.7	+0.7	0.50	-0.40	+1.0	0.8	-0.7	+0.7	0.9	-0.40	+1.0	1.20	-0.7
.71	1.19	+0.8	0.25	-0.50	+1.2	0.40	-0.8	+0.8	0.60	-0.50	+1.2	0.9	-0.8	+0.8	1.1	-0.50	+1.2	1.40	-0.8
1.19	1.97	+1.0	0.30	-0.60	+1.6	0.50	-1.0	+1.0	0.70	-0.60	+1.6	1.1	-1.0	+1.0	1.3	-0.60	+1.6	1.70	-1.0
1.97	3.15	+1.2	0.30	-0.70	+1.8	0.60	-1.2	+1.2	0.80	-0.70	+1.8	1.3	-1.2	+1.2	1.5	-0.70	+1.8	2.00	-1.2
3.15	4.73	+1.4	0.40	-0.90	+2.2	0.70	-1.4	+1.4	1.00	-0.90	+2.2	1.5	-1.4	+1.4	1.9	-0.90	+2.2	2.40	-1.4
4.73	7.09	+1.6	0.50	-1.00	+2.5	0.80	-1.6	+1.6	1.10	-1.00	+2.5	1.7	-1.6	+1.6	2.2	-1.00	+2.5	2.80	-1.6
7.09	9.85	+1.8	0.60	-1.20	+2.8	0.90	-1.8	+1.8	1.40	-1.20	+2.8	2.0	-1.8	+1.8	2.6	-1.20	+2.8	3.20	-1.8
9.85	12.41	+2.0	0.60	-1.20	+3.0	1.00	-2.0	+2.0	1.40	-1.20	+3.0	2.2	-2.0	+2.0	2.6	-1.20	+3.0	3.40	-2.0
12.41	15.75	+2.2	0.70	-1.40	+3.5	1.00	-2.2	+2.2	1.60	-1.40	+3.5	2.4	-2.2	+2.2	3.0	-1.40	+3.5	3.80	-2.2

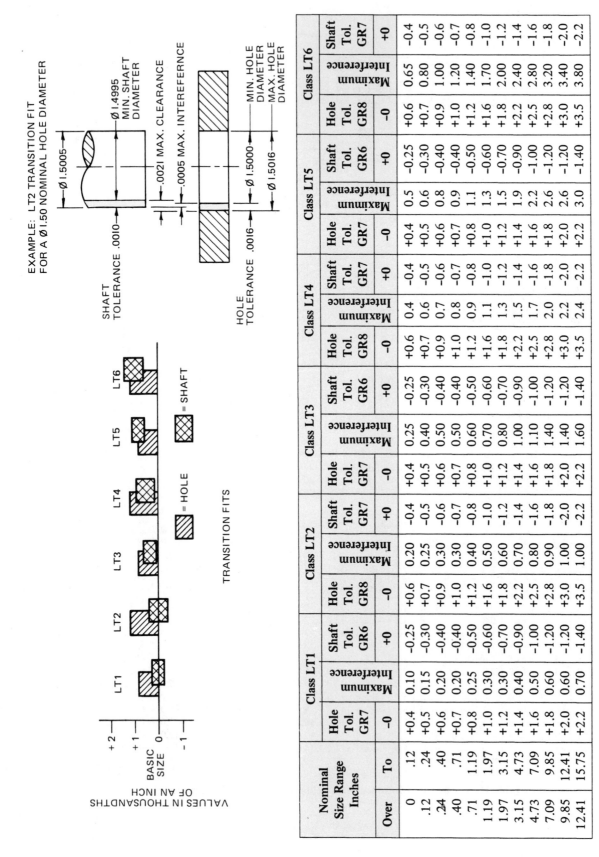

Table 19 Locational Transition Fits (Values in Thousandths of an Inch)

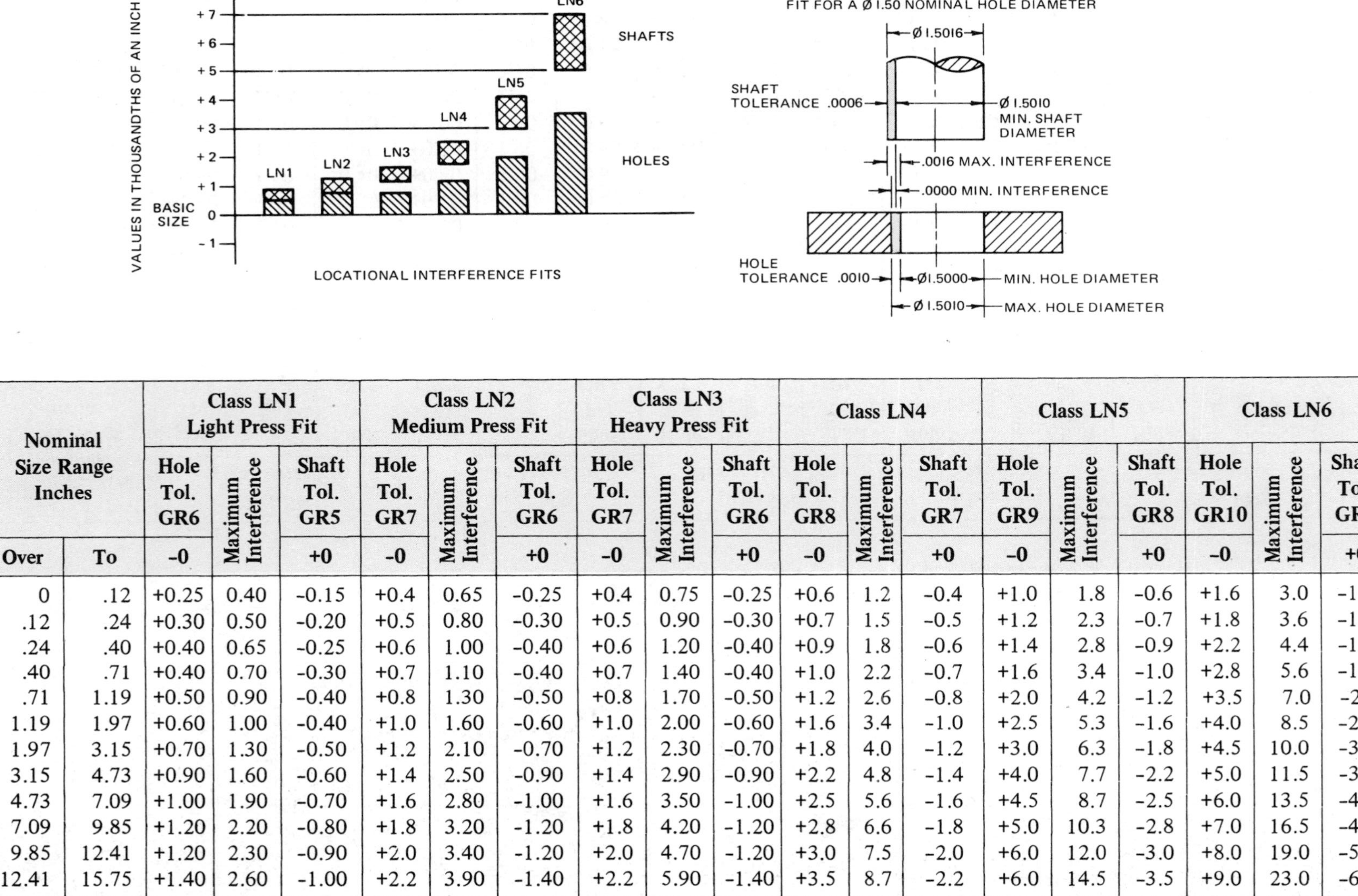

Nominal Size Range Inches		Class LN1 Light Press Fit			Class LN2 Medium Press Fit			Class LN3 Heavy Press Fit			Class LN4			Class LN5			Class LN6		
		Hole Tol. GR6	Maximum Interference	Shaft Tol. GR5	Hole Tol. GR7	Maximum Interference	Shaft Tol. GR6	Hole Tol. GR7	Maximum Interference	Shaft Tol. GR6	Hole Tol. GR8	Maximum Interference	Shaft Tol. GR7	Hole Tol. GR9	Maximum Interference	Shaft Tol. GR8	Hole Tol. GR10	Maximum Interference	Shaft Tol. GR9
Over	To	-0		+0	-0		+0	-0		+0	-0		+0	-0		+0	-0		+0
0	.12	+0.25	0.40	-0.15	+0.4	0.65	-0.25	+0.4	0.75	-0.25	+0.6	1.2	-0.4	+1.0	1.8	-0.6	+1.6	3.0	-1.0
.12	.24	+0.30	0.50	-0.20	+0.5	0.80	-0.30	+0.5	0.90	-0.30	+0.7	1.5	-0.5	+1.2	2.3	-0.7	+1.8	3.6	-1.2
.24	.40	+0.40	0.65	-0.25	+0.6	1.00	-0.40	+0.6	1.20	-0.40	+0.9	1.8	-0.6	+1.4	2.8	-0.9	+2.2	4.4	-1.4
.40	.71	+0.40	0.70	-0.30	+0.7	1.10	-0.40	+0.7	1.40	-0.40	+1.0	2.2	-0.7	+1.6	3.4	-1.0	+2.8	5.6	-1.6
.71	1.19	+0.50	0.90	-0.40	+0.8	1.30	-0.50	+0.8	1.70	-0.50	+1.2	2.6	-0.8	+2.0	4.2	-1.2	+3.5	7.0	-2.0
1.19	1.97	+0.60	1.00	-0.40	+1.0	1.60	-0.60	+1.0	2.00	-0.60	+1.6	3.4	-1.0	+2.5	5.3	-1.6	+4.0	8.5	-2.5
1.97	3.15	+0.70	1.30	-0.50	+1.2	2.10	-0.70	+1.2	2.30	-0.70	+1.8	4.0	-1.2	+3.0	6.3	-1.8	+4.5	10.0	-3.0
3.15	4.73	+0.90	1.60	-0.60	+1.4	2.50	-0.90	+1.4	2.90	-0.90	+2.2	4.8	-1.4	+4.0	7.7	-2.2	+5.0	11.5	-3.5
4.73	7.09	+1.00	1.90	-0.70	+1.6	2.80	-1.00	+1.6	3.50	-1.00	+2.5	5.6	-1.6	+4.5	8.7	-2.5	+6.0	13.5	-4.0
7.09	9.85	+1.20	2.20	-0.80	+1.8	3.20	-1.20	+1.8	4.20	-1.20	+2.8	6.6	-1.8	+5.0	10.3	-2.8	+7.0	16.5	-4.5
9.85	12.41	+1.20	2.30	-0.90	+2.0	3.40	-1.20	+2.0	4.70	-1.20	+3.0	7.5	-2.0	+6.0	12.0	-3.0	+8.0	19.0	-5.0
12.41	15.75	+1.40	2.60	-1.00	+2.2	3.90	-1.40	+2.2	5.90	-1.40	+3.5	8.7	-2.2	+6.0	14.5	-3.5	+9.0	23.0	-6.0

Table 20 Locational Interference Fits (Values in Thousandths of an Inch)

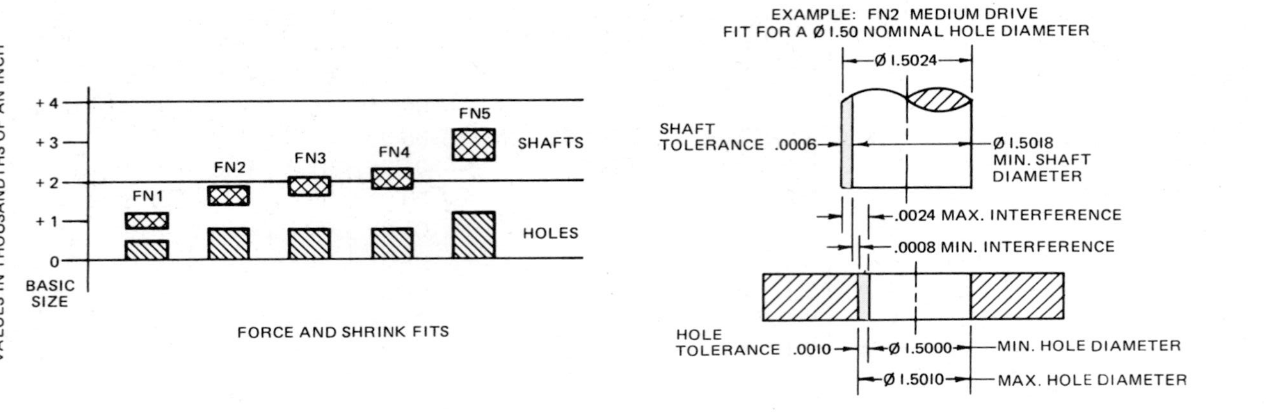

Nominal Size Range Inches		Class FN1 Light Drive Fit			Class FN2 Medium Drive Fit			Class FN3 Heavy Drive Fit			Class FN4 Shrink Fit			Class FN5 Heavy Shrink Fit		
		Hole Tol. GR6	Maximum Interference	Shaft Tol. GR5	Hole Tol. GR7	Maximum Interference	Shaft Tol. GR6	Hole Tol. GR7	Maximum Interference	Shaft Tol. GR6	Hole Tol. GR7	Maximum Interference	Shaft Tol. GR6	Hole Tol. GR8	Maximum Interference	Shaft Tol. GR7
Over	To	-0		+0	-0		+0	-0		+0	-0		+0	-0		+0
0	.12	+0.25	0.50	-0.15	+0.40	0.85	-0.25				+0.40	0.95	-0.25	+0.60	1.30	-0.40
.12	.24	+0.30	0.60	-0.20	+0.50	1.00	-0.30				+0.50	1.20	-0.30	+0.70	1.70	-0.50
.24	.40	+0.40	0.75	-0.25	+0.60	1.40	-0.40				+0.60	1.60	-0.40	+0.90	2.00	-0.60
.40	.56	+0.40	0.80	-0.30	+0.70	1.60	-0.40				+0.70	1.80	-0.40	+1.00	2.30	-0.70
.56	.71	+0.40	0.90	-0.30	+0.70	1.60	-0.40				+0.70	1.80	-0.40	+1.00	2.50	-0.70
.71	.95	+0.50	1.10	-0.40	+0.80	1.90	-0.50				+0.80	2.10	-0.50	+1.20	3.00	-0.80
.95	1.19	+0.50	1.20	-0.40	+0.80	1.90	-0.50	+0.80	2.10	-0.50	+0.80	2.30	-0.50	+1.20	3.30	-0.80
1.19	1.58	+0.60	1.30	-0.40	+1.00	2.40	-0.60	+1.00	2.60	-0.60	+1.00	3.10	-0.60	+1.60	4.00	-1.00
1.58	1.97	+0.60	1.40	-0.40	+1.00	2.40	-0.60	+1.00	2.80	-0.60	+1.00	3.40	-0.60	+1.60	5.00	-1.00
1.97	2.56	+0.70	1.80	-0.50	+1.20	2.70	-0.70	+1.20	3.20	-0.70	+1.20	4.20	-0.70	+1.80	6.20	-1.20
2.56	3.15	+0.70	1.90	-0.50	+1.20	2.90	-0.70	+1.20	3.70	-0.70	+1.20	4.70	-0.70	+1.80	7.20	-1.20
3.15	3.94	+0.90	2.40	-0.60	+1.40	3.70	-0.90	+1.40	4.40	-0.70	+1.40	5.90	-0.90	+2.20	8.40	-1.40

Table 21 Force and Shrink Fits (Values in Thousandths of an Inch)

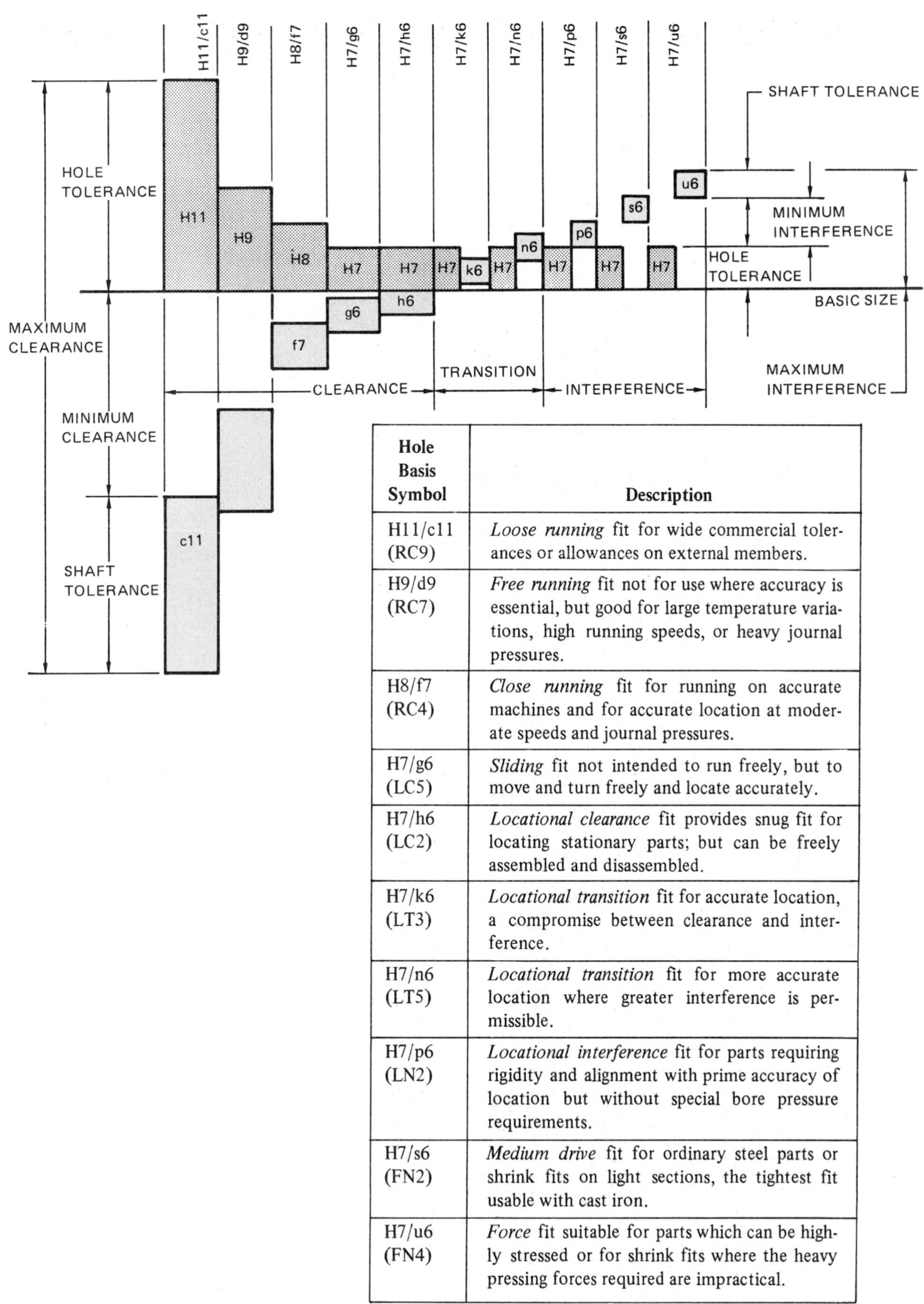

Hole Basis Symbol	Description
H11/c11 (RC9)	*Loose running* fit for wide commercial tolerances or allowances on external members.
H9/d9 (RC7)	*Free running* fit not for use where accuracy is essential, but good for large temperature variations, high running speeds, or heavy journal pressures.
H8/f7 (RC4)	*Close running* fit for running on accurate machines and for accurate location at moderate speeds and journal pressures.
H7/g6 (LC5)	*Sliding* fit not intended to run freely, but to move and turn freely and locate accurately.
H7/h6 (LC2)	*Locational clearance* fit provides snug fit for locating stationary parts; but can be freely assembled and disassembled.
H7/k6 (LT3)	*Locational transition* fit for accurate location, a compromise between clearance and interference.
H7/n6 (LT5)	*Locational transition* fit for more accurate location where greater interference is permissible.
H7/p6 (LN2)	*Locational interference* fit for parts requiring rigidity and alignment with prime accuracy of location but without special bore pressure requirements.
H7/s6 (FN2)	*Medium drive* fit for ordinary steel parts or shrink fits on light sections, the tightest fit usable with cast iron.
H7/u6 (FN4)	*Force* fit suitable for parts which can be highly stressed or for shrink fits where the heavy pressing forces required are impractical.

Table 22 Preferred Hole Basis Metric Fits Description

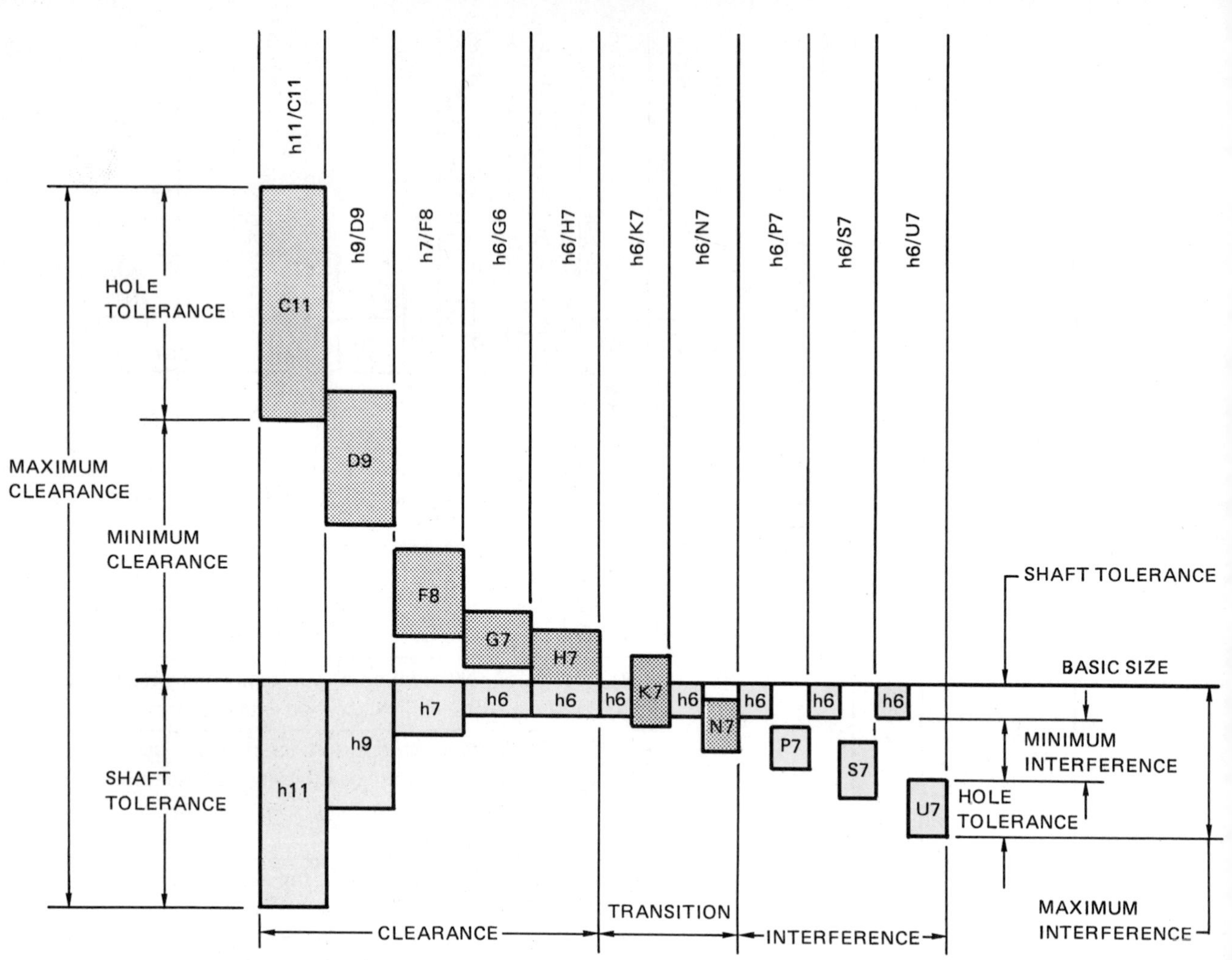

Shaft Basis Symbol	Description	Shaft Basis Symbol	Description
C11/h11	*Loose running* fit for wide commercial tolerances or allowances on external members.	K7/h6	*Locational transition* fit for accurate location, a compromise between clearance and interference.
D9/h9	*Free running* fit not for use where accuracy is essential, but good for large temperature variations, high running speeds, or heavy journal pressures.	N7/h6	*Locational transition* fit for more accurate location where greater interference is permissible.
F8/h7	*Close running* fit for running on accurate machines and for accurate location at moderate speeds and journal pressures.	P7/h6	*Locational interference* fit for parts requiring rigidity and alignment with prime accuracy of location but without special bore pressure requirements.
G7/h6	*Sliding* fit not intended to run freely, but to move and turn freely and locate accurately.	S7/h6	*Medium drive* fit for ordinary steel parts or shrink fits on light sections, the tightest fit usable with cast iron.
H7/h6	*Locational clearance* fit provides snug fit for locating stationary parts; but can be freely assembled and disassembled.	U7/h6	*Force* fit suitable for parts which can be highly stressed or for shrink fits where the heavy pressing forces required are impractical.

Table 23 Preferred Shaft Basis Metric Fits Description

Basic Size		Preferred Hole Basis Clearance Fits														
		Loose Running			Free Running			Close Running			Sliding			Locational Clearance		
		Hole H11	Shaft c11	Fit RC9	Hole H9	Shaft d9	Fit RC7	Hole H8	Shaft f7	Fit RC4	Hole H7	Shaft g6	Fit LC5	Hole H7	Shaft h6	Fit LC2
5	MAX	5.075	4.930	0.220	5.030	4.970	0.090	5.018	4.990	0.040	5.012	4.996	0.024	5.012	5.000	0.020
	MIN	5.000	5.855	0.070	5.000	4.940	0.030	5.000	4.978	0.010	5.000	4.988	0.004	5.000	4.992	0.000
6	MAX	6.075	5.930	0.220	6.030	5.970	0.090	6.018	5.990	0.040	6.012	5.996	0.024	6.012	6.000	0.020
	MIN	6.000	5.855	0.070	6.000	5.940	0.030	6.000	5.978	0.010	6.000	5.988	0.004	6.000	5.992	0.000
8	MAX	8.090	7.920	0.260	8.036	7.960	0.112	8.022	7.987	0.050	8.015	7.995	0.029	8.015	8.000	0.024
	MIN	8.000	7.830	0.080	8.000	7.924	0.040	8.000	7.972	0.013	8.000	7.986	0.006	8.000	7.991	0.000
10	MAX	10.090	9.920	0.260	10.036	9.960	0.112	10.022	9.987	0.050	10.015	9.995	0.029	10.015	10.000	0.024
	MIN	10.000	9.830	0.080	10.000	9.924	0.040	10.000	9.972	0.013	10.000	9.986	0.005	10.000	9.991	0.000
12	MAX	12.110	11.905	0.315	12.043	11.950	0.136	12.027	11.984	0.061	12.018	11.994	0.035	12.018	12.000	0.029
	MIN	12.000	11.795	0.095	12.000	11.907	0.050	12.000	11.966	0.016	12.000	11.983	0.006	12.000	11.989	0.000
16	MAX	16.110	15.905	0.315	16.043	15.950	0.136	16.027	15.984	0.061	16.018	15.994	0.035	16.018	16.000	0.029
	MIN	16.000	15.795	0.095	16.000	15.907	0.050	16.000	15.966	0.016	16.000	15.983	0.006	16.000	15.989	0.000
20	MAX	20.130	19.890	0.370	20.052	19.935	0.169	20.033	19.980	0.074	20.021	19.993	0.041	20.021	20.000	0.034
	MIN	20.000	19.760	0.110	20.000	19.883	0.065	20.000	19.959	0.020	20.000	19.980	0.007	20.000	19.987	0.000
25	MAX	25.130	24.890	0.370	25.052	24.935	0.169	25.033	24.980	0.074	25.021	24.993	0.042	25.021	25.000	0.034
	MIN	25.000	24.760	0.110	25.000	24.883	0.065	25.000	24.959	0.020	25.000	24.980	0.007	25.000	24.987	0.000
30	MAX	30.130	29.890	0.370	30.052	29.935	0.169	30.033	29.980	0.074	30.021	29.993	0.041	30.021	30.000	0.034
	MIN	30.000	29.760	0.110	30.000	29.883	0.065	30.000	29.959	0.020	30.000	29.980	0.007	30.000	29.987	0.000
40	MAX	40.160	39.880	0.440	40.062	39.920	0.204	40.039	39.975	0.089	40.025	39.991	0.050	40.025	40.000	0.041
	MIN	40.000	39.720	0.120	40.000	39.858	0.080	40.000	39.950	0.025	40.000	39.975	0.009	40.000	39.984	0.000
50	MAX	50.160	49.870	0.450	50.062	49.920	0.204	50.039	49.975	0.089	50.025	49.991	0.050	50.025	50.000	0.041
	MIN	50.000	49.710	0.130	50.000	49.858	0.080	50.000	49.950	0.025	50.000	49.975	0.009	50.000	49.984	0.000
60	MAX	60.190	59.860	0.520	60.074	59.900	0.248	60.046	59.970	0.106	60.030	59.990	0.059	60.030	60.000	0.049
	MIN	60.000	59.670	0.140	60.000	59.826	0.100	60.000	59.940	0.030	60.000	59.971	0.010	60.000	59.981	0.000
80	MAX	80.190	79.850	0.530	80.074	79.900	0.248	80.046	79.970	0.106	80.030	79.990	0.059	80.030	80.000	0.049
	MIN	80.000	79.660	0.150	80.000	79.826	0.100	80.000	79.940	0.030	80.000	79.971	0.010	80.000	79.981	0.000
100	MAX	100.220	99.830	0.610	100.087	99.880	0.294	100.054	99.964	0.125	100.035	99.988	0.069	100.035	100.000	0.057
	MIN	100.000	99.610	0.170	100.000	99.793	0.120	100.000	99.929	0.036	100.000	99.966	0.012	100.000	99.978	0.000

Table 24 Preferred Hole Basis Metric Fits (Dimensions in millimeters) **cont'd on page 394**

	Preferred Hole Basis Transition and Interference Fits														
	Locational Transn.			Locational Transn.			Locational Interf.			Medium Drive			Force		
Basic Size	Hole H7	Shaft k6	Fit LT3	Hole H7	Shaft n6	Fit LT5	Hole H7	Shaft p6	Fit LN2	Hole H7	Shaft s6	Fit FN2	Hole H7	Shaft u6	Fit FN4
5 MAX	5.012	5.009	0.011	5.012	5.016	0.004	5.012	5.020	0.000	5.012	5.027	−0.007	5.012	5.031	−0.011
MIN	5.000	5.001	−0.009	5.000	5.008	−0.016	5.000	5.012	−0.020	5.000	5.019	−0.027	5.000	5.023	−0.031
6 MAX	6.012	6.009	0.011	6.012	6.016	0.004	6.012	6.020	0.000	6.012	6.027	−0.007	6.012	6.031	−0.011
MIN	6.000	6.001	−0.009	6.000	6.008	−0.016	6.000	6.012	−0.020	6.000	6.019	−0.027	6.000	6.023	−0.031
8 MAX	8.015	8.010	0.014	8.015	8.019	0.005	8.015	8.024	0.000	8.015	8.032	−0.008	8.015	8.037	−0.013
MIN	8.000	8.001	−0.010	8.000	8.010	−0.019	8.000	8.015	−0.024	8.000	8.023	−0.032	8.000	8.028	−0.037
10 MAX	10.015	10.010	0.014	10.015	10.019	0.005	10.015	10.024	0.000	10.015	10.032	−0.008	10.015	10.037	−0.013
MIN	10.000	10.001	−0.010	10.000	10.010	−0.019	10.000	10.015	−0.024	10.000	10.023	−0.032	10.000	10.028	−0.037
12 MAX	12.018	12.012	0.017	12.018	12.023	0.006	12.018	12.029	0.000	12.018	12.039	−0.010	12.018	12.044	−0.015
MIN	12.000	12.001	−0.012	12.000	12.012	−0.023	12.000	12.018	−0.029	12.000	12.028	−0.039	12.000	12.033	−0.044
16 MAX	16.018	16.012	0.017	16.018	16.023	0.006	16.018	16.029	0.000	16.018	16.039	−0.010	16.018	16.044	−0.015
MIN	16.000	16.001	−0.012	16.000	16.012	−0.023	16.000	16.018	−0.029	16.000	16.028	−0.039	16.000	16.033	−0.044
20 MAX	20.021	20.015	0.019	20.021	20.028	0.006	20.021	20.035	−0.001	20.021	20.048	−0.014	20.021	20.054	−0.020
MIN	20.000	20.002	−0.015	20.000	20.015	−0.028	20.000	20.022	−0.035	20.000	20.035	−0.048	20.000	20.041	−0.054
25 MAX	25.021	25.014	0.019	25.021	25.028	0.006	25.021	25.035	−0.001	25.021	25.048	−0.014	25.021	25.061	−0.027
MIN	25.000	25.002	−0.015	25.000	25.015	−0.028	25.000	25.022	−0.035	25.000	25.035	−0.048	25.000	25.048	−0.061
30 MAX	30.021	30.015	0.019	30.021	30.028	0.006	30.021	30.035	−0.001	30.021	30.048	−0.014	30.021	30.061	−0.027
MIN	30.000	30.002	−0.015	30.000	30.015	−0.028	30.000	30.022	−0.035	30.000	30.035	−0.048	30.000	30.048	−0.061
40 MAX	40.025	40.018	0.023	40.025	40.033	0.008	40.025	40.042	−0.001	40.025	40.059	−0.018	40.025	40.076	−0.035
MIN	40.000	40.002	−0.018	40.000	40.017	−0.033	40.000	40.026	−0.042	40.000	40.043	−0.059	40.000	40.060	−0.076
50 MAX	50.025	50.018	0.023	50.025	50.033	0.008	50.025	50.042	−0.001	50.025	50.059	−0.018	50.025	50.086	−0.045
MIN	50.002	50.000	−0.018	50.000	50.017	−0.033	50.000	50.026	−0.042	50.000	50.043	−0.059	50.000	50.070	−0.086
60 MAX	60.030	60.021	0.028	60.030	60.039	0.010	60.030	60.051	−0.002	60.030	60.072	−0.023	60.030	60.106	−0.057
MIN	60.000	60.002	−0.021	60.000	60.020	−0.039	60.000	60.032	−0.051	60.000	60.053	−0.072	60.000	60.087	−0.106
80 MAX	80.030	80.021	0.028	80.030	80.039	0.010	80.030	80.051	−0.002	80.030	80.078	−0.029	80.030	80.121	−0.072
MIN	80.000	80.002	−0.021	80.000	80.020	−0.039	80.000	80.032	−0.051	80.000	80.059	−0.078	80.000	80.102	−0.121
100 MAX	100.035	100.025	0.032	100.035	100.045	0.012	100.035	100.059	−0.002	100.035	100.093	−0.036	100.035	100.146	−0.089
MIN	100.000	100.003	−0.025	100.000	100.023	−0.045	100.000	100.037	−0.059	100.000	100.071	−0.093	100.000	100.124	−0.146

Table 24 (cont'd) Preferred Hole Basis Metric Fits (Values in millimeters)

	Preferred Shaft Basis Clearance Fits														
	Loose Running			Free Running			Close Running			Sliding			Locational Clearance		
Basic Size	Hole C11	Shaft h11	Fit RC9	Hole D9	Shaft h9	Fit RC7	Hole F8	Shaft h7	Fit RC4	Hole G7	Shaft h6	Fit LC5	Hole H7	Shaft h6	Fit LC2
5 MAX	5.145	5.000	0.220	5.060	5.000	0.090	5.028	5.000	0.040	5.016	5.000	0.024	5.012	5.000	0.020
MIN	5.070	4.925	0.070	5.030	4.970	0.030	5.010	4.988	0.010	5.004	4.992	0.004	5.000	4.992	0.000
6 MAX	6.145	6.000	0.220	6.060	6.000	0.090	6.028	6.000	0.040	6.016	6.000	0.024	6.012	6.000	0.020
MIN	6.070	5.925	0.070	6.030	5.970	0.030	6.010	5.988	0.010	6.004	5.992	0.004	6.000	5.992	0.000
8 MAX	8.170	8.000	0.260	8.076	8.000	0.112	8.035	8.000	0.050	8.020	8.000	0.029	8.015	8.000	0.024
MIN	8.080	7.910	0.080	8.040	7.964	0.040	8.013	7.985	0.013	8.005	7.991	0.005	8.000	7.991	0.000
10 MAX	10.170	10.000	0.260	10.076	10.000	0.112	10.035	10.000	0.050	10.020	10.000	0.029	10.015	10.000	0.024
MIN	10.080	9.910	0.080	10.040	9.964	0.040	10.013	9.985	0.013	10.005	9.991	0.005	10.000	9.991	0.000
12 MAX	12.205	12.000	0.315	12.093	12.000	0.136	12.043	12.000	0.061	12.024	12.000	0.035	12.018	12.000	0.029
MIN	12.095	11.890	0.095	12.050	11.957	0.050	12.016	11.982	0.016	12.006	11.989	0.006	12.000	11.989	0.000
16 MAX	16.205	16.000	0.315	16.093	16.000	0.136	16.043	16.000	0.061	16.024	16.000	0.035	16.018	16.000	0.029
MIN	16.095	15.890	0.095	16.050	15.957	0.050	16.016	15.982	0.016	16.006	15.989	0.006	16.000	15.989	0.000
20 MAX	20.240	20.000	0.370	20.117	20.000	0.169	20.053	20.000	0.074	20.028	20.000	0.041	20.021	20.000	0.034
MIN	20.110	19.870	0.110	20.065	19.948	0.065	20.020	19.979	0.020	20.007	19.987	0.007	20.000	19.987	0.000
25 MAX	25.240	25.000	0.370	25.117	25.000	0.169	25.053	25.000	0.074	25.028	25.000	0.041	25.021	25.000	0.034
MIN	25.110	24.870	0.110	25.065	24.948	0.065	25.020	24.979	0.020	25.007	24.987	0.007	25.000	24.987	0.000
30 MAX	30.240	30.000	0.370	30.117	30.000	0.169	30.053	30.000	0.074	30.028	30.000	0.041	30.021	30.000	0.034
MIN	30.110	29.870	0.110	30.065	29.948	0.065	30.020	29.979	0.020	30.007	29.987	0.007	30.000	29.987	0.000
40 MAX	40.280	40.000	0.440	40.142	40.000	0.204	40.064	40.000	0.089	40.034	40.000	0.050	40.025	40.000	0.041
MIN	40.120	39.840	0.120	40.080	39.938	0.080	40.025	39.975	0.025	40.009	39.984	0.009	40.000	39.984	0.000
50 MAX	50.290	50.000	0.450	50.142	50.000	0.204	50.064	50.000	0.089	50.034	50.000	0.050	50.025	50.000	0.041
MIN	50.130	49.840	0.130	50.080	49.938	0.080	50.025	49.975	0.025	50.009	49.984	0.009	50.000	49.984	0.000
60 MAX	60.330	60.000	0.510	60.174	60.000	0.248	60.076	60.000	0.106	60.040	60.000	0.059	60.030	60.000	0.049
MIN	60.140	59.810	0.140	60.100	59.926	0.100	60.030	59.970	0.030	60.010	59.981	0.010	60.000	59.981	0.000
80 MAX	80.340	80.000	0.530	80.174	80.000	0.248	80.076	80.000	0.106	80.040	80.000	0.059	80.030	80.000	0.049
MIN	80.150	79.810	0.150	80.100	79.926	0.100	80.030	79.970	0.030	80.010	79.981	0.010	80.000	79.981	0.000
100 MAX	100.390	100.000	0.610	100.207	100.000	0.294	100.090	100.000	0.125	100.047	100.000	0.069	100.035	100.000	0.057
MIN	100.170	99.780	0.170	100.120	99.913	0.120	100.036	99.965	0.036	100.012	99.978	0.012	100.000	99.987	0.000

Table 25 Preferred Shaft Basis Metric Fits (Values in Millimeters) **cont'd on page 396**

Basic Size	Locational Transn. Hole K7	Shaft′ h6	Fit LT3	Locational Transn. Hole N7	Shaft h6	Fit LT5	Locational Interf. Hole P7	Shaft h6	Fit LN2	Medium Drive Hole S7	Shaft h6	Fit FN2	Force Hole U7	Shaft h6	Fit FN4
5 MAX	5.003	5.000	0.011	4.996	5.000	0.004	4.992	5.000	0.000	4.985	5.000	−0.007	4.981	5.000	−0.011
MIN	4.991	4.992	−0.009	4.984	4.992	−0.016	4.980	4.992	−0.020	4.973	4.992	−0.027	4.969	4.992	−0.031
6 MAX	6.003	6.000	0.011	5.996	6.000	0.004	5.992	6.000	0.000	5.985	6.000	−0.007	5.981	6.000	−0.011
MIN	5.991	5.992	−0.009	5.984	5.992	−0.016	5.980	5.992	−0.020	5.973	5.992	−0.027	5.969	5.992	−0.031
8 MAX	8.005	8.000	0.014	7.996	8.000	0.005	7.991	8.000	0.000	7.983	8.000	−0.008	7.978	8.000	−0.013
MIN	7.990	7.991	−0.010	7.981	7.991	−0.019	7.976	7.991	−0.024	7.968	7.991	−0.032	7.963	7.991	−0.037
10 MAX	10.005	10.000	0.014	9.996	10.000	0.005	9.991	10.000	0.000	9.983	10.000	−0.008	9.978	10.000	−0.013
MIN	9.990	9.991	−0.010	9.981	9.991	−0.019	9.976	9.991	−0.024	9.968	9.991	−0.032	9.963	9.991	−0.037
12 MAX	12.006	12.000	0.017	11.995	12.000	0.006	11.989	12.000	0.000	11.979	12.000	−0.010	11.974	12.000	−0.015
MIN	11.988	11.989	−0.012	11.977	11.989	−0.023	11.971	11.989	−0.029	11.961	11.989	−0.039	11.956	11.989	−0.044
16 MAX	16.006	16.000	0.017	15.955	16.000	0.006	15.989	16.000	0.000	15.979	16.000	−0.010	15.974	16.000	−0.015
MIN	15.988	15.989	−0.012	15.977	15.989	−0.023	15.971	15.989	−0.029	15.961	15.989	−0.039	15.956	15.989	−0.044
20 MAX	20.006	20.000	0.019	19.993	20.000	0.006	19.986	20.000	−0.001	19.973	20.000	−0.014	19.967	20.000	−0.020
MIN	19.985	19.987	−0.015	19.972	19.987	−0.028	19.965	19.987	−0.035	19.952	19.987	−0.048	19.946	19.987	−0.054
25 MAX	25.006	25.000	0.019	24.993	25.000	0.006	24.986	25.000	−0.001	24.973	25.000	−0.014	24.960	25.000	−0.027
MIN	24.985	24.987	−0.015	24.972	24.987	−0.028	24.965	24.987	−0.035	24.952	24.987	−0.048	24.939	24.987	−0.061
30 MAX	30.006	30.000	0.019	29.993	30.000	0.006	29.986	30.000	−0.001	29.973	30.000	−0.014	29.960	30.000	−0.027
MIN	29.985	29.987	−0.015	29.972	29.987	−0.028	29.965	29.987	−0.035	29.952	29.987	−0.048	29.939	29.987	−0.061
40 MAX	40.007	40.000	0.023	39.992	40.000	0.008	39.983	40.000	−0.001	39.966	40.000	−0.018	39.949	40.000	−0.035
MIN	39.982	39.984	−0.018	39.967	39.984	−0.033	39.958	39.984	−0.042	39.941	39.984	−0.059	39.924	39.984	−0.076
50 MAX	50.007	50.000	0.023	49.992	50.000	0.008	49.983	50.000	−0.001	49.966	50.000	−0.018	49.939	50.000	−0.045
MIN	49.982	49.984	−0.018	49.967	49.984	−0.033	49.958	49.984	−0.042	49.941	49.984	−0.059	49.914	49.984	−0.086
60 MAX	60.009	60.000	0.028	59.991	60.000	0.010	59.979	60.000	−0.002	59.958	60.000	−0.023	59.924	60.000	−0.057
MIN	59.979	59.981	−0.021	59.961	59.981	−0.039	59.949	59.981	−0.051	59.928	59.981	−0.072	59.894	59.981	−0.106
80 MAX	80.009	80.000	0.028	79.991	80.000	0.010	79.979	80.000	−0.002	79.952	80.000	−0.029	79.909	80.000	−0.072
MIN	79.979	79.981	−0.021	79.961	79.981	−0.039	79.949	79.981	−0.051	79.922	79.981	−0.078	79.879	79.981	−0.121
100 MAX	100.010	100.000	0.032	99.990	100.000	0.012	99.976	100.000	−0.002	99.942	100.000	−0.036	99.889	100.000	−0.089
MIN	99.975	99.978	−0.025	99.955	99.978	−0.045	99.941	99.978	−0.059	99.907	99.978	−0.093	99.854	99.978	−0.146

Preferred Shaft Basis Transition and Interference Fits

Table 25 (cont'd.) Preferred Shaft Basis Metric Fits (Values in millimeters)

Quantity	Metric Unit	Symbol	Metric to Inch-Pound Unit	Inch-Pound to Metric Unit
Length	millimeter centimeter meter kilometer	mm cm m km	1 mm = 0.0394 in. 1 cm = 0.394 in. 1 m = 39.37 in. = 3.28 ft 1 km = 0.62 mile	1 in. = 25.4 mm 1 ft. = 30.5 cm 1 yd. = 0.914 m = 914 mm 1 mile = 1.61 km
Area	square millimeter square centimeter square meter	mm^2 cm^2 m^2	1 mm^2 = 0.001 55 sq. in. 1 cm^2 = 0.155 sq. in. 1 m^2 = 10.8 sq. ft. = 1.2 sq. yd.	1 sq. in. = 6 452 mm^2 1 sq. ft. = 0.093 m^2 1 sq. yd. = 0.836 m^2
Mass	milligram gram kilogram tonne	mg g kg t	1 g = 0.035 oz. 1 kg = 2.205 lb. 1 tonne = 1.102 tons	1 oz. = 28.3 g 1 lb. = 0.454 kg 1 ton = 907.2 kg = 0.907 tonnes
Volume	cubic centimeter cubic meter milliliter	cm^3 m^3 m	1 mm^3 = 0.000 061 cu. in. 1 cm^3 = 0.061 cu. in. 1 m^3 = 35.3 cu. ft. = 1.308 cu. yd. 1 mℓ = 0.035 fl. oz.	1 fl. oz. = 28.4 cm^3 1 cu. in. = 16.387 cm^3 1 cu. ft. = 0.028 m^3 1 cu. yd. = 0.756 m^3
Capacity	liter	L	U.S. Measure 1 pt. = 0.473 L 1 pt. = 0.946 L 1 gal. = 3.785 L Imperial Measure 1 pt. = 0.568 L 1 qt. = 1.137 L 1 gal. = 4.546 L	U.S. Measure 1 L = 2.113 pt. = 1.057 qt. = 0.264 gal. Imperial Measure 1 L = 1.76 pt. = 0.88 qt. = 0.22 gal.
Temperature	Celsius degree	°C	$°C = \frac{5}{9}(°F\text{-}32)$	$°F = \frac{9}{5} \times °C + 32$
Force	newton kilonewton	N kN	1 N = 0.225 lb (f) 1 kN = 0.225 kip (f) = 0.112 ton (f)	1 lb (f) = 4.45N = 0.004 448 kN
Energy/Work	joule kilojoule megajoule	J kJ MJ	1 J = 0.737 ft ◦ lb 1 J = 0.948 Btu 1 MJ = 0.278 kWh	1 ft ◦ lb = 1.355 J 1 Btu = 1.055 J 1 kWh = 3.6 MJ
Power	kilowatt	kW	1 kW = 1.34 hp 1 W = 0.0226 ft ◦ lb/min.	1 hp (550 ft ◦ lb/s) = 0.746 kW 1 ft ◦ lb/min = 44.2537 W
Pressure	kilopascal	kPa	1 kPa = 0.145 psi = 20.885 psf = 0.01 ton-force per sq. ft.	1 psi = 6.895 kPa 1 lb-force/sq. ft. = 47.88 Pa 1 ton-force/sq. ft. = 95.76 kPa
	*kilogram per square centimeter	kg/cm^2	1 kg/cm^2 = 13.780 psi	
Torque	newton meter *kilogram meter *kilogram per centimeter	N ◦ m kg/m kg/cm	1 N ◦ m = 0.74 lb ◦ ft 1 kg/m = 7.24 lb ◦ ft 1 kg/cm = 0.86 lb ◦ in	1 lb ◦ ft = 1.36 N ◦ m 1 lb ◦ ft = 0.14 kg/m 1 lb ◦ in = 1.2 kg/cm
Speed/Velocity	meters per second kilometers per hour	m/s km/h	1 m/s = 3.28 ft/s 1 km/h = 0.62 mph	1 ft/s = 0.305 m/s 1 mph = 1.61 km/h

*Not SI units, but included here because they are employed on some of the gages and indicators currently in use in industry.

Table 26 Metric Conversion Tables

Index

Abbreviations, 230–31, 369
structural steel shapes, 230–31
Alignment of parts and holes, 193
foreshortened projection, 193
revolved holes, 194
Allowances, 89
See Tolerances
American Iron and Steel Institute (AISI), 174
Angles
measurement of, 30
Appendix, 369
tables, 369–397

Basis for interpreting drawings, 1
Bearings, 170–71
antifriction, 171–288
ball, 288
plain, types of, 170
journal, 170
sleeve, 170
thrust, 170
roller, 288
shims, 173
Bill of material, 349
Bolts, 378
hexagon head, 378
Boring, 40
Bosses, 131

Cams, types of, 283
displacement, 284
Cap screws, 377
Cast iron, types of, 186
ductile, 186
gray, 186
malleable, 186
nodular, 186
white, 186
Castings, 180–82
design, 186
full mold, 182
sand, 180–81
Casting design, 186
coping down, 188
cored castings, 188
irregular shaped castings, 187
machining lugs, 189
molding, simplicity of, 186
set cores, 187
split pattern, 188
surface coating, 189
Chamfers, 69
Circular features, 37
centerlines, 37
dimension of, 37
dimensioning of cylindrical holes, 39
Clutches, 290–92
Counterbores, 62
Countersinks, 62

Design, 186
Dimensioning, 11, 216
base line, 216
basic rules, 15
chain, 216
choice of, 14
cylindrical features, 37
cylindrical holes, 39
lines, 11
placement of dimensions, 11
rectangular coordinate, 131
Dimensions
not to scale, 55
reference, 49
Dovetails, 47
measuring of, 47
Drafting
computer aided, 21
office, 21
Drawings
abbreviations used on, 24
assembly, 223
bill of material, 223–25
development, 148
dimensioning, 11
identifying parts, 223
lettering, 5
multiple detail, 75
one and two view, 75
phantom outlines, 231
reproduction, 23
revisions, 57
sketching, 5
standards, 5
subassembly, 223
title block, 4
to scale, 45
welding, 236
working, 11
Drawings for numerical control, 176
coordinate system, 176
dimensioning, 176
Drilling, 40
Drills
number and letter size, 373
sizes of, 195
table, metric sizes, 374
Drives, 290
V-belt, 290

Edges, 149

Fasteners, 112–206
pin, 206
threaded, 112
assemblies, 112
conventions, 114
representation, 112
standards, 112
types of, 118
Features
identifying similar sized, 40
Fillets, 40
Fits
description of, 98
force, 392
inch, 98
locational clearance, 389
locational interference, 391
locational transition, 390
metric, 104
metric, preferred hole, 393–96
metric, preferred shaft, 394–98
running, 388
shrink, 392
sliding, 388
standard inch, 100

Gears
bevel, 274
formula, 275
nomenclature, 274
symbol, 275

miter, 274
motor drive, 279
spur, 265
calculation examples, 269
formula, 267
styles of, 266
symbols, 266–67
teeth size, 270
working drawing, 268
trains, 279
center distances, 279

Inches to millimeters
conversion, 372
ISO projection symbol, 4

Joints, 149–225
swivel, 225–26
universal, 225–26

Keys, 129
dimensioning of seats, 130
Woodruff sizes, 384
Knurls, 71

Leaders, 11
Lines
break, 57
center, 37
cutting plane, 60
dimension, 11
extension, 11
hidden, 24
leader, 11
phantom, 231
section, 61
visible, 5

Machining symbols, 54
indicating allowances, 54
Measurements
units of, 13
inch units, 13
SI metric units, 14
Metal
sheet, gauge and thickness, 87, 386
Metric conversion, 397

Nuts
hexagon flanged, 381
hexagon head, 380

Pads, 131
Pins
dowel, 206–07
clevis, 206
cotter, 206–08
machine, 207
radial locking, 209
spring, 209
straight, 206
tapered, 206–07
Pipe
joints and fittings, 159
flanged, 160
screwed, 160
welded, 160
kinds of, 159
brass tubing, 159
cast iron, 159
copper, 159
copper tubing, 159
plastic, 159
steel, 159, 385
wrought iron, 159
valves, 160
check, 161
gate, 160
globe, 161
Piping, symbols, orthographic, 165
flanged, 165
pipe, 165
valve, 165
Projection
ISO symbol, 4
third angle, 1

Reaming, 40
Repetitive features, 39
dimensions, 39
Rings, 290–93
application, 292
O-ring, 290
retaining, 290
types of, 293

Scales
decimal inch, 46
fractional inch, 46
inch and foot, 46
millimeter, 46
Screws
cap screw, 377–78
setscrew, 130, 379
Seals
O-ring, 290–92
Seams, 149
Sections
broken out, 198
partial, 198
revolved and removed, 124
ribs in, 198
spokes in, 200
through shafts, pins, keys, 210
types of, 61
full, 61
half, 62
webs, 198
Shapes, 230–31
structural steel, 230
abbreviations, 230–31
Sketching, 5
Sheet metal
gauge and thickness, 386–87
sizes, 150
Slots
machine, 46
Society of Automotive Engineers (SAE), 174
Spotfaces, 63
Stampings, 151
Steel
alloy effects, 174
specifications AISI, 174
specifications SAE, 174
structural shapes, 230
Surfaces
inclined, 30
Surface texture
control requirements, 83
definitions, 78
ratings, 82
symbol, 79
Symbols
machining, 54
Symmetrical outlines, 49

Tapers
circular, 71
dimensioning, 70
flat, 71
symbols, 70–71
Third angle projection, 1–4
position of views, 2
Threads
inch, 115
metric, 117, 376
right and left handed, 115
table, inch, unified and American, 375
Title block, 4
Tolerances and allowances, 89
definitions, 89
dimension origin symbol, 91
methods
inch, 91
limit dimensioning, 90
millimeter, 91
plus and minus, 91
Tolerancing
coordinate, 336
correlative, 357
coaxiality, 357
concentricity, 361
coplanarity, 357
runout, 358
symmetry, 361
datum, 49, 320
identifying symbol, 321
features by position, 335

form, 14, 304
circularity, 315
cylindricity, 317
flatness, 314
straightness, 304–11
geometric, 301
characteristic symbols, 302
modern engineering, 300
orientation, 327
positional, 337
profile of a line, 353
profile of a surface, 354
projected tolerance zone, 344
symbols, 309
three plane, 320

Undercuts, 70

Valves, 160
check, 160
gate, 160
globe, 160
Views, 194
naming of, 195
partial, 194
primary auxiliary, 138
sectional, 60
secondary auxiliary, 141
selection of, 141

Washers, 230
common sizes, 382–83
Welding
drawings, 236
joints, 236
symbols, 236
basic, 238
fillet, 240
groove, 246
supplementary, 238
Welds
types of
fillet, 240
flange, 259
groove, 246
plug, 254
seam, 258
slot, 255
spot, 256
square, 238
Wheels
kinds of, 296
mechanical advantage of, 296
rachet, 296

8/90(9C1765F)